R00218 99464

FORM 125 M
NATURAL SCIENCES &
USEFUL ARTS DEPT.

The Chicago Public Library

Received April 26, 1972

JUL 0 5 1972

STRATIGRAPHIC BOUNDARY PROBLEMS

Salt Range and greater part of Potwar Plateau (Gemini photograph G5-R4-F19). The Salt Range is visible in lower part of picture as far as some 10 miles east of Khewra in the east (right). Jhelum River is in lower right corner. The rugged area in the west (lower left corner) is the highest part of the Salt Range around Sakesar (4,992 feet). In the far left of the picture the range turns sharply north. Nammal Lake is clearly visible in middle of left side. The narrow river in the upper left corner is the Indus. The continuation of the Salt Range northwest of the Indus is the Surghar Range. Kalabagh lies at the exit of the Indus between the two ranges. The Soan River flows somewhat below the top of the picture and more or less parallel to it. The picture is oriented about 15° W. of N.; the area covered by it is shown in Figure 1 (Kummel and Teichert).

STRATIGRAPHIC BOUNDARY PROBLEMS:
Permian and Triassic of West Pakistan

edited by
Bernhard Kummel and Curt Teichert

Department of Geology · University of Kansas · Special Publication 4

THE UNIVERSITY PRESS OF KANSAS
Lawrence/Manhattan/Wichita/London
1970

© Copyright 1970 by the University Press of Kansas/ Standard Book Number 7006-0064-7/ Library of Congress Catalog Card Number 72-629063/ Printed in the United States of America/ Designed by Fritz Reiber

Preface

The papers assembled in this volume are the outcome of studies initiated by us in Pakistan in 1961 as part of an endeavor to gather and digest data on the stratigraphic and paleontologic relationships at the Permian-Triassic boundary in the Salt Range and adjacent ranges west of the Indus River. Included is our final report on the stratigraphy and paleontology of these "boundary beds," but the bulk of the volume is made up of reports by eight authors who have studied our rock and fossil collections from the point of view of their individual specialization. Only one author, R. E. Grant, bases his contribution in part on material collected by himself. We are fortunate in being able to present all these results in a single volume, instead of having them scattered in perhaps half a dozen, or more, periodicals. We hope that this mode of presentation will greatly facilitate evaluation of the evidence by geologists interested in the exciting subject of era boundaries, both from a stratigraphic and paleontologic point of view.

The bulk of our collections came from the Salt Range where we studied the Permian-Triassic boundary in the greatest detail. The name Trans-Indus ranges is applied informally to several mountain ranges that lie west of the Indus River, and west and southwest of the Salt Range. Of these we have studied only two, Surghar Range and Khisor Range, and stratigraphic relationships in, and fossil material from, all three ranges are described in this volume.

Throughout the volume all geographic distances, elevations, and stratigraphic thicknesses are given in units of the English system of measures, because this is still used in Pakistan and because all previous authors who have worked in the Salt Range, with a single exception, have used English measures. Use of the metric system, although desirable in principle, would have rendered more difficult comparison of our observations with those of earlier workers.

The spelling of geographic names in all contributions is based on usage in the latest available topographic map sheets published by the Survey of Pakistan.

Bernhard Kummel
Curt Teichert

List of Authors

Basil E. Balme
Department of Geology, University of Western Australia, Nedlands, Western Australia

William M. Furnish
Department of Geology, The University of Iowa, Iowa City, Iowa

Brian F. Glenister
Department of Geology, The University of Iowa, Iowa City, Iowa

Richard E. Grant
U.S. Geological Survey, Washington, D.C.

Bernhard Kummel
Museum of Comparative Zoology, Harvard University, Cambridge, Massachusetts

Albert J. Rowell
Department of Geology, The University of Kansas, Lawrence, Kansas

William A. S. Sarjeant
Department of Geology, The University, Nottingham, England

I. Gregory Sohn
U.S. Geological Survey, Washington, D.C.

Walter C. Sweet
Department of Geology, Ohio State University, Columbus, Ohio

Curt Teichert
Department of Geology, The University of Kansas, Lawrence, Kansas

Contents

Preface

Stratigraphy and Paleontology of the Permian-Triassic Boundary Beds, Salt Range and Trans-Indus Ranges, West Pakistan   1
   Bernhard Kummel and Curt Teichert

Lingula from the Basal Triassic, Kathwai Member, Mianwali Formation, Salt Range and Surghar Range, West Pakistan   111
   A. J. Rowell

Brachiopods from Permian-Triassic Boundary Beds and Age of Chhidru Formation, West Pakistan   117
   R. E. Grant

Permian Ammonoid *Cyclolobus* from the Salt Range, West Pakistan   153
   W. M. Furnish and B. F. Glenister

Ammonoids from the Kathwai Member, Mianwali Formation, Salt Range, West Pakistan   177
   Bernhard Kummel

Early Triassic Marine Ostracodes from the Salt Range and Surghar Range, West Pakistan   193
   I. G. Sohn

Uppermost Permian and Lower Triassic Conodonts of the Salt Range and Trans-Indus Ranges, West Pakistan   207
   W. C. Sweet

Acritarchs and Tasmanitids from the Chhidru Formation, Uppermost Permian of West Pakistan   277
   W. A. S. Sarjeant

Palynology of Permian and Triassic Strata in the Salt Range and Surghar Range, West Pakistan   305
   B. E. Balme

Indexes   455
   I. Author Index   455
   II. Systematic Index   458
   III. Subject and Locality Index   467

This is the world's limit that we have come to;
this is the Scythian country,
an untrodden desolation.

Aeschylus, *Prometheus Bound*

Stratigraphy and Paleontology of the Permian-Triassic Boundary Beds, Salt Range and Trans-Indus Ranges, West Pakistan

BERNHARD KUMMEL AND CURT TEICHERT
Harvard University and The University of Kansas

CONTENTS

Abstract   2
Introduction   2
Acknowledgments   3
Salt Range and Trans-Indus Ranges, General Geology and Geography   4
History of Research   5
Present Investigations   15
Stratigraphy of Permian and Triassic Rocks   17
  Permian   17
  Triassic   22
  Summary   27
Detailed Stratigraphy of Beds Near Permian-Triassic Boundary   27
  General Remarks   27
  Chhidru Formation   30
    Introduction   30
    Uppermost Highly Fossiliferous Unit   31
    White Sandstone Unit   31
  Mianwali Formation   36
    Kathwai Member   36
      Introduction   36
      Dolomite Unit   38
        Gross Lithology and Bedding Plane Features   38
        Petrography   43
      Limestone Unit   48
        Gross Lithology   48
        Petrography   49
      Paleontological Content   49
      Depositional and Diagenetic Environments   53
    Basal Mittiwali Member   57
  Comparison with Earlier Observations   58
Nature of Changes at Permian-Triassic Boundary   61
  Definition of Boundary   61
  Lithofacies Changes   61
  Comparative Paleontology of Beds above and below the Boundary   62
  Evaluation of the Evidence   71
Conclusions   77

Contents (continued)

Appendixes   78
   Appendix A: Stratigraphic Sections of the Uppermost Permian and Lowermost Triassic Beds in the Salt, Surghar, and Khisor Ranges   78
   Appendix B: List of Samples Used   97
   Appendix C: Descriptions of Some Rock-Stratigraphic Units of Triassic Age   104
References   106

ABSTRACT

It has been known for about a hundred years that the sedimentary series of the Salt Range and some of the Trans-Indus ranges of West Pakistan encompass a concordant succession of marine, fossiliferous rocks of Late Permian and Early Triassic age. After a brief introduction to geology and geography of the area, the history of research on this sequence and on the problem of the Permian-Triassic boundary is reviewed.

The detailed stratigraphy of the beds immediately above and below the Permian-Triassic boundary is described from nine localities in the Salt Range, two in the Surghar Range, and two in the Khisor Range. The uppermost lithological unit of the Permian Chhidru Formation ("Upper Productus limestone") is a white sandstone unit which is as much as 17 feet thick and in the Khisor Range has yielded brachiopods indicating a Late Permian age. It also contains conodonts, a rich palynological assemblage, and acritarchs.

The white sandstone unit is conformably overlain by a dolomite unit, about 5 to 14 feet thick, containing *Ophiceras, Glyptophiceras, Miocidaris,* and few other megafossils associated with rich assemblages of conodonts, acritarchs, and spores and pollen. This unit forms the base of the Triassic System. It is gradational with an overlying limestone unit that also contains *Ophiceras* and other fossils, including conodonts and palynomorphs. The dolomite and limestone units together form the Kathwai Member which represents the *Ophiceras* Zone, the lowest ammonoid zone of the Triassic System. This is overlain by thin-bedded limestone containing *Gyronites* and other ammonoids (the "Gyronitan" assemblage of Spath)—the "Lower Ceratite limestone" of Waagen, which most previous observers believed to form the base of the Triassic System in the Salt Range.

The nature of the boundary between the Permian and Triassic systems in the Salt Range and the Trans-Indus ranges is discussed from the point of view of lithology, diagenesis, sedimentary structures, and fossil content. The abrupt lithofacies change indicates a sharp change of sedimentary environment. Sedimentary structures, such as mudcracks and capped ripple marks, in combination with dolomitization and dedolomitization patterns, suggest that a period of emergence intervened between the deposition of the white sandstone unit and the dolomite unit, and that the contact between the Permian and Triassic systems in the Salt Range and Trans-Indus ranges, therefore, is a paraconformity which may represent a gap equivalent to as much as, or more than, a stratigraphic stage.

INTRODUCTION

This report is part of a larger effort aimed by us at a restudy of problems connected with evolutionary changes that took place among marine organisms at the passage from the Permian to the Triassic Period. The boundary between these two periods is at the same time the boundary between the Paleozoic and the Mesozoic eras. The nature and significance of era boundaries have presented challenging problems to geologists ever since systems and eras were recognized and named. It does not require deep paleontological insight to recognize that composition and general aspect of life forms in the Paleozoic differ in rather obvious ways from those of the Mesozoic, but exactly how the changeover occurred has been much debated. Most authors have emphasized the factor of suddenness of the changeover, and this plays a considerable role in nearly all hypotheses that have been offered to explain causes of the change.

Obviously, the solution to the problem of extinction of Permian marine faunas and subsequent evolutionary radiation of the rather different Triassic faunas must take its beginning from field observations on stratigraphic sections which encompass this boundary in marine facies. Unfortunately, such stratigraphic sections are extremely rare and, until quite recently, detailed stratigraphic and paleontologic investigations were nonexistent.

In a worldwide review the search for stratigraphic sections for which it has been claimed that marine sedimentation was continuous from the Late Permian into the Early Triassic narrows down to five areas (Teichert, 1968):

1. Southern China
2. Kashmir
3. Northern West Pakistan (Salt Range, Trans-Indus ranges)
4. Azerbeidjan, Armenia (Iran, USSR)
5. Northeast Greenland

Beginning in 1961, we have now completed field investigations in all of these areas except southern China, which for obvious reasons could not be visited by us.

In this report we propose to discuss the geological and paleontological evidence that bears on the problem of the Permian-Triassic boundary in the Salt Range and Trans-Indus ranges. Our purpose is to summarize observations and to present conclusions. We do not wish at this time to offer speculative considerations, or at least not to an extent greater than required for interpretation of the problems encountered. We plan to present a more widely based evaluation of the evidence after all our investigations have been completed.

In 1966 we published a preliminary account of our findings, based essentially on our field work and utilizing a set of preliminary fossil identifications (Kummel and Teichert, 1966). It is now possible to present results in a greatly enlarged and refined manner, thanks, in a large measure, to the efforts of the many paleontological specialists who have worked on material collected by us, whose papers make up the bulk of this volume, and whose names are listed in the Acknowledgments. Much of the chapter on History of Research has been taken almost *verbatim* from our 1966 paper.

The descriptions of stratigraphic sections measured by us, accompanied by locality maps and photographs, are given in Appendix A.

Appendix B lists the exact localities for all of our field numbers cited or described by any of the authors of this volume.

Appendix C contains brief formal descriptions of some rock-stratigraphic units of Triassic age that occur in the Salt Range and Surghar Range.

The illustrations in Appendix A are separately numbered, with prefix A, to facilitate their citation in the main body of the text.

ACKNOWLEDGMENTS

Kummel's field work and travel to Pakistan, as well as his later laboratory work, was supported by National Science Foundation grant No. G-19066. Teichert was a member of an advisory team of the U.S. Agency of International Development (AID) to the Geological Survey of Pakistan from November, 1961, until April, 1964, and his field studies were carried out as part of this assignment. His later laboratory and editorial work was supported by the University of Kansas. The Geological Survey of Pakistan provided logistical support throughout the field project and made available the services of, then Assistant Director, M. R. Khan and, then Deputy Director, A. N. Fatmi, who accompanied us in the field for various periods. During some of his later field trips to the Salt Range, Surghar Range, and Khisor Range, Teichert was accompanied by H. Rahman, then Director, Geological Survey of Pakistan. We are, furthermore, indebted to the Geological Survey of Pakistan for permission to use the unpublished geological maps of the Salt Range, scale one-half mile=one inch, prepared by E. R. Gee in the 1930's.

At all times we received invaluable and indispensable administrative support from John A. Reinemund, then head of the U.S. geological team in Pakistan, his successor, Max G. White, and his administrative assistant, Alan Davis. H. D. Dunkle, who was a member of the same team, gave valuable assistance to Kummel in 1961, before Teichert arrived in Pakistan, and later advised on our finds of fossil fish remains.

We are indebted to many persons for advice, encouragement, and information. First, we want to thank our coauthors in this volume, who have worked on our collections and, collectively, made a vast contribution to the solution of our problem; they are: B. E. Balme, W. M. Furnish, B. F. Glenister, R. E. Grant, A. J. Rowell, W. A. S. Sarjeant, I. G. Sohn, and W. C. Sweet. Others who have helped with identification of various fossil groups at one time or another are G. A. Cooper, L. C. Henbest, Hans Hess (Basel), J. Harlan Johnson, R. L. Kaesler, N. D. Newell, Bobb Schaeffer, and E. L. Yochelson. H.-E. Reineck (Wilhelmshaven) gave valuable advice on the interpretation of bedding-plane features. O. H. Schindewolf (Tübingen) generously made his own Salt Range collections, as well as earlier ones of Koken and Noetling, available to us for study and was ever ready for discussions. Kummel is grateful to Raymond Siever and some of his students for helpful discussions on the petrography of the Kathwai Member. R. C. Moore reviewed the entire manuscript and made innumerable valuable suggestions. A. J. Rowell reviewed the chapter "Stratigraphy of Permian and Triassic Rocks."

Teichert wishes to thank K. N. Kravchenko and I. A. Voskresensky, former advisers to the Oil and Gas Development Corporation of Pakistan, for many helpful discussions and for a guided tour of Khan Zaman Nala in 1963.

Both of us, but especially Kummel, wish to acknowledge the efficient assistance of Victoria Kohler in handling samples, drafting illustrations, and numerous other chores. Roger B. Williams gave valuable assistance in improving several of the previously published illustrations.

Finally, we are indebted to the Harvard University Press and Schweizerbartsche Verlagsbuchhandlung for permission to reproduce illustrations and minor parts of text published by Kummel (1966) and by Kummel and Teichert (1966).

SALT RANGE AND TRANS-INDUS RANGES, GENERAL GEOLOGY AND GEOGRAPHY

The Salt Range forms a belt, composed of rocks of ?Precambrian and Cambrian to Eocene age, running along the southern and southwestern edge of the Potwar Plateau which is underlain predominantly by younger Tertiary rocks of the Siwalik Group (see frontispiece; Hemphill and

Danilchik, 1968). In its eastern part the range is only about 10 miles wide, having heights of 2,200 to 2,500 feet. It widens westward (central Salt Range) to a width of more than 20 miles, with highest altitude, 4,992 feet, at Sakesar. The range then turns into a northwesterly direction and narrows to as little as 4 miles at Nammal Gorge. It reaches the Indus River at Kalabagh.

The rocks in the Salt Range are generally gently folded. There is much faulting, however, both large- and small-scale, as well as local overthrusting with movements toward the south (Gee, 1938, 1945, 1947; also Gansser, 1964, p. 27). The range is essentially treeless except in its highest parts, and it is only sparsely inhabited. Most of it offers patchy and poor grazing land for sheep and goats.

At Kalabagh, a picturesque small town, built around a hill of salt, the Indus River breaks out of the mountains through an impressive gorge, and enters the vast plains to the south. Several ranges west of the Indus in this general area, between about 32° and 35°N. lat., are collectively referred to as Trans-Indus ranges. These are the Surghar Range, Shingarh Range, Khisor Range, and Marwat Range. Only the Surghar Range and Khisor Range have been studied by us. The highest of the Trans-Indus ranges is the Marwat Range reaching a height of 4,516 feet at Sheikh Budin. In their general surface features the Trans-Indus ranges resemble the Salt Range.

The entire area is drained by the Indus River and its mighty tributary, the Jhelum (see frontispiece).

The approach to these ranges is relatively easy, because the area is served by a rather close road net and by some railroads. The nearest airport is at Rawalpindi, less than 60 miles north of the eastern end of the Salt Range, although Lahore, situated 100 miles to the south, provides a more convenient operational base for approaches to the central and western Salt Range and to the Trans-Indus ranges. It also has an airport and is served by frequent flights from Karachi. It is, however, necessary to use jeeps in order to proceed to most of the localities studied by us, and some can be approached only on horseback or by camel or donkey ride.

The entire area has no normal facilities for travelers and tourists. The nearest hotels are in Lahore and Rawalpindi. However, in the Salt Range, Potwar Plateau, and the Jhelum River area, and to a lesser extent west of the Indus, many government rest houses exist which generally are reserved for use by government officials and other persons visiting the area in an official capacity. For most of our field work various rest houses of this type provided convenient operational bases.

HISTORY OF RESEARCH

Literature on the Salt Range begins with its discovery and naming by Mountstuart Elphinstone in 1808 (Elphinstone, 1815). A picture of its stratigraphy began to emerge with the work of Andrew Fleming, a physician who travelled in the Salt Range around the middle of the nineteenth century, observed the principal features of its rock sequence, and made the first fossil collections. Fleming's first report (Fleming, 1848) is largely concerned with coal and other potential economic deposits. In this report he was unsure whether Upper Paleozoic strata were present in the Salt Range. For historical interest it is worth quoting a few sentences in which this question is discussed (Fleming, 1848, p. 514):

> From the former we obtained several specimens of shells of the genera Productus, Terebratula, and probably one Spirifer, associated with Ammonites, Belemnites, etc. The appearance of these fossils, as well as of the limestone in which they are imbedded, is more ancient than that of any of the other fossiliferous strata we have

noticed. Shells of the genera Productus and Spirifer are generally considered characteristic of strata inferior to the Lias, and abound in the magnesian limestone. There are, however, exceptions to this, and at least 3 species of Spirifera, and we think one or two Producti have been found in the Lias itself. Terebratulae are by far the most abundant of all the fossils we noticed in the Limestone, and this genus has been found to occur through all the strata from the chalk formation downwards.

At first sight we were inclined to believe that we were dealing with magnesian limestone, but on subjecting a portion of it to chemical analysis we failed to detect any magnesia in its composition, which earth does not, as far as we can ascertain, exist in any limestone of the salt range.

Fossils of this survey were sent by Fleming to his father and were exhibited at the meeting of the British Association of Science in Edinburgh. De Verneuil and, subsequently, Davidson identified the following species in this collection:

Productus cora d'Orbigny
Productus costatus Sowerby
Productus Flemingii Sowerby
Orthis crenistria? Phillips
Terebratula Royssii L'Eveillé
Terebratula crispata? Sowerby
Several new species of *Terebratula*.

The above list of species first appeared in a publication on the Salt Range that represented abstracts of letters from Fleming to R. I. Murchison (Fleming, 1853a). In the same year Fleming wrote to Murchison on a visit he made to the "Sooliman" (now Suleiman) Range, west of the Indus River (Fleming, 1853b). On this excursion Fleming did not find Permian outcrops, but observed boulders of the Productus limestone. Fleming (1853b, p. 348) explained his reason for not reaching the Paleozoic outcrops as follows:

The difficulties, however, in following out any researches in these hills are very great. One cannot go unarmed or without a party, and hence the explorer cannot follow the by-ways, but must keep to the highways. The Bosders on the Dera Ghazee Khan Frontier, and the Kusranees and Shiovanees on the Dera Ismael Khan Frontier, are all robbers and live by plunder, and would be only too glad to get hold of any of us, as we are here to cut them up if they appear on the plains intent on a fray.

Fleming's full report on his observations in the Salt Range appeared in 1853 (Fleming, 1853c). At that time he still did not have reports on all fossils which he had sent to England, but identification of Carboniferous forms was secure. Strangely, Fleming completely missed identifying the Triassic. The reason is apparent from the following quotation (Fleming, 1853c, p. 266):

Towards the upper part of the lower division of the series, where the limestone becomes argillaceous and thin-bedded and alternates with coarse arenaceous shales, the Brachiopoda become scarce and give place to Cephalopoda, which animals characterize a marine zone of less depth than the Brachiopoda which precede them, and generally occur in seas with muddy bottoms. We have obtained examples of species of the genera Bellerophon, Goniatites (?) and Orthoceras. Associated with these large spiral univalves of the genus Cirrus and Enomphalus are abundant, and, in the slaty limestone at the top of the lower division of the carboniferous series, and also in the middle division, a Cephalopodous shell formerly considered an ammonite, but now constituted into the genus Ceratites abounds, and is generally associated with a small bivalve, probably a species of Passidonia. As Ceratites have hitherto been considered as characteristic of rocks of triassic age and peculiar to the muschelkalk, their occurrence in company with undoubted carboniferous types is highly interesting. We have placed the matter beyond doubt, having in our possession a specimen which we obtained at Moosakhail in which two Orthoceratites and seven Ceratites are lying side by side in a slab 9 in. × 5 in.; Orthoceratites have never been found in strata superior to the carboniferous limestone, but abound throughout the older fossiliferous rocks.

The fossils sent by Fleming to England were described subsequently by Davidson (1862) and de Koninck (1863). A few additional Paleozoic species from the Salt Range were described by de Verneuil (in Verchère, 1866).

Another important paper of this period was one published by Theobald

(1854), who introduced the term "*Productus* limestone" and indicated the thickness of this formation, quite accurately, to be 1,100 feet.

The early history of research on the Permian (then regarded as Carboniferous) and Triassic rocks in the Salt Range has been summarized by Wynne (1878) who recorded 42 papers and notes on the Salt Range published prior to his own report. By this time it was realized that strata of Late Paleozoic ("Carboniferous") age are in close association with rocks of Triassic age and Fleming and others had asserted that Paleozoic brachiopods are present with Triassic ammonites in the same bed.

Wynne carried out his field studies of the Salt Range between 1869 and 1872. During the later stages of his study, after he had completed his mapping, he was joined by W. Waagen, who, with Wynne, made further stratigraphic observations and collected additional fossil faunas.

Wynne's report (1878) includes the first measured stratigraphic sections, which in the light of our recent studies can now be interpreted correctly. These stratigraphic sections formed the background for Waagen's subsequent monographs on the faunas of the "Productus limestone" and of the Ceratite beds, and for his conclusions on the nature and significance of the boundary. For this reason it is pertinent to cite here a few of these authors' more complete and detailed sections across this boundary. At the time of Wynne's report the name "Productus limestone beds" had not been generally accepted, and these beds were referred to by him as the "Carboniferous limestone."

Near the eastern limit of outcrop, of both the Late Paleozoic and Triassic formations in the Salt Range, Wynne (1878, p. 218) and Waagen measured the following section just west of the village of Khoora (now spelled Khura):

		Feet
Trias	Greenish and gray shales, variegated, red, yellow, and blue at top; *Ceratites*	20 - 30
	Ceratite limestone	3 - 4
	Yellow, sandy, calcareous beds, with *Rhynchonellae*	5
Carboniferous	Brown dolomite, like that at Pail	3
	Gray and greenish calcareous and micaceous sandstone, with limestone bands, weathering red in parts and containing *Bellerophons, Productus*, etc.	90 - 100
	Compact carboniferous limestone very rich in fossils, *Strophalosia morrisiana, Productus, Athyris*, etc.	60?

West of the village of Virgal (now spelled Wargal) Waagen measured the following section of "Carboniferous" and Triassic strata (Wynne, 1878, p. 225-226):

		Feet
Trias	Gray limestone with numerous bivalves	2
	Thin-bedded, hard, sandy limestone, no fossils	6
	Ceratite sandstone, thin-bedded, soft, yellow sandstone with gypsum; a *Bellerophon* bed in the upper region	50
	Green *Ceratite* marls	60 - 70
	Thin-bedded limestone with *Ceratites*	8
Carboniferous	Gray sandstone layers	6
	Black coaly, shaly beds, micaceous	3 - 6
	Thick, light gray, concretionary sandstones with nests of fossils, small *Producti, Bellerophon*, and *Gastropoda*	6
	Brown sandstone and limestone with *Bellerophon, Productus costatus*, and a *Dentalium*	200?
	Compact limestone with corals	300

High up in the Amb valley near a small village called Siran-ki-dok (a region we were not able to find or get any information on from the local population) Wynne (1878, p. 241-243) and Waagen measured the following section:

		Feet	Inches
Triassic	Very hard, rusty limestone with numerous sections of *Ceratites* or *Ammonites*, gastropods or bivalves	3	0
	Soft, yellow, sandy beds	3	0
	Hard, rusty-colored layer	1	6
	Gray, cavernous sandstone	3	0
	Very hard, gray limestone, glauconite, and bivalves	6	0
	Soft, yellow sandstone	60	0
	Thin bed of sandstone with indistinct bivalves	0	3
	Hard, brown bed with numerous pebbles of limestone	1	6
	Gray limestone with numerous bivalves	3	0
	Thin-bedded limestone with *Ceratites*	10	0
	Sandstone and limestone with *Ceratites*	20	0
	Ceratite marl, badly seen	30	0
	Brown conglomerate bed	1	6
	Total	142	9
Carboniferous	Rusty dolomite	6	0
	Light colored sandstone, *Bellerophons*, *Athyris subtilita*, *Dentalium*, etc.	about 350	0
	Compact dolomitic, sandy and fossiliferous gray, carboniferous limestone		
	Sandy and ferruginous beds		

Wynne (1878) included a few other sections measured in collaboration with Waagen, but these are very general and not helpful to the problem at hand.

As regards the nature of Permian ("Carboniferous")-Triassic boundary, Wynne (1878, p. 97-98) has this to say:

> These *triassic* rocks, though lithologically distinguishable, present no such marked contrast to the *carboniferous formation* as exists between the *Trias* and succeeding beds. The limestones are more thin-bedded, and the shales or marls of different character and somewhat different colour; but the whole aspect of the group is such that, were it not for the palaeontological evidence, it might pass for a portion of the palaeozoic rocks below with which it was classed by Dr. Fleming.

This view was to be repeated by many subsequent observers.

Wynne (1880) described "Carboniferous" and Triassic rocks from several ranges west of the Indus River, especially from Kalabagh, from the Surghar Range (which he called "Chichali Range"), from the Khisor (Khasor) Range, and from the Sheik Budin area in the Marwat Range. Although his descriptions are rather generalized and he gave few detailed section descriptions, he demonstrated that the Permian-Triassic sequence in these areas was the same as that in the Salt Range.

Waagen, in the introduction to his Productus limestone monograph (Waagen, 1879-88), provisionally recognized Wynne's stratigraphic divisions. He introduced the name Productus limestone for the beds Wynne and previous writers had called "Carboniferous" or Lower Limestone of the Salt Range, and divided it into lower, middle, and upper divisions. The conclusions to the Productus limestone monograph were not written until 1887 at which time, on the basis of his paleontological studies, Waagen proposed a more refined classification of the *Productus* beds:

Upper *Productus* limestone { Topmost beds / Cephalopoda beds / Lower beds

Middle *Productus* limestone { Upper beds / Middle beds / Lower beds

Speckled sandstone group { Lower *Productus* limestone / Speckled sandstone

Waagen concluded from his studies that the Middle and Upper Productus limestone beds which he had previously believed to be Carboniferous were, in fact, Permian in age and contained "a permian fauna that is richer in forms and richer in conspicuous species than anyone that has been described up to the present in the whole world" (Waagen, 1887, p. iv).

The Triassic faunas of the Salt Range were monographed by Waagen in 1895. At that time he adopted the following stratigraphic nomenclature for the Triassic rocks:

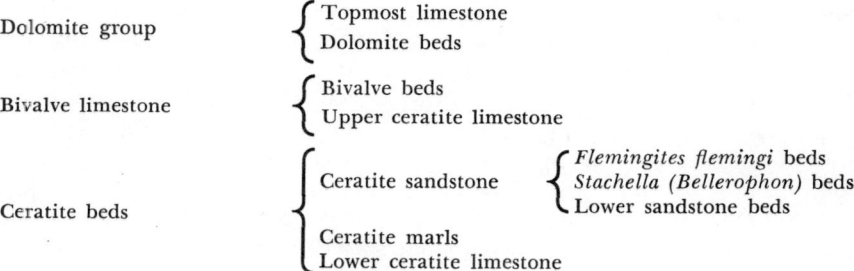

Dolomite group { Topmost limestone / Dolomite beds

Bivalve limestone { Bivalve beds / Upper ceratite limestone

Ceratite beds { Ceratite sandstone { *Flemingites flemingi* beds / *Stachella (Bellerophon)* beds / Lower sandstone beds } / Ceratite marls / Lower ceratite limestone

For the most part, Waagen's stratigraphic data and comments leave much to be desired and are often contradictory and difficult to interpret. In regard to the contact relations between Permian and Triassic, he made the following statement (Waagen, 1895, p. 1):

> According to Mr. Wynne's indications the group forms indeed stratigraphically a member of the palaeozoic series; and it is not on geological, but solely on palaeontological grounds that it can be separated from the Productus-limestone group. As far as my own observations go, the stratigraphical conformity of the palaeozoic rock groups is everywhere obvious. Only on a very close examination one finds, off and on, a section which gives some indications that rather considerable physical changes in the condition of the deposition of rocks must have taken place between the formation of the topmost beds of the Productus-limestone and the strata belonging to the Ceratite formation. At most places some unfossiliferous beds intervene, and sometimes even conglomerates are deposited at the base of the Ceratite group, which show that a certain amount of denudation must have taken place at the commencement of these latter beds.

Thus, to Waagen, the Permian-Triassic boundary was encompassed within an erosional break in deposition. Already in 1879 Waagen had clearly recognized the magnitude of the faunal change from the Productus limestone to the Ceratite beds. We may quote the following short paragraph (Waagen, 1879, p. 2):

> There are two principal boundary lines (in the total Salt Range sequence) which will scarcely escape any observer traveling along the barren valleys of the range and keeping his eyes open to the palaeontological conditions of the strata of which the range is composed; they are the limit between the *Productus*-limestone and the Ceratite beds, and that between the latter and the newer mesozoic formations. There are only a very few genera crossing these limits and extending from the lower formation into the higher, and therefore the boundary lines between those formations mark large stages in the development of life in the country under consideration.

In his monograph of the geological results, Waagen (1891, p. 229) emphasized this faunal change in the following statement summarizing the faunal character of the topmost beds of the Upper Productus limestone:

> There are altogether 63 species, of which 17 are identical with forms occurring already in the preceding divisions of the *productus*-limestone, whilst 46 are peculiar to the sub-division here under consideration. As far as is known to me up to the present, there is not a single species which would extend into the triassic beds of the ceratite formation.

At the same time, Waagen considered nearly 50 percent of the species of his uppermost member to have "mesozoic" affinities, and he concluded that "these beds can barely any more be properly called Permian; they rather seem to form a transitional stage, a sort of passage bed between the palaeozoics and the mesozoics" (Waagen, 1891, p. 230).

At this period the first stratigraphic and paleontological data from the Himalayas had become well established through the studies of Griesbach (1880) who emphasized the presence of *Otoceras* in the lowest Triassic strata of the Himalayas. In reconciling the absence of *Otoceras* in the Salt Range, Waagen (1891, p. 232) came to the conclusion

> . . . that in the Salt Range at most places, some unfossiliferous beds intervene between the fossiliferous topbeds and the Triassic limestone, and it is not impossible that these may represent the *Otoceras* beds in the Salt Range. The absence here of the genus *Otoceras* would then also be explained; as in the Himalayas also it is restricted to a very little extended series of rocks, and is entirely absent in higher strata.

The publication of Waagen's monograph on the fossils of the Ceratite formation in 1895 brought to a close the Wynne-Waagen episode in the long Salt Range controversy. There then appeared upon the stage a most remarkable adversary whose contributions of facts and data were minimal, but who nevertheless in his several papers created an intense controversy that has continued to this date. This person was Fritz Noetling. What is remarkable is that Noetling's interpretation of the Permian-Triassic relations in the Salt Range involved drastic revisions in the then accepted interpretation of the Himalayan and other Triassic sequences and brought him into direct conflict with Diener, Mojsisovics, and others.

Noetling had been delegated by the Geological Survey of India to carry out field investigations in the Salt Range. He records (Noetling, 1901, p. 370) that he spent the winter of 1893-1894, November 1898, and finally the winter of 1899-1900 in the Salt Range. At first, Noetling (1900a) maintained that the Ceratite formation must be included in the Permian System because the gradual passage from the Productus limestone to the Ceratite beds made assignments of these units to two different eras impossible. In addition he reported the presence of *Otoceras* from the Ceratite marls, which he correlated with the Himalayan *Otoceras* beds, assigning a Permian age to all these beds (Noetling, 1900b). This correlation soon proved to be incorrect. Noetling's viewpoint on the interpretation of the stratigraphic position of *Otoceras* was hotly contested by Diener (1900, p. 1) who maintained that the genus characterized the earliest Triassic ammonite zone. Noetling then contributed another paper (1900c) that was highly controversial but added nothing positive to the problem.

Noetling received support in his views on the stratigraphic position of *Otoceras* from A. von Krafft (1900, p. 203) who, in a preliminary report on the stratigraphy of the Spiti region, correlated the *Otoceras* beds there with the Ceratite marls and Lower Ceratite limestone of the Salt Range. In the following year von Krafft (1901) changed his interpretation considerably. He divided the strata exposed between the Kuling shales and the *Hedenstroemia* beds in the Spiti region into three zones (in ascending order): the *Otoceras* bed, the *Ophiceras* bed, and the *Meekoceras* bed. The *Meekoceras* bed he referred to the Triassic, the age of the *Ophiceras* bed was left in doubt, and the *Otoceras* bed was correlated with the Upper Productus limestone on the strength of the identification of *Medlicottia (Episageceras) dalailamae* Diener with *Medlicottia wynnei* Waagen from the zone of *Euphemus indicus* in the Salt Range. Diener (1901a) again refuted this argument, rejecting the identity of these two species of *Medlicottia (Episageceras)*.

In 1901 Noetling published a long paper on the Permian and Triassic formations of the Salt Range and proposed the following zonal scheme:

Triassic	Scythian Stage	Upper Ceratite limestone	Zone of *Stephanites superbus*
		Ceratite sandstone	Zone of *Flemingites flemingianus*
		Ceratite marls	Zone of *Koninckites volutus* Zone of *Prionolobus rotundus*
		Lower Ceratite limestone	Zone of *Celtites(?)* sp.
Upper Permian	Thuringian Stage	Chideru Group	Zone of *Euphemus indicus* Zone of *Medlicottia wynnei* Zone of *Bellerophon impressus* Zone of *Cyclolobus oldhami* Zone of *Derbyia hemisphaerica* Zone of *Productus lineatus*
		Virgal Group	Zone of *Xenodiscus carbonarius* Zone of *Lyttonia nobilis* Zone of *Fusulina kattaensis*
		Amb Group	Zone of *Spirifer marcoui*

This scheme represented a considerable revision of Noetling's previous views, a fact commented upon by Diener (1901b). In regard to the correlation with the Himalayan sequence Noetling correlated the *Otoceras* beds of Spiti with his zone of *Bellerophon impressus* of the Salt Range and the *Ophiceras* beds of Spiti with his *Euphemus indicus* and *Medlicottia wynnei* zones of the Upper Productus limestone. These views on correlation and stratigraphic relations of these strata were maintained in the *Lethaea Palaeozoica* (Noetling, 1902, p. 653) and in the *Lethaea Mesozoica* (Noetling, 1905, p. 71).

Noetling (1900a, p. 179; 1901, p. 451) gave the first description of the section at Mittiwali (misspelled "Mittiali") which corresponds to section 6B of this report. The following description of the boundary beds is excerpted from a more comprehensive section description:

	Thickness	
	Feet	Inches
21. Dark brown, hard and flaggy limestone with numerous ill-preserved fragments of *Ceratites*	3	0
20. Green, unfossiliferous marl	2	0
19. Dark brown, hard and flaggy limestone with numerous ill-preserved fragments of *Ceratites*	5	0
18. Light brown, thin-bedded, calcareous sandstone with argillaceous layers	2	6
17. Limestone of rusty brown color, hard and ringing under the hammer; with indistinct traces of ammonites	0	4
16. Light brown, thin-bedded sandstone, with argillaceous interbeds	3	0
15. Hard, dark, calcareous sandstone, full of fossil fragments, probably *Bellerophon* or *Stachella*	0	6

14. Shaly, calcareous, brown sandstone, apparently unfossiliferous	3	6
13. Hard, dark brown, calcareous sandstone, with shaly interbeds, full of fossil fragments, probably *Bellerophon* or *Stachella*	4	3

Strangely enough, Noetling had failed to find any identifiable fossils in any of these beds, but, after some discussion (1901, p. 453), placed the Permian-Triassic boundary between beds 16 and 17.

Throughout the series of papers on the Permian and Triassic beds of the Salt Range, Noetling maintained that the sequence represented continuous deposition with no erosional breaks. In 1904, he even reported that in the Chua gorge near Virgal (Wargal) he had observed the "most intimate interfingering" of lithological units of the Productus limestone and the Ceratite beds (Noetling, 1904, p. 324-325). In 1901, he offered as explanation for the extinction of the Permian brachiopod fauna the change from deeper water conditions in the Permian to shallow water in the Triassic; the deep water brachiopods died out and were replaced by ammonites, which preferred near-shore, shallow water conditions. At a later date Noetling (1905, p. 128) suggested that an increase in the temperature of the sea was responsible for the extinction of the brachiopods. In fact, to explain the absence of brachiopods in the *Otoceras* beds of the Himalayas (which he considered Permian in age), Noetling suggested that the change in temperature progressed from the north to the south, thus affecting the Himalayan area while the brachiopods persisted in the Salt Range.

The stratigraphic position of *Otoceras* is now no longer in doubt; many new monographs on early Triassic faunas have clearly established its position at the base of the Triassic system in East Greenland (Spath, 1930, 1934), northern Canada (Tozer, 1959), northern Alaska (Kummel, in MS), and in Siberia (Popov, 1958).

A notable contribution was by Tschernyschew (1902, p. 715-728) who discussed age and correlation of the Productus limestone at great length and compared its brachiopod fauna with that of Upper Paleozoic rocks in the Urals and in Timan. These studies led him to conclusions which were at variance with those held at the time by Waagen and by Noetling. Tschernyschew believed the Productus limestone to be encompassed in the Late Carboniferous and Early Permian periods. He believed that this correlation was supported by the existence of a sharp faunal break between the uppermost Paleozoic and the lowermost Triassic beds in the Salt Range. According to Tschernyschew this faunal break corresponded to a stratigraphic hiatus of considerable magnitude which was terminated by transgression of the sea in Scythian time.

The Waagen-Noetling period of the Salt Range Permian-Triassic controversy was followed by a long period of inactivity in either new field observations or in new speculations. The most significant addition to the literature was a splendid review of the boundary problem and the age of *Otoceras* by Diener (1912).

Grabau (1931), although making the "Productus limestone" slightly younger than Tschernyschew had done, nevertheless shared Tschernyschew's view of the existence of a long hiatus between Permian and Triassic rocks in the Salt Range.

In the early 1930's E. R. Gee mapped the entire Salt Range on the scale of 2 miles=1 inch. His maps have not yet been published, but are available for study and inspection in the archives of the Geological Survey of Pakistan in Quetta.

However, Gee did publish occasional observations on the Permian and Triassic rocks, especially a detailed description of the stratigraphic section in Nammal Gorge (Gee, 1947, p. 143-147). The succession immediately below and above the Permian-Triassic boundary was described as follows:

		Thickness Feet
Trias Ceratite Beds (lower part only)	(b) Dull green and flaggy sandstone shales with few limestone bands, shales predominate	177
	(a) Flaggy, gray, Ceratite limestone	4-6
Permian Upper Productus limestone	(c) Brownish-gray, weathering russet-colored, hard, arenaceous limestone	6
	(b) Soft, gray, sandy shale	2.5-3
	(a) Massive, soft, calcareous sandstone (14 feet) passing downward into arenaceous limestones and calcareous sandstone, fossiliferous; shales at base	217

In a work which seems to have been essentially completed in 1944, but which was not published until much later, Pascoe (1959, p. 582) writes that "Gee's work shows that the break in deposition is far greater between the Trias and the Jurassic than it is between the Permian and the Trias." Pascoe suggests that in spite of the conformable junction of the two systems "the time interval between them in fact may have been considerable, for not a single species bridges the two copious faunas, and the lowest stage of the Trias—the *Otoceras* zone—appears to be entirely unrepresented." Pascoe, like his predecessors, regards Waagen's "Lower Ceratite limestone" as the basal beds of the Triassic.

In 1952, O. H. Schindewolf, accompanied by A. Seilacher, visited the Salt Range and measured the Permo-Triassic beds on the west side of Chhidru Nala. The results of this field investigation when published (Schindewolf, 1954) produced the first really detailed stratigraphic section of the critical passage beds. Schindewolf's Chhidru section, which is the same as our locality 6A, is as follows (Schindewolf, 1954, p. 155-157, translated):

Feet

Bed 22. Sandstone, yellow brown, thin-bedded, friable with lenses of light gray limestone; contains *Pseudosageceras multilobatum, Gyronites superior, Pseudoceltites radiosus, P. fortis* and related species. (0.10 m) 0.33

Bed 21. Limestone, light gray, thin-bedded, dolomitic, weathers yellow-brown; very fossiliferous with *Pseudosageceras multilobatum, Gyronites frequens, Koninckites davidsonianus, K. d. truncatus, K. occlusus, K. rotundatus, K.* sp., nautiloids, *Pseudomonotis* sp. sp. (0.40 m) 1.32

Bed 20. Marl, yellow-brown, apparently unfossiliferous. (1.00 m) 3.3

Bed 19. Limestone, light gray, weathers reddish-brown, with occasional layers and veins of gypsum; very fossiliferous with *Pseudosageceras multilobatum, Gyronites frequens, G. undatus, G. atavus, Koninckites davidsonianus, K. d. truncatus, K. occlusus,* nautiloids, *Pseudomonotis* sp. (2.00 m) 6.6

Bed 18. Marl, sandy, fine-grained, alternating with thin-bedded sandstone; poorly fossiliferous, with *Pseudomonotis* sp. (0.80 m) 2.6

Bed 17. Limestone, light gray, contains *Ophiceras connectens, Pseudomonotis* sp., *Lingula* sp., *Orbiculoidea* sp. and numerous rhynchonellids. (0.25 m) 8.2

Bed 16. Sandstone, calcareous with beds of limestone and a marl bed of 0.20 inches in the upper part; poorly fossilifer-

ous, with cidarid spines, *Acrodus* sp., *Colobodus* sp. (2.30 m)	7.6
Bed 15. Sandstone, dolomitic, with numerous green clay nodules; contains *Ophiceras connectens* and *"Bellerophon"* sp. (0.60 m)	2.0
Bed 14. Sandstone, friable, apparently unfossiliferous. (2.50 m)	8.2
Bed 13. Sandstone, light gray, calcareous, friable, apparently unfossiliferous—the transition from the Permian to the Triassic is considered to be within beds 13 and 14. (1.00 m)	3.3
Bed 12. Limestone, sandy, dolomitic, hard, weathers gray-brown; contains *Schizodus (Eoschizodus) pinguis, Pleurophorus* cf. *subovalis, pectinids,* limids, mytilids, *Plagioglypta herculea, Bellerophon jonesianus* (abundant), *Euphemites indicus, Metacoceras* sp., *Enteletes* sp., *Derbyia hemisphaerica, Productus (Dictyoclostus) indicus, P. (Waagenoconcha) purdoni, Spirigerella derbyi, Hemiptychina* sp. (0.40 m)	1.3
Bed 11. Sandstone, calcareous, friable; with *Schizodus* sp., *Bellerophon* sp., *Productus (Dictyoclostus) indicus, P. (Waagenoconcha) purdoni.* (5.80 m)	19.1

Beds 1-10 in Schindewolf's section represent the remainder of most of the Upper Productus limestone and need not be cited here. The most significant feature of this section is the announcement of the discovery of *Ophiceras connectens* in beds 15 and 17 which lie below the Lower Ceratite limestone (beds 19-22). This species of *Ophiceras* is nearly identical with the typical ophiceratids of the Himalayas and other areas, in some of which they are associated with *Otoceras*.

Schindewolf concluded from his study of the Chhidru exposures that these represented an "essentially" unbroken stratigraphic sequence, and he placed the boundary between the Permian and Triassic systems arbitrarily between his unfossiliferous beds 13 and 14. He concluded that terrestrial causes were inadequate to explain the abrupt changes in fauna between beds 12 and 15 and he was, therefore, "inclined to attribute them to cosmic effects, and to associate them with episodic quantitative alterations of penetrating cosmic rays. It is suggested that these may have had a worldwide effect, and may have extinguished some organisms and caused others to undergo rapid mutation" (Schindewolf, 1954, p. 182).

We have published two preliminary accounts of our work done in 1961-64 (Kummel and Teichert, 1964, 1966).[1] In the 1966 paper we described some of our stratigraphic sections and discussed distribution of lithology and faunas of the beds above and below the Permian-Triassic boundary, but this paper was written without the benefit of the results of the several specialist contributions by Balme, Furnish and Glenister, Grant, Rowell, Sarjeant, Sohn, and Sweet, which are now available. In 1966, we discussed the significance of the sharp lithologic change at the boundary; announced the occurrence of Permian-type brachiopods in lowermost Triassic beds without coming to a decision on whether they were survivors from the Permian or redeposited from Permian rocks; and concluded that the boundary between the Permian and Triassic systems in the Salt Range was best regarded "as a paraconformity of undetermined magnitude."

---

[1] A paper submitted by us to the XXII International Geological Congress in 1964 had not yet been published when this report went to press (September, 1969). This paper was based on field observations only and statements contained in it that are in conflict with data presented in the present report should be disregarded. A report entitled *The Permian-Triassic boundary in the Salt Range of West Pakistan*, published as Pre-Publication Issue No. 2 of the Geological Survey of Pakistan in May, 1966, represents a first rough draft, written in 1963, of our 1966 paper and should be disregarded.

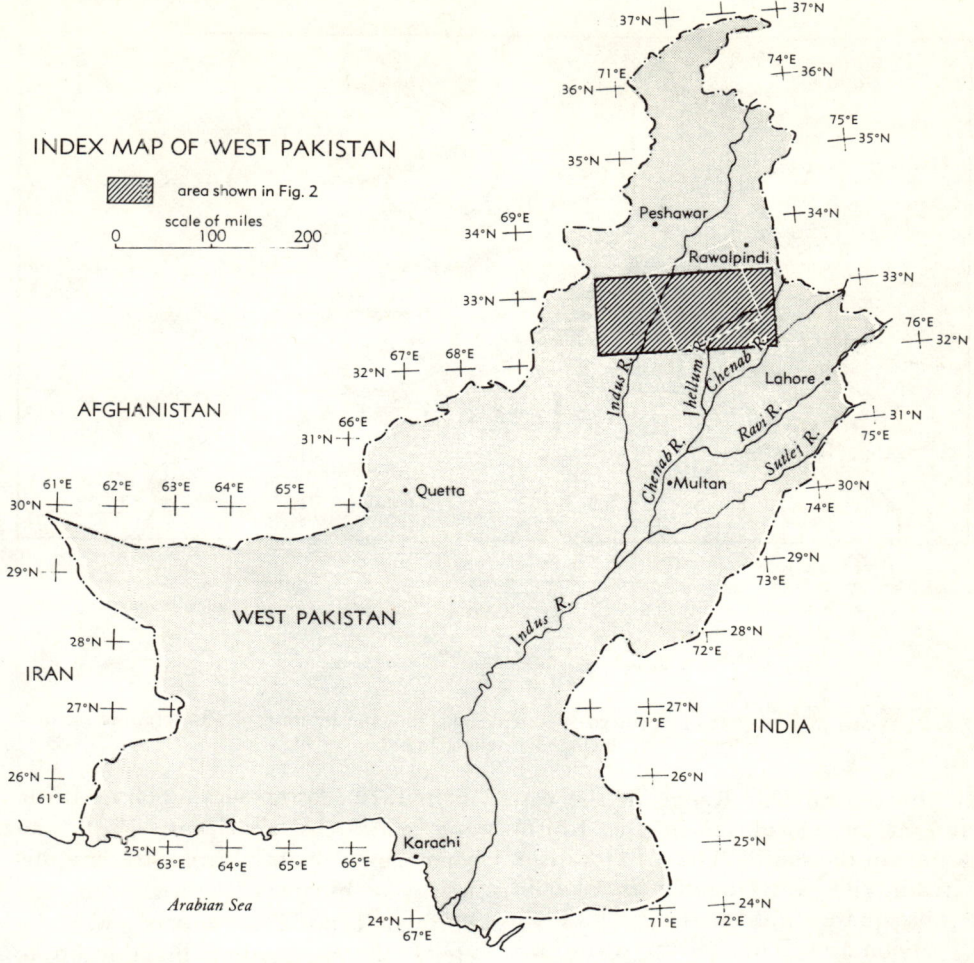

FIG. 1. Index map of West Pakistan. The white frame outlines the area shown in the Gemini photograph (frontispiece).

PRESENT INVESTIGATIONS

Kummel arrived in Pakistan in early October, 1961, and visited selected localities in the western Salt Range and some of the Trans-Indus ranges in the company of H. D. Dunkle. While Dunkle returned to his base in Quetta, Kummel was joined in the field by Teichert in early November, 1961, and during the winter field season of 1961-62 we had opportunity to carry out extensive field studies in the Salt Range and Surghar Range of West Pakistan (Fig. 1, 2). The objectives of this field program were twofold: Teichert made a study of the "Productus limestone," and Kummel of the "Ceratite beds"; the beds near the Permian-Triassic boundary were treated as a joint effort. Teichert did additional field work on the boundary beds in these ranges in November, 1962, and in March and December, 1963, and in the Khisor Range (Fig. 2) in February, 1964.

Detailed field studies and stratigraphic measurements were made at 11 localities between Landa Pusha (lat. 32° 58′ N., long. 71° 12′ E.) in the Surghar Range in the west, and Kathwai (lat. 32° 29′ N., long. 72° 12′ E.)

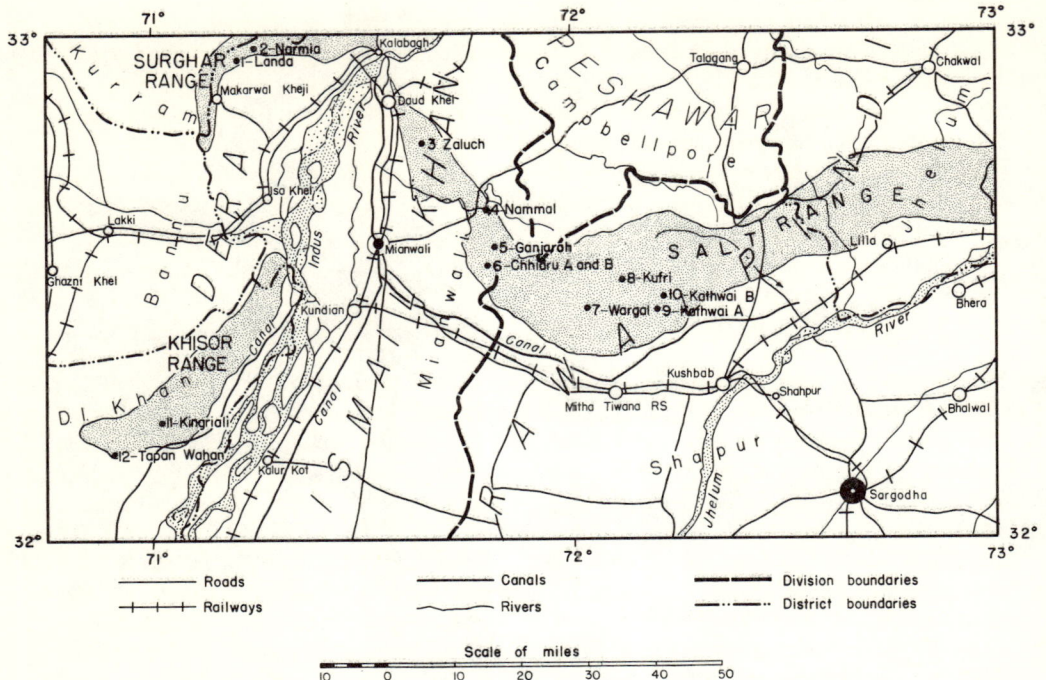

Fig. 2. Sketch map of part of northern West Pakistan, showing location of stratigraphic sections included in chart Fig. 4.

in the central Salt Range in the east and at two localities in the Khisor Range in the south. These 13 localities circumscribe a triangular area about 3,250 square miles in size (Fig. 1, 2).

Soon after Kummel's return to the United States in 1962 our collections were made available to specialists for detailed paleontological investigations. Additional collections were made by Teichert, especially from the Salt Range and Surghar Range in 1963 and from the Khisor Range in 1964. In view of the paucity of megafossils in beds near the Permian-Triassic boundary, news that the rocks yielded a large conodont fauna and rich and varied spore-pollen assemblages was most welcome. These fossil groups are described in two important papers by Sweet (1970) and by Balme (1970). Among other, rarer, groups the following yielded enough material to warrant descriptions in special papers: Lingulids (Rowell, 1970), ammonoids (Furnish and Glenister, 1970; Kummel, 1970), ostracodes (Sohn, 1970), and acritarchs (Sarjeant, 1970). In some papers material from stratigraphic levels at greater distances from the Permian-Triassic boundary was included in the studies, in addition to those from beds immediately above and below the boundary, because the information was essentially new and of considerable intrinsic value.

The bulk of the articulate brachiopods described by Grant (1970) come from only two localities: Narmia Nala in the Surghar Range and Tapan Wahan[1] in the Khisor Range. The Narmia fauna in the basal bed of the Kathwai Member of the Mianwali Formation was discovered by us in 1962. In 1963, Teichert sent a sample weighing about 300 lbs. from this bed to G. A. Cooper at the U.S. National

---

[1] *Wahan* is the Punjabi word, *nala* the Urdu word for a river or stream bed. *Pusha* is the equivalent word in Pashtoo. All three words occur in the names of localities studied by us.

Museum, who had examined our first small collections from it (Kummel and Teichert, 1966, p. 321-322). The work on the larger sample sent in 1963 was entrusted to R. E. Grant who went to Pakistan during the winter of 1963-64, made additional collections at the Narmia locality, and discovered in the Khisor Range a richly fossiliferous pocket in the generally very poorly fossiliferous white sandstone unit at the very top of the Permian sequence. They are the two faunas described in Grant's contribution, with only small additions made by us from other localities.

STRATIGRAPHY OF PERMIAN AND TRIASSIC ROCKS

Purpose of this chapter is to provide a general summary of stratigraphy and paleogeography of the area now covered by the Salt Range and Trans-Indus ranges (Table 1). Paleontologic aspects are mentioned to the extent that they shed light on problems of stratigraphic correlation. Additional information on the fossil content of some of these rocks, especially of beds close to the Permian-Triassic boundary, is given in other sections of this report.

PERMIAN

In the Salt Range, Khisor Range, and Marwat Range, and partly also in the Surghar Range, the Permian System is represented by a rock sequence, about 2,000 feet thick, which is readily subdivided into two roughly equal halves. The lower part is predominantly of continental origin. It was called Nilawan series by Gee (in Pascoe, 1959, p. 746) and referred to as Nilawan Group by Teichert (1966). The upper part of the Permian sequence is entirely marine; it was formerly referred to as the Productus limestone and is now known as Zaluch Group (Teichert, 1966). Each group is approximately 1,000 feet thick, although variations in thickness of the order of 300 to 400 feet occur.

Deposition of the oldest Permian rocks in the area was preceded by a very long interval of nondeposition that embraced the time from about the middle of the Cambrian until the end of the Carboniferous. The Permian rocks rest unconformably on Cambrian and, possibly, Precambrian strata which are only slightly deformed and locally faulted (Gee, 1938, 1945; Teichert, 1967).

Stratigraphic nomenclature for the lower part (Nilawan Group) of the Permian sequence in the area was formalized by the Stratigraphic Commission of Pakistan (H. Rahman, secretary) a few years ago, to comply with the Stratigraphic Code of Pakistan (Stratigraphic Nomenclature Committee of Pakistan, 1962). Some of the new names have recently been published by Husain (1967), and some are included in a "Pre-Publication issue" distributed by the Geological Survey of Pakistan (Rahman, 1968). None of these documents can be said to have established any of the new formational names formally, but they are used here in anticipation that action to introduce them formally will be taken in the near future.

The oldest Permian rocks are boulder beds and conglomerates which have been known, though little studied, since the latter part of the last century. Following Noetling (1901) they have been called "Talchir boulder beds" by many authors and were commonly referred to as "Talchir Stage" in Indian and English publications (e.g., Fox, 1931; Pascoe, 1959). Teichert (1967) pointed out that the occurrence of true Talchir beds closest to the Salt Range is in Jodhpur State, India, 400 miles to the southeast, and

Table 1. Permian and Triassic Formations, Salt Range, West Pakistan

System	Series	Stage	Group	Formation		Old informal names
TRIASSIC	Upper?			Kingriali Dolomite		Kingriali Dolomites (Gee, 1947)
TRIASSIC	Middle			Tredian Formation	Khatkiara Member	Kingriali Sandstones (Gee, 1947)
TRIASSIC	Middle			Tredian Formation	Landa Member	Kingriali Sandstones (Gee, 1947)
TRIASSIC	Lower	Scythian		Mianwali Formation	Narmia Member	Topmost Limestone Dolomite Beds Bivalve Beds (Waagen, 1895)
TRIASSIC	Lower	Scythian		Mianwali Formation	Mittiwali Member	Ceratite Beds (Waagen, 1895)
TRIASSIC	Lower	Scythian		Mianwali Formation	Kathwai Member	Upper Productus Limestone (Waagen, 1879)
PERMIAN	Upper	Chhidruan	ZALUCH GROUP	Chhidru Formation		
PERMIAN	Upper	Guadalupian	ZALUCH GROUP	Wargal Limestone	Kalabagh M.	Middle Productus Limestone (Waagen, 1879)
PERMIAN	Upper	Artinskian	ZALUCH GROUP	Amb Formation		Lower Productus Limestone (Waagen, 1879)
PERMIAN	Lower	Sakmarian	NILAWAN GROUP	Sardi Formation		Lavender Clay (Waagen, 1878)
PERMIAN	Lower	Sakmarian	NILAWAN GROUP	Warchha Sandstone		Speckled Sandstone (Waagen, 1878)
PERMIAN	Lower	Asselian	NILAWAN GROUP			Eurydesma-Conularia beds (Waagen, 1886)
PERMIAN	Lower	Asselian	NILAWAN GROUP	Tobra Formation		Talchir Boulder Beds (Noetling, 1901)

he regarded it as undesirable to carry a stratigraphic name over such long distances across areas where the formation does not exist. Therefore, he used the informal name Salt Range boulder beds. Following a written proposal by R. E. Gee this unit now is known as Tobra Formation, a name first used in print by Husain (1967).

For many decades a general consensus has held that the Tobra Formation was deposited under conditions in which ice-action played some part (see Schindewolf, 1964, 1967; Teichert, 1967, for earlier literature). As far as we know, Lotze (1966a, 1966b) is the only author who in recent years has questioned this idea and who interpreted the boulder beds in the eastern Salt Range as coarse conglomerates of fluviatile origin, noting occurrence in overlying beds of single boulders which could have been ice-rafted.

Teichert (1967), on the basis of

observations on the "boulder beds" in various parts of the Salt Range and Khisor Range, came to the conclusion that three facies are present:

(1) A tillite facies in the easternmost Salt Range, grading upward into marine sandstone containing *Conularia* and *Eurydesma* faunas.

(2) A fresh-water facies of alternating siltstone and shale, containing a spore flora, with only scattered boulders, or none at all, at their base. This facies seems to be characteristic of the central Salt Range.

(3) A mixed facies of diamictite, sandstone, and boulder beds, increasing westward to a thickness of several hundred feet, in the western Salt Range and in the Trans-Indus Khisor Range.

Unfortunately, it is not at all clear to which of these facies the name Tobra Formation should be applied. According to Rahman (1968) the formation is "exposed all along the Salt Range and also in the southern Khisor Range." Thus, it is presumably intended to include all three facies discussed by us and it is so used here.

Transitions between the three facies belts have not yet been studied and the exact location of their boundaries is not known. The ice that deposited the tillites in the eastern Salt Range was probably not part of an extensive inland ice sheet, but of a local glaciation. The ice did not extend into the area now occupied by the central and western Salt Range and by the Khisor Range. In the central Salt Range sediments were laid down in a lake, or lakes, that were only moderately influenced by glacial conditions, as shown by the scarcity of boulders in this area. In the western Salt Range, and probably also in the Khisor Range, there is evidence of periodic intensive ice-rafting of glacigene material alternating with deposits made by meltwater streams, possibly at least partly in a marine or estuarine environment.

In the western Salt Range the Tobra Formation has yielded a small spore assemblage indicating an Early Permian age (Balme in Teichert, 1967).

In the eastern Salt Range the Tobra Formation is overlain by sandstone containing a varied fauna of brachiopods, gastropods, and pelecypods—the so-called *Conularia-Eurydesma* beds (Reed, 1936). This fauna is believed to have close affinities to that of the "Lower Marine series" of New South Wales in Australia which was once regarded as Carboniferous in age, but now is known to be Permian (Teichert, 1944, 1954). Elements of this fauna can be traced westward to the vicinity of Kathwai (our localities 9 and 10), but it is absent farther west, where the marine beds probably grade into freshwater strata containing spores and a small bivalve fauna.

The *Conularia-Eurydesma* beds in the east and the boulder beds in the west are overlain by a sandstone formation, formerly known as "Speckled sandstone," a name derived from the mottled appearance of the sandstone in many places due either to iron staining or discoloration. However, since in many places the sandstone is not speckled at all, the name should be abandoned.

Noetling (1901) proposed the name *Warcha-Gruppe* to include both the Speckled sandstone and the overlying "Lavender series" (see below), but Husain (1967) restricted use of this name, as "Warcha Sandstone," to the Speckled sandstone and it is so used here, with corrected spelling Warchha Sandstone, however.

The Warchha Sandstone is mostly medium- to coarse-grained, conglomeratic in places, and has interbeds of shale. It varies in color from almost white to dark purple, always in reddish hue. Cross-bedding is very common. This and the variable thickness (80 to over 400 feet) suggest that the formation is a fluviatile deposit formed in large alluvial flats.

The Warchha Sandstone is also present in the southern Khisor Range, where it has a maximum thickness of 464 feet (Gee in Pascoe, 1959, p. 754).

The uppermost stratigraphic unit of the Nilawan Group is made up of soft fine-grained sandstone and shale. In the literature it is known as "Lavender series" or "Lavender clay," although its colors are mostly of reddish hue, with greenish-gray colors present in the upper part. It has now been named Sardi Formation by Husain (1967). Its thickness is rather uniformly 150 to 200 feet. It is devoid of megafossils and probably represents a deposit formed in a large lake.

The essentially continental period during which sediments of the Nilawan Group were laid down was terminated by a marine transgression which lasted from late Early Permian through all, or most, of Late Permian time. During this interval the rocks of the Zaluch Group (formerly Productus limestone) were formed (Teichert, 1966).

The Zaluch Group consists of three formations: Amb Formation, Wargal Limestone, and Chhidru Formation. The Amb Formation consists mostly of sandstone and shale. It is 150 to 250 feet thick, richly fossiliferous, containing abundant fusulines, mostly *Parafusulina kattaensis* (Schwager), which determines the age of the formation as Artinskian. Associated with it is a rich fauna in which brachiopods predominate, as they do in the other formations of the Zaluch Group. From various sources, mainly Waagen and Reed, Pascoe (1959, p. 758-761) has listed about 165 species from the Amb Formation, of which only about 28 are mollusks.

In the Khisor Range the lower part of the Amb Formation consists of dark shale containing thin bands of limestone that have yielded some bryozoans and brachiopods. Husain (1967) has named this unit Saiduwali Member of the Amb Formation. It is less than 100 feet thick and seems to be somewhat transitional between the Sardi Formation and the typical Amb Formation.

Although in the Salt Range the Amb Formation itself is lithologically somewhat transitional between the Sardi Formation and the Wargal Limestone, its contacts with both formations are sharp. In the upper part of the Amb Formation shale carrying *Glossopteris* and *Gangamopteris* occurs, possibly indicating temporary return of lacustrine conditions to the area. Many previous observers have included these plant beds in the base of the "Middle Productus limestone" (=Wargal Limestone of this report), but their noncarbonate lithology seems to tie them more closely to the lithology of the Amb Formation (Teichert, 1966).

The overlying Wargal Limestone is essentially a carbonate unit, with only inconsiderable amounts of sandstone and shale. Because of loss of part of the Permian column in the eastern Salt Range due to post-Cretaceous erosion, the Wargal Limestone is known only in the western Salt Range and some of the Trans-Indus ranges. In the western Salt Range it increases in thickness from 360 feet at Chhidru, to 480 feet in Nammal Gorgs (Gee in Pascoe, 1959, p. 757), and to 587 feet in Zaluch Nala (Teichert, 1966). The formation is predominantly limestone, with subordinate amounts of dolomite. It is very richly fossiliferous and yields an abundant brachiopod fauna, but sponges, corals, bryozoans, pelecypods, gastropods, cephalopods, ostracodes, trilobites, and fishes are also present. Pascoe (1959, p. 761-770) listed about 400 fossil species from this formation, roughly two-thirds of them being brachiopods. Not too much confidence should be placed in these figures because the fauna is badly in need of revision, and many fossil names, as pointed out by Branson (1965), are probably invalid, being "forgotten names" under the International Code of Zoological Nomenclature.

Ammonoids are restricted to the upper part of the formation where they are represented by the genus *Xenodiscus* (Schindewolf, 1954) and, probably by *Pseudogastrioceras* (Teichert, 1966). The upper 90 feet of the Wargal Limestone is a nodular limestone with shaly interbeds. This part of the formation has been named Kalabagh Member (Teichert, 1966, p. 12). It is somewhat transitional to the overlying Chhidru Formation, although the contact between the two is quite sharp.

The Chhidru Formation begins with a dark shale unit which is 20 to 40 feet thick and contains ostracodes and brachiopods. The rest of the formation is composed mainly of alternations of calcareous sandstone and arenaceous limestone. As measured by Teichert, the formation is thickest in the central Salt Range, where it is 278 feet thick near Kathwai and 270 feet at Chhidru; thickness decreases northwestward to 210 feet in Nammal Gorge and to 193 feet in Zaluch Nala.

The formation is highly fossiliferous and again brachiopods predominate. However, various species of *Bellerophon* and a large scaphopod, *Plagioglypta*, are extremely common, especially in the upper part of the formation. Pascoe (1959, p. 770-777) listed upward of 300 species from this formation, but the same reservations apply here as in the case of the Wargal Limestone fauna. The following ammonoid genera have been reported from the Chhidru Formation: *Stacheoceras, Cyclolobus, Eumedlicottia, Episageceras,* and *Xenodiscus,* but they are generally rare.

The uppermost bed of the Chhidru Formation is a fine-grained sandstone that originally formed a continuous cover in the area of the present Salt Range and Trans-Indus ranges, varying in thickness from almost nothing to about 15 feet. We have named this the "white sandstone unit" and it is described in greater detail below. In the Salt Range and Surghar Range the unit is poorly fossiliferous, but in the Khisor Range it contains highly fossiliferous pockets (Grant, 1970).

Previously unknown or neglected fossils of the Chhidru Formation, such as ostracodes, conodonts, acritarchs, spores, and pollen, have been studied by Sohn (1970), Sweet (1970), Sarjeant (1970), and Balme (1970) and they are discussed elsewhere in this report.

The age of the rocks here regarded as Permian has been the subject of some discussion. Disregarding earlier erroneous views, it was long customary to classify them as "Permo-Carboniferous" and they were treated in this manner by the Geological Survey of India. However, as mentioned above, the Tobra Formation in the western Salt Range contains a small palynological assemblage indicating an Early Permian age and the tillite in the eastern Salt Range is associated with marine beds of demonstrably Early Permian age. The base of the Nilawan Group is thus approximately correlative with the base of the Permian System.

The Amb Formation which overlies the Nilawan Group is Artinskian in age. The Nilawan Group, therefore, seems to represent the equivalent of the marine Asselian and Sakmarian stages.

The age of the Wargal Limestone is Guadalupian (Teichert, 1966; Furnish, 1966; Grant, 1968, 1970). The exact age of the Chhidru Formation is somewhat controversial at present. It is regarded as Guadalupian by Grant (1970), but as post-Guadalupian by Furnish (1966) and by Furnish and Glenister (1970). This question is discussed at greater length in another chapter (p. 71-76).

In summary, the paleogeography of the Permian Period in the area under consideration offers a picture of considerable diversity and contrast. At the beginning of Permian time conditions were affected by an ice cap, probably situated a short distance south or southeast of the present Salt Range, leading to deposition of tillites and aqueo-glacial and lacustrine sediments. When the ice disappeared the area changed into a large flood plain (Warchha Sandstone), but the source of the alluvial sediments is not known. Toward the end of Sakmarian time a lake covered the area in which up to 200 feet of fine-grained sediments were deposited (Sardi Formation).

During at least most of the remainder of Permian time the area was covered by sea. At first sedimentation was predominantly terrigenous detrital, but during Guadalupian time carbonate deposition prevailed. In post-Guadalupian time carbonate deposition continued, but at the same time large quantities of detrital quartz were brought into the area, perhaps again from a source situated to the south or southeast.

TRIASSIC

The Triassic formations of the Salt Range and Surghar Range have played a particularly important role in the development of our Triassic zonal scheme, especially for the Scythian Stage. They are represented by approximately 1,000 feet of strata in the Surghar Range and thin eastward in the Salt Range, where, in addition, the upper units are truncated by post-Cretaceous erosion. A comprehensive review, including many new data on these formations, has been published by Kummel (1966).

Until recently these formations had been classified according to a scheme proposed by Waagen (1895) as cited on p. 9.

A modern nomenclature for some of these Triassic formations was introduced by Kummel and Teichert (1966) and is shown here in Table 1. Their names were developed in cooperation with the Stratigraphic Commission of Pakistan in 1964 and were to be published formally by the Geological Survey of Pakistan. Kummel and Teichert (1966) and Kummel (1966) used them in anticipation of such publication, but unfortunately the final report of the Stratigraphic Commission has not yet appeared. Very brief, preliminary descriptions are included in the previously mentioned "Pre-Publication issue" (Rahman, 1968). More complete formal descriptions of some units of Triassic age are given in Appendix C.

The name Mianwali was first used by Gee (in Pascoe, 1959, p. 852) for a series unit to include all rocks in the Salt Range believed to be of Triassic age. It was restricted by Kummel (1966, p. 374) to a lithostratigraphic unit of formation rank, equivalent to only the lower part of Gee's Mianwali series. Mianwali is the name of an administrative district that includes the western Salt Range where the Permian and Triassic formations are well developed. The Mianwali Formation includes the fossiliferous units which have yielded all of the Triassic ammonites described from the Salt Range. It comprises all facies units (e.g., "Lower Ceratite limestone" through "Topmost limestone") recognized by Waagen (1895). The formation represents a great wedge of varied facies; it is thickest in the west and thins eastward (Fig. 3).

The greatest thickness of the formation is found in Narmia Nala (loc. 2) in the Surghar Range, where it is 635 feet thick. In the easternmost outcrop area in the Salt Range, at Kathwai (loc. 9, 10), the formation is only 48 feet thick but here the section is truncated by erosion and overlain by the Dhak Pass Formation of Paleocene age. Complete development of the Mianwali Formation is found in the Surghar Range and in the western part of the Salt Range in Zaluch Nala and Nammal Gorge. In Chhidru Nala (loc. 6B) the upper part of the Narmia

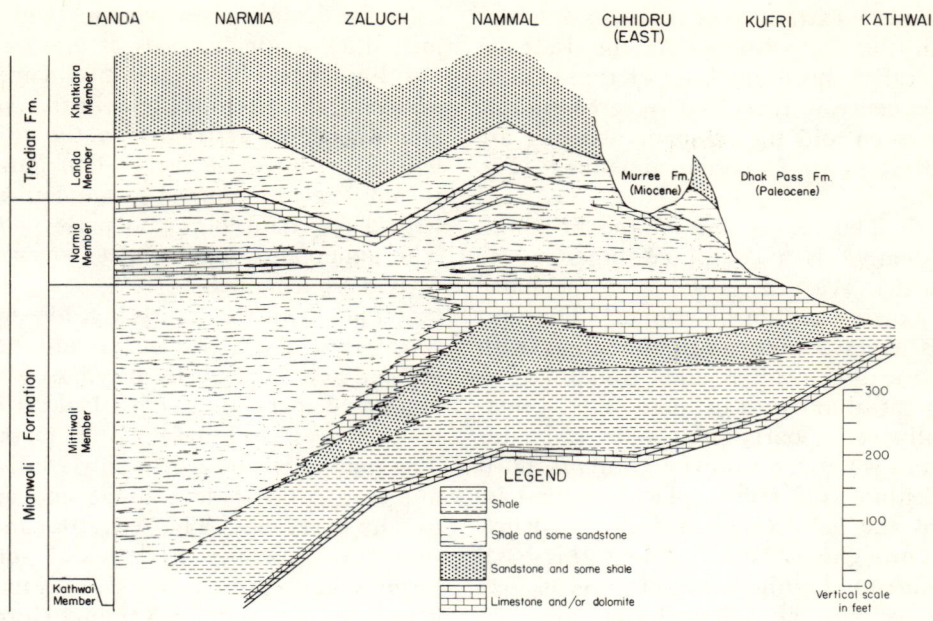

FIG. 3. Diagrammatic reconstruction of facies relationships of Lower and Middle Triassic strata in the Salt Range and Surghar Range (from Kummel, 1966).

Member is truncated and overlain by the Murree Formation of Miocene age, but one mile farther east, in Khan Zaman Nala, Teichert found the Mianwali Formation to be complete, though he was not able to determine its thickness. At Kufri (loc. 8), the Kathwai Member and only 10 feet of the Narmia Member are preserved. From Nammal Gorge westward to the Surghar Range, the Mianwali Formation is overlain conformably by the Landa Member of the Tredian Formation. The contact with the underlying Chhidru Formation is the central theme of this report and is discussed in detail in later chapters.

The lowest unit of the Mianwali Formation, the Kathwai Member, is a remarkably uniform dolomite and limestone bed present throughout the outcrop area of the Triassic formations in the Salt Range and in the Trans-Indus ranges. The type locality of this member is our locality 10, Kathwai B (Fig. 4, A10), where it is 5.5 feet thick. Megafossils are not uncommon in the Kathwai Member but at the same time are neither abundant nor well preserved. The most significant fossils are *Ophiceras connectens* Schindewolf and *Glyptophiceras himalayanum* (Griesbach) (Kummel, 1970). These two species clearly indicate that the Kathwai Member is of earliest Scythian (*Otoceras-Ophiceras* Zone) age. No additional comments are needed here, as this member is discussed in detail in a later chapter.

The Mittiwali Member (see Appendix C) comprises the units Waagen (1895) designated as "Lower Ceratite limestone," "Ceratite marl," "Ceratite sandstone," and "Upper Ceratite limestone." The type locality of this member is on the east side of Chhidru Nala (loc 6B). The site has long been known by the local population as Mittiwali—a place where clay is extracted. The member is 488 feet thick in Narmia Nala (loc. 2), 322 feet in Zaluch Nala (loc. 3), 263 feet at Chhidru B (loc. 6B), 253 feet at Kufri (loc. 8), and only 40 feet at Kathwai (loc. 9, 10) in the eastern extremity of Triassic outcrops in the Salt Range. At the Kathwai locality the upper part of the Triassic sequence is truncated by pre-Cenozoic erosion and the remnants existing are overlain by Paleocene limestone (Fig. 3).

The lowest unit of the Mittiwali Member is a coquinoid limestone to which Waagen (1895) gave the name "Lower Ceratite limestone." The unit is a gray, fine-grained limestone, with little glauconite, and containing a great abundance of ammonites, generally very poorly preserved. It is this unit which most previous authors, until Schindewolf (1954), placed at the base of the Salt Range Triassic sequence. Throughout the Salt Range this basal limestone unit is about 5 to 6 feet thick. In the Khisor Range (loc. 11, 12), west of the Indus River, Teichert measured 25 feet for the unit. It is distinguished from the underlying limestone unit of the Kathwai Member by absence of glauconite and much greater abundance of fossils, almost entirely ammonites, with locally a few pelecypods. The ammonites belong to the Gyronitan age of Spath (1934), the second zone in the sequence of Scythian ammonoid zones. Like the underlying Kathwai Member, the "Lower Ceratite limestone" is remarkably uniform throughout the Salt Range and the Trans-Indus ranges.

Above the "Lower Ceratite limestone" an interesting lithofacies diversity is found in the Mittiwali Member (Fig. 3). Waagen's sequence of "Ceratite marls," "Ceratite sandstone," and "Upper Ceratite limestone" is recognizable in a general way only in the central region of the Salt Range from Nammal Gorge eastward to Kufri. West of Nammal Gorge this portion of the Mittiwali Member is a fairly homogeneous sequence of shale and silty shale, with some thin sandstone and limestone beds.

The "Ceratite marls" are clay shales, greenish- to grayish-black in color, with fairly numerous, 1 to 6 inch thick lenticular beds of argillaceous limestone. The "Ceratite marls" are very fossiliferous, but fossils are almost entirely restricted to the thin lenticular limestone beds. The fauna is completely dominated by ammonites, with a few nautiloids and pelecypods. The ammonoids are being extensively revised by Kummel.

The "Ceratite marls" grade upward into the "Ceratite sandstone" beds. These sandy beds are massive to laminated in bedding, cross bedded in many places, and have ripple marks. The laminated units generally contain numerous thin shale laminae and are usually micaceous. The "Ceratite sandstone" is generally friable and soft, forming low, covered slopes. This unit is well exposed only in Nammal Gorge and along the east side of Chhidru Nala. At all other localities studied in this part of the Salt Range, it is generally covered. The "Ceratite sandstone" is characterized by two very conspicuous fossils: large specimens of *Flemingites*, up to 2 feet in diameter, and the bellerophontid *Stachella*. Waagen (1895) considered this bellerophontid to be confined to the mid-part of his "Ceratite sandstone" and on this basis divided the unit into three divisions and zones. There is one outcrop of the "Ceratite sandstone" in the upper part of Chhidru Nala where *Stachella* does occupy this position, but in another part of the Nala where the "Ceratite sandstone" crops out, *Stachella* is absent. *Stachella* was likewise not found in the "Ceratite sandstone" exposed in Nammal Gorge, nor at any other locality. It is highly dubious that this bellerophontid has any stratigraphic significance. As in the other members of the Mianwali Formation, ammonites dominate the fauna; pelecypods, gastropods, and brachiopods are present but not common.

The "Ceratite sandstone" grades upward into the "Upper Ceratite limestone." This unit is somewhat misnamed as limestone comprises only about 60 percent of it, the remainder consisting of shale and sandstone. Eastward at Chhidru and at Kufri, sandstone and shale comprise an even larger percentage of the unit, and much of the limestone is conspicuously sandy. The limestone is gray, thin- to medium-bedded, fine-grained, and contains abundant fragmental shell remains. The limestone beds are generally quite fossiliferous but unfortunately the preservation is, with few exceptions, very poor. Like the lower units of the Mianwali Formation, the ammonites completely dominate the fossil faunas. In addition, a few pelecypods are seen here and there, but little else. The ammonite fauna includes *Anasibirites, Prionites, Hemiprionites,* etc., indicating the middle zone of the Scythian Stage.

The Narmia Member (see Appendix C) comprises the strata Waagen (1895) called "Bivalve beds," "Dolomite beds," and "Topmost limestone." It has been the least understood part of the "Ceratite Formation," owing mainly to inadequate treatment of these units by Wynne (1878) and Waagen (1889, 1895). The type section of the Narmia Member is in Narmia Nala, our locality 2 (Fig. A1, A2).

Throughout the Salt Range and Trans-Indus ranges the basal bed of the Narmia Member is a thin, approximately 10 feet thick, complex limestone. At Chhidru (loc. 6B) and Kufri (loc. 8) this basal limestone unit is the "Bivalve limestone" of Waagen. It is a hard, light gray, massive limestone, mainly a coquina of pelecypods. At Chhidru B this limestone bed also contains poorly preserved ammonites and nautiloids (*Enoploceras* and orthocerids), in addition to the pelecypods. In Nammal Gorge (loc. 4) it is 7 feet thick and consists of alternating coquinoid beds, like those that make up the whole unit at Chhidru and Kufri, and irregularly bedded limestone containing poorly preserved brachiopods, ammonites, pelecypods, and gastropods. In Zaluch Nala (loc. 3) the unit is similar to outcrops at Narmia, but contains some glauconite. In the Surghar Range, at Narmia Spring and Landa Pusha (loc. 1, 2), this basal limestone unit no longer is a pelecypod coquina but a dark gray to brown, fragmental limestone, sandy in part, containing poorly preserved pelecypods, ammonites, and brachiopods.

Much of the remainder of the Narmia Member consists of olive to grayish-black shales, but in some places thin laminae and beds of fine sandstone occur interbedded with 2 to 10 foot beds of limestone. The limestone beds are highly variable in their lithology. Many of them are dolomitic, glauconitic, pelletal, sandy, and some have thin shale interbeds. The uppermost unit of the Narmia Member in the Surghar Range is a pisolite bed containing *Spiriferina* and other brachiopods, and echinoid spines. The pisolite bed is 7 feet thick in Narmia Nala and 4.5 feet thick in Landa Pusha. In the Salt Range there are much fewer limestone beds in the Narmia Member, and the few hard carbonate beds which do exist are either dolomitic limestone or dolomite. In Zaluch Nala, for instance, the uppermost beds of this member are gray to brown massive dolomite. This may also be observed in Nammal Gorge. In the Salt Range, in Nammal Gorge and Zaluch Nala, the noncarbonate portion of the Narmia Member contains a much higher percentage of sand relative to shale than in the Surghar Range. At Chhidru the Narmia Member, aside from the basal "Bivalve limestone" and an upper dolomite bed, consists of tan sandstone that is micaceous and shaly in part.

The Narmia Member is 128 feet thick in Landa Pusha, 140 feet in Narmia Nala, 75 feet in Zaluch Nala, 189 feet in Nammal Gorge, 118.5 feet in Chhidru Nala, and only 10 feet at Kufri, where most of the member is eroded away. In Chhidru Nala the member is truncated by erosion and overlain by the Miocene Murree Formation. In Khan Zaman Nala, about one mile east of the Chhidru Nala sections, Teichert observed that the Narmia Member is again fully developed, though its thickness was not measured. At Kufri the Narmia Member is overlain by Paleocene limestone. The thinness of the member in Zaluch Nala is believed to be due to minor faulting within the section and has no regional significance.

The limestone beds of the Narmia Member are fossiliferous but, unfortunately, fossils are neither abundant nor well preserved. The fauna includes ammonites, nautiloids, brachiopods, echinoid spines, crinoid remains, and conodonts. The most important of these fossils, at least for dating the member, are are ammonites. While Waagen (1895) had described four species of ammonites from this member, Kummel (1966) was able to increase their number of fifteen. This fauna clearly indicates a late Scythian age (*Prohungarites* Zone) for the bulk of the Narmia Member.

Immediately overlying the Mianwali Formation is a series of strata about which there has been some confusion. Gee (1945) was the first to recognize them and proposed the name Kingriali sandstones for them. Subsequent work revealed that two distinct mappable units can be recognized within the Kingriali sandstones: a lower predominantly shale unit and an upper predominantly sandstone unit. At the time he introduced the name Kingriali sandstones, Gee (1945) also named the overlying strata the Kingriali dolomites. Both are presumably of Triassic age, but to avoid a double usage of the name Kingriali, the name Tredian Formation (Gee in Kummel, 1966) was introduced for the "Kingriali sandstones" (see Table 1). For the lower shaly member of the Tredian Formation, Kummel (1966) proposed the name Landa Member, and for the upper sandy member of the formation, Danilchik and Shah (in press) proposed the name Khatkiara Member (see Kummel, 1966). The Tredian Formation and its members are very briefly described by Rahman (1968, p. 9).

The type section of the Landa Member (see Appendix C) is in Landa Pusha, our locality 1 (Fig. A1), in the Surghar Range, where it is 100 feet thick. The member consists of sandstone and shale, in about equal proportions. The sandstone may be black, pink, or red, and is micaceous, thin to massively bedded, with ripple marks and slump structures. The shale is generally dark, sandy, and micaceous. The member is 72 feet thick in Zaluch Nala (loc. 3), 63 feet in Nammal Gorge (loc. 4). The member is not present in Chhidru Nala, but Teichert found it to be fully developed one mile to the east in Khan Zaman Nala. East of Khan Zaman Nala the member is progressively beveled and it is not present at Kufri and Kathwai.

The only megafossils observed in the Landa Member are poorly preserved and fragmentary plant remains. The samples of fossil plants and shale reported on by Sitholey (1943), Pant (1949), and Pant and Srivastava (1964), from south of Sakesar Ridge in the Salt Range, most probably came from the Landa Member. Kummel (1966) suggested a possible Middle Triassic age for the Tredian Member, and this age assignment appears to receive support from a study of the spore and pollen assemblages (Balme, 1970). Balme reports that the spore-pollen assemblages of the Landa and Khatkiara are essentially similar, but the Landa Member, in addition, contains acritarchs.

The Khatkiara Member was named from Khatkiara Nala in the Surghar Range, where it is 112 feet thick. The member consists mostly of sandstone which is light yellowish-gray to white and purplish, soft, generally thin-bedded, but in places massive and cross-bedded, fine- to very fine-grained, and composed of subangular quartz grains. In places claystone and siltstone layers are intercalated. In Nammal Gorge the member is 140 feet thick. No megafossils have been found in the Khatkiara Member, but Balme (1970) reported from it a rich spore-pollen assemblage which confirms Kummel's (1966) suggestion of a Middle Triassic age.

To round off the discussion of the Triassic, brief mention should be made of the Kingriali Dolomite, first named Kingriali dolomites by Gee (1945). This formation consists of yellowish and gray fine-grained, thick-bedded dolomite and dolomitic limestone and is as much as 400 feet thick, though mostly thinner. Some of the rocks have a brecciated appearance and show honeycomb weathering. The formation contains only indeterminable remains of pelecypods and cephalopods and was assigned a Triassic age by Gee (1945, p. 277) "on stratigraphical grounds," presumably because it is overlain by the Datta Formation ("variegated stage" of Gee) of demonstrably Jurassic age. The Kingriali Dolomite is not considered further in this report.

In a general way, conditions of sedimentation in this area did not change much during the remainder of the Mesozoic and Early Tertiary. Jurassic and Cretaceous are represented by about 1,000 feet of partly continental, partly marine rocks, and Paleocene and Eocene by about 2,000 feet of predominantly marine deposits. After the Cretaceous the entire, conformable, Permian to Cretaceous sequence was gently tilted westward and eroded in such a way that, proceeding in the Salt Range from west to east, progressively older formations are bevelled off and in the eastern Salt Range Paleocene rocks rest on older Permian.

SUMMARY

During the time span represented by the Permian and Triassic periods the area of the Salt Range and Trans-Indus ranges went through one complete environmental cycle—from continental to marine and back to continental conditions. The first continental phase occupied the early part of the Early Permian and may have lasted about 15 million years. The following marine phase began in the Artinskian and extended through at least some of Middle Triassic time for a total of about 60 million years, equal to the duration of the Cenozoic Era. This marine phase was interrupted by an emergence at the end of Permian time which, however, is very poorly and incompletely documented in the rock record (see p. 77). The final continental phase comprises the Late Triassic, perhaps including part of the Middle Triassic, and extends into Jurassic times (continental Datta Formation of Danilchik and Shah, in press). This second continental phase includes the time of deposition of the Kingriali Dolomite which was formed in a greatly restricted sea.

DETAILED STRATIGRAPHY OF BEDS NEAR PERMIAN-TRIASSIC BOUNDARY

GENERAL REMARKS

The main body of our field observations is presented in the stratigraphic correlation chart (Fig. 4) in which the 13 stratigraphic sections measured by us are shown. This chart also contains data on distribution and occurrences of 24 selected fossils and

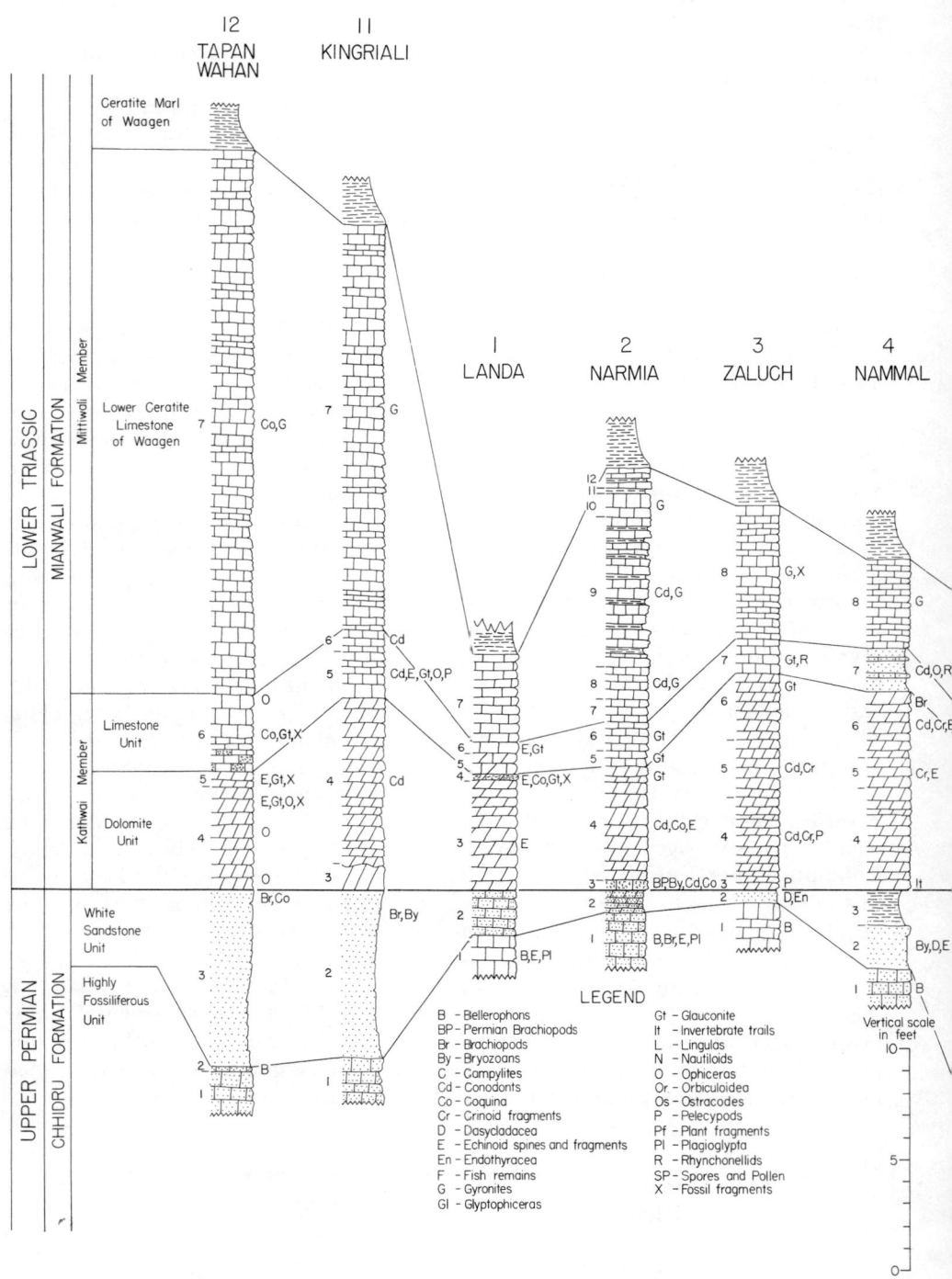

Fig. 4. Stratigraphic sections of latest Permian and earliest Triassic strata in the Salt Range and Trans-Indus ranges. (The section descriptions are found in Appendix A.)

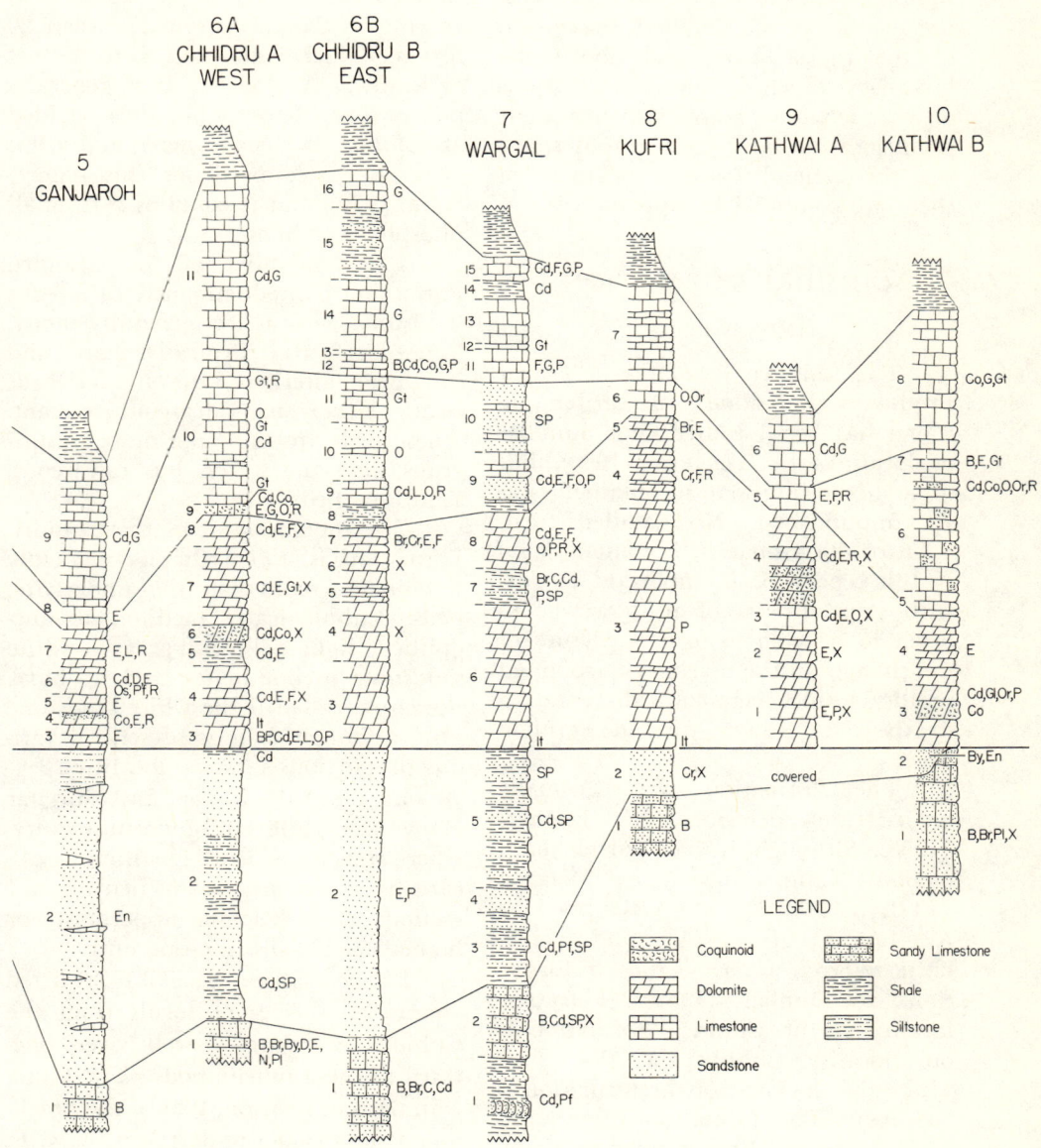

Fig. 4 (CONTINUED)

fossil groups, as well as of glauconite. Considering the general scarcity and poor state of preservation of fossils in the white sandstone unit of the Chhidru Formation and in the dolomite unit of the Kathwai Member, the fossil distribution data may be expected to be incomplete. The chart should be consulted in connection with the reading of the detailed description of the rocks below and above the boundary which follows. Descriptions of the sections, maps showing their exact location, and photographs showing the general features of most of them are compiled in Appendix A.

CHHIDRU FORMATION

Introduction

The Chhidru Formation ("Upper Productus limestone" of earlier authors) has been studied in a number of localities by Teichert (1966), but some of the information obtained is as yet unpublished. No detailed treatment of the formation is contemplated in this report. It is important to deal with some aspects of it, however, in order to be able to form a picture of the changes in the marine environment in the closing stages of sedimentation in the Salt Range area in Permian time.

The formation was formally named and described by Teichert (1966), although it had earlier been named "Chidru beds" by Waagen (1891, p. 243), "Chideru Gruppe" by Noetling (1901, p. 438), and it had been referred to as "Chideru formation" by Dunbar (1962). The type locality of the Chhidru Formation is our locality Chhidru A (loc. 6A), where the thickness of the formation is 268 feet. The formation thins westward, for it is 210 feet thick in Nammal Gorge (loc. 4) and 193 feet in Zaluch Nala (loc. 3). Toward the east the formation thickens only slightly; it is 278 feet thick at Kathwai A (loc. 9). These figures are based on measurements by Teichert. It should be noted that they differ, in some cases significantly, from thicknesses given for the "Upper Productus limestone" by Pascoe (1959, p. 757).

The base of the Chhidru Formation is a shale unit, pale yellowish-gray to medium dark gray in color, and varying in thickness from less than 20 feet in Zaluch Nala (loc. 3) to 45 feet at Kathwai A (loc. 9). It is generally poorly fossiliferous, but has yielded *Mesolobus, Waagenoconcha,* and a few ostracodes (see p. 66 of this report; Sohn, 1970), and it contains rare small phosphate nodules.

The remainder of the Chhidru Formation is predominantly calcareous sandstone with a few, generally sandy, limestone beds. Typically, hard and soft beds alternate, the harder beds being richer in calcareous cement. These beds are as a rule more fossiliferous than the softer, less calcareous, sandstone beds.

The topmost unit of the Chhidru Formation is a bed that is markedly lighter in color than the underlying beds. Lithologically, medium- to fine-grained, light yellowish-gray to white sandstone predominates, and we have, therefore, called it the white sandstone unit, although shale is present in various proportions in some localities.

In the Salt Range and Surghar Range the white sandstone unit everywhere rests on a richly fossiliferous calcareous sandstone bed which in our sections is called the uppermost, or highest, richly fossiliferous unit.

Pascoe (1959, p. 770-777) listed about 330 species of fossils from the Chhidru Formation, well over one-third of them brachiopods. The fauna is in need of revision, that of the brachiopods now being undertaken by R. E. Grant. Relevant aspects of the paleontology of the Chhidru Formation are discussed elsewhere (p. 62-71 of this report).

The age of the Chhidru Formation is Late Permian. However, diverse opinions have been voiced concerning its precise placement in the Upper Permian. These also are discussed elsewhere (p. 71-73).

Uppermost Richly Fossiliferous Unit

In the Salt Range and Surghar Range, this unit is generally a light gray, hard, sandy limestone, or calcareous sandstone, containing varying amounts of quartz and abundant brachiopods, bellerophontids, and *Plagioglypta*, a scaphopod. It is generally 1 to 2 feet thick and is the highest of many similar beds that are present throughout the Chhidru Formation, where they alternate with more sandy, softer, and less fossiliferous layers.

In eight thin sections studied from this unit at different localities the amount of terrigenous material varied widely between less than one percent and about 75 percent, consisting mostly of quartz, and minor amounts of fresh feldspar, muscovite, biotite, and iron oxide. Quartz grains vary widely both in angularity and sphericity, but tend to be subangular. Most of them are in the fine-grained size range. The groundmass consists of very finely crystalline calcite, containing fossils, pellets, and patches of medium to, rarely, coarsely crystalline calcite (Pl. 1, fig. 1, 2).

From this bed in the Chhidru section (our loc. 6A) Schindewolf (1954, p. 156, "Schicht 12") identified a rich fauna which has been quoted in the section "History of Research" (p. 13). We were not able to recognize all these forms, because the rock is very hard and fossils do not weather out easily. However, the cited assemblage is probably fairly typical of the bed over a wider area.

In the Salt Range, the same bed has yielded the following conodonts (Sweet, 1970): *Anchignathodus typicalis, Ellisonia teicherti, E. triassica, Neogondolella carinata*. According to Sweet, all four species are also found in the overlying white sandstone unit and range into the basal part of the Kathwai Member of the Mianwali Formation. Other microfossils, seen in thin sections, include calcispheres and endothyraceans *(Climaccamina?, Colaniella?)*.

In the Khisor Range the uppermost highly fossiliferous unit is absent at Kingriali Mountain (loc. 11) where the white sandstone unit rests on unfossiliferous, yellowish-gray, friable, calcareous sandstone such as is prevalent in the Chhidru Formation. In Tapan Wahan (loc. 12), in the southern Khisor Range, a thin layer with specimens of *Bellerophon* occurs at the top of this yellowish-gray sandstone.

White Sandstone Unit

This unit is predominantly fine- to medium-grained sandstone, white to yellowish-gray, soft and easily eroded, generally thinly laminated and, in places, faintly cross-bedded. It has varying amounts of interbeds of dark shale. At Chhidru A (loc. 6A) the sand/shale ratio is 3:2, at Wargal (loc. 7) probably almost 1:1, but elsewhere it is much higher and in some localities shale is present only as a thin band at the top of the unit. Generally, this top band is not more than 2 inches thick, but in Nammal Gorge it is about 6 inches.

Symmetrical ripple marks were seen in one place, Nammal Gorge (loc. 4), on a bedding plane 4 feet above the base of the unit. The ripples are oriented approximately E-W, have an amplitude of about 6 cm, and a height of about 1.2 cm.

Hardness of the rock increases with calcium carbonate content. Thus, in the Chhidru A section (loc. 6A) a

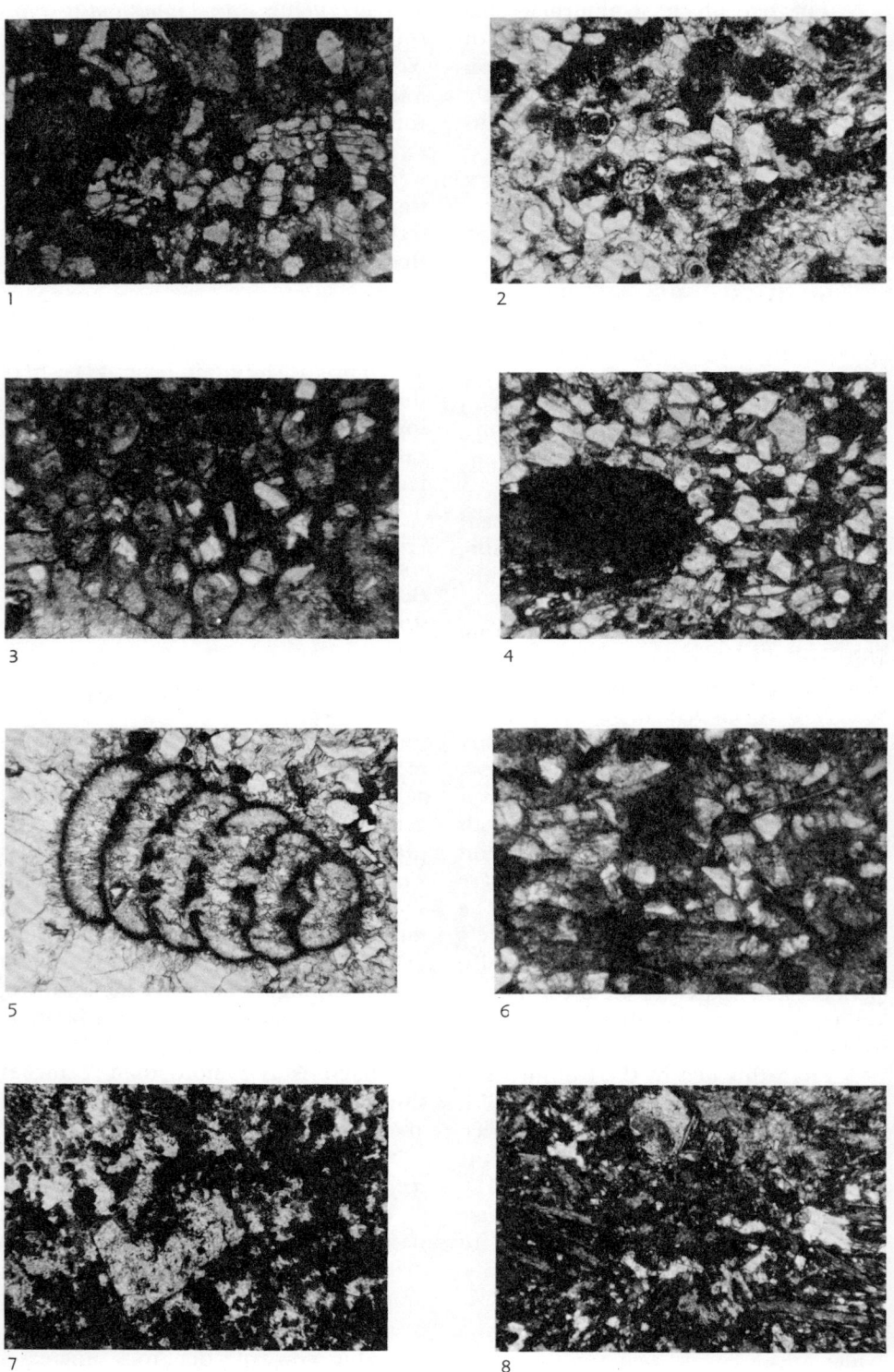

PLATE 1 (EXPLANATION ON FACING PAGE)

hard calcareous band in the middle of the unit caused Schindewolf (1954) to subdivide the unit into two units—his Schichten 13 and 14.

In the softer, more sandy parts terrigenous components may make up as much as 85 percent of the rock (Pl. 1, fig. 4-6). They are predominantly quartz and feldspar, usually present in ratios between 4:1 and 5:1. The feldspars are in roughly equal amounts microcline and plagioclase. There is present usually a small amount (1-2 percent) of biotite, some muscovite, and small amounts of iron oxide. The quartz is in the fine- to medium-grained range, and the grains are mostly angular to subangular. Most grain contacts are tangential and there is little evidence of sutured contacts. Also, little interaction between quartz grains and calcitic groundmass, and little encroachment of calcite on quartz has been observed.

The groundmass consists generally of finely crystalline calcite, which in places is recrystallized into medium, and even coarsely, crystalline patterns. More rarely, it is partly dolomitized. In patches the groundmass makes up as much as 70 percent of the rock, which is then more properly described as sandy limestone. In such rocks the limestone may be pelletal and contain fossil fragments.

The white sandstone unit is thickest, up to 17 feet, in the central part of the western Salt Range, but even here considerable variations in thickness over short distances were observed. In general, the unit thins eastward and westward from this central area. It is only 1 foot thick at Narmia in the Surghar Range, and is absent on the south side of a small creek about one-half mile south of Kathwai, close to our locality 10. In the Khisor Range it is 7.5 feet thick at Kingriali Mountain and 8 feet at Tapan Wahan.

Presence of oscillation ripple marks in the white sandstone unit at Nammal Gorge probably suggests considerable shallowness of the water in this vicinity. Rarity of this kind of ripple marks, however, indicates that the sediment/water interface for the most part must have been below the zone of wave turbulence.

Plate 1

All photomicrographs taken under parallel nichols, ×25; unless otherwise indicated, oriented with top side pointing up stratigraphically.

Figures
1. Calcareous sandstone containing small *Bellerophon*. Uppermost highly fossiliferous unit of Chhidru Formation, Nammal Gorge, Salt Range. Loc. 4, bed 1; field no. T61-55.
2. Calcareous sandstone containing dasycladacean algae and unidentified foraminifers. Uppermost highly fossiliferous unit of Chhidru Formation, Zaluch Nala, Salt Range. Loc. 3, bed 1; field no. K13-A; unoriented.
3. Sandy limestone containing fragment of a bryozoan. White sandstone unit of Chhidru Formation, Kingriali, Khisor Range. Loc. 11, 2-4 inches below top of bed 2; field no. T64-7.
4. Calcareous sandstone containing echinoderm fragment. White sandstone unit of Chhidru Formation, west of Chhidru Nala, Salt Range. Loc. 6A, lower part of bed 2; field no. T61-34.
5. *Climaccamina* sp. in calcareous sandstone. White sandstone unit of Chhidru Formation, Nammal Gorge, Salt Range. Loc. 4, probably bed 2; field no. K13-A; unoriented.
6. Calcareous sandstone containing shell and echinoderm fragments. White sandstone unit of Chhidru Formation, Kingriali, Khisor Range. Loc. 11, bed 2, 4 feet below top of bed; field no. T64-4; top is right.
7. Calcitic dolomite, intimate mixture of clear, crystalline calcite (stained dark) and brownish dolomite with in places lattice-like intertwining of both minerals. Dolomite unit of Kathwai Member, Kathwai A, Salt Range. Loc. 9, bed 4; field no. T61-164.
8. Calcitic dolomite containing shell fragments; calcite (stained dark) occurs in irregular patches; above center dolomite rhombohedron enclosing only partly dolomitized echinoderm fragment. Dolomite unit of Kathwai Member west of Chhidru Nala, Salt Range. Loc. 6A, bed 6; field no. T61-37.

Fig. 5. Contact between white sandstone unit and Kathwai Member, below Kingriali, Khisor Range. The hammer head indicates the contact; the uppermost 2 feet of the white sandstone unit are fossiliferous. (Teichert photograph.)

In the Salt Range and Surghar Range the unit is very poorly fossiliferous. Poorly preserved and indeterminate brachiopod, pelecypod, and crinoid fragments were observed in the field. Some samples have been shown to contain up to 10 percent fragments of Foraminifera, bryozoan fragments, bivalves, and echinoderm plates (Pl. 1, fig. 3, 4). Endothyraceans were identified in thin sections from Kathwai (loc. 10), Chhidru (loc. 6A), and

Zaluch Nala (loc. 3) in the Salt Range, and from Narmia (loc. 2) in the Surghar Range. They represent *Climacammina* sp. (Pl. 1, fig. 5) and *Brunsiina?* sp. The dasycladacean alga *Vermiporella* was identified in a thin section from Chhidru (loc. 6A).

The white sandstone unit is more fossiliferous in the Khisor Range. At Kingriali (loc. 11) lenses of fossiliferous hard, calcareous sandstone occur near the top of the unit (Fig. 5). On top of the ridge about one-half mile south of the measured section in Tapan Wahan (loc. 12) R. E. Grant found a richly fossiliferous lens at the very top of the unit which yielded a fauna of about 22 species of brachiopods (Grant, 1970). The fauna from these two localities has the following composition:

> *Aulosteges* sp.
> *Callispirina* sp.
> *Chonetella* sp.
> Chonetid indet.
> *Cleiothyridina* cf.
>   *C. capillata* (Waagen)
> *Derbyia* cf. *D. plicatella* Waagen
> Dielasmatids indet.
> *Enteletes* sp. 1
> *Hemiptychina* sp.
> *Hustedia* sp.
> *Kiangsiella* sp.
> *Linoproductus* sp.
> *Lyttonia* sp.
> *Martinia?* sp.
> *Neospirifer* sp.
> *Orthotichia* sp. 1
> *Orthotichia* sp. 2
> *Richthofenia* sp.
> *Spiriferella?* sp.
> *Spirigerella* sp.
> *Waagenoconcha?* sp.
> *Whitspakia* sp. 1

Grant concludes that this is an assemblage that is typical for the Chhidru Formation as a whole, which he believes to be Guadalupian in age. Conflicting opinions regarding the age of the Chhidru Formation are being evaluated elsewhere in this report (p. 70-75).

In addition to these fossils the white sandstone unit has yielded the following conodonts (Sweet, 1970):

> *Anchignathodus typicalis* Sweet
> *Ellisonia teicherti* Sweet
> *E. triassica* Müller
> *E.* sp.
> *Neogondollella carinata* (Clark)

All five species are also found in the basal part of the overlying Kathwai Member.

The acritarchs of the white sandstone unit are as yet incompletely known, only one assemblage from one locality having been described so far (Sarjeant, 1970). This consists of the following species, represented by 372 specimens:

> *Micrhystridium inconspicuum* (Defl.) Defl.
> *M. breve* Jansonius
> *M. karamurzae* Sarjeant
> *M. densispinum* Valensi
> *M. setasessitante* Jansonius
> *M. pakistanense* Sarjeant
> *M. circulum* Schön
> *M.* aff. *keratoides* Spode
> *M. microspinosum* Schaarschmidt
> *Veryhachium valensii* (Valensi), Downie and Sarjeant emend.
> *V. irregulare* forma *subtetraedron* Jekhowsky
> *V.? riburgense* Brosius and Bitterli
> *Wilsonastrum colonicum* Jansonius
> *Polyedryxium* sp.
> *Leiofusa stassfurtensis* Schön
> *Deunffia unispinosa* (Schön) comb. nov.

In addition Sarjeant (1970) has described *Tasmanites* sp. from the same locality.

MIANWALI FORMATION

KATHWAI MEMBER

Introduction

The name Kathwai Member was first used by us in 1966 (Kummel and Teichert, 1966; Kummel, 1966), though it was not formally proposed and defined in these publications. As type locality of the member we herewith designate our locality 10, Kathwai B, situated on the north side of a small creek, 200 yards east of the Chakwal-Khushab road, one-quarter mile southeast of Kathwai, central Salt Range. Here the Kathwai Member is 12.5 feet thick. A description of the type section is given in this report, Appendix A.

The Kathwai Member is subdivided into a lower dolomite unit and an upper limestone unit. These two divisions have been recognized by us in every locality studied. Their thicknesses, as well as thickness of the total Kathwai Member, are given in Table 2. The thickness of the member varies between 5.5 feet at Ganjaroh (loc. 5) and 15.8 feet at Chhidru A (loc. 6A). This is remarkable because the two localities are only 2.75 miles apart and elsewhere the member varies less in thickness over much longer distances.

The variations in thickness of the constituent units are interesting, that of the dolomite unit varying from 5 feet at Ganjaroh (loc. 5) to 14 feet at Kufri (loc. 8), that of the limestone unit from 0.5 feet at Ganjaroh (loc. 5) to 7 feet at Kathwai B (loc. 10).

A more meaningful insight into the mutual relationships of the two units of the Kaithwai Member is gained, if their relative proportions are considered as expressed in percentage thickness of total thickness of the member (Table 2). It will be seen that the dolomite unit makes up over 90 percent of the member at Ganjaroh where the member is only 5.5 feet thick, and at Kathwai A where its thickness is 11 feet. In the geographically close locality Kathwai B, however, the dolomite unit makes up less than half of the Kathwai Member.

Thus, it is not only the total thickness of the Kathwai Member which is subject to rapid variations, but also the relative proportions in which the dolomite unit and the limestone unit contribute to the lithologic makeup of the member. The significance of these facts is discussed elsewhere.

TABLE 2. Thickness (in feet) of Kathwai Member and Its Units, Followed by Percentage of Total Thickness for Each Unit

Localities	Limestone unit	Dolomite unit	Kathwai Member
1. Landa Pusha	1.5 (22%)	5.25 (78%)	6.75
2. Narmia Spring	2.0 (26%)	5.7 (74%)	7.7
3. Zaluch Nala	1.6 (15%)	9.8 (85%)	11.4
4. Nammal Gorge	2.0 (18%)	9.0 (82%)	11.0
5. Ganjaroh	0.5 (9%)	5.0 (91%)	5.5
6A. Chhidru A	6.25 (46%)	9.6 (54%)	15.8
6B. Chhidru B	6.5 (48%)	9.2 (52%)	15.7
7. Wargal	5.5 (36%)	9.9 (64%)	15.4
8. Kufri	1.5 (10%)	14.0 (90%)	15.5
9. Kathwai A	1.0 (9%)	10.0 (91%)	11.0
10. Kathwai B	7.0 (56%)	5.5 (44%)	12.5
11. Kingriali	2.8 (25%)	8.2 (75%)	11.0
12. Tapan Wahan	3.5 (40%)	5.4 (60%)	8.9

Fig. 6. Laminated, cross-bedded dolomite in top part of dolomite unit of Kathwai Member, west side of Chhidru Nala, Salt Range (loc. 6A). Length of knife 3 inches. (Teichert photograph.)

Fig. 7. Lower part of Dolomite unit of Kathwai Member, below Kingriali, Khisor Range (loc. 11), showing cross-bedding. (Teichert photograph; from Kummel and Teichert, 1966.)

Fig. 8. Lowermost 6-inch bed of Kathwai Member, west of Chhidru, Salt Range (loc. 6A, bed 3), showing cross-bedding. This is overlain by dolomite unit bed 4 with trails in positive hyporelief on bottom surface. Position of this surface in section is indicated in Figure 9. Length of knife 3 inches. (Teichert photograph.)

Dolomite Unit

Gross Lithology and Bedding Plane Features

The dolomite unit of the Kathwai Member differs sharply lithologically from the white sandstone unit of the Chhidru Formation on which it rests with apparent conformity. A striking feature is the color of its weathered surface which has been described as "rusty" or "rusty brown" by most au-

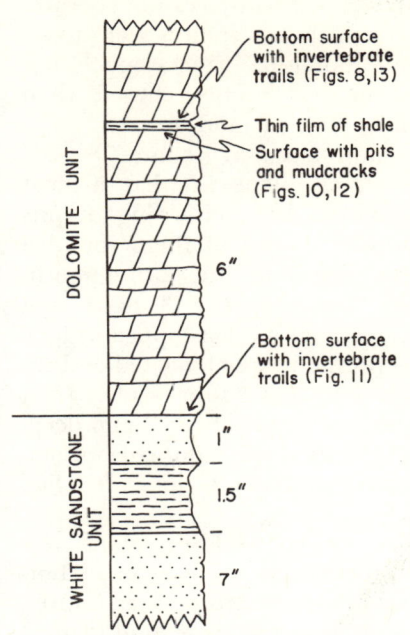

Fig. 9. Top of white sandstone unit and basal dolomite unit to indicate position of various bedding plane features shown on Figures 8, 10, 11, 12, 13. Scale 1:2. West side of Chhidru Nala, Salt Range. (Teichert photograph.)

thors. In terms of the rock-color chart of the National Research Council (Goddard et al., 1948) the color is almost invariably grayish-orange pink, modified in places to grayish-orange or to very pale orange where less weathered. The color of the fresh rock is almost invariably grayish-orange.

The unit is generally of massive appearance, although it can be characterized as being medium-bedded. In places it is laminated and thin-platy (Fig. 6). In other places it is cross-bedded (Fig. 7). A remarkable, rather constant feature of the unit is presence of a basal bed 6 to 12 inches thick, which shows local cross-bedding, and is generally followed by a massive dolomite bed, several feet thick (Fig. 5, 8, A2, A4, and others). The basal 6- to 12-inch bed was studied in detail only at Chhidru A (loc. 6A) and a description follows.

Invertebrate trails were seen in the dolomite unit in several localities, such as Nammal Gorge (loc. 3), Chhidru A (loc. 6A), and Munta Nala (loc. 7), but they were studied in detail only at Chhidru A, where other interesting bedding plane features also occur (Fig.

Fig. 10. Mud cracks and capped cross ripples, surfaces of lowermost 6-inch bed of Kathwai Member, west of Chhidru Nala, Salt Range (loc. 6A, unit 3). Position of bedding plane in the section is indicated in Figure 9. Length of pen about 5.5 inches. Teichert photograph.)

9). In this locality an abundance of trails and markings is seen along a bedding plane only one inch above the base of the Kathwai Member (Fig. 10). There are essentially three types:

(1) Slightly winding trails, with faint parallel lineation, up to 13 mm wide, which appear as positive hyporeliefs. The longest trail seen was 25 cm long.

(2) Smooth trails with midrib, 12-13 mm wide, which appear as negative hyporeliefs. The midrib is about 2 mm wide. Only one fragmentary trail, about 6 cm long, was seen.

(3) Small cones in positive hyporelief, about 5 mm in diameter, a few millimeters high. These are probably casts of shallow diggings.

In addition, there is at this bedding plane evidence of small vertical burrows, a few mm in diameter. All these features provide evidence for presence of an abundant benthonic fauna at the very beginning of deposition of the Kathwai Member.

At Chhidru, as at other localities, as mentioned above, the bottom of the Kathwai Member almost everywhere consists of one bed which is generally 5 to 6 inches thick, but in places reaches 12 inches (Fig. 8). It has not been observed whether or not this bed is separated everywhere from the white sandstone unit by an inch-thick layer of friable dolomite as in the Chhidru West section. The 6-inch bed is cross-bedded (Fig. 8), although this feature is not clearly seen everywhere. At Chhidru A the upper surface of this bed shows mud cracks and a peculiar pitting, features deserving of close study and description.

Mud cracks were observed on the surfaces of two blocks broken from the basal bed (Fig. 11). The polygons are of somewhat irregular shape and size and vary from 8 to 20 cm in diameter. Their general pattern is very similar to mud cracks developed in a tank experiment as illustrated by Twenhofel (1932, p. 687). It is now well known that crack patterns similar to mud cracks formed subaerially may also form on sediment surfaces under water cover, but such occurrences seem to be rare and individual cracks much less straight or less evenly curved than subaerially formed mud cracks.

A most peculiar feature of the same bedding plane is what at first sight seems to be a crowding of pits or depressions, most of them rounded to subrounded in outline, 2 to 5 mm in diameter and generally 3 to 4 mm apart, separated by flat-topped ridges. Some depressions are elongated and up to 10 mm long and 3-4 mm wide. They are rather uniformly about 2 mm deep, and have a shallowly concave bottom. Their interpretation is difficult (Fig. 12).

Very similar bedding plane features in rocks from the Keuper of Lorraine have been interpreted as impressions of blue-green algae and named *Rivularites* by Fliche (1905, pl. 3, fig. 4). White (1929, p. 41) described similar bedding plane features from the Permian Hermit Shale of the Grand Canyon and named them *Rivularites permiensis*. The specimen illustrated by White on plate 6 has an appearance very similar to the pitted surfaces described above, although the pits are larger, ranging from 5 to 10 mm in diameter.

Fenton (1946) found it "difficult to separate *Rivularites* from small symmetrical ripple marks of irregular pattern, so that the existence of this form-genus in early sediments is no more than a possibility." In the *Treatise on Invertebrate Paleontology* Häntzschel (1962, p. W236) placed *Rivularites* among the inorganic "fossils."

It is not possible to entertain seriously the idea of an organic origin for the features described here from the Kathwai Member. Interpretation as somewhat unusual, capped interference ripples is the most probable. These ripples were probably formed in the intertidal zone, because it can be seen

FIG. 11. Trails in positive hyporelief at bottom surface of lowermost 6-inch bed of Kathwai Member west of Chhidru Nala, Salt Range (loc. 6A, unit 3). Position of this surface in section is indicated in Figure 9. Length of knife 3 inches. (Teichert photograph.)

FIG. 12. Pitted bedding plane, surface of lowermost 6-inch bed of Kathwai Member, west of Chhidru Nala, Salt Range (loc. 6A, unit 3). The structure is interpreted as a system of capped crossed ripple marks. Position of bedding plane in section is indicated in Figure 9. Length of knife 3 inches. (Teichert photograph; specimen in Dept. of Geology, Univ. of Kansas.)

on the left side of the surface illustrated in Figure 10 that the pits here are "fading out" as if they were washed away by small waves. The flatness of the areas between the pits can be interpreted as due to capping of ripples by wave action.

Kindle (1914, pl. 8; 1917, pl. 23, fig. A) showed almost identical features from the sediment surface at the shore of a pond. Since no scale was given, it is difficult to judge the size of the pits; they are possibly somewhat larger than the ones described here. Kindle (1914, pl. 8; 1917, pl. 30, fig. B) also showed similar surfaces from the Silurian Lockport Dolomite and from the Mississippian Berea Sandstone. In the former the pits seem to be of a size similar to the ones in the Kathwai Member. The other picture suffers from lack of a scale. Kindle remarked that these features formerly had been interpreted as "tadpole nests."

That similar bedding plane features are indeed due to the activity of tadpole swarms in very shallow water has been more recently reasserted by Maher (1962) and by Dionne (1969), who both described and illustrated sedimentary surface features quite similar to the ones discussed here, though differing in size and outline of the holes. They were in a drainage ditch and in small pools, formed under just a few centimeters of water, and were in all instances associated with tadpoles.

Martinsson (1965, p. 191) showed another similar bedding plane feature in a siltstone of Middle Cambrian age, where the pits, however, are mostly more elongated. Again the picture has no scale. These ripples also seem to be capped, although Martinsson does not refer to that feature in the text. He calls them "Kinneyan ripples" with reference to Walcott's (1914) alleged algal "genus," *Kinneya,* which seems to

Fig. 13. Trails in positive hyporelief on bottom surface of dolomite unit 4, west of Chhidru, Salt Range (loc. 6A), as also shown in Figure 9. Position of this surface in section is indicated in Figure 9. For size compare with Figure 8. (Teichert photograph.)

have been based on a specimen showing capped ripples, not however of the interference type.

The literature contains many descriptions and illustrations of interference ripples that are not capped (e.g., Pettijohn and Potter, 1964, pl. 86A, 87A). By studying such pictures one may visualize how the surfaces might have looked, if the ripples had been capped. The result would have been a pitted surface such as here described from the Kathwai Member.

A latex cast of the bedding plane illustrated in Figure 12 was recently sent to H.-E. Reineck at Wilhelmshaven, Germany, who suggested that the feature described here might be comparable to *Runzelmarken* (wrinkle marks), described and named by Häntzschel and Reineck (1968, pl. 6). Reineck (1969) has plausibly demonstrated that *Runzelmarken* are made by wind blowing across wet sediment, either exposed to the air or covered by not more than 1 cm of water.

Considering all evidence, it seems most likely that the bedding plane features that occur 6 to 7 inches above the base of the Kathwai Member at Chhidru Nala were formed in the intertidal zone.

This surface must then have been covered with a thin mud layer, because the features on the lower surface of the next following dolomite bed are very different (Fig. 8, 13). This surface carries positive hyporeliefs of various kinds of trails and markings:

(1) A mostly straight or only slightly sinuous trail with rounded, almost semicircular cross section.

(2) Mound-shaped eminences, 2 to 3 cm long, and less than 1 cm high. These are casts of resting or digging marks.

(3) Vertical burrows about 5 to 8 mm in diameter.

All features described here are preserved in a rather coarse, saccharoidal dolomite; and if any finer features, such as striations, had been present, they have been destroyed by dolomitization.

Again the trails and markings at this level give evidence of a rich benthonic fauna.

Petrography

The dolomite unit consists of a highly peculiar rock. Although early observers such as Wynne and Waagen (in Wynne, 1878) had, in some sections, correctly identified it as dolomite, Waagen (in Wynne, 1878, p. 226) had described it as a "grey sandstone," 6 feet thick, at the Munta Nala section, west of Wargal (our loc. 7). Gee (1947) called it an arenaceous limestone, and Schindewolf (1954) a partly dolomitic, partly calcareous, sandstone. It is best described as a fine- to coarse-grained dolomite, containing varying amounts of calcite, occasionally minor amounts of detrital mineral grains, and also minor amounts of glauconite.

Usually, as the name we gave it suggests, the unit consists entirely of dolomite (Pl. 1, fig. 7, 8). Only one locality, discovered by Teichert in 1964, is known to us in which dolomitization is somewhat patchy and where parts of the unit are preserved as limestone (Fig. 14). This occurrence is described in detail below.

The original rock, before dolomitization, was essentially a bioclastic limestone, consisting mostly of huge numbers of fragments of echinodermal skeletons and of comminuted shell fragments, frequently not more than 1 mm in size. Dolomitization altered, but in most cases did not completely destroy, the primary sedimentary features.

The basal 2 feet regularly contain an admixture of detrital, terrigenous mineral grains, mostly quartz, but including varying amounts of feldspar and mica. These components may constitute anything from less than 1 to 15 percent of the rock, but tend to be around 10 percent in most places (Pl. 2, fig. 3). Above 2 feet the detrital minerals decrease rapidly in number in most places except Narmia Spring (loc. 2). That part of the unit above the lowermost 3 to 4 feet is either without detrital mineral grains, or has only a minute amount of silt-size particles.

Approximately at the level where the detrital component is diminished, glauconite grains begin to appear and they are fairly characteristic of the upper part of the dolomite unit, as well as of higher beds (Pl. 2, fig. 6).

A major feature of the dolomite unit, and one that is of great significance, is dedolomitization, of which examples can be seen in many thin sections, especially from the upper half of the unit.

Although superficially similar, two texturally different kinds of rock are represented. One consists very largely of rather minute "tabular bodies" which are from one to a few millimeters long and 0.5 to 1 mm wide (Pl. 1, fig. 8; Pl. 2, fig. 2, 4, 5). They are dolomitized or consist of granular or fibrous calcite which almost certainly does not reflect original textures of the fragments. In many thin sections it was seen that the "bodies" consist largely of calcite surrounded by an exceedingly thin rim of dolomite. Quite commonly the edges of the fragments appear frayed owing to invasion of them by dolomite rhombs. Because of their shape the fragments are here interpreted as shell fragments, but what kind of fossil shells they represent is difficult to determine. Brachiopod shells, identifiable by remnants of fibrous texture and usually by wavy outlines, are rarely seen in thin sections. Among molluscs, pelecypods are very rare in the dolomite unit. Probably the fragments are derived from the shells of cephalopods, presumably all *Ophiceras connectens*. This rock type could, therefore, suitably be called a dolomitized coquinoidal microbreccia.

Another basic rock type consists of an aggregate of dolomite rhombohedra, each of which encloses some fragment of an echinoderm, which is usually dolomitized, the dolomite rhombohedron forming an overgrowth of the fragment. Lattice structure is clearly visible in many fragments and here and there the lattices are filled with glauconite. In many cases, however, the lattice structure is obscure because of intense dolomitization and in some instances only a dark smudge is visible in the core of the rhombohedra, which nevertheless can be interpreted confidently as remnant of an echinoderm fragment. The outlines of even the better-preserved fragments are generally fuzzy or quite obscure. This rock type may be characterized as dolomitized echinodermal breccia.

A variety of this type is a rock in which the echinodermal fragments may be either dolomitized or calcitic, but in which the dolomitized fragments show no overgrowth of dolomite. As Lucia (1962) has pointed out for a somewhat similar rock type of Devonian age from Texas, the dolomite here forms a pseudomorphic replacement of the echinoderm fragment.

As contributors to the echinoderm breccia we can identify echinoids, ophiuroids, and very small crinoids. Echinoid spines are not uncommonly seen in thin sections, just as entire spines are seen on bedding planes in many places. All of these probably represent *Miocidaris pakistanensis*. Ophiuroids are known to be present from the unique find of *Aplocoma* cf. *A. torrii*, but no crinoid has yielded identifiable remains. Unmistakable crinoid stems are fairly abundant in many thin sections; and small crinoid ossicles, 1-2 mm wide and 3-4 mm long, are seen rarely on suitably weathered surfaces.

As may be expected, transitional rock types exist between the coquinoidal microbreccia and the echinodermal breccia in which shell and echinoderm fragments occur together in various proportions. In such rocks all combinations of dolomitized and calcitic shell and echinoderm fragments may be present.

The space between the fossil fragments, or overgrowth rhombohedra, is filled with either clear, or turbid, calcite or with turbid, limonitized dolomite, or with a generation of small dolomite rhombohedra, generally significantly smaller than the overgrowth rhombohedra (Pl. 2, fig. 3, 8). In rocks with such bimodal distribution of rhombohedra, the small ones tend to be less than 50 μ in size, the overgrowth rhombohedra several 100 μ up to 1 mm.

In rocks with interstitial cement of calcite or dolomite it is generally impossible to specify the nature of the original sediment, though presumably it was lime mud.

It has been mentioned above that undolomitized portions of the dolomite unit were seen in only one place. This is on the ridge above Tapan Wahan about 0.5 miles east of the measured section (loc. 12). Here the Kathwai Member is underlain by the white sandstone unit containing the highly fossiliferous pocket discovered by Grant (1970). The relations in this locality are shown in Figure 14.

The section runs approximately N-S, parallel to the strike of the rocks. The rocks dip steeply to the west (Fig. A19) and for a short distance across the crest of the ridge they are well exposed along the strike. The locality is of great importance for understanding of the nature of the dolomitization process in the Kathwai Member.

Both the dolomite and limestone units are exposed, although part of the latter is eroded away. The basal 5 feet of the dolomite unit are of greatest interest. This unit is designated number 2 in the section (Fig. 14). In the northern half of the outcrop it is a dolomite of dark yellowish-orange

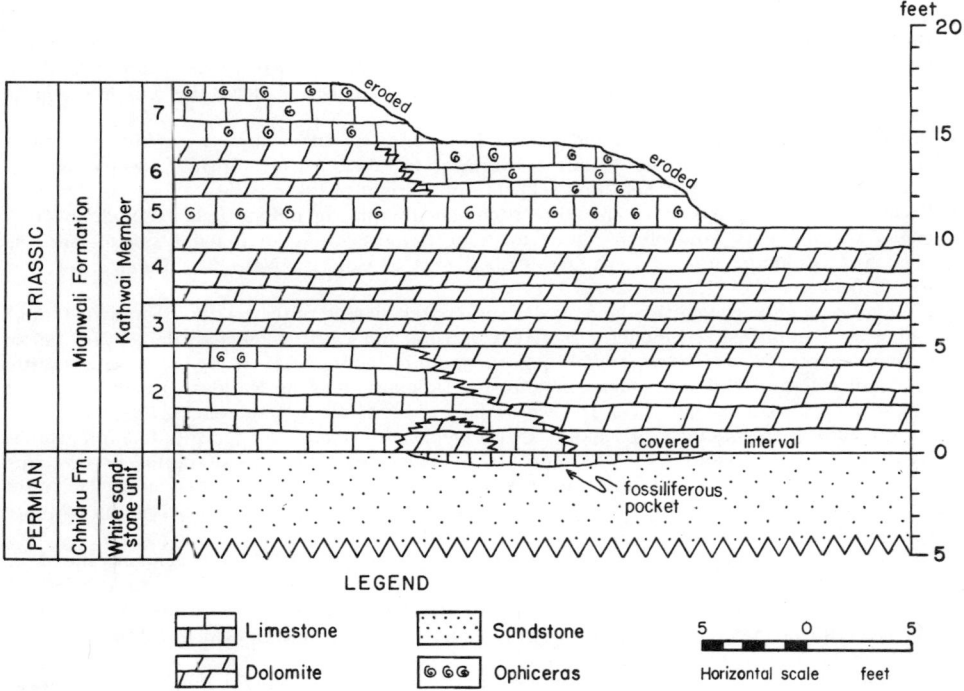

FIG. 14. Relationships between limestone and dolomite in the Kathwai Member, near Tapan Wahan, Khisor Range, about 0.5 miles south of loc. 12. Section oriented approximately N-S in the direction of the strike (see Fig. A19).

color, weathering light brown, which has the features of the dolomitic echinodermal breccia described above. Shell fragments are completely, or almost completely, absent. The rock contains less than one percent of detrital minerals and very scattered pelletal glauconite. Many dolomite rhombohedra have outer calcitic zones, and irregular patches of calcite are also present in the dolomitic ground mass.

In the southern part of the outcrop the same bed 2 consists of yellowish-gray to pinkish-gray limestone, weathering grayish-orange pink. The rock is crowded with minute shell fragments and is identical in thin section with the one described above as dolomitized coquinoidal microbreccia, except that it is not dolomitized. In this rock echinoderm fragments are extremely rare or absent. Glauconite pellets are present in very small numbers. The shell debris consists of fragments that are not longer than 7.5 mm, generally about 5 mm long. Near the top of the bed poorly preserved whole cephalopods were seen, presumably *Ophiceras connectens*. No cephalopods were seen in the dolomitized part.

Of considerable interest is a pocket of dolomitized rock in the limestone near its boundary with the dolomite,

---

PLATE 2

All photomicrographs taken under parallel nichols; unless otherwise indicated, oriented with top side pointing up stratigraphically.

FIGURES

1. Dolomitic limestone; turbid limestone matrix containing many dolomite rhombohedra; large dolomite rhombohedron in center is partly dedolomitized into calcite and has only partly dolomitized echinoderm fragment as core; black spots inside this rhomb are glauconite. Dolomite unit of Kathwai Member, Narmia Spring, Surghar Range. Loc. 2, bed 3; field no. T62-7; top is right. ×75.
2. Dolomite rhombohedron enclosing only partly dolomitized echinoderm fragment and having calcitic laminae parallel to edge; dark patches are irregularly distributed calcite. Dolomite unit of Kathwai Member, Ganjaroh, Salt Range. Loc. 5, bed 7; field no. T62-159; top is right. ×75.
3. Dolomite; aggregate of anhedral and euhedral dolomite crystals, containing about 10 percent angular quartz grains. Dolomite unit of Kathwai Member, Narmia Spring, Surghar Range. Loc. 2, basal part of bed 4; field no. T62-304; top is right. ×25.
4. Dolomite, consisting mostly of dolomite rhombohedra having dolomitized echinoderm fragments as core; dark lamellae in outer zones of rhombs are dolomite, not calcite; slide is unstained. Dolomite unit of Kathwai Member, west of Chhidru Nala, Salt Range. Loc. 6A, bed 5; field no. T63-127; unoriented. ×75.
5. Limestone, highly dolomitic; large dolomite crystal having only partly dolomitized, and somewhat limonitized, echinoderm fragment as core and calcitic lamellae near edge; below it dasycladacean alga (*Vermiporella*? sp.); surrounding gray areas are turbid calcite, mostly recrystallized shell and echinoderm fragments. Dolomite unit of Kathwai Member, east of Chhidru Nala, Salt Range. Loc. 6B, bed 7; field no. T61-48; unoriented. ×75.
6. Groundmass predominantly of small (20-100 μ) calcite crystals; clusters of dolomite rhombohedra near center and right of center; large glauconite pellet left of center broken, but fragments still in close position; this rock is transitional to the limestone unit. Dolomite unit of Kathwai Member, Narmia Spring, Surghar Range. Loc. 2, 4 inches below top of unit; field no. T62-8; unoriented. ×75.
7. Dolomite rhombohedron enclosing a brown core, probably echinoderm fragment; the dark material between the lamellae is calcite (stained) as is the light gray area to the left of the rhomb; the light gray grain cutting across the lamellae at bottom is quartz. Dolomite unit of Kathwai Member, Kingriali, Khisor Range. Loc. 11, bed 4; field no. T64-11; top is right. ×75.
8. Groundmass of very finely crystalline brown dolomite, with many strongly altered fossil fragments and angular to subangular quartz grains; in center, cross-section of ostracode shell; in lower left, dolomite rhombohedron with remnant of echinoderm fragment; the elongated objects are unrecognizable shell fragments. Dolomite unit of Kathwai Member, Ganjaroh, Salt Range. Loc. 5, bed 5; field no. T62-157; unoriented, top either up or down. ×25.

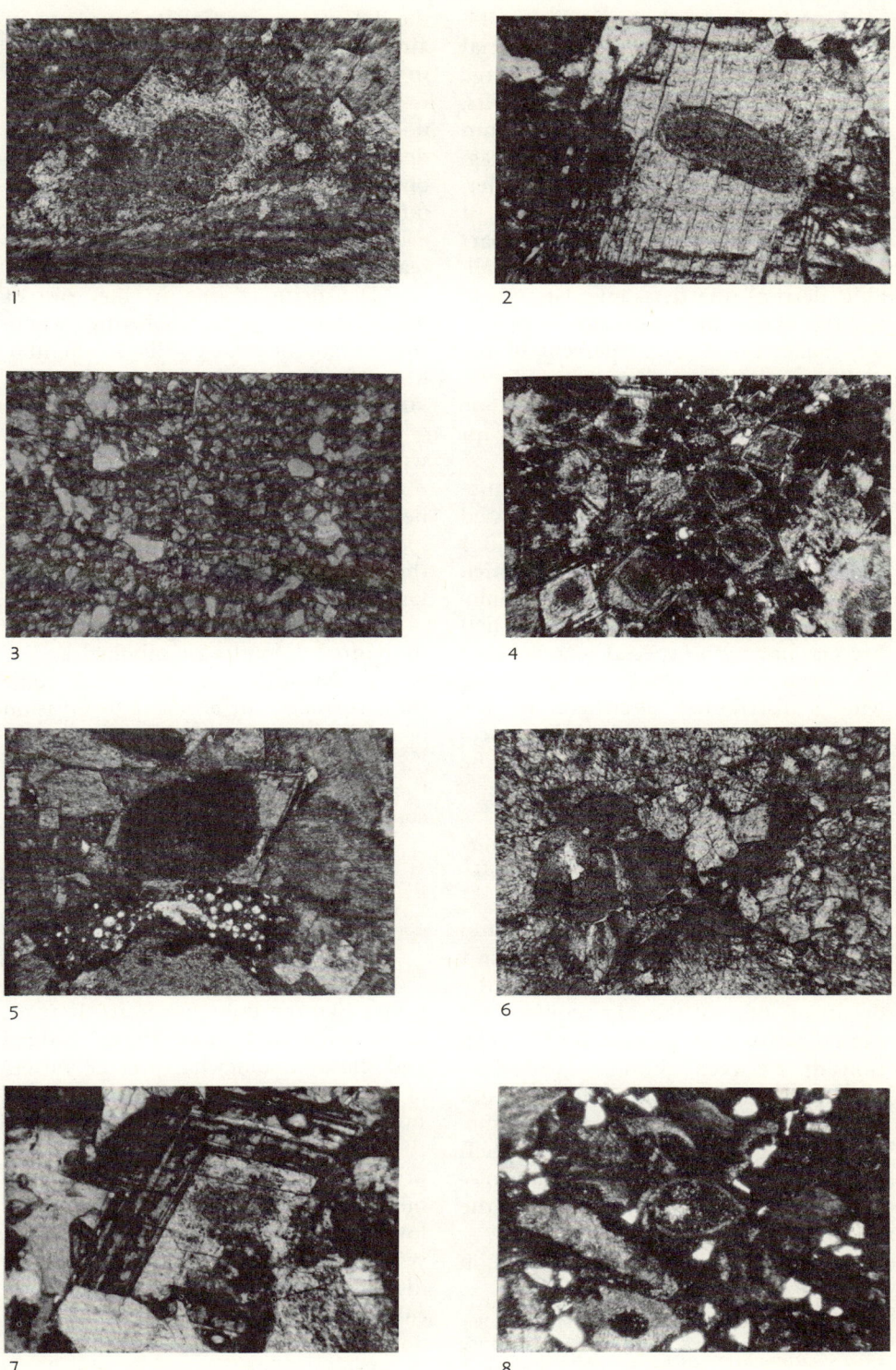

PLATE 2 (EXPLANATION ON FACING PAGE)

yet clearly set off from the main mass of the dolomite, as shown in Figure 14. The rocks in this pocket have special characteristics. They contain large numbers of echinodermal fragments, many of which are identifiable in thin section as echinoid spines. Shell fragments are absent. The rock, further, contains a much greater amount of detrital material than any other part of bed 2. In the one thin section available, detrital quartz, as angular to subangular grains up to 0.2 mm in diameter, makes up 10 to 20 percent of the rock.

The transition between limestone and dolomite of bed 2 is rather abrupt as indicated in Figure 14.

The remainder of the dolomite unit in this locality can be discussed somewhat summarily. Beds 2, 3, and 4 constitute the dolomite unit in which beds 3 and 4 are made up of dolomitized mixed echinodermal and shell breccia, having no special features that set them apart from similar rocks widely distributed elsewhere in the Kathwai Member, except that glauconite seems to be absent in them. The amount of detrital quartz grains is less than one percent.

Beds 5, 6, and 7 make up the limestone unit and are described in the discussion of that unit.

The distribution of limestone and dolomite in bed 2 of the dolomite unit as discussed above throws light on the origin of some of the Mg required for the dolomitization process, for the relationship between abundance of echinodermal remains and dolomitization is obvious. As has been shown, echinodermal fragments are abundant, shell fragments rare or absent in the dolomite, whereas in the limestone, the relative proportions are reversed.

Dedolomitization[1] is a common feature among all dolomitic rocks of the dolomite unit. Calcite is present in almost all rock samples examined; and in some cases, especially where calcite is present in the interstices between dolomite rhombohedra, it is difficult to decide whether it is residual or secondary. That much of the calcite is due to dedolomitization processes, however, is indicated by several observable features:

1) Calcite is seen in the core of many dolomite rhombohedra, where the calcite generally fills a rhombohedral space surrounded by dolomite lamellae.

2) Calcite appears interlayered with dolomite lamellae in the outer zones of many large dolomite rhombohedra (Pl. 1, fig. 7, 8; Pl. 2, fig. 1, 7).

3) Calcite may be distributed in the form of irregular blobs throughout larger dolomite rhombohedra.

4) Calcite may completely replace individual dolomite rhombohedra.

5) Dolomite rhombohedra may show corroded surfaces due to invasion of calcite into the dolomite.

The environmental significance of the dedolomitization process is discussed in a succeeding chapter (p. 56).

Limestone Unit

Gross Lithology

Superficially, the limestone unit is rather like the dolomite unit. It tends to be somewhat more thinly bedded and shows a somewhat greater variety of colors, both on fresh and weathered surfaces. The grayish-orange pink, so characteristic of the dolomite, is still present on many weathered surfaces, but others are light brownish-gray, yellowish-gray, light olive-gray, and pale yellowish-brown. In general, the reddish hue is less predominant as surface color of the limestone unit than of the dolomite.

Freshly broken surfaces of rocks of the limestone unit exhibit a variety of

---

[1] Smit and Swett (1969) have been critical of the use of the term "dedolomitization," branding it, among other things, as ambiguous. Granting that "calcitization" might be more logical, it should be pointed out that, when properly defined, no scientific term can be ambiguous.

colors, though rarely the ubiquitous grayish-orange of the dolomite unit. Colors identified with the help of the rock-color chart include dusky yellow, yellowish-gray, light olive-gray, pale yellowish-orange, pale yellowish-brown, grayish-orange, and light gray.

No cross-bedding has been observed in the rocks of the limestone unit.

In some places, such as Nammal Gorge (loc. 4) and Wargal (loc. 7) the limestone unit contains interbeds of fine-grained sandstone and some shale.

As in the dolomite unit megafossils are uncommon. However, *Ophiceras* is seen in greater number and in better preservation than in the dolomite unit.

Petrography

The rocks of the limestone unit are distinct from those of the dolomite unit in their generally small extent of dolomitization and in the much smaller contribution made by echinodermal fragmental material. The limestone matrix is generally aphanitic, rarely recrystallized. Embedded in the matrix are minute shell fragments in varying abundance and this type of rock is quite similar to the coquinoidal microbreccia of the dolomite unit (Pl. 3, fig. 2). Fragments of echinoderm skeletons may be present or absent, but if present, they are invariably subordinate. No particular pattern in the distribution and relative abundance of the echinoderm fragments could be detected.

Many rocks of the limestone unit contain a greater or smaller amount of dolomite rhombohedra, usually in the size range of 20 to 300 μ (Pl. 3, fig. 1). Generally, they are scattered through the limestone as single individuals; more rarely they form small clusters. Dolomite also occurs in the form of dolomitized echinodermal fragments, but the typical dolomitic overgrowth observed in rocks of the dolomite unit does not occur. Very rarely, shell fragments may be dolomitized, or are invaded by small dolomite rhombohedra.

Dolomitization has proceeded only locally to the point where the rock becomes truly dolomitic. Thus, in the Nammal Gorge section (loc. 4) the limestone unit is mostly a calcitic dolomite (Pl. 3, fig. 3). Elsewhere highly calcitic dolomites occur, along with strongly dolomitic limestone. Since dolomite is rarer in the limestone unit than in the dolomite, dedolomitization also is more rarely observed. In fact, evidence for this process was seen in only one thin section (Pl. 3, fig. 4), from near the top of bed 6 at Kathwai B (loc. 10).

Glauconite is almost invariably present in the limestone unit, though in greatly varying amounts. Most of it consists of pelletal grains ranging in size from 0.1 to 0.5 mm. Some grains are broken, but in general they show little evidence of prolonged transportation (Pl. 3, fig. 6). This type of glauconite generally seems to constitute 2 to 3 percent of the rock, though in some thin sections it was less than one percent. In other thin sections, as much as 5 percent glauconite was seen.

Glauconite is also present in the interstices of some echinodermal skeletal fragments (Pl. 3, fig. 5). For example, in the Ganjaroh section (loc. 5) the limestone unit (bed 8) is less than a foot thick and consists of highly dolomitic limestone which is rich in echinoderm fragments, some dolomitized, some not, but all filled with glauconite. However, it seems that this type of glauconitization of echinodermal skeletons is not as common in the limestone unit as in the dolomite unit.

Paleontological Content

A partial description of the fossil content of the Kathwai Member, based on incomplete and preliminary identifications, was published by us in 1966 (Kummel and Teichert, 1966, p.

320). It is now possible to give a much more comprehensive list based on some of our own identifications, but mostly on the work of Grant (1970), Newell (written communication), Hans Hess (written communication), Sweet (1970), Balme (1970), Schaeffer (written communication), and Sarjeant (1970).

In 1966, we had described separately the faunas of the dolomite and limestone units, but it seems to us now that no essential difference exists between the paleontological contents of these two units, except in composition of the conodont assemblages which, however, is only slight and is discussed elsewhere.

Following is a complete list of fossil forms recovered from and identified in the Kathwai Member:

Bryozoa
  Gen. et sp. indet.
Brachiopoda
  Lingula cf. *L. borealis* Bittner
  Orbiculoidea sp.
  Enteletes sp.
  Orthothetina cf. *O. arakeljani* Sokolskaya
  O. sp.
  Ombonia sp.
  Derbyia? sp.
  Spinomarginifera sp.
  Linoproductus sp.
  Lyttonia sp.
  Spirigerella sp.
  Crurithyris? *extima* Grant
  Martinia sp.

PLATE 3

All photomicrographs taken under parallel nichols; unless otherwise indicated, oriented with top side pointing up stratigraphically.

FIGURES

1. Dolomitic limestone; groundmass of turbid sparry calcite, sprinkled with individual and clustered dolomite rhombohedra. Limestone unit of Kathwai Member, west of Chhidru Nala, Salt Range. Loc. 6A, basal part of bed 10; field no. T63-136; unoriented. ×25.
2. Coquinoid limestone consisting of densely packed shell fragments, presumably of *Ophiceras*. Limestone unit of Kathwai Member, Kathwai A, Salt Range. Loc. 9, bed 5; field no. T61-166. ×25.
3. Dolomite, consisting largely of dolomite rhombohedra enclosing dolomitized echinoderm fragments whose lattices are filled with calcite (the dark, irregular material inside the rhombs); dark material between rhombs and forming marginal lamellae in rhomb below center is turbid calcite. Limestone unit of Kathwai Member, Nammal Gorge, Salt Range. Loc. 4, near top of bed 7; field no. T61-58; top is right. ×75.
4. Limestone, very finely crystalline; rhombic areas have outer lamellae of iron oxide, enclosing calcite,—these are interpreted as calcitized dolomite rhombohedra; small sprinkling of silt-size quartz grains. Limestone unit of Kathwai Member, Kathwai B, Salt Range. Loc. 10, upper part of bed 6; field no. T62-207; top is right. ×75.
5. Limestone, sparry and granular calcite with echinoderm and shell fragments; lattice of the echinoderm fragments is filled with glauconite. Limestone unit of Kathwai Member, Zaluch Nala, Salt Range. Loc. 3, bed 7; field no. K13-F; unoriented (top probably either left or right). ×25.
6. Glauconite grain, fractured, but moved only slightly apart, indicating only weak turbulence or current action. Limestone unit of Kathwai Member, Narmia Spring, Surghar Range. Loc. 2, bed 6; field no. T62-10; top is right. ×25.
7. Shell fragments, poorly oriented, probably of ammonoids, and a small gastropod, in matrix of granular, turbid calcite. "Lower Ceratite limestone" of Mittiwali Member, Munta Nala, Salt Range. Loc. 7, bed 15; field no. K11-14; unoriented (top is either up or down). ×25.
8. Limestone, very finely crystalline; "ghosts" of finely recrystallized shell fragments vaguely seen; dark spots are iron oxide, partly pseudomorphs after dolomite; unidentified fossil fragment in center. "Lower Ceratite limestone" of Mittiwali Member, Chhidru A, Salt Range. Loc. 6A, top 4 inches of bed 11; field no. T63-140. ×25.
9. Coquinoid limestone with granular calcite matrix; shell fragments (probably mostly ammonoids) well oriented. "Lower Ceratite limestone" of Mittiwali Member, east of Chhidru Nala, Salt Range. Loc. 6B, bed 12; field no. K4-14; unoriented (top is either left or right). ×25.

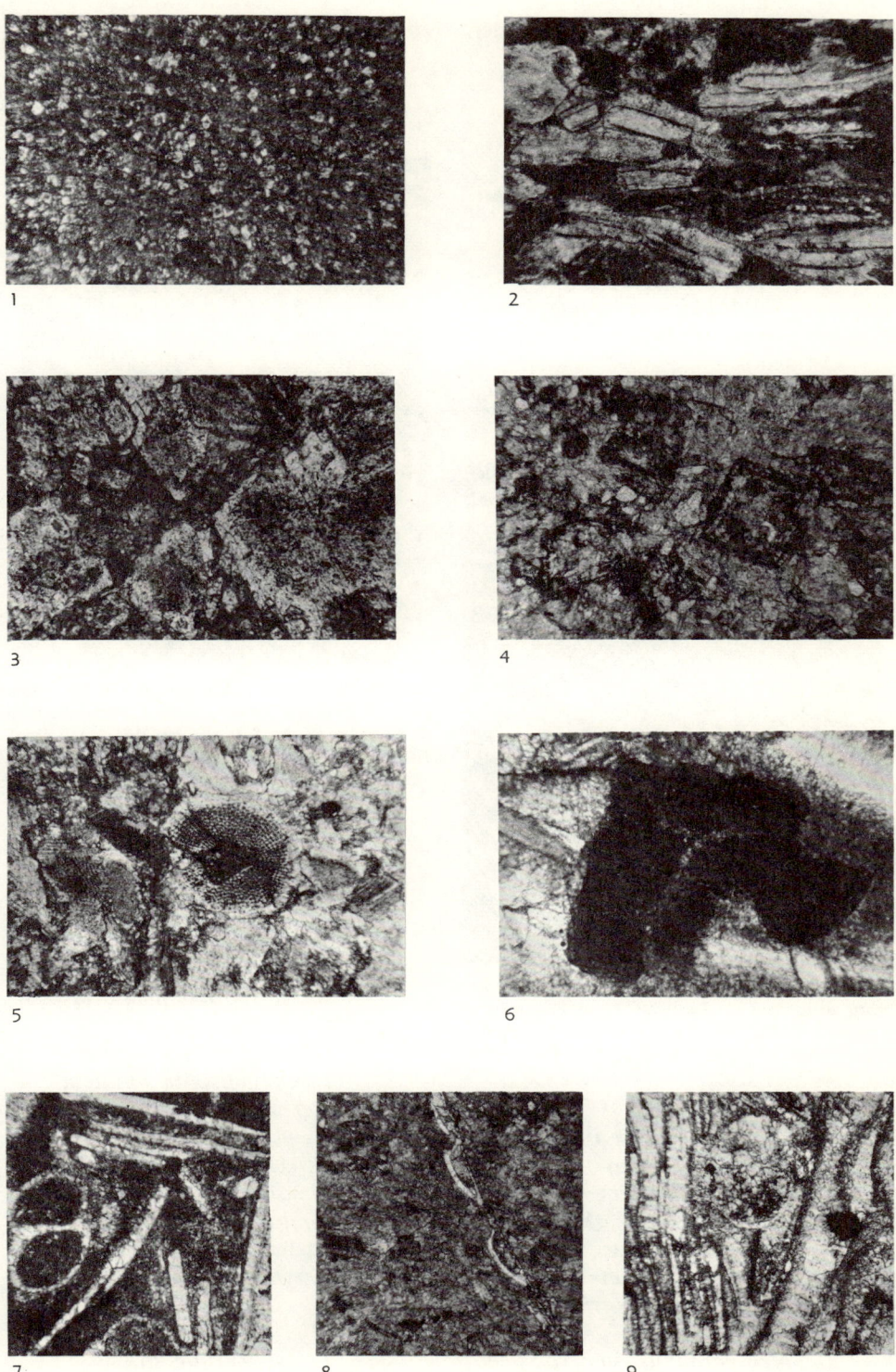

PLATE 3 (EXPLANATION FOR FACING PAGE)

Whitspakia sp.
Dielasmatids, gen. et sp. indet.
Pelecypoda
    Pernopecten? or *Entolium?*
Gastropoda
    Order and family uncertain
Cephalopoda
    Menuthionautilus kieslingeri Collignon
    Ophiceras connectens Schindewolf
    Glyptophiceras himalayanum (Griesbach)
Ostracoda
    Order and family uncertain
Echinodermata
    Crinoid stems
    Aplocoma cf. *A. torrii* (Desio)
    Miocidaris pakistanensis Linck

FIG. 16. *Miocidaris pakistanensis* Linck. Upper part of dolomite unit. Kathwai Member, west of Chhidru Nala, Salt Range (loc. 6A). Geol. Institut, Universität Tübingen, Ech 1058/1. (Coll. O. H. Schindewolf. Photograph courtesy of Dr. Otto Linck.)

FIG. 15. *Aplocoma* cf. *A. torrii* (Desio). Dolomite unit, bed 8, Kathwai Member, west of Chhidru Nala, Salt Range (loc. 6A). Kansas Univ. Mus. Invert. Paleontology, no. 54151.

Conodontophorida
    Anchignathodus isarcicus (Huckriede)
    A. typicalis Sweet
    Ellisonia triassica Müller
    E. gradata Sweet
    E. teicherti Sweet
    Neogondolella carinata (Clark)
    Neospathodus dieneri Sweet
    N. kummeli Sweet
    Xaniognathus curvatus Sweet
    X. deflectens Sweet
Fishes
    Acrodus sp.
    Colobodus sp.
    Saurichthys sp.

Trace fossils
    Invertebrate trails, burrows, and resting marks
Palynomorphs
    Punctatisporites fungosus
    Densoisporites playfordi
    Calmospora landiana
    Osmundacidites senectus
    Lundbladispora obsoleta
    Taeniaesporites novaulensis
    T. pellucidus
    T. cf. *T. transversundatus*
    ?Klausipollenites schaubergeri
    Fimbriaesporites? sp.
Acritarcha
    Veryhachium reductum Deunff
    Micrhystridium sp.
    Leiosphaeridia sp.
Algae
    Dasycladaceae, gen. et sp. ind.

Fragments of dasycladacean algae occur very rarely in the dolomite unit (Pl. 2, fig. 5). They were seen in two thin sections, but the derivation of one of them is in doubt.

Because of their unusual interest as oldest Mesozoic representatives of their respective classes *Aplocoma* cf. *A. torrii* is here illustrated as Figure 15 and the holotype of *Miocidaris pakistanensis* is reillustrated as Figure 16.

Some aspects of the Kathwai fossil assemblages are discussed in the chapter dealing with the comparative paleontology of the Permian-Triassic boundary beds. We here examine only aspects that are relevant to interpretation of the sedimentational history of the Kathwai Member.

When we wrote our discussion of this fauna which was published in 1966, the study of some of the most important fossil groups had scarcely begun (e.g., conodonts) or their presence was not even suspected (palynomorphs, acritarchs). Perhaps the most controversial aspect discussed by us in 1966 was the nature and origin of the articulate brachiopods of definitely Permian affinities found in the Kathwai Member in at least three localities. Our argumentation was based on relatively small collections which had been submitted to G. A. Cooper for study who found them to be "rather Middle Permian than Upper in aspect."

After detailed study of considerably larger collections, Grant (1970) has concluded that the "Permian-type" brachiopod fauna in the Kathwai Member which we list above is decidedly different from that of the underlying Chhidru Formation and he believes it to be of Dzhulfian age, correlative with Dzhulfian brachiopod faunas listed and partly described by Sokolskaya, Sarycheva, Ivanova, and Grunt (in Ruzhentsev and Sarycheva, 1965). The possibility of redeposition of fossils in the Kathwai Member, particularly those of Paleozoic affinities which we discussed in 1966, has been considered by several authors who have studied our collections.

Rowell (1970) emphasizes the fact that the specimens of *Lingula* studied by him preserve fine detail of surface sculpture and are unlikely to have suffered long transportation. Newell (written communication, 1968) states that the bivalve shells examined by him "do not show indication of reworking." Sweet (1970), with reference to our previous discussion (Kummel and Teichert, 1966), paid special attention to the question of reworking of conodonts in the basal Kathwai and came to the following conclusion:

> Conodont elements from the samples studied by Cooper [in Kummel and Teichert, 1966] are not numerous, but they are amber in color, their fragile processes and denticles are mostly intact, their surfaces are smooth and lustrous, and they represent the same fauna found at the base of the Mianwali Formation in all the sections we have studied. Presumably the conodonts, at least, have not been reworked, and they represent the same species recorded in the uppermost part of the Chidru Formation in three Salt Range sections (Fig. 2). In short, brachiopods of Permian aspects in basal Kathwai strata may have been reworked, but it is unlikely that the conodont elements were.

As pointed out in description of the petrography of the Kathwai Member, both the dolomite unit and the limestone unit contain much comminuted shell and skeletal material, giving evidence of considerable water turbulence. Occurrence of well-preserved fossils in such kind of rock seems at first glance anomalous. This matter is considered more fully in the following section.

Depositional and Diagenetic Environments

The rocks of the Kathwai Member possess a number of highly peculiar characteristics which, although they at least in part may suggest seemingly contradictory conclusions, yet if evaluated critically lead to an acceptable and internally consistent interpretation of their mode of formation. The following features are of prime importance:

(1) Bedding plane features and cross-bedding in the dolomite unit.

(2) Extreme fragmentation of shell and echinodermal material forming the bulk of the rock.

(3) Less fragmentary preservation of the "Dzhulfian" brachiopod fauna at Narmia Spring (loc. 2) in the Surghar Range.

(4) Poor preservation of pollen and spores.

(5) Lack of evidence of lengthy transportation of lingulids, pelecypods, and conodonts.

(6) Dolomitization.

(7) Dedolomitization.

It is of interest that the two localities which have yielded some of the most important types of information are also the ones which have been studied most intensely. These are: Narmia Spring (loc. 2), where a "Dzhulfian" brachiopod fauna is present in the Kathwai, and Chhidru A (loc. 6A), where bedding plane features were observed suggesting presence of intertidal conditions. Both localities were visited by us together several times and later several more times by Teichert alone. In all other localities the Kathwai Member received considerably less attention, although it was studied in some detail everywhere and most localities were visited at least twice. It is possible, therefore, that some of the unusual features found in the two localities mentioned might be more common than appears now on the basis of available observations. It must be remembered that the Kathwai Member in the Salt Range, Surghar Range, and Khisor Range forms continuous outcrops that are at least 120 miles long. Our 13 stratigraphic sections, therefore, provide only one narrow control strip for an average of every 10 miles.

The data gathered by us provide basis for certain general conclusions. The first of these is that the lowermost part of the dolomite unit was deposited at least in part in the intertidal zone. This is demonstrated by the occurrence of mud cracks and capped ripple marks. These features, however, were observed only at Chhidru A. Nevertheless, our previous suggestion (Kummel and Teichert, 1966, p. 330) that the contact between the Permian and the Triassic here is a paraconformity, is considerably strengthened, for it is reasonable to assume that formation of an intertidal, or shoreline, deposit was preceded by an episode of emergence.

The second important conclusion is that the remainder of the Kathwai Member was deposited in an infratidal environment. This is confirmed by the fact that there is no evidence for emergent conditions. In addition, glauconite is present everywhere in the upper half of the dolomite unit and in all of the limestone unit. Two types of glauconite occur in both units: (1) pelletal glauconite, and (2) infillings of echinoderm skeletal fragments. Interestingly, these two types also occur in sediments of the Sahul Shelf as reported by van Andel and Veevers (1967, p. 59-60) who found glauconitized skeletal material to be much more abundant than pelletal glauconite. Although glauconitized echinodermal fragments were not mentioned by these authors they stated that glauconite has invaded Bryozoa and even very small and remote chambers of Foraminifera. Replacement glauconite was found to be commonest in water depths between 20 and 60 fathoms, glauconite pellets between 20 and 45 fathoms.

The original sediment from which the rocks of the Kathwai Member were formed was very largely composed of comminuted fragments of shells, probably mostly ammonoid and clam shells, and of equally fragmented echinoid, ophiuroid, and crinoid skeletal parts. To these was added in the earliest part of the sedimentation period a small amount, rarely more than 10 percent, of terrigenous detrital material in the form of angular to subangular quartz grains, feldspar, and mica. The primary cause of this skeletal breakdown is difficult to assess. It seems most likely that fragmentation was the result of a combination of mechanical and biologic agents. The linearity of the narrow Kathwai outcrop belt gives us a two-dimensional picture only. Whereas the Kathwai is composed primarily of skeletal debris, a limited amount of whole fossils are present. These include primarily ammonites, pelecypods, and brachiopods, especially *Lingula*. As already mentioned, Newell, Rowell, and Sweet have observed that shell surface features of pelecypods, lingulids, and conodonts are well preserved. The "Dzhulfian" brachiopods described by Grant in this volume consist mainly of fairly large pieces of broken shells. It should also be pointed out that the seemingly impoverished fauna of the Kathwai Member is, in fact, comparable to earliest Ccythian faunas wherever found throughout the world.

Dolomitization of the Kathwai Member is an intriguing problem, as is perhaps that of all dolomite formations. Again, the nature of the outcrop data for the Kathwai Member limits the range of interpretation and speculation. It is tempting to look to the large amount of echinodermal skeletal fragments as a partial source of the magnesium. The well-known investigations by Vinogradov (1953) and Chave (1954) have demonstrated that in water with a temperature of 22° C or more the skeletons of echinoids, ophiuroids, and crinoids contain around 16 percent by weight of $MgCO_3$. However, we need to note the following statement by Hsu (1966, p. 135): "Magnesium is present in amounts up to 15% $MgCO_3$ in some carbonate sediments (Chave, 1954). Recrystallization of magnesium calcite to calcite during diagenesis could provide a fraction of the magnesium necessary for the dolomitization of a magnesian sediment. Nevertheless, the formation of large, pure dolomite bodies from lime sediments requires an import of magnesium by the movement of subsurface waters." At the same time, the close relationship between echinoderm abundance and dolomitization is striking, as described above (p. 45) for an outcrop near Tapan Wahan. This could reflect "biogenic magnesium enrichment," discussed by Fairbridge (1957, p. 132) as a possible preliminary stage towards dolomitization. On the other hand the relationships displayed at Tapan Wahan could also reflect nothing more than differential porosities.

Recent dolomites are now known from many places, where invariably the dolomite crystals are of small size and have formed in supratidal environments. In the Persian Gulf the dolomite rhombohedra are usually in the 1 to 5 μ size range (Illing, Wells, and Taylor, 1965), in the Bahamas 1 to 2 μ (Shinn, Ginsburg, and Lloyd, 1965), and on Bonaire Island, Netherlands Antilles, about 2 μ (Deffeyes, Lucia, and Weyl, 1965; Lucia, 1968). In contrast, in the Kathwai Member the dolomite rhombohedra range in size from about 20 to 300 or 400 μ, and larger crystals, up to 1 mm in diameter, are not uncommon. Thus the dolomite rhombohedra in the Kathwai Member are in a size range usually associated with later stages of diagenesis as discussed by Teichert (1965, p. 85). Deffeyes *et al.* (1965) considered the Recent dolomite of Bonaire Island, Netherlands Antilles, as comparable to what most authors describe as penecontemporaneous or primary dolomite. Such ancient dolomites are described by Deffeyes *et al.* (1965, p. 87) as "characterized by some or all of the following criteria: very restricted faunas, fine grain size, an association with gypsum, the presence of thin extensive dolomite beds, and dolomite clasts broken from these beds." In addition, Plio-Pleistocene dolomites exist on Bonaire Island which the same authors believe are similar to dolomites considered to be of replacement, or secondary, origin. They describe this suite as "characterized by a normal marine fauna, grain size in the range from 40 to 80 microns, and a pattern of dolomitization which cuts across the depositional bedding." Thus, Deffeyes *et al.* present the following convincing explanation of dolomitization of the Plio-Pleistocene rocks of Bonaire: within flat supratidal areas evaporation of sea water is depositing calcium carbonate and gypsum, producing dense brines having large Mg/Ca ratios; the dense brines flow downward into permeable sediments setting the stage for dolomitization. In addition, Lucia (1969) has pointed out that calcification (dedolomitization) of dolomite is a further stage in these processes.

The diagenetic history of the Kathwai Member could well be comparable to that proposed for the Plio-Pleistocene strata of Bonaire Island. The dolomite rhombs of the Kathwai are much larger than those in any Recent dolomite known; they are in the size range generally found in ancient dolomites, where dolomite forms as an authigenic mineral. The thickness of the dolomite unit of the Kathwai Member is rather variable, suggesting that the dolomitization cuts across depositional bedding. The fauna of the Kathwai Member is "normal" marine. Finally, the paleogeographic setting is compatible. The Salt Range lies on the margin of the Precambrian shield of peninsular India, and the axial area of Tethys was to the north. During earliest Triassic time the shoreline moved progressively in a general southeasterly direction, marking a slow transgression. One needs to hypothesize that evaporitic hypersaline basins were present landward of the shoreline.

The Kathwai Member encompasses the lower Scythian *Ophiceras-Otoceras* zone which had a probable duration of one to two million years. The limestone unit of the Kathwai Member has much less dolomite. In fact dolomite rhombs are usually scattered singly in the limestone but occasionally form small clusters. Dolomitization proceeded only locally to the point where the rocks are truly dolomitic. The overlying lower beds of the Mittiwali Member (the Lower Ceratite limestone of Waagen) contain no dolomite. Thus it would seem plausible that the conditions for dolomitization suggested above may have existed only during the time of deposition of the Kathwai Member and in fact began to break down late in Kathwai time.

Extensive dedolomitization observed in the Kathwai Member supplies further arguments that strengthen this model. It is not easy to account for dedolomitization under conditions that are deducible from the present surface geology. Most authors who have studied dedolomitization in rocks of various ages agree that presence of gypsum-saturated solutions is an essential condition for conversion of dolomite to calcite (Shearman, Khouri, and Taha, 1961; Goldberg, 1967; Evamy, 1967; de Groot, 1967; and earlier literature quoted in these papers). Such solutions must have been derived from evaporites that were contemporaneous with deposition of the Kathwai Member and were later removed by erosion or are now covered by younger rocks.

Derivation of the gypsum from contemporary Early Triassic evaporites is also suggested by the results of experiments by de Groot (1967), who found that dedolomitization in the presence of gypsum-saturated water took place only at very low carbon dioxide partial pressures and at temperatures lower than 50° C. De Groot stated that "carbon dioxide partial pressures in the subsurface are generally much higher than that of the atmosphere," where at the earth's surface it is about $10^{-3.5}$ atm., and he concludes that dedolomitization is a near-surface process. This conclusion had been reached on the basis of field observation by Tatarskiy (1949), Goldberg (1967), Evamy (1967), and others. Goldberg even believed to have found evidence in Jurassic rocks of Israel that dedolomitization took place in the supratidal zone.

We can now set up a tentative model of the paleogeography of, and sedimentation in, the general area of the Salt Range and Trans-Indus ranges at the very beginning of the Triassic Period. Somewhere outside the area of our investigation an environment existed where shells and echinoderm skeletons were broken up into small to very tiny fragments, either by organic action or by surf along a beach. This bioclastic sand and mud was then transported into the area covered by our studies, as shown by widespread presence of cross-bedding in the Kathwai Member. In the beginning the sediments were laid down, at least locally, in the intertidal zone, where populations of soft-bodied benthonic crawlers and burrowers lived. The bioclastic mud carried with it large numbers of pollen grains and spores which are poorly preserved.

As sedimentation proceeded, the sea must have deepened somewhat, for no evidence is seen for emergent conditions anywhere in the Kathwai Member except in its lowermost part. This is confirmed also by the fact that glauconite is present everywhere. At the same time a scarce benthonic fauna existed in the area, consisting of lingulids, few articulate brachiopods and pelecypods, and occasional echinoids and ophiuroids. A small brachiopod assemblage of Dzhulfian affinities seems to have existed in the beginning of Kathwai time, suffering only short transportation, if any. The waters in the sedimentation area were populated by nektonic conodontophorids and fishes and by planktonic acritarchs. Fossil remains of these three groups show excellent preservation.

In summary, we interpret the history of the Kathwai Member as follows: In earliest Triassic time a shallow sea transgressed slowly over the Salt Range and Trans-Indus range area. During the first phase of this transgression, at least part of the area was supratidal. The hinterland was one of very low relief, and the area southeast of the shoreline contained evaporitic basins. Approximately a foot of sediment accumulated during this initial phase which was followed by a lowering of the depositional interface. A limited fauna became widely distributed and glauconite was formed. Brines concentrated in the evaporitic basins southeast of the shoreline were carried down dip into the calcareous sands and muds accumulating offshore, thus triggering the process of dolomitization. The evaporitic basins were present through most of the time during which the Kathwai Member was deposited and disappeared by the close of that interval. Dolomitization and dedolomitization were essentially completed during Kathwai time which, on the basis of ammonite zonation, is believed to be of the order of magnitude of one to two million years.

Basal Mittiwali Member

The basal unit of the Mittiwali Member of the Mianwali Formation is the Lower Ceratite limestone of Waagen (1895). Most previous authors writing on Salt Range geology considered this the basal unit of the Triassic formations. The unit is extremely homogeneous in thickness and lithology throughout most of the Salt Range and Trans-Indus ranges. The thickness is generally between four and seven feet except in the Khisor Range where Teichert measured 25 feet. It is of interest to note that in the Salt Range and Surghar Range this unit, with the Kathwai Member, forms a platelike body. The westward thickening of the Triassic formations takes place in the overlying units of the Mittiwali Member (Fig. 3).

The medium to light gray, fine-grained, thin-bedded limestone is composed almost entirely of shell fragments, and it differs from the underlying limestone unit of the Kathwai Member in the scarcity of glauconite and the great abundance of ammonoids (Pl. 3, figs. 7, 8, 9). The matrix is generally very finely crystalline, containing a minor amount of dolomite, generally in the form of very small rhombohedra, 50-75 μ in diameter. In one thin section, out of five studied, a minute quantity of silt-size detrital quartz was seen.

In many places the unit is a coquina of ammonoids; pelecypods, very rare, tiny, *Soleniscus*-like gastropods, and occasional fish teeth are the only other fossils present. The preservation of the fossils in this unit is generally very poor. A well-preserved fauna was found only in Munta Nala near Wargal (loc. 7). The most common ammonoid genera are *Gyronites, Prionolobus,* and *Proptychites.* These ammonites belong to the Gyronitan age of Spath (1934), the second zone in the sequence of Scythian ammonoid zones.

COMPARISON WITH EARLIER OBSERVATIONS

The detailed stratigraphic and paleontologic data now available lead to a clearer understanding of reports by earlier authors on placement of the Permian-Triassic boundary in our area. One of the clues to such understanding is the fact that many observers have identified the rocks of the Kathwai Member as sandstone or calcareous sandstone and described its color as "rusty" or "rusty brown." Thus, in the Khura section measured by Wynne and Waagen (Wynne, 1878; p. 8 of this report), the 3-foot bed of brown dolomite at the top of the "Carboniferous" sequence is no doubt the dolomite unit with *Ophiceras* which we place in the Lower Triassic. The overlying 5 feet of "yellow, sandy, calcareous beds" is the limestone unit of the Kathwai Member. In the Wargal section of the same authors (p. 8 of this report), the 6-foot "gray sandstone layer" at the top of the "Carboniferous" sequence appears to be the dolomite unit of the Kathwai Member and may include the limestone unit as well.

The section measured by Wynne and Waagen at Siran-i-dok near Amb (p. 8 of this report) furnished the data from which Waagen (1891, p. 227) concluded presence of a marked erosional interval between Permian and Triassic strata in this area. However, we can be certain that the 6-foot bed of "rusty dolomite" at the top of Wynne and Waagen's "Carboniferous" represents the dolomite unit of the Kathwai Member, and that their "basal Triassic" unit, a brown conglomerate bed, is not a basal conglomerate. It is more likely to be a facies of either the limestone unit or the Lower Ceratite limestone of some peculiar texture or structure.

Noetling (1900a, 1901, 1905) published descriptions of a stratigraphic section near Chhidru which he seems to have measured at or close to the site of our locality 6B (Chhidru B). He called this the "Mittiali-Schlucht," an obvious error for Mittiwali (see this report, Appendix C). In an earlier paper (Kummel and Teichert, 1966, p. 323) we attempted to interpret Noetling's section in light of present knowledge, concluding that his bed 17 is equivalent to our lowest bed of the dolomite unit of the Kathwai Member, that is, bed 3 of our locality 6A (Chhidru A), and that Noetling's placement of the Permian-Triassic boundary between his beds 16 and 17 agrees with our own. Renewed scrutiny of Noetling's section description, however, has raised doubts in our minds.

TABLE 3. Geologic Age Assignments of Boundary Beds by Different Authors

Stratigraphic field units (this report)				Geologic age assignments				
				This report	Waagen and Wynne, 1878	Noetling, 1901	Gee, 1947	Schindewolf, 1954
Lowermost part of Mianwali Formation	Lower part of Mittiwali Member	Limestone unit		Triassic	TRIASSIC	TRIASSIC	TRIASSIC	TRIASSIC
		Dolomite unit			?		?	
	Kathwai Member							
Uppermost part of Chhidru Formation	White sandstone unit			Permian	PERMIAN	PERMIAN	PERMIAN	PERMIAN
	Uppermost richly fossiliferous unit							

The most profound change in lithology in the Mittiwali section is that from "Lower Ceratite limestone" to "Ceratite marl" (see Appendix A, loc. 6B). This is the boundary between Noetling's beds 21 and 22. In our section this boundary is placed 24.4 feet above the base of the dolomite unit of the Kathwai Member. In Noetling's section this is the position of the base of his bed 12 (4 inches of "brauner Thon"). Allowing for minor errors in measurements it now seems to us that the correlation of Noetling's section with ours is as follows:

Noetling	*This report*
Bed 22	"Ceratite marl"
Beds 19-21	Beds 12-16 ("Lower Ceratite limestone")
Beds 16-18	Beds 8-11 (limestone unit)
Beds 13-15	Beds 3-7 (dolomite unit)
Beds 10-12	Bed 2 (white sandstone unit)
Bed 9	Bed 1 (highest fossiliferous unit of Chhidru formation)

Bed 9 is the highest in Noetling's section from which he reports abundant Permian fossils. His mention of probable occurrence of *"Bellerophon* or *Stachella"* in beds 13 and 15 could have been based on observations of other fossils, perhaps poor impressions of ammonoid shells. The only discrepancy in this interpretation is in the thickness of the white sandstone unit which we measured as 12 feet, whereas in Noetling's section beds 10-12 have an aggregate thickness of 17 feet 10 inches. Such a discrepancy might be explained by the fact that the white sandstone unit in this locality is rather difficult of access.

In the discussion of his section Noetling expressed doubt whether to assign beds 10-16 to the Permian or the Triassic, but decided in favor of the latter. Thus, Noetling's Permian-Triassic boundary coincides approximately with that between the dolomite unit and the limestone unit of this report, that is, the same position in which Waagen and Wynne had placed it previously.

Noetling also measured a section in Munta Nala, which he called "Muntanar-Schlucht," near Wargal (Noetling, 1901, p. 454), which corresponds to our locality 7. He noted the prevalence of shale in the section and, after some discussion, placed the Permian-Triassic section between his beds 7 and 8, which would seem to correspond to the boundary between beds 5 and 6 in our section (see Appendix A, loc. 7).

In Gee's section of Nammal Gorge (Gee, 1947, p. 147) the dolomite unit of the Kathwai Member is represented by the uppermost bed c of the "Upper *Productus* beds" (see p. 13 of this report). The immediately underlying bed b ("soft, gray sandy shale, 2.5-3 feet") is easily recognizable as the white sandstone unit of the present report.

Our locality 6A (west of Chhidru Nala) is the same as that described by Schindewolf (1954, p. 155-157). His beds 1-12 are the fossiliferous beds of the Chhidru Formation, 12 corresponds to our "highly fossiliferous unit," bed 1. Schindewolf's beds 13 and 14 are the white sandstone unit of the Chhidru Formation (our bed 2); beds 15 and 16 are the dolomite unit of the Kathwai Member (our beds 3-8); beds 17 and 18 are the limestone unit of the Kathwai Member (ours beds 9, 10); and beds 19-22 are the "Lower Ceratite limestone" of the Mittiwali Member (our bed 11).

Most authors before Schindewolf seem to have included the limestone unit of the Kathwai Member in the overlying "Lower Ceratite limestone." However, it is not always possible to be sure.

The geologic age assignments of the various boundary beds by different authors are summarized in Table 3.

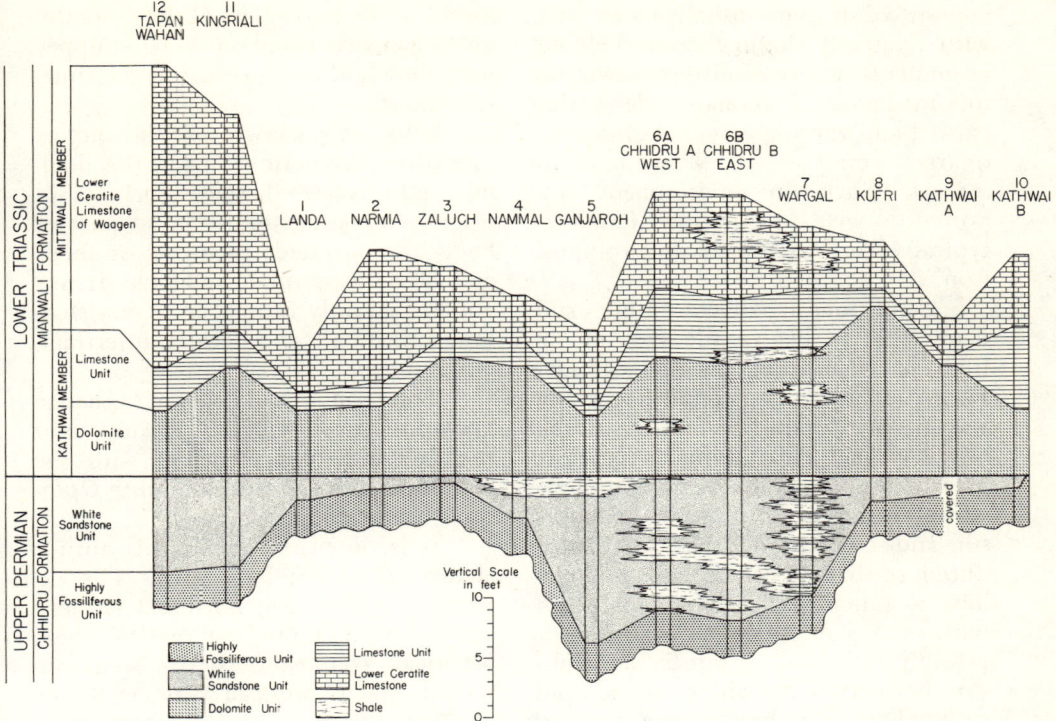

FIG. 17. Facies relationships of latest Permian and earliest Triassic strata in the Salt Range and Trans-Indus ranges of West Pakistan.

NATURE OF CHANGES AT PERMIAN-TRIASSIC BOUNDARY

DEFINITION OF BOUNDARY

Our criterion for definition of the Permian-Triassic boundary is the first appearance of *Ophiceras*. Kummel (1970) has reviewed the arguments in some detail and the reader is referred to his publication. In the Salt Range and Trans-Indus ranges, *Ophiceras* makes its appearance in the Kathwai Member of the Mianwali Formation. Most specimens are poorly preserved, but, with the exception of one specimen assignable to *Glyptophiceras,* all identifiable specimens of ammonoids were found to belong to *Ophiceras connectens* Schindewolf. We conclude, therefore, that poorly preserved specimens of size and shape similar to that species are conspecific. Such specimens have been seen by us at the bottom surface of the Kathwai Member. Thus, we regard the base of this member as marking the base of the Triassic System in the area under consideration.

It has been shown that in our view all previous authors have placed the Permian-Triassic boundary either a little higher or a little lower in the stratigraphic sequence than the datum chosen by us. Although the differences are of the order of only a few feet, the former placements of the boundary obscured the important fact that an abrupt change in lithofacies occurs at the boundary of the two systems as identified by us, namely between the white sandstone unit of the Chhidru Formation and the basal dolomite unit of the Kathwai Member of the Mianwali Formation (Fig. 17).

LITHOFACIES CHANGES

Rocks and fossils of the Chhidru Formation characterize this unit as a deposit of a type usually associated with relatively shallow-water shelf environments, where conditions favorable to moderate carbonate deposition existed and varying amounts of detrital quartz were added from external sources. Such an environment supported a rich invertebrate fauna of typically Permian aspect and composition.

Conditions changed rather suddenly shortly before the end of the time of deposition of the Chhidru Formation, when carbonate formation was sharply diminished and the supply of fine-grained detrital quartz increased greatly, leading to the formation of the white sandstone unit. Also, in places, silt and clay-size material was introduced, so that the white sandstone may have as much as 40 percent shale content, or a sand/shale ratio of 3:2, or possibly 1:1. Shale is rare in the Chhidru Formation anywhere stratigraphically above the basal shale member that has been described above.

The paucity of fossils in the white sandstone unit, especially in the Salt Range and Surghar Range, is in marked contrast to fossil abundance in the immediately underlying beds. However, in spite of these differences the white sandstone unit is lithologically and paleontologically closely linked with the Chhidru Formation and is, therefore, considered to be part of it.

In contrast, the original depositional environment of the overlying basal dolomite unit of the Kathwai Member of the Mianwali Formation was one conducive to the formation of calcarenite, with addition of only small quantities of detrital quartz and a small amount of glauconite formation. The lowest few feet of the dolomite unit may contain as much as 10 percent detrital quartz in silt-size and fine-grained sand range. The upper part of the dolomite unit generally, the limestone unit everywhere, have little, if any, detrital quartz.

Glauconite, which is very rarely found in the lower part of the dolomite unit, is more common in the upper part, and is always present in the limestone unit.

It has been shown that, in spite of intensive diagenetic changes, the dolomite unit is very largely bioclastic in origin, consisting for the most part of finely comminuted fragments of shells and of echinodermal skeletons. It was almost certainly formed in a very shallow, infratidal, and locally intertidal, environment.

The pronounced lithofacies change between the white sandstone unit and the dolomite unit coincides with the first appearance of the ammonite *Ophiceras* in the section.

It is noteworthy that this abrupt change affected a large area. The extreme northwestern, northeastern, and southwestern points at which these identical conditions have been observed, that is, western Surghar Range (Landa Pusha, loc. 1), central Salt Range (Kathwai, loc. 10), and southern Khisor Range (Tapan Wahan, loc. 12), form a triangle covering about 3,250 square miles. Most probably the original size of the area throughout which these conditions prevailed was even larger.

COMPARATIVE PALEONTOLOGY OF BEDS ABOVE AND BELOW THE BOUNDARY

The nature of the faunal changes at the boundary of the Permian and Triassic systems in West Pakistan as defined by us can now be assessed and evaluated in much greater detail than has been possible previously. Here we summarize knowledge of the distribution of phyla and major classes in the beds immediately below and above this boundary, although information on fossil distributions in lower as well as higher strata is given where relevant to our problem. Most of the data are culled from the paleontological papers assembled in this volume, but some are based on our own observations and identifications, or extracted from the literature.

Foraminiferida

In the Zaluch Group Fusulinacea seem to be restricted to the Amb Formation, where *Parafusulina kattaensis* and *Monodiexodina* sp. are abundant (Dunbar, 1933; Douglass, 1968), but few other fusulinid species occur. Endothyracea occur throughout the Zaluch Group up to and including the white sandstone unit of the Chhidru Formation where they are represented by *Climaccamina* sp., *Brunsiina?* sp. and possibly other genera. A common endothyracean genus in the Wargal Limestone, more rarely represented in the Chhidru Formation, is *Colaniella* (identified by L. G. Henbest). This form has not been found in the white sandstone unit.

No fusulinids or other foraminifers have been identified from any of the Triassic formations of the Salt Range and Trans-Indus ranges.

Anthozoa

Corals from the Zaluch Group have been described by Waagen and Wentzel (in Waagen, 1886, 1888) and by Heritsch (1937). They are common locally in the Wargal Limestone from which Waagen and Wentzel described eight species, most of which are in need of revision. To these Heritsch added two new species, *Hapsiphyllum indicum* and *Bradyphyllum indicum*, both from the Wargal Limestone.

According to Heritsch, one Wargal Limestone species, *Waagenophyllum indicum* Waagen and Wentzel, was found in the Chhidru Formation by Koken, identified by him as *Lonsdaleia* n. sp. (Koken, 1907, p. 474). A small colonial form in the Noetling collection, coming from the "uppermost" Productus limestone at Chhidru, was described by Heritsch as *Wentzelella timorica* Gerth.

No corals are known from any Triassic rocks in the Salt Range and Trans-Indus ranges.

Bryozoa (Ectoprocta)

The Bryozoa of the Zaluch Group have not been studied since publication of work by Waagen and Wentzel (in Waagen, 1886). Pascoe (1959, p. 770, 776) listed 15 species of cyclostome, trepostome, and fenestrate Bryozoa from the Chhidru Formation, but generally they are not common, and seem to become rarer in the upper part of the formation. A fragment of a trepostome bryozoan was seen by us in one thin section of calcareous sandstone from the white sandstone unit.

A bryozoan fragment was seen in a thin section from the basal 8 inches of the Kathwai Member at Narmia which we regard as Triassic in age, but which, according to Grant (1970), contains a brachiopod fauna of Dzhulfian affinities.

Brachiopoda

Inarticulata

The only inarticulate brachiopod ever recorded from the Zaluch Group is *Lingula scrutata* Reed (1931), which occurs in the Wargal Limestone. No inarticulate brachiopods are known from the Chhidru Formation.

In the Kathwai Member a lingulid described by Rowell (1970) as *Lingula* cf. *L. borealis* Bittner is reasonably common in both the dolomite and limesone units. Specimens identified in the field as *Orbiculoidea* sp. were recorded by us from the lower part of the dolomite unit as well as from the limestone unit at Kathwai B (loc. 10) and from the limestone unit at Kufri (loc. 8), but were either not collected or lost later. Schindewolf (1954) reported *Orbiculoidea* sp. from his Schicht 17 (=our bed 9 of limestone unit at Chhidru A, loc. 6A).

Articulata

The Zaluch Group is characterized by a great abundance of articulate brachiopods. An apparently complete, though uncritical, list of all described species and varieties was published by Pascoe (1959, p. 758-768, 771-773, 777-778). From the Chhidru Formation alone Pascoe listed 133 species of brachiopods. Without doubt, a revision of this fauna, now being undertaken by R. E. Grant, will result in reduction of the number of recognized species.

It seems that, in the Salt Range at least, the number of brachiopod species decreases toward the top of the Chhidru Formation. From his Schicht 12 at Chhidru, which corresponds to our bed 1 in locality 6A (Chhidru A), Schindewolf recorded *Enteletes* sp., *Derbyia hemisphaerica* Waagen, *Dictyoclostus indicus* Waagen, *Waagenoconcha purdoni* Davidson, *Spirigerella derbyi* Waagen, and *Hemiptychina* sp. Although we did not recognize all of these species at this locality, the assemblage seems to be representative of the uppermost, highly fossiliferous bed immediately below the white sandstone unit.

Highly fossiliferous pockets were found in the white sandstone unit in the Khisor Range only. From these Grant has described the 22 species of articulate brachiopods which have been listed on p. 50 (Grant, 1970; this paper, Appendix A, loc. 11, 12).

Grant (1970) concludes that this is a typical Chhidru fauna which is closely related to that of the Wargal Limestone of Guadalupian age. Our views concerning the correlation of the Chhidru formation are stated below.

The Kathwai Member of the Mianwali Formation generally contains rare and poorly preserved rhynchonellid-type brachiopods which cannot be identified. However, better preserved brachiopods were found in the member in three localities: Narmia Spring, Surghar Range (loc. 2), and Chhidru A (loc. 6A), and Khan Zaman Nala (no loc. number, see Fig. 2), Salt Range. From these localities Grant (1970) described the following brachiopods.

Narmia Spring (loc. 2), lowermost 6-12 inches of Kathwai Member: *Enteletes* sp. 2, *Orthothetina* cf. *O. arakeljani* Sokolskaya, *Orthothetina* sp., *Ombonia* sp., *Derbyia?* sp., *Spinomarginifera* sp., *Linoproductus* sp., *Lyttonia* sp., *Spirigerella* sp., *Crurithyris? extima* Grant, *Martinia* sp., *Whitspakia* sp. 2, Dielasmatids, gen. et sp. indet.

Chhidru A (loc. 6A), 6 inches above base of member: *Spinomarginifera* sp.

Khan Zaman Nala, 5-6 feet above base of member: *Crurithyris? extima* Grant.

Grant concludes that the fauna contained in the lowermost 6-12 inches of the Kathwai Member is of Dzhulfian age. Our views on this remarkable assemblage and its possible derivation are set forth below (p. 76-77).

Bivalvia

Bivalves are fairly common in the Zaluch Group. Pascoe (1959, p. 773-774, 777) listed 54 species from the Chhidru Formation alone. Schindewolf recorded *Eoschizodus pinguis* Waagen, *Pleurophorus* cf. *P. subovalis* (author?), as well as unspecified pectinids, limids, and mytilids, from the uppermost highly fossiliferous bed of the Chhidru Formation at our locality 6A (Chhidru A). The fact that bivalves have not been recorded from the white sandstone unit may be explained by the general scarcity of fossils in this unit.

In the Kathwai Member we found remains of bivalves in 6 out of our 13 localities, but all of them were extremely poorly preserved. The collections were submitted for study to Norman D. Newell, who reported: "I have been unable to isolate any forms that could be objectively recognized save one ammusioid scallop quite similar to

Pernopecten or *Entolium*. Both are long-ranging genera, the former Carboniferous and Permian, the latter, Jurassic. The Salt Range species can as well be assigned to one as the other. The shells certainly do not show indication of reworking" (N. D. Newell, written communication, February 7, 1968).

Bivalves occur also in the Mittiwali Member and Narmia Member of the Mianwali Formation. A coquina limestone bed, about 5 feet thick, forms the base of the Narmia Member in the Salt Range and Khisor Range, but its fauna has never been studied.

Gastropoda

Gastropods are reasonably common in rocks of the Zaluch Group. Pascoe (1959, p. 774-775, 777) listed 50 species as occurring in the Chhidru Formation, among them 21 species of bellerophontids, including species assigned to *Bellerophon*, *Stachella*, *Warthia*, and *Euphemus (recte Euphemites)*. Bellerophontids are indeed common to abundant in many beds of the Chhidru Formation, including the highly fossiliferous beds immediately below the white sandstone unit.

No gastropods have so far been found in the white sandstone unit of the Chhidru Formation.

In the Kathwai Member of the Mianwali Formation numerous small high-spired gastropods, 3-4 mm long, were seen in a thin section from the limestone unit at Narmia Spring. In no other locality were gastropods noted in this member.

The bellerophontid *Stachella* has long been recorded from the upper part of the Mittiwali Member. Specimens were seen by Kummel only in the "Ceratite sandstone" (for discussion see Kummel, 1966, p. 371, 379). Peculiarly, despite its unusual Triassic occurrence, the species seems never to have been described.

Cephalopoda

The Chhidru Formation has yielded about a dozen species of nautiloids which belong to five families: Tainoceratidae *(Tainoceras, Pleuronautilus, Parametacoceras)*; Koninckioceratidae *(Temnocheilus, Planetoceras)*; Grypoceratidae *(Pselioceras, Stearoceras, Virgaloceras)*; Solenochilidae *(Solenochilus)*; and Liroceratidae *(Liroceras)*. The tainoceratid genera *Tainoceras* and *Pleuronautilus* are also known from Triassic rocks, but only *Pleuronautilus* has also been recorded from the Triassic of the Salt Range. Of the other families, only the Grypoceratidae and Liroceratidae are known to continue on into the Triassic, but none of the genera by which they are represented in the Chhidru Formation does so. The nautiloid fauna of the Chhidru Formation thus has a decidedly Paleozoic aspect, containing only two genera (out of ten) that survived into the Triassic.

No nautiloids have been found in the white sandstone unit nor in the immediately underlying highly fossiliferous bed.

The Chhidru Formation contains the following ammonoids (Teichert, 1966; Furnish and Glenister, 1970): *Stacheoceras antiquum* (Waagen), *Cyclolobus oldhami* (Waagen), *C. teicherti* Furnish and Glenister, *C.* cf. *C. walkeri* Diener, *Eumedlicottia primas* (Waagen), *Episageceras wynnei* (Waagen), *Xenodiscus carbonarius* (Waagen), *X. plicatus* Waagen. All are rare to very rare. Only a few specimens of *Stacheoceras*, *Eumedlicottia*, and *Episageceras* have ever been found, whereas *Cyclolobus* and *Xenodiscus*, though far from common, are not quite so rare. Grant (1968) believed that the holotype of *Cyclolobus oldhami* came from the Kalabagh Member of the Wargal Formation, but the highest authentically recorded occurrences of all three species of *Cyclolobus* are from 36 to 60, or possibly 77, feet below the top of the Chhidru Formation. The genus thus is characteristic of the uppermost Chhidru Formation, though it may range downward into the uppermost Wargal Limestone. Its significance for problems of correlation of the Chhidru Formation is discussed elsewhere in this report.

The only known cephalopods in the Kathwai Member are *Menuthionautilus kieslingeri* Collignon, *Ophiceras connectens* Schindewolf, and *Glyptophiceras himalayanum* (Griesbach) (Kummel, 1966, 1970). The first and last species are rare, but *Ophiceras connectens* is locally abundant, though mostly very poorly preserved. Parts of the Kathwai Member, both the dolomite and limestone units, consist of a coquina of comminuted shell fragments. It is reasonable to assume that the shells of these species, especially *Ophiceras,* contributed significantly to the construction of such rock types.

The ammonoids of the higher parts of the Mianwali Formation (Mittiwali Member, Narmia Member) have been listed and in part described by Kummel (1966) and these are not dealt with here. In the Mittiwali Member the nautiloids *Taenionautilus trachyceras* Frech, *Menuthionautilus kieslingeri* Collignon, *Grypoceras bidorsatoides* Kummel, *G. aemulans* Kummel, and *Pleuronautilus kokeni* Frech occur (Kummel, 1953). The Narmia Member has yielded representatives of *Enoploceras* (Kummel, 1966) and fragmentary orthoconic shells that probably belong to *Michelinoceras.*

Scaphopoda

Scaphopods are common in the Chhidru Formation where they are represented by a large species, *Plagioglypta herculea* (de Koninck), also reported from Bokhara and Japan. No scaphopods are known from any of the Triassic rocks in the Salt Range and Trans-Indus ranges.

Trilobitae

Trilobites have only recently become known from the Permian of the Salt Range, when Grant (1966) described *Ditomopyge fatmii* and *Kathwaia capitorosa* from the Kalabagh Member of the Wargal Limestone. They are rare and among the latest known trilobites in the world.

No trilobites have been found in the Chhidru Formation and none in Triassic rocks.

Ostracoda

Ostracodes are found in moderate numbers in all formations of the Zaluch Group. Since it appears that only one ostracode has ever been described from Permian rocks of the Salt Range and Trans-Indus ranges (Waagen, 1879, p. 22), we have compiled the following list of identifications made by I. G. Sohn (written communication, 1966), based on collections made by Teichert in 1961-1962 and by R. E. Grant and A. Fatmi in 1964.

Amb Formation

　　Youngiellacea?, gen. indet.
　　Geisiinidae, gen. indet.
　　Bairdiacypris? sp.
　　Pseudobythocypris? sp.
　　Cavellina sp.

Kalabagh Member of Wargal Limestone

　　Amphissites n. sp. 1
　　Roundyella? sp.
　　Youngiella? sp.
　　Geisina? sp.
　　Hypotetragona sp.
　　Bairdia sp.
　　Acratia? sp.
　　Bairdiacypris? sp.
　　Pseudobythocypris? sp.
　　Cavellina sp.
　　Cytherelloidea? sp.
　　Healdia spp.
　　Microcheilinella sp.
　　Silenites? sp.

[In addition, Sohn (1970) reports the occurrence of *Ceratobairdia* from an unspecified locality of the "Middle Productus limestone."]

Chhidru Formation (mostly in basal shale member)

 Youngiellacea, gen. indet.
 Geisina? sp.
 Bairdia spp.
 Cavellina sp.
 Graphiadactyllis? sp. aff.
  Basslerella australae
  Crespin, 1945
 Cytherelloidea? sp.
 Silenites sp.

The many uncertainties in identification of the species are mainly due to poor preservation of the fossils. In addition to the above-mentioned species Sohn (1970) lists *Glyptopleuroides* and *Carboprimitia* from unspecified horizons in the "Productus limestone" (=Zaluch Group).

Ostracodes were seen in the Kathwai Member of the Mianwali Formation in only one thin section (Pl. 2, fig. 8), but no identifiable specimens were recovered from this unit. However, Kummel's collections from the Mittiwali and Narmia members have yielded a small ostracode fauna of highly interesting composition which has been described by Sohn (1968, 1970):

Mittiwali Member

 Carinaknightina carinata
  Sohn
 C. discarinata Sohn
 Judahella? sp.
 Monoceratina? sp.
 Lutkevichinella? ornata Sohn
 Hungarella? sp.

Narmia Member

 Kirkbyidae gen. indet., sp.
  or spp.
 Bairdia? sp.
 Bairdiacypris? sp.
 Microcheilinella sp.
 Reubenella? sp.

Of these ostracodes *Lutkevichinella? ornata*, *Hungarella?* sp., *Judahella?* sp., and *Reubenella?* sp. "are Mesozoic and younger in aspect" (Sohn, 1970). Kirkbyids are typically Late Paleozoic, although a few records of Early Triassic occurrences exist. *Carinaknightina* Sohn (1970) also is of kirkbyid affinities. *Bairdiacypris* and *Microcheilinella* have not previously been reported from Triassic rocks.

ECHINODERMATA

Echinodermal fragments are common in the Permian and Triassic rocks of the Salt Range and Trans-Indus ranges, but identifiable specimens are extremely rare.

Crinoidea. Waagen (1885) described several species of crinoids from the Salt Range, almost all from the Wargal Limestone. These need not concern us here, because no identifiable crinoids are known with certainty from either the Chhidru or Mianwali formations. In places the Kathwai Member is composed of debris of echinoderm skeletons, including very small crinoid stems, and these can also be seen on weathered surfaces at many places.

Ophiuroidea. No ophiuran remains are known from rocks of the Zaluch Group. However, ophiuroids contribute to the echinodermal debris found in the Kathwai Member, as shown by the find of an arm fragment which Dr. Hans Hess of Basel (written communication, 1968) identified as *Aplocoma* cf. *A. torrii* (Desio). This species, described by Desio (1951) as *Ophioderma,* from the Upper Triassic (Rhaetic) of the Bergamo Alps in northern Italy, was redescribed by Hess (1965). The only other record of occurrence of ophiuroids in rocks of Scythian age appears to come from Kueichou, China, where Yang (1960) described a new species under the name of *Ophioderma schistovertebrata*.

Echinoidea. The only echinoid known from the Zaluch Group is

Archaeocidaris forbesianus (de Koninck) from the Wargal Limestone, variously assigned to *Cidaris, Eocidaris,* and *Permocidaris*. We here follow Fell (1966, p. *U*317) in assigning the species to *Archaeocidaris*.

Some beds of the Kathwai Member of the Mianwali Formation are rich in echinoid spines and cross sections of them are seen in many rock thin sections. However, only one somewhat damaged test with spines attached has been found. It was described as *Miocidaris pakistanensis* by Linck (1955). From the rather uniform appearance of the detached spines found in many localities it seems reasonable to conclude that only this single species of echinoids occurs in the Kathwai Member.

According to Fell (1966), the range of *Archaeocidaris* is Lower Carboniferous to Permian, the Salt Range occurrence being the only record in the Permian. The range of *Miocidaris* is ?Lower Carboniferous and Permian to Lower Jurassic. *Miocidaris* is the only genus of Paleozoic echinoids known to survive in post-Paleozoic time.

Conodontophorida

Although Teichert collected several suites of samples from the entire Zaluch Group, W. C. Sweet was unable to recover identifiable conodont assemblages from these samples, except in the uppermost few feet of the Chhidru Formation where they are relatively common. Conodonts are very abundant in the Kathwai Member and higher members of the Mianwali Formation. Sweet (1967, 1970) reported that 99 samples taken by us from these rocks yielded 21,459 identifiable conodont elements, separable into 52 morphologic categories, or form species.

In the uppermost Permian and Lower Triassic sequence of the Salt Range and Trans-Indus ranges, Sweet was able to establish presence of nine biostratigraphic zones based on conodont assemblages. These are named:

9. Zone of *Neospathodus timorensis*
8. Zone of *Neogondolella jubata*
7. Zone of *Neospathodus waageni*
6. Zone of *Neospathodus pakistanensis*
5. Zone of *Neospathodus cristagalli*
4. Zone of *Neospathodus dieneri*
3. Zone of *Neospathodus kummeli*
2. Zone of *Neogondolella carinata*
1. Zone of *Anchignathodus typicalis*

The distribution of the conodont zones in relation to lithostratigraphic divisions shows some interesting features (Sweet, 1970, fig. 2, 3).

Zone 1, which yielded 4,998 conodont elements, embraces the uppermost part of the Chhidru Formation and the entire dolomite unit of the Kathwai Member, except at Kathwai A (loc. 9), where at least the upper half of the dolomite unit belongs to conodont zone 2, and at Kingriali (loc. 11), where the base of the limestone unit belongs in zone 1. The lower boundary of zone 1 has not been defined, but the characteristic zone assemblage has been found both in the white sandstone unit (Chhidru A, Tapan Wahan), and in the uppermost highly fossiliferous unit of the Chhidru Formation (Chhidru B, Wargal). Zone 1 thus straddles the Permian-Triassic boundary as defined in this paper. Sweet has studied conodonts from the basal 6-12 inches of the Kathwai Member at Narmia Spring (loc. 2) and concluded that it is unlikely that they were reworked. This is the bed from which Grant (1970) describes a brachiopod fauna that he believes to be Dzhulfian in age. These conflicting conclusions are discussed in more detail below.

Conodont zone 2, which yielded 2,518 conodont elements, is essentially identical with the limestone unit of the Kathwai Member; only at Chhidru A (loc. 6A) does it embrace at least the lowermost bed of the "Lower Ceratite limestone" of the Mittiwali Member. Unfortunately, this is the only section for which sufficiently closely spaced sample control is available. Therefore, it is possible that zone 2 includes the base of the Mittiwali Member elsewhere.

As far as available sample control allows one to judge, zone 3, which has yielded 638 conodont elements, seems to lie entirely in the lower part of the "Lower Ceratite limestone," though samples are available from only five localities.

Zone 4, which has yielded 3,823 conodont elements, is generally coincident with the upper part of the "Lower Ceratite limestone," but anomalously at Nammal Gorge (loc. 4), it includes the basal part of the "Ceratite marl."

It is most likely that zones 5, 6, and 7 are confined within the limits of the Mittiwali Member, and zones 8 and 9 within those of the Narmia Member, but samples are relatively few and widely spaced. Thus, while there is no doubt about the reality of the conodont zones themselves, the exact positions of their boundaries in relation to the lithostratigraphic framework is somewhat uncertain and may differ somewhat from those shown by Sweet (1970) which were drawn in consultation with us. The number of conodont elements recovered from rocks of zone 5 is 6,934, of zone 6, 809, of zone 7, 268, of zone 8, 1,159, and of zone 9, 312.

In summary, it may be stated that some conodont zone boundaries are transgressive across lithostratigraphic boundaries, but the extent of such discrepancies is insufficiently documented.

Chondrichthyes, Osteichthyes

From fragmentary material in the "Productus limestone," mostly teeth and spines, Waagen (1879, 1880) described nine species of fossil fish, most of them of holocephalian or elasmobranch affinities, including some bradyodonts. Of these, at least two, *Xystracanthus gigas* and *Thaumatacanthus blanfordi*, an ichthyodorulite, are described as coming from the "uppermost" or "highest" beds of the "Productus limestone." Since all of Waagen's species are almost certainly in need of revision, they will not be considered further.

Branson (1935) described an edestid, *Helicampodus kokeni*, from rocks that seem to belong to the Chhidru Formation and included in this species a specimen that Koken (1901) had described as *Helicoprion* sp. *Helicampodus* was not found by us in rocks younger than Permian, but it is interesting to note that its occurrence in rocks regarded as earliest Triassic has been recorded in the Dzhulfa section of the Armenian SSR by Obruchev (*in* Ruzhentsev and Sarycheva, 1965).

From various levels of his "Ceratite formation" Waagen (1895) described several species of fishes which he assigned to the genera *Saurichthys?*, *Colobodus*, *Gyrolepis*, and *Acrodus*. Included was one *Acrodus* sp. indet. from "a limestone at Chidroo" which might well have come from the Kathwai Member.

Schindewolf (1954) listed *Acrodus* sp. and *Colobodus* sp. from the lower part of the Kathwai Member at Chhidru in a stratigraphic position corresponding approximately to our locality 6A, beds 4-7. We collected fish teeth and spines from the top of the dolomite unit (our bed 8) at the same locality. Elsewhere we found fish remains in the upper dolomite unit at Chhidru B (loc. 6B), in the upper dolomite unit, the lower limestone unit, and in the "Lower Ceratite limestone" at Wargal (loc. 7), and in the upper dolomite unit at Kufri (loc. 8).

Acritarcha

Discovery of acritarchs in our samples is due to Balme (1970) who found them to be abundant in both Permian and Triassic rocks. He reports that in the Chhidru Formation leiospheres and spinose acritarchs constitute 10 to 30 percent of total microfloral assemblages and that in the Kathwai Member acritarchs appear in huge numbers.

Sarjeant (1970) has so far described only the acritarchs of the white sandstone unit where 16 species are present, belonging to the genera *Micrhystridium, Veryhachium, Wilsonastrum, Polyedryxium, Leiofusa,* and *Deunffia. Micrhystridium* is represented by nine species, *Veryhachium* by three, and the other genera by one species each. In addition, *Tasmanites* occurs. Two species are left unnamed, two are new. The remaining nine species are known elsewhere to range from Permian to Triassic or even younger Mesozoic beds; one species was previously known only from the Jurassic, one from the Triassic; and only one species is elsewhere confined to the Permian. Thus, the acritarch assemblage has a decidedly Mesozoic aspect.

The acritarchs from Triassic rocks in our collections have not yet been studied. Balme (1970) noted a decline in acritarch assemblages in the Narmia Member of the Mittiwali Formation and in the Landa Member of the Tredian Formation. They are absent, of course, in the nonmarine Khatkiara Member.

Spores and Pollen

Sixty-one species of spores and pollen have been found in Permian rocks of the Salt Range (Balme, 1970), only 25 of which range throughout the Zaluch Group. Most important are 1) haploxylonoid, striatitid, and 2) nontaeniate, disaccate pollen grains. Among the first group Balme regards *Guttalapollenites hannonicus* as most unusual, because it was previously known only from the Permian and Triassic of Madagascar. Among the second group are the morphologically unusual *Pinuspollenites thoracatus* and the typically Zechstein species, *Klausipollenites schaubergeri*. These two groups, believed to be of gymnospermous origin, constitute 75 to 90 percent of the total spore-pollen assemblages in most samples.

The greatest diversity of spore-pollen assemblages is found in the upper part of the Chhidru Formation, due mainly to an increase in the variety of trilete and monolete forms, some of which are believed to be of lycopodiaceous origin. Curiously, a few of these are elsewhere more typical of Early Mesozoic strata. Two species, *Nevesisporites fossulatus* and *Tigrisporites playfordi,* are characteristic of the Triassic of Australia, *Guthoerlisporites cancellosus* of the Triassic, including Rhaetic, of Europe, and there are others. The total of spore-pollen form species in the Chhidru Formation is 57.

In the early Scythian (Kathwai Member, Mittiwali Member) the number of species was reduced to 18 (nine of spores, nine of pollen grains) of which 14 were survivors from the Chhidru Formation where 12 of them are extremely rare. Apart from *Taeniaesporites,* disaccate pollen grains are rare in these earliest Triassic rocks. The environmental significance of these changes in spore-pollen associations is considered below.

Another break in continuity of development of the spore-pollen associations occurs either in the upper part of the Mittiwali Member or near the base of the Narmia Member of the Mianwali Formation, as discussed by Balme (1970). From the prevalence of *Falcisporites,* Balme concludes that corystosperms were predominating forms in the Middle Triassic of the Salt Range, resembling in this respect contemporaneous floras of Australia.

On the whole, Balme finds the succession of spore assemblages in the Salt Range Triassic to be strikingly similar to that of the Perth Basin of Western Australia. An important link with Triassic floras in both the northern and southern hemisphere is provided by the genus *Atatrisporites* of lycopodiaceous origin.

Algae

Algae are comparatively rare in the Chhidru Formation. Fragments of dasycladaceans were seen in the uppermost highly fossiliferous bed near Narmia Spring (loc. 2) and calcispheres also occur here. *Permocalculus* (identified by J. Harlan Johnson) was found two feet below the top of the Chhidru Formation near Kathwai B (loc. 10) at a place where the white sandstone unit is not present. *Vermiporella* and perhaps other forms occur rarely in the white sandstone unit.

Remains of dasycladacean algae are extremely rare in the Kathwai Member, where they are probably restricted to the dolomite unit. The fragment illustrated on Plate 2, figure 5 is in a rock close to the top of the dolomite unit and is, thus, unlikely to be "derived."

Trace Fossils

No trace fossils have been seen in the Chhidru Formation, but they are not uncommon in the Kathwai Member of the Mianwali Formation, especially in the dolomite unit. Most are invertebrate trails, of which four or five different types can be distinguished, but burrows and resting marks also occur. The occurrences at locality 6A (west of Chhidru Nala) are described in some detail elsewhere in this report (p. 38-43). Invertebrate trails were also observed, but not studied in detail, in the dolomite unit in Nammal Gorge (loc. 4), in Munta Nala near Wargal (loc. 7), and near Kufri (loc. 8).

EVALUATION OF THE EVIDENCE

The observations and conclusions presented in this report, and in the companion papers by Balme, Furnish and Glenister, Grant, Rowell, Sarjeant, Sohn, and Sweet, make possible a drastic reevaluation of the stratigraphic and paleontologic evidence bearing on the problem of the Permian-Triassic boundary in West Pakistan. In 1966, we proposed, as "best suggestion," to "regard the contact between the Permian and Triassic systems in the Salt Range and Trans-Indus ranges as a paraconformity" (Kummel and Teichert, 1966, p. 330). We now judge that this suggestion is supported by a very large body of evidence.

Our reasons for placing the boundary between the Permian and Triassic systems in the Salt Range and adjacent ranges at the base of the first occurrence of *Ophiceras*, which coincides with the beginning of the deposition of the Kathwai Member (p. 36), are presented in detail by Kummel (1970). Here we undertake to evaluate the bearing of work by our specialist collaborators on the boundary problem with reference to the datum recognized by us. It is probably important to point out that the specialists who worked on our collections did so independently of one another and, very largely, without knowledge of their fellow workers' results and conclusions. We believe that, thus, a rather unique occasion has been created in which a diversity of talents is applied to the solution of a difficult geological problem.

Analysis of lithologic and diagenetic features of the earliest Triassic rocks, the Kathwai Member of the Mianwali Formation (p. 36-58), led us to conclude that early sedimentation of the dolomite unit took place, at least locally, in an intertidal environment and that an emergent area with evaporitic basins existed at an unknown, though not very great, distance from the Salt Range and neighboring areas during the time of deposition of the dolomite unit. At the same time it is possible that the bioclastic material of which the Kathwai Member is composed was brought into the area from the outside, suggesting existence of a shore line at a distance which may not have been great.

Our conclusions receive strong support from the palynological investigations carried out by Balme (1970), who points out (p. 431) that the abundance of lycopodiaceous spores in the Chhidru Formation may "reflect the colonization of coastal marshes by lycopodiaceous elements, which filled the ecological niche occupied today by halophytic families such as the Chenopodiaceae and Plumbaginaceae." He further states that "Taken in its entirety the palynological evidence points towards a strongly regressive phase at the top of the Chhidru Formation . . . ." The Permian-Triassic boundary, as defined in this report, is characterized by "a strong and complete palynological break," documented not only in West Pakistan, but also in Western Australia, Madagascar, and Arctic Russia (Balme, 1970, p. 431). For the earliest Triassic, Balme again stresses colonization of coastal areas, probably by lycopodiaceous forms, in an environment in which spores and pollen of upland floras "were effectively screened out of marine deposits."

The principal problem facing us now is determination of the magnitude of the stratigraphic break between the Chhidru and Mianwali formations, a question which is most intimately tied to determination of the age of the Chhidru Formation.

It has long been considered, on the basis of the presence of *Cyclolobus,* that the Chhidru Formation is of latest Permian age. In their review of the genus *Cyclolobus,* Furnish and Glenister (1970) come to the conclusion that the Salt Range cyclolobids "are referable to the middle Dzhulfian Chhidruan stage, intermediate between the Araksian and the Changhsingian." These authors extend the term Dzhulfian to embrace all post-Guadalupian Upper Permian rocks. They regard the unit as a series divided into three stages: Araksian, Chhidruan, and Changhsingian. The newly named Araksian corresponds to the type Dzhulfian of Ruzhentsev and Sarycheva (1965). The Changhsingian contains the Changhsing Limestone of Chao (1965) as type unit.

The thesis of Furnish and Glenister is that *Cyclolobus* is part of an evolutionary series containing *Timorites* of Guadalupian age as its immediate ancestor and *Changhsingoceras* (Chao, 1965), characteristic of the Changhsingian Stage, as an immediate descendant. This evolutionary series cannot be demonstrated in any one locality. Each cyclolobid occurrence consists essentially of a single species complex. Thus the assumed relative stratigraphic position of the various *Cyclolobus* species, found in different parts of the world, is based primarily on phylogenetic considerations. Accordingly, Furnish and Glenister are led to conclude that the Chhidru Formation is older than the Changhsing Limestone of South China but younger than strata assigned by Ruzhentsev and Sarycheva (1965) to the Dzhulfian.

On the other hand, it seems to us that the evidence presented by Furnish (1966) and by Furnish and Glenister (1970) might possibly be consistent with the existence in the Late Permian of one or two worldwide species of *Cyclolobus,* each represented by several local geographic races.

Grant (1968, 1970) believes to have recognized the locality from which Waagen obtained the first specimen and holotype of *Cyclolobus oldhami* and is convinced that the rocks here belong to the Kalabagh Member of the Wargal Limestone, not to the Chhidru Formation as generally believed. While, admittedly, the Kalabagh Member has a distinct nodular limestone lithology that is not easily confused with that of other parts of the Permian stratigraphic column in the Salt Range, nodular limestone beds do occur in the Chhidru Formation, though observed only in small thickness (Teichert, 1966, p. 15, 18). Furthermore, certain features of Waagen's (1872; 1879, p. 25) description of the type locality, not quoted by Grant, should be considered. After the description of the locality, which has been quoted in full by Grant (1968, p. 5), Waagen (1872, p. 352) continues:

> Above this thin-bedded calcareous series, with the fossils just mentioned, follows, at Jabbi, a brown sandy dolomitic rock, rather thick-bedded in which *Dentalium Herculeum*, Koninck, and *Bellerophon Jonesianum*, Koninck, are found in abundance. The series with these two species includes, all along the Salt Range, the highest beds in which *Producti* and *Athyris* sometimes occur. Above them follow everywhere limestones and green marls, with the *Ceratites* described by Koninck . . . .

In our opinion this passage is important, for the description refers without any doubt to the Chhidru Formation and with great probability to its upper part, where bellerophons and *Plagioglypta ("Dentalium") herculea* are very common. If the Chhidru Formation here rests on the Kalabagh Member, it would have been almost impossible for Waagen to miss the shale member at the base of the Chhidru which nowhere in observed outcrops is found to be less than 25 feet thick.

That strata next above the cephalopod beds were uppermost "Upper Productus limestone" must also have been Waagen's impression, for in 1891 (p. 241) he placed the Cephalopoda beds (=Jabbi beds) in the middle of that unit. In 1891 (p. 208) he indicated that the top of the Cephalopoda beds was 10 feet below the top of the Upper Productus limestone.

It should also be mentioned that all specimens of *Cyclolobus* from Pakistan studied by Furnish and Glenister (1970) for which accurate stratigraphic occurrence data are available come from the Chhidru Formation and were found in beds 36 to 60, or possibly 77, feet below its top. No authentic find of *Cyclolobus* has been reported from the lower half of the Chhidru Formation.

In striking contrast to the conclusion just stated based on ammonoids, Grant (1970), in his study of the brachiopods, concludes that "the Chhidru Formation is not Dzhulfian in age, but is lower to middle Guadalupian, and that links with the Dzhulfian begin in the Salt Range only in the base of the overlying Mianwali Formation. The further implication is that most, if not all, of the Dzhulfian is absent from the Salt Range, and that the basal part of the Kathwai Member of the Mianwali Formation represents the minimal deposit of what was essentially a depositional hiatus in this region" (Grant, 1970, p. 117).

In the opinion of Grant (1970, p. 120) the brachiopod fauna of the white sandstone unit of the Chhidru Formation is essentially a typical, though impoverished, Chhidru fauna. Argumentation for a Guadalupian age of this brachiopod assemblage is detailed and complex as shown by Grant's paper in this volume (Grant, 1970). Difficulties in age assignment of the brachiopod assemblage and philosophies followed are summarized by Grant (1970, p. 129), thus:

> Brachiopods are notoriously difficult to use in correlation, and Permian ones especially so because of their provinciality (Cooper in Dunbar *et al.*, 1960; Ruzhentsev, in Ruzhentsev and Sarycheva, eds., 1965, p. 102). Estimation of the age of the Chhidru Formation must depend partly upon comparisons of the general "aspect" of the fauna as well as with particular details of this and other faunas that are considered approximately synchronous, and partly upon absence from this fauna of elements that are present in demonstrably younger faunas.

The above conclusions on the age of the brachiopod assemblage from the white sandstone unit are difficult to accept. It appears that the direction of comparisons has been downward stratigraphically with no attention paid to the possibility of longer ranges of

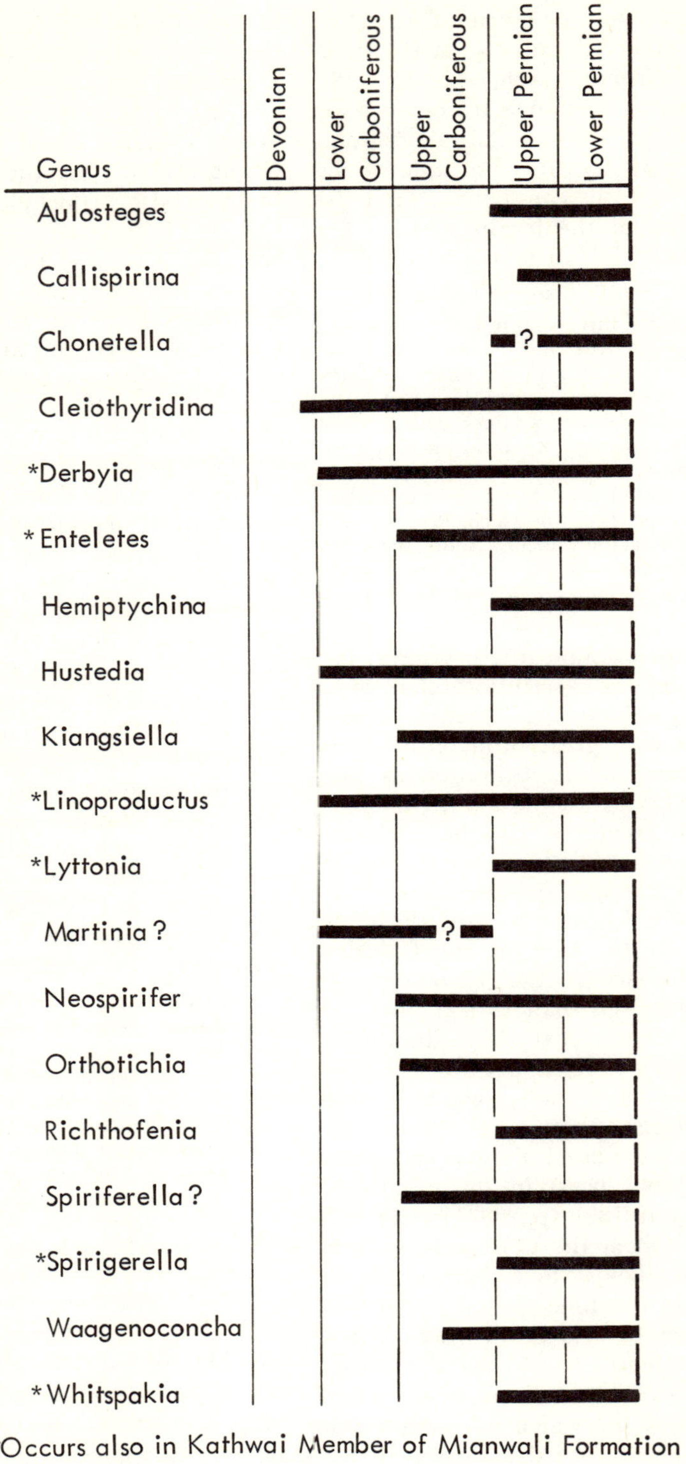

Genus	Devonian	Lower Carboniferous	Upper Carboniferous	Upper Permian	Lower Permian
Aulosteges				■■■■	■■■■
Callispirina				■■■	■■■■
Chonetella				■? ■	■■■■
Cleiothyridina		■■■■	■■■■	■■■■	■■■■
*Derbyia		■■■■	■■■■	■■■■	■■■■
*Enteletes			■■■■	■■■■	■■■■
Hemiptychina				■■■■	■■■■
Hustedia		■■■■	■■■■	■■■■	■■■■
Kiangsiella			■■■■	■■■■	■■■■
*Linoproductus		■■■■	■■■■	■■■■	■■■■
*Lyttonia				■■■	■■■■
Martinia?		■■■	? ■		
Neospirifer			■■■■	■■■■	■■■■
Orthotichia			■■■■	■■■■	■■■■
Richthofenia				■■■	■■■■
Spiriferella?			■■■■	■■■■	■■■■
*Spirigerella				■■■■	■■■■
Waagenoconcha			■■■■	■■■■	
*Whitspakia				■■■	■■■■

*Occurs also in Kathwai Member of Mianwali Formation

FIG. 18. Range zones of genera of articulate brachiopods occurring in the white sandstone unit of the Chhidru Formation (data from Williams *et al.*, 1965).

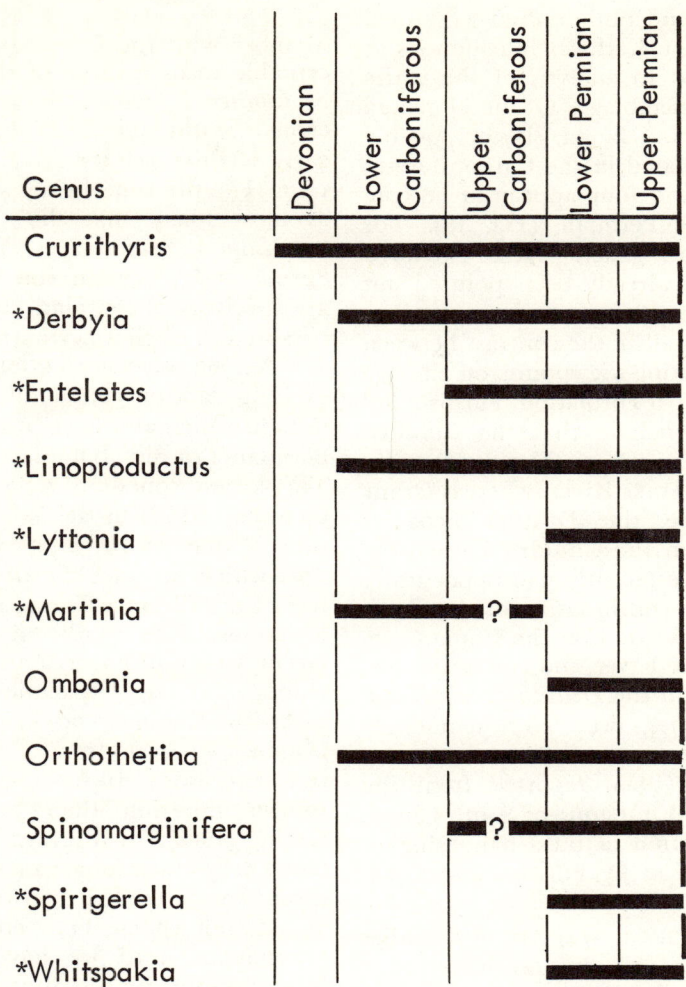

*Occurs also in white sandstone unit of Chhidru Formation

Fig. 19. Range zones of genera of articulate brachiopods occurring in the Kathwai Member of Mianwali Formation (data from Williams et al., 1965).

the taxa concerned. None of the species have been definitely identified and named and most of the 19 genera recognized in the white sandstone assemblage have long ranges (Fig. 18). Eleven of them originated in the Carboniferous or Devonian; eight genera range through most or all of the Permian. Attempting to make refined correlations on such long-ranging genera is indeed a difficult task.

The "Permian" brachiopod fauna of the Kathwai Member Grant (1970) considers to be quite different from that of the underlying white sandstone unit. It should be noted, however, that of the 11 genera recorded from the Kathwai Member seven are also present in the white sandstone unit and that only one species, a new one, has been definitely named. As to the age of the brachiopod assemblage from the dolomite unit of the Kathwai Member, Grant (1970, p. 125) concludes that it "contains some genera that point to a latest Guadalupian (Lamar equivalent) age, and others that point to a Dzhulfian, or even early Triassic age." The same techniques and philosophies were used to analyze the Kathwai brachiopods as were applied to analysis of the white sandstone assemblage. Of the 11 genera in the Kathwai fauna, six, or possibly seven, originated in the Carboniferous, or earlier, and four appear to be confined to the Permian (Fig. 19). For some of the genera these extended ranges have already been pointed out by Stepanov (1967).

To crystallize the contrast between these conflicting viewpoints on the age of the Chhidru Formation, Furnish and Glenister conclude that the Chhidru Formation is younger than the Dzhulfa beds of the Araks River, whereas Grant would place the Dzhulfa beds as younger than the Chhidru Formation.

We have recently had opportunity to study the continuation of the classical Dzhulfa section on the Iranian side of the Araks River and, in strata corresponding to the *Dzhulfites* and *Paratirolites* zones, we have tentatively identified three genera of ammonoids that Chao (1965) reported from the Changhsing Limestone of South China, the type section of the Changhsingian Stage proposed by Furnish and Glenister.

In a general way Grant's studies lend support to the postulate of a stratigraphic break between the Chhidru and Mianwali formations. His paleogeographic interpretations are as follows (Grant, 1970, p. 149):

> Regression began as the middle of the Wargal [Limestone] was being deposited. The shore moved closer to the area, terrigenous sediments were introduced, and the sequence from Kalabagh Member through the Chhidru Formation records the steady regression. The sea shallowed to a shallow basin having minimal deposition during the Dzhulfian [=Araksian of Furnish and Glenister, 1970]. The dolomite unit of the Kathwai Member was deposited during this period, and there was stillstand or some erosion. Transgression began again and the limestone unit of the Kathwai Member was deposited, followed by deposition of the argillaceous limestones of the upper member of the Mianwali Formation.

The correlation of the Kathwai Member with the Dzhulfian is not acceptable to us because of the presence of *Ophiceras,* which by generally accepted world standards indicates an early Scythian age for this stratigraphic unit. For the time being we regard it as most likely that the "Dzhulfian" brachiopod assemblages at Narmia Spring and places in the Salt Range are survivors of Permian faunas which persisted in earliest Scythian time.

Among other fossil groups new and exciting data on conodonts have come to light which also have bearing on the Permian-Triassic boundary problem. The lowest conodont zone recognized by Sweet (1970) in the Salt Range section is that of *Anchignathodus typicalis* which includes the uppermost 14 feet of the Chhidru Formation and the lowermost 3 to 9 feet of the superjacent Mianwali Formation. Sweet concludes, "If the systemic boundary picked by Kummel and Teichert (1966) is indeed a paraconformity, the interruption it records must have been of much shorter duration than proposed by Grant (1968), or the *typicalis* fauna had a very much longer range than did those that succeeded it in the Triassic. We cannot choose between these two alternatives until we have more information on the distribution of conodonts in Upper Permian rocks." Recently, Sweet (written communication) has recognized the *A. typicalis* assemblage in the Ervay Tongue of the Upper Permian Phosphoria Formation in Wyoming. Further, on the basis of collections made by us in 1967, Sweet (written communication) has found that the Zone of *A. typicalis* also straddles the Permian-Triassic boundary in the famous Kap Stosch section of northeastern Greenland.

No identifiable ostracodes have been found in any beds close to the Permian-Triassic boundary, but the fact that Sohn (1970) describes several ostracodes of Paleozoic aspect occurring in higher parts of the Mianwali Formation is significant.

Finally, the acritarchs, an important constituent of the Salt Range fossil assemblages, are yet insufficiently known, but are under study (Sarjeant, 1970). From an investigation of one selected assemblage found in the white sandstone unit Sarjeant concludes that the acritarchs indicate an intermediate marine environment, "not far offshore, but not, apparently, enclosed." Of 15 named species three were known previously only from Mesozoic rocks, nine from both Permian and Mesozoic deposits, and one only from the Permian, Two species are new. Thus, Sarjeant concludes, this is "very definitely an intermediate assemblage, with characteristics both of Permian and of Triassic." Balme (1970) has emphasized the appearance of huge swarms of acritarchs in the Kathwai Member, but these assemblages have not yet been studied.

CONCLUSIONS

We believe that the evidence is now overwhelming that the Permian-Triassic boundary in the Salt Range and Trans-Indus ranges is a paraconformity reflecting a recession of the sea and an emergent condition in latest Permian time followed by a transgression in the earliest Triassic as defined by the first appearance of *Ophiceras*. Previous interpretations of the faunal change at this boundary assumed an almost complete change in composition of the faunas. Intensified new studies have shown that, while the change in faunal (and floral) composition is still significant, it is by no means as great as previously thought. Permian-type brachiopods apparently survived briefly into the Triassic. The uppermost beds of the Chhidru Formation and the lower most few feet of the Kathwai Member of the Mianwali Formation both occur within one conodont zone, the Zone of *Anchignathodus typicalis*. The main evolutionary lines of Late Permian nautiloids continue into the Triassic without significant changes. Similarly, Paleozoic-type ostracodes are now known from the Lower Triassic. The acritarchs of the uppermost Permian beds have a strongly Mesozoic aspect. On the other hand, a strong break in the succession of palynological assemblages is indicated at the boundary as defined by us.

While we have defined the Permian-Triassic boundary in the Salt Range and Trans-Indus ranges on the basis of evidence furnished by ammonoid occurrences, it is obvious that, if other groups had been chosen as guides, alternative positions of the boundary might have been suggested, although these would have differed only slightly from the one chosen by us.

One of the principal problems is the exact age of the Chhidru Formation. The problem as to what constitutes the Upper Permian in West Pakistan is obviously unsettled. In so far as the Salt Range is concerned, an earliest Triassic age of the Kathwai Member, based on the presence of *Ophiceras* and *Glyptophiceras,* we regard as established. In regard to the underlying Chhidru Formation, Furnish and Glenister, on the basis of ammonoids, recognize the absence of the uppermost Permian stage (their Changhsingian). Grant, on the other hand, postulates an even greater stratigraphic gap between the Permian and Triassic formations. We tend to favor the interpretation of Furnish and Glenister, though by no means in all its aspects, over that of Grant. Regardless of which of these specialists one follows, their conclusions imply a significant stratigraphic break, of the magnitude of a stratigraphic stage or, possibly, more, between the uppermost Permian and the lowermost Triassic beds in the Salt Range and Trans-Indus ranges.

APPENDIXES

APPENDIX A: STRATIGRAPHIC SECTIONS OF THE UPPERMOST PERMIAN AND LOWERMOST TRIASSIC BEDS IN THE SALT, SURGHAR, AND KHISOR RANGES

Most of the fossil identifications are by persons to whom acknowledgments have been made in the Introduction and in appropriate places in the text.

One of us (Teichert) has consistently used the National Research Council's Rock-color Chart (Goddard et al., 1948) for designation of rock colors, whereas Kummel used conventional color designations which include such terms as buff, tan, and rusty brown, not utilized by Goddard et al. This explains some inconsistencies in rock color designations in the descriptions of stratigraphic sections.

1. Landa Pusha, 3 miles northeast of the village of Malla Khel, Surghar Range (lat. 32°58′ N., long. 71°12′ E.), just south of loading area of coal mine (Fig. A1).

Thickness
(in feet)

LOWER TRIASSIC
MIANWALI FORMATION
Mittiwali Member (basal part= "Lower Ceratite limestone")
  7. Limestone, light gray, thin-bedded, poorly exposed 4.0

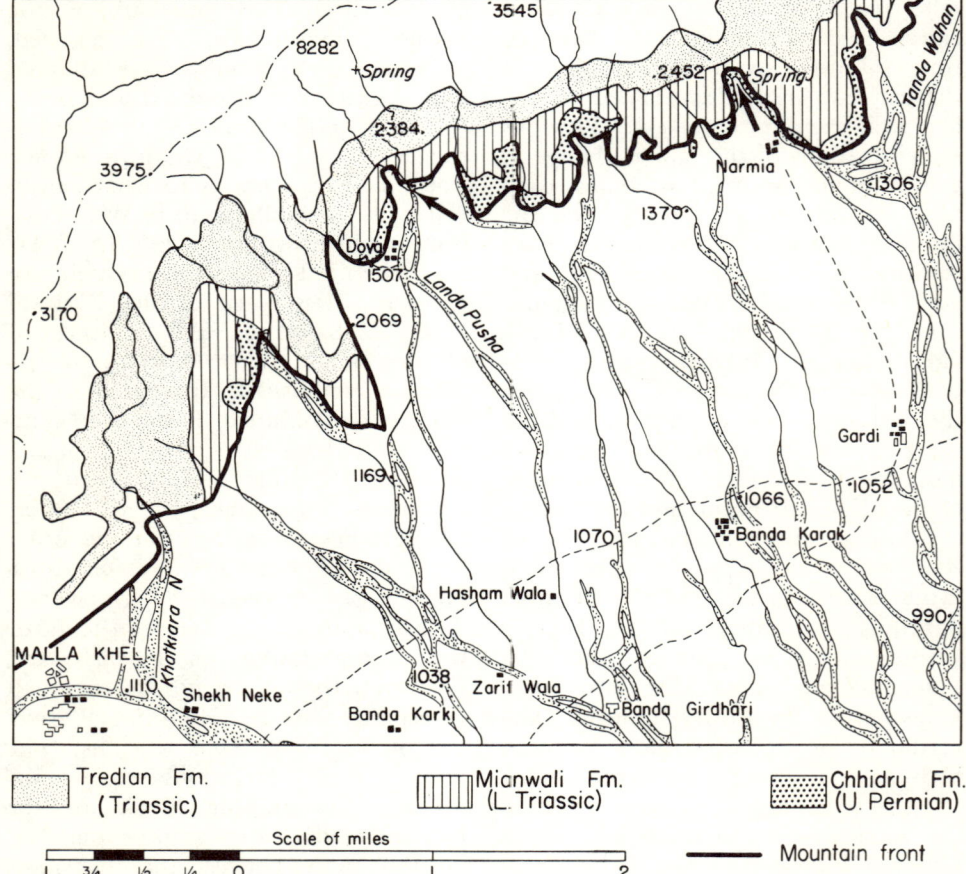

FIG. A1. Locality map of part of Surghar Range, showing location of Landa Pusha (loc. 1) and Narmia Spring (loc. 2), indicated by arrows. Outcrops of Late Permian and Triassic formations adapted from unpublished map by W. Danilchik and A. Shah.

FIG. A2. The Permian-Triassic contact as exposed at Narmia Spring, Surghar Range. The massive ledge is the Kathwai Member. The brachiopods described by Grant (1970) from this locality occur in the lower 1 foot of the member. Below it the white sandstone unit is visible. (Kummel photograph; from Kummel and Teichert, 1966.)

2. Narmia Nala at spring, Surghar Range (lat. 31°58′ N., long. 71°13.5′ E.) (Fig. A1, A2).

Kathwai Member
  Limestone unit
    6. Limestone, light gray, glauconitic, forming one bed rich in echinodermal debris 0.5
    5. Limestone, dolomitic, glauconitic 1.0
  Dolomite unit
    4. Dolomite, yellowish-gray; coquina of shell and echinoderm fragments, glauconitic 0.25
    3. Dolomite, calcitic, yellowish-gray to grayish-orange pink, medium-bedded; composed predominantly of dolomitized echinodermal fragments 5.0

UPPER PERMIAN
CHHIDRU FORMATION
  White sandstone unit
    2. Limestone, sandy medium-light gray; contains some echinodermal debris 2.0
  Highly fossiliferous unit
    1. Limestone finely crystalline, medium-light gray, medium-bedded; contains *Bellerophon*, *Plagioglypta*, and echinodermal debris; <2% detrital quartz particles up to 0.2 mm in diameter ....

Thickness (in feet)

LOWER TRIASSIC
MIANWALI FORMATION
  Mittiwali Member (basal part= "Lower Ceratite limestone")
    12. Limestone, gray, weathers tan, beds 2-4 inches thick, with a few thin shale interbeds 1.0
    11. Shale, tan and gray, silty 0.2
    10. Limestone, gray, argillaceous, massive, weathers tan; contains poorly preserved ammonites 1.0
    9. Limestone, gray, fine-grained, weathers tan, with some thin beds of black, silty shale; contains poorly preserved ammonites 6.8
    8. Limestone, gray, medium-bedded; contains typical "Gyronitan" assemblage of poorly preserved ammonoids and conodonts (*Ellisonia gradata* Sweet, *Neospathodus dieneri* Sweet, *N. kummeli* Sweet) 1.5
    7. Limestone, gray, like bed 8 .... 1.0
Kathwai Member
  Limestone unit

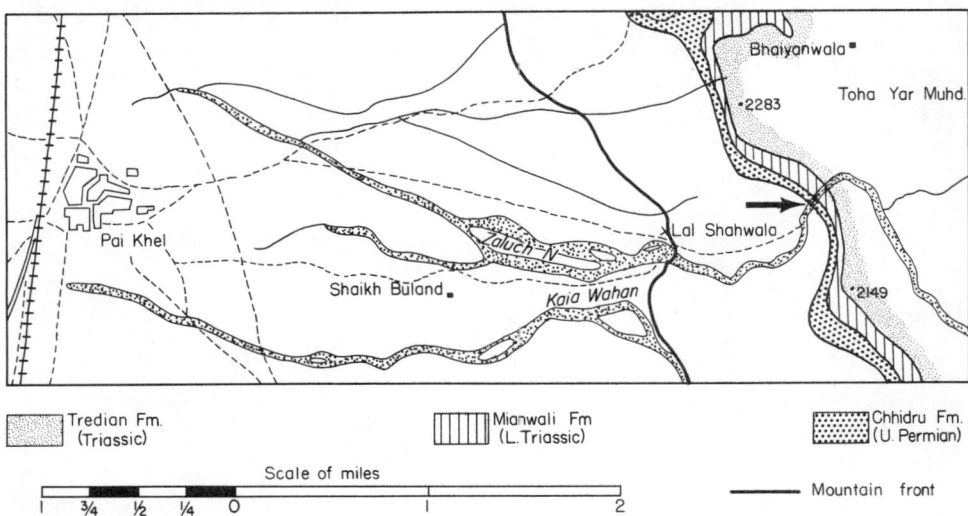

Fig. A3. Locality map of part of western Salt Range, showing location of Zaluch Nala and of locality 3, indicated by arrow. Outcrops of Late Permian and Triassic formations adapted from unpublished map by E. R. Gee.

Fig. A4. The Permian-Triassic contact as exposed in Zaluch Nala. Hammer head rests on top of white sandstone unit. (Teichert photograph.)

6. Limestone, light gray, hard, massive glauconitic; no fossils seen 1.3
5. Limestone, very light gray, speckled brown, glauconitic; contains conodonts (*Anchignathodus typicalis* Sweet, *Ellisonia teicherti* Sweet, *E. triassica* Müller) 0.7

Dolomite unit

4. Dolomite, generally grayish-orange, pinkish-gray in uppermost part, fine-grained; coquinoidal, consisting mainly of echinoderm fragments; uppermost foot contains glauconite; contains conodonts (*Anchignathodus typicalis* Sweet, *Ellisonia teicherti* Sweet, *Xaniognathus curvatus* Sweet) 5.2
3. Limestone, very pale orange to yellowish-gray, coquinoidal, partly dolomitized; contains P e r m i a n-type brachiopods (*Crurithyris? extima, Derbyia?* sp., dielasmatid undet., *En-*

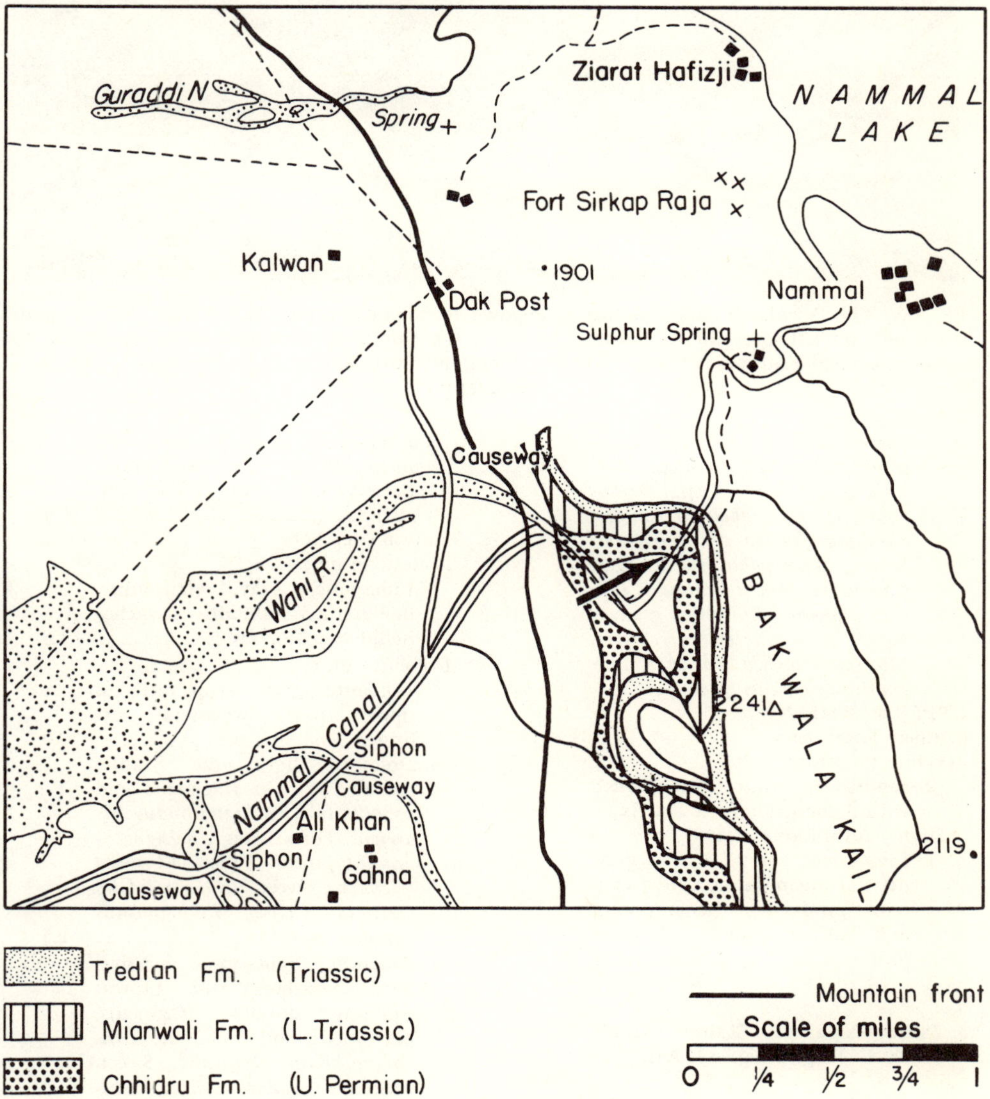

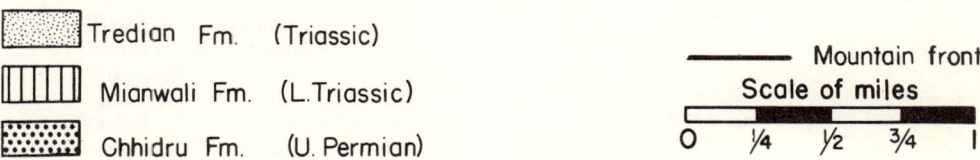

Fig. A5. Locality map of part of western Salt Range, showing location of Nammal Gorge and of locality 4, indicated by arrow. Outcrops of Late Permian and Triassic formations adapted from unpublished map by E. R. Gee.

Fig. A6. The Permian-Triassic contact as exposed in Nammal Gorge, Salt Range. White sandstone unit on left, overlain by Kathwai Member; the thin-platy limestone above the thick- to medium-bedded Kathwai is "Lower Ceratite limestone," overlain by "Ceratite marl." (Teichert photograph.)

teletes sp. 2, *Lingula* sp., *Linoproductus* sp., *Lyttonia* sp., *Martinia* sp., *Ombonia* sp., *Orthotetina* cf. *O. arakeljani*, *O.* sp., *Spinomarginifera* sp., *Spirigerella* sp., *Whitspakia* sp. 2), and conodonts (*Anchignathodus typicalis* Sweet, *Ellisonia gradata* Sweet, *Ellisonia teicherti* Sweet, *Ellisonia triassica* Müller, *Xaniognathus curvatus* Sweet) 0.5

UPPER PERMIAN
CHHIDRU FORMATION
 White sandstone unit
  2. Sandstone, white, fine-grained with dolomitic calcitic matrix 1.0
 Highly fossiliferous unit
  1. Limestone, very light gray, detrital organogenic, sandy; contains productids, bellerophons, and other typical Permian forms ....

3. Zaluch Nala, Salt Range (32°49′ N. lat., 71°38′ E. long.) (Fig. A3, A4).

Thickness
(in feet)

LOWER TRIASSIC
MIANWALI FORMATION
 Mittiwali Member (basal part= "Lower Ceratite limestone")

  8. Limestone, gray, in beds 2-3 inches thick; very fossiliferous, preservation poor (*Gyronites*, etc.) 6.0

Kathwai Member
 Limestone unit
  7. Limestone, gray, glauconitic, fine-grained; contains rhynchonellid brachiopods 1.6
 Dolomite unit
  6. Dolomite, gray, glauconitic in upper part, weathers rust brown 3.0
  5. Dolomite, light gray, friable; contains crinoid fragments and conodonts [*Anchignathodus isarcicus* (Huckriede), *Anchignathodus typicalis* Sweet, *Ellisonia teicherti* Sweet, *Neogondolella carinata* (Clark), *Xaniognathus curvatus* Sweet] 2.6
  4. Dolomite, light gray, massive, hard, weathers rust brown; contains crinoid fragments, pelecypods, and conodonts (*Anchignathodus typicalis* Sweet, *Ellisonia gradata* Sweet, *Ellisonia teicherti* Sweet, *Ellisonia triassica* Müller) 3.6
  3. Dolomite, grayish-tan, fine-grained 0.6

UPPER PERMIAN
CHHIDRU FORMATION
 White sandstone unit
  2. Sandstone, grayish-white 0.6
 Highly fossiliferous unit
  1. Limestone with abundant bellerophons ....

4. Nammal Gorge, Salt Range (32°40′ N. lat., 71°48′ E. long.) (Fig. A5, A6).

Thickness
(in feet)

LOWER TRIASSIC
MIANWALI FORMATION
 Mittiwali Member (basal part= "Lower Ceratite limestone")
  8. Limestone, gray, fine-grained, in beds 1-3 inches thick; contains abundant poorly preserved ammonites (*Gyronites*, etc.) 4.0
Kathwai Member
 Limestone unit
  7. Dolomite, highly calcitic, yellowish-gray, thin-bedded with some interbeds of sandy limestone; contains poorly preserved ammonites (*Ophiceras*), rhynchonellid brachiopods, and conodonts [*Neogondolella carinata* (Clark)] 2.0
 Dolomite unit
  6. Dolomite, gray, sandy, weathers brown, in beds 1-3 inches thick; lower two-thirds of unit contains cidarid spines and crinoid remains, upper part contains

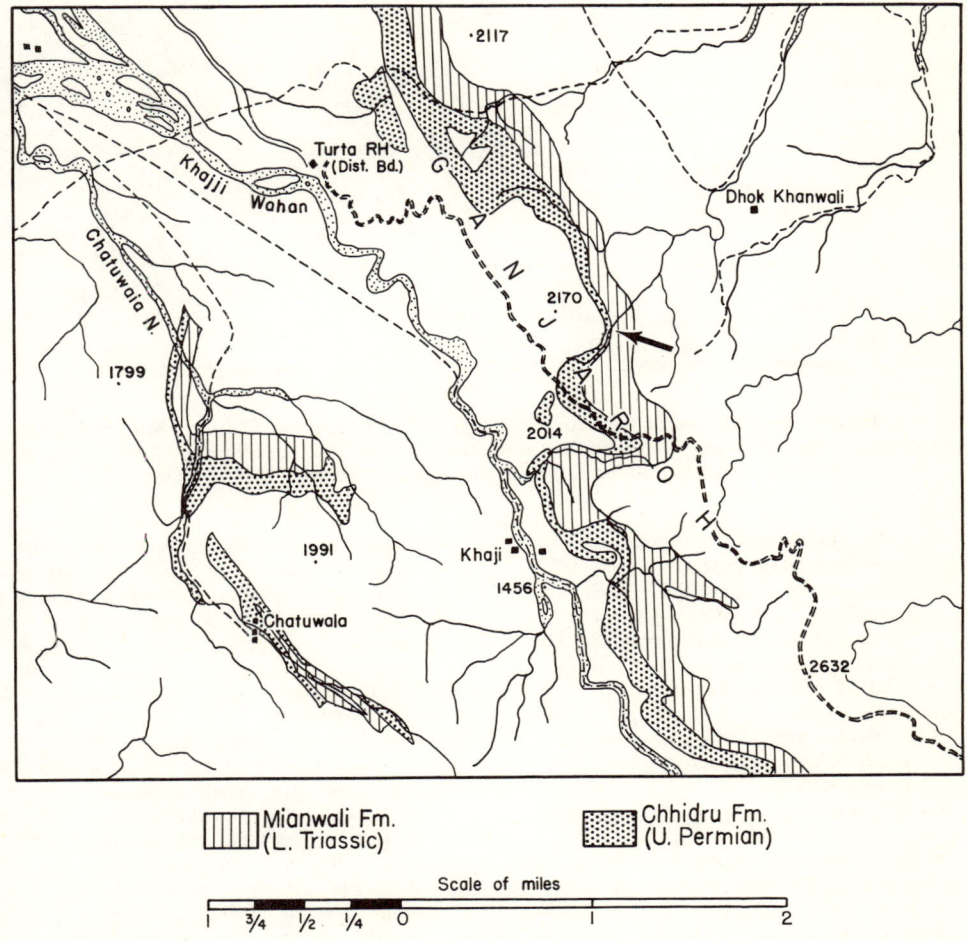

FIG. A7. Locality map of part of western Salt Range, showing location of Ganjaroh Hill and of locality 5, indicated by arrow. Outcrops of Late Permian and Triassic formations adapted from unpublished map by E. R. Gee.

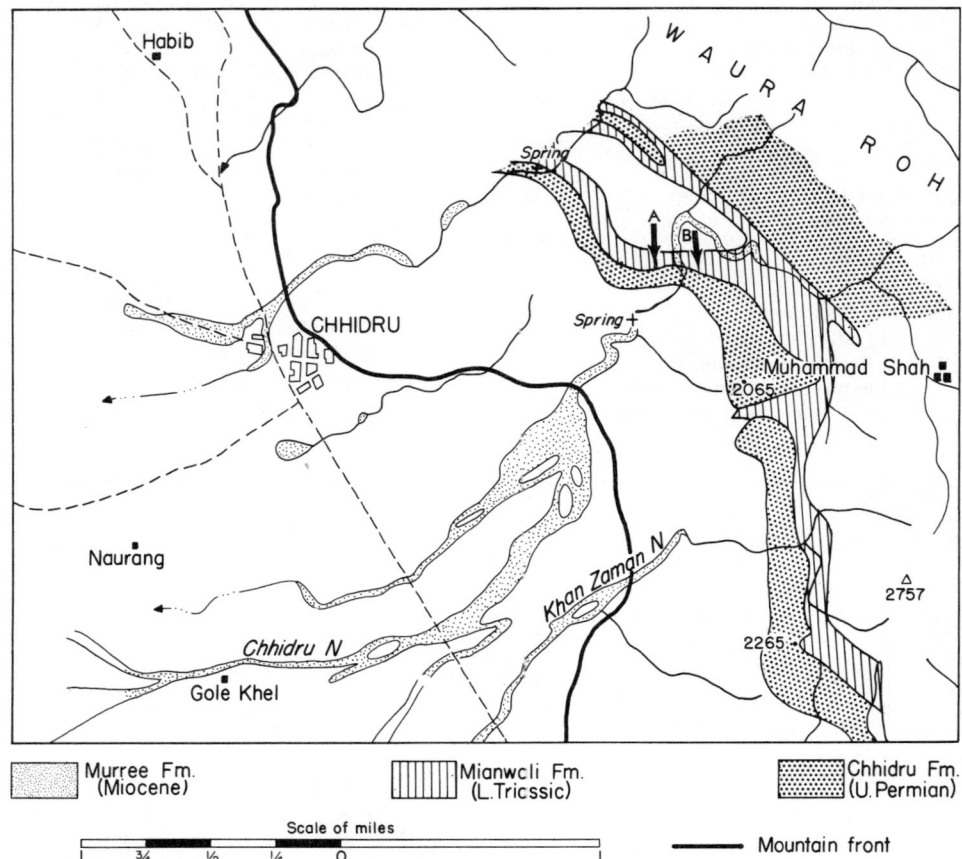

Fig. A8. Locality map of part of western Salt Range, showing location Chhidru Nala and Khan Zaman Nala and of localities 6A and 6B, indicated by arrows. Outcrops of Late Permian, Triassic, and Miocene formations from unpublished map by E. R. Gee.

small brachiopods; conodonts [*Anchignathodus typicalis* Sweet, *Ellisonia triassica* Müller, *Neogondolella carinata* (Clark), *Xaniognathus curvatus* Sweet] .. 3.0
5. Dolomite as bed 4; contains cidarid spines and crinoid remains 1.4
4. Dolomite, gray, sandy, massive, weathers brown; no fossils seen .. 4.6

UPPER PERMIAN
CHHIDRU FORMATION
  White sandstone unit
    3. Shale, black, micaceous, with laminae of tan micaceous siltstone 1.5
    2. Sandstone, white, massive, friable 2.0
  Highly fossiliferous unit
    1. Limestone, white, very sandy, massive; contains bellerophons ..

5. Dry tributary heading on jeep trail from Mianwali to Sakesar, Ganjaroh, Salt Range, 6.5 miles east of road to Nammal Gorge 32°35.5′ N. lat., 71°50′ E. long.) (Fig. A7).
Thickness
(in feet)

LOWER TRIASSIC
MIANWALI FORMATION
  Mittiwali Member (basal part= "Lower Ceratite limestone")
    9. Limestone, gray argillaceous in part, beds 1-2 inches thick; rich in poorly preserved ammonoids (*Gyronites*, etc.) and conodonts [*Ellisonia gradata* Sweet, *Ellisonia triassica* Müller, *Neogondolella carinata* (Clark), *Neospathodus dieneri* Sweet, *Neospathodus kummeli* Sweet, *Xaniognathus curvatus* Sweet, *Xaniognathus deflectens* Sweet] 6.0

Fig. A9. West side of Chhidru Nala, Salt Range. The Permian-Triassic contact is just below the crest of the sharp-crested hill in center of photograph. (Teichert photograph; from Kummel and Teichert, 1966.)

Fig. A10. White sandstone unit and Kathwai Member, west side of Chhidru Nala, Salt Range. This is the far end of the ridge shown in Figure 14, viewed from the other side. (Teichert photograph; from Kummel and Teichert, 1966.)

Kathwai Member
Limestone unit
8. Limestone, pale yellowish-brown, fine-grained, highly dolomitic; rich in echinodermal remains; base firmly welded to unit 7 .... 0.2-0.8

Dolomite unit
7. Dolomite, slightly calcitic, grayish-orange, weathering grayish-orange pink, medium-bedded, coarse-grained; contains abundant echinoderm fragments, some shell debris, rhynchonellids, *Lingula* cf. *L. borealis* Bittner, and conodonts [*Anchignathodus typicalis* Sweet, *Ellisonia triassica* Müller, *Neogondolella carinata* (Clark), *Xaniognathus curvatus* Sweet] 1.8
6. Dolomite, highly calcitic, grayish-orange, weathering same, finely crystalline; contains echinoderm fragments, rhynchonellids, ostracodes, and dasycladacean algae(?); some detrital quartz 0.9
5. Dolomite, grayish-orange, weathering grayish-orange pink, coarsely crystalline; contains echinoderm fragments 0.8
   Dolomite, slightly calcitic, grayish-orange, weathering light brown, coquinoid in upper half; contains echinoderm fragments, including *Miocidaris* spines, rhynchonellids; about 15% detrital quartz 0.5
3. Dolomite, calcitic, grayish-orange, weathering light brown; contains few echinoderm remains; about 20% detrital quartz 1.0

UPPER PERMIAN
CHHIDRU FORMATION
White sandstone unit
2. Sandstone, white to light tan, parts friable, other parts hard with calcareous cement, unit generally incompetent, massive; upper 1.5 feet thin-bedded and shaly 14.0

Highly fossiliferous unit
1. Limestone, gray, very sandy; contains bellerophons 10.0

6A. West side of Chhidru Nala, Salt Range (32°33′ N. lat., 71°48′ E. long.) (Fig. A8-A10).

Thickness
(in feet)

LOWER TRIASSIC
MIANWALI FORMATION
Mittiwali Member (basal part= "Lower Ceratite limestone")

11. Limestone, pale yellowish-brown, hard, thin-bedded, slightly dolomitic and glauconitic; contains abundant poorly preserved ammonoids (*Gyronites* and others), and conodonts [*Ellisonia gradata* Sweet, *E. triassica* Müller, *Neogondolella carinata* (Clark), *Neospathodus cristagalli* (Huckriede), *N. kummeli* Sweet, *Xaniognathus curvatus* Sweet, *Xaniognathus deflectens* Sweet] 8.0

Kathwai Member
Limestone unit
10. Limestone, dolomitic, friable, laminated; contains echinoderm fragments and, at 3.5 to 4 feet, poorly preserved ammonoids (*Ophiceras*?); conodonts [*Ellisonia gradata* Sweet, *E. triassica* Müller, *Neogondolella carinata* (Clark), *Xaniognathus curvatus* Sweet, *X. deflectens* Sweet] 5.75
9. Limestone, light gray, finely crystalline, glauconitic, coquinoidal; abundant shell fragments, rhynchonellid brachiopods, subordinate echinoderm remains, *Ophiceras connectens*, and conodonts [*Ellisonia triassica* Müller, *Neogondolella carinata* (Clark), *Xaniognathus deflectens* Sweet] 0.5

Dolomite unit
8. Dolomite, calcitic, grayish-orange, upper part dolomitic limestone; rich in shell and echinoderm fragments, *Miocidaris* spines, *Aplocoma* cf. *A. torrii* (Desio), fish teeth, and conodonts [*Anchignathodus typicalis* Sweet, *Ellisonia gradata* Sweet, *E. triassica* Müller, *Neogondolella carinata* (Clark), *Xaniognathus curvatus* Sweet] .. 1.25
7. Dolomite, calcitic, dark, yellowish-orange, thin-bedded, laminated; <1% detrital quartz; abundant shell and echinoderm fragments, very little glauconite; *Miocidaris* spines; conodonts [*Anchignathodus isarcicus* (Huckriede), *A. typicalis* Sweet, *Ellisonia gradata* Sweet, *E. teicherti* Sweet, *E. triassica* Müller, *Neogondolella carinata* (Clark), *Xaniognathus curvatus* Sweet] .. 3.25
6. Dolomite, calcitic, dark yellowish-orange, one massive bed, coquinoidal; contains shell fragments and conodonts [*Anchignathodus isarcicus* (Huckriede), *Anchignathodus typicalis*

Sweet, *Ellisonia gradata* Sweet, *E. teicherti* Sweet, *E. triassica* Müller, *Neogondolella carinata* (Clark), *Xaniognathus curvatus* Sweet] 0.7

5. Shale, olive-green, upper 4.5 inches interbedded with thin layers of grayish-orange dolomite with abundant echinodermal fragments; conodonts [*Anchignathodus typicalis* Sweet, *Ellisonia teicherti* Sweet, *Neogondolella carinata* (Clark), *Xaniognathus curvatus* Sweet] .. 1.0

4. Dolomite, grayish-orange to dark yellowish-orange, finely crystalline; rich in shell and echinoderm fragments, including *Miocidaris* spines, fish teeth, and conodonts [*Anchignathodus typicalis* Sweet, *Ellisonia teicherti* Sweet, *E. triassica* Müller, *Neogondolella carinata* (Clark), *Xaniognathus curvatus* Sweet] .. 2.5

3. Dolomite, grayish-orange to dark yellowish-orange, one bed, finely crystalline, laminated, locally cross-bedded; contains *Lingula* cf. *L. borealis* Bittner, *Spinomarginifera* sp., *Pseudomonotis*-like bivalves, *Ophiceras connectens*, echinoderm fragments, and conodonts [*Anchignathodus typicalis* Sweet, *Ellisonia teicherti* Sweet, *E. triassica* Müller, *Neogondolella carinata* (Clark)] [Between units 3 and 4 the bedding plane features (mud cracks, pitted surface, and invertebrate trails) described on p. 39 occur.] 0.9

UPPER PERMIAN
CHHIDRU FORMATION
White sandstone unit

2. Alternating shale and soft sandstone as follows:
   l. Friable, platy, laminated sandstone, locally with 2-inch shale layer at top—9 inches
   k. Soft, white sand—34 inches
   j. Dark shale and laminated medium-gray, brownish, and light gray siltstone, and very fine-grained sandstone—18 inches
   i. Soft light gray sand—6 inches
   h. Dark shale with thin, white sand layers at 12-14 inches—18 inches
   g. Soft, light gray sand—8 inches
   f. Friable sandstone—4 inches
   e. Hard, calcareous sandstone, laminated—7 inches
   d. Soft, light gray sand—7 inches
   c. Dark shale—13 inches
   b. Soft, light gray sandstone—10 inches
   a. Dark shale—2 inches
   Unit contains conodonts [*Anchignathodus typicalis* Sweet, *Neogondolella carinata* (Clark)] 11.3

Highly fossiliferous unit
1. Limestone, light gray, mottled, sandy, massive; contains productids, bellerophons, and other typical Permian fossils ....

6B. East side of Chhidru Nala, Mittiwali, Salt Range (lat. 32°33′ N., long. 71°48′ E.) (Fig. A8).

Thickness (in feet)

LOWER TRIASSIC
MIANWALI FORMATION
Mittiwali Member (basal part= "Lower Ceratite limestone")

16. Limestone, gray, hard; very fossiliferous (*Gyronites*, etc.) 1.6
15. Shale, black, with thin beds of fine-grained, tan sandstone, mostly covered 3.0
14. Limestone, gray, fine-grained, hard, thin- to medium-bedded; very fossiliferous (*Gyronites*, etc.), but preservation poor 3.0
13. Sandstone, yellowish-gray, fine-grained, laminated 0.2
12. Limestone, gray, coquinoidal consisting of brachiopod, pelecypod, and gastropod fragments; contains poorly preserved ammonites (*Gyronites*, etc.), pelecypods, and conodonts [*Neogondolella carinata* (Clark), *Xaniognathus deflectens* Sweet] 0.9

Kathwai Member
Limestone unit
11. Limestone, grayish-tan, fine-grained, thin-bedded, scoured in part, dolomitic, with minor amount of glauconite 2.0
10. Sandstone, grayish-tan, fine-grained, thin-bedded, with 2-inch bed of limestone in center containing *Ophiceras connectens* Schindewolf 2.5
9. Limestone, light gray, hard, fragmental; contains *Lingula*, rhynchonellids, *Ophiceras connectens* Schindewolf, *Menuthionautilus kieslingeri* Collignon, and conodonts [*Ellisonia gradata* Sweet, *Neogondolella carinata* (Clark), *Xaniognathus curvatus* Sweet] 1.0
8. Shale, olive-gray, with some thin sandstone beds 1.0

Dolomite unit
7. Dolomite, gray, hard; contains

	Thickness (in feet)
brachiopods, crinoid remains, cidarid spines, teeth and bone fragments	1.0
6. Dolomite, like bed 5, but harder; contains fossil fragments	1.3
5. Dolomite, grayish-orange, fine-grained, thin-bedded, shaly, unit soft	0.8
4. Dolomite, grayish-orange, medium- to massively-bedded, sandy in lower part; contains fossil fragments	2.5
3. Dolomite, grayish-orange, mottled, sandy, hard, massive	3.6

UPPER PERMIAN
CHHIDRU FORMATION
  White sandstone unit

2. Sandstone, yellowish-gray, fine-grained, friable, with some shaly interbeds, contains echinoderm fragments and poorly preserved pelecypods	12.0

Highly fossiliferous unit

1. Sandstone, very pale orange, calcareous; abundant bellerophons, productids, and other typical Permian forms, and conodonts [*Anchignathodus typicalis* Sweet, *Ellisonia teicherti* Sweet, *Neogondolella carinata* (Clark)]

7. Munta Nala, 1.5 miles west of Wargal, Salt Range (lat. 32°27' N., long. 72°02' E.) (Fig. A11, A12).

Thickness (in feet)

LOWER TRIASSIC
MIANWALI FORMATION
  Mittiwali Member (basal part= "Lower Ceratite limestone")

15. Limestone, gray, fine- to medium-grained; very fossiliferous, with small pectinid pelecypods, ammonoids (*Gyronites*, etc.) and fish remains	1.0
14. Shale, gray; contains conodonts [*Neospathodus cristagalli* (Huckriede), *N. dieneri* Sweet]	0.8
13. Limestone, yellowish-gray, thin-bedded; contains numerous ammonoids (*Gyronites*, etc.)	1.9
12. Limestone, light gray, dense, hard, glauconitic, weathers brown	0.2

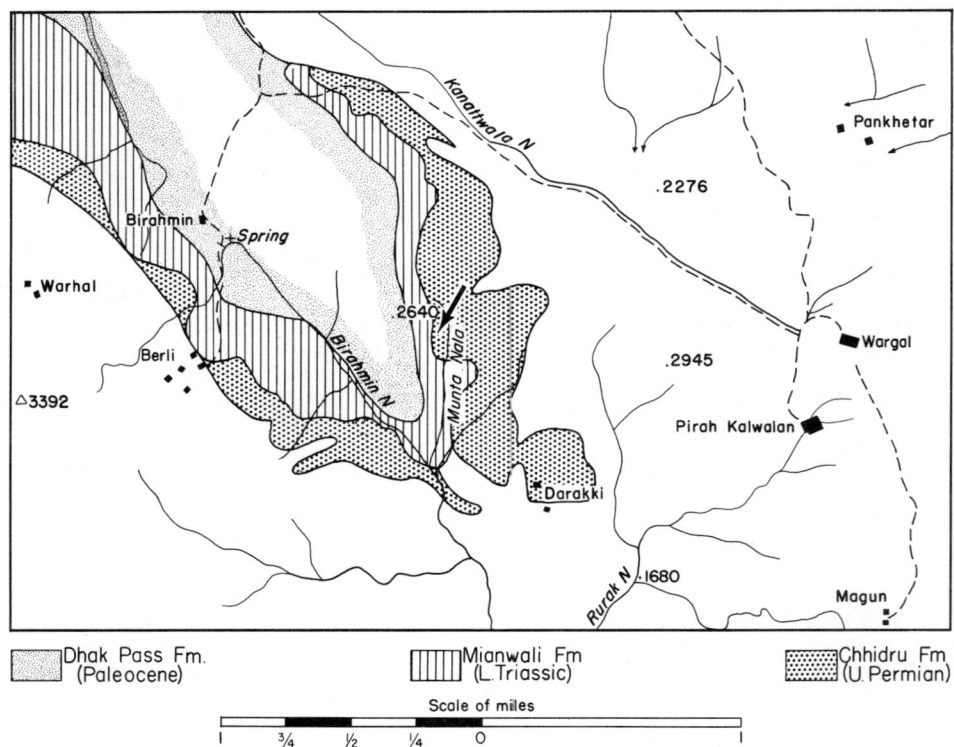

FIG. A11. Locality map of part of central Salt Range, showing location of Wargal and Munta Nala and of locality 7, indicated by arrow. Outcrops of Late Permian, Early Triassic, and Paleocene formations from unpublished map by E. R. Gee.

Fig. A12. The Permian-Triassic contact as exposed in Munta Nala, 1.5 miles west of Wargal, Salt Range. The dark beds below are shale forming the top of the white sandstone unit, overlain by massive strata of the dolomite unit of the Kathwai Member. (Kummel photograph; from Kummel and Teichert, 1966.)

11. Limestone, very light gray, irregular texture; contains large pelecypods, fragments of ammonoids, fish teeth 1.4

Kathwai Member
Limestone unit

10. Sandstone, gray, massive, with a few 1-inch calcareous beds; contains spores and pollen (*Cycadopites follicularis* Wilson and Webster, *Punctatisporites fungosus* Balme, *Striatopodocarpites pantii* Jansonius, *Taeniaesporites noviaulensis* Leschik) 3.0

9. Limestone, pinkish-gray, fine-grained glauconitic; unit begins with 4 inches of grayish-black shale, followed by 3 inches of lenticular limestone, then 1 foot of laminated to thin-bedded sandstone, then limestone; contains small pectinids, echinoderm fragments, ammonoids (*Ophiceras?*), fish teeth, and conodonts [*Neogondolella carinata* (Clark)] 2.5

Dolomite unit

8. Dolomite, light gray, massive, weathers iron-brown; contains shell fragments, rhynchonellid brachiopods, eumorphotid pelecypods, ammonoids (*Ophiceras?*), cidarid spines, fish teeth, and conodonts [*Neogondolella carinata* (Clark)] 2.4

7. Shale, dark gray, with laminae of fine, tan sandstone; contains poorly preserved brachiopods and pelecypods 1.5

6. Dolomite, tan, fine-grained, massive, upper 1 foot laminated, weathers brown 6.0

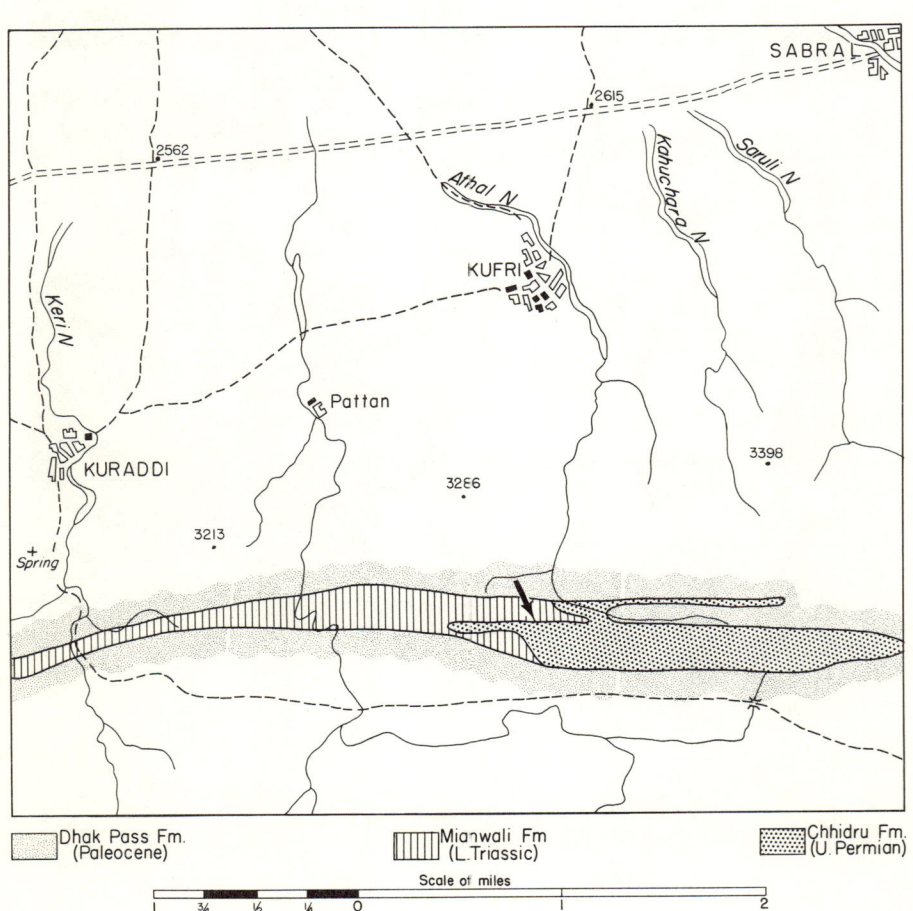

FIG. A13. Locality map of part of central Salt Range, showing location of Kufri and of locality 8, indicated by arrow. Outcrops of Late Permian, Early Triassic, and Paleocene formations from unpublished map by E. R. Gee.

UPPER PERMIAN
CHHIDRU FORMATION
  White sandstone unit
   5. Shale, dark gray, micaceous, with thin beds of cross-bedded sandstone and laminae of sandstone; contains spores and pollen [*Acanthotriletes tereteangulalus* Balme and Hennelly, *Allisporites tenuicorpus* Balme, *Cedripites prisus* Balme, *Densoisporites complicatus* Balme, *Ephedripites* sp., *Klausipollenites schaubergeri* Potanié and Klaus, *Kraeuselisporites rallus* Balme, *Leavigatosporites callosus* Balme, *Leiotriletes* sp., cf. *L. adnatus* (Kosanke), *Lueckisporites singhii* Balme, *L. virkkiae* Potanié and Klaus, *Lunulasporites vulgaris* Wilson, *Marsupipollenites triradiatus* Balme and Hennelly, *Nevesisporites fossulatus* Balme, *Paravittatina lucifer* Bharadwaj and Salujha, *Peltacystia venosa* Balme and Segroves, *Plicatipollenites indicus* Lele, *Pretricolpipollenites bharadwaji* Balme, *Protohaploxypinus limpidus* (Balme and Hennelly), *P. microcorpus* (Schaarschmidt), *Protohaploxypinus varius* Bharadwaj, *Striatopodocaripites cancellatus* (Balme and Hennelly), *Tigrisporites playfordi* de Jersey and Hamilton, *Triquitrites proratus* Balme] 5.9
   4. Sandstone, white, massive, friable, lenticular, bed ranges in thickness from 2 inches to 1 foot 1.0
   3. Shale, dark gray, clay, with laminae of fine sandstone and thin beds of micaceous sandstone; contains plant fragments, and spores and pollen [*Densipollenites indicus* Bharadwaj, *Inapeturopollenites nebulosus* Balme, *Vitreisporites pallidus* (Reissinger)] 3.0
  Highly fossiliferous unit
   2. Sandstone, yellowish-gray, calcareous, massive, weathers tan, unit lenticular ranging in thickness from 4 inches to 3 feet; contains bellerophons and other typical Permian fossils, and conodonts (*Ellisonia triassica* Müller) 3.0
   1. Shale, dark gray, clay, micaceous, with laminae of fine-

FIG. A14. Country about 1.5 miles south of Kufri, Salt Range, near locality 8. The hills in the right half of photograph are made up of Chhidru Formation; the dip slopes of these hills consist of Kathwai Member and "Lower Ceratite limestone"; the saddle is underlain by "Ceratite marl," and the slope of the hill just left of center consists of "Ceratite sandstone" and "Upper Ceratite limestone"; the crest of this hill is "Bivalve Limestone" of the Narmia Member of the Mianwali Formation. (Kummel photograph.)

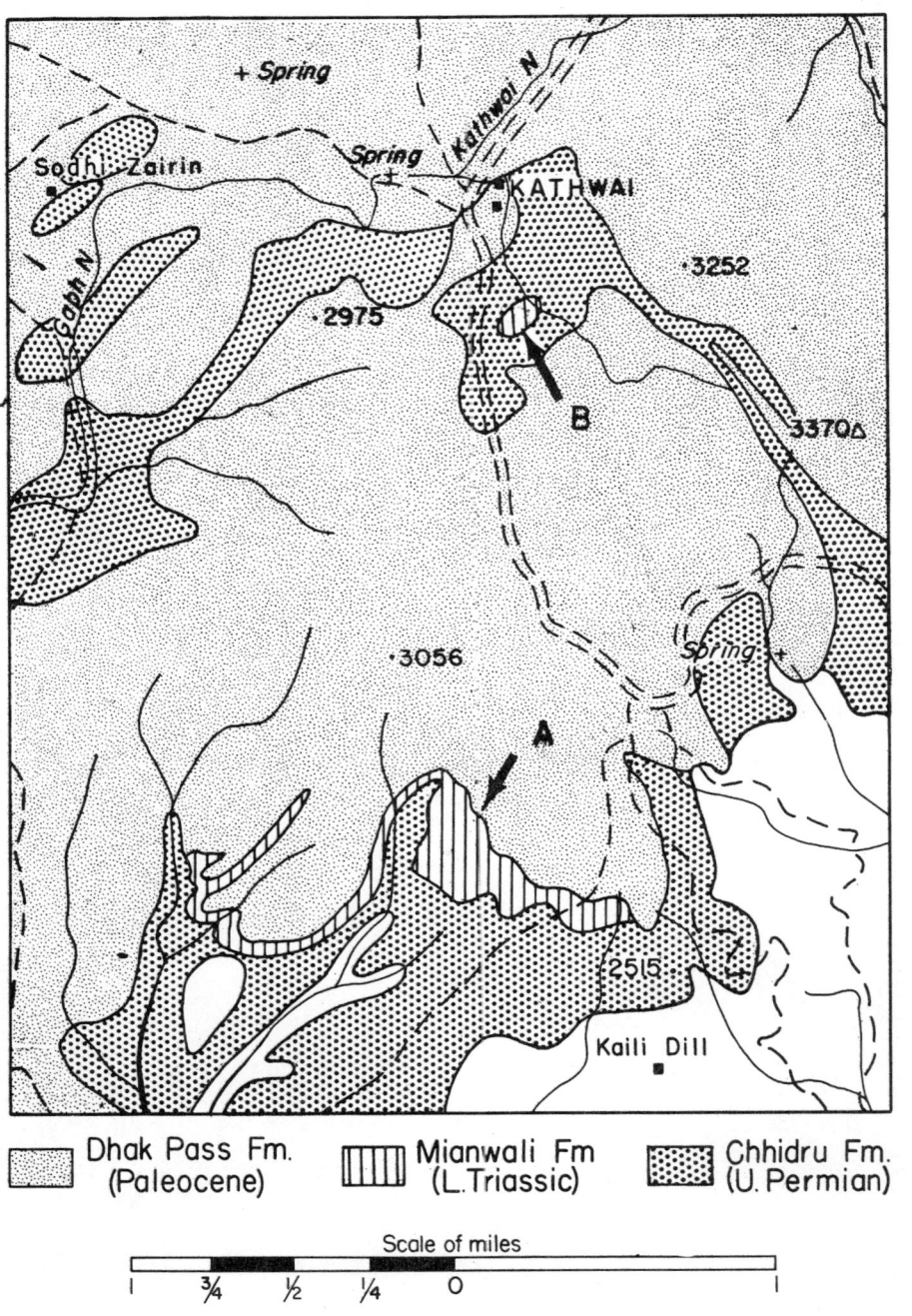

Fig. A15. Locality map of part of central Salt Range, showing location of Kathwai and of localities 9 (A) and 10 (B), indicated by arrows. Outcrop of Late Permian, Early Triassic, and Paleocene formations from unpublished map by E. R. Gee.

grained, very pale orange sandstone and 1-inch to 1-foot beds of cross-bedded sandstone; contains plant fragments, and spores and pollen [*Iraqispora labrata* Singh, *Kraeuselisporites vargalensis* Balme, *Lophotritetes novius* Singh, *Polypodiisporites mutabilis* Balme, *Polypodiidites* sp., *Punctatosporites* sp. cf. *P. minutus* Ibrahim, *Striatopodocarpites rarus* Bharadwaj and Salujha, *Sulcatisporites ovatus* (Balme and Hennelly)] 6.0

8. 1.5 miles southeast of Kufri, 0.5 miles north of highway marker 31 on road from Kuraddi to Kathwai, Salt Range (32°31.5′ N. lat., 72°6.5′ E. long.) (Fig. A13, A14).

Thickness
(in feet)

LOWER TRIASSIC
MIANWALI FORMATION
Mittiwali Member (basal part= "Lower Ceratite limestone")
7. Limestone, gray, thin-bedded; contains numerous poorly preserved ammonoids (*Gyronites*, etc.) 4.0

Kathwai Member
Limestone unit
6. Limestone, gray, thin-bedded; contains rare, poorly preserved ammonoids *(Ophiceras?)* and *Orbiculoidea* 1.5

Dolomite unit
5. Dolomite, gray-brown, massive, hard, weathers brown; contains cidarid spines and small brachiopods 1.0
4. Dolomite, yellow-brown, thin- to medium-bedded; contains rhynchonellid brachiopods, fish teeth, crinoid fragments, and burrows 3.0
3. Dolomite, yellow-brown, weathers brown; contains poorly preserved pelecypods 10.0

UPPER PERMIAN
CHHIDRU FORMATION
White sandstone unit
2. Sandstone, white, friable, massive; contains crinoid remains and shell fragments 2.0

Highly fossiliferous unit
1. Limestone, gray, sandy, massive; contains bellerophons 2.0

9. Beds exposed at head of unnamed creek 1.25 miles south of Kathwai, Salt Range (32°27.5′ N. lat., 72°11.5′ E. long.) (Fig. A15, A16).

Thickness
(in feet)

LOWER TRIASSIC
MIANWALI FORMATION
Mittiwali Member (basal part= "Lower Ceratite limestone")
6. Limestone, gray, hard; contains abundant, poorly preserved am-

FIG. A16. View of unnamed valley 1.75 miles south of Kathwai, near locality 9. The slope opposite consists of Chhidru Formation; the Kathwai Member is visible at top of slope on right side. (Teichert photograph.)

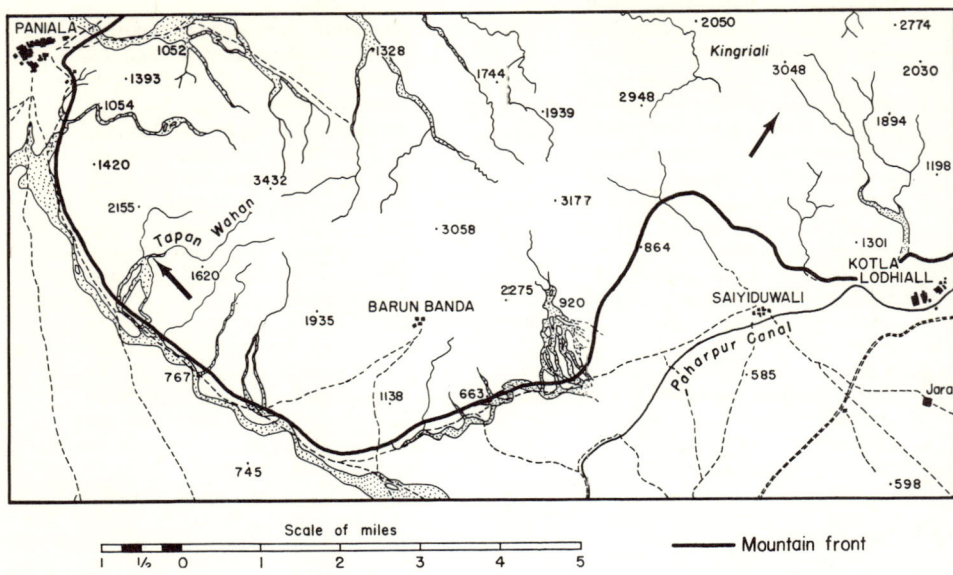

Fig. A17. Locality map of part of southern Khisor Range showing location of Kingriali and Tapan Wahan and of localities 11 and 12, indicated by arrows.

monoids (*Gyronites*, etc.), and conodonts [*Ellisonia gradata* Sweet, *E. triassica* Müller, *Neogondolella carinata* (Clark), *Neospathodus dieneri* Sweet, *Xaniognathus curvatus* Sweet, *Xaniognathus deflectens* Sweet] 3.0

Kathwai Member
Limestone unit
5. Limestone, yellowish-gray, yellowish-brown in part, massive, hard; contains abundant rhynchonellid brachiopods, cidarid spines and other echinodermal fragments, and smooth pectinids 1.0

Dolomite unit
4. Dolomite, calcitic, grayish-orange, thick- to medium-bedded, coquinoid in part, weathering grayish-orange pink; contains small rhynchonellid brachiopods, echinoid spines, much echinodermal and shell debris, and conodonts [*Ellisonia gradata* Sweet, *E. triassica* Müller, *Neogondolella carinata* (Clark), *Xaniognathus curvatus* Sweet] 4.0

3. Limestone, light gray, hard, massive, weathering light olive-gray; contains rare ammonoids (*Ophiceras*?) and echinodermal fragments, abundant shell fragments, and conodonts [*Neogondolella carinata* (Clark)] 1.0

2. Dolomite, grayish-orange, hard, massive, porous; contains echinodermal and shell fragments .. 2.0
1. Dolomite, dark yellowish-orange, massive weathering grayish-orange; contains poorly preserved bivalves, and much echinodermal and shell debris; about 2% detrital quartz 3.0

UPPER PERMIAN
CHHIDRU FORMATION
Uppermost 25 feet of this unit covered

10. North side of small creek 200 yards east of road, one-quarter mile southeast of Kathwai, Salt Range (lat. 32°29′ N., long. 72°12′ E.) (Fig. A15).

Thickness
(in feet)

LOWER TRIASSIC
MIANWALI FORMATION
Mittiwali Member (basal part= "Lower Ceratite limestone")
8. Limestone, light brownish-gray, fine-grained, coquinoidal, slightly glauconitic; abundant poorly preserved ammonoids (*Gyronites*, etc.) 5.9

Kathwai Member
Limestone unit
7. Limestone, yellowish-gray, fine-grained, faintly glauconitic; contains brachiopod and echinoderm fragments 1.0

6. Limestone, grayish-orange, thin-bedded, soft, coquinoidal; contains *Orbiculoidea*, rhynchonellids, cidarid spines, other echinoid and crinoid fragments, ammonoids *(Ophiceras)*; and conodonts [*Ellisonia gradata* Sweet, *E. triassica* Müller, *Neogondolella carinata* (Clark)]; most fossils are in a 6-inch bed 1 foot from top of unit 5.0
5. Limestone, grayish-orange, highly dolomitic; echinoderm fragments 1.0

Dolomite unit
4. Dolomite, grayish-orange, medium to coarsely crystalline, in beds 1-2 inches thick; echinoderm fragments 2.9
3. Dolomite, pale to dark yellowish-orange, massive, fine-grained; contains *Orbiculoidea*, large smooth pectinids, *Glyptophiceras himalayanum* (Griesbach) and conodonts *(Anchignathodus typicalis* Sweet, *Ellisonia teicherti* Sweet, *E. triassica* Müller); fossils come from upper 6 inches 2.6

UPPER PERMIAN
CHHIDRU FORMATION
White sandstone unit
2. Sandstone, white, soft, friable .. 0-1.0

Highly fossiliferous unit
1. Limestone, yellowish-gray, sandy, massive, with abundant Permian fossils; on west side of creek, 50 yards off this section this unit immediately underlies bed 3 and the white sandstone unit is absent. Here unit 1 contains abundant dasycladacean algae *(Permocalculus)*, bryozoans, brachiopods, nautiloids *(Domatoceras, Metacoceras)*, gastropods, including *Euphemites indicus*, several species of *Bellerophon*, and *"Strobeus"* of *"S." avellenoides* (de Koninck) (E. L. Yochelson, written communication, 1963), and large scaphopods, probably *Plagioglypta herculea* (de Koninck).

11. Southeast side of Kingriali Mountain, Khisor Range (lat. 32°14′ N., long. 71°02.5′ E.) (Fig. A17, A18).

Thickness (in feet)

LOWER TRIASSIC
MIANWALI FORMATION
Mittiwali Member (basal part= "Lower Ceratite limestone")
7. Limestone, gray; rich in ammonoids, not studied in detail 25

FIG. A18. Section of Triassic rocks at Kingriali, Khisor Range. Cliff in foreground is Kathwai Member, with base of "Lower Ceratite limestone" (platy beds) on top; slope in middle is "Ceratite marl," followed by another cliff, the "Upper Ceratite limestone"; the cliff at the top is Kingriali dolomite, and the slope in between is made up of the Narmia Member of the Mittiwali Formation and the Landa and Khatkiara Members of the Tredian Formation. (Teichert photograph.)

FIG. A19. Tapan Wahan, Khisor Range, on right, viewed from half-a-mile to the south. Rocks in foreground are Kathwai Member; relationships of limestone and dolomite are shown diagrammatically in Fig. 14. (Teichert photograph.)

Kathwai Member
Limestone unit
6. Limestone, pale yellowish-brown to medium light gray, hard, weathering pale yellowish-brown, non-glauconitic; contains conodonts [*Ellisonia gradata* Sweet, *E. triassica* Müller, *Neogondolella carinata* (Clark), *Xaniognathus curvatus* Sweet] .. 0.5-1.0
5. Limestone, light olive-gray, hard, medium-bedded, highly glauconitic, cross-bedded; near the top are found pelecypods, *Miocidaris* spines, *Ophiceras*, and conodonts [*Anchignathodus typicalis* Sweet, *Ellisonia gradata* Sweet, *E. teicherti* Sweet, *E. triassica* Müller, *Neogondolella carinata* (Clark), *Xaniognathus curvatus* Sweet] .. 2.1

Dolomite unit
4. Dolomite, grayish-orange to pale yellowish-orange, medium crystalline, irregularly thick-bedded, coarsely cross-bedded, weathering light brown; contains conodonts [*Anchignathodus isarcicus* (Huckriede), *Anchignathodus typicalis* Sweet, *Ellisonia gradata* Sweet, *E. teicherti* Sweet, *E. triassica* Müller, *Neogondolella carinata* (Clark), *Xaniognathus curvatus* Sweet] .. 7.0-7.6

3. Dolomite, grayish-orange, one massive bed; upper contact with unit 4 undulating 0.6-1.2

UPPER PERMIAN
CHHIDRU FORMATION
White sandstone unit
2. Sandstone, white to light yellowish-gray, calcareous, fine- to medium-grained, without distinct bedding, generally friable with the exception of thin, hard, more calcareous beds in the middle and near the top which contain bryozoans and brachiopods. The latter include *Cleiothyridina* cf. *C. capillata*, *Derbyia* cf. *D. plicatella*, *Enteletes* sp. 1, ?*Orthotichia* sp., *Spirigerella* sp. (abundant and well preserved), *Waagenoconcha* sp., *Whitspakia* sp. 1 .... 7.5

Highly fossiliferous unit
1. Sandstone, yellowish-gray, friable, highly calcareous, indistinctly bedded; unfossiliferous

12. Bottom of Tapan Wahan, southern Khisor Range (lat. 32°12.5′ N., long. 70°54.5′ E.) (Fig. A17, A19).

Thickness (in feet)

LOWER TRIASSIC

MIANWALI FORMATION
Mittiwali Member (basal part= "Lower Ceratite limestone")
7. Limestone, thin-bedded to flaggy, gray, coquinoidal; containing abundant ammonoids 25.0

Kathwai Member
Limestone unit
6. Limestone, lowermost 10 inches pale yellowish-orange, friable, thin-bedded, in part slightly dolomitized; remainder of unit is light gray, hard, thin- to medium-bedded, in part coquinoidal; contains abundant shell fragments, *Ophiceras* common in upper part; entire unit glauconitic 3.5

Dolomite unit
5. Dolomite, grayish-orange, thin-bedded to flaggy; consists of finely comminuted shell and echinoderm fragments; contains some glauconite and traces of silt-size detrital quartz grains 0.7
4. Dolomite, weakly calcitic, grayish-orange, massive; consists mostly of comminuted echinoderm fragments, with an admixture of shell fragments in the upper part; fine angular to subangular detrital quartz grains, decreasing in bulk from about 5% in the lower part to about 1% in the upper part; the upper part is slightly glauconitic 4.7

UPPER PERMIAN
CHHIDRU FORMATION
White sandstone unit
3. Sandstone, white, friable, no studied in detail; one-half mile south of section contain brachiopods (*Aulosteges* sp., *Callispirina* sp., *Chonetella* sp., chonetid undet., *Cleiothyridina* cf. *C. capillata*, *Derbyia* cf. *D. plicatella*, dielasmatids indet., *Enteletes* sp. 1, *Hemiptychina* sp., *Hustedia* sp., *Kiangsiella* sp., *Linoproductus* sp., *Lyttonia* sp., *Martinia?* sp., *Neospirifer* sp., *Orthotichia* sp. 1, *Orthotichia* sp. 2, *Richthofenia* sp., *Spiriferella?* sp., *Spirigerella* sp., *Waagenoconcha?* sp., *Whitspakia* sp. 1) 8.0

Highly fossiliferous units
2. Calcareous sandstone with *Bellerophon* 0.2
1. Sandstone, calcareous, orange-gray, fine-grained .... 15+

APPENDIX B: LIST OF SAMPLES USED

This Appendix contains a list giving precise occurrence data for all samples used by Balme (1970), Rowell (1970), Sohn (1970), Sweet (1970), and by ourselves. Included are references to maps and figures in this report [indicated by K & T] and in earlier publications.

K1-12	Loc. 2, bed 4, Narmia Spring, Surghar Range (K & T, fig. A1, A2; bed 2 in Kummel, 1966, p. 415); dolomite unit of Kathwai Member, Mianwali Formation.
K1-16	Loc. 2, bed 8, Narmia Spring, Surghar Range (K & T, fig. A1, A2; bed 6 in Kummel, 1966, p. 415); basal unit of Mittiwali Member, Mianwali Formation=Lower Ceratite limestone of Waagen.
K1-17	Loc. 2, Narmia Spring, Surghar Range (K & T, fig. A1, A2; bed 7 in Kummel, 1966, p. 415); basal unit of Mittiwali Member of Mianwali Formation=Lower Ceratite limestone of Waagen.
K1-21	Loc. 2, Narmia Spring, Surghar Range (K & T, fig. A1, A2; bed 11 in Kummel, 1966, p. 415); Mittiwali Member of Mianwali Formation.
K1-36	Loc. 2, Narmia Spring, Surghar Range (K & T, fig. A1, A2; bed 26 in Kummel, 1966, p. 414); Narmia Member of Mianwali Formation.
K1-37	Loc. 2, Narmia Spring, Surghar Range (K & T, fig. A1, A2; bed 27 in Kummel, 1966, p. 414); Narmia Member of Mianwali Formation.
K1-42	Loc. 2, Narmia Spring, Surghar Range (K & T, fig. A1, A2; bed 32 in Kummel, 1966, p. 413); Narmia Member of Mianwali Formation.
K1-50	Loc. 2, Narmia Spring, Surghar Range (K & T, fig. A1, A2; bed 39 in Kummel, 1966, p. 413); Narmia Member of Mianwali Formation.
K2-4	Loc. 5, beds 3-7, Ganjaroh, Salt Range (K & T, fig. A7); dolomite unit of Kathwai Member, Mianwali Formation.
K2-5	Loc. 5, bed 9, Ganjaroh, Salt Range (K & T, fig. A7); basal unit of Mittiwali Member, Mianwali Formation=Lower Ceratite limestone of Waagen.
K3-2A	Loc. 6A, bed 2, Chhidru A, Salt Range (K & T, fig. A7; unit 11 in Teichert, 1966, p. 17); white sand-

K3-3	Loc. 6A, bed 3, Chhidru A, Salt Range (K & T, fig. A8, A9, A10; bed 1 in Kummel, 1966, p. 409); dolomite unit of Kathwai Member, Mianwali Formation. stone unit of Chhidru Formation, top 6 inches.
K3-5	Loc. 6A, bed 5, Chhidru A, Salt Range (K & T, fig. A8, A9, A10; bed 2 in Kummel, 1966, p. 409); dolomite unit of Kathwai Member, Mianwali Formation.
K3-6	Loc. 6A, beds 6, 7, Chhidru A, Salt Range (K & T, fig. A8, A9, A10; beds 3, 4 in Kummel, 1966, p. 409); dolomite unit of Kathwai Member, Mianwali Formation.
K3-7	Loc. 6A, bed 8, Chhidru A, Salt Range (K & T, fig. A8, A9, A10; bed 5 in Kummel, 1966, p. 409); dolomite unit of Kathwai Member, Mianwali Formation.
K3-8	Loc. 6A, beds 9, 10, Chhidru A, Salt Range (K & T, fig. A8, A9, A10; beds 6-11 in Kummel, 1966, p. 409); limestone unit of Kathwai Member, Mianwali Formation.
K3-10	Loc. 6A, lower 3 feet of "Ceratite marl," Chhidru A, Salt Range (K & T, fig. A8, A9, A10; bed 15 in Kummel, 1966, p. 408); Mittiwali Member of Mianwali Formation.
K3-12	Loc. 6A, 10 feet above "Lower Ceratite limestone," Chhidru A, Salt Range (K & T, fig. A8, A9, A10; bed 15 in Kummel, 1966, p. 408); Mittiwali Member of Mianwali Formation.
K3-14	Loc. 6A, about 15 feet above "Lower Ceratite limestone," Chhidru A, Salt Range (K & T, fig. A8, A9, A10; bed 15 in Kummel, 1966, p. 408); Mittiwali Member of Mianwali Formation.
K3-15	Loc. 6A, 20 feet above "Lower Ceratite limestone," Chhidru A, Salt Range (K & T, fig. A8, A9, A10; bed 15 in Kummel, 1966, p. 408); Mittiwali Member of Mianwali Formation.
K4-2	Loc. 6B, bed 1, Chhidru B, Salt Range (K & T, fig. A8, top of unit 10 in Teichert, 1966, p. 17); uppermost highly fossiliferous unit of Chhidru Formation.
K4-11	Loc. 6B, bed 9, Chhidru B, Salt Range (K & T, fig. A8); limestone unit of Kathwai Member, Mianwali Formation.
K4-14	Loc. 6B, bed 12, Chhidru B, Salt Range (K & T, fig. A8); basal unit of Mittiwali Member, Mianwali Formation=Lower Ceratite limestone of Waagen.
K4-LCM	Loc. 6B, Chhidru B, Salt Range (K & T, fig. A8); Mittiwali Member of Mianwali Formation (lower part of Waagen's "Ceratite marl").
K4-CM	Loc. 6B, Chhidru B, Salt Range (K & T, fig. A8); Mittiwali Member of Mianwali Farmation (from Waagen's "Ceratite marl").
K4-UCS	Loc. 6B, Chhidru B, Salt Range (K & T, fig. A8); Mittiwali Member of Mianwali Formation (from Waagen's "Ceratite sandstone").
K4-UCL	Loc. 6B, Chhidru B, Salt Range (K & T, fig. A8); Mittiwali Member of Mianwali Formation (from Waagen's "Upper Ceratite limestone").
K6-1A	Loc. 4, bed 2, Nammal Gorge, Salt Range (K & T, fig. A5, A6), uppermost highly fossiliferous unit of Chhidru Formation.
K6-5	Loc. 4, bed 6, Nammal Gorge, Salt Range (K & T, fig. A5, A6; bed 3 in Kummel, 1966, p. 411); dolomite unit of Kathwai Member, Mianwali Formation.
K6-6	Loc. 4, bed 7, Nammal Gorge, Salt Range (K & T, fig. A5, A6; bed 4 in Kummel, 1966, p. 411); limestone unit of Kathwai Member, Mianwali Formation.
K6-A	Loc. 4, Nammal Gorge, Salt Range (K & T, fig. A5, A6; bed 6, 6 feet above base, in Kummel, 1966, p. 411); Mittiwali Member of Mianwali Formation.
K6-B	Loc. 4, Nammal Gorge, Salt Range (K & T, fig. A5, A6; bed 6, 40 feet above base, in Kummel, 1966, p. 411); Mittiwali Member of Mianwali Formation.
K6-8A	Loc. 4, Nammal Gorge, Salt Range (K & T, fig. A5, A6; bed 6, 4 feet above base, in Kummel, 1966, p. 411); Mittiwali Member of Mianwali Formation.
K6-8B	Loc. 4, Nammal Gorge, Salt Range (K & T, fig. A5, A6; bed 6, 15 feet above base, in Kummel, 1966, p. 411); Mittiwali Member of Mianwali Formation.
K6-8C	Loc. 4, Nammal Gorge, Salt Range (K & T, fig. A5, A6; bed 6, 30 feet above base, in Kummel, 1966, p. 411); Mittiwali Member of Mianwali Formation.
K6-8D	Loc. 4, Nammal Gorge, Salt Range (K & T, fig. A5, A6; bed 6, 50 feet above base, in Kummel, 1966, p. 411); Mittiwali Member of Mianwali Formation.
K6-8E	Loc. 4, Nammal Gorge, Salt Range

	(K & T, fig. A5, A6; bed 6, 80 feet above base, in Kummel, 1966, p. 411); Mittiwali Member of Mianwali Formation.		Narmia Member of Mianwali Formation.
K6-8F	Loc. 4, Nammal Gorge, Salt Range (K & T, fig. A5, A6; bed 6, 95 feet above base, in Kummel, 1966, p. 411); Mittiwali Member of Mianwali Formation.	K6-43A	Loc. 4, Nammal Gorge, Salt Range (K & T, fig. A5, A6; bed 41, 3 feet above base, in Kummel, 1966, p. 409); Landa Member of Tredian Formation.
K6-8G	Loc. 4, Nammal Gorge, Salt Range (K & T, fig. A5, A6; bed 6, 10 feet from top of unit, in Kummel, 1966, p. 411); Mittiwali Member of Mianwali Formation.	K6-43B	Loc. 4, Nammal Gorge, Salt Range (K & T, fig. A5, A6; bed 41, from middle of unit, in Kummel, 1966, p. 409); Landa Member of Tredian Formation, about 14 feet above base of member.
K6-9	Loc. 4, Nammal Gorge, Salt Range (K & T, fig. A5, A6; bed 7 in Kummel, 1966, p. 411); Mittiwali Member of Mianwali Formation, about 123 feet above base of member.	K6-44	Loc. 4, Nammal Gorge, Salt Range (K & T, fig. A5, A6; bed 42 in Kummel, 1966, p. 409); Landa Member of Tredian Formation.
K6-27A	Loc. 4, Nammal Gorge, Salt Range (K & T, fig. A5, A6; bed 25 in Kummel, 1966, p. 410); Mittiwali Member of Mianwali Formation.	K6-45A	Loc. 4, Nammal Gorge, Salt Range (K & T, fig. A5, A6; bed 43, upper foot of unit, in Kummel, 1966, p. 409); Landa Member of Tredian Formation.
K6-30	Loc. 4, Nammal Gorge, Salt Range (K & T, fig. A5, A6; bed 28 in Kummel, 1966, p. 410); Narmia Member of Mianwali Formation.	K7-2	Loc. 8, bed 2, Kufri, Salt Range (K & T, fig. A13, A14; bed 00 in Kummel, 1966, p. 404), white sandstone unit of Chhidru Formation.
K6-31A	Loc. 4, Nammal Gorge, Salt Range (K & T, fig. A5, A6; bed 29, in Kummel, 1966, p. 410); Narmia Member of Mianwali Formation, about 35 feet above base of member.	K10-1	Loc. 9, bed 1, Kathwai A, Salt Range (K & T, fig. A15, A16), dolomite unit of Kathwai Member, Mianwali Formation.
		K10-2	Loc. 9, bed 3, Kathwai A, Salt Range (K & T, fig. A15, A16); dolomite unit of Kathwai Member, Mianwali Formation.
K6-31B	Loc. 4, Nammal Gorge, Salt Range (K & T, fig. A5, A6; bed 29, 5 feet below top of unit, in Kummel, 1966, p. 410); Narmia Member of Mianwali Formation.	K10-3	Loc. 9, bed 4, Kathwai A, Salt Range (K & T, fig. A15, A16); dolomite unit of Kathwai Member, Mianwali Formation.
K6-33	Loc. 4, Nammal Gorge, Salt Range (K & T, fig. A5, A6; bed 31 in Kummel, 1966, p. 409); Narmia Member of Mianwali Formation, about 77 feet above base of member.	K10-5	Loc. 9, bed 6, Kathwai A, Salt Range (K & T, fig. A15, A16); basal unit of Mittiwali Member, Mianwali Formation=Lower Ceratite limestone of Waagen.
K6-35	Loc. 4, Nammal Gorge, Salt Range (K & T, fig. A5, A6; bed 33 in Kummel, 1966, p. 409); Narmia Member of Mianwali Formation, about 110 feet above base of member.	K10-6	Loc. 9, Kathwai A, Salt Range (K & T, fig. A15, A16); Lower part of Mittiwali Member of Mianwali Formation.
K6-36	Loc. 4, Nammal Gorge, Salt Range (K & T, fig. A5, A6; bed 34 in Kummel, 1966, p. 409); Narmia Member of Mianwali Formation.	K11-LCL	Loc. 7, Munta Nala, 1.5 mi. west of Wargal, Salt Range (K & T, fig. A11, A12); basal unit of Mittiwali Member, Mianwali Formation=Lower Ceratite limestone of Waagen.
K6-37	Loc. 4, Nammal Gorge, Salt Range (K & T, fig. A5, A6; bed 35 in Kummel, 1966, p. 409); Mittiwali Member of Mianwali Formation.	K11-CL	Loc. 7, Munta Nala, 1.5 mi. west of Wargal, Salt Range (K & T, fig. A11, A12); basal unit of Mittiwali Member, Mianwali Formation=Lower Ceratite limestone of Waagen.
K6-41A	Loc. 4, Nammal Gorge, Salt Range (K & T, fig. A5, A6; bed 39, 5 feet above base, in Kummel, 1966, p. 409); Narmia Member of Mianwali Formation.	K11-2C	Loc. 7, bed 1, Munta Nala, 1.5 mi. west of Wargal, Salt Range (K & T, fig. A11, A12; bed 2 in Kummel, 1966, p. 406); uppermost highly fossiliferous unit of Chhidru Formation, 2 feet below top.
K6-41B	Loc. 4, Nammal Gorge, Salt Range (K & T, fig. A5, A6; bed 39, upper 2 feet, in Kummel, 1966, p. 409);		

K11-2D Loc. 7, bed 1, Munta Nala, 1.5 mi. west of Wargal, Salt Range (K & T, fig. A11, A12; bed 2 in Kummel, 1966, p. 406); uppermost highly fossiliferous unit of Chhidru Formation, uppermost foot.

K11-2F Loc. 7, bed 1, Munta Nala, 1.5 mi. west of Wargal, Salt Range (K & T, fig. A11, A12; bed 2 in Kummel, 1966, p. 406); uppermost highly fossiliferous unit of Chhidru Formation, lower 6 inches.

K11-3 Loc. 7, bed 2, Munta Nala, 1.5 mi. west of Wargal, Salt Range (K & T, fig. A11, A12; bed 3 in Kummel, 1966, p. 406); uppermost highly fossiliferous unit of Chhidru Formation.

K11-4A Loc. 7, bed 3, Munta Nala, 1.5 mi. west of Wargal, Salt Range (K & T, fig. A11, A12; bed 4 in Kummel, 1966, p. 406); white sandstone unit of Chhidru Formation, bottom of unit.

K11-4B Loc. 7, bed 3, Munta Nala, 1.5 mi. west of Wargal, Salt Range (K & T, fig. A11, A12; bed 4 in Kummel, 1966, p. 406); white sandstone unit of Chhidru Formation, 8 feet below top of unit.

K11-4C Loc. 7, bed 3, Munta Nala, 1.5 mi. west of Wargal, Salt Range (K & T, fig. A11, A12; bed 4 in Kummel, 1966, p. 406); white sandstone unit of Chhidru Formation, 7 feet below top of unit.

K11-6A Loc. 7, bed 5, Munta Nala, 1.5 mi. west of Wargal, Salt Range (K & T, fig. A11, A12; bed 6 in Kummel, 1966, p. 406); white sandstone unit of Chhidru Formation, from lower 3 inches.

K11-6B Loc. 7, bed 5, Munta Nala, 1.5 mi. west of Wargal, Salt Range (K & T, fig. A11, A12; bed 6 in Kummel, 1966, p. 406); white sandstone unit of Chhidru Formation, from middle of bed.

K11-6C Loc. 7, bed 5, Munta Nala, 1.5 mi. west of Wargal, Salt Range (K & T, fig. A11, A12; bed 6 in Kummel, 1966, p. 406); white sandstone unit of Chhidru Formation, 2 feet below top.

K11-6D Loc. 7, bed 5, Munta Nala, 1.5 mi. west of Wargal, Salt Range (K & T, fig. A11, A12; bed 6 in Kummel, 1966, p. 406); white sandstone unit of Chhidru Formation, 3 feet from top.

K11-6E Loc. 7, bed 5, Munta Nala, 1.5 mi. west of Wargal, Salt Range (K & T, fig. A11, A12; bed 6 in Kummel, 1966, p. 406); white sandstone unit of Chhidru Formation, from upper 1 foot.

K11-8 Loc. 7, bed 7, Munta Nala, 1.5 mi. west of Wargal, Salt Range (K & T, fig. A11, A12; bed 9 in Kummel, 1966, p. 405); dolomite unit of Kathwai Member of Mianwali Formation.

K11-9 Loc. 7, bed 8, Munta Nala, 1.5 mi. west of Wargal, Salt Range (K & T, fig. A11, A12; bed 9 in Kummel, 1966, p. 405); dolomite unit of Kathwai Member, Mianwali Formation.

K11-10 Loc. 7, bed 9, Munta Nala, 1.5 mi. west of Wargal, Salt Range (K & T, fig. A11, A12; bed 10 in Kummel, 1966, p. 405); limestone unit of Kathwai Member of Mianwali Formation.

K11-14 Loc. 7, bed 13, Munta Nala, 1.5 mi. west of Wargal, Salt Range (K & T, fig. A11, A12; bed 14 in Kummel, 1966, p. 405); basal unit of Mittiwali Member of Mianwali Formation (=Lower Ceratite Limestone of Waagen).

K11-15 Loc. 7, bed 14, Munta Nala, 1.5 mi. west of Wargal, Salt Range (K & T, fig. A11, A12; bed 15 in Kummel, 1966, p. 405); basal unit of Mittiwali Member, Mianwali Formation =Lower Ceratite limestone of Waagen.

K12-1 Loc. 1, Landa Pusha, Surghar Range (K & T, fig. A1; bed 1 in Kummel, 1966, p. 416); basal unit of Narmia Member, Mianwali Formation.

K12-2 Loc. 1, Landa Pusha, Surghar Range (K & T, fig. A1; bed 2 in Kummel, 1966, p. 416); lower part of Narmia Member, Mianwali Formation.

K12-5 Loc. 1, Landa Pusha, Surghar Range (K & T, fig. A1; bed 5 in Kummel, 1966, p. 416); Narmia Member of Mianwali Formation.

K12-6 Loc. 1, Landa Pusha, Surghar Range (K & T, fig. A1; bed 6 in Kummel, 1966, p. 416); Narmia Member of Mianwali Formation.

K12-7 Loc. 1, Landa Pusha, Surghar Range (K & T, fig. A1; bed 7 in Kummel, 1966, p. 416); Narmia Member of Mianwali Formation.

K12-9 Loc. 1, Landa Pusha, Surghar Range (K & T, fig. A1; bed 9 in Kummel, 1966, p. 416); Narmia Member of Mianwali Formation.

K12-10 Loc 1, Landa Pusha, Surghar Range (K & T, fig. A1; bed 10 in Kummel, 1966, p. 416); Narmia Member of Mianwali Formation.

K12-12 Loc. 1, Landa Pusha, Surghar Range

	(K & T, fig. A1; bed 12 in Kummel, 1966, p. 416); Narmia Member of Mianwali Formation.
K12-14	Loc. 1, Landa Pusha, Surghar Range (K & T, fig. A1; bed 14 in Kummel, 1966, p. 416); Narmia Member of Mianwali Formation.
K12-18A	Loc. 1, Landa Pusha, Surghar Range (K & T, fig. A1; bed 18, 5 feet above base, in Kummel, 1966, p. 416); Landa Member of Tredian Formation.
K12-21	Loc. 1, Landa Pusha, Surghar Range (K & T, fig. A1); 135 feet below bed 1 in Kummel, 1966, p. 416); Mittiwali Member of Mianwali Formation.
K13-A	Loc. 3, bed 1, Zaluch Nala, Salt Range (K & T, fig. A3, A4; Teichert, 1966, fig. 3, top of unit 47); uppermost fossiliferous unit of Chhidru Formation.
K13-9	Loc. 3, Zaluch Nala, Salt Range (K & T, fig. A3, A4; bed 15 in Kummel, 1966, p. 413); Mittiwali Member of Mianwali Formation.
K13-11	Loc. 3, Zaluch Nala, Salt Range (K & T, fig. A3, A4; bed 17 in Kummel, 1966, p. 412); Mittiwali Member of Mianwali Formation.
K13-13	Loc. 3, Zaluch Nala, Salt Range (K & T, fig. A3, A4; bed 19 in Kummel, 1966, p. 412); basal unit of Narmia Member, Mianwali Formation.
K13-24	Loc. 3, Zaluch Nala, Salt Range (K & T, fig. A3, A4; bed 30 in Kummel, 1966, p. 412); Landa Member of Tredian Formation, 44 feet above base of member.
K14-3	Loc. 10, bed 3, Kathwai B, Salt Range (K & T, fig. A15; bed 1 in Kummel, 1966, p. 402); dolomite unit of Kathwai Member, Mianwali Formation.
K14-6	Loc. 10, bed 6, Kathwai B, Salt Range (K & T, fig. A15; bed 4 in Kummel, 1966, p. 402); limestone unit of Kathwai Member of Mianwali Formation.
K14-9	Loc. 10, Kathwai B, Salt Range (K & T, fig. A15, bed 7 in Kummel, 1966, p. 402); Mittiwali Member of Mianwali Formation.
T61-37	Loc. 6A, bed 6, Chhidru A, Salt Range (K & T, fig. A8, A9, A10); dolomite unit of Kathwai Member of Mianwali Formation, 5 feet above base of member.
T61-48	Loc 6B, bed 7, Chhidru B, Salt Range (K & T, fig. A8); dolomite unit of Kathwai Member of Mianwali Formation, near top of unit.
T61-55	Loc. 4, bed 1, Nammal Gorge, Salt Range (K & T, fig. A5, A6); uppermost highly fossiliferous unit of Chhidru Formation, immediately below white sandstone unit.
T61-58	Loc. 4, bed 7, Nammal Gorge, Salt Range (K & T, fig. A5, A6); limestone unit of Kathwai Member of Mianwali Formation, upper part of unit.
T61-61	Loc. 4, bed 2, Nammal Gorge, Salt Range (K & T, fig. A5, A6); white sandstone unit of Chhidru Formation, just below top.
T61-64	Loc. 4, Nammal Gorge, Salt Range (K & T, fig. A5, A6); basal part of Wargal Limestone.
T61-81	Loc. 4, Nammal Gorge, Salt Range (K & T, fig. A5, A6); Wargal Limestone, 130 feet above T61-64.
T61-164	Loc. 9, bed 4, top of bed, Kathwai A, Salt Range (K & T, fig. A15); dolomite unit of Kathwai Member, Mianwali Formation.
T61-166	Loc. 9, bed 5, Kathwai A, Salt Range (K & T, fig. A15, A16); limestone unit of Kathwai Member of Mianwali Formation, 3 inches below top of unit.
T61-178	Loc. 9, Kathwai A, Salt Range (K & T, fig. A15, A16); Chhidru Formation, about 80 feet below top.
T61-206	Loc. 7, bed 5, Munta Nala, 1.5 mi. west of Wargal, Salt Range (K & T, fig. A11, A12); white sandstone unit of Chhidru Formation, 3.5 feet below top of unit.
T61-207	Loc. 7, bed 5, Munta Nala, 1.5 mi. west of Wargal, Salt Range (K & T, fig. A11, A12); Chhidru Formation, about 15 feet below top.
T61-208	Loc. 7, bed 5, Munta Nala, 1.5 mi. west of Wargal, Salt Range (K & T, fig. A11, A12); white sandstone unit of Chhidru Formation, 3.5 feet below top of unit.
T61-208A	Loc. 7, bed 5, Munta Nala, 1.5 mi. west of Wargal, Salt Range (K & T, fig. A11, A12); Chhidru Formation, 6 inches below top of unit 1; 14 feet below top of formation.
T61-209	Loc. 7, bed 5, Munta Nala, 1.5 mi. west of Wargal, Salt Range (K & T, fig. A11, A12); white sandstone unit of Chhidru Formation, 5.5 feet below top of unit.
T61-217	Loc. 7, bed 5, Munta Nala, 1.5 mi. west of Wargal, Salt Range (K & T, fig. A11, A12); white sandstone unit of Chhidru Formation, 6 inches below top of unit.
T61-218	Loc. 7, bed 7, Munta Nala, 1.5 mi. west of Wargal, Salt Range (K & T,

	fig. A11, A12); dolomite unit of Kathwai Member of Mianwali Formation, 7.5 feet above base.
T61-226	Loc. 7, bed 10, Munta Nala, 1.5 mi. west of Wargal (K & T, fig. A11, A12); limestone unit of Kathwai Member, Mianwali Formation, 1.5 feet below top of unit.
T62-7	Loc. 2, bed 3, Narmia Spring, Surghar Range (K & T, fig. A1, A2); dolomite unit of Kathwai Member of Mianwali Formation, basal 4 inches of unit.
T62-8	Loc. 2, bed 4, Narmia Spring, Surghar Range (K & T, fig. A1, A2); dolomite unit of Kathwai Member of Mianwali Formation, 4 inches below top of unit.
T62-10	Loc. 2, bed 6, Narmia Spring, Surghar Range (K & T, fig. A1, A2); limestone unit of Kathwai Member of Mianwali Formation, near top of unit.
T62-73	Loc. 4, Nammal Gorge, Salt Range (K & T, fig. A5, A6); Chhidru Formation, basal shale member, 2-3 feet above base.
T62-87	Loc. 4, Nammal Gorge, Salt Range (K & T, fig. A5, A6); Chhidru Formation, shale 148 feet above base.
T62-157	Loc. 5, bed 5, Ganjaroh, Salt Range (K & T, fig. A7); dolomite unit of Kathwai Member of Mianwali Formation, 2.5 feet above base of unit.
T62-159	Loc. 5, bed 7, Ganjaroh, Salt Range (K & T, fig. A7); dolomite unit of Kathwai Member of Mianwali Formation, top 16 inches of unit.
T62-173	Loc. 3, Zaluch Nala, Salt Range (K & T, fig. A3, A4; unit 42 in Teichert, 1966, p. 6); Chhidru Formation, basal shale unit.
T62-191	Loc. 3, bed 4, Zaluch Nala, Salt Range (K & T, fig. A3, A4); dolomite unit of Kathwai Member, Mianwali Formation, 2 feet above base of unit.
T62-192	Loc. 3, bed 5, Zaluch Nala, Salt Range (K & T, fig. A3, A4); dolomite unit of Kathwai Member, Mianwali Formation, 4.5 feet above base of unit.
T62-193	Loc. 3, bed 5, Zaluch Nala, Salt Range (K & T, fig. A3, A4); dolomite unit of Kathwai Member, Mianwali Formation, 5 feet above base of member.
T62-207	Loc. 10, bed 6, Kathwai B, Salt Range (K & T, fig. A15); limestone unit of Kathwai Member of Mianwali Formation, 4 feet above base of unit.
T62-284	Loc. 3, Zaluch Nala, Salt Range (K & T, fig. A3, A4); Landa Member of Tredian Formation, 73 feet above base of member.
T62-286	Loc. 3, Zaluch Nala, Salt Range (K & T, fig. A3, A4); Khatkiara Member of Tredian Formation, 50 feet above base.
T62-287	Loc. 3, Zaluch Nala, Salt Range (K & T, fig. A3, A4); Khatkiara Member of Tredian Formation, 55 feet above base.
T62-300	Loc. 2, bed 3, Narmia Spring, Surghar Range (K & T, fig. A1, A2); dolomite unit of Kathwai Member, Mianwali Formation, lowermost 6 inches of unit (same as T63-19, T63-162).
T62-304	Loc. 2, basal part of bed 4, Narmia Spring, Surghar Range (K & T, fig. A1, A2); dolomite unit of Kathwai Member of Mianwali Formation, 18-21 inches above base of unit.
T62-320	Dhodha Wahan, Salt Range (Teichert, 1966, fig. 4); Amb Formation, 67-70 feet above base.
T62-347	Warchha Water Tank, near Warchha, Salt Range (Teichert, 1966, p. 8-9); Amb Formation, 126 feet above base.
T62-350	Warchha Water Tank, near Warchha, Salt Range (Teichert, 1966, p. 8-9); Amb Formation, 160 feet above base.
T63-19	Loc. 2, bed 3, Narmia Spring, Surghar Range (K & T, fig. A1, A2); dolomite unit of Kathwai Member, Mianwali Formation, lowermost 6 inches of unit (same as T62-300, T63-162).
T63-22	200 yards east of Loc. 2, Narmia Spring, Surghar Range, from equivalent of bed 3; dolomite unit of Kathwai Member, Mianwali Formation, 3 inches above base of unit.
T63-122	Loc. 6A, bed 3, Chhidru A, Salt Range (K & T, fig. A8, A9, A10); dolomite unit of Kathwai Member, Mianwali Formation, 1-5 inches above base of unit.
T63-123	Loc. 6A, bed 3, Chhidru A, Salt Range (K & T, fig. A8, A9, A10); dolomite unit of Kathwai Member, Mianwali Formation, 1-6 inches above base of unit.
T63-124	Loc. 6A, bed 4, Chhidru A, Salt Range (K & T, fig. A8, A9, A10); dolomite unit of Kathwai Member, Mianwali Formation, 20-24 inches above base of bed 4.
T63-127	Loc. 6A, bed 5, Chhidru A, Salt Range (K & T, fig. A8, A9, A10); dolomite unit of Kathwai Member

T63-128   Loc. 6A, bed 6, Chhidru A, Salt Range (K & T, fig. A8, A9, A10); dolomite unit of Kathwai Member, Mianwali Formation, 4 feet above base of unit.

T63-130   Loc. 6A, bed 7, Chhidru A, Salt Range (K & T, fig. A8, A9, A10); dolomite unit of Kathwai Member, Mianwali Formation, 8-12 inches above base of bed 7.

T63-131   Loc. 6A, bed 7, Chhidru A, Salt Range (K & T, fig. A8, A9, A10); dolomite unit of Kathwai Member, Mianwali Formation, 32-38 inches above base of bed 7.

T63-132   Loc. 6A, bed 8, Chhidru A, Salt Range (K & T, fig. A8, A9, A10); dolomite unit of Kathwai Member, Mianwali Formation, 2-6 inches above base of bed 8.

T63-136   Loc. 6A, bed 10, Chhidru A, Salt Range (K & T, fig. A8, A9, A10); limestone unit of Kathwai Member, Mianwali Formation, 16-20 inches above base of bed 10.

T63-139   Loc. 6A, bed 11, Chhidru A, Salt Range (K & T, fig. A8, A9, A10); basal unit of Mittiwali Member, Mianwali Formation (=Lower Ceratite limestone of Waagen), 0-3 inches above base.

T63-140   Loc. 6A, bed 11, Chhidru A, Salt Range (K & T, fig. A8, A9, A10); "Lower Ceratite limestone" unit of Mittiwali Member of Mianwali Formation, top 4 inches of unit.

T63-144   Khan Zaman Nala (K & T, fig. A8); Kathwai Member of Mianwali Formation, basal part of limestone unit or uppermost part of dolomite unit.

T63-145   Khan Zaman Nala (K & T, fig. A8); Kathwai Member of Mianwali Formation, same as T63-144.

T63-151   Loc. 6B, bed 3, Narmia Spring, Surghar Range (K & T, fig. A8); dolomite unit of Kathwai Member, Mianwali Formation.

T63-153   Loc. 6A, bed 11, Chhidru A, Salt Range (K & T, fig. A8, A9, A10); basal unit of Mittiwali Member, Mianwali Formation (=Lower Ceratite limestone of Waagen), 14-24 inches above base.

T63-155   Loc. 6A, bed 11, Chhidru A, Salt Range (K & T, fig. A8, A9, A10); basal unit of Mittiwali Member, Mianwali Formation (=Lower Ceratite limestone of Waagen), 24-37 inches above base.

T63-157   Loc. 6A, bed 11, Chhidru A, Salt Range (K & T, fig. A8, A9, A10); basal unit of Mittiwali Member, Mianwali Formation (=Lower Ceratite limestone of Waagen), 59-74 inches above base.

T63-158   Loc. 6A, bed 11, Chhidru A, Salt Range (K & T, fig. A8, A9, A10); basal unit of Mittiwali Member, Mianwali Formation (=Lower Ceratite limestone of Waagen); 74-91 inches above base.

T63-162   Loc. 2, bed 3, Narmia Spring, Surghar Range (K & T, fig. A1, A2); dolomite unit of Kathwai Member, Mianwali Formation, lowermost 6 inches (same as T62-300, T63-19).

T63-163   Loc. 2, bed 3, Narmia Spring, Surghar Range (K & T, fig. A1, A2); dolomite unit of Kathwai Member, Mianwali Formation, 6-12 inches above base of unit.

T63-165   Loc. 2, bed 4, Narmia Spring, Surghar Range (K & T, fig. A1, A2); dolomite unit of Kathwai Member, Mianwali Formation, 6-9 inches above base of unit.

T63-167   Loc. 2, Narmia Spring, Surghar Range (K & T, fig. A1, A2; bed 39 in Kummel, 1966, p. 413); Narmia Member of Mianwali Formation.

T64-4   Loc. 11, bed 2, 2-4 inches below top of bed, Kingriali, Khisor Range (K & T, fig. A17, A18); white sandstone unit of Chhidru Formation, 4 feet below top of unit.

T64-7   Loc. 11, bed 2, 2-4 inches below top of bed, Kingriali, Khisor Range (K & T, fig. A17, A18); white sandstone unit, Chhidru Formation.

T64-11   Loc. 11, bed 4, Kingriali, Khisor Range (K & T, fig. A17, A18); dolomite unit of Kathwai Member of Mianwali Formation, 2 feet above base of unit.

T64-13   Loc. 11, bed 4, Kingriali, Khisor Range (K & T, fig. A17, A18); dolomite unit of Kathwai Member, Mianwali Formation, 44-46 inches above base of unit.

T64-14   Loc. 11, bed 4, Kingriali, Khisor Range (K & T, fig. A17, A18); dolomite unit of Kathwai Member, Mianwali Formation, 60-64 inches above base of unit.

T64-15   Loc. 11, bed 4, Kingriali, Khisor Range (K & T, fig. A17, A18); dolomite unit of Kathwai Member, Mianwali Formation, top 4 inches of unit.

T64-17   Loc. 11, bed 5, Kingriali, Khisor Range (K & T, fig. A17, A18); limestone unit of Kathwai Member, Mianwali Formation, 8-10 inches above base.

T64-18 Loc. 11, bed 5, Kingriali, Khisor Range (K & T, fig. A17, A18); limestone unit of Kathwai Member, Mianwali Formation, 10-12 inches above base.

T64-19 Loc. 11, bed 5, Kingriali, Khisor Range (K & T, fig. A17, A18); limestone unit of Kathwai Member, Mianwali Formation, 20-24 inches above base.

T64-21 Loc. 11, bed 6, Kingriali, Khisor Range (K & T, fig. A17, A18); limestone unit of Kathwai Member, Mianwali Formation, 25-37 inches above base of unit.

APPENDIX C: DESCRIPTIONS OF SOME ROCK-STRATIGRAPHIC UNITS OF TRIASSIC AGE

Following are formal descriptions of some Triassic units which were named and discussed by us in previous publications (Kummel, 1966; Kummel and Teichert, 1966), but which have not before been described in a manner satisfying the Stratigraphic Code of Pakistan (Stratigraphic Nomenclature Committee of Pakistan, 1962).

MIANWALI FORMATION

Type locality: Mianwali District of Salt Range area. The type section is Nammal Gorge, within this district (lat. 32° 40′ N., long. 71° 48′ E.). This is our locality 4.

Description: In the type section this formation consists of a varied sequence of dolomite, limestone, shale, and sandstone divided into the Kathwai, Mittiwali, and Narmia members. The Kathwai Member is a dolomite and limestone unit and is described elsewhere in this report (p. 36); the Mittiwali and Narmia members are primarily of clastic facies with minor carbonate beds.

Distribution and thickness: The formation represents a great wedge of varied facies; it is thickest in the west and thins eastward. The greatest thickness of the formation is found in Narmia Nala in the Surghar Range, where it is 635 feet thick. In the easternmost outcrop areas in the Salt Range, at Kathwai, the formation is only 48 feet thick but here it is truncated by erosion and overlain by the Dhak Pass Formation of Paleocene age. Complete development of the Mianwali Formation is found in the Surghar Range and in the western part of the Salt Range in Zaluch Nala and Nammal Gorge.

Topographic expression: Generally slope-forming.

Stratigraphic relationships: The Mianwali Formation is paraconformable with the underlying Upper Permian Chhidru Formation and conformable with the overlying Tredian Formation.

Fossils and age: The abundant ammonoid faunas indicate that the formation comprises the whole of the Scythian (Lower Triassic) stage.

Synonyms: The name Mianwali was first used by Gee (*in* Pascoe, 1959, p. 852) for a series unit. It was restricted by Kummel (1966, p. 374) to a lithostratigraphic unit of formation rank equivalent to only the lower part of Gee's Mianwali series. This formation comprises all the Triassic facies units (e.g., "Lower Ceratite limestone" through "topmost limestone") recognized by Waagen (1895).

Kathwai Member[1]

Type locality: Head of an unnamed creek, 1.25 miles south of Kathwai and 200 yards east of road leading south from Kathwai (32° 27.5′ N. lat., 72° 11.5′ E. long.). This is our locality 9.

Description: The member consists of a lower dolomite unit and an upper limestone unit. Both have been described in detail elsewhere in this report (p. 36-57).

Distribution and thickness: See p. 36 of this report.

Topographic expression: Forms conspicuous cliff, 10-20 feet high, of grayish-orange to grayish-orange pink color.

Fossils and age: The member contains *Ophiceras* and *Glyptophiceras,* few other megafossils, but abundant conodonts, palynomorphs, and acritarchs. The age of the member is Early Triassic (early Scythian).

Synonym: Sandstone containing *Ophiceras* (Schindewolf, 1954).

Mittiwali Member

Type locality: East side of Chhidru Nala, Salt Range, at a site long known by the local population as Mittiwali—a place where clay is extracted (lat. 32° 33′ N., long. 71° 48′ E.). This is our locality 6B.

---

[1] It should be noted that the name "Kathwai shales" was used informally by Venkatachala and Kar (1967) for a shale unit "25 feet above the Talchir boulder bed" in the eastern Salt Range. It would seem that this unit belongs either to the Tobra Formation or to the Warchha Sandstone as here understood. It is hoped that, to avoid confusion, use of the name Kathwai shales will be discontinued.

Description: The lowest unit of the member is a coquinoid limestone to which Waagen (1895) gave the name "Lower Ceratite limestone." This limestone is gray, fine-grained, and contains an abundance of ammonoids, generally poorly preserved. This lower limestone is followed by a diversity of facies. In the Surghar Range, where the member is thickest, this portion of the Mittiwali Member is a fairly homogeneous sequence of shale and silty shale with some thin sandstone and limestone interbeds. The member thins to the east. In the central Salt Range the lowest unit is a shale, followed by a sand unit, and finally by a sandy limestone unit making up the Ceratite marls, Ceratite sandstone, and Upper Ceratite limestone of Waagen (1895).

Distribution and thickness: Widely distributed in Salt Range and Trans-Indus ranges. The thickness at the type locality is 263 feet. In Narmia Nala the member is 488 feet thick. East of Chhidru at Kathwai the member is only 40 feet thick, but here the upper part has been truncated by pre-Cenozoic erosion.

Topographic expression. Mainly slope-forming.

Stratigraphic relationships: The member is conformably underlain by the Kathwai Member and overlain by the Narmia Member of the Mianwali Formation.

Fossils and age: The abundant ammonoid fauna of this member indicates an age covering the lower half of the Scythian except for the lowest zone *(Otoceras-Ophiceras).*

Synonyms: This member comprises the "Lower Ceratite limestone," "Ceratite marls," "Ceratite sandstone," and "Upper Ceratite limestone" of Waagen (1895).

Narmia Member

Type locality: Narmia Nala, Surghar Range (lat. 31° 58' N., long. 71° 13.5' E.). This is our locality 2.

Description: The basal beds of the formation are approximately 9 feet of dark gray and brown limestone, in part coquinoidal and containing poorly preserved ammonoids. The remainder of the formation in the type section consists of gray shale with interbedded sandstone and sandy limestone, and five prominent beds of limestone and dolomite, 2 to 14 feet thick, with characteristics similar to those of the basal limestone. In the Surghar Range and Khisor Range, the top of the formation is formed by a pisolitic limestone which is dolomitic in places and by a massive carbonate unit in the Salt Range.

Distribution and thickness: Widely distributed in the western Salt Range and in the Trans-Indus ranges. The thickness in the type locality is 140 feet. The greatest thickness, 189 feet, was measured in Nammal Gorge. In Chhidru Nala 120 feet are present, but here the top of the formation is eroded by pre-Murree (late Miocene) erosion. Eastward of Dhodha Wahan the formation is progressively beveled by post-Cretaceous erosion and is overlain by the Dakh Pass Formation of Paleocene age. In the Kufri area only the lowest coquinoidal limestone bed remains.

Topographic expression: Slope-forming.

Stratigraphic relationships: The member conformably overlies the Mittiwali Member of the Mianwali Formation and conformably underlies the Landa Member of the Tredian Formation.

Fossils and age: A sparse ammonoid fauna clearly indicates a latest Scythian (*Prohungarites* zone) age.

Synonyms: This member comprises the "Bivalve beds," "Dolomite beds," and the "Topmost limestone" of Waagen (1895).

TREDIAN FORMATION

Landa Member

Type locality: Landa Pusha, 3 miles northeast of the village of Malla Khel, Surghar Range (lat. 32° 58' N., long. 71° 12' E.). This is our locality 1.

Description: This member consists of sandstone and shale, in about equal proportions. The sandstone may be black, pink, or red, and is micaceous, thin- to massively-bedded, with ripple marks and slump structures. The shale is generally dark, sandy, and micaceous.

Distribution and thickness: At the type locality this member is 100 feet thick, in Zaluch Nala 72 feet, and 63 feet in Nammal Gorge. The member is not present in Chhidru Nala but is fully developed 1 mile to the east in Khan Zaman Nala. East of Khan Zaman Nala the member is progressively beveled and it is not present at Kufri and Kathwai.

Topographic expression: Slope-forming.

Stratigraphic relationships: The formation conformably overlies the Narmia Member of latest Scythian age and is conformably overlain by the Khatkiara Member of the Tredian Formation.

Fossils and age: The only fossils observed in the Landa Member are poorly preserved and fragmentary plant remains. Spore and pollen assemblages studied by Balme (1970) suggest a Middle Triassic age.

Synonyms: The name Tredian Formation (Gee in Kummel, 1966) was introduced to supplant Kingriali sandstones of Gee (1947). The Landa Member is the lower shaly portion of the Tredian Formation.

REFERENCES

Andel, T. H. van, and Veevers, J. J., 1967, Morphology and sediments of the Timor Sea: Australia Bur. Mineral Resources, Geology, Geophysics, Bull. no. 83, x + 173 p., 5 pl.

Balme, B. E., 1970, Palynology of Permian and Triassic strata in the Salt Range and Surghar Range, West Pakistan, *in* Kummel, B., and Teichert, C., eds., Stratigraphic Boundary Problems: Permian and Triassic of West Pakistan: Dept. Geology, Univ. Kansas Spec. Publ. 4, p. 305-453.

Branson, C. C., 1935, A Labyrinthodont from the Lower Gondwana of Kashmir and a new edestid from the Permian of the Salt Range: Mem. Conn. Acad. Arts Sci., v. 9, Art. 2, p. 19-26, 2 pl.

———, 1965, Selective effect of fifty-year rule of names of Late Paleozoic fossils in the Indian region: Dr. D. N. Wadia Commem. Volume, Mining and Metallurgical Inst. India, p. 226-230.

Chao King-Koo, 1965, The Permian ammonoid-bearing formations of South China: Scientia Sinica, v. 14, p. 1814-1825, 2 pl.

Chave, K. E., 1954, Aspects of the biogeochemistry of magnesium. 1. Calcareous marine organisms: Jour. Geology, v. 62, p. 266-283.

Danilchik, W., and Shah, A., in press, Stratigraphic nomenclature of formations in the Trans-Indus mountains, Mianwali District, West Pakistan: Mem. Geol. Survey, Pakistan.

Davidson, Thomas, 1862, On some Carboniferous Brachiopoda collected in India by A. Fleming, M.D., and W. Purdon, Esq., F.G.S.: Geol. Soc. London, Quart. Jour., v. 18, p. 23-35.

Deffeyes, K. S., Lucia, J. F., and Weyl, P. K., 1965, Dolomitization of Recent and Plio-Pleistocene sediments by marine evaporite waters on Bonaire, Netherlands Antilles: Soc. Econ. Paleontologists Mineralogists, Spec. Publ. no. 13, p. 71-88.

Desio, Ardito, 1951, *Ophioderma torrii*, nuova specie di Ofiura nel Retico del M. Albenza (Prealpi Bergamasche): Riv. Ital. Paleontologia Stratigrafia, v. 57, p. 1-10, pl. 3.

Diener, Carl, 1900, Ueber die Grenze des Perm- und Triassystems im ostindischen Faunen-Gebiet: Centralbl. Mineralogie, Geologie, Palaeontologie, 1900, p. 1-5.

———, 1901a, Ueber das Alter der *Otoceras* beds des Himalaya: Same, 1901, p. 513-518.

———, 1901b, Zur Frage des Alters der *Otoceras* beds im Himalaya: Same, 1901, p. 655-657.

———, 1912, The Trias of the Himalayas: Geol. Survey India, Mem. 36, p. 202-360, pl. 3.

Dionne, J.-C., 1969, Tadpole holes: a true biogenic sedimentary structure: Jour. Sed. Petrology, v. 39, p. 358-360.

Douglass, R. C., 1968, Morphologic studies of fusulinids from Lower Permian of West Pakistan: Am. Assoc. Petroleum Geologists, Bull., v. 52, p. 525-526.

Dunbar, C. O., 1933, Stratigraphic significance of the fusulinids of the Lower Productus limestone of the Salt Range: Geol. Survey India, Rec., v. 66, p. 405-413, pl. 22.

———, 1962, Faunas and correlation of the Late Paleozoic rocks of northeast Greenland. Part III. Brachiopoda: Medd. om Grønland, v. 167, no. 6, 14 p., 2 pl.

———, *et al.*, 1960, Correlation of the Permian formations of North America: Geol. Soc. America, Bull., v. 71, p. 1763-1806, 1 pl.

Elphinstone, Mountstuart, 1815, An Account of the Kingdom of Caubul and Its Dependencies in Persia, Tartary, and India Comprising a View of the Afghaun Nation and a History of the Dooraunee Monarchy: Longman, Hurst, Rees, Orm, and Brown, London, 675 p.

Evamy, B. D., 1967, Dedolomitization and the development of rhombohedral pores in limestones: Jour. Sed. Petrology, v. 37, p. 1204-1215.

Fairbridge, R. W., 1957, The dolomite question, *in* Regional Aspects of Carbonate Sedimentation, R. J. Blanc and J. G. Breeding, eds., Soc. Econ. Paleontologists Mineralogists, Spec. Publ. no. 5, p. 125-178.

Fell, H. B., 1966, Cidaroids, *in* Part U, Echinodermata 3, *of* Moore, R. C., ed., Treatise on Invertebrate Paleontology: Geol. Soc. America and Univ. Kansas Press, p. $U312$-$U339$, fig. 235-254.

Fenton, C. L., 1946, Algae of the Pre-Cambrian and Early Paleozoic: Am. Midland Naturalist, v. 36, p. 259-263.

Fleming, Andrew, 1848, Report on the Salt Range, and on its coal and other minerals: Asiatic Soc. Bengal, Jour., v. 17, pt. 2, p. 500-526.

———, 1853a, On the Salt Range of the Punjab: Geol. Soc. London, Quart. Jour., v. 9, p. 189-200.

———, 1853b, On the geology of part of the Sooliman Range: Same, v. 9, pt. 1, p. 346-349.

———, 1853c, Report on the geological structure and mineral wealth of the Salt Range in the Punjab: Asiatic Soc. Bengal, Jour., v. 22, p. 229-279, 333-368, 444-462.

Fliche, P., 1905, Flore fossile du Trias en Lorraine et en Franche-Comté: Soc. Sci.

de Nancy, Bull. des séances, sér 3, v. 6, no. 3, p. 1-66, pl. 1-5.
Fox, C. S., 1931, The Gondwana system and related formations: Geol. Survey India, Mem., v. 58, p. 241, 8 pl.
Furnish, W. M., 1966, Ammonoids of the Upper Permian *Cyclolobus*-Zone: Neues Jahrb. Geologie Palaeontologie, Abh., v. 125, p. 265-296, pl. 23-26.
Furnish, W. M., and Glenister, B. F., 1970, Permian ammonoid *Cyclolobus* from the Salt Range, West Pakistan, *in* Kummel, B., and Teichert, C., eds., Stratigraphic Boundary Problems: Permian and Triassic of West Pakistan: Dept. Geology, Univ. Kansas Spec. Publ. 4, p. 153-175.
Gansser, Augusto, 1964, Geology of the Himalayas: Interscience Publ., London, New York, Sydney, xvi + 289 p., 95 phot., 4 pl.
Gee, E. R., 1938, The economic geology of the northern Punjab, with notes on adjoining portions of the North-West Frontier Province: Mining Geol. Metall. Inst. India Trans., v. 38, p. 263-354, pl. 9-19.
———, 1945, The age of the saline series of the Punjab and of Kohat: Natl. Acad. Sci. India, Sec. B Proc., v. 14 (1944), pt. 6, p. 269-310.
———, 1947, Further note on the age of the saline series of the Punjab and of Kohat: Same, v. 16 (1946), p. 95-154, 19 pl.
Goddard, E. N., Trask, P. D., DeFord, R. K., Rove, O. N., Singewald, J. T., and Overbeck, R. M., 1948, Rock-color chart: Natl. Research Council, Washington, D.C. (reprinted 1963).
Goldberg, Moshe, 1967, Supratidal dolomitization and dedolomitization in Jurassic rocks of HaMakhtesh HaQatan, Israel: Jour. Sed. Petrology, v. 37, p. 760-773.
Grabau, A. W., 1931, The Permian of Mongolia: Amer. Museum Natural History, Natural History of Central Asia, v. 4, 665 p., pl. 1-35.
Grant, R. E., 1966, Late Permian trilobites from the Salt Range, West Pakistan: Palaeontology, v. 9, pt. 1, March 1966, p. 64-73.
———, 1968, Structural adaptation in two Permian brachiopod genera, Salt Range, West Pakistan: Jour. Paleontology, v. 42, p. 1-32, pl. 1-9.
———, 1970, Brachiopods from Permian-Triassic Boundary Beds and Age of Chhidru Formation, West Pakistan, *in* Kummel, B., and Teichert, C., eds., Stratigraphic Boundary Problems: Permian and Triassic of West Pakistan: Univ. Press Kansas, Dept. Geology, Univ. Kansas Spec. Publ. 4, p. 117-151.
Griesbach, C. L., 1880, Palaeontological notes on the Lower Trias of the Himalayas: Geol. Survey India, Rec., v. 13, p. 94-113, pl. 1-4.
Groot, K. de, 1967, Experimental dedolomitization: Jour. Sed. Petrology, v. 37, p. 1216-1240.
Häntzschel, Walter, 1962, Trace fossils and problematica, *in* Part W, Miscellanea, *of* Moore, R. C., ed., Treatise on Invertebrate Paleontology: Geol. Soc. America and Univ. Kansas Press, p. $W177$-$W245$.
———, and Reineck, H.-E., 1968, Fazies-Untersuchungen im Hettangium von Helmstedt (Niedersachsen): Geol. Staatsinst. Hamburg, Mitt., Heft 37, p. 5-39, pl. 1-16.
Hemphill, W. R., and Danilchik, Walter, 1968, Geologic interpretation of a Gemini photo: Photogrammetric Engineering, v. 34, p. 150-154.
Heritsch, Franz, 1937, Rugose Korallen aus dem Salt Range, aus Timor und aus Djoulfa mit Bemerkungen über die Stratigraphie des Perms: Akad. Wiss., Wien, Math.-naturwiss. Kl., Abt. 1, Bd. 146, Heft 1 and 2, p. 1-16, pl. 1, 2.
Hess, Hans, 1965, Trias-Ophiuren aus Deutschland, England, Italien und Spanien: Mitt. Bayer. Staatssamml. Palaeontologie, hist. Geologie, v. 5, p. 151-177, 20 fig., pl. 13-16.
Hsu, K. J., 1966, Origin of dolomite in sedimentary sequences: a critical analysis: Mineralium Deposita, v. 2, p. 133-138.
Husain, B. R., 1967, Saiduwali Member, a new name for the lower part for the Permian Amb Formation, West Pakistan: Univ. Studies [Karachi], Sci. and Technol. No., v. 4, no. 3, p. 88-95.
Illing, L. V., Wells, A. J., and Taylor, J. C. M., 1965, Penecontemporaneous dolomite in the Persian Gulf, *in* Pray, L. C., and Murray, R. C., eds., Dolomitization and Limestone Diagenesis, A Symposium: Soc. Econ. Paleontologists Mineralogists, Spec. Publ. no. 13, p. 89-111.
Kindle, E. M., 1914, An inquiry into the origin of *Batrachioides* the *antiquor* of the Lockport Dolomite of New York: Geol. Mag., Dec. 6, v. 1, p. 158-161, pl. 8, 9.
———, 1917, Recent and fossil ripple marks: Geol. Survey Canada, Mus. Bull. no. 25, Geol. Ser. no. 34, 56 p., 33 pl.
Koken, E., 1901, Helicoprion im Productus-Kalk der Saltrange: Centralbl. Mineralogie, Geologie, Palaeontologie, 1901, p. 225-227.
———, 1907, Indisches Perm und die permische Eiszeit: Neues Jahrb. Mineralogie, Geologie, Palaeontologie, Festband, p. 446-546, pl. 19.
Koninck, L. de, 1863, Descriptions of some fossils from India discovered by Dr. A. Fleming of Edinburgh: Geol. Soc. London, Quart. Jour., v. 19, p. 1-19.

Krafft, A. von, 1900, Stratigraphical notes on the Mesozoic rocks of Spiti: Gen. Rep., Geol. Survey India, 1 April 1899 to 31 March 1900, p. 199-229.

———, 1901, Über das permische Alter der *Otoceras*-Stufe des Himalaya: Centralbl. Mineralogie, Geologie, Palaeontologie, 1901, p. 275-279.

Kummel, Bernard, 1953, Lower Triassic Salt Range nautiloids: Breviora, Mus. Comp. Zoology, no. 20, p. 1-8.

———, 1966, The Lower Triassic formations of the Salt Range and Trans-Indus ranges, West Pakistan: Mus. Comp. Zoology, Bull., v. 134, no. 10, p. 361-429, 22 fig., 4 pl.

———, 1970, Ammonoids from the Kathwai Member, Mianwali Formation, Salt Range, West Pakistan, in Kummel, B., and Teichert, C., eds., Stratigraphic Boundary Problems: Permian and Triassic of West Pakistan: Dept. Geology, Univ. Kansas Spec. Publ. 4, p. 177-192.

———, and Teichert, Curt, 1964, The Permian-Triassic boundary in the Salt Range of West Pakistan: Rept. 22nd Sess., Internat. Geol. Congress India, Vol. of Abstracts, p. 120.

———, and ———, 1966, Relations between the Permian and Triassic formations in the Salt Range and Trans-Indus ranges, West Pakistan: Neues Jahrb. Geologie Paläontologie, Abh., v. 125, p. 297-333, pl. 27-28.

Linck, Otto, 1955, Ein bemerkenswerter Seeigel-Rest (*Miocidaris pakistanensis* n. sp.) aus der Unter-Trias der Salt Range (Pakistan): Neues Jahrb. Geologie Paläontologie, Monatsh., 1955, p. 489-495.

Lotze, Franz, 1966a, Zur permokarbonischen Eiszeit auf dem indischen Subkontinent: Neues Jahrb. Geologie Palaeontologie, Monatsh., 1966, p. 751-752.

———, 1966b, Permokarbonische Vereisung im indischen Raum: Akad. Wiss. u. Literatur Mainz, Jahrb. 1965, p. 155.

Lucia, F. J., 1962, Diagenesis of a crinoidal sediment: Jour. Sed. Petrology, v. 32, p. 848-865.

———, 1968, Recent sediments and diagenesis of South Bonaire, Netherlands Antilles: Same, v. 38, p. 845-858.

———, 1969, Recognition of evaporite-carbonate shoreline sedimentation: Am. Assoc. Petroleum Geologists, v. 53, p. 729-730.

Maher, S. W., 1962, Primary structures produced by tadpoles: Jour. Sed. Petrology, v. 32, p. 138-139.

Makhlaev, V. G., 1957, Dedolomitized rocks in the Dankova-Lebedyansk beds: Akad. Nauk SSSR, Dokl., v. 117, p. 1011-1014 (Consultants Bureau Translation).

Martinsson, Anders, 1965, Aspects of a Middle Cambrian thanatotope on Öland: Geol. Fören. Stockholm Förh., v. 67, p. 181-230.

Noetling, Fritz, 1900a, Note on the relationship between the Productus limestone and the Ceratite formation of the Salt Range: Gen. Rep. Geol. Survey India, April 1899 to 31 March 1900, p. 176-183.

———, 1900b, Ueber die Auffindung von *Otoceras* sp. in der Salt Range: Neues Jahrb. Mineralogie, Geologie, Palaeontologie, v. 1, p. 139-141.

———, 1900c, Die *Otoceras* beds in Indien: Centralbl. Mineralogie, Geologie, Palaeontologie, 1900, p. 216-217.

———, 1901, Beiträge zur Geologie der Salt Range, insbesondere der permischen und triassischen Ablagerungen: Neues Jahrb. Mineralogie, Geologie, Palaeontologie, v. 14, p. 369-471.

———, 1902, Die Dyas in Indien: in Frech., F., Lethaea geognostica. I. Palaeozoicum, p. 639-658.

———, 1904, Ueber das Verhältniss zwischen Productuskalk und Ceratitenschichten in der Salt-range (Indien): Centralbl. Mineralogie, Geologie, Palaeontologie, 1904, p. 321-327.

———, 1905, Die asiatische Trias.: in Frech, F., Lethaea geognostica. II. Mesozoicum, p. 107-221, pl. 9-33.

Pant, D. D., 1949, On some Triassic plant remains from the Salt Range in the Punjab: Nature, v. 163, p. 914.

———, and Srivastava, G. K., 1964, Further observations on some Triassic plant remains from the Salt Range, Punjab: Palaeontographica, Abt. B., v. 114, p. 79-93.

Pascoe, E. H., 1959, A Manual of the Geology of India and Burma, vol. II, 3rd ed., Govt. India Press, Calcutta, IX-XXII, p. 485-1343, 1 map.

Pettijohn, F. J., and Potter, P. E., 1964, Atlas and Glossary of Primary Sedimentary Structures: Springer, New York, xv + 370 p., 117 pl.

Popov, Y. N., 1958, Nakhodka *Otoceras* v nizhnem Triase vostochnogo Verkhoyan'ya: Izvest. Akad. Nauk SSSR, Geol. Ser., no. 12, p. 105-109.

Rahman, H., Sec., 1968, Lithostratigraphic units of Potwar-Kohat Region. A summary of recommendations of the Stratigraphic Committee of Pakistan: Geol. Survey Pakistan, Pre-Publ. Issue, no. 84, 20 + vi p., 2 charts.

Reed, F. R. C., 1931, New fossils from the Productus limestones of the Salt Range, with notes on other species: Geol. Survey India, Palaeontologia Indica, n. s., v. 17, 56 p., 8 pl.

———, 1936, Some fossils from the *Eurydesma* and *Conularia* beds (Punjabian) of the

Salt Range: Same, n. s., v. 23, Mem. 1, p. 1-36, 5 pl.
Reineck, Hans-Erich, 1969, Die Entstehung von Runzelmarken: Natur u. Museum, v. 99, p. 386-388.
Rowell, A. J., 1970, Lingula from the basal Triassic Kathwai Member, Mianwali Formation, Salt Range and Surghar Range, West Pakistan, in Kummel, B., and Teichert, C., eds., Stratigraphic Boundary Problems: Permian and Triassic of West Pakistan: Univ. Press Kansas, Dept. Geology, Univ. Kansas Spec. Publ. 4, p. 111-116.
Ruzhentsev, V. E., and Sarycheva, T. G., eds., 1965, Razvitie i smena morskikh organismov na rubezhe Paleozoya i Mezozoya: Akad. Nauk SSSR, Paleont. Inst., Trudy, v. 108, 431 p., 58 pl.
Sarjeant, W. A. S., 1970, Acritarchs and tasmanitids from the Chhidru Formation, uppermost Permian of West Pakistan, in Kummel, B., and Teichert, C., eds., Stratigraphic Boundary Problems: Permian and Triassic of West Pakistan: Univ. Press Kansas, Dept. Geology, Univ. Kansas Spec. Publ. 4, p. 277-304.
Schindewolf, O. H., 1954, Über die Faunenwende vom Paläozoikum zum Mesozoikum: Zeitschr. Deutsche Geol. Ges., v. 105, p. 154-183.
———, 1964, Über die jungpaläozoische Vereisung der Salt Range (W.-Pakistan): Neues Jahrb. Geologie Palaeontologie, Abh., v. 121, p. 55-66, 4 pl.
———, 1967, Zur permischen Vereisung der Salt Range (W.-Pakistan): Geol. Rundschau, v. 56, p. 914-918, 2 pl.
Shearman, D. J., Khouri, J., and Taha, S., 1961, On the replacement of dolomite by calcite in some Mesozoic limestones from the French Jura: Geologists' Assn., Proc., v. 72, p. 1-12.
Shinn, E. A., Ginsburg, R. N., and Lloyd, R. M., 1965, Recent supratidal dolomite from Andros Island, Bahamas, in Pray, L. C., and Murray, R. C., eds., Dolomitization and Limestone Diagenesis, A Symposium: Soc. Econ. Paleontologists Mineralogists, Spec. Publ. no. 13, p. 112-123.
Sitholey, R. V., 1943, Plant remains from the Triassic of the Salt Range of the Punjab: Natl. Acad. Sci. India, Sec. B., v. 13, p. 300-325.
Smit, D. E., and Swett, Keene, 1969, Devaluation of "dedolomitization": Jour. Sed. Petrology, v. 39, p. 379-380.
Sohn, I. G., 1968, Relict Paleozoic ostracodes in Early Triassic time: Geol. Soc. America, Progr. 1968 Ann. Meetings, p. 286-287 (abs.).
———, 1970, Early Triassic marine ostracodes from the Salt Range and Surghar Range, West Pakistan, in Kummel, B., and Teichert, C., eds., Stratigraphic Boundary Problems: Permian and Triassic of West Pakistan: Univ. Press Kansas, Dept. Geology, Univ. Kansas Spec. Publ. 4, p. 193-206.
Spath, L. F., 1930, The Eo-Triassic invertebrate fauna of East Greenland: Medd. om Grønland, v. 83, no. 1, p. 1-90, pl. 1-12.
———, 1934, Catalogue of the Fossil Cephalopoda in the British Museum (Natural History) Part IV. The Ammonoidea of the Triassic: Trustees British Mus., London, xvi + 521 p., 17 pl.
Stepanov, D. L., 1967, Pravilnyy put k resheniyu sloznoy problemy: Paleont. Zhurnal, 1967, no. 4, p. 144-150. (AGI Translation: The right way to solve a complicated problem, Paleont. Journal, p. 119-125, 1968).
Stratigraphic Nomenclature Committee of Pakistan, 1962, Stratigraphic Code of Pakistan: Geol. Survey Pakistan, Mem., v. 4, pt. 1, ii + 8 p.
Sweet, W. C., 1967, Sequence of Lower Triassic (Scythian) conodonts in West Pakistan: Geol. Soc. America, Progr. 1967 Ann. Meetings, p. 218 (abs.).
———, 1970, Uppermost Permian and Lower Triassic conodonts of the Salt Range and Trans-Indus ranges, West Pakistan, in Kummel, B., and Teichert, C., eds., Stratigraphic Boundary Problems: Permian and Triassic of West Pakistan: Univ. Press Kansas, Dept. Geology, Univ. Kansas Spec. Publ. 4, p. 207-276.
Tatarskiy, V. G., 1949, O rasprostranennosti razdolomichennykh porod: Akad. Nauk SSSR, Dokl., v. 69, no. 6, p. 849-851.
Teichert, Curt, 1944, Two new ammonoids from the Permian of Western Australia: Jour. Paleontology, v. 18, p. 83-89, 1 pl.
———, 1954, A new ammonoid from the eastern Australian Permian province: Royal Soc. New South Wales, Jour. Proc., v. 87, p. 46-50, pl. 7.
———, 1965, Devonian rocks and paleogeography of central Arizona: U.S. Geol. Survey Prof. Pap. 464, 181 p., 32 pl.
———, 1966, Stratigraphic nomenclature and correlation of the Permian "Productus limestone," Salt Range, West Pakistan: Geol. Surv. Pakistan, Rec. 15, pt. 1, p. 1-19.
———, 1967, Nature of Permian glacial record, Salt Range and Khisor Range, West Pakistan: Neues Jahrb. Geologie Palaeontologie, Abh., v. 129, p. 167-184, pl. 15.
———, 1968, Paleontology: McGraw-Hill Yearb. Sci. Technology, 1968, p. 286-287.
Theobald, W., 1854, Notes on the geology of the Panjab Salt Range: Asiatic Soc. Bengal, Jour., v. 23, p. 651-678.
Tozer, E. T., 1959, Triassic stratigraphy and faunas, Queen Elizabeth Islands, Arctic

Archipelago: Geol. Surv. Canada, Mem. 316, p. 1-116, pl. 1-30.

Tschernyschew, T., 1902, Die obercarbonischen Brachiopoden des Urals und des Timan: Comité géol. Russie, Mém. 316, p. 1-116, pl. 1-30.

Twenhofel, W. H., 1932, Treatise on sedimentation: Williams and Wilkins Co., Baltimore, xix + 926 p.

Venkatachala, B. S., and Kar, R. K., 1967, Palynology of the Kathwai Shales, Salt Range, West Pakistan. 1. Shales 25 ft. above the Talchir Boulder Bed: The Palaeobotanist, v. 16, no. 2, p. 156-166, 6 pl.

Verchère, Albert M., 1866, Geology of Kashmir, the Western Himalaya and the Afghan Mountains, a geological paper; with a note on the fossils by M. Edouard de Verneuil: Asiatic Soc. Bengal, Jour., v. 36, pt. 2, p. 201-229.

Vinogradov, A. P., 1953, The elementary chemical composition of marine organisms: Sears Found. Marine Research, Mem. 2, 647 p.

Waagen, William, 1872, On the occurrence of *Ammonites,* associated with *Ceratites,* and *Goniatites* in the Carboniferous deposits of the Salt Range: Geol. Survey India, Mem., v. 9, p. 351-358, 1 pl.

———, 1879-1888, Salt Range fossils. I. Productus-limestone fossils: Palaeontologia Indica, ser. 13, 1. Pisces-Cephalopoda, p. 1-72, pl. 1-6 (23 May, 1879); 2. Pisces-Cephalopoda: Supplement. Gastropoda, p. 73-183, pl. 7-16 (4 June, 1880); 3. Pelecypoda, p. 185-328, pl. 17-24 (28 Sept., 1881); 4. (fasc. 1) Brachiopoda, p. 329-390, pl. 25-28 (22 Dec., 1882); 4. (fasc. 2) Brachiopoda, p. 391-546, pl. 29-49 (30 Aug., 1883); 4. (fasc. 3) Brachiopoda, p. 547-610, pl. 50-57 (1 May, 1884); 4. (fasc. 4) Brachiopoda, p. 611-728, pl. 58-81 (10 Dec., 1884); 4. (fasc. 5) Brachiopoda, p. 729-770, pl. 82-86 (2 July, 1885); 5. Bryozoa-Annelida-Echinodermata, p. 771-834, pl. 87-96 (1 Oct., 1885); 6. Coelenterata, p. 835-924, pl. 97-116 (10 Dec., 1886); 7. Coelenterata-Amorphozoa-Protozoa, p. 925-998, pl. 117-128 (Feb., 1888).

———, 1889-1891, Salt Range fossils. IV. Geological results: Palaeontologia Indica, ser. 13, v. 4, pt. 1, p. 1-88, 4 pl. (1889); v. 4, pt. 2, p. 89-242, pl. 1-8 (1891).

———, 1895, Salt Range fossils. II. Fossils from the Ceratite formation: Same, ser. 13, v. 2, p. 1-323, pl. 1-40.

Walcott, C. D., 1914, Cambrian geology and paleontology, III, No. 2, Pre-Cambrian Algonkian algal flora: Smithsonian Miscell. Coll., v. 64, no. 2 (Publ. 2271), 156 p., 26 pl.

White, David, 1929, Flora of the Hermit Shale, Grand Canyon, Arizona: Carnegie Inst. Washington, Publ. 405, 18 p., 51 pl.

Williams, Alwyn, *et al.,* 1965, Part H, Brachiopoda, *of* Moore, R. C., ed., Treatise on Invertebrate Paleontology: Geol. Soc. America and Univ. Kansas Press, xxxiii + 927 p.

Wynne, A. B., 1878, On the geology of the Salt Range in the Punjab: Geol. Surv. India, Mem., v. 14, p. 1-313, 30 pl.

———, 1880, On the Trans-Indus extension of the Punjab Salt Range: Same, v. 17, p. 211-305, 6 pl.

Yang Tsun-Yi, 1960, On the discovery of a Scythic ophiuroid from Kueichou, China: Acta Palaeont. Sinica, v. 8, no. 2, p. 162-163, pl. 1.

Lingula from the Basal Triassic Kathwai Member, Mianwali Formation, Salt Range and Surghar Range, West Pakistan

A. J. ROWELL
University of Kansas

ABSTRACT

Lingulids from the Mianwali Formation are referred to *Lingula* sp. cf. *L. borealis* Bittner. The distribution of the valves and preservation of their surface ornament suggest that although not in their life position, they have suffered little transportation. It is considered most unlikely that they are reworked specimens from substantially older rocks.

PREVIOUS WORK

Lingulid brachiopods are not conspicuous elements of the fauna contained in the Permian and Triassic rocks of the Salt Range. In spite of the extensive collections that were made towards the end of the last century, they were seemingly unknown in the region until 1931, when Reed (1931) described *Lingula scrutata* from sandstones in the Productus Limestone (probably from the Amb Formation [Teichert, pers. communication, 1969]). Subsequently, Schindewolf (1954, p. 155) recorded *Lingula* sp. from the section at Chhidru. Schindewolf's material came from his Bed 17, a bed which Kummel and Teichert (1966, p. 324) have recognized as forming part of the limestone unit of the Kathwai Member, the basal member of the Kathwai Formation. Underlying the limestone are a few feet of strata, consisting predominantly of dolomite, which the latter authors (1966) term the dolomite unit of the Kathwai Member, a unit which rests on the upper beds of the Chhidru Formation.

PRESENT INVESTIGATIONS

More recently, Kummel and Teichert (1966, p. 314) have recorded *Lingula* sp. from the basal 12 inches of the dolomite unit exposed on the western side of Chhidru Nala. Cooper (in Kummel and Teichert, 1966, p. 322) also identified the genus in collections made from the lowermost 6 inches of the unit at Narmia, a locality some 50 miles northwest of Chhidru.

The collections available for the present study were made by Kummel and Teichert and came from both units of the Kathwai Member. However, the limestone unit yielded only a small lingulid fauna of four specimens, all from the east side of Chhidru Nala (loc. 6B). Material from the underlying dolomite unit came from three localities: west side of Chhidru Nala (loc. 6A), three miles north at Ganjaroh (loc. 5), and one at Narmia Spring (loc. 2). At both Chhidru Nala and Narmia, the specimens are from the basal foot of the unit, but they are relatively rare. The material from Ganjaroh is more abundant and comes from slightly higher in the sequence, about 4 feet above the base of the formation.

PALEOECOLOGY

A few paleoecological comments may be made of the specimens from Ganjaroh. Although these apply only to the relatively thin band containing lingulids, they add to the sum of our knowledge of the depositional environment of the unit. The shells of *Lingula* occur here as disassociated valves in a thin band, some 0.25 inches thick. The block of dolomite from this locality was not oriented and did not contain any obvious bedding planes, but it is considered probable that the band of lingulids was at least approximately parallel to the bedding. Within the band, the valves are haphazardly arranged, both with regard to their convexity, some being convex up and others convex down, and also with regard to the orientation of their plane of symmetry (Pl. 1, fig. 1).

All modern lingulids inhabit burrows which they form in relatively soft sediments. The burrows are typically vertical or subvertical, and the animal lives in the upper part with its beak pointing downwards. The lingulid is able to move vertically within the tube-like structure, but is attached to the sediment at the base of the burrow by the distal end of its long, extensile pedicle. Recent forms are not known to voluntarily leave their burrow, but the bond between pedicle tip and sediment is not very strong, being effected only by mucus. This mode of life is not a recent adaptation, for Carboniferous lingulids are frequently found with both valves associated and disposed normal to the bedding planes (Craig, 1952) and are reasonably interpreted as being in their life position. Lingulids have been occasionally reported in similar positions in rocks as old as the Ordovician (Teichert, 1930; Öpik, 1930). Consequently, it appears that this way of life is a characteristic of the family, and it is probable that the Triassic forms from Ganjaroh also lived in burrows.

Clearly the distribution of the *Lingula* valves at Ganjaroh is not consistent with their being in a life position. The shells have been removed from their burrows, or more probably the sediment has been scoured from around them and the valves separated. The latter occurs readily in lingulids, for only adhesion of outer epithelium to the valves holds them together; this epithelium is thin and rots easily. However, in view of the possibility that some of the brachiopods from the Kathwai Member may be reworked Permian forms (Kummel and Teichert, 1966, p. 327), some estimate of the extent of their transportation is desirable.

The principal inorganic component of a lingulid shell is an apatite, and fortunately this is less readily altered than calcite during diagenesis. Consequently, lingulids tend to retain their fine surface ornament even in a rock that has been extensively dolomitized, whereas the same features may be obliterated in any associated calcareous shells, removing one of the lines of evidence which might be used to assess the amount of abrasion they have suffered. The Ganjaroh *Lingula* specimens retain details of their growth lines, and their surface is not severely eroded or corroded. It appears extremely unlikely that they are reworked forms; and it is considered probable that, although they are not in their life position, they have moved only a very short distance. This viewpoint is supported by the lack of evidence for strong directional currents; the valves, although elongate, are not aligned, neither have they consistently assumed the convex upwards position which is the stable orientation in moderately high water velocities. All of the evidence is consistent with the notion that the Ganjaroh lingulids lived essentially in the area in which they are now found.

SYSTEMATIC DESCRIPTIONS

Order LINGULIDA Waagen, 1885
Superfamily LINGULACEA Menke, 1828
Family LINGULIDAE Menke, 1828
Genus LINGULA Bruguière, 1797

Lingula sp. cf. *L. borealis* Bittner, 1899

Pl 1, fig. 1-10

Description: Elongate oval in outline, lateral margins subparallel, length of valve about twice maximum width. Anterior margin of both pedicle and brachial valves rounded; in outline two valves differ only slightly, the posterior and posterolateral margins of the brachial valve are rounded, whereas those of the pedicle valve tend to be more pointed at the beak. Both valves relatively convex for genus, maximum height being about one tenth of length: the posterior profile is more uniformly convex in the brachial than the opposing valve. In the pedicle valve, there is an abrupt change of curvature along the midline, the lateral flanks being very gently convex; this is particularly noticeable in the posterior third of the valve. Both valves ornamented by fine growth lines.

Digesting the dolomite in 10 percent formic acid produced several free fragments of valves. Unfortunately the specimens were usually cracked and no complete valves were recovered.

Internally, the posterior margin of the brachial margin is typical of the genus, the pseudointerarea being represented by a slight thickening (Pl. 1, fig. 5). The central muscles form a pair of faint, posterolaterally directed scars slightly in front of the midlength of the valve. Between these scars the shell is thickened and a low ridge extends forwards to about two-thirds the valve length, bearing the anterior oblique muscles on its anterior end.

There are two conspicuous features on the internal surface of the pedicle valve, the pseudointerarea and a pair of ridges. The pseudointerarea

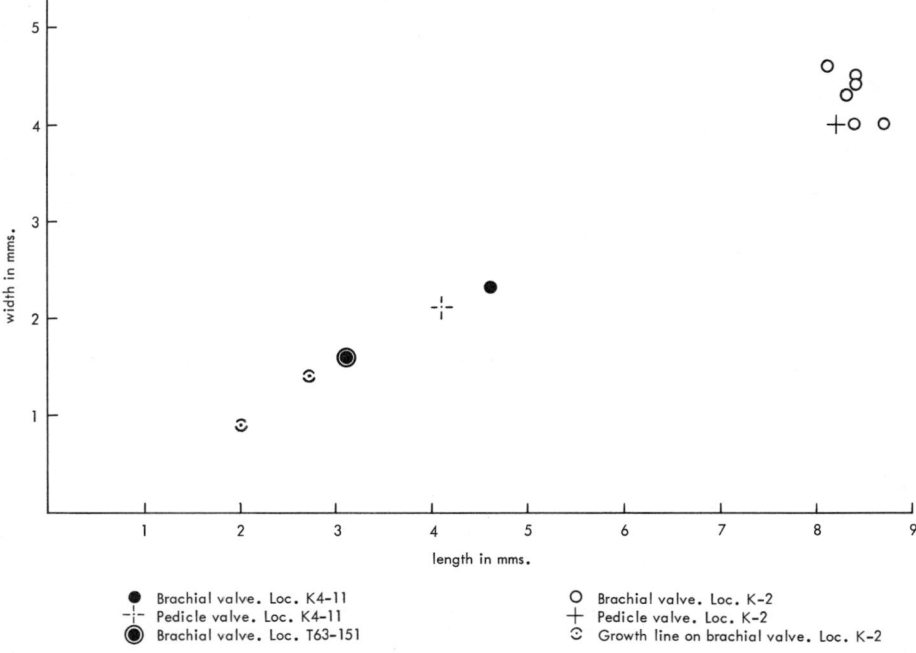

FIG. 1. Bivariate scatter diagram of length and width of *Lingula* cf. *L. borealis* from the Kathwai Member, Mianwali Formation, Salt Range, West Pakistan.

is relatively strong; no specimens revealing the structure were sufficiently complete to allow measurement of its relative length, but it would seem to extend forwards nearly one-tenth of the valve length. The propareas are solid and formed by marked thickening of the posterolateral margins of the valve, they are separated by a deep pedicle groove. Arising a short distance in front of the propareas are a pair of narrow ridges, almost low blades, which diverge slightly anteriorly, they seemingly extend forwards slightly more than half the valve length (Pl. 1, fig. 4).

Discussion: The species is referred to *Lingula,* but internally the pedicle valve differs conspicuously from *Lingula anatina,* the type species of the genus. The latter has a relatively diminutive ventral pseudointerarea and lacks the pair of long, medially located ridges. The ventral pseudointerarea of the material from Pakistan recalls that of *Barroisella* Hall and Clark, but this genus also lacks the paired median ridges and the internal markings of the brachial valve are quite unlike those of our specimens. It could be argued that the morphological differences are of sufficient magnitude to merit the creation of a new lingulid genus, but the systematics of Upper Paleozoic and Mesozoic members of the family are so poorly understood that there is seemingly little to be gained by such action until more is known about them.

All of the lingulid material from the Kathwai Member of the Salt Range is referred to the same species. There are differences between the material from different localities, but these are predominantly in size (Fig. 1). As there are very few specimens from any locality other than Ganjaroh, the possible significance of these differences cannot be rigorously evaluated. The similarity of the proportions of the smaller individuals from both the limestone unit and lower part of the dolomite unit with the larger specimens from Ganjaroh suggests that they may be juveniles of the same species.

In external form, the specimens from the Salt Range agree closely with *Lingula borealis* Bittner as recently revised by Dagis (1965, p. 10), but even the largest of our specimens is little more than half the size of the material figured by Dagis. There are seemingly internal differences, notably in the presence of a single median ridge in the brachial valve of this species. It is not known to what extent these differences are more apparent than real, particularly in the pedicle valve. The internal features of *L. borealis* are known primarily from internal molds, and the available illustrations are at too small a scale to show all the features clearly.

EXPLANATION OF PLATE 1

Fig. 1-10. *Lingula* sp. cf. *L. borealis* Bittner.

1. Slab showing haphazard orientation of lingulids on presumed bedding plane. ×1.5. Dolomite unit, Kathwai Member. Locality K-2. (56990).——2. Partly exfoliated brachial valve, external. ×8. Limestone unit, Kathwai Member. Locality K4-11. (56991).——3. Partly exfoliated pedicle valve, external. ×8. Limestone unit, Kathwai Member. Locality K4-11. (56992).——4. Fragment of pedicle valve, internal view, showing propareas and pair of ridges. ×8. Dolomite unit, Kathwai Member. Locality K-2. (56993).——5. Fragment of brachial valve, internal view showing thickened posterior margin. ×12. Dolomite unit, Kathwai Member. Locality K-2. (56994).——6. Fragment of brachial valve showing central muscle scars. ×12. Dolomite unit, Kathwai Member, Locality K-2. (56995).——7. Brachial valve, external. ×4. Dolomite unit, Kathwai Member. Locality K-2. (56989).——8. Incomplete pedicle valve, external. ×4. Dolomite unit, Kathwai Member. Locality K-2. (56988).——9. Brachial valve, external. ×4. Dolomite unit, Kathwai Member. Locality K-2. (56987).——10. Same specimen as 9, photographed under water, showing track of anterior oblique scars. ×4. Specimens in fig. 2-9 coated with ammonium chloride. Repository and numbers of figured material Museum of Invertebrate Paleontology, University of Kansas.

PLATE 1

Comparable material to the Salt Range specimens has been described from the Lower Triassic Dinwoody Formation of Wyoming by Newell and Kummel (1942), who identified their lingulids as *L. borealis*. There are minor differences in outline between the Dinwoody and Mianwali specimens, and also differences in the depth of the central muscle scars. Larger collections of the Pakistan material are needed to assess the significance of the differences.

Unfortunately, the lingulids contribute little to our understanding of the relative age of the Kathwai Member. *L. borealis* and forms very similar to it are known from the Scythian in several areas, but our knowledge of the stratigraphic range of these forms and the extent to which they occur in the Upper Permian is extremely limited.

Occurrences: Kathwai Member of Mianwali Formation, Salt Range, West Pakistan: Limestone unit (Bed 9) at Chhidru East (loc. 6B, K4-11); Dolomite unit at Chhidru West (loc. 6A, Bed 3, T63-151, K3-3); Ganjaroh (loc. 5, Bed 7, K2-4); Narmia Spring (loc. 2, Bed 3, T63-162) (see Kummel and Teichert, 1970).

ACKNOWLEDGMENTS

I am indebted to Bernhard Kummel and Curt Teichert for making their material available to me, and to my wife for translating the relevant passages of Dagis (1965).

REFERENCES

Bittner, Alexander, 1899, Versteinerungen aus den Trias-Ablagerungen des Süd-Ussuri-Gebietes in der ostsibirischen Küstenprovinz: Mémoires du Comité Géologique [St. Petersburg], v. 7, no. 4, 35 p., 4 pl.

Craig, G. Y., 1952, A comparative study of the Ecology and Palaeoecology of Lingula: Edinburgh Geol. Soc. Trans., v. 15, p. 110-120.

Dagis, A. S., 1965, Triasovye brakhiopody Sibiri: Akad. Nauk SSSR, Sibirskoe Otdelenie, Institut Geologii i Geofiziki, 186 p., 26 pl.

Kummel, Bernhard, and Teichert, Curt, 1966, Relations between the Permian and Triassic Formations in the Salt Range and Trans-Indus ranges, West Pakistan: Neues Jahrb. Geologie Paläontologie Abh., v. 125, p. 297-333, pl. 27, 28.

———, and ———, 1970, Stratigraphy and paleontology of the Permian-Triassic boundary beds, Salt Range and Trans-Indus ranges, West Pakistan: in Kummel, B., and Teichert, C., eds., Stratigraphic Boundary Problems: Permian and Triassic of West Pakistan: Univ. Press Kansas, Dept. Geology, Univ. Kansas Spec. Publ. 4, p. 1-110.

Newell, N. D., and Kummel, Bernhard, 1942, Lower Eo-triassic stratigraphy, western Wyoming and Southeast Idaho: Geol. Soc. Amer. Bull., v. 53, p. 937-995, 3 pl.

Öpik, A. A., 1930, Brachiopoda Protremata der estländischen ordovizischen Kukruse-Stufe: Acta et Comment. Univ. Tartuensis (Dorpatensis), A, v. 17, p. 1-261, 22 pl.

Reed, F. R. C., 1931, New fossils from the Productus Limestones of the Salt Range, with notes on other species: Mem. Geol. Survey India, Palaeontologica Indica, New Series, v. 18, p. 1-56, 8 pl.

Schindewolf, O. H., 1954, Über die Faunenwende vom Paläozoikum zum Mesozoikum: Deutsche Geol. Gesellsch. Zeitsch., v. 105, p. 154-183.

Teichert, Curt, 1930, Biostratigraphie der Poramboniten. Eine entwicklungsgeschichtliche, paläogeographische und vergleichendgeotektonische Studie: Neu. Jahrb. Mineralogie Geologie, Paläontologie, v. 63, p. 177-246.

Brachiopods from Permian-Triassic Boundary Beds and Age of Chhidru Formation, West Pakistan

RICHARD E. GRANT
U.S. Geological Survey, Washington, D.C.

ABSTRACT

Comparison of the brachiopod fauna of the uppermost beds of the Chhidru Formation with that in the dolomite unit at the base of the Kathwai Member of the Mianwali Formation suggests that 1) both are Permian in age, 2) the faunas are different from one another, 3) the uppermost Chhidru brachiopods represent the same suite as that found lower in the formation and also in the Kalabagh Member of the Wargal Formation, and most likely are Guadalupian in age, 4) the Kathwai brachiopods resemble those from the Dzhulfa area of Armenia, and most likely are Dzhulfian in age.

One new species, *Crurithyris? extima,* ranges higher in the Kathwai Member than the other brachiopods, spanning the Permo-Triassic boundary. A brief discussion of each taxon is presented, and most are illustrated.

INTRODUCTION

The age and correlation of the Zaluch Group (formerly "Productus Limestone") have been based traditionally on fossils from two zones. The lowermost lithic unit, the Amb Formation, contains fusulinids that are considered Artinskian in age, approximately equivalent to the Leonard Series of western North America (Dunbar, 1933; Dunbar et al., 1960). The Amb contains no ammonoids. The uppermost unit, the Chhidru Formation, has been considered Late Permian, largely on the basis of the presence of the ammonoid *Cyclolobus;* no fusulinids are recorded from the Chhidru. Brachiopods of the Salt Range Permian have been dated primarily by their relation to these two reference horizons and only secondarily by comparison and correlation with brachiopod faunas of other regions.

The lower level of reference, based upon *Parafusulina kattaensis* (Schwager) from the upper part of the Amb Formation, has not been studied critically for this report. However, the Artinskian age assignment does not conflict with data from other animal groups, and it seems reasonable to accept that age at least provisionally at present.

The upper reference zone, based upon ammonoids, particularly *Cyclolobus,* has been taken to indicate a very late Permian age for the Chhidru Formation, equivalent at least in part to the Ochoa of West Texas and the Dzhulfian of Soviet Armenia (Miller and Furnish, 1940, p. 30; Glenister and Furnish, 1961, p. 680-684; Ruzhentsev and Sarycheva, eds., 1965, p. 103). The thesis of this paper is that the Chhidru Formation is not Dzhulfian in age but is lower to middle Guadalupian, and that links with the Dzhulfian begin in the Salt Range only in the base of the overlying Mianwali Formation. The further implication is that most, if not all, the Dzhulfian is absent from the Salt Range, and that the basal part of the Kathwai Member of the Mianwali Formation represents the minimal deposit of what was essentially a depositional hiatus in this region.

ACKNOWLEDGMENTS

I wish to thank the many people and institutions who have helped my work in the Salt Range. The Walcott Fund of the Smithsonian Institution, through the offices of G. Arthur Cooper, made my work there possible. The U.S. Geological Survey and the Geological Survey of Pakistan provided important logistical support; Ali N. Fatmi, G.S.P., accompanied me in the field, contributing substantially to the fossil collection, and to understanding of the structure and stratigraphy of the area; Curt Teichert, then U.S.G.S., now University of Kansas, acquainted me with problems of the Permo-Triassic boundary, led Mr. Fatmi and me to several key localities, and contributed his collections of brachiopods. In working on this particular problem, I benefited from discussions with G. A. Cooper, U.S. National Museum; Curt Teichert, University of Kansas; Bernhard Kummel, Harvard University; N. D. Newell, Columbia University; and W. M. Furnish and B. F. Glenister, University of Iowa. Publication was authorized by the Director, U.S. Geological Survey.

BRACHIOPODS OF THE CHHIDRU FORMATION

Lower part of formation: The former "Upper Productus Limestone" (Waagen, 1889) is now defined lithostratigraphically as the Chhidru Formation (Teichert, 1966). Brachiopods of this unit were described by Waagen (1882-1885), with additions by Reed (1944). Lists of the brachiopods from each of the three units of the Zaluch Group, including the Chhidru Formation, are available in these two monographs and also in Waagen (1889) and Pascoe (1959, p. 754-780). These lists are satisfactory for comparison on the generic level with faunal lists from other pertinent areas, although some of the generic names must be up-dated by reference to recent monographic studies of certain brachiopod groups (*e.g.*, Muir-Wood and Cooper, 1960) and Part H of the *Treatise on Invertebrate Paleontology* (Williams *et al.*, 1965).

Fossils are not distributed uniformly in the Chhidru Formation. Brachiopods are rare to absent in the lowermost 20-40 feet of rock, which normally is arenaceous shale or argillaceous sandstone. Above this basal unit the fossils occur in beds that are individually recognizable to some extent in the field. These are beds of sandy limestone or calcareous sandstone; certain beds are characterized by abundant specimens of *Derbyia,* others by *Linoproductus,* and others by relatively varied and diverse faunas containing many of the diagnostic genera of the top beds of the Wargal Limestone (the Kalabagh beds of Waagen, 1889; Kalabagh Member of Wargal Limestone: Strat. Comm. Pakistan, Teichert, 1966). The number of ammonoids increases upward in the section, and as stated by Pascoe (1959, p. 779), the fauna of the Chhidru Formation is distinguishable from that of the Wargal Limestone "almost solely by the greater development of ammonoids which, however, are by no means plentiful."

Ammonoids are most abundant in the upper third or fourth of the Chhidru Formation (see Schindewolf, 1954, p. 156, 158), although all the genera present there have their origins in older beds (Teichert, 1966, p. 328). Only *Cyclolobus* was considered to be confined to the Chhidru Formation or beds of its age or younger (Teichert, 1966), and its presence was taken to indicate an age for the Chhidru fauna significantly younger than that of the fauna of the Kalabagh Member of the

Wargal. Indeed, Furnish (1966, tab. 1, 2) advocated recognition of a Chhidruan Stage above the Guadalupian and below the Dzhulfian, containing a Jabian Substage in the upper part based largely upon occurrence of *Cyclolobus* in the upper part of the Chhidru Formation. Biostratigraphic and lithostratigraphic evidence presented by Grant (1968, p. 4, 17) indicates that Waagen's "Jabi beds" from which *Cyclolobus* was first obtained are not high in the Chhidru Formation, but are the Kalabagh Member of the underlying Wargal Limestone. This conclusion extends the range of *Cyclolobus,* thus weakening the case for an age of the Chhidru Formation that is significantly younger than that of the Wargal, itself conceded to be Guadalupian (Furnish, 1966, p. 268).

Recognition of the "Jabi beds" as the Kalabagh Member obviates them as a unit of the Chhidru, or as a Jabian Substage within the Chhidruan. A central conclusion of the present paper, to be documented below, is that the faunas of the Kalabagh and the Chhidru are correlative with much or all of the Guadalupian, and that no subsequent Chhidruan Series or Stage should be recognized.

Chao (1965) cited evidence based on study of ammonoids in South China, that Permian rocks younger than Dzhulfian exist. Data from the Salt Range cast no light on this problem, but the possibility of post-Dzhulfian Permian supports the argument for a hiatus of some magnitude between the latest Permian rocks and the earliest Triassic rocks that are exposed in the Salt Range area.

Topmost beds: The white sandstone that uniformly makes up the topmost few feet of the formation (Kummel and Teichert, 1966, p. 318) is commonly friable and unfossiliferous. At two places in the Trans-Indus Khisor Range, however, the topmost bed of the white sandstone unit was found to be sufficiently well cemented to preserve identifiable fossils. One was about one-half mile east of the narrow gorge of Tapan Wahan (Survey of Pakistan map sheet 38 L/16; U.S. National Museum locality number 9049) in a lens about 6 inches thick, about 6 feet long, and about 2 feet wide. The other was a similar but smaller lens on the southeast side of Kingriali Mountain (Survey of Pakistan map sheet 38 P/4; U.S. National Museum locality number 9175). The top bed of the white sandstone unit is overlain by the dolomite unit of the Kathwai Member of the Mianwali Formation (Kummel and Teichert, 1966, p. 319) which contains Permian brachiopods and Triassic ammonoids in other parts of the Salt Range and the Trans-Indus ranges, but only Triassic ammonoids *(Ophiceras)* at Tapan Wahan and Kingriali Mountain.

Teichert and Kummel examined and sampled the white sandstone unit over most of its outcrop area, and I searched it for fossils at several places, but no other similar fossiliferous lenses were found. Either of two hypotheses might explain the presence of fossils in the two lenses in the Khisor Range and their apparent absence elsewhere. Preservation of fossils in the two lenses may be the result of fortuitous cementation of the sandstone to an extent sufficient to hold the shells intact. On the other hand, the somewhat greater amount of calcareous cement in the lenses here than elsewhere may be the result of the partial leaching of calcite from a concentration of shells. The first hypothesis implies that the sea in which the top beds were deposited contained a normal complement of shells, but that only a few were preserved. The second hypothesis could mean that brachiopods were scarce and patchy in their distribution during the last phases of deposition of the Chhidru Formation. The scarcity of such lenses, and the fact that the fauna of the overlying dolomite unit of the Kathwai Member is strikingly different, lend force to the idea that the fauna of the Chhidru Formation was becoming impoverished and patchy as deposition ceased, or that regression of the sea brought conditions to the area that were only locally favorable to brachiopod survival.

Brachiopods of the white sandstone unit: Most of the brachiopods in the fossiliferous lenses at Tapan Wahan and Kingriali Mountain are preserved as single valves; the only complete shells are *Spirigerella* and a small *Hustedia*. Many specimens are molds, although sufficient shell material is preserved on some specimens to retain delicate ornamentation, and some molds are filled with secondary calcite that simulates the original shell. Possibly the ones that appear to be secondary fillings are merely weathered shells whose structure has been destroyed. Brachiopods of the white sandstone unit are as follows, in alphabetical order:

Aulosteges sp.
Callispirina sp.
Chonetella sp.
chonetid undet.
Cleiothyridina cf. *C. capillata* (Waagen)
Derbyia cf. *D. plicatella* Waagen
dielasmatids undet.
Enteletes sp. 1
Hemiptychina sp.
Hustedia sp.
Kiangsiella sp.
Linoproductus sp.
Lyttonia sp.
Martinia? sp.
Neospirifer sp.
Orthotichia sp. 1
Orthotichia sp. 2
Richthofenia sp.
Spiriferella? sp.
Spirigerella sp.
Waagenoconcha? sp.
Whitspakia sp. 1

This list is essentially an abbreviated list of the fauna of the major part of the Chhidru Formation. The Chhidru fauna in turn reflects the fauna of the Kalabagh Member at the top of the Wargal Limestone (cf. Waagen, 1882-1885, 1889; Reed, 1944; Pascoe, 1959). No elements of the faunule of the white sandstone unit indicate an age significantly younger than that of the lower beds of the Chhidru Formation. Although the ammonoid *Cyclolobus* was not obtained in this part of the Range, it occurs 10 to 40 feet below the top of the formation and, by inference, that far below the lenses in the white sandstone unit. Considering only the brachiopods, the entire Chhidru Formation and the Kalabagh Member of the Wargal Limestone would be regarded as a faunal unit of essentially one age; that is, of deposition within one stage of the Permian. Presence of supposedly latest Permian ammonoids in the upper third of the Chhidru Formation formerly was taken to indicate that this part of the formation is younger than the lower part.

Age of the Chhidru brachiopods: Brachiopods are notoriously difficult to use in correlation, and Permian ones especially so because of their provinciality (Cooper in Dunbar *et al.,* 1960; Ruzhentsev in Ruzhentsev and Sarycheva, eds., 1965, p. 102). Estimation of the age of the Chhidru Formation fauna must depend partly upon consideration of the general "aspect" of the fauna as well as of particular details of this and other faunas that are considered approximately synchronous, and partly upon absence from this fauna of elements that are present in demonstrably younger faunas.

The total impression given by the brachiopods of the Chhidru Formation is not that of a latest Permian (Ochoa or Dzhulfian) age. The abundance of *Enteletes* in the Kalabagh Member and in the white sandstone fossiliferous lenses would argue for an age not younger than early Guadalupian (Word) if correlation were made with the southwestern United States. However, *Enteletes* occurs in several Permian faunas of Asia that are thought mian faunas of Asia that are thought to be as young or younger than the fauna of the Chhidru (Grabau, 1931). In addition, *Enteletes* occurs in the lowest bed of the Kathwai Member of the Mianwali Formation which overlies the Chhidru Formation in the Salt Range, and recently a small species of *Enteletes* was reported from the Indus (i.e., Scythian) Stage of the Lower Triassic in Soviet Armenia (Ruzhentsev and Sarycheva, eds., 1965).

Waagenoconcha in the Salt Range is confined to the Kalabagh Member of the Wargal Limestone and the Chhidru Formation; it is locally very abundant in the upper part of the Kalabagh. This genus is characteristic of Lower Permian beds in Bolivia (Kozlowski, 1914) and the Ural and Timan regions of the USSR. (Tschernyschev, 1902). It also occurs in the Loping Series of the Yangtze Valley of China. This series has been considered to be very late Permian, although this correlation seems to be based largely on faunal similarity to the Chhidru Formation. Correlation with the southwestern United States would indicate an age not younger than early Guadalupian (Wordian) on the basis of the occurrence of *Waagenoconcha* (Girty, 1908, p. 237; King, 1931, p. 80; Muir-Wood and Cooper, 1960, p. 253). Boreal occurrences of *Waagenoconcha* also indicate an age for the genus no younger than early Guadalupian (Dunbar, 1955; Harker and Thorsteinsson, 1960; Gobbett, 1963), although evidence from these distant and distinct faunas is perhaps less relevant to south Asian correlations. The only latest Permian reference to a species now assigned to *Waagenoconcha* is by Merla (1931, p. 86), who mentioned but did not illustrate a dorsal valve of "*Productus abichi* Waagen" from the Bellerophon Limestone. Recent research in Dzhulfian strata (Ruzhentsev and Sarycheva, eds., 1965) suggests the probability that this was a valve of *Tschernyschewia*. Heritsch (1934b, p. 50) also questioned the identification of *Waagenoconcha* in the Bellerophon Limestone and suggested the likelihood that it was *Tschernyschewia*, although he correlated at least the lower part of the Bellerophon Limestone with the Chhidru Formation (Heritsch, 1934a, tab. 2).

Stenoscisma is common in the Kalabagh Member and the lower half of the Chhidru Formation in the Salt Range. Similarly, it is abundant in the Lower and early Upper Permian of the southwestern United States, northern Mexico (Cooper in Cooper et al., 1953), and Oregon (Cooper, 1957). It is rare in beds younger than early Guadalupian (Wordian) in North America and the Arctic, although a distinctive species was reported from the Dzhulfian of Soviet Armenia by Sokolskaya (in Ruzhentsev and Sarycheva, eds., 1965), The related genus *Cyrolexis* also occurs in the Kalabagh Member in the Salt Range, and appears not to occur in strata younger than Kazanian elsewhere (Grant, 1965, p. 91).

Heretofore, *Kiangsiella* was known in the Salt Range only from the Kalabagh Member (Pascoe, 1959, p. 762) and the lower half of the Chhidru Formation (Waagen, 1884, p. 589), but specimens in the fossiliferous lens at Tapan Wahan extend its range to the top of the Chhidru. It is common in Lower and lower Upper Permian rocks of Asia and Australia (Thomas, 1958, p. 68). *Kiangsiella* has been reported from the Late Permian Bellerophon Limestone of the Dolomites (Merla, 1931, p. 78, pl. 7, fig. 12) on the basis of one distorted imprint of a dorsal valve, conceded by Merla to be unreliable. Heritsch (1934b, p. 27) described "*Streptorhynchus pectiniformis* Davidson" (i.e., *Kiangsiella*) from the Bellerophon Limestone of the Dolomites and West Serbia, but unfortunately did not illustrate it. He cited faunal lists from various localities in the Bellerophon Limestone to demonstrate that the parts sampled are not all of the same age; he considered the West Serbian and South Tyrolian parts to be somewhat older than those at other Alpine localities. *Kiangsiella* was not reported from the Dzhulfian (Ruzhentsev and Sarycheva, eds., 1965), so, except for the ambiguous occurrence of the genus in the lower part of the Bellerophon Limestone, most of the evidence indicates that it does not extend into the latest Permian.

Derbyia is another genus whose range extends only tenuously beyond the Guadalupe. It has been reported from the West Serbian (i.e., lower) Bellerophon Limestone, and Walter (1953, p. 686) reports a species from the lower part of the Rustler Formation of west Texas, assigned to the Ochoa Series by Dunbar and others (1960, chart 7). *Derbyia* is abundant in the Carboniferous and Early and early Late Permian, but could be considered abundant in late and post-Guadalupian beds only if the Chhidru Formation is correlated with the Dzhulfian or the Ochoan. Walter (1953, p. 688) emphasized the resemblance of elements in the impoverished Rustler fauna to forms found in the Guadalupe and the Upper Productus Limestone, although the similarity is slight at best.

Ruzhentsev (in Ruzhentsev and Sarycheva, eds., 1965, p. 102-107) has summarized very neatly the biostratigraphic evidence based upon brachiopods of the Guadalupian Stage of most of the world. His conclusion is that the brachiopods of the Middle Productus limestone are Guadalupian. The implication is that those of the Upper Productus, that is, the Chhidru Formation, are younger. This is also implied in his table 14 where the *Cyclolobus* horizon is placed definitely in the Dzhulfian, in full accordance with his previously stated opinions as to its age (e.g., Ruzhentsev, 1955, p. 123).

The brachiopod fauna of the so-called Middle Productus limestone (=Wargal Limestone, Teichert, 1966) is derived almost entirely from the fossiliferous Kalabagh Member just above the top of the massive cliff formed by the Wargal Limestone. Preliminary comparison of the fauna of the Kalabagh Member with that of the Chhidru Formation confirms what is apparent from study of faunal lists by Waagen (1882-1885, 1889), Reed (1944), and Pascoe (1959), when allowance is made for Reed's "splitting" of species and some ambiguity in the stratigraphic placement of some collections. The conclusion is that the Chhidru fauna begins in the Kalabagh Member and continues well into the Chhidru Formation. The fossiliferous lenses in the white sandstone unit in the Khisor Range confirm that the Chhidru fauna continues to the top of the formation, although some elements of it were not recovered from those lenses. Similarity of the faunas of the Kalabagh Member and the Chhidru Formation is striking when examined in the field. Some beds in the middle part of the Chhidru contain nearly complete repetitions of the Kalabagh fauna, even to such genera as *Oldhamina*, *Waagenoconcha*, *Stenoscisma*, and *Kiangsiella* which are not found below the top of the Wargal Limestone cliff.

Authorities agree (e.g., Pascoe, 1959) that the two faunas are highly similar and closely related. Disagreement is largely in interpretation of the significance of ammonoids of the upper part of the Chhidru Formation where *Cyclolobus* formerly was believed confined. The Kalabagh Member contains a definitely Guadalupian ammonoid (Glenister and Furnish, 1961, p. 684) and coral fauna (Heritsch, 1937, p. 13), and its age is not in contention. It seems from the literature that the Late Permian age ascribed to *Cyclolobus* is based to some extent upon its occurrence in the Chhidru Formation, which many have assumed to be latest Permian, as well as upon its advanced suture pattern (Miller and Furnish, 1940, p. 25; Schindewolf, 1954, p. 153, 163). However, Schindewolf (1954, p. 166) admits that, in view of the rarity and apparent stratigraphic isolation of *Cyclolobus* in the Chhidru Formation, it is a poor zone fossil, and that the brachiopods upon further study should prove to be more satisfactory.

The foregoing analysis points to a Guadalupian age for the brachiopod fauna of the Kalabagh Member and the Chhidru Formation. However, a considerable suite of Permian fossils was discovered by Teichert and Kummel in the base of the overlying dolomitic unit of the Kathwai Member of the predominantly Triassic Mianwali Formation. Comparison and contrast of this brachiopod fauna with that of the Chhidru Formation seem to indicate that the Chhidru fauna not only is not latest Permian, but that it is not even latest Guadalupian.

BRACHIOPODS OF THE KATHWAI MEMBER, MIANWALI FORMATION

Occurrence: The basal unit of the Kathwai Member is a rusty brown dolomite 5 to 8 feet thick which has been described in detail by Kummel and Teichert (1966, p. 319). Preliminary examination of brachiopods from the lower part of this unit by Cooper (in Kummel and Teichert, 1966, 1970) showed that the brachiopods definitely are Permian, although Cooper warned that their fragmentary preservation should keep alive the hypothesis that they were reworked from the underlying Chhidru Formation. One brachiopod of Paleozoic aspect, a species of *Crurithyris*(?), is abundant in the upper part of the dolomite unit as well as in the basal part, and although *Spinomarginifera* occurs only in the basal part of the dolomitic unit, it was found at one locality above the level of the lowest occurrence of the Triassic ammonoid *Ophiceras connectens* Schindewolf (Kummel and Teichert, 1966).

Brachiopods of the dolomite unit: As in the fossiliferous lens at the top of the Chhidru Formation at Tapan Wahan, most of the brachiopods of the dolomite unit of the Kathwai Member are preserved as single valves. Only the small and subglobular species *Crurithyris? extima* n. sp. occurs as complete shells; most of these are filled with sparry calcite and much of the shell material is missing. Of the others, some are molds of interiors or exteriors, many appear to retain their original shells although in fragmentary condition, and a few seem to be calcite fillings in molds that simulate original shell material. Further detailed study of the brachiopods, and the fortunate coincident appearance of the study of the Dzhulfian fauna, edited by Ruzhentsev and Sarycheva (1965), have allowed the list to be somewhat expanded from Cooper's preliminary list (in Kummel and Teichert, 1966). Detailed occurrences are given following the descriptions of each taxon; the composite list for the dolomite unit is as follows:

> *Crurithyris? extima* Grant, n. sp.
> *Derbyia?* sp.
> dielasmatid undet.
> *Enteletes* sp. 2
> *Lingula* sp.
> *Linoproductus* sp.
> *Lyttonia* sp.
> *Martinia* sp.
> *Ombonia* sp.
> *Orthothetina* cf. *O. arakeljani* Sokotskaya
> *Orthothetina* sp.
> *Spinomarginifera* sp.
> *Spirigerella* sp.
> *Whitspakia* sp. 2

This rather short list contains some noteworthy differences from the list obtained from the lens at the top of the Chhidru Formation at Tapan Wahan. Many of the Tapan Wahan genera are absent from the dolomite unit. Most significantly absent are *Cleiothyridina,* one of the most abundant in the

Tapan Wahan lens; *Kiangsiella,* which is virtually a guide to the fauna of the Chhidru and the Kalabagh Member of the Wargal; *Orthotichia* which is abundant low in the Chhidru as well as in the fossiliferous lens at Tapan Wahan; and *Neospirifer,* a common element throughout the Zaluch Group. Two others warrant mention: the species of *Enteletes* in the dolomite unit is so strikingly different from the one in the white sandstone unit that it is undoubtedly a different species, and the questioned *Derbyia* in the dolomite unit is so doubtfully identifiable that its presence on the list almost certainly can be discounted.

Fully as significant are the elements of the fauna of the dolomite unit that are absent from the white sandstone and from other beds of the Chhidru. *Crurithyris* has not been reported previously from the Salt Range, although I collected two small specimens from the Kalabagh Member in Warchha Gorge. The species in the dolomite unit is greatly different, however, and internal differences cast doubt on its placement in *Crurithyris*. The *Martinia* of the dolomite unit is different from that in the Tapan Wahan lens and, as stated by Cooper (in Kummel and Teichert, 1966, p. 322), it differs from any described previously from the Salt Range. *Ombonia* is unreported from the Zaluch Group, as are *Orthothetina* and *Spinomarginifera*.

The question of possible reworking: As mentioned previously, Cooper suggested in his report to Kummel and Teichert (1966) the possibility that the Permian elements in the base of the dolomite unit of the Kathwai Member have been reworked, presumably from the Chhidru Formation. Several lines of evidence argue against these elements having been reworked from the Chhidru. First is the disparity in the two faunal lists. The fossiliferous lenses at the top of the white sandstone unit show that the Chhidru brachiopod fauna persisted throughout the deposition of the Chhidru Formation and that radically different elements were not added at the top. It might be argued that the shells in the dolomite unit were reworked from some now absent deposit that existed previously above the Chhidru Formation, but the new elements in it seem to preclude reworking from the Chhidru Formation.

The relatively poor state of preservation of some of the shells was mentioned (Cooper in Kummel and Teichert, 1966, p. 321) along with the detrital aspect of a polished surface of the dolomite (p. 322) as indicating possible reworking. Study of etched surfaces as well as polished surfaces of samples from T63-162 shows that the "clay balls" mentioned are weathered patches of slightly rounded detrital dolomite in a biosparite matrix with euhedral rhombic dolomite. The shells are not severely abraded, and spine bases are present on several. Undisturbed spines are rare in any bed, regardless of whether the shells were reworked or simply agitated by contemporaneous wave action. Curved shells in the etched and polished sections contain sparry calcite under the concave surfaces, indicating that they were deposited as shells, and not as pieces from some pre-existing, partially indurated bed. The samples show the kind of bedding that would be expected in a shell heap, with most specimens lying parallel or subparallel to the bedding surfaces, and enough at other angles to suggest lack of severe current action. Finally, the samples are nearly devoid of sand-size quartz grains (the insoluble residue is a micaceous quartz silt). The upper beds of the Chhidru Formation at most localities are fine- to medium-grained quartz sandstone with calcite and dolomite cement. These factors, when considered together, seem to preclude the possibility that the brachiopods of the dolomite unit were reworked from the Chhidru Formation. Reworking from a pre-existing formation above the Chhidru remains a possibility but is irrelevant to the question of the post-Chhidru age of the brachiopods of the basal Mianwali Formation.

Age of the Kathwai Member brachiopods: The number of brachiopod genera recovered from the Kathwai Member is small, but fortunately some of the genera are sufficiently distinctive to provide an indication of their age relationships. The small species of *Enteletes* is similar to *E. dzhagrensis* Sokolskaya (in Ruzhentsev and Sarycheva, eds., 1965) which was found in the Lower Triassic of the Dzhulfa region of Armenia. This hints at a late age for the fauna of the dolomite unit, as does the new species *Crurithyris? extima* which was found with Triassic ammonoids in the dolomite unit of the Salt Range. *Enteletes* sp. 2 is not the same as *E. dzhagrensis,* however, and its strong plication is normal for Permian species of *Enteletes,* as pointed out by Cooper (in Kummel and Teichert, 1966). Furthermore, the species of *Crurithyris?* occurs low in the dolomite unit with definitely Permian genera, as well as at the top with Triassic genera, and may possibly span the Permian-Triassic boundary.

Other elements of the brachiopod fauna provide firmer links with the Permian. *Linoproductus,* although too poorly preserved for specific identification, is definitely Permian; *Lyttonia* must be considered Permian also, although it provides no closer determination. The species of *Martinia* in the dolomite unit is well represented by many identifiable fragments and by a few nearly complete valves that show its rather bizarre form. It is so nearly identical to the adult form of *M. rhomboidalis* Girty, which is abundant in the upper Guadalupian Lamar Limestone of Texas, that if the specimens were found together they undoubtedly would be considered conspecific. This species of *Martinia* can be taken as evidence that the dolomite unit is late Guadalupian in age.

Ombonia is represented in the Kathwai Member by only one definitely identifiable ventral valve with the characteristic spondylium. Its presence is a strong indication of a Late Permian age: its greatest abundance in the West Texas fauna is in the upper Guadalupian Lamar Limestone, although its range begins lower in rocks of Word age. This genus is a prominent element of the Upper Permian Bellerophon Limestone (Merla, 1931, p. 79) and is considered by Ruzhentsev (in Ruzhentsev and Sarycheva, eds., 1965, p. 105) to be an indicator of Capitan age and a link with Mediterranean faunas.

Spirigerella is an abundant element of the fauna of the Chhidru, especially in the upper part, but is poorly represented in the Kathwai Member by only one certainly identifiable valve, and a few doubtful ones. Were the Kathwai fauna reworked from the Chhidru, one could expect this genus to be better represented in the Kathwai.

Orthothetina cf. *O. arakeljani* Sokolskaya, *Orthothetina* sp., and *Spinomarginifera* sp. are elements of the Dzhulfian fauna described by Sokolskaya (in Ruzhentsev and Sarycheva, eds., 1965). They point to a possible Dzhulfian or younger age for the fauna of the basal dolomite unit of the Kathwai. *Spinomarginifera* is said to extend into the Triassic in the Dzhulfa area and was found in beds above the level of the lowest occurrence of Triassic ammonoids in the Kathwai. Moreover, absence of these genera from the fauna of the lens at Tapan Wahan and from beds lower in the Chhidru Formation provides further evidence that the Kathwai fauna is not derived from the Chhidru fauna.

In summary, the dolomite unit of the Kathwai Member contains some genera that point to a latest Guadalupian (Lamar equivalent) age, and others that point to a Dzhulfian, or even Early Triassic age. Possibly the Dzhulfian as now constituted (Ruzhentsev and Sarycheva, eds., 1965) in-

TABLE 1. Comparative Lists of Brachiopod Genera*

	Soviet Armenia and Nakhichevanskaya S.S.R. (Dzhulfa region)	Central Elburz Range, Iran	Salt Range and Trans-Indus Ranges, West Pakistan
TRIASSIC — Scythian Stage	Araxathyris, Enteletes, Haydenella, Orthothetina	Elikah Formation	Manwali Formation
UPPER PERMIAN — Dzhulfan Stage	Orthotichia, Spinomarginifera, Terebratuloidea — Comelicania, Janiceps — Araxathyris, Alexenia, Chonopectoides, Compressoproductus, Doroshamia, Gubleria, Haydenella, Lyttonia, Martiniopsis, Ogbinia, Pseudowellerella, Spinomarginifera, Spiriferellina, Stenoscisma, Tyloplecta, Wellerella — Araxathyris, Compressoproductus, Crenispirifer, Edriosteges, Gubleria, Haydenella, Lyttonia, Neophricadothyris, Notothyris, Oldhamina, Orthothetina, Orthotichia, Parenteletes, Poikilosakos, Spinomarginifera, Tschernyshewia, Uncinunellina, Wellerella — Araxilevis, Araxathyris, Compressoproductus, Gubleria, Lyttonia, Notothyris, Neochonetes, Orthothetina, Spinomarginifera, Tschernyschewia, Wellerella	Upper Nesen Formation — Araxathyris, Orthothetina, Enteletes, Neophricadothyris, Orthotichia, Tyloplecta, Spinomarginifera (Level 4) — Cancrinella, Crurithyris, Dielasma, Linoproductus, Marginifera, Orthothetina, Orthotichia, Spinomarginifera (Level 3) — Dielasma, Reticulatia?, Streptorhynchus, Tyloplecta (Level 2) — Cleiothyridina, Dielasma, Haydenella, Neophricadothyris, Spinomarginifera	Dolomite unit of Kathwai Member — Crurithyris?, Spinomarginifera (Ophiceras and other Triassic ammonoids) — Hiatus? — Crurithyris?, Derbyia?, dielasmatid, Enteletes, Lingula, Linoproductus, Lyttonia, Martinia, Ombonia, Orthothetina, Spinomarginifera, Spirigerella, Whitspakia — Hiatus?

TABLE 1 (continued)

		Kutch			White sandstone unit / Chhidru Formation / Guadalupian Stage
		Level 5			
"Composita"	Orthotichia	Costiferina	Linoproductus	Aulosteges	Linoproductus
Chonostegoides	Rhipidomella	Derbyia	Marginifera	Callispirina	Lyttonia
Dielasma	Richthofenia	Dielasma	Neochonetes	Chonetella	Martinia?
Krotovia	Septospirigerella	Echinoconchus	Spinomarginifera	chonetid	Neospirifer
Marginifera	Spiriferellina			Cleiothyridina	Orthotichia
Neochonetes	Spinomarginifera			Derbyia	Richthofenia
Ogbinia	Terebratuloidea			Enteletes	Spiriferella?
Orthothetina				Hemiptychina	Spirigerella
		Level 1		Hustedia	Waagenoconcha
Chonostegoides	Notothyris	Cancrinella	Orthotichia	Kiangsiella	Whitspakia
Composita	Ogbinia	Cleiothyridina	Spiriferellina		
Dielasma	Orthotichia	Compresso-	Streptorhynchus		
Edriosteges	Orthothetina	productus	Stepanoviella		
Geyerella	Parenteletes	Cyrolexis	Whitspakia		
Keyserlingina	Reticularina	Dielasma			
Krotovia	Richthofenia	Linoproductus			
Linoproductus	Septospirigerella	Marginifera			
Lyttonia	Spinomarginifera				
Martiniopsis	Spiriferellina				
Marginifera	Terebratuloidea				
Neochonetes	Uncinunellina				
Neophrica-	Vediproductus				
dothyris	Wellerella				

UPPER PERMIAN — Guadalupian Stage

* From the Armenian region of the Soviet Union (Ruzhentsev and Sarycheva, eds., 1965), the Elburz Range of Iran (Fantini Sestini, 1965; Fantini Sestini and Glaus, 1966), and the Salt Range and Trans-Indus Ranges of West Pakistan (original identifications in this paper).

cludes rocks equivalent in age to the Lamar Limestone. Nevertheless, this demonstrably younger fauna in the Kathwai Member constitutes additional evidence for the Guadalupian, rather than Dzhulfian, age of the Chhidru Formation.

At least two interpretations are possible. One is that Dzhulfian (and possibly late Guadalupian) beds were deposited in the area but later eroded and their brachiopods incorporated into the basal Triassic bed (the lower 1 foot of the dolomite unit). Another interpretation derives from the fact that the Permian brachiopods, although occurring in the lowermost bed of the dolomite unit, rarely were found associated with Triassic ammonoids. Permian brachiopods and Triassic ammonoids appear to occur at the same level, but only upon the assumption that the basal bed of the dolomite unit is continuous. The alternative interpretation is that the basal foot of the Kathwai Member is Permian (Dzhulfian) at some localities (e.g., Narmia) and Triassic at other localities (e.g., Munta Nala, Kathwai), with apparent mixing at few places, such as at the west side of Chhidru Nala (Kummel and Teichert, 1966, bed 4; 1970, loc. K3-3). This alternative would fit a picture of "starved" Dzhulfian deposition, and subsequent shallow erosion cutting out the Dzhulfian elements at most places before the Triassic transgression, but leaving some fragments to be incorporated into the basal bed at some places.

Thus the Permian-Triassic boundary would be drawn at the base of the dolomite unit of the Kathwai Member by the first interpretation. The alternative interpretation would cause the boundary to be drawn at the base at some localities, and about 1 foot higher at others, the choice depending upon the fossils contained in the basal foot. Where mixing occurs, the boundary would depend on the lowest Triassic ammonoid, Permian brachiopods being assumed to have been reworked.

Comparison of Dzhulfian faunas of adjacent areas: Recent studies of the Late Permian faunas of the Elburz Mountains of Iran (Fantini Sestini, 1965) and the Dzhulfa region of Soviet Armenia (Ruzhentsev and Sarycheva, eds., 1965) bear upon the correlation of the brachiopod fauna of the Kathwai Member in the Salt Range.

Fantini Sestini (1965, p. 18) correlates the fauna of the Ruteh Limestone of the central Elburz in Iran with the Middle Productus Limestone, which is to say, the fauna of the Kalabagh Member. She assigns a "Murgabian" age (in the terminology of Miklukho-Maklai, 1963), which is the lower of two subdivisions of the Upper Permian, roughly equivalent to the Guadalupian. The Upper Productus Limestone (i.e., Chhidru Formation) is assigned to the "Pamirian," the latest subdivision of the Upper Permian in Miklukho-Maklai's regional terminology for Central Asia. Comparison of the lists given for the several levels within the Ruteh Limestone with those from the Kathwai Member in the Salt Range and those by Sarycheva *et. al.* (in Ruzhentsev and Sarycheva, eds., 1965) suggests that the Ruteh spans the boundary between Guadalupian and Dzhulfian. The composite list from the Ruteh (Fantini Sestini, 1965, p. 26) masks the rather significant difference between the fauna of the lowest collections (level 1) and the higher collections. Only the list from level 5 seems anomalous, and Fantini Sestini (1965, p. 24) questions its stratigraphic position. The genera listed at level 5 seem to be those of a lower level, correlative with the Chhidru fauna.

Lists of brachiopods from the three areas, Salt Range, Elburz Range, and Soviet Armenia, are summarized on Table 1. The list of genera from the Guadalupian of Armenia contains several names that do not appear in the list from the lenses in the white sandstone unit, and which also are missing from lists of the Chhidru fauna by

TABLE 2. Distribution List of Brachiopod Genera*

| | Permian | | | | | | Triassic | | |
| | Guadalupian | | | Dzhulfian | | | | | |
	A	E	SR	A	E	SR	A	E	SR
Araxathyris				×	×				
Cleiothryidina		×	×		×				
Compressoproductus		×	×	×					
Costiferina		×	×						
Crurithyris			×		×				×
Cyrolexis		×	×						
Derbyia		×	×						
Dielasma	×	×	×		×				
Enteletes			×		×	×	×		
Haydenella			×	×	×		×		
Linoproductus	×	×	×		×	×			
Lyttonia	×		×	×	×	×			
Marginifera	×	×	×		×				
Martiniopsis	×		×	×					
Neochonetes	×	×	×	×					
Neophricodathyris	×		×	×					
Notothyris	×		×	×					
Oldhamina			×	×					
Orthothetina	×			×	×	×	×		
Orthotichia	×	×	×	×	×		×		
Rhipidomella	×	×							
Richthofenia	×	×							
Septospirigerella	×	×	×						
Spiriferellina	×	×	×	×					
Spinomarginifera	×	×		×	×	×	×		×
Stenoscisma			×	×					
Terebratuloidea	×		×				×		
Tyloplecta				×	×				
Uncinunellina	×		×	×					
Whitspakia		×	×			×			

* Common to at least two of the following areas: A—Armenian region of Soviet Union, data from Ruzhentsev and Sarycheva (1965); E—Elburz Range of northern Iran, data from Fantini Sestini (1965), Fantini Sestini and Glaus (1966); SR—Salt Range and Trans-Indus ranges, West Pakistan, data from Waagen (1889), Pascoe (1959), and original observations. [*Septospirigerella* is not listed by Fantini Sestini; its occurrence in Iran is recorded by Grunt (in Ruzhentsev and Sarycheva, eds., 1965, p. 238); it also occurs in the Salt Range where it has been identified as *Spirigerella gigas* Reed (1944). *Lyttonia* is recorded from the Late Permian of the Elburz Range by Glaus (1964, p. 499). *Permophricodothyris* of Fantini Sestini and Glaus (1966) is retained as *Neophricadothyris*.]

Waagen (1889), Reed (1944), and Pascoe (1959).

The similarities and differences among the Armenian, Elburz Range, and Salt Range faunas are summarized in Table 2. Names that appear in the table are those of genera that occur in more than one of the three areas, so some that are significant but occur in only one area are omitted, thus reducing the meaningfulness of this kind of comparison. The table shows that Guadalupian faunas are more similar to one another than to Dzhulfian faunas and more similar to Dzhulfian faunas than to Triassic faunas. This is to be expected if the correlations are correct. It also shows a fairly high similarity between the Guadalupian and Dzhulfian faunas, which, in absence of other evidence, tends to obscure the differences.

Ombonia and *Martinia* are omitted from Table 2 because they occur in the Salt Range but not in Armenia or the Elburz Range. These two genera are important to the correlation of the Kathwai Member in the Salt Range, linking it to the late Guadalupian or early Dzhulfian of Southern Europe and the southwestern United States. This suggests that the latest Guadalupian is not present in the Zaluch Group.

The Dzhulfian fauna from Armenia contains the largest number of genera, the Dzhulfian faunas of the Elburz and Salt ranges being relatively small. This suggests that the last two areas lay nearer the margins of the regressing Tethyan sea.

SYSTEMATIC DISCUSSION OF BRACHIOPODA

Nomenclature and classification: Most of the brachiopods are identified only to the generic level. For some, this is because the specimens are poorly preserved, but for many it is a reflection of the poor state of knowledge of the species of the "Productus Limestone." The number of species established by Reed (1944) is excessive, and restudy of the fauna is now under way to attempt to bring a more reasonable order to the specific designations.

Supergeneric categories are confined to Class, Order, and Family, although admittedly Suborder or Superfamily would be more relevant to some groups. The classification follows the *Treatise on Invertebrate Paleontology* (Williams *et al.,* 1965), and the full taxonomic position of each genus can be found there.

Repositories: Types and illustrated specimens are deposited in the U.S. National Museum, Washington, D.C.

Localities: Each locality is assigned a U.S. National Museum locality number (USNM). The localities that were sampled by Teichert and by Kummel also are designated by their field numbers (beginning with T or K) in order that the discussions here may be coordinated with those by Kummel and Teichert (1966, 1970).

White sandstone unit of Chhidru Formation.

USNM loc. 9049: Top bed of white sandstone unit 0.5 mi. east of the narrow gorge of Tapan Wahan, Khisor Range, Survey of Pakistan map sheet 38 L/16. The faunal list for this locality is given on page 96. [Near loc. 12, Kummel and Teichert, 1970.]

USNM loc. 9175: Lens in top bed of white sandstone unit on southeast side of Kingriali Mountain, Khisor Range, Survey of Pakistan map sheet 38 P/4; Section F. of Kummel and Teichert (1966, p. 315): *Cleiothyridina* cf. *C. capillata, Derbyia* cf. *D. plicatella, Enteletes* sp. 1, *Orthotichia?* sp., *Spirigerella* sp. (abundant and well preserved), *Waagenoconcha* sp., *Whitspakia* sp. 1. [Loc. 11, Kummel and Teichert, 1970.]

Dolomite unit of Kathwai Member, Mianwali Formation.

USNM loc. 9156 (K3-3): West side of Chhidru Nala, Salt Range, Survey of Pakistan map sheet 38 P/16, 6 inches above base of dolomite unit: *Spinomarginifera* sp. [Loc. 6A, Kummel and Teichert, 1970.]

USNM loc. 9157 (T63-141): Khan Zaman Nala, Salt Range, Survey of Pakistan map sheet 38 P/14, about 5 feet above base of dolomite unit: *Crurithyris? extima* (abundant).

USNM loc. 9158 (T63-144): Khan Zaman Nala, Salt Range, Survey of Pakistan map sheet 38 P/14, top of dolomite unit (or base of overlying limestone unit): *Crurithyris? extima* (abundant).

USNM loc. 9159 (K1-12, T62-2, T62-300, T63-19): Narmia Spring, Surghar Range (Trans-Indus) Survey of Pakistan map sheet 38 P/1, lowermost 6 inches of dolomite unit, just above creek bottom: *Derbyia?* sp., dielasmatid indet., *Enteletes* sp., *Lingula* sp., *Linoproductus* sp., *Lyttonia* sp., *Martinia* sp., *Orthothetina* sp., *Spinomarginifera* sp., *Spirigerella* sp. [Loc. 2, Kummel and Teichert, 1970.]

USNM loc. 9160 (T63-20): Same locality as 9159, but 6-12 inches above base of dolomite unit: *Crurithyris? extima* (common).

USNM loc. 9161 (T63-162): Narmia Nala, Surghar Range, 20 yards east of locality 9159, lowermost 6 inches of dolomite unit: *Derbyia?* sp., *Enteletes* sp., *Lingula* sp., *Linoproductus* sp., *Lyttonia* sp., *Martinia* sp., *Ombonia* sp., *Orthothetina* sp., *O.* cf. *O. arakeljani* Sokolskaya, orthotetid indet., *Orthotichia* sp., *Spinomarginifera* sp., *Spirigerella* sp., *Whitspakia* sp. 2. [Near loc. 2, Kummel and Teichert, 1970.]

USNM loc. 9162 (T63-165): Same locality as 9161, but 6-9 inches above base of dolomite unit: *Crurithyris? extima* (common).

USNM loc. 9163 (T63-163): Same locality as 9161, but 6-12 inches above base of dolomite unit: *Enteletes* sp., *Linoproductus* sp., orthotetid indet.

USNM loc. 9164 (T63-22): Narmia Nala, Surghar Range, Survey of Pakistan map sheet 38 P/1, about 200 yards west of locality 9159, 3 inches above base of dolomite unit: *Crurithyris? extima* (abundant): [200 yards west of loc. 2, Kummel and Teichert, 1970.]

Class ARTICULATA Huxley, 1869

Order ORTHIDA Schuchert and Cooper, 1932

Family ENTELETIDAE Waagen, 1884

Genus ENTELETES Fischer, 1825

Diagnosis: Strongly convex and strongly ribbed uniplicate schizophoriids with narrowly divergent dental plates and a median septum in the ventral valve, brachiophores of dorsal valve supported by thin divergent plates.

Enteletes sp. 1

Plate 1, figures 1-5

Remarks: This species is similar in form to Reed's (1944) *E. socialis* or *E. conjunctus*, although its adult size is smaller. Those two species occur in the Kalabagh beds at the top of the Wargal Formation. It also resembles the Himalayan species *E. tschernyscheffi* Diener (1897) from the Chitichun Limestone which is considered to be equivalent in age to the Kalabagh beds or perhaps to part of the Chhidru Formation. *Enteletes* sp. 1 is somewhat small for a Salt Range *Enteletes;* it has three or four costae on each side of the median dorsal fold, and the mature shell has a compressed appearance due to thickening by growth along the valve margins that does not produce proportional widening. The ornament is typical of *Enteletes,* with fine costellae, many of which terminate in small, anteriorly opening pores that may have contained setae in life. The radial ornament is crossed by strong concentric growth laminae.

Occurrence: Enteletes sp. 1 was found in the top of the white sandstone unit of the Chhidru Formation at Tapan Wahan, Khisor Range (USNM loc. 9049). It is one of the most abundant constituents of the faunule there, individuals and fragments numbering about the same as those of *Linoproductus* and *Derbyia.* A single specimen was found at the same level at Kingriali Mountain, Khisor Range (USNM loc. 9175).

Enteletes sp. 2

Remarks: Specimens of *Enteletes* from the basal part of the Kathwai member are smaller, less abundant, and more fragmentary than those in the top of the Chhidru Formation. They are too poorly preserved for effective illustration, although they retain enough of the characteristics of the genus for unequivocal generic identification. They may be small because they are juveniles, but the costation begins so near the beaks that they seem to be representatives of a small species similar to *E. kayseri* Waagen or *E. dzhagrensis* Sokolskaya.

Occurrence: This small *Enteletes* occurs in the base of the dolomite unit of the Kathwai Member, Mianwali Formation, Narmia Spring, Surghar Range (USNM loc. 9161, 9163, 9159).

Genus ORTHOTICHIA Hall and Clarke, 1892

Diagnosis: Biconvex, uniplicate to bisulcate; delthyrium large, triangular, open; teeth supported by dental plates that outline sides of muscle area; median septum present; brachiophores supported in dorsal valve by basal plates outlining posterior part of muscle area; cardinal process bilobed, in apex of notothyrium.

Orthotichia sp. 1

Plate 1, figures 9-11, 13

Remarks: A species of rather low convexity is represented by about 20 specimens, the largest being a ventral valve 16 mm long. Most are small and intermediate-size single valves. This species most nearly resembles *O. corallina* (Waagen, 1884) and contrasts with the larger and more convex "species 2" with which it occurs.

Occurrence: See below.

Orthotichia sp. 2

Plate 1, figure 12

Remarks: Two strongly convex and rather large dorsal valves (18 and 20 mm long) are sufficiently different from the smaller and flatter "species 1" to warrant separate mention. They resemble *O. derbyi* (Waagen), which occurs in the Kalabagh beds and in the upper part of the Chhidru Formation.

Occurrence: The two forms of *Orthotichia* were found in the white sandstone unit of the Chhidru Formation, Tapan Wahan, Khisor Range (USNM loc. 9049); a doubtful specimen of species 1 at Kingriali Mountain, Khisor Range (USNM loc. 9175).

Order STROPHOMENIDA Öpik, 1934

Family MEEKELLIDAE Stehli, 1954

Genus ORTHOTHETINA Schellwien, 1900

Diagnosis: Shell conical or subconical, commonly distorted; plications absent, weak, or irregular; dental plates long, extending along floor of ventral valve without joining.

Orthothetina cf. O. arakeljani Sokolskaya

Plate 1, figure 6

Cf. *Orthothetina arakeljani* Sokolskaya, in Ruzhentsev and Sarycheva, eds., 1965, p. 205, pl. 30, fig. 3, text-fig. 27.

Remarks: One nearly complete dorsal valve is identified tentatively with *O. arakeljani* on the basis of the distinct but weak radial plications which begin about 11 mm anterior to the beak, about midlength on the specimen. The costae on the Salt Range specimen as well as on the type from Soviet Armenia are strong enough for them to be considered morphologically intermediate between *Orthothetina* and *Meekella*. Apparently this costation is secondarily acquired in *Orthothetina*, rather than having been derived directly from *Meekella*, for two reasons. One is that five of the seven species described by Sokolskaya (in Ruzhentsev and Sarycheva, eds., 1965) have some degree of weak costation on shells that otherwise resemble *Meekella* only slightly. The other reason is that *Meekella* is more typical of the Carboniferous and Lower Permian, becoming relatively rare in Guadalupian and younger rocks, whereas noncostate and weakly costate forms ascribed to *Orthothetina* are abundant in the late Guadalupian and the Dzhulfian.

Occurrence: Lower part of dolomite unit of Kathwai Member, Mianwali Formation, Narmia Nala, Surghar Range (USNM loc. 9161).

Orthothetina sp.

Remarks: Numerous fragments that are too poor to illustrate can be assigned to *Orthothetina* on the basis of the characteristic meekelloid costellation, the conical shape of the ventral valves, the pair of dental plates visible in the beaks of several of the ventral valves, and the long and divergent brachiophore bases of the dorsal valves (also made visible by weathering or by transparency of the shell). These fragments are among the most numerous constituents of the faunule of the dolomitic unit. They provide an important link with the late Guadalupian and Dzhulfian faunas of the Dzhulfa region where several species of *Orthothetina* are reported by Sokolskaya (in Ruzhentsev and Sarycheva, eds., 1965).

Occurrence: Lower part of dolomitic unit of Kathwai Member, Mianwali Formation, Narmia Spring, Surghar Range (USNM loc. 9159, 9161).

Genus OMBONIA Caneva, 1906

Diagnosis: Shell conical, commonly distorted, finely costellate but not plicate; dental plates joined above ventral valve floor, forming Y-shaped spondylium.

Ombonia sp.

Plate 1, figure 7

Remarks: A single, finely costellate and slightly distorted ventral valve has the fused dental plates forming the large spondylium that characterizes *Ombonia*. The beak was cut and polished to reveal the spondylium unambiguously, and a slice was cut into the shell near the anterior part of the spondylium to confirm that it retains the Y shape for its entire length.

Ombonia is characteristic of Upper Permian rocks in Sicily, the Carnic Alps, the Northern Caucasus, and the Guadalupe Mountains of Texas. It is common in the Lamar Limestone of late Capitan age in Texas, and, along with a strongly folded species of *Martinia* with which it also occurs in the Salt Range, it provides a possible correlation between the Kathwai Member and the Lamar.

Occurrence: Lower part of dolomitic unit of Kathwai Member, Mianwali Formation, Narmia Nala, Surghar Range (USNM loc. 9161).

Family SCHUCHERTELLIDAE
Williams, 1953

Genus KIANGSIELLA Grabau and Chao, 1927

Diagnosis: Strongly costellate and costate Streptorhynchinae with elongate ventral beak, large bilobed cardinal process, recurved socket plates; dental or other septal plates absent from both valves.

Kiangsiella sp.

Plate 1, figure 8

Remarks: Many badly broken fragments show the characteristic costation of *Kiangsiella,* and a few also retain the typical costellae and strong growth lines. The generic identification is certain, but the material is too fragmentary to allow specific identification or comparison.

Species of *Kiangsiella* occur only in the upper part of the Zaluch Group: in the Kalabagh beds at the top of the Wargal Limestone and in the Chhidru Formation. Heretofore it has not been found higher than about the middle of the Chhidru Formation.

Occurrence: Fragments of species of *Kiangsiella* are fairly abundant in the white sandstone of the Chhidru Formation at Tapan Wahan, Surghar Range (USNM loc. 9049).

Family ORTHOTETIDAE Waagen, 1884

Genus DERBYIA Waagen, 1884

Diagnosis: Shell rectimarginate to shallowly sulcate or coarsely and irregularly wrinkled; ventral valve moderately deep, subconical, irregular, capped by convex dorsal valve; costellae fine, distinct, intercalated; ventral interior with high median septum and no dental plates; dorsal interior with socket plates anteriorly divergent, not recurved.

Remarks: This diagnosis is meant to include wrinkled specimens that have been called *"Plicatoderbyia"* Thomas. Study of *Derbyia* from the Wargal and Chhidru Formations of the Salt Range and from the Permian of the Glass Mountains of Texas indicates that the wrinkled form occurs sporadically within many species. Irregular, random wrinkles, no matter how strong, do not constitute a valid generic distinction from *Derbyia*.

Derbyia cf. *D. plicatella* Waagen

Plate 1, figures 14-19

Cf. *Derbyia plicatella* Waagen, 1884, p. 601, pl. 55, fig. 3.

Remarks: This small to moderate-size *Derbyia* is the largest species of any genus found in the white sandstone unit at the top of the Chhidru Formation, and in addition it is the most abundantly represented species. The smaller specimens are less markedly plicated, but the largest specimen is strongly wrinkled, not only by irregularly radial plications but also by concentric rugae.

Waagen reported *D. plicatella* as a rare constituent of the upper beds of the Chhidru Formation, and the specimen that he illustrated is rather weakly and regularly plicated. I have collected numerous specimens with greater and lesser degrees of wrinkling, not only from the Chhidru Formation but also from the Kalabagh beds at the top of the Wargal Formation. Wrinkled specimens clearly belong to more than one species, so the identity of the specimens from the white sandstone with *D. plicatella* is doubtful at best.

Occurrence: White sandstone, top of Chhidru Formation at Tapan Wahan (USNM loc. 9049) and Kingriali Mountain (USNM loc. 9175), Khisor Range.

Derbyia? sp.

Remarks: Fragmentary small valves of orthotetids in the lower part

---

EXPLANATION OF PLATE 1

FIGURE

1-5. *Enteletes* sp. ×2, USNM loc. 9049.——1. Small dorsal valve, USNM no. 153691.——2. Small dorsal valve, side view, USNM no. 153692.——3. Small dorsal valve, USNM no. 153693.——4, 4a. Average size ventral valve, USNM no. 153694; 4, ventral view; 4a, side view.——5. Dorsal valve, somewhat larger than average, USNM no. 153695.

6. *Orthothetina* cf. *O. arakeljani* Sokolskaya, dorsal valve, ×2, USNM no. 153696, USNM loc. 9161 (T63-162).

7. *Ombonia* sp., beak of ventral valve cut and etched to show spondylium ×4, USNM no. 153697, USNM loc. 9161 (T63-162).

8. *Kiangsiella* sp., side view of ventral valve, ×2, USNM no. 153698, USNM loc. 9049.

9-13. *Orthotichia* spp., USNM loc. 9049.——9. *Orthotichia* sp. 1, ventral valve ×4, USNM no. 153699.——10, 10a. *Orthotichia* sp. 1, ventral valve ×3, USNM no. 153700; 10, ventral view; 10a, side view.——11. *Orthotichia* sp. 1, large ventral valve ×1.5, USNM no. 153701.——12. *Orthotichia* sp. 2, large inflated dorsal valve ×1.5, USNM no. 153702.——13. *Orthotichia* sp. 1, ventral interior showing plates, ×1.5, USNM no. 153703.

14-19. *Derbyia* cf. *D. plicatella* Waagen, USNM loc. 9049.——14. Large fragment of partly exfoliated dorsal valve, ×1.5, USNM no. 153704.——15. Large narrow fragment of ventral valve with broad flat umbonal cicatrix of attachment, ×1.5, USNM no. 153705.——16. Fragment of ventral interior, showing median septum, ×2, USNM no. 153706.——17. Large ventral valve with concentric rugae and irregular radial plications, ×1, USNM no. 153707.——18. Partly exfoliated dorsal valve, ×1.5, USNM no. 153708.——19. Interarea of small conical ventral valve, ×2, USNM no. 153709.

20. *Chonetella* sp., ×2, USNM no. 153710, USNM loc. 9049.——20, 20a. Calcite cast of posterior part of ventral valve; 20, interior view; 20a, exterior view.——20b, Mold that formed cast shown in 20 and 20a.

21. *Aulosteges* sp. ×2, USNM no. 9049, USNM loc. 153711 partly exfoliated fragment of ventral valve.——21. Side view showing beak and pustulose ornamentation.——21a. Anterior oblique view showing beak.

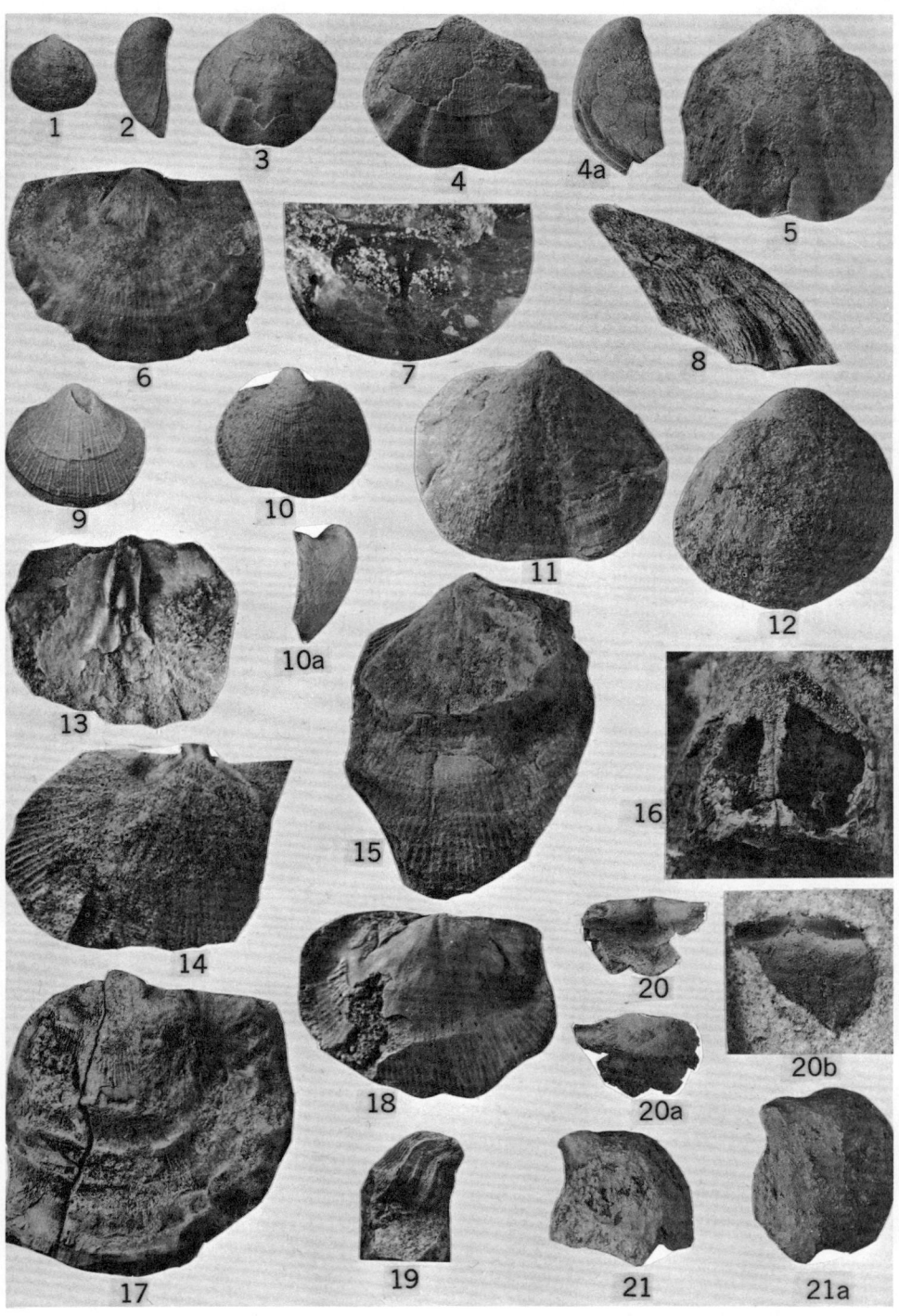

PLATE 1

of the dolomitic unit of the Kathwai Member of the Mianwali Formation may belong to one or more species of *Derbyia*. Admittedly, most are dorsal valves that could belong to *Orthothetina*, which also occurs in this unit, but the transverse outline, low convexity, and weak, irregular wrinkles suggest a tentative identification with *Derbyia*. One wrinkled fragment of a ventral valve can be identified more certainly as a specimen of a species of *Derbyia*, establishing presence of that genus in the Kathwai Member.

Occurrence: Rare in the dolomite unit at Narmia Spring, Surghar Range (USNM loc. 9159, 9161, 9163).

Family CHONETIDAE Bronn, 1862

Chonetid gen. et sp. indet.

Plate 2, figure 19

Remarks: A small, flat fragment of a valve appears to belong to a species of chonetid. It has a shallow sulcus, no costae, low pustules, and apparently a median septum. It is impossible to determine whether the pustules are internal or external. It may belong to a species of *Lissochonetes* or *Neochonetes*.

Occurrence: White sandstone unit of Chhidru Formation, Tapan Wahan, Khisor Range (USNM loc. 9049).

Family AULOSTEGIDAE Muir-Wood and Cooper, 1960

Genus AULOSTEGES von Helmersen, 1847

Diagnosis: Strophalosiacea with relatively high interarea; beak commonly twisted or otherwise distorted by attachment; trail recurved; spines rather short, prostrate, and suberect; cardinal process trilobate; median septum long; alveolus triangular.

Aulosteges sp.

Plate 1, figures 21-21a

Remarks: Three small specimens are identified as belonging to a species of *Aulosteges* on the basis of high and distorted beaks, characteristic spine bases, and typically aulostegid outline and convexity. They are much smaller than *A. dalhousi* Davidson, the species that occurs most typically in the Kalabagh Member at the top of the Wargal Limestone, although they are similar in external form. They may be stunted representatives of the species, but more probably belong to a separate species.

Occurrence: Rare in white sandstone unit of Chhidru Formation at Tapan Wahan, Khisor Range (USNM loc. 9049).

Family CHONETELLIDAE Likharev, 1960

Genus CHONETELLA Waagen, 1884

Diagnosis: Small, strongly concavo-convex strophalosiaceans with alate hinge; costae low, rounded; trail extended at midline to form narrow trough; spines few, proportionately stout, projecting laterally in row along posterior margin, three or four on each side; cardinal process short, blunt, simple.

Chonetella sp.

Plate 1, figures 20-20b

Remarks: One ventral valve has the shape, size, convexity, and low costae of *Chonetella*. The valve appears to be a secondary filling of calcite in a mold from which the original shell material was dissolved; therefore it is not clear whether the concave surface is the inner surface of the ventral valve, or perhaps the outer surface of part of the dorsal valve.

Occurrence: White sandstone unit of Chhidru Formation at Tapan Wahan, Khisor Range (USNM loc. 9049).

Family RICHTHOFENIIDAE Waagen, 1885

Genus RICHTHOFENIA Kayser, 1881

Diagnosis: Coraliform productoids cemented by beak and anchored by spines; miocoelidium outlined by two septa and bisected by one; apertural spines on anterior inner surface of ventral valve, simple, short; endospines of dorsal valve short, abruptly tapered, few, and widely spaced.

Remarks: This diagnosis is based upon silicified specimens from the Wargal and basal Chhidru formations.

Richthofenia sp.

Plate 2, figure 18

Remarks: Four fragments are identified as belonging to a species of *Richthofenia* on the basis of shape and external ornament. They are small; the longest cone is only 12 mm high. The miocoelidium is not preserved, so identity with *Richthofenia* proper cannot be demonstrated, but the fact that *Richthofenia* is the only member of the Richthofeniacea that has been found in the Salt Range Permian makes it most probable that these specimens belong to that genus.

Occurrence: White sandstone unit of Chhidru Formation, Tapan Wahan, Khisor Range (USNM loc. 9049).

Family MARGINIFERIDAE Stehli, 1954

Genus SPINOMARGINIFERA Huang, 1932

Diagnosis: Marginiferidae with numerous spines scattered rather densely and evenly over visceral disc of ventral valve, shallow pits on dorsal valve, row of spines on each flank lacking; profile rather evenly curved, with only slight geniculation; costae and rugae weak; sulcus shallow or absent; cardinal process short, bilobed; marginal diaphragm of dorsal valve finely striated.

Spinomarginifera sp.

Plate 2, figures 1-3

Remarks: Fragmentary or decorticated specimens of this genus, perhaps belonging to more than one species, are characterized by even convexity, a very shallow sulcus that is essentially absent at the anterior of larger specimens, and numerous low spine bases distributed evenly over the visceral disc of the ventral valve. Part of the marginal diaphragm is visible on an external impression of a dorsal valve. The best preserved of the specimens most nearly resemble those identified by Sarycheva (in Ruzhentsev and Sarycheva, eds., 1965, pl. 37) as *S. helica* (Abich).

Occurrence: Dolomite unit of Kathwai Member, Mianwali Formation, Narmia Spring, Surghar Range (USNM loc. 9159, 9161); west side of Chhidru Nala, Salt Range (USNM loc. 9156). One specimen at 9156 was collected several inches above the lowest occurrence of the Triassic ammonoid *Ophiceras* (Kummel and Teichert, 1970, loc. 6A, bed 4).

Family ECHINOCONCHIDAE Stehli, 1954

Genus WAAGENOCONCHA Chao, 1927

Diagnosis: Echinoconchid productoids with weak lamellae of growth, long, thin, quincuncially arranged spines over ventral visceral disc, forming dense corona of spines around edge of visceral disc, and short, hairlike spines on trail and dorsal valve.

Waagenoncha? sp.

Plate 2, figures 4-5

Remarks: A large fragment and several smaller fragments bear the characteristic ornament of spine bases that normally indicates *Waagenoconcha*. The largest fragment is part of a dorsal valve with a low median fold. The specimen may be part of the dorsal valve of *Tschernychewia* Stoyanov, although the fold is low enough to be well within the range for *Waagenoconcha*.

Occurrence: White sandstone unit of Chhidru Formation, Tapan Wahan (USNM loc. 9049) and Kingriali Mountain (USNM loc. 9175), Khisor Range.

Family LINOPRODUCTIDAE Stehli, 1954

Genus LINOPRODUCTUS Chao, 1927

Diagnosis: Strongly convex, finely costellate productids with few spines; rugae only on ears of ventral valve, on most of visceral disc of dorsal valve; cardinal process trilobate, with alveolus.

Linoproductus spp.

Plate 2, figures 7-9

Remarks: Generically identifiable fragments of one or more species of *Linoproductus* are among the most abundant constitutents of faunules of the white sandstone and the dolomitic unit. The most frequently preserved fragments are the umbonal and visceral part of the ventral valve, although several nearly complete dorsal valves also were obtained from the white sandstone. The trail area of the ventral valve was not preserved on any of the specimens.

Specimens from both stratigraphic units are smaller than the typical "Productus Limestone" linoproductoids. They all resemble one another in size, convexity, and ornamentation, so it is not possible to determine whether more than one species is present.

Occurrence: Abundant fragments in white sandstone unit of Chhidru Formation, Tapan Wahan, Khisor Range (USNM loc. 9049), and in dolomite unit of Kathwai Member, Mianwali Formation, Narmia Spring, Surghar Range (USNM loc. 9159, 9161, 9163).

Family LYTTONIIDAE Waagen, 1883

Genus LYTTONIA Waagen, 1883

Diagnosis: Flattish oldhaminoids with broad posterior flap, symmetrical muscle and septal systems, median anterior incision of dorsal valve shallow or absent.

Nomenclature: The name *Leptodus* was introduced by Kayser (1883) for a fossil from China that he interpreted as part of a fish. Waagen (1883) shortly thereafter correctly identified this form as an aberrant brachiopod, for which he substituted the name *Lyttonia*, on the grounds that Kayser's name perpetuated a misidentification. This procedure was invalid by the rules of nomenclature, and the name *Leptodus* gradually has come to supplant *Lyttonia* (e.g., Williams *et al.*, 1965). However, Likharev (1965, p. 149) pointed out that the oldhaminoids now are known to be a complex group comprising several genera and that Kayser's specimens probably are distinct from the Salt Range form. Casts of the holotype of *Leptodus richthofeni* Kayser in the U.S. National Museum support Likharev's contention that the species from China is generically distinct from those of the Salt Range that were called *Lyttonia* by Waagen. The generic identity of the Chinese form is uncertain at present, but it is sufficiently different from the Salt Range form to warrant use of the name *Lyttonia* for a generic entity that is something other than *Leptodus*. Therefore, Likharev's suggestion is followed here, and the name *Lyttonia* is used.

Lyttonia spp.

Plate 2, figure 6

Remarks: Numerous fragments can be identified as belonging to species of *Lyttonia* by their pinnate structure and by the granular ornament of the dorsal valves. The muscle and septal patterns are symmetrical, and most of the specimens occur in the same formation as undoubted good specimens of *Lyttonia,* so the generic assignment seems fairly certain.

Occurrence: Fragments of one or more species of *Lyttonia* are rather abundant in the fossiliferous lens at the top of the white sandstone, Chhidru Formation, Tapan Wahan, Khisor Range (USNM loc. 9049). Fragments of the genus are rare in the dolomite unit of the Kathwai Member of the Mianwali Formation at Narmia Spring (USNM loc. 9159, 9161). In view of the characteristic irregularity of these shells, it is impossible to tell whether the specimens from the Chhidru Formation are the same or a different species from those in the Kathwai Member. All the specimens are well below the median size for species of *Lyttonia* from farther down in the Zaluch Group.

Order SPIRIFERIDA Waagen, 1883

Family RETZIIDAE Waagen, 1883

Genus HUSTEDIA Hall and Clarke, 1893

Diagnosis: Strongly costate, punctate spiriferids without obvious fold and sulcus; beak elongate with conjunct deltidial plates, circular foramen; dental and septal plates absent from ventral valve; cardinal process large, recurved; spiralia transverse, united by elaborate jugum.

Hustedia sp.

Plate 2, figures 10-10b

Remarks: One complete shell is similar in shape although somewhat smaller than *H. indica* (Waagen) var. *chittidilensis* Reed (1944), which typically occurs in the Kalabagh Member at the top of the Wargal Limestone. It is more elongate and has sharper costae than *H. indica* proper, which occurs in the so-called Katta beds of Waagen (1889) at the top of the Amb Formation. Specimens of *Hustedia* are found at all levels in the Zaluch Group, but they are not abundant.

Occurrence: White sandstone unit of Chhidru Formation, Tapan Wahan, Khisor Range (USNM loc. 9049).

Family ATHYRIDIDAE M'Coy, 1844

Genus SPIRIGERELLA Waagen, 1883

Diagnosis: Uniplicate, without costae or radial ornament; growth laminae distinctive; valve edges beveled; foramen minute or absent; umbonal region thickened, normally burying rudimentary dental plates; cardinal plate large, with two deep muscle marks; spire similar to *Composita.*

Remarks: Spirigerella is one of the most abundant shells in the Permian of the Salt Range and spans a wide range of variability. Therefore a great number of species have been described, especially by Reed (1944). Until the large collections from the main part of the Zaluch Group can be studied, it seems best not to attempt to identify the species in the white sandstone and the dolomite unit. These specimens are few and fragmentary, although they retain enough of the form and concentric ornament to be identified confidently as belonging to species of *Spirigerella.*

Spirigerella spp.

Plate 2, figures 11-13

Remarks: One or more species of *Spirigerella* are represented by about a dozen fragments in the white sandstone unit, mostly ventral valves. These were found at Tapan Wahan (USNM loc. 9175) and Kingriali Mountain (USNM loc. 9049), Khisor Range.

One ventral valve from the dolomitic unit of the Kathwai Member of the Mianwali Formation is certainly identifiable as belonging to a species of *Spirigerella.* Four others in the same

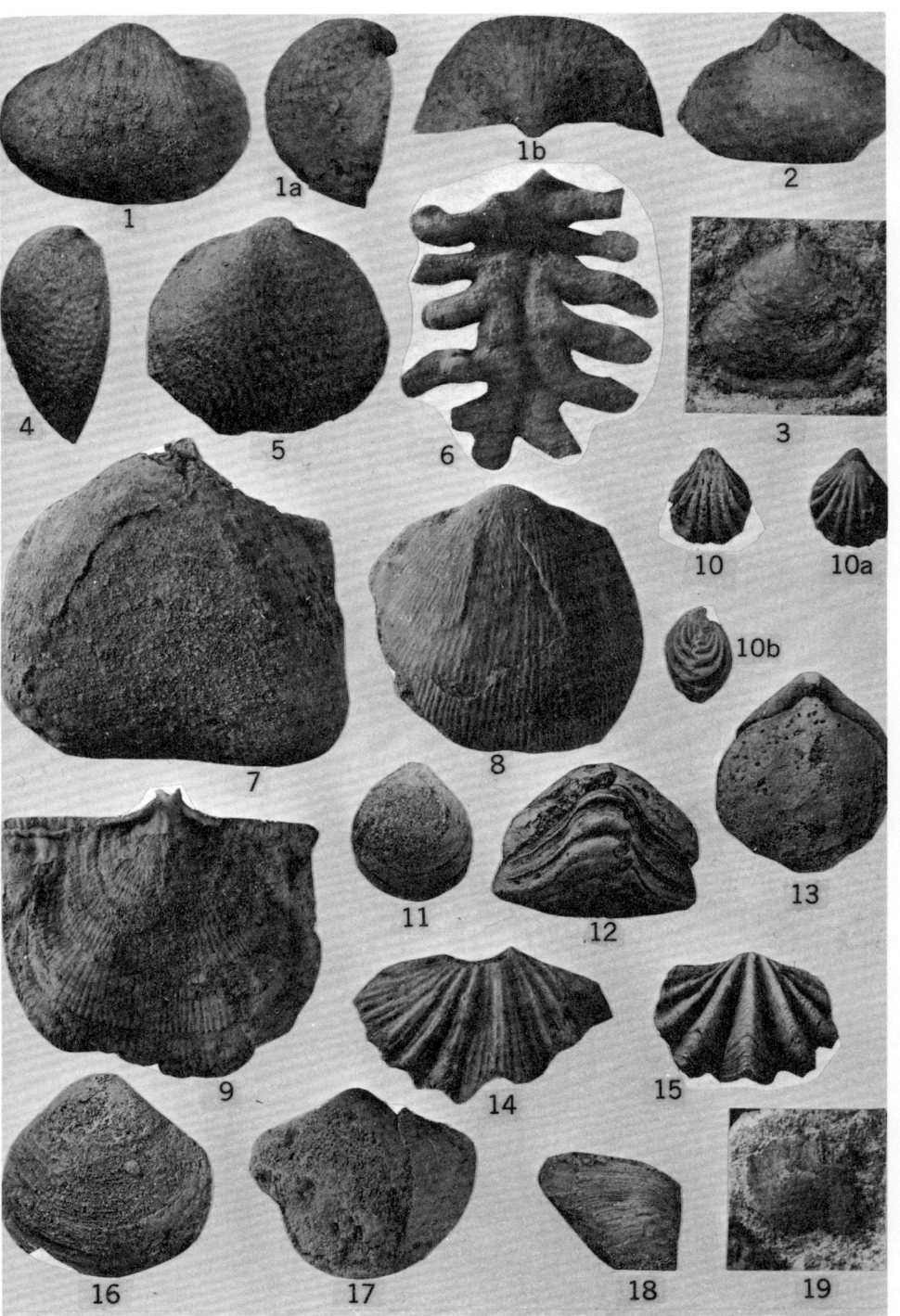

Plate 2

sample (T63-162) are tentatively assigned to the genus. *Spirigerella* does, therefore, extend into the Kathwai Member, but in view of its superabundance in the Zaluch Group it must be considered exceedingly rare in the Kathwai.

Genus CLEIOTHYRIDINA Buckman, 1906

Diagnosis: Thick-shelled athyridinids with small pedicle foramen; growth lamellae numerous, distinct, evenly spaced, projecting as serrated or fringed laminations, simulating numerous fine spines; cardinal plate large, thick, apically perforated; dental plates small, typically buried in secondary shell material of mature specimens; spiralia and jugum as in *Athyris*.

Cleiothyridina cf. *C. capillata* (Waagen)

Plate 2, figures 16-17

Cf. *Athyris capillata* Waagen, 1883, p. 479, pl. 39, figs. 6-9; pl. 40, figs. 1-5; pl. 42, figs. 1-5.

Remarks: Specimens of *Cleiothyridina* resemble the small and rather strongly convex *C. capillata* from the upper part of the Chhidru Formation. However, the few with the anterior margin preserved show no evidence of plication, and the external surface is so badly exfoliated that it cannot be determined whether the growth laminae are as closely spaced as described for *C. capillata* by Waagen.

Occurrence: Moderately abundant fragments in the white sandstone unit of the Chhidru Formation, Tapan Wahan (USNM loc. 9049) and Kingriali Mountain (USNM loc. 9175), Khisor Range.

Family AMBOCOELIIDAE George, 1931

Genus CRURITHYRIS George, 1931

Diagnosis: Noncostate, weakly

EXPLANATION OF PLATE 2

FIGURE

1-3. *Spinomarginifera* sp., ×2.——1, 1a, 1b. Exfoliated ventral valve showing spine channels, USNM no. 153712, USNM loc. 9156 (K3-3); 1, ventral view; 1a, side view; 1b, posterior view.——2. Ventral valve with spine bases, ventral view, USNM no. 153713, USNM loc. 9161 (T63-162).——3. Dorsal valve with pustules and marginal diaphragm, USNM no. 153714, USNM loc. 9159 (T63-19).

4, 5. *Waagenoconcha?* sp., USNM loc. 9175, USNM no. 153715.——4, Side view of ventral valve, ×2.——5, Ventral view of ventral valve, showing shallow sulcus and spine bases, ×2.

6. *Lyttonia* sp., exterior of dorsal valve, ×2, USNM no. 153717, USNM loc. 9049.

7-9. *Linoproductus* sp., USNM loc. 9049.——7. Large ventral valve, partly exfoliated, ×1.5, USNM no. 153718.——8. Smaller, rather evenly convex ventral valve, ×1.5, USNM no. 153719.——9. Interior of large dorsal valve, ×1.3, USNM no. 153720.

10. *Hustedia* sp., complete shell, ×2, USNM no. 153721, USNM loc. 9049; 10, ventral view; 10a, dorsal view; 10b, side view.

11-13. *Spirigerella* sp.——11. Exterior of small ventral valve, showing characteristic growth lines, ×2, USNM loc. 9049, USNM no. 153722.——12. Anterior of weathered shell, showing strong growth laminae, ×2, USNM loc. 9175, USNM no. 153723.——13. Dorsal view of small shell, ×2, USNM loc. 9175, ×2, USNM no. 153724.

14. *Neospirifer* sp., fragment of dorsal valve, ×2, USNM no. 153725, USNM loc. 9049.

15. *Callispirina* sp., exterior of nearly complete dorsal valve, ×2, USNM no. 153726, USNM loc. 9049.

16, 17. *Cleiothyridina* cf. *C. capillata* ×2, USNM loc. 9049.——16. Exterior of ventral valve showing growth lines, USNM no. 153727.——17. Partly exfoliated exterior of rather wide dorsal valve, USNM no. 153728.

18. *Richthofenia* sp., side view of fragment of ventral valve, ×2, USNM no. 153729, USNM loc. 9049.

19. Chonetid undet., fragment of ventral valve showing pustules, ×2, USNM no. 153730, USNM loc. 9049.

folded, biconvex to planoconvex ambocoeliinids with minute spinose ornament, elongate narrow dorsal adductor muscles located in the posterior part of the valve; rodlike crura cemented to the posterior floor of the dorsal valve become slightly elevated anteriorly and give rise to an elaborate pair of transversely coiled spiralia without jugum.

Crurithyris? extima Grant n. sp.

Plate 3, figures 1-3d, Text-fig. 1

Description: Shell small to moderate size for *Crurithyris;* outline subelliptical, width slightly greater than length; commissure rectimarginate to almost imperceptibly uniplicate in small shells or sulcate in larger shells; hinge about half as wide as shell; posterolateral margins strongly curved, not angular or alate; ornament consisting of weak growth lamellae; spine bases or other ornamentation not preserved.

Ventral valve strongly and evenly convex, without sulcus or other median depression; beak short for *Crurithyris,* moderately strongly curved to suberect position; interarea low, concave, only indistinctly set off from curve of valve near beak, more sharply delimited laterally; delthyrium narrowly triangular, with rudimentary deltidial plates along edges, uniting under beak to form short deltidium.

Dorsal valve flatly convex, but relatively strongly convex for *Crurithyris,* slightly swollen near umbo, slightly flattened near anterior margin; interarea very low; notothyrium open: chilidial plates thicker than deltidial plates, observed in section only.

Ventral interior without plates or median ridge; slight thickening in posterior part of valve; hinge teeth blunt, one on each side of delthyrium.

Dorsal interior with low rounded cardinal process; sockets elevated on thin supporting plates that continue anteriorly beyond sockets, becoming crural plates, remaining attached to floor for more than one-fourth length of shell, about one-sixth length of valve, then becoming detached, reduced to pair of thin rods that give rise anteriorly to paired spiralia. Muscle area poorly observed, apparently consisting of paired, elongate, narrow marks beginning near where crura detach from floor.

Comparisons: Crurithyris? extima differs in several features from other species of *Crurithyris* and may be in fact generically distinct. Species of *Crurithyris* normally have the dorsal valve nearly flat and the commissure obviously plicated, either uniplicate as in normal juveniles, slightly sulcate as in many adults, or emarginate as a result of shallow sulcation of both valves. In contrast, *C.? extima* has no hint of a sulcus on the ventral valve, and the commissure is so weakly, sporadically, and occasionally asymmetrically undulose as to suggest fortuitous deformation rather than genetically determined median folding. The dorsal valve is much more strongly convex than that of other Late Paleozoic species of *Crurithyris.* The umbonal part of the ventral valve is shorter than typical for *Crurithyris,* the interarea consequently is lower and less strongly concave, and the interarea of the dorsal valve is extremely low for a species of *Crurithyris.*

Two more important differences from typical *Crurithyris* are present in the interior of the dorsal valve. The sockets are not low and essentially cemented directly to the valve floor, but instead are raised on short plates (Fig. 1) which, when combined with the thickness of the socket plates themselves, serve to elevate the socket considerably. The other difference is the relatively great length of the parts of the crura that extend along the floor, and their anterior continuance as rods rather than as thin plates.

On the other hand, many features argue for inclusion of *C.? extima* with *Crurithyris.* The general aspect, both external and internal, is crurithyridid.

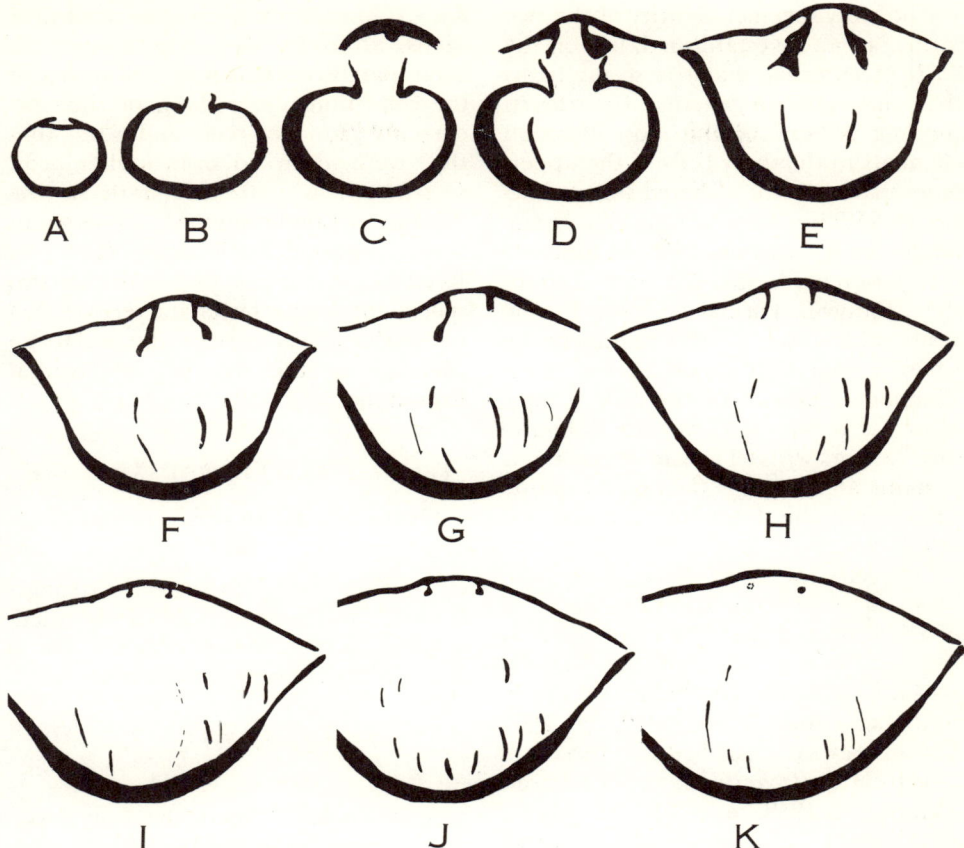

Fig. 1. Serial sections ×6 of *Crurithyris? extima* (USNM no. 153741) from the dolomite unit of the Kathwai Member, Mianwali Formation, locality T63-22 (USNM loc. 9164). Original dimensions: Length, 7.9 mm; width, 8.8 mm; thickness, 5.0 mm. Numbers below indicate distance of each section from end of the umbo, just behind the beak; sections are perpendicular to the plane of commissure. A. 0.7 mm, small deltidium in apex of delthyrium; B. 1.1 mm, low deltidial plates; C. 1.4 mm, high deltidial plates: apex of dorsal valve; D. 1.5 mm, beginnings of sockets and socket plates; E. 1.6 mm, sockets, socket plates, teeth, and mesial whorls of spiralia; F. 1.7 mm, anterior ends of sockets and teeth, socket supporting plates remain; G. 1.8 mm; H. 2.0 mm, plates low, becoming crural plates; I. 2.2 mm, crura supported by low thin plates; J. 2.4 mm; K. 2.6 mm, crura unsupported, extending anteriorly just above floor of dorsal valve.

The hinge is narrower than the widest part of the shell, a feature that distinguishes *Crurithyris* from *Ambocoelia*, and the muscle marks of the dorsal valve appear to be in the normal position for *Crurithyris*. In view of the paucity and relatively poor preservation of the material, it seems best to assign the species tentatively to *Crurithyris*, while emphasizing its peculiar differences from other species of the genus as well as its anomalous stratigraphic position.

Externally *C.? extima* bears some resemblance to *Crurithyris speciosa* Wang (1956) from the Upper Permian Changhsing Limestone of Kweichow Province, China, especially in the convexity of the dorsal valve and the rectimarginate anterior commissure. The outline of *C.? extima* appears to be widest somewhat farther posterior to the midline, and the beak less strongly curved. Growth lines are not preserved on *C.? extima*, and nothing is known of the interior of *C. speciosa*, so generic (or possibly specific) identity of the two forms cannot be established, but on the basis of external shape it seems likely that they are congeneric. *Crurithyris speciosa* is "one of the most common elements in the shaly beds of the uppermost part of the Changhsing Limestone" (Wang, 1956, p. 581). Its resemblance to *C.? extima*, and the possible correlation with the dolomite unit of the Mianwali Formation that it suggests, adds force to the argument by Chao (1965) that much of the latest Permian is absent in the Salt Range. How *C.? extima* got into the dolomite unit along with Triassic ammonoids remains an enigma: *C.? extima* occurs throughout the dolomite unit, not just at the base.

Species of *Crurithyris* have also been reported from the Late Permian of the central Elburz in Iran (Fantini Sestini, 1965, p. 62) and the Dzhulfa area of Armenia (Arthaber in Frech, Arthaber, and Frech, 1900, p. 266). Fantini Sestini identified her species as *Crurithyris? tschernyschewi* Likharew (Likharev, 1939, p. 114) and placed Arthaber's "*Martinia planoconvexa* Shumard" in synonymy with it. *C.? extima* differs from that species in its shorter ventral beak with consequently lower delthyrium, its more convex dorsal valve, and possibly also by the peculiar (for *Crurithyris*) internal structures of *C.? extima*; the interior of *C.? tschernyschewi* is unknown.

Occurrence: Crurithyris? extima occurs in the dolomite unit of the Kathwai Member, Mianwali Formation, Narmia Spring, Surghar Range (USNM loc. 9160, 9162); Khan Zaman Nala, Salt Range (USNM loc. 9157, 9158). Sixteen specimens were freed by cracking the rock, and about another twelve in more fragmentary condition were visible in rock samples. Numerous specimens are visible in the rock samples from T63-141 and 144, which were taken from 5-6 feet above the base of the dolomitic unit at Khan Zaman Nala, well into the Triassic *Ophiceras*

Zone. Most specimens are complete shells, although they appear to have been partly dolomitized. Nothing in the condition of the shells or their occurrence in the rock indicates that they were reworked from earlier beds. *Crurithyris* is extremely rare in the Productus Limestone itself. The genus is not reported by Waagen (1884) or Reed (1944), although I collected two specimens with the characteristic flattened dorsal valve from the Kalabagh Member at the top of the Wargal Limestone.

Family SPIRIFERIDAE King, 1846

Genus NEOSPIRIFER Fredericks, 1924

Diagnosis: Transverse, strongly folded, fascicostate, fold and sulcus costate; dental plates present; crural plates absent; hinge denticulate.

Neospirifer sp.

Plate 2, figure 14

Remarks: A small, broken, dorsal valve bears the strong fasciculate costation of a species of *Neospirifer*. Unfortunately not enough of the outline is present to indicate its shape, but the costation most nearly resembles that of *N. warchensis* Reed from the Kalabagh Member.

Occurrence: White sandstone unit of Chhidru Formation, Tapan Wahan, Khisor Range (USNM loc. 9049).

Family BRACHYTHYRIDIDAE Fredericks, 1924

Genus SPIRIFERELLA Chernyshev, 1902

Diagnosis: Hinge narrow; costae simple, strong; fold and sulcus weakly costate; dental plates well developed.

Spiriferella? sp.

Remarks: A fragmentary ventral valve has the strong and nonfasciculate costae that characterize the *"Elivina"* (Frederiks, 1924) kind of *Spiriferella* that was illustrated by Reed (1944, pl. 29) from the Kalabagh Member at the top of the Wargal Limestone. The beak was cut and etched, revealing strong dental plates, but no other internal structures were made visible. The sulcus appears to be smooth, but only the posterior part is preserved, and costae could have been present farther forward. The identification is uncertain, but the characters revealed are sufficient to indicate presence in the faunule of an impunctate spiriferid other than *Neospirifer*.

Occurrence: White sandstone of Chhidru Formation, Tapan Wahan, Khisor Range (USNM loc. 9049).

Family SPIRIFERINIDAE Davidson, 1884

Genus CALLISPIRINA Cooper and Muir-Wood, 1951

Diagnosis: Strongly convex, punctate spiriferinids with strong angular plications; growth lines distinct, numerous, regularly spaced.

Callispirina sp.

Plate 2, figure 15

Remarks: One dorsal plate has the strong plications, numerous, distinct, and regular growth laminae, and the rounded lateral outline that identify it as belonging to a species of *Callispirina*. It is less convex than typical for *C. ornata* (Waagen), and its plications are more angular than those of the other two Salt Range species identified by Reed (1944, p. 250).

The shape of the valve and the angularity of the plications of this specimen resemble the Dzhulfian species *Crenispirifer dzhulfensis* Ivanova (in Ruzhentsev and Sarycheva, eds., 1965, p. 236). However, the ornament of fine and regular growth lines on the Salt Range specimen identifies it as belonging to a species of *Callispirina* rather than *Crenispirifer* which has closely spaced low pustules.

Occurrence: One dorsal valve and its external mold found in white sandstone unit of Chhidru Formation at Tapan Wahan, Khisor Range (USNM loc. 9049).

Family MARTINIIDAE Waagen, 1883

Genus MARTINIA M'Coy, 1844

Diagnosis: Unribbed, strongly uniplicate reticulariids with shagreen texture on lamellar shell layer, no internal septa or plates, and with axes of spiralia nearly directly transverse.

Martinia sp.

Plate 3, figures 4-6

Remarks: This species is characterized by its exaggerated fold that produces the anterior margin of the ventral valve into a long tongue. Weak concentric growth lines are visible, and the shell is decorticated in such a manner that radial pallial lines show through, simulating costellae. The dorsal valve is nearly straight along the ridge of the fold, the beak being only slightly convex. The strong fold of this species recalls that of most specimens of *M. rhomboidalis* Girty from the upper Guadalupian Lamar Limestone of the Guadalupe Mountains, Texas, although Girty's (1908, pl. 13) illustrated specimens of that species are immature and do not show the strong fold.

Occurrence: Well-preserved specimens were found in the base of dolomite unit of Kathwai Member, Mianwali Formation, Narmia Spring and vicinity, Surghar Range (USNM loc. 9159, 9161).

Martinia? sp.

Remarks: A fragmentary dorsal valve from the fossiliferous lens of the white sandstone unit of the Chhidru

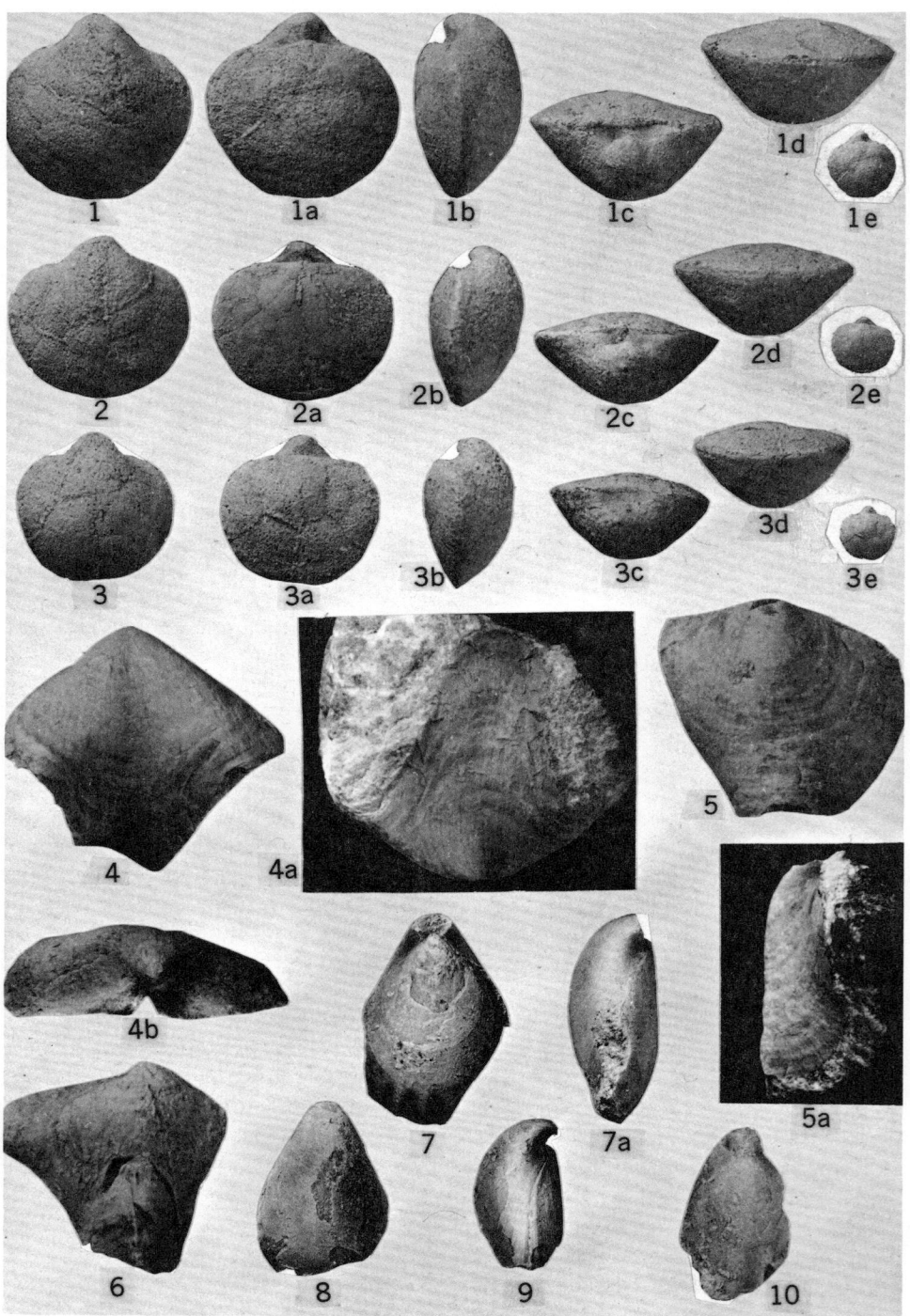

PLATE 3

Formation, Tapan Wahan, Khisor Range, appears to belong to *Martinia* (USNM loc. 9049). The longitudinal convexity along the crest of the fold is greater than on the specimens from the overlying Kathwai Member, and the fold apparently is not as high.

Order TEREBRATULIDA Waagen, 1883

Family DIELASMATIDAE Schuchert, 1913

Genus HEMIPTYCHINA Waagen, 1882

Diagnosis: Anterior commissure rectimarginate, with several low, short costae; cardinal plate divided, each half reaching floor mesially and inner socket ridge distally; dental plates absent.

Hemiptychina sp.

Plate 3, figures 7-7a

Remarks: This small dielasmatid has the generic characters listed above for *Hemiptychina,* but does not seem to belong to any of the species described by Waagen (1882). It most nearly resembles *H. himalayensis* (Davidson) and *H. sparsiplicata* Waagen in its low, few, and short anterior plications. It is not as convex as either of those species, is more elongate and not as broadly ovate, and occurs stratigraphically much higher than either of them.

Occurrence: Four fragmentary specimens from white sandstone unit of Chhidru Formation, Tapan Wahan, Khisor Range (USNM loc. 9049).

Genus WHITSPAKIA Stehli, 1964

Diagnosis: Anterior commissure sulciplicate; cardinal plate divided, each half joined to floor mesially, to inner socket ridge distally; dental plates present.

Whitspakia sp. 1

Plate 3, figure 10

Remarks: One shell consisting of a nearly complete dorsal valve and the beak of the ventral valve shows the characteristic low convexity and "biplicate" anterior margin of *Whitspakia.* The mesial edges of the divided hinge plate and the lines of junction of the inner socket ridges with the floor can be seen through the transparent punctate shell. The ventral valve is broken in such a manner as to display one dental plate. The specimen is smaller than *W. biplex* (Waagen) or *W. acu-*

EXPLANATION OF PLATE 3

FIGURE

1-3.   *Crurithyris? extima* Grant, n. sp., complete shells, USNM loc. 9164 (T63-22).——1, 1a, 1b, 1c, 1d. Shell of average size, holotype, ×3, USNM no. 153731, ventral, dorsal, side, posterior, anterior views; 1e, dorsal view, ×1.——2, 2a, 2b, 2c, 2d. Slightly smaller shell, ×3, USNM no. 153732, ventral, dorsal, side, posterior, anterior views; 2e, dorsal view, ×1.——3, 3a, 3b, 3c, 3d. Small shell ×3, USNM no. 153733, ventral, dorsal, side, posterior, anterior views; 3e, dorsal view, ×1.

4-6.   *Martinia* sp., ×1.5, USNM loc. 9161 (T63-162).——4, 4a, 4b. Complete exterior of ventral valve with sulcus and long anterior tongue, USNM no. 153734. 4, view of large part of valve; 4a, specimen inverted, showing long tongue, sulcus; 4b, posterior view showing interarea and delthyrium.——5, 5a. Dorsal valve, USNM no. 153735. 5, dorsal view; 5a, side view showing height of fold.——6. Fragmentary dorsal valve with high fold, dorsal view, USNM no. 153736.

7.   *Hemiptychina* sp. fragmentary shell, ×2, USNM no. 153737, USNM loc. 9049. 7, dorsal view showing plications; 7a, side view.

8, 9.   Dielasmatids indet., ×2, USNM loc. 9049.——8. Ventral view of fragmentary ventral valve, USNM no. 153738.——9. Side view of fragmentary shell, USNM no. 153739.

10.   *Whitspakia* sp. 1, dorsal view of fragmentary shell, ×2, USNM no. 153740, USNM loc. 9049.

tangula (Waagen) from lower in the section; it is also somewhat smaller than *Whitspakia* sp. 2 from the overlying dolomite unit of the Mianwali Formation.

Occurrence: White sandstone of Chhidru Formation, Tapan Wahan (USNM loc. 9049) and Kingriali Mountain (USNM loc. 9175), Khisor Range.

Whitspakia sp. 2

Remarks: A dorsal valve has the low convexity and anterior biplication of a species of *Whitspakia*. The traces of the divided cardinal plate and the socket ridges are visible through the shell, but in the absence of the ventral valve, the presence of dental plates cannot be ascertained. This specimen is larger than the specimens from the top of the Chhidru Formation, and the plication is more gentle. It most nearly resembles *W. acutangula* (Waagen), although it is smaller and somewhat less strongly plicated.

Occurrence: Lower part of dolomite unit of Kathwai Member, Mianwali Formation, Narmia Spring, Surghar Range (USNM loc. 9161).

Dielasmatid gen. et sp. indet.
Plate 3, figures 8, 9

White sandstone unit: Five specimens lacking anterior margins belong to some species of dielasmatid. They have the characteristic shape, punctate shell, absent dental plates, and divided cardinal plates, and probably belong to species of *Hemiptychina*. Tapan Wahan, Khisor Range (USNM loc. 9049).

Dolomite unit: An incomplete ventral valve shows punctate shell structure and the absence of dental plates but cannot be identified generically. Narmia Spring, Surghar Range (USNM loc. 9159).

CONCLUSIONS

Analysis of the brachiopod faunas of the Chhidru Formation and of the Kathwai Member of the Mianwali Formation suggests that:

1. The fauna of the Chhidru Formation begins in the Kalabagh Member of the Wargal Limestone and is Guadalupian in age. Furthermore, its youngest beds are not equivalent to the highest Guadalupian, but only to the part below the uppermost Capitanian.

2. The fauna of the dolomite unit of the Kathwai Member is younger than that of the Chhidru and contains latest Guadalupian, Dzhulfian, and Triassic elements.

3. The Kathwai Member fauna is depauperate in comparison with the Dzhulfian fauna of Armenia, indicating that some of it was eroded away, or that conditions were marginal for the growth of brachiopods in the Salt Range area during the latest stage of the Permian.

4. Because much of the Armenian Dzhulfian fauna is absent from the Kathwai fauna, and the fossiliferous part of the dolomite unit is only a few feet thick, very slow deposition or perhaps erosion was taking place during the late Dzhulfian in the Salt Range area, and there is a definite hiatus between the latest Permian and the earliest Triassic there. This conclusion agrees with that of Grabau (1931, p. 424) and disagrees with that of Noetling (1901, p. 415, 422) and Schindewolf (1954, p. 153 *et seq.*).

5. The Zaluch Group in the Salt Range and nearby Trans-Indus ranges is almost a classic example of a transgressive-regressive sequence. Setting aside the question of a possible disconformity between the Amb and the Wargal Formation (Grabau, 1924, p. 328), which would represent only a short hiatus at best, the generalized lithic sequence is as follows:

Lowermost is the Amb Formation, with shale and siltstone at the base and bituminous shales at many localities, followed by sandstone with calcareous cement. The amount of sand decreases upward in the section and the amount of calcareous cement increases, but the formation remains sandy to the top (with local argillaceous patches such as that at Katha).

The Wargal Limestone follows, beginning with rather thin-bedded limestone, indicating numerous small shale breaks. Bedding increases in massiveness, although close inspection shows that the microbedding of the limestone is very thin, but without terrigenous partings. Most of the Wargal is a tightly cemented biosparite, but toward the top of the main part that forms the high cliff, the gross bedding again becomes thinner, with progressively more numerous shaly partings. The cliff is capped by the Kalabagh Member of argillaceous limestone or calcareous shale.

Above the Kalabagh Member the basal beds of the Chhidru Formation are shale, or locally siltstone or fine sandstone. Grain size of terrigenous particles increases sporadically upward in the formation; and the uppermost beds, the "white sandstone unit" of Kummel and Teichert (1966), are poorly cemented by lime, although the sand is only medium grained.

Above the terminal sands of the Chhidru Formation lies the dolomite unit of the Kathwai Member of the Mianwali Formation, only 5-8 feet thick, with Permian fossils sporadically present in the base and Triassic fossils sporadically present throughout. This is euhedral dolomite, distinctly bedded, with generous admixtures of calcareous cement.[1]

The sequence of events can be interpreted as follows: Swamps and near-shore deposits of the basal Amb Formation were flooded by the transgressing sea, conditions becoming progressively more marine and the shore-line retreating from the area. Great deposits of limesand accumulated far from shore on a shallow shelf, producing the Wargal Limestone. Regression began as the middle of the Wargal was being deposited. The shore moved closer to the area, terrigenous sediments were introduced, and the sequence from Kalabagh Member through the Chhidru Formation records the steady regression. The sea shallowed to a shallow basin having minimal deposition during the Dzhulfian. The dolomite unit of the Kathwai Member was deposited during this period, and there was still-stand or some erosion. Transgression began again and the limestone unit of the Kathwai Member was deposited, followed by deposition of the argillaceous limestones of the upper member of the Mianwali Formation.

REFERENCES

Chao, K-K, 1965, The Permian ammonoid-bearing formations of South China: Scientia Sinica, v. 14, no. 12, p. 1813-1826, 2 pl., 4 fig., 2 tab.

Cooper, G. A., 1957, Permian brachiopods from central Oregon: Smithsonian Misc. Coll., v. 134, no. 12, 79 p., 12 pl., 2 fig.

———, et al., 1953, Permian fauna at El Antimonio, western Sonora, Mexico: Smithsonian Misc. Coll., v. 119, no. 2, 106 p., 25 pl., 3 fig.

Diener, Carl, 1897, The Permocarboniferous fauna of Chitichun No. 1: Geol. Survey India, Palaeontologia Indica, ser. 15, v. 1, pt. 3, 105 p., 13 pl.

Dunbar, C. O., 1933, Stratigraphic significance of the fusulinids of the Lower Productus Limestone of the Salt Range: Geol. Survey India Recs., v. 66, pt. 4, p. 405-413, pl. 22.

———, 1955, Permian brachiopod faunas of central East Greenland: Medd. om Grønland, Bd. 110, nr. 3, 169 p., 32 pl., 22 fig.

[1] For detailed petrographical description of dolomite unit, see Kummel and Teichert, 1970, p. 43-48.—Eds.

———, et al., 1960, Correlation of the Permian formations of North America: Geol. Soc. America Bull., v. 71, p. 1763-1806, 1 pl., 2 fig.

Fantini Sestini, N., 1965, The geology of the upper Djadjerud and Lar Valleys (North-Iran), pt. 2, Palaeontology; Bryozoans, brachiopods and molluscs from Ruteh Limestone (Permian): Rivista Ital. Paleontologia, Stratigrafia, v. 71, no. 1, p. 13-108, pl. 2-8, 6 fig.

———, and Glaus, M., 1966, Brachiopods from the Upper Permian Nesen Formation (North Iran): Rivista Ital. Paleontologia, Stratigrafia, v. 72, no. 4, p. 887-930, pl. 63-66, 4 fig.

Frech, F., Arthaber, G. von, and Frech, V., 1900, Neue Forschungen in den kaukasischen Ländern. II. Abteilung. Über das Paläozoikum in Hocharmenien und Persien: Beiträge Paläontologie und Geologie Österreich-Ungarns u. Orients, Bd. 12, H. 4, p. 161-308, pl. 15-22, 23 fig., 3 maps.

Frederiks, G., 1924, O Verkhne-Kamennougolnykh spiriferidakh Urala [On the Upper Carboniferous spiriferids of the Urals]: Russia, Geol. Komitet, Izvestiya, v. 38 (for 1919), no. 2, p. 295-324, 7 fig.

Furnish, W. M., 1966, Ammonoids of the Upper Permian *Cyclolobus*-Zone: Neues Jahrb. Geologie Paläontologie, Abh., v. 125 (Festband Schindewolf), p. 265-296, pl. 23-26, 4 fig., 2 tab.

Girty, G. H., 1908, The Guadalupian fauna: U.S. Geol. Survey Prof. Paper 58, 651 p., 31 pl. [1909].

Glaus, Martin, 1964, Trias und Oberperm im zentralen Elburs (Persien): Eclogae Geol. Helvetiae, v. 57, no. 2, p. 497-508, pl. 1-3.

Glenister, B. F., and Furnish, W. M., 1961, The Permian ammonoids of Australia: Jour. Paleontology, v. 35, no. 4, p. 673-736, pl. 78-86, 17 fig.

Gobbett, D. J., 1963, Carboniferous and Permian brachiopods of Svalbard: Norsk Polarinst., Oslo, Skr., no. 127, 201 p., 25 pl., 27 fig.

Grabau, A. W., 1924, Stratigraphy of China, pt. 1; Paleozoic and older: Peking, Geol. Survey China, 528 p., 6 pl., 306 fig.

———, 1931, Natural History of Central Asia, v. 4; The Permian of Mongolia: New York, Am. Mus. Nat. History, 665 p., 35 pl., 72 fig.

Grant, R. E., 1965, The brachiopod superfamily Stenoscismatacea: Smithsonian Misc. Coll., v. 148, no. 2, 192 p., 24 pl., 34 fig.

———, 1968, Structural adaptation in two Permian brachiopod genera, Salt Range, West Pakistan: Jour. Paleontology, v. 42, no. 1, p. 1-32, pl. 1-9, 21 fig.

Harker, Peter, and Thorsteinsson, Raymond, 1960, Permian rocks and faunas of Grinnell Peninsula, Arctic Archipelago: Canada Geol. Survey Mem. 309, 89 p., 25 pl., 9 fig., 7 tab.

Heritsch, Franz, 1934a, Die Stratigraphie von Oberkarbon und Perm in den Karnischen Alpen: Geol. Gesell. Wien, Mitteil. Bd. 26, p. 162-190, 2 tab.

———, 1934b, Die oberpermische Fauna von Zazar und Vrzdenec in den Savefalten: Yugoslavia, Serv. Géol. Bull. (Vesnik), t. 3, f. 1, p. 6-61, 2 pl., 10 fig.

———, 1937, Rugose Korallen aus dem Salt Range, aus Timor und aus Djoulfa mit Bemerkungen über die Stratigraphie des Perms: Akad. Wiss. Wien, Math.-Naturwiss. Klasse, Sitzungsber., Abt. 1, Bd. 146, H. 1-2, p. 1-16, pl. 1-2.

Kayser, E., 1883, Obercarbonische Fauna von Loping: in F. F. von Richthofen, China: v. 4, pt. 8, p. 160-206, Berlin.

King, R. E., 1931, Geology of the Glass Mountains, Texas; Part II, Faunal summary and correlation of the Permian formations with description of Brachiopoda: Texas Univ. Bull. 3042, 245 p., 44 pl., 5 fig.

Kozlowski, R., 1914, Les brachiopodes du Carbonifère Supérieur de Bolivie: Ann. Paléontologie, v. 9, 100 p., 11 pl., 24 fig.

Kummel, Bernhard, and Teichert, Curt, 1966, Relations between the Permian Triassic formations in the Salt Range and Trans-Indus ranges, West Pakistan: Neues Jahrb. Geologie Paläontologie, Abh., v. 125 (Schindewolf Festschrift), p. 297-333, pl. 27-28, 4 fig.

———, and ———, 1970, Stratigraphy and paleontology of the Permian-Triassic boundary beds, Salt Range and Trans-Indus ranges, West Pakistan: in Kummel, B., and Teichert, C., eds., Stratigraphic Boundary Problems: Permian and Triassic of West Pakistan, Univ. Press Kansas, Dept. Geology, Univ. Kansas Spec. Publ. 4, p. 1-110.

Likharev, Boris K., 1939, Klass Brachiopoda, in Likharev, Boris K., et al., Atlas rukovodyashchikh form iskopaemikh faun SSSR [Atlas of the leading forms of the fossil fauna of the USSR 1: v. 6, Permian]: USSR Tsentralnyi Nauchnogo-issledovatelskii Geologo-razvedochnii Inst., Leningrad, Moscow, p. 76-121, pl. 16-29.

———, 1965, Neskolko zamechanii po povodu stati T. G. Sarycheva "Oldhaminoidnie brachiopodi is Permi Zakavkazya" [Some comments on the article "Oldhaminoid brachiopods of the Permian of the Transcaucasus" by T. G. Sarycheva]: Paleont. Zhurnal, no. 2, p. 149-150.

Merla, Giovanni, 1931, La fauna del Calcare a *Bellerophon* della Regione Dolomitica: Padua, Univ. Istituto Geologico Mem., v. 9, no. 2, 221 p., 11 pl.

Miklukho-Maklay, A. D., 1963, Verkhnii Paleozoi Srednei Azii [Upper Paleozoic of Central Asia]: Leningrad Gosudar. Univ., 329 p., 8 pl., 8 tab.
Gosudar. Univ., 329 p., 8 pl., 8 tab.
Miller, A. K., and Furnish, W. M., 1940, Permian ammonoids of the Guadalupe Mountain region and adjacent areas: Geol. Soc. America Spec. Paper no. 26, 242 p., 44 pl., 59 fig., 6 tab.
Muir-Wood, H., and Cooper, G. A., 1960, Morphology, classification and life habits of the Productoidea (Brachiopoda): Geol. Soc. America Mem. 81, 447 p., 135 pl., 8 fig.
Noetling, F., 1901, Beiträge zur Geologie der Salt Range, inbesondere der permischen und triassischen Ablagerungen: Neues Jahrb. Mineralogie, Geologie, u. Palaeontologie, Beilage-Band 14, p. 369-471, 4 fig.
Pascoe, E. H., 1959, A manual of the geology of India and Burma, v. 2, (3rd ed.): Calcutta, Govt. India Press, 1343 p.
Reed, F. R. C., 1944, Brachiopoda and Mollusca from the Productus Limestone of the Salt Range: Geol. Survey India Mem., Palaeontologia Indica, new ser., v. 23, no. 2, 678 p., 65 pl.
Ruzhentsev, V. E., 1955, Osnovnye stratigraficheskie kompleksy ammonoidei permskoi sistemy: Akad. Nauk. SSSR Isvest., Ser. Biol. no. 4, p. 120-132.
———, and Sarycheva, T. G., eds., 1965, Razvitie i smena morskikh organizmov na rubezhe Paleozoya i Mesozoya [Development and succession of marine organisms at the boundary of the Paleozoic and Mesozoic]: Akad. Nauk SSSR, Paleont. Inst., Trudy, v. 108, 430 p., 58 pl., 59 fig.
Schindewolf, O. H., 1954, Über die Faunenwende vom Paläozoikum zum Mesozoikum: Deut. Geol. Gesell. Zeitschr., v. 105, p. 153-182, 3 pl., 5 fig., 2 tab.

Teichert, Curt, 1966, Stratigraphic nomenclature and correlation of the Permian "Productus limestone" of the Salt Range, West Pakistan: Geol. Survey Pakistan, Records, v. 15, pt. 1, 20 p., 6 fig.
Thomas, G. A., 1958, The Permian Orthotetacea of Western Australia: Australia Bur. Mineral Resources, Geology and Geophysics, Bull. 39, 158 p., 22 pl., 13 fig., 17 tab.
Tschernyschev, T. [Chernychev, T.], 1902, Verkhnekamennougolniya brakhiopodi Urala i Timana [Die obercarbonischen Brachiopoden des Ural und des Timan]: Russia, Geol. Komitet, Trudy, v. 16, no. 2, pt. 1, text: 749 p., 85 fig.; pt. 2, atlas: 63 pl. (in Russian; German summary).
Waagen, W., 1882-1885, Salt Range fossils, Productus Limestone fossils: Geol. Survey, India Palaeontologia Indica, ser. 13, v. 1, 998 p., 128 pl., 34 fig. [See Kummel and Teichert, 1970, for complete bibliographic reference.]
———, 1889, Salt Range fossils, Geological results: Same, ser. 13, v. 4, pt. 1, 242 p., 8 pl.
Walter, J. C., Jr., 1953, Paleontology of Rustler Formation, Culberson County, Texas: Jour. Paleontology, v. 27, no. 5, p. 679-702, pl. 70-73, 3 fig.
Wang, Y., 1956, New species of brachiopods II: Scientia Sinica, v. 5, no. 5, p. 577-601, pl. 6-10. [Kraus Reprint Limited, Nendeln-Niechtenstein, 1967: first published in Chinese in Acta Palaeontologia Sinica, v. 4, no. 3, p. 385-405, 1956.]
Williams, Alwyn, et al., 1965, Part H, Brachiopoda, in Moore, R. C., ed., Treatise on Invertebrate Paleontology: Geol. Soc. America and Univ. Kansas Press, 2 vols.: 927 p., 746 fig.

Permian Ammonoid *Cyclolobus* from the Salt Range, West Pakistan

W. M. FURNISH AND BRIAN F. GLENISTER
Department of Geology, University of Iowa

ABSTRACT

A century of collecting has yielded 15 specimens of *Cyclolobus* from the Upper Permian Chhidru Formation of the Salt Range, West Pakistan. The genus may also occur in the subjacent Kalabagh Member of the Wargal Limestone. Most representatives are referable to the type species, *C. oldhami* (Waagen), a form which occurs as a rare element in the Kuling Shale of the central Himalayas. *C. insignis* Diener, described originally from the Himalayas, is suppressed as a junior synonym of *C. oldhami*. *C. teicherti* Furnish and Glenister, n. sp., is an extremely rare Chhidru taxon which is related to *C. kullingi* (Frebold) from the Foldvik Creek Formation of East Greenland and the lower Dzhulfian of Armenia and Iran. A third Salt Range species, *C. walkeri* (Diener), was first described from the Himalayas, and occurs abundantly in the Ambilobé beds of northern Madagascar. *Krafftoceras kraffti* Diener and *K. haydeni* Diener, from the Kuling Shale of the Himalayas, and *Cyclolobus madagascariensis* Besairie and *C. astrei* Besairie, from the Ambilobé beds of Madagascar, are suppressed as junior synonyms of *C. walkeri*.

Direct comparison of virtually all known specimens of *Cyclolobus*, and of the ancestral *Timorites*, indicates that the Salt Range occurrences of *Cyclolobus* are younger than Guadalupian. They are referable to the middle Dzhulfian Chhidruan Stage, intermediate between the Araksian and the Changhsingian.

INTRODUCTION

A century ago, the discovery of a single true ammonite in direct association with Carboniferous-type productid brachiopods (Waagen, 1872) was an occasion of great interest to paleontologists. The similarity of this fossil to Middle and Upper Triassic arcestaceans has since been demonstrated to be superficial, but some early workers such as Joachim Barrande questioned the authenticity of discovery in the Permian. Waagen (1879, p. 25) defended his observation, still based upon a solitary specimen, and the unique nature of this rare ammonoid caused it to receive, meanwhile, a disproportionate amount of attention in standard reference texts such as those of Zittel. Cephalopod authorities such as Mojsisovics (1873) helped provide a perspective. The lack of recognizable progeny in the Lower Triassic served to justify erection of a new Paleozoic genus *Cyclolobus* by Waagen (1879), but that author observed "how little we yet know" concerning its developmental history. Waagen's assignment was substantiated to some extent by Gemmellaro (1887, 1888) and Rothpletz (1892), who recognized a relationship with less complicated Permian species. Diener (1903) described a series of *Cyclolobus* specimens from the Himalayan Permian; later, others were recorded from Greenland (Frebold, 1932; Miller and Furnish, 1940b) and from Madagascar (Treat, 1933). The genus still remains a great rarity in the type region of the Salt Range, but intensive search within recent years has provided a representative collection for study, about a dozen specimens (Schindewolf, 1954; Teichert, 1966; Kummel and Teichert, 1966; Grant, 1968).

SYSTEMATIC PALEONTOLOGY

Family CYCLOLOBIDAE Zittel, 1895

The cyclolobids are globular to discoidal ammonoids characterized by multilobed ammonitic sutures. The typical genus *Cyclolobus* Waagen appears to represent a final stage in complexity of the lineage. *Glassoceras* Ruzhentsev, 1960 (type: *Stacheoceras normani* Miller and Furnish, 1957; orig. desig.) has been identified as the late Early Permian link with more primitive Vidrioceratidae; all other representatives of the Cyclolobidae are Upper Permian. *Mexicoceras* Ruzhentsev, 1955 (type: *Waagenoceras cumminsi guadalupense* Girty, 1908; orig. desig.), *Waagenoceras* Gemmellaro, 1887 (type: *W. mojsisovicsi;* orig. desig.), and *Timorites* Haniel, 1915 (type: *T. curvicostatus;* subs. desig., Diener, 1921) are intermediate stages that occur widely in the Guadalupian Series.

Various members of the Cyclolobidae have been recorded as occurring in the Permian of South China (Chao, 1965). Perhaps the most significant form is *Changhsingoceras* Chao, 1965 (*C. meishanense, nomen nudum,* attributed to Chao and Liang, from the Changhsing Limestone of Chekiang). According to Chao, this fossil occurs with a diverse Late Permian ammonoid fauna (100 species in 30 genera and 7 families) being studied for a comprehensive monograph under joint authorship. From available information, this faunal assemblage appears comparable to that in the *Paratirolites*-beds of Armenia, which are younger than the Salt Range Permian; the unit may logically be named "Changhsingian Stage." Some of the taxa proposed by Chao (1965) are too sketchily presented to establish the names, but the sole illustration of *Changhsingoceras,* a suture drawing, is clearly a complex cyclolobid. In general configuration and development of tertiary elements, the suture is like *Cyclolobus;* however, in Chao's genus there are only about seven lateral lobes outside the umbilical shoulder at 50 mm diameter, instead of the usual nine or ten in *Cyclolobus.* The principle of retrograde evolution in number of lateral lobes can be assumed to have applied in other members of the family; i.e., *Mexicoceras.*

To a large degree, an understanding of age relationships in the Upper Permian is based upon ammonoid faunas. A phylogenetic sequence of cyclolobids constitutes the primary reference, although there are a number of other less diagnostic families ranging through this portion of the system. A tabulation of apparent relationship in the Cyclolobidae has been tentatively arranged in Table 1.

There is little unanimity of opinion regarding Permian stage names, particularly in the upper portion of the system. Ruzhentsev and Sarycheva (1965) have reviewed terminology, and Furnish and Glenister are preparing a manuscript to elaborate on the names listed in the accompanying table. The Amarassian Stage from Timor has already been presented briefly (Furnish and Glenister, 1968) to resignate uppermost Guadalupian. New data within the last few years suggest relationships for the post-Guadalupe sequence not visualized previously (Furnish, 1966). The best example is that found in the Upper Permian section of Araks River gorge near Dzhulfa Soviet Armenia where seven ammonoid zones have been identified (Ruzhentsev and Shevyrev in Ruzhentsev and Sarycheva, 1965). The lowest of these units at Dzhulfa contains a primitive *Cyclolobus,* and *Paratirolites* marks uppermost Permian, according to our present interpretation. Dzhulfian, first used in such a broad sense by Glenister and Furnish (1961), thus seems to be most appropriate for a series term. Further, it is here proposed that Araksian Stage, from the Araks River, be used for the lower portion of the Dzhulfian Series.

TABLE 1. Stratigraphic Ages of Important Species of Cyclolobidae.

Stage	Species	Localities
Changhsingian	*Changhsingoceras meishanense* Chao and Liang	Chekiang
Chhidruan	*Cyclolobus oldhami* (Waagen) et affines	Punjab, North India, Tibet, Madagascar
Araksian	*Cyclolobus kullingi* (Frebold)	Greenland, Armenia
Amarassian	*Cyclolobus persulcatus* Rothpletz *Timorites curvicostatus* Haniel	Coahuila, Timor
Capitanian	*Timorites striatus* Haniel *Mexicoceras guadalupense* (Girty) *Waagenoceras karpinskyi* Miller	Timor, Coahuila, Japan, Siberia
Capitanian	*Timorites uddeni* Miller and Furnish *Mexicoceras guadalupense* (Girty) *Waagenoceras richardsoni* Plummer and Scott	West Texas, Coahuila
Wordian	*Mexicoceras guadalupense* (Girty) *Waagenoceras girtyi* Miller and Furnish	West Texas, Coahuila, South China, Canada
Wordian	*Waagenoceras mojsisovicsi* Gemmellaro et affines	Sicily, West Texas, Timor, Coahuila
Roadian	*Glassoceras normani* (Miller and Furnish)	West Texas, Coahuila, Northern Rockies

This new name is thought to be largely synonymous with Godthaabian (Furnish, 1966) from East Greenland, but it is introduced because it now has the advantage of a diverse fauna in sequence. Seven species of the characteristic *Araxoceras* have been named; and related genera include *Prototoceras*, *Pseudotoceras*, *Vescotoceras*, *Urartoceras*, and *Rotaraxoceras*. These indices are poorly known elsewhere. However, strata in the type area also contain *Cyclolobus*, *Stacheoceras*, *Pseudogastrioceras*, and *Syrdenites*, all found commonly in other Upper Permian sections. In type Araksian there are nautiloids as well as fusulinids, tabulate and rugose corals, bryozoans, brachiopods, and ostracodes. As a stage, Chhidruan is an historic name based upon the Salt Range locality in Pakistan (Teichert, 1966); this element contains typical *Cyclolobus* and a variety of other ammonoids. Changhsingian, from a locality in Chekiang of South China, has not been formally designated previously and is relatively poorly known. According to Chao (1965) the Changhsing Limestone and equivalent Talung Formation constitute the upper part of the Loping Series and contain a varied ammonoid fauna: *Changhsingoceras, Pseudotirolites, Pleuronodoceras, Pseudostephanlites,* and *Tapashanites* (listed *nomina nuda,* though illustrated) together with *Pseudogastrioceras*.

Genus CYCLOLOBUS Waagen, 1879

[=*Krafftoceras* Diener, 1903; *Godthaabites* Frebold, 1932.]

Type species: Phylloceras Oldhami Waagen, 1872; original designation, p. 24.

Diagnosis: All known representatives of *Cyclolobus* are closely similar. The genus includes relatively large subdiscoidal cyclolobids, normally 100 to 200 mm in diameter at maturity, with pronounced biconvex flexures in the apertural margin. There are fine ribs on the shell surface of most representatives to a diameter in excess of 100 mm. Constrictions, likewise, are generally retained in moderately large conchs. Whorls are involute, with an umbilicus/diameter ratio ranging from 10 to 20 percent in large specimens. Sutures represent the ultimate complexity for Paleozoic cephalopods; externally, there are a dozen to 14 pairs of lateral lobes, subequal but diminishing in size toward the umbilicus. One distinguishing sutural feature is a tertiary element which divides the first lateral saddle. There is an arcuate alignment of the sutural trace more pronounced than in other members of the family.

Remarks: Definition of *Cyclolobus* involves a progression of evolutionary stages within the family in which *Timorites* Haniel, 1915 (type: *T. curvicostatus;* subs. desig. Diener, 1921) is the immediate predecessor. Separation of these two genera is arbitrary, for several levels of development are found in the Timor Upper Permian. Miller and Furnish (1940a) revised the concept of *Timorites* to give the genus a broader meaning than that of the original author, and they used the taxon as a primary index for the Capitanian Stage. However, a problem still exists in the typology of *T. curvicostatus,* for the lectotype of that species (Haniel, 1915, pl. 7, fig. 9a-c) portrays a simple cyclolobid suture at a diameter of only 13.5 mm, far too immature for critical features to have been developed. To a diameter of nearly 30 mm the whorls are evolute and strongly ribbed; larger specimens may be presumed to have existed, but none of larger size has been recognized in the Timor collections. Reasonable doubt concerning this species, and consequently the generic concept, will exist until larger specimens of *T. curvicostatus,* or strictly comparable species, have been studied.

In practice, the concept of *Timorites* is based largely upon *T. striatus* Haniel, a form which occurs in somewhat older strata and may differ significantly from the generotype. *T. striatus* exemplifies a group characteristic of Capitanian strata in West Texas, Coahuila, Siberia, and Japan. It is now known that mature conchs in this gens normally attained a diameter in excess of 200 mm (e.g., the type of *Waagenoceras intermedium* Wanner), and only the inner whorls resemble the types of *T. curvicostatus*.

At the type locality of *T. curvicostatus* near Soefa, Amarassi, another quite different cyclolobid occurs in association; this is *Cyclolobus persulcatus* Rothpletz (1892), a primitive representative of the more complex genus. The two species also occur in direct association at Ruasnain (previously known as Kuafeu, Kunafeu, or Koeafeoe) and other localities in Amarassi Province. *C. persulcatus,* together with normally associated ammonoids, has also been found farther east in the vicinity of Basleo at Nipol, Sumpek (Niipol Soempek) along Noe (Noil) Bunu. Collectively designated the "neodyadische Amarassifauna" by Haniel (1915), these beds contain, in addition to *T. curvicostatus* and *C. persulcatus,* the following: *Sundaites levis, Parapronorites* sp., *Episageceras noetlingi, Strigogoniatites angulatus, Hyattoceras subgeinitzi, Stacheoceras tridens, Syrdenites* n. sp., *Propinacoceras?* n. sp., and a variety of adrianitids. On the basis of several ammonoid relationships, mostly within the Las Delicias sequence, Coahuila, this fauna can be correlated with the upper Guadalupian Series; whereas *Timorites striatus* and affines appear to constitute an index for middle Guadalupian.

There had been some uncertainty regarding the exact nature of *Cyclolobus persulcatus,* established by Rothpletz upon a single specimen from western Timor; there is now a series of well-preserved growth stages (Pl. 4A-G) in the large collection available for study. Early whorls of 10 to 20 mm diameter are involute and subglobular; larger specimens are subdiscoidal. At about 125 mm diameter (Fig. 1C) the suture is nearly as complex as that in Salt Range *Cyclolobus oldhami,* but differs in certain details. For example, the tertiary element dividing the first lateral saddle is relatively small and lies distinctly in a flank position, rather than being of equal size with other subdivisions and central in position. Such a minor difference is locally consistent but gradational within known *Cyclolobus;* it is therefore regarded as taxonomically significant only at a specific level.

Godthaabites Frebold, 1932 (type: *G. kullingi;* orig. desig.) was based upon an immature cyclolobid from East Greenland. The species is still imperfectly known (Nassichuk, Furnish, and Glenister, 1966), but it definitely represents a species of *Cyclolobus* more primitive than typical forms of the genus. If a separate taxonomic name becomes useful for this stage in the lineage, Frebold's name can be revalidated. Somewhat similarly, Diener proposed *Krafftoceras* in 1903 [type: *Cyclolobus (Krafftoceras) haydeni* Diener; subs. desig., Diener, 1921] for material from the Kuling Shale in the Himalayas. This proposal was made conditionally on supposed differences in the sutures, but with the realization that Waagen's type for *Cyclolobus* from the Salt Range was poorly preserved and inaccurately figured. Reexamination of all types, as well as topotype material from both localities, substantiates the belief (Furnish, 1966) that *Krafftoceras* is a synonym. The several variants known to Diener occur in association. In 1903, it seems that Diener logically proposed his generic name with *Cyclolobus krafti* in mind. That species was the only one illustrated to portray critical features, and it has been assumed by several authors to be the type. Nevertheless, Diener's only true designation (1921) specified *C. (K.) haydeni.* From a practical viewpoint, there is no difference; *Krafftoceras* falls in synonymy.

Dimensions and proportions of species of *Cyclolobus,* recognized by us, are stated in Table 2.

Occurrence: Waagen originally reported *Cyclolobus* from the Chhidru Formation (=Upper Productus Limestone). The restricted Salt Range zonation of Noetling (1901) has been largely discredited, for this ammonoid has been secured from various levels of this general part of the section (Grant, 1968). Elsewhere, occurrences at a comparable stratigraphic level have been well documented, particularly in the Himalayas and in Madagascar.

Several authors, in addition to those at the turn of the century, have recognized *Cyclolobus* as a zonal index for the uppermost Permian. Böse (1919) was probably the first to define a sequence of Permian ammonoid "zones" including the genus. Haniel (1915) had already utilized these same index fossils for determining a faunal sequence in Timor, and he recognized their equivalencies elsewhere. Spath (1934) amplified the discussion of age relationships, based upon ammonoids. Wedekind (1935) then used *Perrinites, Waagenoceras,* and *Cyclolobus* to designate Permian "Stufen" in a sense conforming with modern concepts. Coincident with a restudy of *Timorites,* this genus was placed in the lineage and three cyclolobid "zones" were listed by Miller and Furnish (1940a). Such an alignment for Upper Permian ammonoids has been followed in much of the literature (Dunbar *et al.,* 1960). The stratigraphic terms Wordian, Capitanian, and Ochoan have been used as equivalent "stage" names; Dzhulfian is a substitute for Ochoan (e.g., Glenister and Furnish, 1961).

Although the fund of knowledge

TABLE 2. Dimensions (in mm) and Proportions of Species of *Cyclolobus* and *Timorites*.

	D	H/D	W/D	U/D
C. oldhami (Waagen)				
Hypotype (Tübingen), Wargal	140	.50	.32	.15
Hypotype (Brit. Mus. C10459), Wargal	124	.52	.31	.14
Holotype (G.S. India 3110), Jabbi	107	.50	.33	.19?
Hypotype (Tübingen), Wargal	103	.53	.33	.16
Hypotype (G.S. India 7420), Spiti	95	.54	.32?	.17
Type *insignis* (G.S. India 7419), Spiti	64	.55	.38	.17
C. walkeri Diener				
Holotype (G.S. India 7318), Tibet	85	.53	.39	.12
Type *kraffti* (G.S. India 7423), Spiti	61	.57	.43	.10
Type *haydeni* (G.S. India 7421), Spiti	56	.55	.39	....
Hypotype (Zurich), Kumaon	80	.55	.42	.09
Hypotype (Copenhagen), Spiti	44	.57	.45	.11
Hypotype (G.S. India 7422), Spiti	27	.56	.52	.17
Hypotype (Copenhagen), Spiti	19	.58	.58	.16
Hypotype (Sorbonne), Ambilobé	70	.57	.37?	.06
Type *astrei* (Sorbonne), Ambilobé	79	.52	.41	.16
Type *madagascariensis* (Sorbonne), Ambilobé	44	.52?	.48	.19
C. teicherti Furnish and Glenister, n. sp.				
Holotype (Univ. Kansas), Kathwai	125	.50	.40	.19
C. kullingi (Frebold)				
Hypotype (Copenhagen), East Greenland	100	....	....	.17
Hypotype (Copenhagen), East Greenland	50	....	.50?	....
Holotype (Copenhagen), East Greenland	19	.42	.63	.32
Hypotype (Moscow), Dzhulfa	55	.46	.55	.20
Hypotype (Geol. Survey Iran), Julfa	55	.48	.57	.22
C. persulcatus Rothpletz				
Hypotype (Iowa 12646), Basleo	192	.48	.36	.19
Hypotype (Amsterdam T766), Basleo	125	.52	.48	.17
Hypotype (Amsterdam T766) (inner volution)	78	.51	.57?	.17
Hypotype (Iowa 12344), Basleo	35	.49	.51	.16
Hypotype (Iowa 12338), Ruasnain	20	.50	.75	.20
Hypotype (Iowa 12338), Ruasnain	8	.38	.63	.31
Timorites striatus Haniel				
Topotype (Iowa 12647), Basleo	200	.43	.45	.28
Topotype (Iowa 33572), Basleo?	135	.48	.48	.27
Topotype (Iowa 31898), Basleo	33	.36	.58	.45
Lectotype (Delft 12754), Basleo	21	.33	.50	.55

D—conch diameter, H—height of whorl, W—conch width, U—width of umbilicus.

concerning Permian ammonoids has been greatly enlarged within the past century, and many refinements can be made in faunal relationships, the general concept of *Cyclolobus* as a post-Guadalupian Upper Permian index remains. There appears to be no conflict with evidence from the fusulinids (Ross, 1967). Some authorities (e.g., Grant, 1968) have suggested that brachiopod occurrences indicate a Guadalupian age for the *Cyclolobus*-beds.

Evidence from the ammonoids refutes such a correlation. However, one primitive representative of the genus, *C. persulcatus,* does occur in the upper portion of the Timor Permian, in association with *Timorites* and other elements of an assemblage characteristic of highest Guadalupian elsewhere. Such an occurrence can be regarded merely as a reflection of the arbitrary nature of systematics, while strengthening the idea of continuous lineages.

Cyclolobus oldhami (Waagen, 1872)

Plates 1*A*, 2*E-G*, 3*F*, 4*K*

Phylloceras Oldhami Waagen, 1872, p. 353, 354, pl. 1, fig. 1, 1a.
Arcestes Oldhami (Waagen), Mojsisovics, 1873, 1875, p. 72, 83.
Cyclolobus oldhami (Waagen), Waagen, 1879, p. 24-26, pl. 1, fig. 9a-c; Miller and Furnish, 1940b, p. 5, 6; Branson, 1948, p. 773; Schindewolf, 1954, p. 156; Miller and Furnish, 1957, p. 54, text-fig. 65; Jeannet, 1959, p. 9-12, pl. 1, fig. 1-5, text-fig. 1; Nassichuk, Furnish, and Glenister, 1966, p. 35, text-fig. 9; Furnish, 1966, p. 270-272, pl. 23, fig. 7, text-fig. 3b; Teichert, 1966, p. 18 (part); Grant, 1968, p. 5, 6.
Cyclolobus Oldhami (Waagen), Gemmellaro, 1888, p. 6, pl. A, fig. 8; Noetling, 1902, p. 646, fig.; Diener, 1921, p. 27; Haniel, 1915, p. 119, 120; Böse, 1919, p. 169.
Cyclolobus cf. *Oldhami* (Waagen), Diener, 1903, p. 162-164, pl. 6, fig. 6a-d.
Cyclolobus insignis Diener, 1903, p. 164, 165, pl. 6, fig. 5a-c.
Cyclolobus Oldhamianus Koken, 1907, p. 474.

Diagnosis: A species of *Cyclolobus* characterized by a combination of complex suture and conspicuous shell ornament. Constrictions are retained on the internal mold to diameters greater than 100 mm; fine ribs and prominent constrictions appear on the shell surface to at least that size. The twelfth lateral lobe is commonly situated on the umbilical shoulder at 125 mm conch diameter.

Description: Ten representatives of *Cyclolobus oldhami,* virtually all the specimens assembled in a century of collecting, are available for direct study. Seven are from the general type area, the Salt Range, whereas the remaining three are from the Himalayas. Actual preservation of the holotype (Pl. 2*F, G*) contrasts impressively with the highly idealized representations presented previously (Waagen, 1872, pl. 1, fig. 1,1a; 1879, pl. 1, fig. 9a-c). In fact, this specimen is so poorly preserved and extensively prepared and abraded that form of the surface sculpture on the internal mold is uncertain. In addition, although the general trace of the suture is available, the details of the taxonomically critical first lateral saddle remain obscure. However, one mature hypotype from the Salt Range (Pl. 1*A*) appears to be uncrushed and displays all necessary sutural details. An additional specimen from the same general area (Pl. 3*F*) is immature and lacks sutural information, but retains the shell and displays details of sculpture. Comparable information is available from a similar Himalayan specimen (Pl. 4*K*), the holotype of *C. insignis* Diener. Perhaps the most nearly complete representation of shell morphology is available from a Himalayan specimen described by Jeannet (1959). In practice, our understanding of *C. oldhami* rests largely on these hypotypes.

The largest known representative of *C. oldhami* (Pl. 1*A*) is fully septate to the ultimate end of the specimen where it has a diameter of 140 mm. Neither constrictions nor ribs are discernible on the internal mold at diameters larger than 100 mm. However, smaller specimens, including the holotype and the representative described by Jeannet (1959), display six highly arched constrictions per volution, each tracing a rounded hyponomic sinus and a shallow reentrant across the crest of the lateral salient. Constrictions are shallowly incised on the internal mold, but are deeply entrenched on the shell surface of smaller specimens (Pl. 3*F*). Numerous fine ribs appear on the shell surface, increasing in number by intercalation toward the venter, where 35 are present

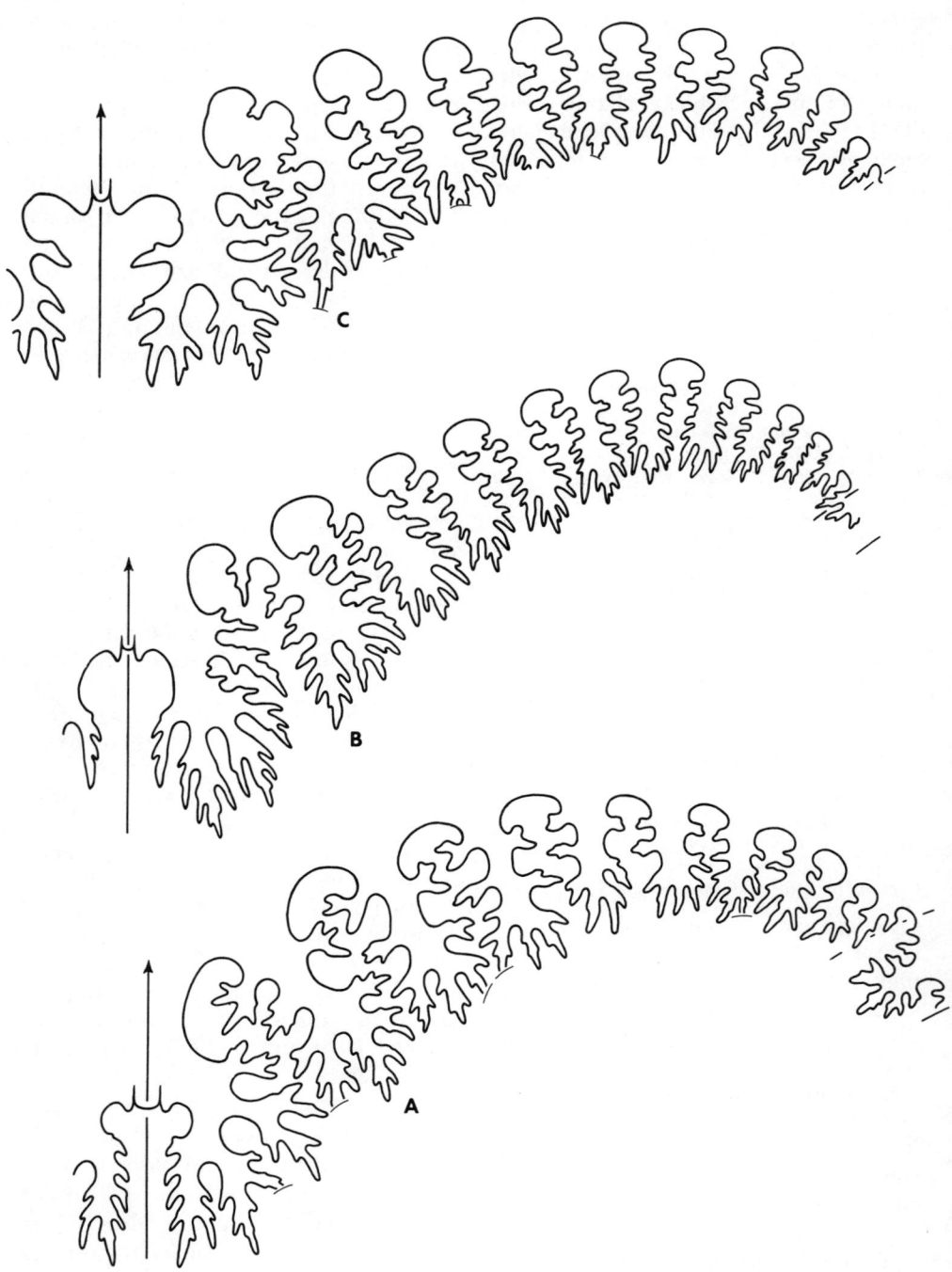

Fig. 1 (EXPLANATION ON FACING PAGE)

between adjacent constrictions in specimens of moderate size (50 mm).

Conch dimensions and proportions are given in Table 2. The convergent flanks are only slightly convex, whereas the ventrolateral shoulder and venter are narrowly and uniformly rounded. The rounded umbilical wall extends with uniform curvature onto the dorsolateral flank.

Details of the mature suture have been presented previously (Furnish, 1966, fig. 3b; Jeannet, 1959, fig. 1).

Comparisons: Conch form, sculpture, and sutural pattern permit differentiation of *Cyclolobus oldhami* from other species of the genus in all but juvenile growth stages. Dimensions and ratios of the conch may be compared by reference to Table 2. The smaller ratio of W/D in *C. oldhami* permits differentiation of the broader species (*C. persulcatus, C. kullingi,* and *C. teicherti*), but *C. walkeri* intergrades with the type species in this ratio. In comparison with other species, the flanks of large *C. oldhami* are less strongly curved, and they converge at a higher angle toward a proportionally narrower venter. Umbilical ratios (U/D) are difficult to determine in some specimens, but the umbilicus of *C. walkeri* is appreciably narrower than that of the type species.

Fine ribs probably characterize the shell surface of all species of *Cyclolobus,* with the exception of *C. walkeri,* to a conch diameter of at least 100 mm. Both *C. persulcatus* and *C. teicherti* differ from the type species in retaining broad conspicuous umbilical ribs on the internal mold to a diameter of at least 125 mm. Comparable umbilical sculpture is unknown in any growth stage of *C. oldhami* or *C. walkeri.*

Simple counts of sutural elements are probably not specifically definitive in *Cyclolobus.* However, specimens of *C. oldhami* appear to have the twelfth lateral lobe situated on the umbilical shoulder at approximately 100 mm conch diameter, whereas the eleventh lobe is so situated in *C. teicherti* and *C. walkeri,* and the tenth occurs on the umbilical shoulder of similar-sized *C. persulcatus.* Both *C. persulcatus* and *C. teicherti* differ from other species of the genus in the small size and asymmetric situation of the tertiary element which divides the first lateral saddle.

Occurrence: The holotype of *Cyclolobus oldhami* (Geol. Surv. India Mus. 3110, Pl. 2F, G, herein; Waagen, 1872, pl. 1, fig. 1, 1a) was collected about 1870 by W. Waagen at Jabbi at the top of a 20- to 25-foot alternation of limestone and marl. Waagen stated that, although the area was faulted,

FIG. 1. Sutures of *Timorites* and *Cyclolobus* from the Upper Permian, at a conch diameter of about 100 mm, ×1.5. A, *T. schucherti* Miller and Furnish based upon two topotypes (Univ. Iowa 32855, 33507) from black shales of the Capitanian Stage (Bed 5 of King, 1944) on the south side of Arroyo Difunta, southwestern Coahuila, Mexico; B, *C. teicherti* Furnish and Glenister, n. sp., based upon the holotype (Univ. Kansas Mus. Invertebrate Paleontology, 58929) from the Chhidru Formation near Kathwai, Salt Range, Pakistan; C, *C. persulcatus* Rothpletz based upon a hypotype (Geol. Inst. Univ. Amsterdam, T766) from the Amarassian in the vicinity of Basleo, western Timor, Indonesia. [The original figure of *T. schucherti* (Miller and Furnish, 1940a, fig. 26B) was based upon the lectotype (Yale Peabody Museum, 16694) at a diameter of 70 mm, and certain of the details had been destroyed by weathering. The present figure is traced from a fresh surface, but it became necessary to employ two specimens found at exactly the same location and stratigraphic level in order to show the entire external suture. In such ammonoids, there is normally some variation in the precise nature of tertiary and quarternary sutural elements between specimens and even on opposite sides of the same specimen. Also, sutural details were added throughout growth of the shells; *T. schucherti* attained a mature conch diameter of at least 200 mm. The suture of typical *Cyclolobus* is characterized by a distinct segmentation of the ventral prongs into only three complex elements and a nearly symmetrical subdivision of the first lateral saddle. The ancestral *Timorites* differs in both these respects and also possesses a large umbilical complex instead of small "auxiliary" lobes. *C. teicherti* and *C. persulcatus* may be regarded as "primitive" representatives of *Cyclolobus* in terms of these comparisons.]

origin of the specimen was unequivocally in the "upper division of the Productus-limestone," now known as the Chhidru Formation. Grant (1968, p. 5-6) was able to recognize the precise locality (U.S. National Mus. Loc. 9207 and 9222) from Waagen's description. However, he concluded that both the stratigraphic succession and brachiopod fauna of these "Cephalopoda beds" indicate assignment to the Kalabagh Member of the uppermost Wargal Limestone (Teichert, 1966) rather than the Chhidru. Other data would tend to substantiate this correlation, for *C. oldhami* has been recorded throughout the directly overlying Chhidru Formation. Also, Noetling (1901) and others have noted local abundance of ammonoids in the Kalabagh Member, a thin unit which has traditionally been placed with the underlying massive limestone of the Wargal instead of the Chhidru Formation above. All other *Cyclolobus* known from the Salt Range Permian have been identified from various levels within the Chhidru Formation.

Arcestes priscus Waagen (Holotype, Geol. Surv. India Mus. 3117) from Khúra, illustrated (1879, pl. 2, fig. 6c) to have an ammonitic suture, might be supposed to represent *Cyclolobus*; but this is not the case. A sutural trace was created erroneously by the delineator; the species should be placed in synonymy with *Stacheoceras antiquum* (Waagen). Apparently, no additional Salt Range *Cyclolobus* were recovered until about 1900 when E. Koken found two large specimens of *C. oldhami* (Brit. Mus. Nat. Hist. C10459 and Inst. Mus. Geologie-Paläontologie, Univ. Tübingen collections) from the vicinity of Wargal. A poorly preserved third specimen was secured by Koken and F. Noetling from the same locality. Considerably the best study specimen (Pl. 1*A*) was found in 1952 by O. H. Schindewolf during a study of the rocks at Wargal. Both the latter specimens are deposited at Tübingen. R. E. Grant and A. N. Fatmi, in 1965, secured a specimen with shell (Pl. 3*F*) from type Chhidru Formation (U.S. National Mus. Loc. 9239) about 36 feet below the top of the unit. Also, a specimen questionably referable to this species was collected due south of Sodhi Zairin (USNM Loc. 9237) from approximately 77 feet below the top of the Chhidru Formation.

A few specimens identifiable as *C. oldhami* have come from the Kuling Shale in the central Himalayas. One representative collected about 70 years ago by H. H. Hayden north of Kágá (Geol. Surv. India Mus. 7420, Pl. 2*E*; Diener, 1903, pl. 6, fig. 6a-d) has been referred to the species. The type of *Cyclolobus insignis* (Geol. Surv. India Mus. 7419, Pl. 4*K*; Diener, 1903, pl. 6, fig. 5a-c) was secured by A. von Krafft from Lilang, Spiti. In Kumaon, at Col Lebong, the 1936 Heim-Gansser Expedition found an exceptionally well-preserved specimen (Inst. Géol. École Polytechnique Féd. Zurich; Jeannet, 1959, pl. 2, fig. 1-5).

Cyclolobus teicherti Furnish and Glenister, n. sp.

Figure 1*B*, Plate 1*B-D*

Cyclolobus oldhami (Waagen), Teichert, 1966, p. 18 (part).

Diagnosis: A species of *Cyclolobus* characterized by retention in large specimens (conch diameter of 125 mm) of a broad whorl section (W/D 40 percent), large umbilicus (U/D 20 per-

---

PL. 1. *Cyclolobus* from the Upper Permian Chhidru Formation of the Salt Range, ×0.75. *A*, *C. oldhami* (Waagen), mature hypotype (Inst. Mus. Geologie-Paläontologie, Univ. Tübingen 1376/1), retaining portion of the body chamber; collected by O. H. Schindewolf from vicinity of Wargal. *B-C*, *C. teicherti* Furnish and Glenister, n. sp., holotype (Univ. Kansas Mus. Invertebrate Paleontology, 58929), entirely septate; constrictions and umbilical ribs preserved on internal mold, and a portion of shell displays subdued ribs across venter; collected by Curt Teichert from the upper Chhidru Formation, vicinity of Kathwai (Loc. 9).

Plate 1 (explanation on facing page)

cent), and prominent ribs and constrictions. A tertiary subdivision is situated asymmetrically on dorsal flank of the first lateral saddle.

Description: A single fully septate specimen of approximately 140 mm diameter represents this new species of *Cyclolobus*. Preservational differences in either side of the last septum suggest that the terminal end of the specimen is the base of the body chamber. The conch remains symmetrical, without trace of crushing or distortion. Most of the outer volution comprises internal mold. Weathering on one side is sufficiently slight that most sutural details are preserved. Portions of the shell are retained, especially across the venter.

Conch dimensions and proportions are presented in Table 2. Six strongly arched constrictions appear in the final volution (Pl. 1*B-D*). They are shallowly impressed in the internal mold and are reflected on the shell surface. Near the umbilical shoulder, the internal mold exhibits two broad ribs between adjacent constrictions, but these ribs disappear on the dorsolateral flank. Approximately 40 fine ribs appear on the shell surface between adjacent constrictions; the number increases from the dorsolateral shoulder by ventral intercalation of additional ribs. Both the ribs and constrictions outline a deep angular hyponomic sinus and a shallow reentrant across the axis of the lateral salient.

The mature suture is represented in Fig. 1*B*.

Comparisons: Closest overall affinities of *C. teicherti* are with the most primitive species of *Cyclolobus, C. persulcatus* Rothpletz (1892) from the Amarassi beds of Timor. Conch form in the two species is closely similar, particularly in respect to the relatively large mature umbilicus, and both retain comparable ribs and constrictions to diameters exceeding 100 mm. However, *C. teicherti* has a conspicuously more narrowly rounded venter than the Indonesian form. Differences in the ventrolateral suture (Fig. 1*B, C*) are judged to have greater taxonomic significance. Details of denticulation in the ventral prongs of *C. teicherti* differ markedly from those of *C. persulcatus,* and the tertiary element dividing the first lateral saddle is deeper and much more pronounced in the Salt Range species. All other species of *Cyclolobus* differ in the position of this tertiary element within the crest of the first lateral saddle, whereas in even the largest representatives of *Timorites* the element is lodged on the dorsal flank of the saddle.

Another species, *Cyclolobus kullingi* (Frebold, 1932), is closely related to *C. teicherti,* but definitive comparison is precluded by paucity of material in both species. *C. cullingi* was described originally from the Foldvik Creek Formation of central East Greenland, and rare new occurrences are now known from Iran and Soviet Armenia.

Fine ribs of the Greenland representatives of *Cyclolobus kullingi* are known to persist to a conch diameter of at least 50 mm, where they resemble those of *C. teicherti* at slightly greater size. However, larger specimens from Greenland seem to lack the prominent umbilical ribs which characterize the Salt Range species. Additionally, the umbilicus of *C. kullingi* is probably somewhat smaller than that of *C. teicherti* (Table 2). Reexamination of the larger crushed Greenland specimen of *C. kullingi* (Miller and Furnish, 1940, pl. 1, fig. 5) reveals that the umbilicus is distinctly larger (approximately 15 to 20 percent of diameter) than suggested by the existing illustration. The suture of this specimen also differs significantly from published interpretations (Miller and Furnish, 1940, text-fig. 3); some details are obscure, but the tertiary subdivision in the first lateral saddle is situated asymmetrically rather than at the crest. *C. teicherti* and *C. kullingi* are approximately the same in sutural complexity.

It is noteworthy that the intermediate-sized specimen illustrated from the Dzhulfian *Araxoceras* beds of Soviet Armenia as *Krafftoceras* n. sp. (Ruzhentsev in Ruzhentsev and Sarycheva, 1965, pl. 17, fig. 3a,b) is virtually indistinguishable from similar-sized *C. kullingi* now available to us (Nassichuk, Furnish, and Glenister, 1966). We are referring the Soviet specimen (Table 2) to *C. kullingi*. Within the past several years, personnel of the Geological Survey of Iran have conducted extensive investigations of the Permian in the Dzhulfa area but to the south of the Araks Gorge. One excellent specimen of *Cyclolobus* has been found by them in strata correlated with the *Araxoceras* beds to the north. This specimen is the same size as that studied by Ruzhentsev and agrees with the Soviet specimen in all particulars. Still another critical specimen of the same species was found, in association with the *Araxoceras* fauna, near Abadeh in central Iran (Taraz, 1969). This Abadeh specimen is partially silicified and only moderately well preserved; the camerate portion is about 55 mm in diameter.

Occurrence: The holotype of *Cyclolobus teicherti* (Univ. Kansas, KU 58929) was secured by Curt Teichert from an unnamed gorge, a tributary of Gabh Nala, one mile southwest of Kathwai (Kummel and Teichert, 1970, loc. 9). It was found loose on the slope, 60 feet below the top of the Chhidru Formation where that unit is 276 feet thick. Occurrence is consistent with possible derivation from somewhat higher in the section. The lithology, brown calcarenite with quartz detritus and shell debris, is comparable to that at several levels.

Cyclolobus walkeri (Diener, 1903)

Figure 2A-D, Plates 2A-D, 3A-E, 4H-J

Cyclolobus (Krafftoceras) Walkeri Diener, 1903, p. 12-14, 162, pl. 1, fig. 3a-d; Diener, 1927, p. 74, pl. 14, fig. 1, text-fig. 7e.
Cyclolobus (Krafftoceras) Kraffti Diener, 1903, p. 165-167, pl. 6, fig. 9a-c; Diener, 1921, p. 27; Besairie, 1930, p. 185-186, pl. 5, fig. 3, text-fig. 11.
Cyclolobus (Krafftoceras) Haydeni Diener, 1903, p. 167-169, pl. 6, fig. 7a, b, 8a, b; Diener, 1904, p. 56-58, text-fig. 5; Diener, 1921, p. 27.
Cyclolobus Walkeri Diener, Diener, 1921, p. 27; Treat, 1926, p. 1093; Treat, 1933, p. 25-29, pl. 1, fig. 1-7, pl. 2, fig. 1-6a, text-fig. 4-6; Jeannet, 1959, p. 12-14, pl. 2, fig. 1-5, text-fig. 2.
Cyclolobus Haydeni Diener, Besairie, 1930, p. 186, pl. 5, fig. 4.
Cyclolobus walkeri Diener, Besairie, 1936, p. 105, pl. 4, fig. 6-9; Branson, 1948, p. 773; Miller and Furnish, 1940b, p. 4, 5.
Cyclolobus walkeri madagascariensis Besairie, 1936, p. 105, pl. 4, fig. 5; Branson, 1948, p. 773.
Cyclolobus astrei Besairie, 1936, p. 106, pl. 4, fig. 4; Branson, 1948, p. 772.
Cyclolobus cf. *walkeri* Diener, Reed, 1944, p. 369, pl. 64, fig. 1, 1a.
Cyclolobus oldhami (Waagen), Schindewolf, 1954, p. 156, 158.
Cyclolobus kraffti (Diener), Nassichuk, Furnish, and Glenister, 1966, p. 35, text-fig. 11; Furnish, 1966, p. 284-285, pl. 23, fig. 1-4, pl. 24, fig. 1, 2, text-fig. 2.
Cyclolobus sp. Grant, 1968, p. 31.

Diagnosis: A species of *Cyclolobus* characterized by prominent constrictions at conch diameters less than 50 mm and by fine growth lines in all growth stages. Suture is relatively simple; generally, the ninth lateral lobe is situated on the umbilical shoulder in specimens of 50 mm diameter, and the eleventh lobe is normally in that position at a fully mature phragmocone diameter of 100 mm.

Description: Specimens of *Cyclolobus walkeri* are known from three general areas, and virtually all materials assembled in three-quarters of a century of collecting are available for study. The holotype is one of approximately 35 representatives of *C. walkeri* known from the Himalayas. Its preservation is indifferent, but details of the suture are portrayed satisfactorily. Additional conspecific representatives from the same general area comprise the types of *C. (Krafftoceras) kraffti*, *C. (K.) haydeni* (Diener, 1903), and the relatively well-preserved specimens described by

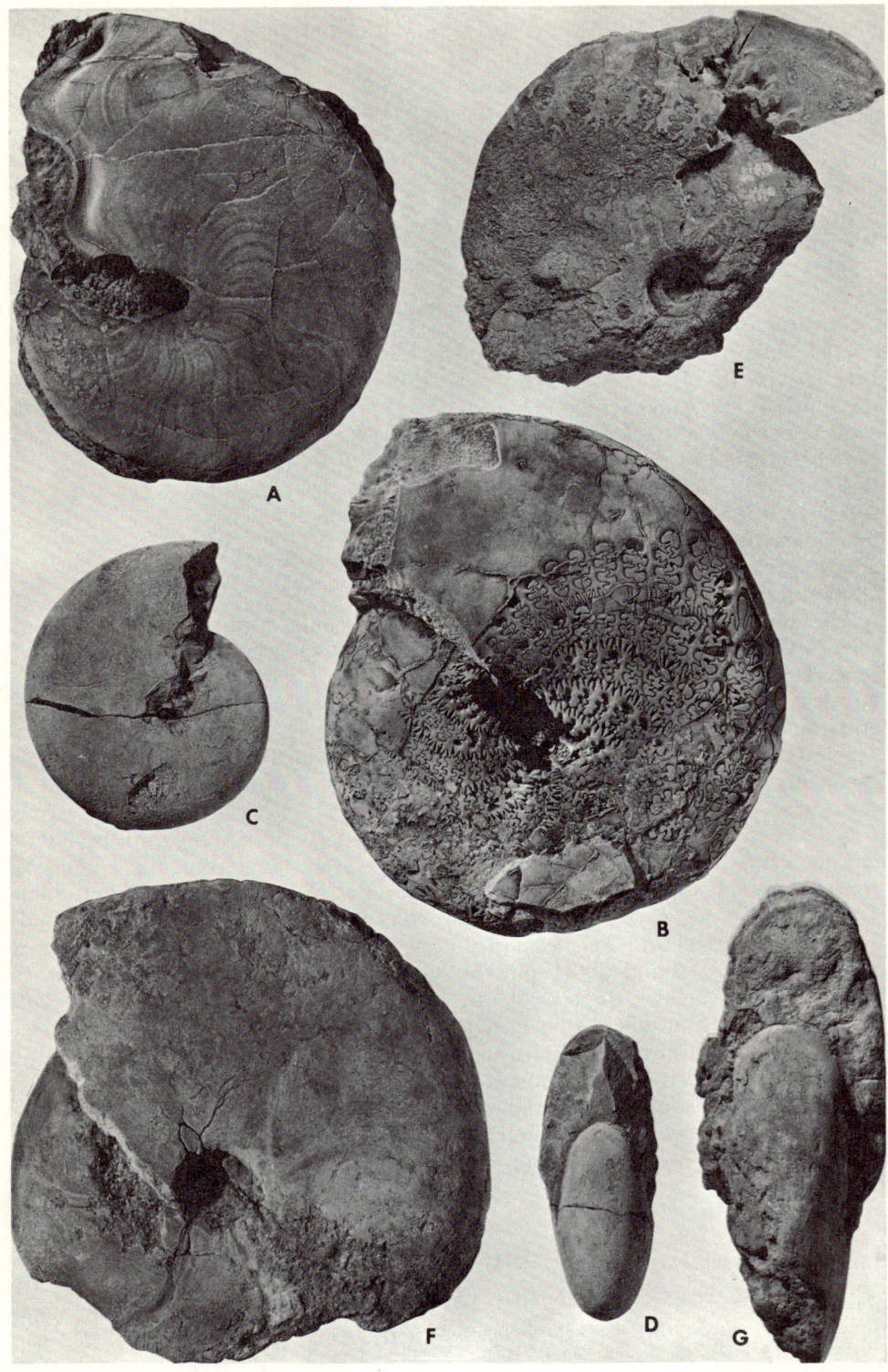

PLATE 2 (EXPLANATION ON FACING PAGE)

Jeannet (1959), Nassichuk, Furnish, and Glenister (1966), and Furnish (1966). Six specimens probably referable to the species are included in recent collections from the Salt Range; all are imperfect, but one retains exceptionally well-preserved sutures. Finally, in excess of 150 representatives from northern Madagascar have been examined. Although commonly crushed, these latter materials display all details of ontogeny, including modifications at full maturity.

The Himalayan holotype of *Cyclolobus walkeri* (Pl. 3E) is an abraded phragmocone of 85 mm diameter. No trace of constrictions is discernible on the well-preserved portion of the internal mold, although surface weathering is so extensive in most areas that any original ornament would have been destroyed. Conch proportions and dimensions are presented in Table 2. Sutural details are available near the base of the ultimate volution, and are represented in Fig. 2B. Jeannet (1959) gave detailed information on one additional Himalayan fragment of *C. walkeri*. There is no evidence that this specimen differs appreciably from the holotype in either conch form, ornamentation, or sutural pattern. The supposed angularity of the venter in the body chamber of Jeannet's specimen is regarded as a function of lateral compression; similar distortion occurs in many conspecific forms from Madagascar.

Three Himalayan representatives of *C. walkeri* were described by Diener (1903) as the types of *C. (Krafftoceras) krafti*. Diener's holotype (Pl. 3A-C) is a well-preserved internal mold of 61 mm diameter. Constrictions are absent from the ultimate volution, and preserved fragments of the shell display only fine growth lines. Conch dimensions and proportions are given in Table 2, and details of the suture are presented in Fig. 2C.

The seven Himalayan specimens for which Diener (1903) erected *C. (K.) haydeni* range in diameter from 26-57 mm. Conch form (Table 2) is closely similar to that of other representatives of *C. walkeri*, as are the fine shell ornament and absence of constrictions in larger specimens (Pl. 2C, D). Conspicuous constrictions are present on the shell surface to a diameter of at least 26 mm. Preserved portions of the suture are undiagnostic. An additional Himalayan hypotype of *C. haydeni* (Diener, 1904; Pl. 3G, H, herein) achieves a diameter of only 24 mm. It agrees in all morphologic detail with the primary types of that species, and should probably be referred to *C. walkeri*.

Twenty exceptionally well-preserved specimens of *Cyclolobus walkeri* are known from the central Himalayas (Min. Geol. Mus. Univ. Copenhagen) and seven of them have been illustrated previously as *C. krafti* (Furnish, 1966). Sutural pattern, conch form, and shell features are displayed to a maximum diameter of 100 mm. Table 2 lists relevant conch dimensions and proportions, and sutural ontogeny was presented by Nassichuk, Furnish, and Glenister (1966, fig. 11). Conspicuous constrictions appear on both the internal mold and shell surface up to the diameter of

---

PL. 2. Representative Upper Permian *Cyclolobus* from Madagascar, the Himalayas, and the Salt Range, ×0.75. *A, B, C. walkeri* Diener. *A*, hypotype, Ambilobé beds, Ankitohazo, northern Madagascar (Waterlot Collection, Mus. national Hist. natur. B-7520; Treat, 1933, pl. 1, fig. 4), retaining fully mature apertural margin and modifications. *B*, specimen from same locality (Besairie Collection, Brit. Mus. Nat. Hist. C-36641), showing significantly larger phragmocone. *C, D, C. haydeni* Diener, holotype, type Kuling Shale, Spiti area (Geol. Surv. India Mus. 7421; Diener, 1903, pl. 6, fig. 7a-c), collected by A. von Krafft. *E-G, C. oldhami* (Waagen). *E*, (Geol. Surv. India Mus. 7420; Diener 1903, pl. 6, fig. 6a, b), Kuling Shale, north-northwest of Kala, Spiti area, collected by H. H. Hayden. *F, G*, holotype (Geol. Surv. India Mus. 3110; Waagen, 1879, pl. 1, fig. 9a-c), presumably from the Kalabagh Member of Wargal Formation at Jabbi, collected by Waagen.

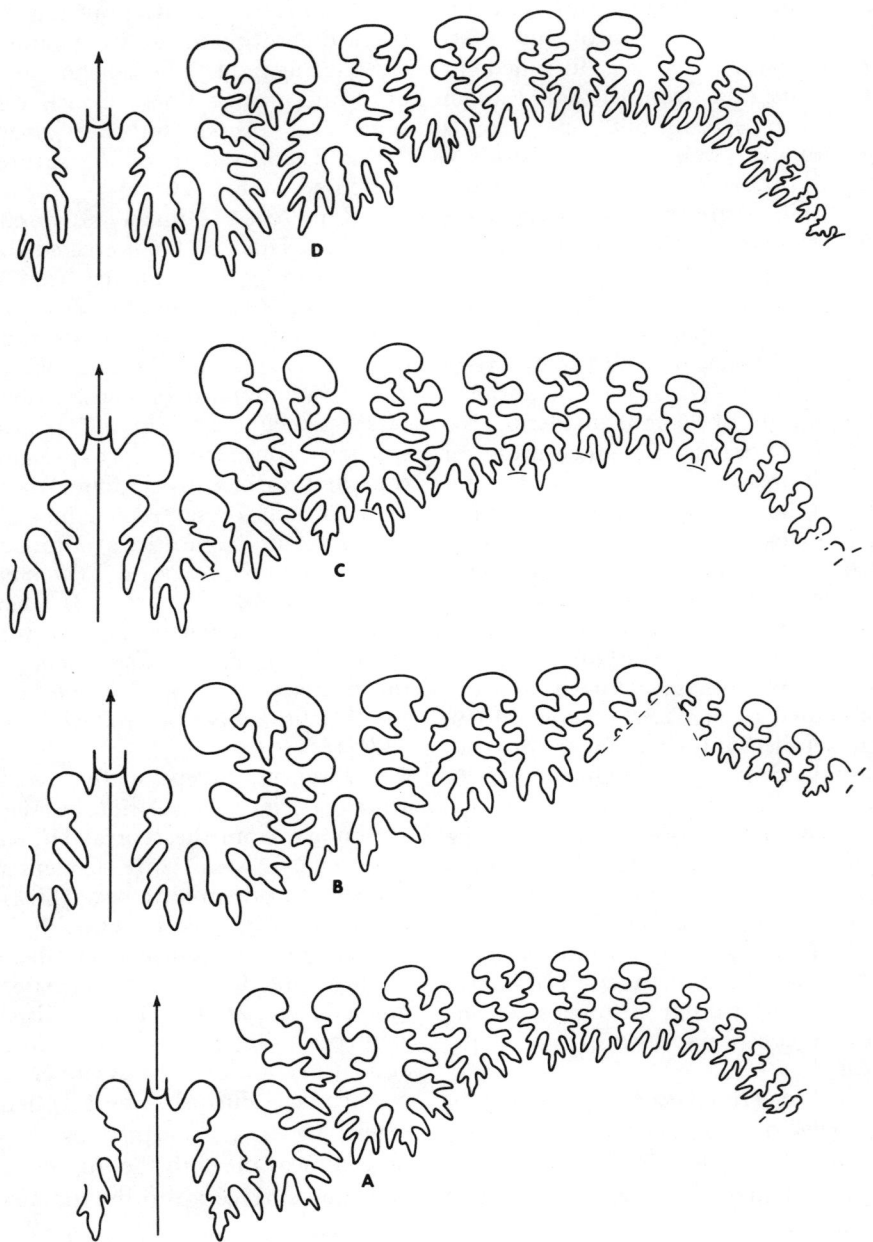

Fig. 2. Representative sutures of *Cyclolobus walkeri* Diener from the Late Permian; all ×3, at conch diameter of approximately 50 mm. [The exact lobe count is not regarded as specifically definitive in *Cyclolobus*. For example, at approximately the same diameter the holotype of *C. krafti (C)* possesses two more lateral lobes than the other figured representatives of *C. walkeri (A, B, D)*.] *A*, hypotype (Inst. Mus. Geologie-Paläontologie, Univ. Tübingen) from the type locality of the Chhidru Formation, Salt Range; *B*, holotype (Geol. Surv. India Mus. 7318), Chitichun-I, Tibet; *C*, holotype of *C. krafti* (Geol. Surv. India Mus. 7423) from type Kuling Shales at Lilang, Spiti; *D*, hypotype (Treat Coll., Mus. national d'Hist. nat., Paris) from Ambilobé beds, northern Madagascar.

PL. 3. Representative Upper Permian *Cyclolobus* from the Himalayan area and the Salt Range, ×0.75, except G, H, which are ×1.5. A-E, C. walkeri Diener. A-C, holotype of *C. krafftii* Diener (Geol. Surv. India Mus. 7423; Diener, 1903, pl. 6, fig. 9a-c), Kuling Shale, Lilang, Spiti area, collected by A. von Krafft; D, a specimen (Inst. Mus. Geologie-Paläontologie, Univ. Tübingen 1376/2) from the Chhidru Formation, about 50 feet below top, Chhidru Nala, collected by A. Seilacher; E, holotype (Geol. Surv. India Mus. 7318; Diener, 1903, pl. 1, fig. 3a-c), limestone at Chitichun-I, Hundés region, Tibet, collected by T. L. Walker. F, *C. oldhami* (Waagen), specimen (U.S. National Mus. Loc. 9239), upper Chhidru Formation, Chhidru Nala, collected by R. E. Grant and A. N. Fatmi. G, H, ?C. walkeri Diener, immature hypotype of *C. haydeni* Diener (Geol. Surv. India Mus. 7873; Diener, 1904, fig. 5), Kuling Shale, south of Pomerang, Spiti area, collected by H. H. Hayden.

35 mm, and fainter varices occur in slightly larger representatives (Furnish, 1966, pl. 23, fig. 1-4, pl. 24, fig. 1, 2). Fine growth lines characterize the shell surface.

Four of six specimens known from the Salt Range are in the collections of the U.S. National Museum. They are crushed internal molds, ranging in diameter from 65 to 105 mm, which display the general form of the whorl section and suture. Specific reference is uncertain, but referral to *C. walkeri* is suggested by the relatively small number of sutural elements, the tenth lateral lobe being situated on the umbilical shoulder at a conch diameter of 100 mm. One additional specimen (Pl. 3D; Inst. Mus. Geologie Paläontologie, Univ. Tübingen) is a phragmocone of 65 mm diameter which preserves the details of the suture (Fig. 2A) and the internal mold. No constriction is discernible on the internal mold at a diameter as small as 50 mm. At that diameter, the suture is strikingly similar to those of the holotype and specimens from Madagascar, the ninth lateral lobe being lodged on the umbilical shoulder. The remaining specimen (Reed, 1944) is so poorly preserved that positive specific identification is precluded.

The numerous specimens from Madagascar (Fig. 2D; Pl. 2A, B; 4H-J) exhibit a full ontogenetic range of all morphological details, with the exception of whorl section. Many specimens display clear evidence of lateral crushing, so that width measurements are suspect in all cases. Conch measurements and proportions are presented in Table 2. Maximum width occurs on the dorsolateral flanks, with gently curved convergence toward the broadly rounded venter. The umbilical wall is inclined, with an indistinct shoulder.

Constrictions are deeply impressed on the internal mold of the Madagascar specimens to a conch diameter of about 25 mm, but become progressively shallower at larger sizes. Even the best-preserved specimens generally lack constrictions at diameters greater than 50 mm, although one representative of *C. walkeri* (the lectotype of *C. astrei* Besairie, designated herein; Besairie, 1936, pl. 4, fig. 4) retains constrictions on the internal mold to a diameter of 75 mm. The number of constrictions per volution increases progressively from 4 at 10 mm diameter to 5 in the largest varicose whorls. Forward arching of the constrictions becomes stronger with increase in diameter; in the larger whorls, constrictions trace a deep, rounded, hyponomic sinus and a shallow reentrant across the crest of the high lateral salient. Preserved fragments of the shell exhibit fine growth lines but lack prominent ribs.

Some ten specimens from Madagascar are known to preserve the fully mature peristome. Ultimate size is variable: ranging from a minimum of 90 mm, averaging approximately 100 mm, and reaching an estimated maximum of 150 mm. The mature body chamber is slightly less than one volution in length (Pl. 2A). Weakly geniculate coiling results in slight temporary reduction in the umbilical diameter.

Pl. 4. Representative Upper Permian *Cyclolobus* from Timor, Madagascar, and the Himalayas, *A, B*, ×1.5; *C, D*, ×1; *E-K*, ×0.75. *A-G, C. persulcatus* Rothpletz from the Amarassi beds of Timor. *A, B*, two views of small topotype with shell (Inst. Paläontologie, Univ. Bonn; Haniel Coll. 33a, Haniel, 1915, pl. 53, fig. 6a, b) from Soefa; *C, D*, moderately sized internal mold (Univ. Iowa 12344), collected by Soejono Martodjojo from the vicinity of Basleo; *E*, crushed specimen retaining the shell (Univ. Iowa 12339), collected by Soejono Martodjojo from Ruasnain (Kuafeu); *F, G*, large septate specimen (Geol. Inst. Univ. Amsterdam T766) from the Basleo area, inner volution retaining the ornate shell as well as traces of the internal suture from succeeding whorl. *H-J, C. walkeri* Diener, three views of a specimen (Besairie Coll., Univ. Paris, Sorbonne) from the Ambilobé beds, Ankitohazo, northern Madagascar. *K, C. oldhami* (Waagen), the holotype of *C. insignis* Diener (Geol. Surv. India Mus. 7419; Diener, 1903, pl. 6, fig. 5a-c), Kuling Shale at Lilang, Spiti area, collected by A. von Krafft.

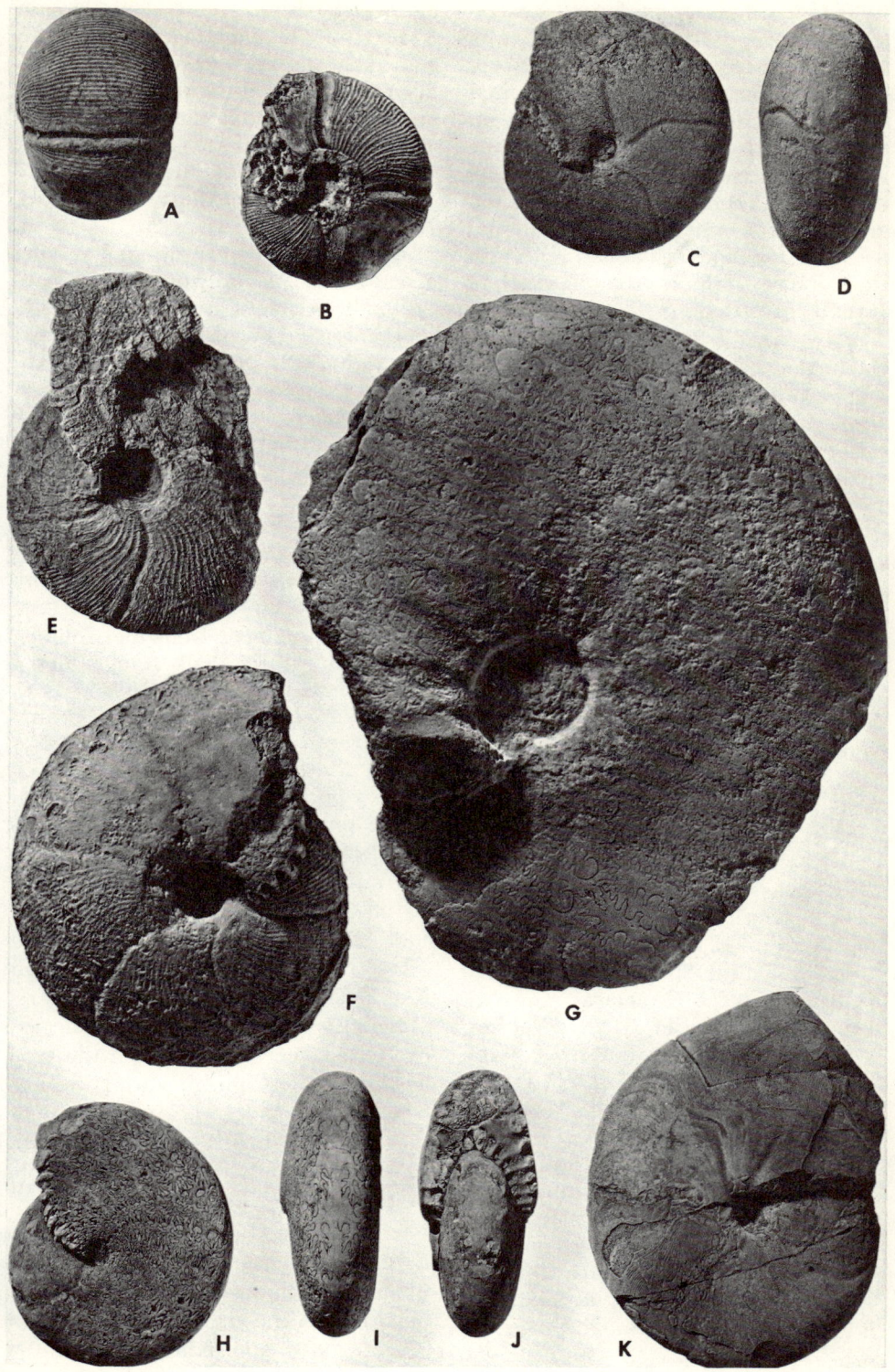

PLATE 4 (EXPLANATION ON FACING PAGE)

Indistinct sinuous undulations characterize the internal mold, and the fully mature peristome is marked by a conspicuous constriction and terminal flare. A deep, elongate pit occurs in the terminal constriction across the ventrolateral shoulder (Pl. 2A). Both the terminal constrictions and the undulations encountered on the internal mold of the body chamber reflect in exaggerated fashion the outline of juvenile constrictions.

The suture of an intermediate-sized Madagascar specimen is represented in Fig. 1D. At full maturity, the eleventh lateral lobe lies on the umbilical shoulder, and two additional elements occur on the umbilical wall.

Comparisons: Detailed comparisons of *Cyclolobus walkeri* are given under the appropriate heading of *C. oldhami*. All other species of *Cyclolobus* differ from *C. walkeri* in possession of pronounced ribs up to a diameter of at least 50 mm.

Occurrence: The holotype of *Cyclolobus walkeri* (Geol. Surv. India Mus. 7318, Fig. 2B, Pl. 3E; Diener, 1903, pl. 1, fig. 3a-d) was collected by T. L. Walker from Chitichun No. 1, Hundés region, Himalayas. The hypotype collected by the A. Heim and A. Gansser 1936 Swiss Expedition (Inst. Géol. École Polytechnique Féd. Zurich; Jeannet, 1959, pl. 2, fig. 1-5) was secured from the Kuling Shale at Col Lebong, Kumaon, central Himalayas. Two of the original types of *C. (Krafftoceras) krafftí*, including the holotype (Geol. Surv. India Mus. 7423, Fig. 2C, Pl. 3A-C; Diener, 1903, pl. 6, fig. 9a-c) were secured by A. von Krafft from the Kuling Shale at Lilang, Spiti area, Himalayas; Diener's remaining type was collected by H. H. Hayden from the nearby Gyundi River. All primary types of *C. (K.) haydeni* are from the Kuling Shale of the Spiti area; the holotype (Geol. Surv. India Mus. 7421, Pl. 2C, D; Diener, 1903, pl. 6, fig. 7a, b) and figured paratype (Geol. Surv. India Mus. 7422; Diener, 1903, pl. 6, fig. 8a, b) were secured by A. von Krafft from Kuling, together with an additional specimen from Lilang; H. H. Hayden assembled the remaining types, one from Rátang River and the three additional primary types from the vicinity of Po. The hypotype of *C. (K.) haydeni* (Geol. Surv. India Mus. 7873, Pl. 3G, H; Diener, 1904, fig. 5) is from near Pomarang, in the same general area and horizon. Twenty additional specimens, referred originally to *C. krafftí* (Furnish, 1966, fig. 2a-f; pl. 23, fig. 1-4; pl. 24, fig. 1, 2), were secured from the Kuling Shale of Muth and Lilang by the Eigil Nielsen party in 1950 (Mineral. Geol. Mus., Univ. Copenhagen). The holotype of *C. walkeri*, from Tibet, is preserved in white fine-grained limestone, whereas the remaining forms from the Kuling Shale were secured from dark calcareous nodules.

Four specimens from the Chhidru Formation of the Salt Range, designated herein as *Cyclolobus* cf. *C. walkeri*, are from the 1965 R. E. Grant–A. N. Fatmi collection of the U.S. National Museum (loc. 9236, approximately 50 feet below top of formation where unit is 191 feet thick, 3 specimens; loc. 9237, approximately 77 feet below top of formation, 1 specimen) about 0.8 miles south-southeast of Sodhi Zairin. A comparable specimen (Reed, 1944, pl. 64, fig. 1, 1a) is from Amb. The remaining Salt Range representative of *C. walkeri* was collected by A. Seilacher in 1952 from the type locality of the Chhidru Formation, approximately 50 feet below the top of that unit (Schicht 7 of Schindewolf, 1954, p. 156; unit 10 of Teichert, 1966, p. 15; Tübingen 1376/2).

More than 150 specimens of *Cyclolobus walkeri* have been collected from nodules which weather from argillaceous strata in the vicinity of Ankitohazo (Ankitokazo) approximately 25 miles south of Ambilobé, northern Madagascar. Major collections have been assembled by G. Waterlot, Henri Besairie, Maurice Collignon, F. G. Stehli, and Bernhard Kummel; they are reposited in Brit. Mus. Nat. Hist., Univ. Paris Sorbonne, Mus. national Hist. nat., Acad. Moirans, Case-Western Reserve Univ., Harvard Univ., and Univ. Iowa. Included are the lectotypes of *Cyclolobus walkeri madagascariensis* Besairie (1936, p. 105, pl. 4, fig. 5) and *C. astrei* Besairie (1936, p. 106, pl. 4, fig. 4). It is noteworthy that Besairie did not designate the figure representing his new "variety"; however, his original label on this specimen specifies the type in question.

CONCLUSIONS

Phylogenetic analysis of Salt Range representatives of *Cyclolobus* confirms age assignments derived from study of associated ammonoid groups, particularly *Stacheoceras* and the more abundant family Xenodiscidae. It is concluded that the Chhidru Formation and the subjacent Kalabagh Member of the Wargal Limestone are younger than both the fossiliferous type Guadalupian and the uppermost Guadalupian Amarassian beds of Timor and Mexico. Similarly, the Salt Range occurrences are older than the late Dzhulfian Changhsingian Stage of southern China. Assignment of the Salt Range occurrences of *Cyclolobus* should be to the middle Dzhulfian Chhidruan Stage.

ACKNOWLEDGMENTS

It is a pleasure to acknowledge the many individuals and institutions who have contributed study material or information bearing on the present research. Curt Teichert (University of Kansas), R. E. Grant (U.S. Geological Survey), and A. N. Fatmi (Geological Survey of Pakistan) provided specimens from the Salt Range and relevant stratigraphic data. O. H. Schindewolf and A. Seilacher (Universität Tübingen) secured two critical specimens from the Salt Range, and these and older Tübingen collections were lent for study through the courtesy of Schindewolf, Jürgen Kullmann, and Frank Westphal. Eigil Nielsen (University of Copenhagen) provided critical reference specimens from the Himalayas. The single cyclolobid from Soviet Armenia was studied in Moscow through the courtesy of V. E. Ruzhentsev (Akademiya Nauk SSSR). Study of large collections from Madagascar was facilitated by Maurice Collignon (Académie Moirans), Éliane Basse de Ménorval (Centre national de la Recherche scientifique), J. P. Lehman (Muséum national d'Histoire naturelle), M. K. Howarth and Dennis Phillips (British Museum of Natural History), and F. G. Stehli (Case-Western Reserve University). Types and other cyclolobids from Timor were made available by H. A. Brouwer, H. J. MacGillavry, and M. v. d. Boogaard (Universiteit van Amsterdam), J. Dufour, J. L. H. Bemelmans, and Eva Buschmann (Technische Hogeschool Delft), and H. K. Erben and K. J. Müller (Rhein. Friedrich-Wilhelm-Universität, Bonn). Soejono S. Martodjojo (Institut Teknologi Bandung) served as research associate in the project and assisted substantially by collecting reference faunas from Indonesian Timor. M. V. A. Sastry, S. C. Shah and P. R. Chandra (Geological Survey of India, Calcutta) enabled us to study critical type collections from the Salt Range and the Himalayas. Reference specimens from Greenland were provided by Svend Bendix-Almgreen (Grønlands Geologiske Undersøgelse). Iranian ammonoids were studied through the courtesy of J.

Stöcklin and Hushang Taraz (United Nations Geological Survey Institute and Geological Survey of Iran).

R. J. Beinert, J. M. Cocke, and W. B. Saunders (University of Iowa) aided in the preparation of specimens and illustrations. Financial support was provided by the National Science Foundation (GB-5530) and the Graduate College, University of Iowa.

REFERENCES

Besairie, Henri, 1930, Recherches géologiques à l'étude des ressources minérales, 5e Partie, Notes paléontologiques sur les principaux fossiles caractéristiques des terrains étudiés; fossiles permiens: Soc. Hist. Nat. Toulouse Bull., v. 60, p. 185-186, pl. 5, fig. 11.

———, 1936, Recherches géologiques à Madagascar, lre Suite: La Géologie du Nord-Ouest; Chapitre 3, Les fossiles: Acad. Malgache Mém., fasc. 21, p. 105-207, pl. 4-24, fig. 6-15.

Böse, Emil, 1919, The Permo-Carboniferous ammonoids of the Glass Mountains, west Texas, and their stratigraphical significance: Texas Univ. Bull. 1762 (1917), 241 p., 11 pl.

Branson, C. C., 1948, Bibliographic index of Permian invertebrates: Geol. Soc. America, Mem. 26, 1049 p.

Chao, K.-K., 1965, The Permian ammonoid-bearing formations of South China: Scientia Sinica, v. 14, p. 1813-1826, 4 fig., 2 pl.

Diener, Carl, 1903, Permian fossils of the central Himalayas: Geol. Survey India Mem., Palaeontologia Indica, ser. 15, v. 1, pt. 5, 204 p., 10 pl.

———, 1904, Note on *Cyclolobus Haydeni*, Diener: Geol. Survey India, Rec., v. 31, p. 56-58, fig. 5.

———, 1921, Ammonoidea permiana: Fossilium Catalogus, I, Animalia, pars 14, p. 1-36.

———, 1927, in G. Gürich, Leitfossilien des marinen Perm: Leitfossilien, Lief. 5, Borntraeger, Berlin, 84 p., 14 pl., 10 fig.

Dunbar, C. O., et al., 1960, Correlation of the Permian formations of North America: Geol. Soc. America Bull., v. 71, p. 1763-1806.

Frebold, Hans, 1932, Marines Unterperm in Ostgrönland und die Frage der Grenzziehung zwischen dem pelagischen Oberkarbon und Unterperm: Medd. Grønland, v. 84, no. 4, p. 1-35, pl. 1, 4 fig.

Furnish, W. M., 1966, Ammonoids of the Upper Permian *Cyclolobus*-zone: Neues Jahrb. Geologie Palaeontlogie, Abh., v. 125, p. 265-296, pl. 23-26, 4 fig., 2 tables.

———, and Glenister, B. F., 1968, The Guadalupian Series: Geol. Soc. America, Program 1968 Ann. Meet., p. 105-106.

Gemmellaro, G. G., 1887, La Fauna dei Calcari con Fusulina della Valle del Fiume Sosio nella Provincia di Palermo: Gior. Sci. Nat. ed Econ., v. 19, p. 1-106, pl. 1-10. App., 1888; ibid., v. 20, p. 9-36, pl. A-D.

Glenister, B. F., and Furnish, W. M., 1961, The Permian ammonoids of Australia: Jour. Paleontology, v. 35, p. 673-736, pl. 78-86, 17 fig., 3 tables.

Grant, R. E., 1968, Structural adaptation in two Permian brachiopod genera, Salt Range, West Pakistan: Same, v. 42, p. 1-32, pl. 1-9, 21 text-fig.

Haniel, C. A., 1915, Die Cephalopoden der Dyas von Timor: Pälaontologie von Timor, Lief. 3, Abh. 6, p. 1-153, pl. 46-56, 38 fig.

Jeannet, A., 1959, Ammonites permiennes et faunes triassiques de l'Himalaya Central: Geol. Survey India Mem., Palaeontologia Indica, new ser., v. 34, mem. 1, 189 p., 21 pl., 173 fig.

King, R. E., et al., 1944, Geology and paleontology of the Permian area north of Las Delicias, southwestern Coahuila, Mexico: Geol. Soc. America Spec. Paper 52, 172 p., 45 pl., 29 fig.

Koken, E., 1907, Indisches Perm und die permische Eiszeit: Neues Jahrb. Mineralogie, Geologie Palaeontologie, Festband, p. 446-546, pl. 19.

Kummel, Bernhard, and Teichert, Curt, 1966, Relations between the Permian and Triassic formations in the Salt Range and Trans-Indus ranges, West Pakistan: Neues Jahrb. Geologie Palaeontologie, Abh., v. 125, p. 297-333, pl. 27, 28, 4 fig., 2 tables.

——— and ———, 1970, Stratigraphy and paleontology of the Permian-Triassic boundary beds, Salt Range and Trans-Indus ranges, West Pakistan: in Kummel, B., and Teichert, C., eds., Stratigraphic Boundary Problems: Permian and Triassic of West Pakistan, Univ. Press Kansas, Dept. Geology, Univ. Kansas Spec. Publ. 4, p. 1-110.

Miller, A. K., and Furnish, W. M., 1940a, Permian ammonoids of the Guadalupe Mountain region and adjacent areas: Geol. Soc. America Spec. Paper 26, 242 p., 44 pl., 59 fig., 6 tables.

——— and ———, 1940b, *Cyclolobus* from

the Permian of eastern Greenland: Medd. Grønland, v. 112, p. 1-10, pl. 1, 3 fig.

─── and ───, 1957, Paleozoic Ammonoidea, *in* Part L, Mollusca 4, *of* Moore, R. C., ed., Treatise on Invertebrate Paleontology, p. 11-36, 47-79, fig. 1-37, 46-123.

Mojsisovics, E. von, 1873, 1875, Das Gebirge um Hallstatt; Thiel 1, Die Mollusken-Faunen der Zlambach-und Hallstätter-Schichten, Heft 1: K. K. Geol. Reichsanst., Abh., Bd. 6, p. i-vii, 1-82, pl. 1-32, 1873; Heft 2: Same, p. 83-174, pl. 33-70, 1875.

Nassichuk, W. W., Furnish, W. M., and Glenister, B. F., 1966, The Permian ammonoids of arctic Canada: Geol. Survey Canada, Bull. 131 (1965), 56 p., 5 pl., 17 fig.

Noetling, Fritz, 1901, Beiträge zur Geologie der Salt Range, insbesondere der permischen und triassischen Ablagerungen: Neues Jahrb. Mineralogie, Geologie, Palaeontologie, v. 14, p. 369-471, 4 fig.

───, 1902, Die Dyas in Indien, *in* Frech, F., Lethaea geognostica I. Palaeozoicum, p. 639-658.

Reed, F. R. C., 1944, Brachiopoda and Mollusca from the Productus Limestone of the Salt Range: Geol. Survey India Mem., Palaeontologia Indica, n. ser., v. 23, mem. 2, p. 353-379, pl. 60-65.

Ross, C. A., 1967, Development of fusulinid (Foraminiferida) faunal realms: Jour. Paleontology, v. 41, p. 1341-1354, 9 fig.

Rothpletz, A., 1892, Die Perm-, Trias-, und Jura- Formation auf Timor und Rotti im indischen Archipel.: Palaeontographica, v. 39, p. 57-106, pl. 9-14.

Ruzhentsev, V. E., and Sarycheva, T. G., eds., 1965, Razvitie i smena morskikh organismov na rubezhe Paleozoya i Mezozoya: Akad. Nauk. SSSR, Paleont. Inst., Trudy, v. 108, 431 p., 58 pl., 59 fig., 16 tables.

Schindewolf, O. H., 1954, Über die Faunenwende vom Paläozoikum zum Mesozoikum: Zeitschr. Deutsch. Geol. Gesellsch., Bd. 105, Teil 2, p. 153-183, pl. 5, 6, 5 fig., table 1-3.

Spath, L. F., 1934, Catalogue of the fossil Cephalopoda in the British Museum (Natural History), Part 4, The Ammonoidea of the Trias: 521 p., 18 pl., 160 fig.

Taraz, Hushang, 1969, Permo-Triassic section in Central Iran: Amer. Assoc. Petroleum Geologists Bull., v. 53, p. 688-693, 2 fig.

Teichert, Curt, 1966, Stratigraphic nomenclature and correlation of the Permian "Productus limestone," Salt Range, West Pakistan: Geol. Survey Pakistan, Rec. 15, pt. 1, p. 1-20, 5 fig.

Treat, Ida Vaillant-Couturier, 1926, Note sur le Permienne marin de Madagascar: C. R. Acad. sci. (Paris), v. 182, p. 1092-1094.

───, 1933, Paléontologie de Madagascar, 19, Le Permo-Trias Marin: Ann. Paléont., t. 22, fasc. 2, p. 37-59, pl. 5-10, 17 fig.

Waagen, William, 1872, On the occurrence of *Ammonites*, associated with *Ceratites*, and *Goniatites* in the Carboniferous deposits of the Salt Range: Geol. Survey India Mem., v. 9, p. 351-358, 1 pl.

───, 1879, Salt Range fossils. I. Productus-limestone fossils: Palaeontologia Indica, ser. 13, 1. Pisces-Cephalopoda, p. 1-72, pl. 1-6.

Wedekind, Rudolf, 1935, Einführung in die Grundlagen der historischen Geologie; Band I, Die Ammoniten-, Trilobiten- und Brachiopodenzeit: Ferdinand Enke, Stuttgart, 109 p., 27 pl., 19 fig.

Ammonoids from the Kathwai Member, Mianwali Formation, Salt Range, West Pakistan

BERNHARD KUMMEL
Harvard University

ABSTRACT

Ophiceras connectens Schindewolf and *Glyptophiceras* sp. indet. are described from the Kathwai Member of the Mianwali Formation in the Salt Range of West Pakistan. These species are directly correlative with species of the *Otoceras-Ophiceras* fauna of the Himalayas. The formal composition of the earliest Scythian ammonoid zone is discussed with particular attention on the genus *Glyptophiceras*.

INTRODUCTION

The most significant fossil discovery resulting from Professor O. H. Schindewolf's visit in 1952 to Chhidru Nala in the central Salt Range was his discovery of *Ophiceras connectens* in the dolomite unit of the Kathwai Member. In field studies of the Kathwai Member carried out by Dr. Curt Teichert and myself, we recognized impressions and fragments that probably are this species at a number of additional localities, but in most cases the preservation left much to be desired. A number of fairly well-preserved specimens were obtained, however, from the limestone unit of the Kathwai Member on the east side of Chhidru Nala (loc. 6B, bed 9, Chhidru B of fig. 3 in Kummel and Teichert, 1970). This bed also contains *Menuthionautilus kieslingeri* Collignon. I agree with Schindewolf (1954) that *Ophiceras connectens* is closely related to the ophiceratids of the Spiti region described by Diener (1897).

In addition a specimen assigned to *Glyptophiceras* sp. indet. was obtained from the dolomite unit of the Kathwai Member at Kathwai (loc. 10, bed 3, Kathwai B of fig. 3 in Kummel and Teichert, 1970). This specimen cannot be separated specifically from the specimen (on a slab with *Otoceras woodwardi*) that Griesbach (1880) described as *Ophiceras himalayanum*.

GENERAL COMMENTS ON *OTOCERAS-OPHICERAS* ZONE

The *Otoceras-Ophiceras* beds of the Himalayas have long been recognized as including the basal Triassic zone. Diener (1912) thoroughly reviewed the data and evolution of thought on the stratigraphy and fauna of this zone. His concluding remarks summarize quite well the accepted view (Diener, 1912, p. 33): "Thus the fauna of the *Otoceras* stage represents one single palaeontological zone only, which, from its most conspicuous types, should be called zone of *Otoceras Woodwardi* and *Ophiceras Sakuntala*."

The *Otoceras-Ophiceras* beds of the Spiti and Painkhanda regions in the Himalayas are approximately one meter thick (Diener, 1912). The lower and upper thirds of this unit are abundantly fossiliferous. At Spiti, *Otoceras* is confined to the lower third of this unit; but in Painkhanda, 130 miles to the southeast, *Otoceras* is present in both the lower and upper thirds of the unit. Species of *Ophiceras* are the predominant element in the lower third and upper third of this unit. This difference in representation of *Otoceras* and *Ophiceras* in the lowest Scythian unit of the Himalayas is well demonstrated in the numbers of specimens of each genus from nine collections studied by Diener (1897):

Localities	Numbers of specimens*	
	Ophiceras	*Otoceras*
Shalshal	221	52
Kiunglung	79	2
Gaichund	11	x
Khar	7	x
Tengdi	5	0
Kaga	x	0
Kuling	4	x
Ensa	x	x
Muth	x	x

* An x indicates no specific number of specimens was listed, but presumably there were few.

It is clear from these data that *Ophiceras* is by far the predominant element in the ammonoid fauna, and *Otoceras* the second ranking element. The five other genera of this fauna are represented by only nine specimens. In the lowest Scythian strata of East Greenland, *Otoceras* is likewise a minor element of the fauna in contrast to *Ophiceras* (Tove Birkelund, personal communication). My own observations over a period of approximately one month at Kap Stosch have confirmed this. Data are not available on the relative abundance of *Otoceras* and *Ophiceras* from northern Alaska (Kummel in Reeside, 1957), the Arctic Islands of Canada (Tozer, 1961), Spitsbergen (Petrenko, 1963), or Siberia (Popov, 1961).

The beds with *Otoceras* in Siberia also contain the following genera of ammonoids: *Ophiceras, Glyptophiceras, Tompophiceras,* and *Episageceras* (Yu. N. Popov, personal communication). In East Greenland *Otoceras* and *Ophiceras* are associated in the *Ophiceras* beds (Spath, 1935). The underlying *Glyptophiceras* beds contain only small glyptophicerids and *Otoceras boreale*. Excellent specimens of *Otoceras boreale* were collected from the *Glyptophiceras* beds at Kap Stosch in the summer of 1967. The overlying *Vishnuites* beds of Spath (1935) contain species of *Ophiceras* but *Otoceras* is not present. Thus in this case the range of *Ophiceras* extends beyond that of *Otoceras*. The same situation prevails in the Arctic Islands of Canada (Tozer, 1967). It is of special interest to note that there is considerable uncertainty in correlation of the early Scythian between the Arctic Islands of Canada and East Greenland (Tozer, 1967).

Thus on the basis of available data the lowest Scythian ammonoid zone is characterized by the predominance of *Ophiceras,* and much lesser presence of *Otoceras*. In most places the two genera occur together and thus cannot be used to mark two distinct biostratigraphic zones for intercontinental correlation. Separation into two distinct zones for purposes of provincial correlation as has been done in East Greenland and Arctic Canada is quite acceptable, but this scheme breaks down when extended beyond these particular regions. The fact that even in the two principal areas of Triassic outcrops in the Himalayas (Spiti and Painkhanda) *Otoceras* is present throughout the range of *Ophiceras* in one locality but is confined to the lowermost bed in the other clearly demonstrates that the absence of *Otoceras* needs to be evaluated very carefully. More data are needed, but it appears that the present distributional pattern of *Otoceras* suggests that this genus became extinct in the Arctic region shortly after the beginning of the Triassic but persisted slightly longer in Tethys. In regard to the two species of ammonoids in the Kathwai Member, they clearly are extremely close to comparable forms in the *Otoceras-Ophiceras* beds of the Himalayas and justify stratigraphic correlation between the two areas, the absence of *Otoceras* in the Salt Range notwithstanding.

THE GENUS *GLYPTOPHICERAS*

Lowermost Triassic ammonoid faunas of the *Otoceras-Ophiceras* zone consist of approximately 10 genera of ammonoids. Only three localities have yielded anything approaching a fairly representative fauna in terms of numbers of specimens. These localities are the Himalayas, East Greenland, and Arctic Canada (Ellesmere and Axel Heiberg Islands). The largest number of genera among these faunas are found in the Himalayas. The Arctic faunas have less than half the number of genera than the faunas of the Himalayas. Nearly all Arctic genera are present in the Himalayas.

The ammonoid genera of the lowermost Scythian are, for the most part, characterized by smooth conchs. There is only one group of ornamented ammonoids, characterized by ribs, and this has generally been assigned to the genus *Glyptophiceras*. The specimen described here I find impossible to distinguish from *Ophiceras himalayanum* (Griesbach, 1880, p. 111, pl. 3, fig. 8), discussed in more detail by Diener (1897, p. 41, pl. 14, fig. 14). Griesbach's species was based on a single specimen and is embedded in a slab of rock adjacent to a specimen of *Otoceras woodwardi* (Pl. 2, fig. 3). A single specimen from a ridge between the Dharma and Lissar valleys, Kumaon, was assigned by von Krafft and Diener (1909, p. 92) to *Xenodiscus himalayanus,* but it appears there is uncertainty as to the exact stratigraphic position of this specimen. It was presumed to be from the "*Otoceras* stage." I have had the opportunity of examining the Triassic collections made by Dr. Eigil Nielsen in 1950 in the Muth and Lilang areas of Spiti and found that there are no specimens of *Glyptophiceras* in these collections. I know of no other specimens of *Glyptophiceras* from the *Otoceras-Ophiceras* beds in this region of the Himalayas, where I would judge the genus to be quite rare.

Diener (1913) described a fauna from Pastannah, Kashmir, which he considered to be of *Otoceras* zone age. This fauna included species of *Ophiceras, Glyptophiceras,* and *Vishnuites.* Diener's conclusion that this was a fauna comparable to, and of the same age as, that of the *Otoceras-Ophiceras* faunas of the Spiti and adjoining areas of the Himalayas was contested by Bion (1914). The crux of this difference of interpretation lies in the interpretation of the stratigraphic sequence of the Scythian strata in the Pastannah region. The geological study of the Pastannah area was carried out by Middlemiss (1910), and his collections were reported on by Diener (1913). Because of the geologic setting and Bion's subsequent observations it is worthwhile to quote Middlemiss' statements regarding the *Ophiceras* fauna (Middlemiss, 1910, p. 241-243):

> Whilst working out the excellent sections in the Muschelkalk at this place, I was unduly detained by a spell of rainy weather, and my attention was then drawn to numerous blocks of clear, dark, blue-grey limestone among the gravel and detritus fans near the village, and which were seen to contain a very well preserved *Ophiceras* fauna. It was some time before I was able to track these to their home *in situ,* where they occupy . . . the secondary little spurs (mostly forest-covered) W. by S. of Pastannah village, referred to on p. 240. The outcrops are very imperfectly shown because of the prevailing forest and soil-covered dip slopes, but their position as overlying the Zéwan (Permo-Carboniferous) and underlying the Muschelkalk, admitted of no doubt. They consisted of blocks *in situ,* and occasional definite layers of dark grey, compact limestone, protruding from the soil-cap of the occasionally-found barer slopes.
>
> The apparent dip follows that of the beds below and above and is approximately 40° N.N.W. Owing to the downhill dip, it is not easy to estimate the thickness, but there must be at least 300 feet, of which some 50 feet are specially fossiliferous.
>
> The commonest form of ammonite in these lowermost Trias limestones is present in great abundance, single blocks often yielding a large number of specimens compacted together. Occasionally individ-

uals may be broken out from the rock in a very fair state of completeness and preservation. A large percentage seem to belong to Diener's amplification of Griesbach's genus *Ophiceras* (Pal. Ind., Series XV, Vol. II, Pt. 1, p. 100) although in no single instance is there visible any of the delicate concentric or spiral striation that is supposed to characterize the innermost layer of the shell. Other differences are to be found in the transverse section which is never cordate, only occasionally lanceolate, and generally roughly oval with no well-marked umbilical margin or wall.

Nevertheless, the general similarity of the forms in size, shape, ornamentation, sutures and range of variation to those united by Diener into his group *Ophiceras sakuntala* cannot be denied, and with this group therefore I am compelled provisionally to classify them.

Ophiceras, according to Diener, is entirely restricted to the lowest division of the Trias, the *Otoceras* beds. As already remarked, the latter genus is apparently wanting in this part of the Kashmir area, so that it is quite possible that the forms discovered by me may be slightly different in horizon and exact affinities.

The two species of this group to which most of my specimens most nearly correspond are those of *O. sakuntala*, Dien., and *O. ptychodes*, Dien., but the two seem to pass into each other by transitional forms, whilst there are others, *Ophiceras* sp., with characteristics which seem to connect *Ophiceras* and *Xenodiscus (Danubites).*

An almost equally large percentage of ammonites in this bed belong undoubtedly to the genus *Xenodiscus (Danubites)*. Several species seem indicated, but I have not attempted to sort them out specifically. All or most have a general similarity to the species described by Diener and von Krafft (Pal. Ind., Series XV, Vol. II, pt. 1, and Vol. VI, pt. 1). There are also a few other undetermined fragments of ammonites.

The observations by Bion on the Triassic strata of this region are contained in the annual report of the Director of the Geological Survey of India for 1914. In this report the Director (H. H. Hayden) considered Bion's remarks on the Lower Triassic of such importance that he quoted in full his remarks, which are as follows (Bion, 1914):

About 20 feet above the base of the black shales there is a layer of calcareous nodules from which many specimens of *Otoceras* have been obtained, associated with almost all the other members of the fauna of the *Otoceras* beds of the central Himalayas. Good collections have been obtained from Nagaberan in the Dachhigam State Rakh and from the Pahlgam-Aru basin. Some thirty feet above the *Otoceras* layer there is another fossiliferous horizon characterized by *Ophiceras* from which one specimen of *Otoceras* was also procured, but the rest of the black shale division seems to be barren. A surprising element in the fauna of the basal *Otoceras* layer is furnished by the presence of the genus *Productus*, of which three specimens have been obtained from near Pahlgam. In spite of this Permian element I consider that the fauna of the *Otoceras* beds of Kashmir has a decided Triassic aspect.

The black shale division passes up, by a gradual increase in the calcareous intercalations, into some 300 feet of thin-bedded blue limestones which almost invariably break into well-marked crags. The lower of these crags is absolutely barren, but the upper one, which contains subordinate shales, has yielded fossils. The fauna obtained by C. S. Middlemiss from the Guryul ravine came from these upper beds, and while examining the section at Pastannah from which the same observer obtained his *Ophiceras* fauna, I was fortunate enough to ascertain that this horizon also occurs in the upper part of these limestones. It now becomes evident that the Guryul ravine fauna is on practically the same horizon as the *Ophiceras* horizon of Pastannah, a fact strongly at variance with the conclusions arrived at by Professor Diener as a result of his examination of C. S. Middlemiss' collections. Professor Diener referred the Pastannah fauna to the horizon of the *Otoceras* beds of Spiti and the Guryul ravine fauna to that of the *Hedenstroemia* stage of the same area (Pal. Ind., New Series, Vol. V, Mem. 1, p. 120). It is now evident that the Pastannah fauna occurs at an horizon some two to three hundred feet above the *Otoceras* bed proper, and that there is very little difference of horizon between it and that of the Guryul ravine.

At the end of this quotation from Bion's report the Director added the following comments: "In conclusion Mr. Bion is strongly of the opinion that the Pastannah fauna is very nearly allied to that of the *Otoceras* beds but a slightly different and younger variant, while the Guryul ravine fauna is most nearly allied to that of the *Meekoceras* beds, as was originally supposed by Mr. Middlemiss."

I had the opportunity of a brief visit to the Pastannah area in June of 1968. Unfortunately I was not able to locate any fossiliferous outcrops or float blocks of the *Ophiceras* beds. The area is heavily forested and outcrop conditions very poor. I found that the sketch profiles by Middlemiss (1910, pl. 32, 35) reflect well the physiographic conditions of the region. Looking up the valley from the village of Pastannah the left-hand side of the valley is framed by the dip slope of Permian strata, on the right side of the valley Middle Triassic strata crop out. The difficulty of establishing the precise horizon of the *Ophiceras* fauna is clearly understandable under the circumstances. However, Bion is ambiguous in his use of the term "horizon." For him to say that "there is very little difference of horizon between it and that of the Guryul ravine" is completely unwarranted. The detailed stratigraphic succession of the Triassic formations of Kashmir is still not sufficiently well known to justify such a statement.

Spath (1934) also came to the conclusion that the Pastannah fauna was slightly younger than the *Ophiceras* fauna of the Spiti region. This could well be the case but it should be emphasized that in spite of Bion's comments the actual stratigraphic level of the Pastannah fauna is still not known. The absence of *Otoceras* could after all be merely a collection failure. If the Pastannah fauna is not exactly synchronous with the *Ophiceras* beds of Spiti, it appears at most to be a transitional fauna to the next higher zone.

The only other region where *Glyptophiceras* is a predominant element is in the *Otoceras-Ophiceras* beds of East Greenland. In his first report on the East Greenland lower Scythian faunas Spath (1930) recognized four species of *Glyptophiceras*. In his second report on the East Greenland fauna Spath (1935) recognized 11 species of *Glyptophiceras*. Spath's discussion and description of these species reads much like those of the Kashmir fauna by Diener. Spath used the same criteria to distinguish his species as Diener did for his Kashmir fauna: the results were predictably the same, resulting in a proliferation of species names for a highly variable group. Measurements of most of the more complete specimens of *Glyptophiceras* studied by Spath are tabulated on Table 1 and plotted on the graph of Figure 1. It can readily be seen that variation in umbilical diameter and whorl width are within the range of *Glyptophiceras himalayanum*. Variation in rib pattern appears likewise to be completely gradational. The following species of East Greenland ophiceratids I believe to be synonymous:

Glyptophiceras extremum Spath, 1935
G. gracile Spath, 1930
G. minimum Spath, 1935
G. minor Spath, 1930
G. nielseni Spath, 1930
G. pascoei Spath, 1930
G. pseudellipticum Spath, 1930
G. serpentinum Spath, 1935
G. subextremum Spath, 1935

I retain *G. gracile* as the species name for this group. *Glyptophiceras minor* has page priority in Spath's 1930 paper, but that species has very weakly developed ribs and represents one extreme in the total range of variation in rib pattern. *Glyptophiceras gracile* occupies a more central position in terms of rib pattern. There are two species of *Glyptophiceras* (*G. triviale* and *G. polare*) that occur in the lowest Scythian strata of East Greenland that I exclude from the above synonymy. These are very small species, as yet poorly known. New collections made in July, 1967, from the Kap Stosch area of East Greenland will make possible a more thorough study of these species.

TABLE 1. Measurements (in mm) and Proportions of Specimens of *Glyptophiceras* from the Lower Scythian of East Greenland Described by Spath (1930, 1934).*

	D	W	H	U	W/D	H/D	U/D
1.	61.1	?	10.6	33.2	?	17.3	54.3
2.	49.9	?	13.7	26.2	?	27.5	52.5
3.	47.0	12.5	15.5	22.0	26.6	32.9	46.8
4.	46.5	13.5	13.8	23.9	29.0	29.7	51.4
5.	44.8	11.7	14.0	21.3	26.1	31.3	47.5
6.	43.8	11.3	12.7	20.1	25.8	28.9	45.9
7.	43.7	10.3	11.9	21.0	23.6	27.2	48.1
8.	41.7	8.2	11.1	22.7	19.7	26.6	54.4
9.	36.0	?	12.1	15.0	?	33.6	41.7
10.	34.7	8.4	10.9	15.8	24.2	31.4	45.5
11.	32.2	7.6	8.5	17.4	23.6	26.4	54.0
12.	29.6	10.1	10.1	13.2	34.1	34.1	44.6
13.	29.3	?	10.6	11.8	?	36.2	40.3
14.	27.8	7.3	9.3	11.8	26.3	33.5	42.4
15.	23.4	7.0?	7.0	11.3	29.9?	29.9	48.3
16.	18.2	4.6	6.0	8.0	25.3	32.9	43.9
17.	17.0	4.6	6.0	7.3	27.1	35.3	42.9
18.	14.6	4.0	4.7	7.0	27.4	32.2	47.9

1. Paratype, *Glyptophiceras extremum* Spath (1935, pl. 11, fig. 4), MMH 8003.
2. Holotype, *G. nielseni* Spath (1935, pl. 19, fig. 1), MMH 8073.
3. Plesiotype, *G. gracile* var. *robusta* Spath (1935, pl. 18, fig. 5), MMH 8069.
4. Holotype, *G. extremum* Spath (1935, pl. 19, fig. 7), MMH 8079.
5. Holotype, *G. gracile* Spath (1930, pl. 7, fig. 5), MMH 8656.
6. Paratype, *G. pascoei* Spath (1930, pl. 8, fig. 5), MMH 8665.
7. Plesiotype, *G. nielseni* var. *modesta* Spath (1935, pl. 9, fig. 4), MMH 7993.
8. Paratype, *G. nielseni* Spath (1935, pl. 9, fig. 5), MMH 7994.
9. Plesiotype, *G. gracile* Spath (1935, pl. 11, fig. 9), MMH 8008.
10. Holotype, *G. pseudellipticum* Spath (1930, pl. 8, fig. 8), MMH 8668.
11. Plesiotype, *G. nielseni* var. *modesta* Spath (1935, pl. 5, fig. 2), MMH 7959.
12. Plesiotype, *G. pascoei* var. *rotunda* Spath (1935, pl. 6, fig. 5), MMH 7966.
13. Plesiotype, *G.?* sp. ind. cf. *minor* Spath (1935, pl. 7, figs. 8a-c), MMH 7978.
14. Plesiotype, *G.* sp. ind. (Spath, 1930, pl. 8, fig. 9), MMH 8669.
15. Paratype, *G. pascoei* Spath (1930, pl. 8, fig. 7), MMH 8667.
16. Plesiotype, *G.* sp. ind. Spath (1930, pl. 7, fig. 10), MMH 8660.
17. Plesiotype, *G.* sp. ind. Spath (1930, pl. 7, fig. 9), MMH 8659.
18. Paratype, *G. minimum* Spath (1935, pl. 8, fig. 3), MMH 7984.

* Meaning of D, W, H, and U same as in Table 1.
MMH=Museum Mineralogicum Hafniense (Mineralogical Museum, University, Copenhagen, Denmark).

What are the relationships between the East Greenland *G. gracile* and the Himalayan *G. himalayanum?* Spath (1934, p. 83) included the specimen Diener (1913, p. 3, pl. 2, fig. 4) described as *Xenodiscus himalayanus* and a specimen from Pastannah in the British Museum (Natural History) in *Glyptophiceras pascoei,* a species first established on specimens from East Greenland. Figure 1 shows clearly that in regard to umbilical diameter and whorl width there is essentially no difference between the East Greenland and Kashmir forms. However, in the kinds of rib patterns and the range of variations in rib pattern the East Greenland and Kashmir species are different. This is a subjective evaluation and possibly could be established on a rigorous basis.

The early Scythian glyptophiceratids with radial, or nearly radial, ribs are remarkably similar to the Middle and Late Permian ribbed *Xenodiscus*. In fact, both Griesbach (1880) and Diener (1913) assigned these species of ribbed ammonites from the *Otoceras-Ophiceras* fauna of the Himalayas to the genus *Xenodiscus*. The relationship of *Xenodiscus* to *Glyptophiceras* warrants some attention to clarify the phylogenetic position of *Glyptophiceras*.

There are approximately 30 genera of Late Permian ammonoids (Chhidruan and Dzhulfian stages, and the Changhsing and Talung formations of Chao, 1965). Most of these genera are specialized either in shape, ornamentation, or suture. There are two genera, however, that are highly generalized forms: these are *Xenodiscus* and *Xenaspis*. The type species of *Xenodiscus* is *X. plicatus* Waagen (1879) from the Chhidru Formation of the Salt Range (Pl. 2, fig. 1,2); the type species of *Xenaspis* Waagen (1895) is *Ceratites*

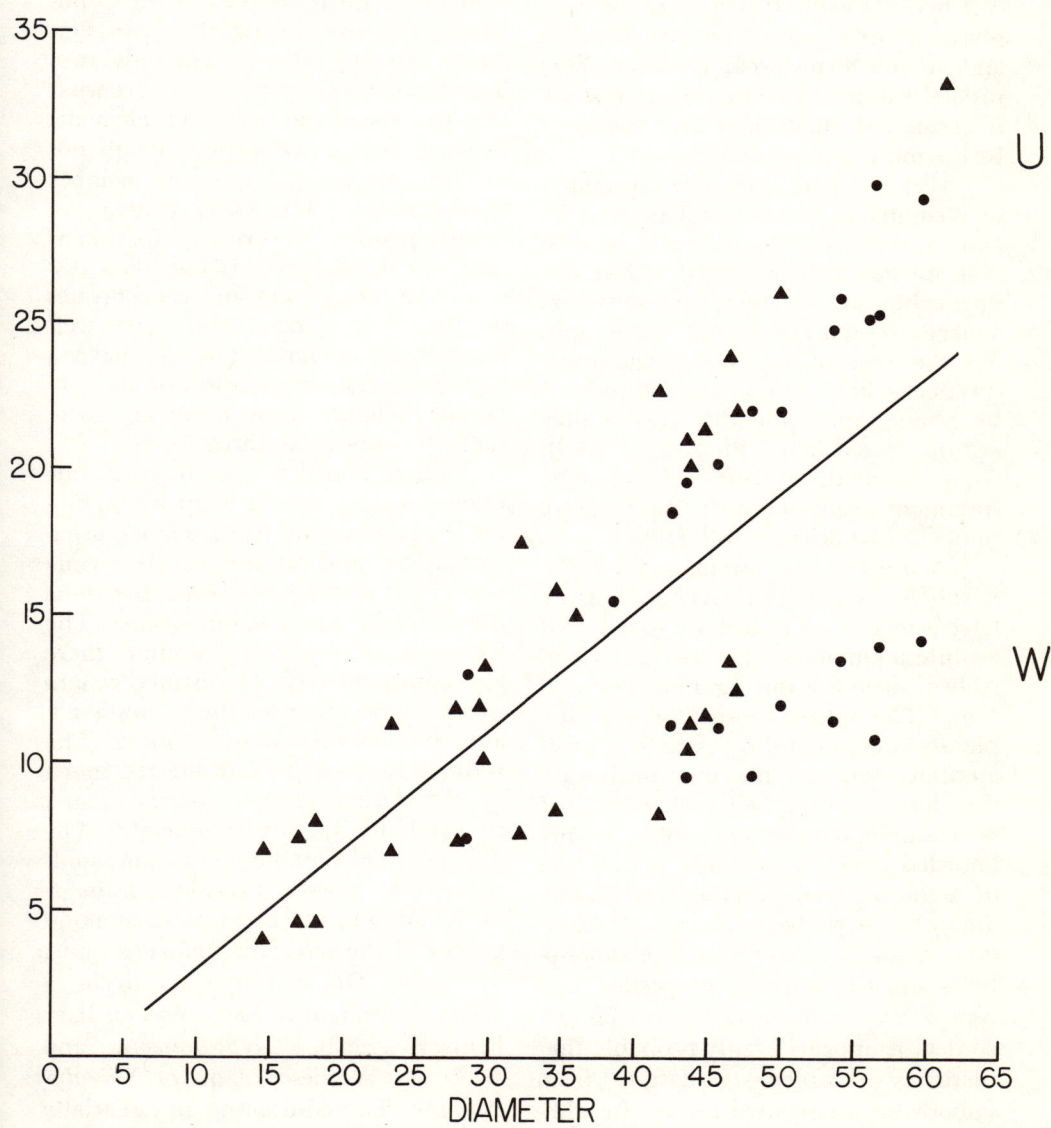

Fig. 1. Variation in umbilical diameter and whorl width of glyptophicerids from Spiti, Kashmir (dots), and East Greenland (triangles). Data are from Tables 1 and 2.

carbonarius Waagen (1872), also from the Chhidru Formation of the Salt Range (Pl. 2, fig. 4-9). These two genera were differentiated on the basis of ornamentation, *plicatus* having radial ribs, *carbonarius* being essentially smooth. Both genera have generally been accepted. Recently Schindewolf (1954) has presented convincing arguments for placing *Xenaspis* in synonymy of *Xenodiscus*, and this is the procedure followed here as was done by Teichert (1966). It needs to be emphasized that both the ornamented and smooth forms occur together. Furnish (1966, p. 291) stated that the relatively smooth shells have been found to be the more common form.

Waagen had only one specimen of *Xenodiscus plicatus* and four of *X. carbonarius*. These specimens in general are not well preserved, a fact not noticeable in the fine line drawings Waagen reproduced in his monograph. To the best of my knowledge these specimens have never been illustrated by photographs; for this reason they are illustrated here (Pl. 2, fig. 1,2,4-9). Some good illustrations of topotype specimens from the Salt Range were published by Schindewolf (1954).

Accepting the premise that *Xenaspis* is a synonym of *Xenodiscus*, we have a picture of a stock of generalized evolute ammonoids that are smooth to ribbed, evolving during Late Permian time. The suture consists of two simple, serrated, lateral lobes with a small auxiliary lobe on the umbilical wall, the dorsal suture being characterized by a simple, narrow, bifid lobe. To my knowledge no single large population of well-preserved specimens of *Xenodiscus* has as yet been discovered. However, from an analysis of the descriptions and illustrations of specimens of *Xenodiscus* and *Xenaspis* so far recorded, it appears highly probable that there is complete gradation from smooth to ornamented forms. In addition, when ribbing does occur, its pattern is highly variable. Admitting that the Late Permian has its share of correlation problems, the records of *Xenodiscus* to date do not show any significant evolutionary change through that period of time, possibly 15 million years. Though there were changes in total numbers of genera at each moment in the Late Permian, the total is still fairly large.

As mentioned above, the lowest Scythian horizons are characterized by approximately 10 genera of ammonoids, which is in great contrast to the populations existing during the Late Permian. Among these genera only two, *Ophiceras* and *Glyptophiceras*, are what can be considered common elements of these faunas. *Otoceras*, though not as common, is an important member of the fauna. The other genera are known mainly by very few specimens from the Himalayas. If one then discards the rare genera and concentrates on the three genera which are well represented or important, we have a highly interesting situation of the ammonite adaptive zone occupied essentially by only these three forms.

There is still a vast chasm in our understanding of the adaptive significance of ammonoid conch shapes, ornamentation, and suture. At the same time, most researchers accept that these features have adaptive significance. The fact that in the Late Permian there are approximately 30 distinct genera suggests that many of the niches available to ammonites were occupied. The wave of extinctions that affected much of the Late Permian marine fauna changed this picture completely. The only genus of Late Permian ammonoids to survive into the Triassic is *Episageceras*. Among the three most common genera of the *Otoceras-Ophiceras* zone, the genus *Otoceras* appears to be a direct descendant of *Pseudotoceras* Ruzhentsev (1962). *Glyptophiceras* and *Ophiceras* are descendants of *Xenodiscus*, one line maintaining an essentially smooth conch with minor elaboration of the suture, as pointed out by Schindewolf (1954). The other line incorporated ribbing as a fundamental conch feature. *Glyptophiceras, Ophiceras,* and *Otoceras* of the lowermost Scythian are extremely plastic stocks. Nearly every researcher who has worked with these genera has testified to this fact. All three genera show a very large amount of variation in nearly all conch parameters and yet evolved very few species. Throughout its long history *Xenodiscus* remained a fairly stable stock, at the same time it was but a single element in rather diverse faunas.

SYSTEMATIC PALEONTOLOGY*

Genus GLYPTOPHICERAS Spath, 1930

Type species: Xenodiscus aequicostatus Diener, 1913 [=*Ophiceras himalayanum* Griesbach, 1880 (subj. syn.)].

Glyptophiceras himalayanum (Griesbach)

Plate 1, figure 1

Ophiceras himalayanum Griesbach, 1880, p. 111, pl. 3, fig. 8.
Danubites himalayanus, Diener, 1897, p. 41, pl. 14, fig. 14.
Danubites sp. ind. ex. aff. *himalayanus,* Diener, 1897, p. 44, pl. 14, fig. 10.
Xenodiscus himalayanus, von Krafft and Diener, 1909, p. 92, pl. 13, fig. 2; Diener, 1913, p. 3, pl. 2, fig. 3, 4; Diener, 1915, p. 312.
Xenodiscus aequicostatus Diener, 1913, p. 6, pl. 2, fig. 10; Diener, 1915, p. 311.
Xenodiscus althothae Diener, 1913, p. 8, pl. 2, fig. 6, 11; Diener, 1915, p. 311.
Xenodiscus comptoni Diener, 1913, p. 10, pl. 2, fig. 7; Diener, 1915, p. 312.
Xenodiscus cf. *ellipticus* Diener, 1913, p. 9, pl. 3, fig. 1.
Xenodiscus cf. *lissarensis* Diener, 1913, p. 5, pl. 1, fig. 11.
Xenodiscus cf. *ophioneus,* Diener, 1913, p. 12, pl. 2, fig. 8, 9; Diener, 1915, p. 314.
Xenodiscus cf. *rotula,* Diener, 1913, p. 11, pl. 3, fig. 2; Diener, 1915, p. 315.
Xenodiscus salomonii Diener, 1913, p. 7, pl. 2, fig. 5; Diener, 1915, p. 315.
Xenodiscus cf. *sitala* Diener, 1913, p. 14, pl. 3, fig. 3.
Xenodiscus ellipticus Diener, 1915, p. 312.
Xenodiscus lissarensis Diener, 1915, p. 313.
Xenodiscus sitala Diener, 1915, p. 315.
Glyptophiceras aequicostatum, Spath, 1930, p. 35; Spath, 1934, p. 81, fig. 15.
Glyptophiceras himalayanum, Spath, 1930, p. 33; Spath, 1934, p. 82.
Glyptophiceras kashmiricum Spath, 1930, p. 37; Spath, 1934, p. 83, pl. 1, fig. 3a, b.
Glyptophiceras pascoei Spath, 1930, p. 36, pl. 7, fig. 1-7, 16; Spath, 1934, p. 83.
Glyptophiceras ophioides Spath, 1934, p. 82, pl. 12, fig. 1.

Description: This species is represented by a single, fragmentary, and poorly preserved specimen which is embedded in dolomite matrix. Only one side of one half of the specimen is exposed. Its principal features are the evolute nature of the conch, the compressed whorl section—the whorls being higher than wide—the rounded venter and the lateral, straight ribs. The ribs are most prominent in the dorsal half of the whorl side, decreasing in prominence toward the venter. No trace of a suture is preserved. The specimen has a diameter of approximately 53 mm, a whorl height of 14 mm, and an umbilical diameter of 24 mm.

Discussion: The *Ophiceras* fauna collected by Middlemiss contained 96 specimens from which Diener (1913) assigned 44 to the species of *Xenodiscus* included in the list of synonyms above. Subsequently Spath (1930, p. 37) established *Glyptophiceras kashmiricum* for *X.* cf. *lissarensis* (Diener, 1913, p. 5, pl. 1, fig. 11) and assigned one of the specimens Diener (1913, p. 3, pl. 2, fig. 4) included in *X. himalayanus* to *Glyptophiceras pascoei* Spath, the type of which comes from the *Otoceras* beds of Clavering Island, East Greenland. In addition, Spath (1934, p. 82) introduced the name *Glyptophiceras ophioides* for *X.* cf. *ophioneus* Diener (1913, p. 12, pl. 2, fig. 8, 9; non Waagen).

---
* The abbreviations used here are as follows: MCZ= Museum of Comparative Zoology, GIT=Geological Institute Tübingen.

TABLE 2. Measurements (in mm) and Proportions of Specimens of *Glyptophiceras himalayanus* (Griesbach) from Kahmir and Spiti described and illustrated by Diener (1897, 1913) and Spath (1934).*

	D	W	H	U	W/D	H/D	U/D
1.	59.5	16.4	14.2	29.4	27.6	23.9	49.4
2.	56.6	17.4	14.0?	25.3	30.7	24.7?	44.7
3.	56.4	15.2	10.8	29.7	26.9	19.1	52.7
4.	56.0	16.2	?	25.2	28.9	?	45.0
5.	54.0	17.8	13.5	25.9	32.9	25.0	47.9
6.	53.7	15.9	11.4	24.8	29.6	21.2	46.2
7.	50.0	16.0	12.0	22.0	32.0	24.0	44.0
8.	48.0	16.3	9.7	22.0	33.9	20.2	45.8
9.	45.7	14.6	11.2?	20.2	31.9	24.5?	44.2
10.	43.5	13.5	9.6	19.6	31.0	22.1	45.1
11.	42.5	14.2	11.3	18.6	33.4	26.6	43.8
12.	38.5?	13.2?	?	15.5	34.3?	?	40.3?
13.	28.5	9.1	7.9	13.1	31.9	27.7	45.9

1. Holotype, *Xenodiscus althothae* Diener (1913, pl. 2, fig. 11), GSI 11271.
2. Plesiotype, *Xenodiscus himalayanus* Griesbach (Krafft and Diener, 1909, pl. 23, fig. 2), GSI 9485.
3. Holotype, *Xenodiscus salomonii* Diener (1913, pl. 2, fig. 5), GSI 11265.
4. Plesiotype, *Xenodiscus himalayanus* Griesbach (Diener, 1913, pl. 2, fig. 4), GSI 11264.
5. *Glyptophiceras pascoei* Spath (1934, p. 83), BMNH C28541.
6. Holotype, *Xenodiscus aequicostatus* Diener (1913, pl. 2, fig. 10), GSI 11270.
7. Holotype, *Glyptophiceras ophioides* Spath (1934, p. 82), BMNH C28539.
8. Plesiotype, *Xenodiscus* cf. *ophioneus* Waagen (Diener, 1913, pl. 2, fig. 8), GSI 11268.
9. Plesiotype, *Danubites* sp. ind. ex aff. *himalayano* (Diener, 1897, pl. 14, fig. 10), GSI 6015.
10. Plesiotype, *Xenodiscus himalayanus* Griesbach (Diener, 1913, pl. 2, fig. 3), GSI 11263.
11. Holotype, *Xenodiscus comptoni* Diener (1913, pl. 2, fig. 7), GSI 11267.
12. Plesiotype, *Xenodiscus himalayanus* Griesbach (1880, pl. 3, fig. 8), GSI 6019.
13. Holotype, *Glyptophiceras kashmiricum* Spath (1934, p. 83, pl. 1, fig. 3), BMNH C28540.
* D—diameter, W—width, H—height of last whorl, U—width of umbilicus.
GSI=Geological Survey of India; BMNH=British Museum (Natural History).

I assume that all these "species" occur together, as Middlemiss (1910, p. 242) writes that "The commonest form of ammonite in these lowermost Triassic limestones is present in great abundance, single blocks often yielding a large number of specimens compacted together." The species recognized by Diener (1913) and Spath (1930, 1934) in this Kashmir fauna were distinguished by differences in ornamentation, involution, and whorl section. Only the specimens illustrated by Diener are still preserved in the Geological Survey of India. Measurements of the illustrated specimens

PLATE 1. GLYPTOPHICERAS AND OPHICERAS.

FIGURES

1. *Glyptophiceras* sp. indet. Side view of specimen from dolomite unit of Kathwai Member at Kathwai (loc. 10, bed 3, of figure 4 in Kummel and Teichert, 1970), MCZ 9668, ×1.
2-9. *Ophiceras connectens* Schindewolf. 2, side view of holotype, GIT 1050/4, ×1; 3, side view, MCZ 9661, ×1.5; 4, side view, MCZ 9662, ×1.5; 5, side view, MCZ 9663, ×3; 6, 7, side and ventral view, MCZ 9664, ×4; 8, side view, MCZ 9665, ×3; 9, side view, MCZ 9666, ×4. Specimen of figure 2 from dolomite unit of Kathwai Member on west side of Chhidru Nala; specimens of figures 3-9 from limestone unit of Kathwai Member on the east side of Chhidru Nala (loc. 6B, bed 9, of figure 4 in Kummel and Teichert, 1970).

MCZ=Museum of Comparative Zoology, Harvard University.
GIT=Geologisches Institut, Universität Tübingen.

PLATE 1

PLATE 2

Diener assigned to various species of *Xenodiscus* are given in Table 2, and data on whorl width and umbilical diameter are plotted on the graph of Figure 1. At least with regard to these two parameters there do not appear to be any significant differences but rather a gradational range. Evaluation of the ribbing is more difficult as the differences are largely a matter of degree. In rib orientation, the pattern varies from radial as in *G. himalayanum* to forward projected as in *G. aequicostatum*. Ontogenetic change in rib pattern is notable in some specimens and has been used for species definition. I believe all these "differences" in rib pattern are no more than one should expect within a population of such ornamented ammonoids. The number of specimens is admittedly small; but from experience with other species of ornamented Scythian ammonoids, e.g., *Columbites parisianus*, this range of variation in rib pattern is common. Rejection of this thesis leaves one with little alternative than to name each individual, a procedure almost adopted by Diener (1913) and Spath (1930, 1934, 1935). I thus come to the conclusion that all of the Spiti and Kashmir species of "*Xenodiscus*" are conspecific. Thus during the earliest Scythian in the Himalayas there are two especially predominant genera, *Ophiceras* and *Glyptophiceras*. Both are highly variable in nearly all conch parameters, representing highly plastic stages in evolutionary development.

Occurrence: Dolomite unit of Kathwai Member at Kathwai (loc. 10, bed 3, Kathwai B, Kummel and Teichert, 1970, fig. 3).

Repository: MCZ 9668.

Genus OPHICERAS Griesbach, 1880

Type species: Ophiceras tibeticum Griesbach, 1880.

Ophiceras connectens Schindewolf

Plate 1, figures 2-9

Ophiceras connectens Schindewolf, 1954, p. 178, pl. 6, fig. 4; text-fig. 4a, b.

Description: The original descrip-

TABLE 3. Measurements (in mm) of Specimens of *Ophiceras connectens* from the Salt Range, West Pakistan.*

	D	W	H	U
Holotype (Pl. 1, fig. 2)	69.6	?	29.5?	29.0?
MCZ 9661 (Pl. 1, fig. 3)	27.5	?	10.5?	9.0?
MCZ 9662 (Pl. 1, fig. 4)	27.1	5.6	10.8	6.4
MCZ 9666 (Pl. 1, fig. 9)	14.0	?	?	?
MCZ 9665 (Pl. 1, fig. 8)	12.3	?	?	?
MCZ 9663 (Pl. 1, fig. 5)	12.0	?	4.7?	4.6
(unfig.)	11.6	?	?	?
MCZ 9664 (Pl. 1, figs. 6,7)	9.5	?	?	?

* Meaning of D, W, H, and U as in Table 1.

PLATE 2. XENODISCUS AND GLYPTOPHICERAS.

FIGURES

1, 2. *Xenodiscus plicatus* Waagen (1879, pl. 2, fig. 1). Side and ventral view of holotype, GSI 3112, ×1.
3. *Glyptophiceras himalayanum* (Griesbach, 1880, pl. 3, fig. 8). Side view of holotype, GSI 6019. On same slab is specimen of *Otoceras woodwardi* Griesbach. ×1.
4-9. *Xenodiscus carbonarius* Waagen. 4, Syntype (Waagen, 1879, pl. 2, fig. 2), GSI 3113, ×1; 5, 6, syntype (Waagen, 1879, pl. 2, fig. 4), GSI 3115; 7, 8, syntype not figured by Waagen (1879) GSI 3116; 9, syntype (Waagen, 1879, pl. 2, fig. 3) GSI 3114. All specimens ×1.
GSI=Geological Survey of India.

tion and discussion of this species by Schindewolf (1954, p. 178) is quite adequate, and after restudy of the holotype I have nothing new to offer. Poor impressions and fragments of ammonites are not uncommon in the Kathwai Member, and presumably most of these are remains of *Ophiceras connectens*. I was able to find only one locality where the preservation could be considered as adequate. This was in the limestone unit of the Kathwai Member on the east side of Chhidru Nala.

This new collection consists of ten specimens and a number of fragments. The measurements of the holotype and seven specimens of the new material are given in Table 3.

The smallest of the specimens (Pl. 1, fig. 6, 7), at a diameter of 9.5 mm, has all the essential features of the adult specimen represented by the holotype. That is, the venter is narrowly rounded and the flanks are broadly arched. The suture likewise has all the essential features of the mature suture (Fig. 2).

Discussion: The early Scythian ophiceratids are an extremely plastic stock, as suggested by Diener (1897) in his monograph on the Himalayan fauna. Spath (1930, 1935), in his study of the East Greenland faunas, recognized this plasticity but yet ended up establishing a large number of species for both faunas, even though both authors freely recognized gradational forms between many of their species. Restudy of the Himalayan and East Greenland ophiceratids monographed by Diener (1897) and Spath (1930, 1935) convinces me that there are relatively few species in these faunas. This is not the place to discuss this problem in detail, but it should be mentioned that on completion of my study of the Himalayan ophiceratids it is probable that *Ophiceras connectens* will be demonstrated to be a synonym.

Occurrence: Schindewolf's holotype came from the dolomite unit of the Kathwai Member on the west side of Chhidru Nala, Salt Range (loc. 6A, bed 4, in Kummel and Teichert, 1970, fig. 4). Schindewolf records additional specimens from his bed 17 (=bed 8, loc. 6A, in Kummel and Teichert, 1970, fig. 4). We have specimens of poorer preservation from the same section in bed 3 (loc. 6A, Kummel and Teichert, 1970, fig. 4). The specimens described and illustrated in this paper are from the limestone unit of the Kathwai Member on the east side of Chhidru Nala (bed 9, loc. 6B, in Kummel and Teichert, 1970, fig. 4). Less well-preserved specimens are in the collections from the dolomite unit of the Kathwai Member at Wargal, Salt Range (bed 8, loc. 7, in Kummel and Teichert, 1970, fig. 4), from Kathwai A, Salt Range

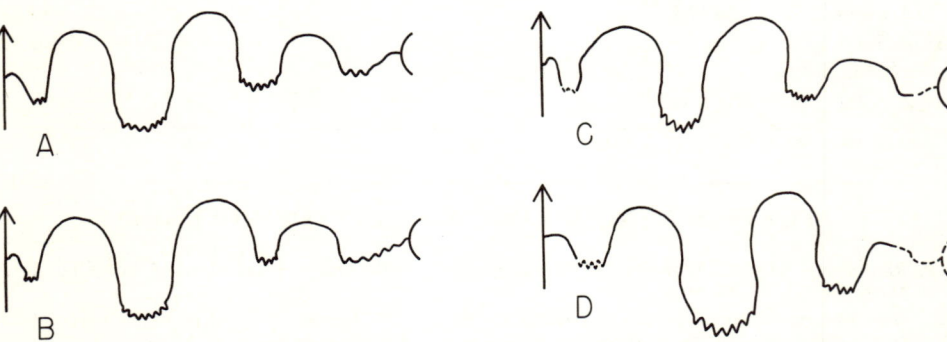

Fig. 2. Diagrammatic representation of sutures of (A) *Ophiceras chamunda* Diener (from Schindewolf, 1954, fig. 2b), (B-D) *Ophiceras connectens* Schindewolf, B—holotype, GIT 1050/4 (from Schindewolf, 1954, fig. 4b); C, MCZ 9661 (Pl. 1, fig. 3) at a whorl height of 7.8 mm; D, MCZ 9664 (Pl. 1, fig. 6, 7) at a whorl height of 3.9 mm.

(bed 3, loc. 9, in Kummel and Teichert, 1970, fig. 4), and from Tapan Wahan, Khisor Range (bed 4, loc. 12, in Kummel and Teichert, 1970, fig. 4). In addition poorly preserved specimens which presumably belong to this species were observed in the limestone unit at Landa Pusha, Surghar Range (bed 6, loc. 1), Nammal (bed 7, loc. 4), Kufri (bed 6, loc. 8), Kathwai B (bed 6, loc. 10), Salt Range, Kingriali, Khisor Range (bed 5, loc. 11), all in Kummel and Teichert, 1970, fig. 4.

Repository: Holotype GIT 1050/4 (Pl. 1, fig. 2); plesiotypes MCZ 9661 (Pl. 1, fig. 3), MCZ 9662 (Pl. 1, fig. 4), MCZ 9663 (Pl. 1, fig. 5), MCZ 9664 (Pl. 1, fig. 6, 7), MCZ 9665 (Pl. 1, fig. 8), MCZ 9666 (Pl. 1, fig. 9), unfigured specimen 9667.

ACKNOWLEDGMENTS

It is a pleasure to acknowledge the competent assistance of Miss Victoria Kohler in preparation of this manuscript. The field studies for this report were supported by National Science Foundation grant G-19066, the laboratory studies by GB-5109. Through the courtesy and help of Mr. M. V. A. Sastry I was able to study all the types of Lower Triassic ammonoids deposited in the offices of the Geological Survey of India in Calcutta. Dr. Tove Birkelund of the Universitetets Mineralogisk-Geologiske Institut in Copenhagen offered the same hospitality to facilitate study of the collections from East Greenland studied by L. F. Spath (1930, 1935).

REFERENCES

Bion, H. S., 1914, Tract N. of Srinagar and Pahlgam, in General report for 1913: Geol. Survey India, Rec., v. 44(1), p. 39-40.

Chao King-koo, 1965, The Permian ammonoid-bearing formations of South China: Scientia Sinica, v. 14, p. 1814-1845.

Diener, Carl, 1897, Himalayan fossils, the Cephalopoda of the Lower Trias: Geol. Survey India, Mem., Palaeontological Indica, ser. 15, v. 2(1), p. 1-191.

———, 1912, The Trias of the Himalayas: Geol. Survey India, Mem., v. 36(1), p. 202-347.

———, 1913, Triassic faunae of Kashmir: Geol. Survey India, Mem., Palaeontological Indica, n. ser., v. 5(1), p. 1-33.

———, 1915, Fossilium Catalogus. I Animalia. Pt. 8, Cephalopoda Triadica: W. Junk, Berlin, 369 p.

Furnish, W. M., 1966, Ammonoids of the Upper Permian *Cyclolobus* Zone: Neues Jahrb. Geologie Palaeontologie, Abhandl., v. 125, p. 265-296.

Griesbach, C. L., 1880, Palaeontological notes on the Lower Trias of the Himalayas: Geol. Survey India, Rec., v. 13(2), p. 94-112.

Krafft, A. von, and Diener, Carl, 1909, Himalayan Fossils. Lower Triassic Cephalopoda from Spiti, Malla Johar, and Byans: Geol. Survey India, Mem., Palaeontologia Indica, ser. 15, v. 6(1), p. 1-186.

Kummel, Bernhard, and Teichert, Curt, 1970, Stratigraphy and paleontology of the Permian-Triassic boundary beds, Salt Range and Trans-Indus ranges, West Pakistan: in Kummel, B., and Teichert, C., eds., Stratigraphic Boundary Problems: Permian and Triassic of West Pakistan, Univ. Press Kansas, Dept. Geology, Univ. Kansas Spec. Publ. 4, p. 1-110.

Middlemiss, C. S., 1910, A revision of the Silurian-Trias sequence in Kashmir: Geol. Survey India, Rec., v. 40(3), p. 206-260.

Petrenko, V. M., 1963, Nekotorye vazhnye nautiloidi Rannetriasovoi fauny na Ostrove Shpitsbergei. Uchenye Zapiski Paleontologie i Biostratigrafiya. (Some important nautiloids in the Lower Triassic fauna on the island of Spitsbergen: Scientific Records of Paleontology and Biostratigraphy.) Nauchno-issledovatel'skii Inst. Geol. Arktiki Gosudarstvennogo Geologicheskogo Komiteta SSSR, no. 3, p. 50-54.

Popov, Yu. N., 1961, Triasovye ammonoidei severo-vostoka SSSR (Triassic ammonoids of northeastern USSR): Trudy Nauchno-issledovatel'skii Inst. Geol. Arktiki, no. 79, p. 1-178.

Reeside, J. B., et al., 1957, Correlation of the Triassic formations of North America exclusive of Canada: Geol. Soc. America, Bull., v. 68, p. 1451-1514.

Ruzhentsev, V. E., 1962, Klassifikatsiya semeistva Araxoceratidae. (Classification of the family Araxoceratidae.): Paleont. Zhurnal, no. 4, p. 88-104.

Schindewolf, O. H., 1954, Über die Faunenwende vom Paläozoikum zum Mesozoikum: Deutsch. Geol. Ges., Zeitschr., v. 105, p. 154-183.

Spath, L. F., 1930, The Eo-Triassic invertebrate fauna of East Greenland: Medd. om Grønland, v. 83, p. 1-90, pl. 1-12.

———, 1934, Catalogue of the fossil Cephalopoda in the British Museum (Natural History). Part IV, The Ammonoidea of the Trias: London, p. 1-521, pl. 1-17.

———, 1935, Additions to the Eo-Triassic invertebrate fauna of East Greenland: Medd. om Grønland, v. 98, p. 1-115, pl. 1-23.

Teichert, Curt, 1966, Nomenclature and correlation of the Permian "Productus Limestone," Salt Range, West Pakistan: Geol. Survey Pakistan, Rec., v. 15 (1), p. 1-20.

Tozer, E. T., 1961, Triassic stratigraphy and faunas, Queen Elizabeth Islands, Arctic Archipelago: Geol. Survey Canada, Mem., v. 316, p. 1-116.

———, 1967, A standard for Triassic time: Geol. Surv. Canada, Bull. 156, p. 1-103, pl. 1-10.

Waagen, W., 1872, On the occurrence of Ammonites, associated with Ceratites, and Goniatites in the Carboniferous deposits of the Salt Range: Geol. Survey India, Mem. 9, p. 351-358, pl. 1.

———, 1879, Salt Range Fossils. I. Productus-limestone fossils: Geol. Survey India, Mem., Palaeontologia Indica, ser. 13, v. 1 (1), p. 1-72, pl. 1-6.

———, 1895, Salt Range fossils. Fossils from the Ceratite Formation. Pt. I, Pisces-Ammonoidea: Geol. Survey India, Mem., Palaeontologia Indica, Ser. 13, v. 2, p. 1-323, pl. 1-40.

Early Triassic Marine Ostracodes from the Salt Range and Surghar Range, West Pakistan

I. G. SOHN

U.S. Geological Survey, Washington, D.C. 20242

ABSTRACT

Fragile, poorly preserved ostracodes obtained by digesting Lower Triassic limestones from West Pakistan with acetic acid are described and illustrated. Typically shallow-water, marine taxa of both Paleozoic and of Mesozoic affinities occur in the same collections, suggesting that some Paleozoic marine ostracodes survived the end of the Permian. The Knightinidae, n. fam., is established in the Kirkbyacea for *Carinaknightina*, n. gen., and *C. carinata*, n. sp., *C.* aff. *C. carinata*, and *C. discarinata*, n. sp., are described. Additional taxa are: *Lutkevichinella? ornata*, n. sp., *Judahella?* sp., *Bairdia?* sp., *Bairdiacypris?* sp., *Microcheilinella* sp., *Hungarella?* sp., and *Reubenella?* sp.

Amphissites n. sp. 1, from the Upper Permian Wargal Formation is described and illustrated, and genera found in collections from the Upper, Middle, and Lower Productus Limestone are listed. *Knightina? cuestaforma* Sohn, 1954, from the Permian of Texas is transferred to *Tenebrion* Zanina, 1956, and *Basslerella australae* Crespin, 1945, from the Permian of Australia is transferred to *Graphiadactyllis* Roth, 1929.

ACKNOWLEDGMENTS

I am grateful to Dr. B. Kummel, Harvard University, and Dr. C. Teichert, University of Kansas, for sharing their collections to be processed for ostracodes. Dr. R. E. Grant, U.S. Geological Survey, contributed parts of some of his collections from the Permian of the Salt Range. Mr. R. H. McKinney and Mr. D. H. Massie, U.S. Geological Survey, prepared the photographs, and Mrs. Elinor Stromberg, U.S. Geological Survey, prepared the plate. Mr. P. J. Jones, Bureau of Mineral Resources, Canberra, Australia, made available the types and duplicates of *Basslerella australae* Crespin, 1945.

Publication was authorized by the Director, U.S. Geological Survey.

INTRODUCTION

Marine Late Permian and marine Early Triassic ostracodes are rare and poorly known. A survey of the literature through 1966 disclosed only six papers in which marine Early Triassic ostracodes were described (Sohn, 1968a, p. 6). Consequently, the ostracodes from West Pakistan described in this paper, although poorly preserved and numerically scarce, are of special interest. Their importance is enhanced by the fact that both Paleozoic types and Mesozoic through Recent types are represented.

The condition of the specimens is due in part to the fact that the ostracodes were obtained by dissolving limestones in acetic acid. The results are extremely fragile and poorly preserved fossils of which only a few were recovered. Many of the recovered specimens consist of two layers with a void between. Evidently, the chemical composition of only the outer and inner surfaces of the valves was insoluble in dilute acetic acid. Some of these specimens were damaged in the process of photography and study.

Crushing of the limestone did not yield any ostracodes. The small number of specimens obtained by etching suggests that the ostracodes represent only a minor fraction of the rock.

The possibilities that the Paleozoic types represent either contamination or were reworked into the Triassic are discounted for the following reasons: Clean pieces of limestones were etched in order to insure against field or laboratory contamination. The Paleozoic Kirkbyidae occur as steinkerns (Pl. 1, fig. 10) created in the laboratory by the acid. Reworking of Paleozoic fossils probably would have disassociated the carapaces into valves. *Carinaknightina* is present as fragile valves that do not show any abrasion caused by reworking. Furthermore, all the specimens in a given collection have approximately the same color.

The material available to me represents but a fraction of the forms that lived during the Early Triassic in the area where they were collected. It is hoped that presentation of these meager results will stimulate future collecting of ostracodes, so that additional forms will become available.

In the systematic part of the paper the locality numbers are those of Kummel and Teichert (1970, fig. 4).

COLLECTION LOCALITIES

Triassic

USGS Mesozoic loc. 29613. Loc. 2, Narmia, Surghar Range, Narmia Member of Mianwali Formation, bed 32 in Kummel, 1966, p. 413. Collected by Bernhard Kummel, field no. K1-42.

USGS Mesozoic loc. 29614. Loc. 6B, east side of Chhidru Nala, Salt Range limestone in lower part of Mittiwali Member of Mianwali Formation, lower part of bed 15 in Kummel, 1966, p. 408. Collected by Bernhard Kummel, field no. K4-CM.

USGS Mesozoic loc. 29615. Loc. 1, Landa Pusha, Surghar Range, limestone in Narmia Member of Mianwali Formation, bed 9 in Kummel, 1966, p. 416. Collected by Bernhard Kummel, field no. K12-9.

Permian

USNM loc. 9210. Kalabagh Member of Wargal Formation, 1 mile due north of mosque at west end of Jabbi; Survey of Pakistan sheet 43 D/3. Collected by R. E. Grant and A. N. Fatmi, Jan. 25, 1964.

HISTORICAL REVIEW

The first paper to describe Early Triassic marine ostracodes is that of Méhes (1911). He described and illustrated eight taxa from the Balaton Mountains, Hungary, that have been reassigned as follows (Sohn, 1968a): *Bairdia hagenowi* Reuss = gen. and sp. indet., *B. subglobosa* Bosquet = *Hungarella?* sp., *Cythere tenera* Brady = gen. and sp. indet. (Podocopida), *Loxoconcha pusilla* Brady and Robertson = gen. and sp. indet. (Cytheracea), *Bairdia* sp., *Xestoleberis?* sp. = gen. indet., *Cythereis* sp. = gen. indet., and *Cytherella* sp.

Bielecka (1956) illustrated *Bairdia carinthiaca* Gümbel = *Clinocypris?* sp. from the Bunter Sandstone, Holy Cross Mountains, Poland. Shneider (1956) described the following marine forms from the Transcaspian Lowland of the USSR: *Gemmanella schweyeri* and *Lutkevichinella bruttanae*. The following year she (Shneider, 1957) described *Pulviella ovalis* and *Renngartenella ovata* from the Astrakhan District of the USSR. Styk (1958) identified *Cythere tubulifera* Gümbel = *Judahella* n. sp. Sohn (1968a) and *Bairdia carinthiaca* Gümbel = *Clinocypris?* sp. Sohn (1968a) from the Upper Bunter Sandstone taken from a borehole in Poland. Shneider (1960) added to her previous list *Gemmanella parva, Renngartenella avdusini,* and *Lutkevichinella involuta*. Belousova (1965) described and illustrated from the Induan (Seisian) Stage of Transcaucasia in the USSR, *Bairdia armenica, B. intermedia* = *Cryptobairdia intermedia* (Belousova), *B.? pseudoo-*

TABLE 1. Stratigraphic Ranges of Genera and Families of Early Triassic Species Discussed in This Paper.

	Pre-Triassic	Triassic Lower	Triassic Middle	Triassic Upper	Post-Triassic
Kirkbyidae gen. indet. sp. or spp.	————	————		?	
Carinaknightina n. gen. n. spp.		————			
Microcheilinella sp.		————			
Bairdiacypris? sp.		————		?	
Bairdia? sp.		————			————
Monoceratina? sp.	?	————			————
Lutkevichinella? n. sp.		————			?
Hungarella? sp.		————			
Judahella? sp.		————			
Reubenella? sp.		————			

buncus = *Cryptobairdia pseudoobunca* (Belousova), *Fabilicypris oboncus* (misspelling of *Fabalicypris*), *F. subgeinitziana*, *Healdia incognita*, *Healdianella doraschamensis*, and *H. splendida*.

An abstract of the findings represented in the present paper was published previously (Sohn, 1968b).

SIGNIFICANCE OF THE TRIASSIC OSTRACODES

The coexistence of ostracode taxa having Paleozoic affinities with those having Mesozoic to Recent affinities indicates that some Paleozoic marine ostracodes survived the end of the Permian. Table 1 shows the known stratigraphic ranges of the genera and families discussed in this paper. This table demonstrates that the Early Triassic ostracode assemblage contains elements of both Paleozoic and post-Paleozoic (Mesozoic and younger) affinities. The fact that the overall makeup of the ostracodes is that of shallow-water rather than deep-water inhabitants is significant. One of the postulated explanations for the repopulation of marine animals after the Permian is that elements that were able to survive in deep water immigrated during the Triassic into shallow-water environments. This assemblage from the Salt Range indicates that such need not be the case.

Terry and Tucker (1968) suggested that cosmic radiation caused by the explosions of supernovae at the end of the Permian could have caused the extinction of many animals, including shallow-water marine organisms. The data on hand show that at least some of the Ostracoda, including the Kirkbyacea, were not affected.

The presence of definite Kirkbyacea in the assemblage is significant, because it sheds light on the classification of that group. Prior to this study only three recorded occurrences of Kirkbyidae, none documented by illustrations, were known from the Triassic: Kollmann (1963, p. 144-146) recorded "*Kirkbyidae?* indet." from the Upper Triassic of the Alps; Gerke (1957, p. 99) mentioned the new genus *Nordvikia (nomen nudum)* as "very similar to Paleozoic Kirkbyids" from the Carnian of Siberia, and Plöchinger (1963, p. 63, 64) mentioned "Kirkbyidae ind." from the Rhaetian of Salzburg. The specimens illustrated on Plate 1, figures 4-11, are definitely Kirkbyidae, although they are steinkerns. There is no question that the new genus *Carinaknightina* is of kirkbyid affinities. These specimens verify the tentative extension of the stratigraphic range of the Palaeocopida into the Triassic (Sohn, 1965, p. 41).

Lutkevichinella? ornata, *Hungarella?* sp., *Judahella?* sp., and **Reuben-**

ella? sp. are Mesozoic and younger in aspect. Of these the first three occur in the same sample as *Carinaknightina*, while *Reubenella?* sp. is from the same collection as the Kirkbyidae. This indicates that during the Early Triassic there were still Paleozoic types of ostracodes living with Mesozoic and younger types.

Except for the one species of *Amphissites* from the Upper Permian, none of the Late Permian taxa are illustrated. Three collections from the lower part of the basal Upper Productus Limestone at Chhidru Nala contain well-preserved specimens of a new species of *Graphiadactyllis* Roth, 1929, that is closely related to the Middle Permian *Graphiadactyllis australae* (Crespin) (=*Basslerella australae* Crespin, 1945) described from the Lower Permian of eastern Australia. The recorded stratigraphic range of *Graphiadactyllis* has been only from the Mississippian, with a few questionable identifications from the Devonian.

Collections from the Lower, Middle, and Upper Productus limestone contain fair to poorly preserved representatives of the following genera: *Glyptopleuroides* Croneis and Gale, 1938, *Youngiella?* Jones and Kirkby, 1895, *Roundyella* Bradfield, 1935, *Hypotetragona* Morey, 1935, *Carboprimitia* Croneis and Funkhouser, 1939, *Bairdia* McCoy, 1844, *Bairdiacypris* Bradfield, 1935, *Silenites* Coryell and Booth, 1933, *Microcheilinella* Geis, 1933, *Acratia?* Delo, 1930, *Healdia* Roundy, 1926, *Cavellina* Coryell, 1928, and a new genus probably in the Platycopida.

A collection of shale from the Middle Productus Limestone from the Salt Range, the exact locality of which is in doubt, housed in the U.S. National Museum, contains specimens of *Bairdia*, *Ceratobairdia* Sohn, 1954, and *Amphissites* Girty, 1910.

SYSTEMATIC DESCRIPTIONS

Order PALAEOCOPIDA
Henningsmoen, 1953

Superfamily KIRKBYACEA
Ulrich and Bassler, 1906

Diagnosis: Straight-backed, reticulated, with kirkbyan pit, with or without lobes, nodes, and carinae; ridge and groove hingement, with or without terminal dentition; valves subequal, overlap slight, free margin of one valve rabbeted to receive opposing valve; one or more marginal rims; dimorphism unknown.

Discussion: Gründel (1965) revised the Kirkbyacea and proposed a classification that differs from that used in the *Treatise* (Sohn in Moore, 1961, p. Q163). The difference is due in part to Gründel's hypothesis that the "kirkbyan pit" is not a valid criterion for suprageneric classification. Gründel's postulated reduction of the kirkbyan pit with time as the group became younger (1965, p. 51) is contradicted by the Triassic specimens on hand.

Earlier (Sohn, 1954, p. 5; in Moore, 1961, p. Q167) I put into the Kellettinidae those genera in the Kirkbyacea that lack a true kirkbyan pit. Furthermore, I noted (Sohn, 1954, p. 11) that the genus *Knightina* Kellett, 1933, included species with and without kirkbyan pits. *Knightina* was provisionally referred to the Kirkbyidae, with the understanding that, should the genus be split, *Knightina* s.s. would more properly belong to the Kellettinidae. In this paper I am removing those taxa without a kirkbyan pit to the new family, the Knightinidae.

Family KIRKBYIDAE
Ulrich and Bassler, 1906

Diagnosis: Elongated, lobed or unlobed, without nodes or carinae.

TABLE 2. Measurements of Kirkbyidae gen. indet. sp. or spp. (in mm).

	Length	Height	Width
Figured specimen (Pl. 1, fig. 4, 5)	0.51	0.25	0.19
Figured specimen (Pl. 1, fig. 6-8)	0.72	0.36	0.27
Figured specimen (Pl. 1, fig. 9)	0.55	0.28	0.23
Figured specimen (Pl. 1, fig. 10)	0.80+	0.41	0.26
Figured specimen (Pl. 1, fig. 11)	0.75	0.37	0.31

Stratigraphic range: Lower Mississippian to Lower Triassic.

Discussion: The following genera are referred to this family: *Kirkbya* Jones, 1859; *Aurikirkbya* Sohn, 1950; *Coronakirkbya* Sohn, 1954; and *?Reviya* Sohn, 1962. Both I (Sohn in Moore, 1961, p. Q164) and Gründel (1965, p. 57) had referred *Knightina* Kellett, 1933, to this family; however, upon restudy of the type species, *Amphissites allerismoides* Knight, 1928, I now remove this genus from the Kirkbyidae.

Kirkbyidae gen. indet. sp. or spp.
Pl. 1, fig. 4-11

The specimens on hand are too poorly preserved for generic designation. The illustrations show the family characters.

The above measurements do not represent the actual sizes of the specimens because they are for steinkerns, and Plate 1, figure 10, clearly shows that the shell was larger. The measurements do, however, indicate relative proportions of the measured parameters.

Occurrence: Landa Pusha (loc. 1, K12-9) and Narmia Spring, Surghar Range (loc. 2, K1-42). Narmia Member of Mianwali Formation.

Age: Early Triassic.

Family AMPHISSITIDAE
Knight, 1928

Genus AMPHISSITES Girty, 1910

See Sohn (1962, p. 115) for a discussion of this genus. A single valve from the Permian Kalabagh Beds of the Salt Range (USNM loc. 9210) is illustrated in order to show the differences between this genus and the new genus *Carinaknightina* described in this paper.

Amphissites n. sp. 1
Pl. 1, fig. 18, 19

Only one right valve of this species is available, consequently the species is not named. It differs from *A. centronotus* (Ulrich and Bassler), 1906, and from *A. truncatus* Sohn, 1962, in having an elongated subcentral node.

Measurements of figured specimen: length 0.65 mm, height 0.38 mm.

Occurrence: One mile due north of mosque at west end of Jabbi, Salt Range, Survey of Pakistan sheet 43 D/3 (USNM loc. 9210); Kalabagh Member of Wargal Formation (top part of "Middle Productus limestone").[1]

Age: Late Permian.

Family KNIGHTINIDAE Sohn,
new family

Diagnosis: Straight-backed, elongated, reticulated, rimmed ostracodes, without a well-developed kirkbyan pit, without nodes or lobes.

Discussion: The absence of a "kirkbyan pit" differentiates this family from the Kirkbyidae. The type species of the nominate genus has a subcentral smooth area that is larger than a single reticulation on the surface of the valve. In addition to the nominate genus, the following are here assigned to this family: *Tenebrion* Zanina, 1956, *Villo-*

---
[1] For an alternative interpretation of the age of this sample, see Kummel and Teichert (1970, p. 73).—Eds.

zona Gründel, 1965, and *Carinaknightina* Sohn, n. gen.

Stratigraphic range: Devonian through Lower Triassic.

Genus KNIGHTINA Kellett, 1933

Type species (original designation): *Amphissites allerismoides* Knight, 1928, upper Fort Scott Limestone, Missouri.

Diagnosis: With obtuse cardinal angles; large, pitlike reticulations, subdued to well-defined posterior shoulder, and subequal ends; posterior margin more truncated; without any lobes, nodes, or carinae; two marginal rims.

Discussion: I previously (Sohn in Moore, 1961, p. Q164) considered *Tenebrion* Zanina, 1956, as a synonym of *Knightina,* but now recognize *Tenebrion* as a distinct genus. The species originally described as *Knightina? cuestaforma* Sohn, 1954, is a *Tenebrion.* The "kirkbyan pit" illustrated in the holotype of *K.? cuestaforma* (Sohn, 1954, pl. 3, fig. 30) is an artifact due to poor preservation. A true kirkbyan pit has a smooth rim on the lateral surface of the valve: the holotype (USNM 118386) and an unfigured paratype (USNM 118387a) do not have smooth rims around the adductor muscle-scar area. The stratigraphic range of *Tenebrion* is here extended from the Upper Devonian through the Permian.

Genus CARINAKNIGHTINA
Sohn, n. gen.

Type species: C. carinata Sohn, n. sp.

Diagnosis: Differs from *Knightina* Kellett, 1933, in having a rim subparallel to the dorsal margin, with or without carinae crossing the lateral surface of the valves. Hinge weak ridge and groove, with or without terminal teeth and sockets.

Description: The carapace is subquadrate to suboval in lateral outline, the cardinal angles are obtuse, ends subequal with a narrower posterior margin. The surface is covered by reticulations with wide partitions. A thin ridge parallels the dorsal margin while the lateral surface may bear variously oriented costae.

Discussion: The ridge subparallel to the dorsal margin is reminiscent of the dorsal shield in *Amphissites* Girty, 1910 (compare Pl. 1, fig. 16, 21, and 24 with 18), from which this genus differs in lacking a "kirkbyan pit" and a subcentral node. *Villozona* Gründel, 1965, differs from the new genus in larger size and absence of dorsal carina. In addition to the species illustrated here, the Permian *Amphissites tscherdynzevi* Shneider (1948, p. 42) probably belongs to this genus.

Carinaknightina carinata Sohn, n. sp.
Pl. 1, fig. 12-17

Kirkbyacea, n. gen., n. sp., Sohn, 1968, p. 12, pl. 3, figs. 9, 10.

Name: Carina rib.

Holotype: USNM 147202.

Paratypes: USNM 160774, 160775, and 160787.

Material: Five valves including juveniles, some broken.

Type locality: East side of Chhidru Nala, Salt Range (loc. 6B, K4-CM).

Stratigraphic occurrence: Lower part of Mittiwali Member of Mianwali Formation, Lower Triassic.

Diagnosis: With well-developed inner rim and subcentral horizontally trending, sinuous or discontinuous carina.

Description: The valves are small, less than 1 mm in greatest length, subovate in lateral outline, the diagnostic dorsal ridge curves downward near the posterior cardinal angle resulting in an apparent sinuous dorsal margin. The hinge, as seen from the inside of the valve is, however, straight. The inner rim is removed from the outer rim, it is as well developed as the dorsal ridge which it joins at about 90° near the anterior cardinal angle, it then continues subparallel with the free margins and joins the dorsal ridge at the posterior cardinal angle. The position of this inner ridge on the surface of the valve varies with individuals; a left valve (USNM 160787) has a wider distance between this rim and the inner rim than the holotype. A juvenile, about half the size of the presumed adults, on which the posterior portion is missing, has the same conformation of ridges and subcentral carina. The position and shape of the subcentral carina also varies with individuals; it consists of slightly raised partitions. On the holotype this carina appears to consist of two parts; one branch angles from the posterior cardinal area to the anteroventral area without reaching the inner rim, the second branch intersects this carina behind midlength about halfway between the dorsal and inner rims from where it trends towards the approximate midanterior margin, but does not quite reach the inner rim. The carina is represented by the concave upward subcentral portion that is slightly more than the combined width of four reticules. Except for the two pits at the cardinal angles of the holotype (Pl. 1, fig. 15), the hingement cannot be discerned. The muscle-scar area cannot be discerned on the outside of the valves; it is reflected on the inside by a low node. Measurements of representative specimens are given in Table 3.

TABLE 3. Measurements of *Carinaknightina carinata* (in mm).

	Length	Height
Holotype	0.41	0.26
Paratype, (Pl. 1, fig. 13, 14)	0.47 (approx.)	0.30
Paratype, broken (Pl. 1, fig. 12)	0.20	0.14
Paratype, USNM 160787	0.40	0.25

Discussion: This species differs from *C.? tscherdynzevi* (Shneider, 1948) from the Upper Permian of Kazan and Timan in smaller size and in that on the Permian species the inner rim joins the dorsal rim farther from the cardinal angles. The Permian species has a convex upward carina that starts at the midpoint of the posterior curved portion of the rim and ends at the anteroventral portion of that rim.

Occurrence: Restricted to type locality.

Age: Early Triassic.

Carinaknightina aff. *C. carinata* Sohn

Pl. 1, fig. 20-22

One left valve that is more elongate and that has a longer sinuous subcentral carina is in the collection. Because of poor preservation the specific identity is uncertain.

Measurements of figured specimen: length 0.55 mm, height 0.30 mm.

Occurrence: Same as nominate species. East side of Chhidru Nala (K4-CM).

Age: Early Triassic.

Carinaknightina discarinata Sohn, n. sp.

Pl. 1, fig. 23-25

Name: Discarinata—without ribs.
Holotype: USNM 160778.
Paratype: USNM 160790.
Material: Two left valves.
Type locality: East side of Chhidru Nala (loc. 6B, K4-CM).
Stratigraphic occurrence: Lower part of Mittiwali Member of Mianwali Formation.
Diagnosis: Without inner rim or subcentral costa.
Description: The valve is suboval in lateral outline; the end margins subequal with the posterior margin shorter. The lateral surface does not have an inner rim or subcentral costa, it is covered with porelike reticules separated by relatively thick partitions.

Measurements of holotype: Length 0.55 mm, height 0.35 mm.

Plate 1

Discussion: This species resembles Pennsylvanian and Permian species of *Knightina* in lateral outline and reticulations, but is readily distinguished by the dorsal rim. Although only two valves are available, the specimens are sufficiently distinct to warrant naming.

Occurrence: Restricted to type locality.

Age: Early Triassic.

Superfamily unknown

Family JUDAHELLIDAE Sohn, 1968

Genus JUDAHELLA Sohn, 1968

Judahella Sohn, 1968a, p. 14.

Type species (original designation): *Judahella tsorfatia* Sohn, 1968.

Upper Muschelkalk (Ladinian), railroad cut northwest of Faulquemont, France.

Diagnosis (from original): Small, approximately equivalved, straight-backed; lateral surface reticulated or pitted; with three or more nodes along the dorsal margin, and horizontal lobe that may be broken into two or more nodes along the ventral margin; overlap slight along free margins; hinge simple, dorsal edge of larger valve fits into smooth groove of smaller valve, shallow cardinal sockets below dorsal edge of larger valve. Internal marginal structure and muscle-scar pattern unknown. May be dimorphic in shape of lateral outline.

PLATE 1

[Magnifications approximately as noted.]

FIGURES

1-3. *Microcheilinella* sp. (p. 204). Right, posterior, and dorsal views of a steinkern of a carapace ×30. Figured specimen USNM 160768.

4-11. Kirkbyidae gen. indet. sp. or spp. (p. 195). [6-8 from Narmia Spring (loc. 2), the others from Landa Pusha (loc. 1)].——4, 5. Dorsal and lateral views of a steinkern ×30. Figured specimen USNM 160769.——6-8. Right, ventral, and left views of a steinkern ×30; note trace of duplicature on figure 7. Figured specimen USNM 160770.——9. Left view of steinkern ×30. Figured specimen USNM 160771.——10. Right view of steinkern with part of the shell preserved ×30. Figured specimen USNM 160772.——11. Left view of broken steinkern, posterior missing ×30. Figured specimen USNM 160773.

12-17. *Carinaknightina carinata* Sohn, n. gen., n. sp. (p. 198).——12. Lateral view of right valve of instar ×60. Paratype USNM 160774.——13, 14. Inside ×30, and outside ×60 of right valve. Paratype USNM 160775.——15-17. Inside, ventral, and outside views of right valve ×60. Holotype USNM 147202.

18, 19. *Amphissites* n. sp. 1 (p. 197). Dorsal and outside views of right valve ×30. Figured specimen USNM 160776.

20-22. *Carinaknightina* aff. *C. carinata* Sohn (p. 198). Inside, dorsal, and outside views of left valve ×30. Figured specimen USNM 160777.

23-25. *Carinaknightina discarinata* Sohn, n. sp. Outside, dorsal, and inside views of left valve ×30. Holotype USNM 160778.

26, 27. *Judahella?* sp. (p. 202). Inside and outside views of right valve ×60. USNM 160779. The apparent node directly below the middle dorsal node and at the end of the ridge separating the posterior dorsal node and the ventral ala is an artifact of photography.

28-31. *Monoceratina?* sp. (p. 202). Dorsal, lateral, ventral, and inside views of a left valve ×60. Figured specimen USNM 160780.

32, 33. *Bairdiacypris?* sp. (p. 204). Ventral and right views of a steinkern ×30. Note the trace of the duplicature. Figured specimen USNM 160781.

34-37. *Lutkevichinella? ornata* Sohn, n. sp. (p. 202).——34, 35. Inside ×30 and outside ×60 views of left valve, broken during photography. Paratype USNM 160782.——36, 37. Outside and inside views of left valve ×30. Holotype USNM 160783.

38-40. *Bairdia?* sp. (p. 203). Inside, dorsal, and outside views of right valve ×30. Figured specimen USNM 160784.

41, 42. *Hungarella?* sp. (p. 204). Inside and outside views of left valve ×30; note broken outer preserved layer (fig. 42). Figured specimen USNM 160785.

43, 44. *Reubenella?* sp. (p. 205). Ventral and right views of a steinkern ×30. Figured specimen USNM 160786.

Judahella? sp.

Pl. 1, fig. 26, 27

Only one broken right valve that probably represents this genus is available. It differs from all the known species of *Judahella* by having only three nodes and in that the ventral node is inflated behind midlength into a lateral ala. A thin ridge separates the dorsal nodes from the ventral lobe, resembling the Cytheracea, from which group the species on hand is distinguishable by having a simple hinge. The specimen broke after photography.

Measurements of figured specimen: length, 0.38 mm, height 0.21 mm.

Occurrence: East side of Chhidru Nala, Salt Range (loc. 6B, K4-CM), lower part of Mittiwali Member of Mianwali Formation.

Age: Early Triassic.

Order PODOCOPIDA Sars, 1865

Suborder CYTHEROCOPINA Gründel, 1967

Gründel (1967) replaced the suborder Podocopida as used in Moore (1961) by the new suborders Bairdiocopina and Cytherocopina. His suborder Cytherocopina is accepted in this paper.

Superfamily CYTHERACEA Baird, 1850

Family BYTHOCYTHERIDAE Sars, 1926

Subfamily MONOCERATININAE Szczechura, 1964

Genus MONOCERATINA Roth, 1928

See Szczechura (1964) for a discussion of this genus.

Monoceratina? sp.

Pl. 1, fig. 28-31

One specimen, a left valve, is present in the collections. Unfortunately, the terminal part of the caudal process was broken during examination, but after photography.

Measurements of figured specimen: length 0.38 mm, height 0.20 mm.

Discussion: Herrig (1967) described and illustrated a morphological form group around the Late Cretaceous *Monoceratina umbonatoides* Kaye, 1964, and referred the taxa to *Bythocytherina* Hornibrook, 1952. The specimen on hand resembles that group of species in having a deep sulcus and extended ventrolateral ala. It differs, however, in lateral outline and lack of reticulations.

Occurrence: East side of Chhidru Nala, Salt Range (loc. 6B, K4-CM), lower part of Mittiwali Member of Mianwali Formation.

Age: Early Triassic.

Family CYTHERISSINELLIDAE Kashevarova, 1958

Genus LUTKEVICHINELLA Shneider, 1956

Lutkevichinella? ornata Sohn, n. sp.

Pl. 1, fig. 34-37

Name: Ornata—ornamented with strong ribs.

Holotype: USNM 160783.

Paratypes: USNM 160782 and 160788.

Material: Three left valves, one broken in handling.

Type locality: East side of Chhidru Nala, Salt Range (loc. 6B, K4-CM).

Stratigraphic occurrences: Lower part of Mittiwali Member of Mianwali Formation, Lower Triassic.

Diagnosis: Posterior margin about half the height of the anterior margin, surface ornamented by distinct anastomosing, wavy ridges between which are deep pits. Subcentral sulcus weakly developed.

Description: The valve is suboval; the anterior margin is about twice as high as the posterior margin, it is curved and truncated ventrally; the posterior margin is more rounded than the anterior margin, commences near the hingeline and terminates at approximate midlength. The ventral margin is gently curved, it deflects backwards and upwards from approximately directly below the anterior cardinal angle, making the greatest height of the valve at that line. The dorsal margin is straight. Dorsal outline lanceolate, more sharply truncated in the posterior than the anterior, and with the greatest width at approximately posterior quarter of greatest length. The hinge is apparently slightly incised; the venter is concave so that the ventral margin is obscured by the ventral inflation of the lateral surface. The hinge of the left valve is a subdued, straight, smooth ridge terminated by shallow sockets; the hinge of the right valve is unknown. Muscle-scar pattern and inner margin not discernable. Measurements of available specimens are given in Table 4.

TABLE 4. Measurements of *Lutkevichinella? ornata* (in mm).

	Length	Height
Holotype	0.51	0.29
Paratype	0.51	0.31
Paratype (unfigured, USNM 160788)	0.46	0.26

Discussion: Although the type species of *Lutkevichinella*, *L. bruttanae* Shneider, 1956, from the Lower Triassic of Russia has a distinct subcentral vertical sulcus, the only additional known species, *L. involuta* Schneider, 1960, also from the Lower Triassic, does not have a distinct sulcus; consequently, I am tentatively referring the new species to this genus. The two previously described species have subconcentric reticulations with deep pits, while the new species has a ridged ornament that is reminiscent of the family Progonocytheridae Sylvester-Bradley, 1948, (Jurassic to Recent) from which, however, it differs by having a simple hinge.

Occurrence: Restricted to type locality.
Age: Early Triassic.

Suborder BAIRDIOCOPINA Gründel, 1967

Gründel (1967, p. 326) established this suborder to include the Bairdiacea, Cypridacea, and ?Darwinulacea. The Darwinulacea is a primitive group that probably does not belong in the Bairdiocopina.

Superfamily BAIRDIACEA Sars, 1887
Family BAIRDIIDAE Sars, 1887
Subfamily BAIRDIINAE Sars, 1887

See Sohn (1961, p. 12) and Kollmann (1963, p. 160-165) for a discussion of this group.

Genus BAIRDIA McCoy, 1844

In an earlier study (Sohn, 1961, p. 16) I had questioned the stratigraphic range of *Bairdia* in sediments younger than the Paleozoic. Subsequent discovery of poorly preserved specimens of *"Bairdia"* in sediments of Carnian age in Israel indicates that the genus does range into the Mesozoic.

Bairdia? sp.
Pl. 1, fig. 38-40

A few steinkerns and one poorly preserved valve of what appears to be *Bairdia* are in the collections. There is no point in illustrating the steinkerns because Kollmann (1960, 1963) established several Triassic genera based on lateral surface ornament, and there is no way of determining the surface ornament of the steinkerns. The valve, although the outer surface may be partly abraded, and the end margins may be partly missing, appears to be a true *Bairdia*.

Measurements of figured specimen: length 0.84 mm, height 0.41 mm.

Occurrence: Landa Pusha, Surghar

Range (loc. 1, K12-9), Narmia Member of Mianwali Formation.
Age: Early Triassic.

Genus BAIRDIACYPRIS Bradfield, 1935

See Sohn (1961, p. 57) for a discussion of this genus. When that paper was prepared, I questioned the stratigraphic range of this genus into the Permian. The presence of a steinkern in the collections on hand that probably belongs to this genus, and of *B.? pannonica* (Méhes), 1911, in the Carnian of Hungary, extend the stratigraphic range of *Bairdiacypris* into the Triassic.

Bairdiacypris? sp.

Pl. 1, fig. 32, 33

Only one steinkern that probably belongs to this genus is present in the collections. The specimen could have been an immature individual because of its small size, but the unmistakable trace of a wide inner lamella on the free margins suggests that the steinkern is of an adult individual.

Measurements of figured specimen: length 0.70 mm, height 0.28 mm.

Occurrence: Narmia Spring, Surghar Range (loc. 2, K1-42), Narmia Member of Mianwali Formation.

Age: Early Triassic.

Family uncertain

Microcheilinella Delo, 1932

Type species (original designation): *Microcheilus distortus* Geis, 1932, Salem Limestone, Indiana.

Diagnosis: Small, less than 0.75 mm, elongate, asymmetrical, smooth ostracodes. Larger valve overlaps on all free margins, overreaches along incised hinge margin where smaller valve apparently overlaps. Inner lamella and duplicature well developed. Muscle-scar pattern unknown.

Discussion: The suprageneric affinities of this genus are unknown because single valves are rare due to the small size and strong overlap. More than 50 species of Silurian to Permian age have been recorded in *Microcheilinella*. Many of the lower Paleozoic species referred to this genus do not belong in it, and three were removed (Sohn, 1961, p. 74-75). Because poorly silicified rare specimens of *Microcheilinella* are present also in a collection of Scythian age from the Island of Chios, Greece, the stratigraphic range of the genus is as young as the Lower Triassic.

The genus is easily recognized, even in steinkerns, by its asymmetrical end views. Species have been differentiated on subtle differences in outline, and, if present, the position of postero-ventral spines.

Microcheilinella sp.

Pl. 1, fig. 1-3

Four steinkerns of fat, small carapaces belonging to this genus were recovered.

Measurements of figured specimen: length 0.33 mm, height 0.18 mm, width 0.21 mm; those of unfigured specimen (USNM 160789): length 0.38 mm, height 0.21 mm, width 0.22 mm.

Occurrence: Landa Pusha, Surghar Range (loc. 1, K12-9), Narmia Member of Mianwali Formation.

Age: Early Triassic.

Suborder METACOPINA Sylvester-Bradley (Moore, 1961)

Family HEALDIACEA Harlton, 1933

Genus HUNGARELLA Méhes, 1911

See Sohn (1968a, p. 28) for discussion of this genus. The healdiid muscle-scar pattern suggests that the genus belongs in the Healdiacea.

Hungarella? sp.

Pl. 1, fig. 41, 42

Poorly preserved valves and fragments of valves that probably represent this genus are present in the collections. These specimens are interesting because of their unusual form of preservation, two layers separated by a void. The illustrated specimen has complete preservation on the inside of the valve, including the inner lamella (Pl. 1, fig. 41), and part of the outside of the valve (Pl. 1, fig. 42). A subcentral healdiid muscle-scar pattern is visible in the area where the outside layer is missing, consequently the genus is referred to the Healdiacea.

Measurements of figured specimen: length 0.88 mm, height 0.54 mm.

Occurrence: East side of Chhidru Nala, Salt Range (loc. 6B, K4-CM), Mittiwali Member of Mianwali Formation.

Age: Early Triassic.

Order PLATYCOPIDA Sars, 1865

Suborder PLATYCOPINA Sars, 1865

Superfamily CAVELLINACEA Egorov, 1950

See Sohn (1968a, p. 17) for a discussion of this group.

Family CAVELLINIDAE Egorov, 1950

Genus REUBENELLA Sohn, 1968

Reubenella? sp.

Pl. 1, fig. 43, 44

One steinkern that probably belongs to *Reubenella* is available in the collections. It differs from the type-species, *R. avnimelechi* Sohn, 1968, from the Ladinian of Israel in having a more lanceolate ventral outline.

Measurements of figured specimen: length 0.62 mm, height 0.32 mm, width 0.19 mm.

Because the specimen is a steinkern, the measurements are useful only to indicate the ratios of the parameters. The size of the specimen is well within the range of size of females in the type species of *Reubenella*.

Occurrence: Landa Pusha, Surghar Range (loc. 1, K12-9), Narmia Member of Mianwali Formation.

Age: Early Triassic.

REFERENCES

Belousova, Z. D., 1965, Ostracoda, *in* Ruzhentsev, V. E., and Sarycheva, T. G., eds., Razvitie i smena morskikh organizov na rubezhe Paleozoia i Mezozoia: Akad. Nauk SSSR, Paleont. Inst. Trudy, v. 108, p. 254-265, pl. 47-50.

Bielecka, Wanda, 1956, Note on Triassic foraminifers of the north-west periphery of the Swiety Krzyz Mountains: Poland Inst. Geol. Bull. 102, p. 81-95, 1 pl., 2 fig.

Crespin, Irene, 1945, Permian Ostracoda from eastern Australia: Queensland Royal Soc., Proc., v. 56, no. 4, p. 31-36, pl. 4.

Gerke, A. A., 1957, O mikrofaune mezozoiskikh otlozhenii severnoi chasti Eniseisko-Lenskogo kraia i ee stratigraficheskom znachenii: Akad. Nauk SSSR, Mezhvedomstv. soveshchaniia po razrabotke unifitsirovannykh stratigraficheskikh skhem Sibiri 1956 g., Trudy Leningrad, p. 98-103.

Gründel, Joachim, 1965, Zur Kenntnis der Kirkbyacea (Ostracoda): Freiberger Forschungshefte, C 182, Palaeontologie, p. 49-61, 7 fig.

———, 1967, Zur Grossgliederung der Ordnung Podocopida G. W. Müller, 1894 (Ostracoda): Neues Jahrb. Geologie Palaeontologie, Monatsh. 1967, no. 6, p. 321-332.

Herrig, Ekkehard, 1967, Zur Phylomorphogenese von *Bythocytherina umbonatoides* (Kaye, 1964), Ostracoda, Crustacea, aus der nordostdeutschen Oberkreide: Geologie, Jahrg. 16, Heit 5, p. 598-614, 7 fig.

Kollmann, Kurt, 1960, Ostracoden aus der alpinen Trias Österreichs. I. *Parabairdia* n. g. und *Ptychobairdia* n. g. (Bairdiidae): [Austria] Geol. Bundesanst., Jahrb., Sonderband 5, p. 79-105, pl. 22-27, fig. 1-30.

———, 1963, Ostracoden aus der alpinen Trias II. Weitere Bairdiidae: [Austria] Geol. Bundesanst. Jahrb., Band 106, p. 121-203, pl. 1-11, 3 tables, 8 fig.

Kummel, Bernhard, 1966, The Lower Triassic formations of the Salt Range and Trans-Indus ranges, West Pakistan: Mus. Comp. Zoology, v. 134, no. 10, p. 361-429, 4 pl.

———, and Teichert, Curt, 1970, Stratigraphy

and paleontology of the Permian-Triassic boundary beds, Salt Range and Trans-Indus ranges, West Pakistan: *in* Kummel, B., and Teichert, C., eds., Stratigraphic Boundary Problems: Permian and Triassic of West Pakistan, Univ. Press Kansas, Dept. Geology, Univ. Kansas Spec. Publ. 4, p. 1-110.

Méhes, Gyula, 1911, Über Trias-Ostrakoden aus dem Bakony, *in* Resultate der wissenschaftl. Erforschung des Balatonsees. Abh. Palaeontologie der Umgebung des Balatonsees, Band 3, Teil 6, 38 p., 4 pl.

Moore, R. C., ed., 1961, Treatise on Invertebrate Paleontology, Part 9, Arthropoda 3: Geol. Soc. America and Univ. Kansas Press, 442 p., 334 fig.

Plöchinger, B., 1963, *in* Grill, R., Kollmann, K., Küpper, H., and Oberhauser, R., eds., D II. Exkursion in den Grünbachgraben am Untersberg-Ostfuss (Salzburg). Exkursionsführer für das Achte Europäische Mikropaläontologische Kolloquium in Österreich: [Austria] Geol. Bundesanst., Verh., Sonderheft F, p. 57-61, pl. 3.

Shneider [Schneider], G. F., 1948, Fauna ostrakod verkhnepermskikh otlozhenii (Tatarskii i Kazanskii yarusy) neftenosnikh raionov SSR: Vses. Neft. Nauch-Issled. Geologo-Razv. Inst. (VNIGRI), n. ser., no. 31, p. 21-48, pl. 1-4.

———, 1956, [New genera], *in* Kiparisova, L. D., Markovskoy, B. P., and Radchenko, G. P., eds., Novye semeistva i rody. Materialy po paleontologii: Vses. Nauch.-Issled. Geol. Inst. (VSEGEI), Ministerstva Geologii i Okhrany Nedr SSSR, n. ser., Paleont., no. 12, Moscow, p. 120-127, pl. 19, 20, 22.

———, 1957, *in* Mandelshtam, M. I., Shneider, G. F., Kuznetsova, Z. V., and Kats [Katz], F., Novye rody ostrakod v semeistvakh Cypridae i Cytheridae: Vses. Paleontologicheskogo Obshchestva, Ezhegodnik, v. 16, p. 181-183, pl. 3, 4.

———, 1960, Fauna ostrakod nizhnetriasovykh otlozhenii Prikaspiiskoi nizmennosti, *in* Brod, I. O., ed., Geologiya i neftegazonosnost yuga SSSR, Turkmenistan i Zapadnyi Kazakhstan: Akad. Nauk SSSR, Kompleksnaia yuzhnaya Geologicheskaya Ekspeditsya, Trudy, no. 5, p. 287-303, 3 pl.

Sohn, I. G., 1954, Ostracoda from the Permian of the Glass Mountains, Texas: U.S. Geological Survey Prof. Paper 264-A, 24 p., 5 pl.

———, 1961, Paleozoic species of *Bairdia* and related genera: U.S. Geol. Survey Prof. Paper 330-A, 105 p., 6 pl. [has a 1960 imprint].

———, 1962, *Aechminella, Amphissites, Kirkbyella*, and related genera: U.S. Geol. Survey Prof. Paper 330-B, p. 107-160, pl. 7-12 [has a 1961 imprint].

———, 1965, Significance of Triassic ostracodes from Alaska and Nevada: U.S. Geol. Survey Prof. Paper 501-D, p. 40-42, 1 fig.

———, 1968a, Triassic ostracodes from Makhtesh Ramon, Israel: Israel Geol. Survey Bull. no. 44, 71 p., 4 pl.

———, 1968b, Relict Paleozoic ostracodes in Early Triassic time: Geol. Soc. America, Progr. 1968 Meetings, p. 286-287 (abs.).

Styk, Olga, 1958, Triassic microfauna in the neighborhood of Chrznow and in the northwestern part of the Mesozoic periphery of the Swiety Krzyz Mts.: Poland Inst. Geol. Bull. 121, Micropaleontological Researches in Poland, v. 3, p. 163-176, 3 fig.

Szczechura, Janina, 1964, *Monoceratina* Roth (Ostracoda) from the upper Cretaceous and lower Paleocene of north and Central Poland: Acta Paleont. Polonica, v. 9, no. 3, p. 357-406, 11 pl.

Terry, K. D., and Tucker, W. H., 1968, Biologic effects of supernovae: Science, v. 159, no. 3813, p. 421-423.

Zanina, I. E., 1956, Ostrakody vizeyskogo yarusa Podmoskovnogo basseyna: Mikrofauna SSSR, pt. 8, Vses. Neft. Nauch-Issled. Geologorazv. Inst., Trudy, new ser., v. 98, Leningrad, p. 185-310, 8 pl.

Uppermost Permian and Lower Triassic Conodonts of the Salt Range and Trans-Indus Ranges, West Pakistan

WALTER C. SWEET
The Ohio State University

ABSTRACT

Ninety-nine samples from uppermost Permian and Lower Triassic rocks in the Salt Range and Trans-Indus range, West Pakistan, yielded more than 21,000 conodont elements, which are referred to 28 species of six conodont genera. The vertical distribution of these species defines a sequence of nine zones, five of which are probably recognizable on at least three continents. Two new genera *(Anchignathodus* and *Xaniognathus)* and 18 new species are established.

INTRODUCTION

The presence of well-preserved conodont elements in Lower Triassic rocks exposed in Chhidru Nala, in the Salt Range, West Pakistan, was noted by Schindewolf (1954), and specimens assigned to 13 form species were described and illustrated from the same section by Huckriede (1958). Neither Schindewolf nor Huckriede, however, made a detailed study of Salt Range Triassic conodonts, and their reports are based entirely on material from the lowermost part of a single section. Nevertheless, their observations indicated an abundance of conodonts in this important Triassic section and served as a stimulus for the more comprehensive study reported here.

Between 1961 and 1964, in connection with their studies of Upper Permian and Lower Triassic rocks in West Pakistan, Bernhard Kummel and Curt Teichert collected and sent to the author 213 samples from rocks of Upper Permian (Chhidruan) and Lower Triassic (Scythian) age. After routine processing, only five of the 100 Permian samples yielded conodont elements, but 94 of the 113 derived from Triassic strata were productive. Consequently, this report is based on material from five Permian and 94 Triassic samples, collected at more or less irregular intervals from nine sections in the Salt Range and from four in the Trans-Indus Surghar Range and Khisor Range of West Pakistan (Figure 1). A report in abstract form has been published earlier (Sweet, 1967).

ACKNOWLEDGMENTS

The 213 samples on which this study is based were collected between 1961 and 1964 by Bernhard Kummel and Curt Teichert, to whom the author is indebted for the opportunity to study them. Samples were processed for conodonts in the Micropaleontological Laboratories of the Department of Geology, at The Ohio State University, by the author, with the capable assistance of Misses Elsa Rubin, Rashel Nikravesh, and Frances Mullins, and Mr. Nelson Ford, whose efforts were supported financially by Ohio State and Harvard universities and by the Federal Work-Study program of the United States Government. During the progress of this study, the author has benefited greatly from discussions with Professor Maurits Lindström and Dr. Reinhold Huckriede, of Phillips University, Marburg, Germany; Dr. Marinus van den Boogaard, of the University of Amsterdam, Netherlands; Drs. Stig M. Bergström and James W. Collinson, of The Ohio State University; Mr. György Hamar, of the University of Oslo, Norway; Dr. D. L. Clark, of the University of Wisconsin; and Dr. L. C. Mosher, of Florida State University. Dr. W. M. Furnish, of the University of Iowa, generously made available on an extended loan the type specimens of Triassic conodont species described by Müller and Clark, and Professor O. H. Schindewolf and Dr. Frank Westphal arranged for a study of the Triassic conodonts assembled by Staesche during a short visit at the University of Tübingen, Germany, in June, 1966. Sayed Afaq Ali, a graduate student at The Ohio State University, drafted Figures 2 and 3 from copy supplied by Kummel.

STRATIGRAPHY

The Permian and Triassic rocks of the Salt Range and Trans-Indus ranges, West Pakistan, have been described most recently by Teichert (1966), Kummel (1966), and Kummel and Teichert (1966). The detailed comments of these authors are not repeated here, and the following summary is included only to place the conodonts described in a later part of this report in intelligible stratigraphic context (see also Kummel and Teichert, 1970).

Marine Permian rocks of the Salt Range and Trans-Indus ranges (Fig. 1) have an aggregate thickness of more than 1,000 feet, are largely sandstones and limestones, and are divided, in ascending order, into the Amb, Wargal, and Chhidru formations. Samples from all three formations were processed for conodonts, but only a few were productive. One fragmentary specimen was derived from the Wargal Formation. All the rest came from rocks in, or just below, the uppermost part of the Chhidru Formation, which is a friable white sandstone, 1 to 17 feet thick.

The Chhidru Formation is succeeded paraconformably by 48 to 635 feet of fossiliferous carbonates, shales, and sandstones and rocks in this interval have produced nearly all the conodonts described in this report. Kummel (1966) includes them all in the Mianwali Formation, a complex lithostratigraphic unit that is divided into three members. The lower 7 to 15 feet of the Mianwali Formation, consisting of a basal dolomite and a superjacent fine-grained limestone, contains abundant conodonts as well as Triassic ammonoids [*Ophiceras connectens* Schindewolf, *Glyptophiceras himalayanum* (Griesbach)] and is set aside as the Kathwai Member. The 40 to 487 feet of rock above the Kathwai Member constitute the Mittiwali Member, which includes Waagen's (1895) "Lower Ceratite Limestone," "Ceratite Marls," "Ceratite Sandstone," and "Upper Ceratite Limestone," and their dominantly shaly lateral equivalents in the Trans-Indus ranges. The Narmia Member, next above the Mittiwali, is 75 to about 150 feet of olive to gray-black shale and interbedded limestone and sandstone, which lie between a 10-foot basal pelecypodal limestone and a 5- to 7-foot pisolite bed that caps the section and defines the top of the Mianwali Formation.

Although they are not numerous in much of the section, ammonoids are the most diagnostic megafossils of the Mianwali Formation. They indicate a Lower Triassic (Scythian) age for the

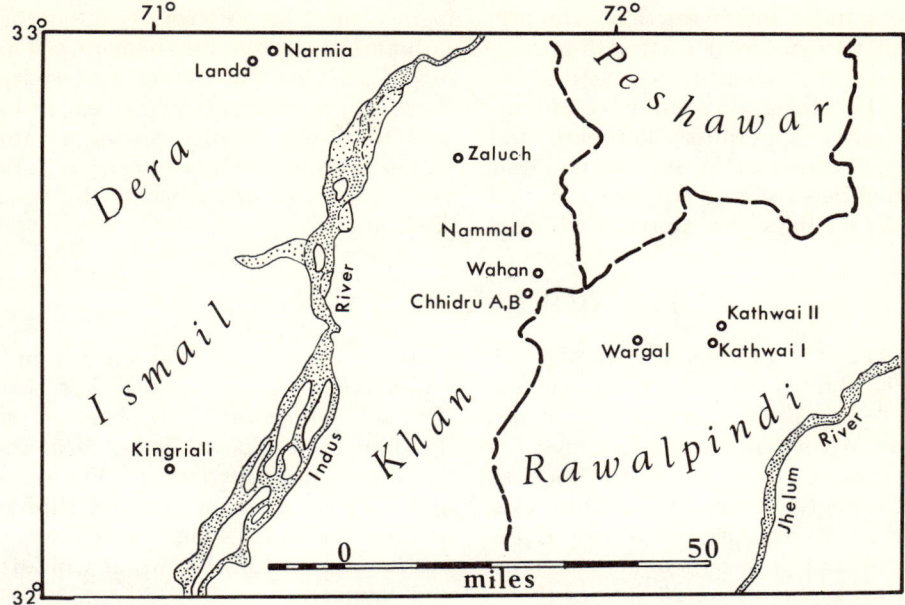

Fig. 1. Outline map of a part of West Pakistan, showing the location of sections from which the conodonts described in this report were collected. Sections at these localities are shown graphically in Figures 2 and 3.

unit (Kummel, 1966), and this fact was recognized by Waagen and Diener (in Mojsisovics, Waagen, and Diener, 1895), who chose the "Ceratite Beds" (=Mittiwali Member) as typical of the Scythian in its pelagic facies. Schindewolf (1954) discovered *Ophiceras* in rocks below the "Ceratite Beds" now included in the Kathwai Member, and demonstrated their Scythian age, and Kummel (1966) has described ammonoids from the Narmia Member that show it also to be Scythian and to be part of the "Prohungaritan" division of that Series. In short, the Mianwali Formation probably represents an almost complete Scythian sequence, and this gives added significance to the conodonts described on later pages of this report.

PREPARATION OF COLLECTIONS

Limestone and dolomite samples, averaging 500 grams in weight, were crushed mechanically to fragments about an inch in diameter and leached in 15 percent acetic or formic acid until all carbonates were removed. Acid-insoluble residues were washed on a 100-mesh screen, and the contents of that screen were then dried and further concentrated in tetrabromoethane or in a Frantz Isodynamic Magnetic Separator. Concentrated residues were searched beneath a binocular stereoscopic microscope, and conodonts found in them were removed to suitably numbered micropaleontologic slides, which were filed for later study.

Shales were crushed, leached in acid, and dried in an oven. Dry, leached samples, averaging 1,000 grams in weight, were then disaggregated by immersion in kerosene and subsequent flushing in water. Dispersed mud was washed through a 100-mesh screen, and the fraction retained on that screen was treated in the same way noted for carbonate residues.

For all samples, the insoluble or dispersed residue finer than 100-mesh was collected in plastic bags along with some water. These fractions are preserved in cardboard cartons for eventual studies of smaller microfossils.

The slides containing conodonts from the uppermost Permian and Lower Triassic of West Pakistan bear the numbers listed in Appendix A, and all these slides are stored and catalogued in the Micropaleontological Laboratories of the Department of Geology, at The Ohio State University. Type and figured specimens bear numbers that refer to the catalog of the Orton Museum of Geology, at The Ohio State University, where the types are housed.

THE CONODONT FAUNAS

The 99 productive samples on which this report is based yielded 21,388 identifiable conodont elements. These are separable into 52 morphologic categories (or form species), which singly, or in combination with others, are interpreted as the representatives of 28 species of six conodont genera. Eight of the 28 species recognized are regarded as "multielement" species. That is, their skeletal apparatuses apparently included elements of several morphologic types. All these species are assigned to *Ellisonia* Müller, which is interpreted very broadly. The remaining 20 species seem to have formed skeletal elements of a single morphologic type, and these species are distributed among *Neogondolella* Bender and Stoppel, *Neospathodus* Mosher, *Prioniodella* Bassler, and the new genera *Anchignathodus* and *Xaniognathus*, which are established in this report.

After the vertical distribution of the major elements of the conodont faunas had been established in each of the 11 sections from which we received sequential samples, these sections were zoned and aligned with the results shown in Figures 2 and 3. From these figures, it is clear that productive samples are widely spaced in most sections and permit only broad correlations. For this reason, we attach no special significance to the exact placement of any section with respect to any of the others. However, from the three sections for which we have the most information (Chhidru A, Chhidru B, Nammal), it is possible to synthesize a more or less complete sequence from a point below the presumed Permian-Triassic contact up to the base of the Tredian Formation. Alignment of the remaining nine sections of Figures 2 and 3 is consistent with, and supplements, the succession of conodont faunas observed in sections at Chhidru and Nammal.

From the correlated sections of Figures 2 and 3, we compiled the data on vertical range shown schematically in Figure 4. In that figure, the uppermost Permian and Lower Triassic rocks of the Salt Range and Trans-Indus ranges are divided into a sequence of nine zones, based entirely on the distribution of conodonts and named either for the species whose compiled range defines the boundaries of the unit, or for the species that is most abundantly represented in it. Because of the erratic spacing of productive samples, the zones recognized are based on data of unequal quality. However, all but the topmost zone are distinguishable in sequence in three to nine of the sections studied, and this suggests that they have value at least as a first approximation and as a framework for the following remarks on conodont faunas and correlation.

1. ZONE OF *ANCHIGNATHODUS TYPICALIS*

The oldest conodont fauna recognized in the collections at hand is composed of *Anchignathodus isarcicus, A. typicalis, Ellisonia triassica, E. gradata,*

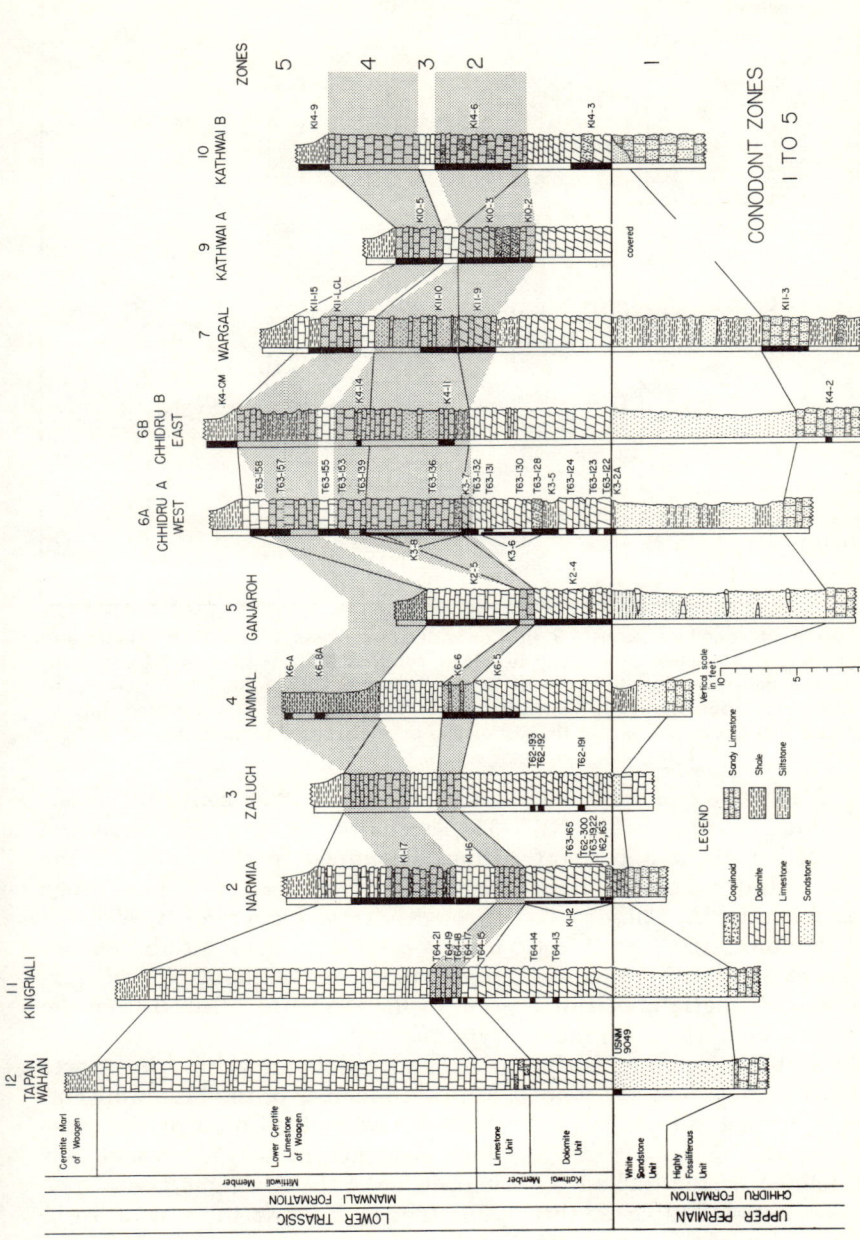

Fig. 2. Distribution of conodont zones 1-5 in uppermost Permian and lowermost Triassic rocks of the Salt Range and Trans-Indus ranges. The lithologic columns are taken from Kummel and Teichert, 1970, fig. 4. The black areas adjacent to the columns indicate positions of conodont-bearing samples. Lithology and paleontologic content of the samples is described in Kummel and Teichert, 1970, Appendix A.

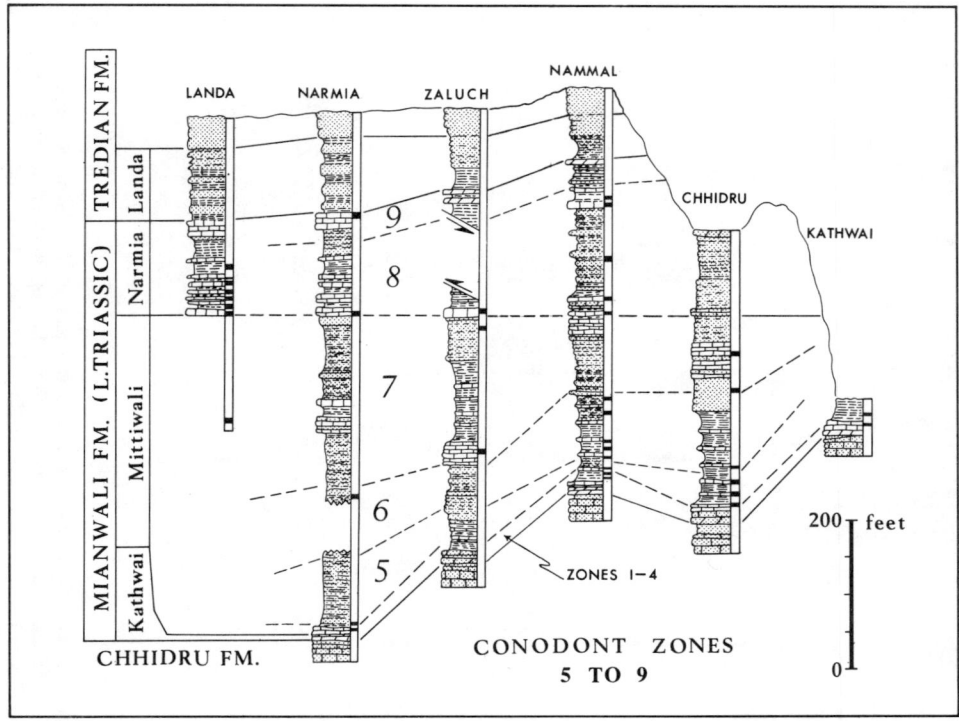

Fig. 3. Distribution of conodont zones 5-9 in Lower Triassic rocks of the Salt Range and Trans-Indus ranges. Black squares adjacent to lithologic columns indicate position of conodont-bearing samples, the field numbers and content of which can be determined by reference to Appendix A. Fault separation shown in Zaluch section is conventional and is intended only to indicate that the upper part of the section has been disrupted by faulting.

E. teicherti, Neogondolella carinata, and *Xaniognathus curvatus* (Fig. 4), which occur together in the uppermost 14 feet of the Chhidru Formation in three sections (Chhidru A, Chhidru B, Wargal) and the lowermost 3 to 9 feet of the superjacent Mianwali Formation in seven sections (Kingriali, Narmia, Zaluch, Nammal, Ganjaroh, Chhidru A, and Kathwai B). The zone distinguished by this aggregation of conodont species is named for *Anchignathodus typicalis,* and the upper zonal boundary is defined by the highest occurrence of elements referable to this species. The position of the zonal base is not known, for the oldest productive sample examined includes elements of the *typicalis* fauna.

Anchignathodus isarcicus, A. typicalis, and *Ellisonia teicherti* are not found in rocks above the *typicalis* zone and, collectively, they constitute about 35 percent of the elements recovered from 30 samples in this zone. Of the remaining four species, which range through several higher zones, *Neogondolella carinata* is the most abundantly represented, and accounts for 56 percent of the elements collected from the *typicalis* zone.

Sandy dolomites in the lower Kathwai Member of the Mianwali Formation have yielded most of the conodont elements on which we base our knowledge of the *Anchignathodus typicalis* fauna. Because these same rocks also contain specimens of *Ophiceras* and *Glyptophiceras* (Kummel and Teichert, 1966, 1970), there can be little doubt that the upper boundary of the *typicalis* zone is in rocks of lowermost Triassic age. The lowest samples yielding elements of the *typicalis* fauna are from limestone with bellerophons and other "typical Permian fossils," and these rocks are just below a white sandstone unit that Kummel and Teichert regard as the uppermost deposit of Permian age in the Salt Range and Trans-Indus ranges. Thus it seems likely that the Zone of *Anchignathodus typicalis* straddles the Permian-Triassic boundary. If the systemic boundary picked by Kummel and Teichert (1966) is indeed a paraconformity, the interruption it records must have been of much shorter duration than proposed by Grant (1968), or the *typicalis* fauna had a very much longer range than did faunas that succeeded it in the Triassic. We cannot choose between

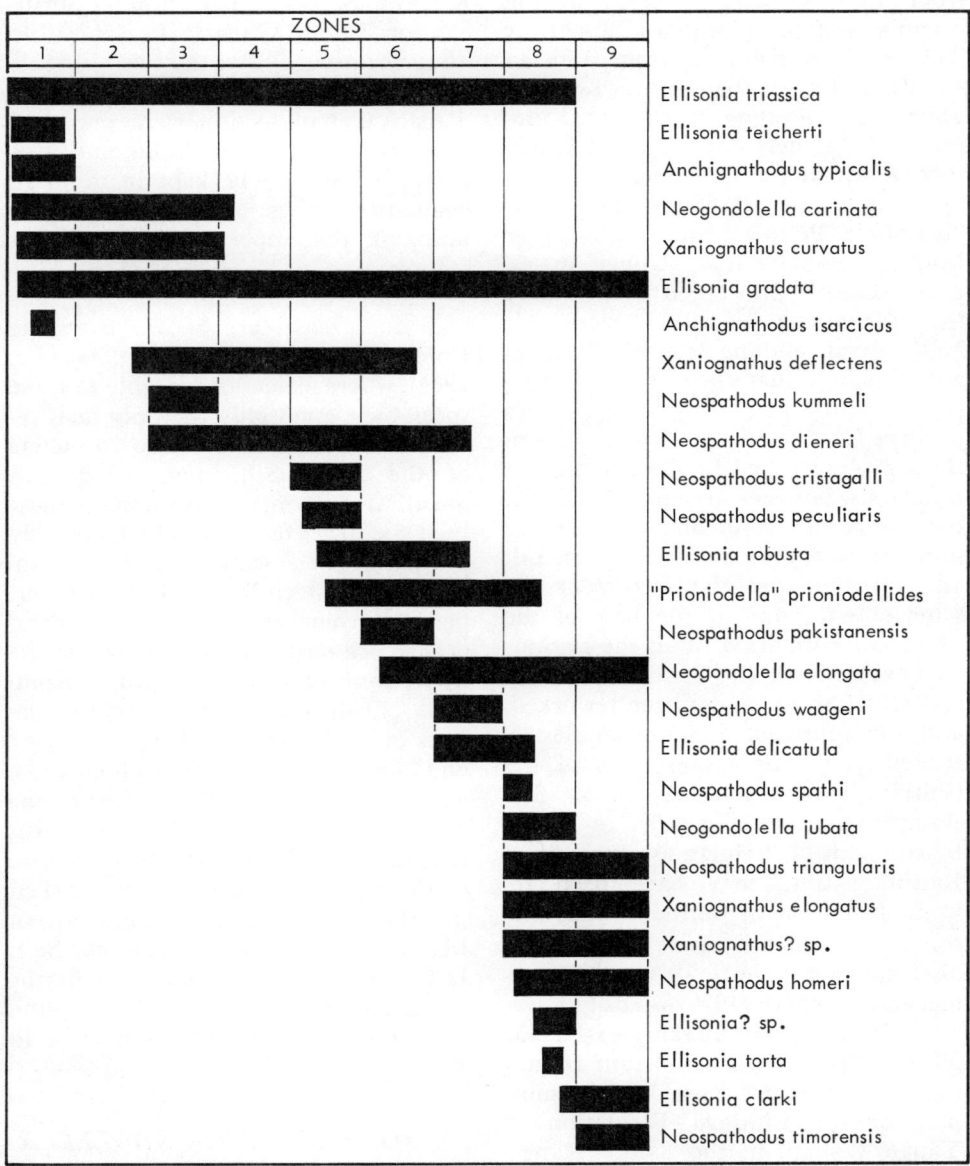

FIG. 4. Ranges of conodont species by zones. The contact between Permian and Triassic rocks is at the base of the range of *Ellisonia gradata* and *Xaniognathus curvatus,* in the lower fourth of Zone 1.

these two alternatives until we have more information on the distribution of conodonts in Upper Permian rocks.

In evaluating Grant's (1968) conclusion that the youngest beds in the Zaluch Group (=Amb, Wargal, and Chhidru formations) are no later than Guadalupian in age, it may be useful to note that we have recovered elements of the *Anchignathodus typicalis* fauna from seven samples collected in the Narmia section (Surghar Range) by Teichert and Kummel from a 6- to 12-inch bed at the base of the Kathwai Member. From these same rocks, Kummel and Teichert also collected brachiopods, which G. A. Cooper (in Kummel and Teichert, 1966) reported to be "definitely Permian and . . . rather Middle Permian than Upper in aspect." Cooper also noted that most of the brachiopods he studied were very badly worn, and he warned Kummel and Teichert that they "must be prepared for the possibility that they are reworked." Conodont elements from the samples studied by Cooper are not numerous, but they are amber in color, their fragile processes and denticles are mostly intact, their surfaces are smooth and lustrous, and they represent the same fauna found at the base of the Mianwali Formation in all the sections we have studied. Presumably the conodonts, at least, have not been reworked, and they represent the same species recorded in the uppermost part of the Chhidru Formation in three Salt Range sections (Fig. 2). In short, brachiopods of Permian aspect in basal Kathwai strata may have been reworked, but it is unlikely that the conodont elements were. [For further discussion see Grant (1970) and Kummel and Teichert (1970).—Eds.]

We also have one sample (USNM 9049) collected by R. E. Grant from a calcareous lens in the sandy uppermost part of the Chhidru Formation at Tapan Wahan, in the Khisor Range. This sample produced two dark-colored conodont elements, which are probably L-elements of an *Ellisonia* but cannot be identified as components of any of the species of *Ellisonia* described in this report. These two specimens are not now stratigraphically diagnostic, but they are distinctly different in morphology and color from the well-preserved amber-colored specimens of the *typicalis* fauna that we have recovered in great abundance from the basal part of the Mianwali Formation in the Kingriali Section, some 10 miles northeast of Tapan Wahan. It is possible (but not proved) that the two elements from the uppermost Chhidru of the Tapan Wahan section were reworked from older Permian strata, and this possibility should be kept in mind in evaluating other components of the fauna in the uppermost part of that section.

Both *Anchignathodus typicalis* and *A. isarcicus* are known from the Dolomites of northern Italy (Staesche, 1964), where elements referable to these species are essentially the only ones recorded from the lower 17 to 20 meters of the Seiser Schichten (=Strati di Siusi). *A. typicalis* is also present in at least the lower 85 feet of the Dinwoody Formation in a section in Horse Canyon, southeastern Idaho (L. C. Mosher, personal communication). The Idaho occurrence seems to be well below the upper limit of a unit from which Kummel (1954) has reported representatives of *Ophiceras* and *Glyptophiceras*, and ammonoids referable to both these genera also occur in the *typicalis* zone of the Salt Range and Trans-Indus ranges. As in West Pakistan, the base of the *typicalis* zone is not established in either northern Italy or southeastern Idaho. However, in at least northern Italy, that boundary could be in the upper part of the Permian Bellerophon Limestone, as seems to be the case in West Pakistan.

2. ZONE OF *NEOGONDOLELLA CARINATA*

Eighteen samples, from the *Ophiceras*-bearing Kathwai dolomites and limestones just above the *typicalis* zone, have yielded 2,518 conodont elements, 81 percent of which are referable to *Neogondolella carinata* (Clark). Other elements recovered from these samples represent *Ellisonia triassica, E. gradata, Xaniognathus curvatus,* and *X. deflectens* (Fig. 4). All of these species but *X. deflectens* are known from the subjacent *typicalis* zone, and all of them (including *Neogondolella carinata*) range upward into rocks above the *carinata* zone. Consequently, the *carinata* zone is based solely on the high relative abundance of *N. carinata,* and it can be distinguished with certainty only in those sections in which overlying or underlying zones, or both, can also be recognized.

Neogondolella is not yet known from the European Lower Triassic, and *N. carinata* (which may include Clark's *N. nevadensis* and *N. planata*) has been reported previously only from rocks in and below the "*Meekoceras* beds" of Nevada (Clark, 1959; 1960), and from a single sample of Lower Triassic limestone from the Anantnag District of the Kashmir Himalayas (Srivastava and Mandwal, 1966). Clearly, too little is known of the distribution of *N. carinata* and its associates to make them very useful stratigraphically at this time, and the Zone of *N. carinata* may be of use in correlation only in West Pakistan.

3. ZONE OF *NEOSPATHODUS KUMMELI*

Four samples, from *Gyronites*-bearing limestones in the basal part of the Mittiwali Member of the Mianwali Formation in three sections (Narmia, Ganjaroh, and Chhidru A), have yielded 638 conodont elements, 76 percent of which are referable to *Neospathodus kummeli* and the balance to *Ellisonia triassica, E. gradata, Neogondolella carinata, Neospathodus dieneri, Xaniognathus curvatus,* and *X. deflectens* (Fig. 4). The range of *Neospathodus kummeli* defines the zone, and specimens of this species and *N. dieneri,* which makes its first appearance there, distinguish the zone from those above and below. *N. kummeli,* the zonal index, is described for the first time in this report and is not known at present from sections in any other region. It is in this zone, however, that the first distinctively Triassic species appear, and they set the pattern by which we recognize subsequent Lower Triassic zones in West Pakistan.

4. ZONE OF *NEOSPATHODUS DIENERI*

From four samples derived from the upper 3 feet of the *Gyronites*-bearing "Lower Ceratite Limestone" (basal Mittiwali Member) at Chhidru A, Kathwai A, and Wargal, and from two samples from the lower 7 feet of the "Ceratite Marls" in the section at Nammal, we have collected 3,823 conodont elements, of which 3,183 (or 83 percent) represent *Neospathodus dieneri*. The remaining 640 elements are referable to *Ellisonia triassica, E. gradata, Neogondolella carinata, Xaniognathus curvatus,* and *X. deflectens* (Fig. 4). All of these species are represented in older rocks, and all but *N. carinata* and *X. curvatus* continue into younger zones. Consequently, the *dieneri* zone must be regarded as a local peak zone and, because it can be recognized only in biostratigraphic context, it may be of use only in local correlations in West Pakistan.

5. ZONE OF *NEOSPATHODUS CRISTAGALLI*

Ten samples, derived from the lower 25 feet or so of the basal "Ceratite Marls" (Mittiwali Member, Mianwali Formation) in six sections (Narmia, Chhidru A and B, Kathwai A and B, and Wargal) have yielded 6,934 conodont elements, 62 percent of which are referable to *Neospathodus dieneri* and 18 percent to *N. cristagalli*. Other elements represent *"Prioniodella" prioniodellides* (Tatge), *Neospathodus peculiaris, Ellisonia robusta, Xaniognathus deflectens, E. triassica,* and *E. gradata* (Fig. 4). Representatives of *"Prioniodella" prioniodellides, Ellisonia robusta, Neospathodus peculiaris,* and *N. cristagalli* make their first appearance in West Pakistan in this interval, and the local range of *N. cristagalli* defines a zone in rocks just above the *N. dieneri* peak zone. Valid representatives of *N. cristagalli* are known only from rocks of the *cristagalli* zone in West Pakistan, and those elements of the *cristagalli* fauna that have been reported from other places have such long ranges that their biostratigraphic utility is limited. In short, we cannot now identify the *cristagalli* zone or its possible correlatives outside of West Pakistan.

6. ZONE OF *NEOSPATHODUS PAKISTANENSIS*

Eight samples, from at least 80 feet of shaly strata above the *cristagalli* zone and below the "Ceratite Sandstone" (Mittiwali Member, Mianwali Formation) in three sections (Narmia, Chhidru A, and Nammal), have yielded 809 conodont elements, 562 (or 60 percent) of which represent *Neospathodus pakistanensis*. The remainder are referable to *Ellisonia gradata, E. robusta, E. triassica, Neogondolella elongata, Neospathodus dieneri, "Prioniodella" prioniodellides,* and *Xaniognathus deflectens*. In the sections we have studied, the known range of *N. pakistanensis* defines a zone immediately above the *cristagalli* zone, and this zone may be very widely recognizable. That is, we have seen conodont elements that may be referable to *N. pakistanensis* in small collections derived by Dr. M. van den Boogaard, of the University of Amsterdam, from the matrix of Lower Triassic ammonoids from Timor, and similar specimens occur in the Spitzbergen Triassic, the conodonts of which are under study by György Hamar, at Oslo University, Norway.

7. ZONE OF *NEOSPATHODUS WAAGENI*

Five samples, collected in three sections (Chhidru B, Landa, and Zaluch) from various parts of the 100 to 185 feet of rock just beneath the Narmia Member of the Mianwali Formation, have yielded 268 conodont elements, 189 (or 70 percent) of which represent *Neospathodus waageni* and *Ellisonia triassica*. Elements of *N. waageni* occur in all samples, and the range of this distinctive species defines a thick *waageni* zone in the section above the *pakistanensis* zone. Other elements of the *waageni* fauna are referable to *Neogondolella elongata; Ellisonia delicatula* (which makes its first appearance in this zone); *E. robusta* and *Neospathodus dieneri* (whose highest occurrence is in the *waageni* zone); and *E. gradata,* which ranges through the entire Lower Triassic of West Pakistan.

The *waageni* zone, which is controlled by only a few specimens from widely separated samples, is very thick, and the lowest of the samples we include in it is from rocks almost 50 feet above the highest ones we can include in the subjacent *pakistanensis* zone. Consequently, we suspect that further collecting in this interval will result in its subdivision into several more restricted units of greater biostratigraphic utility. We predict this because we have seen *Neospathodus waageni* and other components of the *waageni* fauna in a small Lower Triassic collection from Kotal-e-Tera, Afghanistan, and in a similar collection from the Lower Triassic of Timor shown to us by Dr. M. van den Boogaard of the University of Amsterdam. Furthermore, studies in progress by Professor J. W. Collinson and Mr. Walter Hasenmueller, of The Ohio State University, indicate that the *waageni* fauna is represented in at least some parts of the "*Meekoceras* beds" at a number of localities in Nevada. In Afghanistan, Timor, and Nevada, *N. waageni* occurs with distinctive elements that have not been found in the *waageni* zone of West Pakistan, but whose position might be established there by further examination of the strata above the *pakistanensis* zone.

8. ZONE OF *NEOGONDOLELLA JUBATA*

From 15 samples, derived from the lower 130 feet or so of the Narmia Member of the Mianwali Formation in sections at Narmia, Nammal, and Landa, we have collected 1,159 conodont elements, which are referable to 14 species and represent a fauna that is readily distinguishable from that of underlying conodont zones. Nine of the species we recognize in this interval appear for the first time in it (Fig. 4), and four of these are not known to be represented in younger rocks. One of the latter, *Neogondolella jubata*, is especially distinctive (and was apparently quite widespread), and its range in the sections studied defines a *jubata* zone in strata above the *waageni* zone.

In 1955 the late Professor A. K. Miller, of the University of Iowa, sent the author photographs of a few conodont elements that had been collected by Youngquist (1952) from the Thaynes Formation of Bear Lake County, southeastern Idaho. These elements, which were derived from the matrix and fillings of *Columbites*- and *Sibirites*-like ammonoids, are referable to *Neogondolella jubata*, which defines the *jubata* zone, and *Neospathodus homeri*, which makes its initial appearance in the *jubata* zone but ranges upward into strata of the *timorensis* zone. The common occurrence in southeastern Idaho of specimens representing these two conodont species indicates that the zone of *Neogondolella jubata* is contained in some part of the Thaynes Formation, and that further studies in Idaho may make it possible to establish more precise correlations with the section in West Pakistan. The *jubata* zone may also be represented in Nevada, for Mosher (1968) illustrates representatives of *Neospathodus homeri* and *N. triangularis* from strata in that state that are assigned by Silberling to the upper Scythian *Subcolumbites* zone.

Neospathodus homeri and *N. triangularis* occur together in limestones on Chios, Greece, that also produce upper Scythian ammonoids (Renz and Renz, 1948; Bender, 1967?), and representatives of *N. homeri* are known from near the top of the Campiler Schichten, in a section near St. Vigil, in northern Italy (Staesche, 1964). Neither of these species is restricted to the *jubata* zone, and *N. jubata* has not been reported from either Chios or north Italy. However, the presence of *N. triangularis* and/or *N. homeri* in those areas indicates that rocks of the *jubata* and/or *timorensis* zones may be identifiable there and suggests that these elements of the *jubata* and *timorensis* faunas may be of considerable utility in intercontinental correlations.

9. ZONE OF *NEOSPATHODUS TIMORENSIS*

The youngest fauna distinguishable in our collections from West Pakistan includes representatives of eight conodont species (Fig. 4) but is known from only two samples, both of which were taken from a pisolitic limestone at the top of the Narmia Member of the Mianwali Formation in the section at Narmia. Of the 312 specimens collected from these samples, 189 represent species that occur in at least one underlying zone. However, 123 are referable to *Neospathodus timorensis* (Nogami), which is not represented in older rocks and is chosen as the index for a distinctive, if poorly known, Zone of *Neospathodus timorensis*. Little is known about the distribution of the *timorensis* zone. However, *N. timorensis* was founded by Nogami (1968) on material from near Ue Lacan, Portuguese Timor, and conodonts were collected from dark gray limestone containing the ammonites *Leiophyllites timorensis* Bando, *L.* sp., and *Procarnites* aff. *kokeni* (Arthaber). Nogami (1968) expresses some uncertainty as to whether these rocks are uppermost Scythian or lowermost Anisian. An Anisian age is suggested (but by no means established) by noting that conodonts from Anisian (="Hydaspian") rocks on Chios, Greece, that Bender (1967?) named *Spathognathodus gondolelloides* are probobly closely related to the ones herein referred to *N. timorensis*. If this is so, and Nogami's uncertainty as to the age of the rocks yielding the types of *N. timorensis* is well founded, it may be that the uppermost bed of the Narmia (and the Zone of *N. timorensis*) are Anisian in age.

SUMMARY

The uppermost Permian and Lower Triassic rocks of the Salt Range and Trans-Indus ranges of West Pakistan are divided into a succession of nine zones, each distinguished by a characteristic association of conodonts. The oldest zone apparently spans the Permian-Triassic boundary and is recognizable on three continents. Intermediate zones are known mostly at present only in West Pakistan, but elements of the *pakistanensis*, *waageni*, *jubata*, and *timorensis* zones are present in described or undescribed collections from such widely separated places as Timor, Nevada and Idaho, Spitzbergen, northern Italy, and Greece. This suggests that, as Scythian conodont sequences become better known elsewhere, the section in the Salt Range and Trans-Indus ranges of West Pakistan may come into its own as the international reference standard envisioned for it by Mojsisovics, Waagen, and Diener in 1895.

SYSTEMATIC PALEONTOLOGY

The collections on which this study is based are divisible into 52 morphologic groups, or form species, only 12 of which have been named in previous studies of Permian and Triassic conodonts. Form species were assigned numbers (1 through 52), and information with respect to their frequency of occurrence in each of the 99 productive samples was punched on data cards. Using a FORTRAN IV program written for the 7094 computer by Professor Charles E. Corbató, of The Ohio State University, an index of affinity was determined for the two members of every possible pair of form species represented in our collections. Next, using grouping procedures described by Fager (1957) and applied to conodonts by Kohut (1969), the recurrent groups of form species and their associates shown in Figure 5 were extracted at the 0.4-level of affinity by Miss Nancy Ramsey and Mr. Walter Hasenmueller, students at The Ohio State University.

Group I of Figure 5, which includes all the longest-ranging form species in our collections, maintains its identity at the 0.5 and 0.6-levels of affinity, but dissolves into two subgroups (11-15, and 30-33) at the 0.7-level. Each subgroup is composed of a similar array of morphologically intergradational form species, each yields a significant coefficient of rank concordance, and subgroup 30-33 contains the form species *Ellisonia triassica* Müller. Consequently, we interpret Group I as a pair of multielement species of *Ellisonia*, *E. gradata* (subgroup 11-15), and *E. triassica* (subgroup 30-33), which have similar but slightly different ranges in the uppermost Permian and Lower Triassic rocks of West Pakistan.

At the 0.4-level of affinity, Group II includes form species 16-20, 39, and

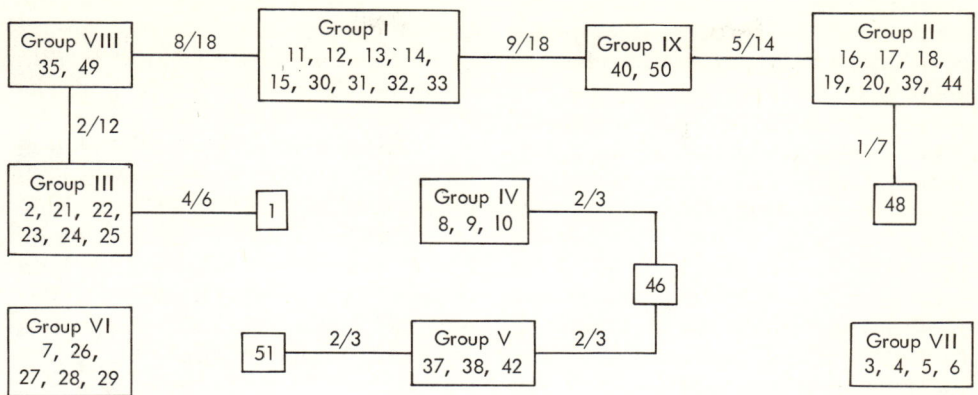

Fig. 5. Recurrent groups of uppermost Permian and Lower Triassic form species formed at the 0.4-level of affinity (Fager, 1957). All members of a group have affinities with every other member of the same group at 0.4 or above. Bonding between groups, and form species associated with groups, are indicated by fractions on the lines joining groups or joining groups and associates. Numbers stand for form species, most of which do not have names.

44. At the 0.5-level, form species 30 disappears from Group II, and, form species 16 and 44 leave the group at the 0.7-level. These facts, and disparities in stratigraphic range, indicate that Group II is divisible into three subgroups (16-20; 39; and 44). Subgroup 16-20 is similar in elemental composition to the two subgroups of Group I, and, for that reason, is referred to a new species of *Ellisonia*, *E. robusta*. Disappearance at the 0.7-level of affinity of form species 16 is attributed to the fact that it is a bilaterally symmetrical, or unpaired, element; hence, because the probability of its occurrence is distinctly lower than the probability of co-occurrence of the paired components of subgroup 16-20, it has an expectably lower affinity with other elements of the subgroup. Subgroup 39 is *Neospathodus cristagalli*, and specimens of this type were apparently the only components of the skeletal apparatus of that species. Subgroup 44 includes only 12 specimens, all of which were derived from the Zone of *N. cristagalli*. These specimens are morphologically similar to *N. cristagalli*, hence they are referred to a new species, *N. peculiaris*.

Group III maintains its identity at the 0.5- and 0.6-levels of affinity, dissolves at the 0.7-level, and can be separated into two subgroups (2; 21-25) on the basis of slight disparities in stratigraphic range and distinct differences in morphology. Subgroup 2 includes the longest-ranging elements in Group III and is interpreted as the single-element species *Anchignathodus typicalis*. Form species 1, which is an associate of Group III (Fig. 5), is morphologically similar to *A. typicalis*, but has a more limited range. It is described as the single-element species *Anchignathodus isarcicus* (Huckriede). Subgroup 21-25 includes a series of element types that are structurally similar and completely gradational morphologically. Subgroup 21-25 is described as a multielement species of *Ellisonia*, *E. teicherti*.

Form species 8, 9, 10, 37, 38, 42, 46, and 51, which form a closely interrelated plexus at the 0.4-level of affinity (Fig. 5), are best arranged as Group IV (form species 8, 9, 10) and Group V (form species 37, 38, 42). Form species 46 is an equally-bonded associate of both groups, whereas form species 51 is associated only with Group V. Elements of Group IV, which maintains its identity at the 0.5-level, form a morphologically intergradational series similar in plan to that of the two subgroups of Group I. We regard the latter subgroups as species of *Ellisonia;* hence we also interpret the form species of Group IV as a species of *Ellisonia, E. delicatula.*

At the 0.5-level, form species 37 leaves Group V to form a group with 46 that is weakly associated with groups IV and V, and form species 51 loses its two bonds with Group V to become independent. At the 0.6-level, form species 46 is independent, and Group V, consisting of form species 38 and 42, has form species 37 as an associate. The changing composition of Group V and its associates at various levels of affinity suggests that, although these form species are closely related in occurrence, they have no other type of relationship. This suggestion is borne out by the fact that the elements of Group V and its associates are morphologically quite distinct and have somewhat different stratigraphic ranges. They are described as the single-element species *Neogondolella jubata* (37), *Neospathodus homeri* (38), *N. triangularis* (42), *N. spathi* (46), and *Xaniognathus elongatus* (51).

Group VI (form species 7; 26-29), which is distinct at the 0.4- and 0.5-levels, has no associates or relationships above 0.4 with other groups, and dissolves completely just above 0.5. On the basis of intergradations in morphology and their similarity in plan to those of *Ellisonia gradata* (a subgroup of Group I), we describe elements 26-29 as *E. torta.* However, it should be noted that the elements of this species occur in just one sample, and, because that sample is also the only one in which we have identified form species 7 (the unpaired element of *Ellisonia delicatula*), that form species groups with the elements of *E. torta* rather than with Group IV, which has numerous representatives in this sample.

Group VII, which includes form species 3-6, is stable at the 0.4-level and 0.5-level of affinity, loses form species 4 at the 0.6-level, and dissolves at the 0.7-level. Because elements of this group have the same stratigraphic range and form an intergradational series similar in plan to others we include in *Ellisonia,* we describe it as the multielement species *E. clarki.*

Group VIII consists of two form species, 35 and 49, which have a high index of mutual affinity (0.72) and are less strongly bonded to the more abundantly represented elements of both Group I and Group III (Fig. 5). The hypothesis that this pair of elements represents a single multielement species is weakened, if not disproved, by noting that form species 35 is Clark's (1959) "*Gondolella carinata*" and form species 49 is an ozarkodina-like form with a short posterior process, which has not been reported to occur in collections that yielded the type material of "*G. carinata.*" Other gondolella-like elements in our collections (e.g., form species 36 and 37) seem to have been the only elements in their respective apparatuses, and other ozarkodina-like elements either fail to group at all or group for reasons of similar range with elements that are only distantly related in form to "*G. carinata.*" For these reasons, and because form species 35 and 49 have somewhat different stratigraphic ranges, we describe them as the single-element species *Neogondolella carinata* (Clark) (35), and *Xaniognathus curvatus. Xaniognathus* and *Neogondolella* probably include closely related species, as we note in discussions of these two genera on later pages, but we cannot make a case for uniting them in a single multielement genus.

Group IX (form species 40 and 50) is similar to Group VIII in that it contains an ozarkodina-like element (50) in combination with another element type (40). These two elements represent species whose ranges overlap in an interval from which we have many samples, but whose range extremes are different. Further, these two species are different morphologically, even though they seem to represent species that were rather closely related. We describe element type 40 as *Neospathodus dieneri* and 50 as *Xaniognathus deflectens*.

Eight of the 52 form species we recognize in collections from West Pakistan are distributed through the section in such a way that they fail to group with even one other form species at any of the levels of significance we have employed. By analogy with better represented groups, one of these form species (34) is probably an L-element of an undescribed species of *Ellisonia*, so we describe it as *E.? sp.* Another (52) is probably a *Xaniognathus*, and all the rest are referred to single-element species of *Neogondolella* (36) and *Neospathodus* (41, 42, 43, 45, 46, and 47).

Genus ANCHIGNATHODUS Sweet, 1970

Diagnosis: Anchignathodus includes conodonts with a skeletal apparatus composed solely of paired, individually asymmetric elements that are more or less conspicuously arched, straight, or slightly bowed blades. A series of high, laterally compressed denticles, which are discrete, fused, or overgrown, forms a distinctive anterior crest: denticles posterior to the apex of this crest decrease in size and length to the posterior extremity of the element. The attachment surface is enclosed by a broadly flaring sheath that is lachrymiform in plan and tapers toward, but extends to, the posterior extremity of the element. Beneath the anterior blade, however, the attachment surface is confined to a narrow groove, which extends nearly to the anterior end of the element. In elements of at least one species, nodelike denticles occur on the upper surface of the laterally flaring sheaths.

Type species: Anchignathodus typicalis Sweet, 1970.

Remarks: The conodonts here assembled in *Anchignathodus*, but included by previous writers in the form genus *Spathognathodus*, constitute a closely interrelated group that ranges from the Lower Carboniferous (uppermost Kinderhookian) into the basalmost Triassic. The group was recognized as a distinctive one by Rexroad and Collinson (1961), who compared it with *Gnathodus* Pander, but concluded that its members are morphologically like *Pandorinellina* Hass, hence, following Müller and Müller (1957), should be retained in *Spathognathodus*. Our comparisons suggest, however, that similarity of these forms and Upper Devonian *Pandorinellina* is only superficial, and we doubt that *Anchignathodus* and *Pandorinellina* are even very closely related. That is, the basal excavation of typical representatives of *Pandorinellina* is a narrow groove, expanded slightly to form a navel beneath the cusp, and bordered posteriorly by narrow flangelike brims that project laterally in a plane almost perpendicular to that of the blade. The attachment surface of elements representing species here included in *Anchignathodus*, on the other hand, is enclosed posteriorly in a capacious, cuplike structure and is not bordered by flangelike brims. In our view, elements typical of *Pandorinellina* are most closely related to those Müller and Müller (1957) included in *Ctenopolygnathus*, whereas *Anchignathodus* elements are morphologically closest to those of the several form genera assembled by Hass (1959, 1962) in the Idiognathodontinae.

As Rexroad and Scott (1964) and Rexroad and Collinson (1965) have noted, the lineage here included in *Anchignathodus* almost certainly began with *A. regularis* (Branson and Mehl, 1938), in which we include *Spathognathodus pulcher* (Branson and Mehl, 1938) and the upper Keokuk and Warsaw specimens Rexroad and Collinson described as *S. sp.* cf. *S. pulcher*. *A. regularis* was probably the progenitor of *A. sp.* of Rexroad and Scott (1964) and the somewhat younger

A. coalescens and *A. penescitulus* of Rexroad and Collinson (1965), and the latter was undoubtedly ancestral to *A. scitulus* (Hinde, 1900) of the St. Louis Limestone and its North American and European equivalents. *A. spiculus* (Youngquist and Miller, 1949), in which we include *Spathognathodus bidens* Youngquist and Miller, 1949, and *A. cristulus* (Youngquist and Miller, 1949), which includes the specimens from the Upper Limestone at Linn Spout, Dalry, that Clarke (1960) identified tentatively as *Spathognathodus minutus?* (Ellison), are ubiquitous Chesterian species and the probable ancestors of *A. minutus* (Ellison, 1941), which ranges from the uppermost Chester (Kinkaid of Illinois) into the Lower Permian. *Anchignathodus* is also represented in the Upper Carboniferous (Desmoinesian) of Colorado by *A. coloradoensis* (Murray and Chronic, 1965).

In the few small Permian faunas that have been described, *Anchignathodus* is clearly represented by elements from the Bone Springs Limestone of Texas and the Minnekhata Limestone of Wyoming that Clark and Ethington (1962) described as *Spathognathodus* n. sp., and by at least some of the specimens from the Middle Permian of Sicily described as *Gnathodus sicilianus* by Bender and Stoppel (1965).

The youngest known representatives of *Anchignathodus* are referable to *A. typicalis* (Sweet) and *A. isarcicus* (Huckriede, 1958), which have been reported from basal Triassic strata in Pakistan and northern Italy (Huckriede, 1958; Staesche, 1964). *A. typicalis* is also present in undescribed collections from the lower 85 feet of the Dinwoody Formation of southeastern Idaho (L. C. Mosher, personal communication), which suggests that *Anchignathodus* was widely distributed as late as the time of its extinction in the earliest Scythian.

As here conceived, *Anchignathodus* includes the majority of those post-Kinderhook (Lower Carboniferous) species previously referred to *Spathognathodus*. That is, *S. campbelli* Rexroad, *S. pusillus* Clarke, and *S. pellaensis* Youngquist and Miller, are Lower Carboniferous species that probably belong in *Gnathodus*; *S. exodentatus* Clarke, of the Scottish Lower Carboniferous, and *S. whitei* Rhodes, of the basal Permian in Wyoming, almost certainly represent undescribed gnathodontid genera; and *S. divergens* Bender and Stoppel, of the German Zechstein, and other species related to Scythian *S. cristagalli* Huckriede are referred in a later part of this study to *Neospathodus* Mosher, 1968. In short, it is concluded that *Spathognathodus* does not range above the Kinderhook and is thus primarily a Lower Paleozoic genus.

The skeletal apparatus of *Anchignathodus* is thought to have included elements of only one morphologic type, because representatives of the several species included in the genus are commonly the only ones in Late Paleozoic collections or have ranges and occur in proportions that are independent of those of elements representing associated species. In this respect *Anchignathodus* differs from gnathodontids like *Gnathodus, Streptognathodus,* and *Idiognathodus,* which are known to be components of multielement skeletal apparatuses (Schmidt, 1934; Scott, 1942; Rhodes, 1952; Schmidt and Müller, 1964).

Anchignathodus typicalis Sweet, 1970
Pl. 1, fig. 13, 20

Spathognathodus cf. *minutus* (Ellison) Huckriede, 1958, p. 162, 167, pl. 10, fig. 8.
Spathognathodus isarcicus Huckriede, Staesche, 1964 (partim), p. 288-289, fig. 60, 61 (non fig. 6, 62, 63, 64).
Anchignathodus typicalis Sweet, 1970, p. 7, pl. 1, fig. 13, 22.

Diagnosis: A species of *Anchignathodus* characterized by slightly arched and bowed bladelike elements, 2 to 2.5 times as long as wide, in which denticles posterior to the cusp tend to diminish regularly in length to the top of the unit.

Material: 1,130 discrete elements; 568 dextral, 562 sinistral.

Description: Elements of *Anchignathodus typicalis* from the uppermost Permian and lowermost Triassic of West Pakistan are laterally expanded, slightly arched, and faintly bowed bladelike structures that range from 0.2 to 1.1 mm in length. In early developmental stages each element consists of an anterior process surmounted by 2 or 3 laterally compressed denticles; an erect to suberect cusp that is longer than any process denticle and is situated above the apex of the basal cavity; and a posterior process that bears 6 or 7 denticles, is about 3 times as long as the anterior process, and is broadly expanded basally.

With growth, denticles are added to the distal extremities of both processes; but, in later developmental stages, proximal denticles of the posterior series and all in the anterior series are incorporated into the cusp, which, in all but the earliest growth stages, is a laterally compressed structure, which is subtriangular in lateral view and makes up the anterior third of the unit. The 7 to 12 denticles of the posterior series decrease in length more or less regularly away from the cusp, and in the distal third of its length, the upper edge of the posterior curves downward more or less abruptly to its junction with the basal margin.

Beneath the posterior process, the attachment surface is enclosed in a deep, asymmetrically subconical, basal cavity, the transverse diameter of which is slightly less than half the length of the unit. Beneath the bladelike anterior process, the attachment surface is enclosed in a narrow groove that is less than a third the depth of the basal cavity under the posterior process.

Remarks: Elements characteristic of *Anchignathodus typicalis* are readily distinguishable from those of other species we include in *Anchignathodus*, but are morphologically closest to the ones from the Winterset Limestone of Missouri on which Ellison (1941) based *Spathodus minutus*, a species herein referred to *Anchignathodus*. We have not seen the type elements of *A. minutus*, but we have compared our Triassic material with obviously conspecific specimens from the Upper Brush Creek Limestone of Ohio, which is about the same age as the Winterset. These elements of *A. minutus* are about 3 times as long as wide, whereas those of *A. typicalis* are more expanded laterally and range from 2 to 2.5 times as long as wide. In addition, denticles of *A. typicalis* elements tend to diminish gradually in length between the cusp and posterior end of the unit; whereas in *A. minutus* elements there tends to be an abrupt offset in the lateral profile immediately posterior to the cusp, and the 6 or 8 denticles following the cusp are all about the same length.

Occurrence: Uppermost highly fossiliferous unit of Chhidru Formation at Chhidru B; White Sandstone unit of Chhidru Formation at Chhidru A; Dolomite unit of Kathwai Member at Kingriali, Narmia, Zaluch, Nammal, Ganjaroh, Chhidru A, Chhidru B, and Kathwai B. The distribution and frequency of occurrence in West Pakistan is summarized in Appendix A. The species is also represented in the Seiser Schichten of northern Italy (Staesche, 1964) and in samples from the Dinwoody Formation in the lower 85 feet of a section in Horse Canyon, southeastern Idaho (L. C. Mosher, personal communication).

Type: Holotype, OSU 28017; paratype, OSU 28016.

Anchignathodus isarcicus (Huckriede), 1958

Pl. 1, fig. 18, 19

Spathognathodus isarcica Huckriede, 1958, p. 162, 167, pl. 10, fig. 6, 7a-c.——Staesche, 1964 *(partim)*, p. 288-289, fig. 6, 61, 63, 64 *(non* fig. 60, 61).

(not) *Spathognathodus isarcicus* Huckriede, Pomesano Cherchi, 1967, p. 228, pl. 16, fig. 4 (=*Neospathodus?* sp.).

Remarks: Anchignathodus isarcicus (Huckriede) was founded on only five specimens from the Seiser Schichten at Pufelsbach, in the Dolomites of northern Italy, and is represented in our collections from West Pakistan by exactly the same number of elements, none of which is especially well preserved. Staesche (1964) assembled an additional 209 specimens from northern Italy and, since his figures indicate all the pertinent features of the species, our few specimens are not described.

Staesche (1964) regarded the Tyrolean specimens Huckriede (1958) referred to *Spathognathodus* cf. *S. minutus* (Ellison) as representatives of the "Anfangsstadium der Entwicklung von *Sp. isarcicus*," hence he included those specimens and others like them from his own collections in *S. isarcicus*. We do not agree with this interpretation, although the developmental patterns indicated in Staesche's detailed figures are convincing evidence of very close relationship between the elements of *Anchignathodus isarcicus* and those of its probable progenitor, *A. typicalis*.

Our decision to follow Huckriede (1958), rather than Staesche (1964), in the interpretation of *Anchignathodus isarcicus* is based largely on the stratigraphic distribution in northern Italy and West Pakistan of elements of these two types. That is, at the time of our visit to Tübingen, we tabulated elements of these two kinds in Staesche's collections and, using the procedure outlined by Shaw (1964), we subsequently compiled into a single composite section the sections from which Staesche collected his material. In this composite section, it is clear that laterally denticulate elements like the type of *A. isarcicus* occur only in the upper half of the range Staesche gives for *A. isarcicus*; laterally undenticulate elements that we refer to *A. typicalis* have a much longer range. The same distribution may also obtain in West Pakistan, where *A. isarcicus* defines a very thin zone a few feet below the top of the much longer range of *A. typicalis*. Further, Pakistan specimens of *A. typicalis* are almost all much larger than any of the five specimens of *A. isarcicus* with which they occur, and this suggests that elements of the former type can hardly be regarded as initial ontogenetic stages of the latter.

Occurrence: Dolomite unit of Kathwai Member at Kingriali, Zaluch, and Chhidru A. The known distribution of specimens in West Pakistan is summarized in Appendix A.

Type: Figured hypotype, OSU 28015.

Genus ELLISONIA Müller, 1956

Type species: Ellisonia triassica Müller, 1956.

Remarks: In collections from the Lower Triassic of West Pakistan, we recognize seven multielement species characterized by a skeletal apparatus that includes some combination of hibbardella- (or ellisonia-), ligonodina-, lonchodina-, prioniodina-, cladognathodus-, and enantiognathus-like elements. By referring these seven species to *Ellisonia,* we indirectly expand the scope of a genus that was originally erected to include only hibbardella-like conodont elements; and, ordinarily, we would prepare an emended diagnosis of *Ellisonia* to take account of this expansion in scope. We do not do this, however, because a large number of conodont species, ranging in age from the Ordovician through at least the Triassic, had skeletons composed of combinations of the intergradational forms that distinguish the multielement groups here assigned to *Ellisonia,* and we do not know how most of these species should be assembled into genera or what the names for those genera should be. The plan common to all these species was apparently a persistent one, and it may ultimately provide the basis for recognizing a suprageneric category. In this study, however, we have devoted our principal attention to the discrimination of Triassic species, and these are described as representatives of *Ellisonia* because the type of the form genus is Lower Triassic and was almost certainly a skeletal component of one of the species we are describing.

In description of skeletal components of the following species of *Ellisonia*, we have attempted to avoid confusion by referring to various types of elements with capital letters. In all these descriptions, the letter U stands for bilaterally symmetrical, unpaired elements of the sort that have been included in the form genera *Ellisonia, Hibbardella, Roundya,* and *Trichonodella*. All other elements are individually asymmetrical, were probably paired in the original skeleton, and are designated "L elements." In those skeletal apparatuses in which there were apparently several different kinds of L-elements, we distinguish LA, LB, LC, LD, LE, and LF compounds. We have attempted to employ the same type of "L" denomination for elements that seem to have been homologous structures, but it is difficult to be certain that this was the case in all combinations.

Ellisonia clarki Sweet, n. sp.
Pl. 4, fig. 15-19

Diagnosis: A species of *Ellisonia* with a skeletal apparatus characterized by U-elements with a short cusp and conspicuously arched posterior process; LA- and LC-elements with no distinguishable short lateral process in late growth stages; and LB-elements with a prominent cusp and two conspicuous processes. In late growth stages, all elements tend to develop a swollen midlateral rib, and the distribution of "white matter" in these elements is such that they are almost completely and uniformly opaque.

Material: 83 discrete elements, representing the four morphologic types whose distribution is shown in Table 1.

TABLE 1. Distribution and Frequency of Elements of *Ellisonia clarki*, n. sp.

Sample number	Frequency of element types*			
	U	LA	LB	LC
T63-167	2	17	40	17
K6-37	1	1	2	1
K6-36	–	2	–	–
Totals ..	3	20	42	18

* Symbols for element types are explained in the text.

Description: The three U-elements at hand are all small, fragile, and incomplete posteriorly. Processes of the anterior arch are short, directed anteriorly, curved gracefully laterally, and surmounted by 3 or 4 discrete, slightly recurved denticles. The cusp, at the junction of the two anterior processes, is similar to, and only slightly longer than, process denticles. The posterior process is laterally compressed, bladelike, and conspicuously arched at midlength. Immediately posterior to the cusp, the posterior process bears one or two short, delicate denticles, which are succeeded posteriorly by a large, conspicuous denticle above the apex of the posterior arch. Posterior to the major denticle of the posterior process is a succession of 3 or more delicate denticles that are much smaller and shorter than the cusp. With growth, the major denticle of the posterior series overgrows minor denticles on both its anterior and posterior sides. The undersurface is narrow, flat, or faintly grooved beneath the cusp and just posterior to the junction of the lateral elements of the anterior arch.

LA-elements are paired, individually asymmetric, bladelike structures that are prominently arched and sinuously bowed. In the smallest forms at hand, a very short lateral process, bearing one or two short, discrete, needlelike denticles, grades proximally into an erect to slightly recurved cusp, which is flanked on its opposite side by a long laterally bowed lateral process bearing 14 to 16 denticles, which are closely similar to, but slightly longer than, denticles of the shorter lateral process. In successively larger elements, denticles of the short lateral process are overgrown by, and largely included within, the cusp; a conspicuous longitudinal rib is formed at element midheight; and denticles of the longer lateral process develop into stout, discrete structures that curve apically toward the cusp and incorporate two or more early-stage denticles. The undersurface of LA-elements is sharp, bluntly wedge-shaped in transverse section, and bears only a faint indentation immediately below the cusp.

LB-elements are paired, individually asymmetric, bladelike structures that are closely similar in all comparable respects to U-elements except that they lack one of the lateral processes of the anterior arch. Elements of this type are arched and consist of an anterolaterally bowed anterior process surmounted by 4 or 5 discrete, recurved denticles; an erect to somewhat reclined cusp that enlarges basally through incorporation of adjacent anterior- and posterior-process denticles; and a long posterior process that bears as many as 15 reclined denticles, the posteriormost 3 or 4 of which are the largest and most conspicuously reclined. The undersurface is transversely flat beneath the processes and faintly indented beneath the cusp in most elements. In a few, however, the attachment surface forms a narrow strip that faces laterally beneath the cusp and posterior process.

LC-elements are strikingly similar to LA-elements in that most of them consist of a short lateral process that bears one or two short, needlelike denticles; a long erect or reclined cusp; and a much longer, sinuously bowed, lateral process with denticles that are similar to, but longer and more numerous than those of the shorter lateral process. In LC-elements, however, the short lateral process and the adjacent margin of the cusp are deflected sharply to one side so that the element is L-shaped in superior or inferior view. As in LA-elements, larger LC-units bear a distinct longitudinal rib at midheight, and denticles of the short lateral process tend to be incorporated in the base and edge of the cusp.

Remarks: *Ellisonia clarki*, which is named for Professor David L. Clark of the University of Wisconsin, is similar to *E. gradata* and *E. delicatula*, and it may have developed from one of those species. *E. clarki* is distinguished from *E. gradata* by the lack of L-elements with a bifid anterior process; by the posteriorly arched U-elements; by LC-elements of strikingly different conformation; and by the tendency for all elements to develop a distinctive longitudinal rib at midheight in late stages of growth. Although U- and L-elements of *E. clarki* and *E. delicatula* are somewhat alike, conspicuous differences in the distribution of "white matter" in elements of the two species distinguish them readily. U-elements of *E. clarki* are reminiscent of the ones Mosher and Clark (1965) included in the form species *Hibbardella acroforme* Mosher and Clark, but denticles between the cusp and the major denticle of the posterior process are much larger in *H. acroforme* than in the U-elements of *E. clarki*.

Occurrence: Narmia Member of Mianwali Formation at Narmia and Nammal. The distribution of specimens from West Pakistan is summarized in Appendix A and Table 1.

Types: Syntypes, OSU 28018-28021, inclusive.

Ellisonia delicatula Sweet, n. sp.
Pl. 4, fig. 9-14

Diagnosis: A species of *Ellisonia* with a skeletal apparatus that includes posteriorly-arched U-elements, LB-elements with a long anterior and a very short posterior process, and LA- and LC-elements with two distinct processes

TABLE 2. Distribution and Frequency of *Ellisonia delicatula*, n. sp.

Sample number	Frequency of element types*			
	U	LA	LB	LC
K12-9	2	5	12	3
K12-7	–	–	1	1
K12-5	–	2	5	5
K1-37	–	–	2	1
K1-36	–	–	1	1
K13-13	–	2	7	2
K12-1	–	–	1	1
K6-30	–	–	1	–
K13-11	–	–	3	–
K4-UCS	–	7	6	2
K13-9	–	8	13	2
Totals	2	24	52	18

* Symbols for element types are explained in the text.

of unequal length. "White matter" is confined to the tips and axial regions of denticles in all elements.

Material: 96 discrete elements, representing the four morphologic types whose distribution is shown in Table 2.

Description: As herein conceived, the skeletal apparatus of *Ellisonia delicatula* consists of four intergradational, but morphologically distinct, elements that exhibit similarities in size, denticulation, and distribution of "white matter." The two small, posteriorly incomplete U-elements in our collections have a posterior process of unknown length and two short anterolateral processes that join to form a bilaterally symmetrical anterior arch. The two processes of the anterior arch project slightly to the anterior then curve abruptly laterally from their point of union at the proximal end of the posterior process. Each anterolateral process bears four distally discrete, erect to slightly recurved denticles, and the crest of the anterior arch is surmounted by a cusp of similar size and conformation. The posterior process is arched, laterally compressed, bladelike, and, at least proximally, it bears a series of more than four narrow, needlelike denticles, the longest and broadest of which is at the crest of the posterior arch and is separated from the cusp by two or three denticles of lesser length. The undersurface of the three processes is transversely flat or slightly channeled, and there is an indistinct indentation beneath the cusp.

LA-elements are paired, individually asymmetric units, with a long, distally pointed, laterally costate cusp that is posteriorly recurved, subtriangular in transverse section, and flanked laterally by denticulated processes of conspicuously different length. The longer lateral process, which projects laterally from the cusp and curves slightly to the posterior, bears 10 short, narrow needlelike denticles of similar length and breadth. The shorter process, which projects laterally and curves slightly anteriorly, bears four short denticles that are similar to those of the longer process. At its base, the posterior face of the cusp swells slightly to form a liplike structure that may be regarded as a short, undenticulated, posterior process. The undersurface of the lateral processes is transversely flat or slightly channeled, and there is a shallow cuplike indentation beneath cusp and posterior process.

LB-elements of *Ellisonia delicatula* are paired, individually asymmetric, bladelike structures that are prominently arched and conspicuously bowed. The anterior margin of the erect, sharp-pointed cusp grades into an anterior process that curves downward and laterally, and bears on its upper edge a series of as many as 9 sharp-pointed, compressed, distally discrete denticles, the anterior few of which tend to be incorporated in the cusp in elements representing successively larger growth stages. The posterior process, which is a continuation of the sharp posterior edge of the cusp, is short, compressed, straight, and bears four or more needlelike denticles that fuse in large specimens into a single blade but are recognizable by axial strips of "white matter" preserved in that blade. The undersurface is transversely flat beneath processes and faintly indented beneath the cusp.

LC-elements are similar to LB-elements, but in these forms the short posterior process and the posterior cusp margin are deflected abruptly to one side to form a short, undenticulated posterolateral process. Elements of this type have been included by previous authors in the form-genera *Apatognathus* or, more recently, *Enantiognathus*.

Remarks: *Ellisonia delicatula* is most closely related morphologically to *E. gradata*, but it differs from that species in two major respects. That is, U-elements of *E. delicatula* have a conspicuously arched posterior process, whereas the comparable structure in U-elements of *E. gradata* is straight and none of the LB-elements of *E. delicatula* have the bifid anterior process that characterizes LB2-elements of *E. gradata*. Differences in the distribution of "white matter" in elements of *E. delicatula* and *E. gradata* are difficult to describe accurately, but they are apparent in the illustrations of the types of those two species, which are mounted side-by-side on plate 4.

Occurrence: Mittiwali Member of Mianwali Formation at Zaluch and Chhidru B; Narmia Member of Mianwali Formation at Landa, Narmia, Zaluch, and Nammal. Distribution in West Pakistan is summarized in Appendix A and Table 2.

Types: Syntypes, OSU 28022-28026, inclusive.

TABLE 3. Distrbution and Frequency of Elements of *Ellisonia gradata* Sweet

Sample number	U	LA	LB1	LB2	LC	Sample number	U	LA	LB1	LB2	LC
T63-167	4	19	16	10	25	T63-157	4	22	15	4	8
K6-37	–	–	1	–	–	K10-5	–	2	–	1	3
K6-36	–	3	1	–	–	K1-16	–	–	2	–	–
K6–31B	–	–	–	3	–	K2-5	–	–	2	1	2
K12-9	–	4	–	2	–	T63-155	–	2	3	1	3
K12-5	6	17	17	–	9	K3-3	2	3	3	–	2
K1-37	1	5	–	1	2	T63-153	1	10	9	2	10
K1-36	1	8	2	4	10	T63-145	1	–	–	1	–
K13-13	–	1	–	–	–	T63-139	1	–	–	–	1
K12-1	–	–	–	1	–	K10-3	–	1	3	–	3
K6-27A	1	12	6	3	2	K14-6	–	–	1	–	–
K4-UCL	–	1	–	–	–	K4-11	–	–	–	–	2
K4-UCS	1	3	–	1	5	T63-136	–	–	–	–	1
K13-9	1	5	4	4	2	K3-7	2	3	2	–	4
K6-B	–	1	–	1	–	T64-21	2	3	4	–	6
K3-15	4	8	2	1	9	T64-19	–	–	2	–	3
K3-14	17	52	24	24	18	T64-18	4	5	2	2	7
K1-17	–	18	5	1	5	T63-132	4	8	–	1	1
K3-12	6	34	6	7	11	T64-17	1	5	1	1	2
K10-6	10	13	9	4	11	T63-131	3	3	2	–	3
K3-10	12	28	12	7	13	T63-130	–	1	–	–	2
K4-CM+LCM	7	8	6	6	10	T64-14	–	–	–	–	1
T63-158	2	10	4	1	3	T62-191	–	–	–	1	–
K6-A	8	8	5	1	1	T63-19	–	1	–	–	–
K11-CL+LCL	2	5	–	–	3	Totals	108	332	171	97	203

* Symbols for element types are explained in the text.

Ellisonia gradata Sweet, 1970

Pl. 4, fig. 1-8

Lonchodina latidentata (Tatge) Huckriede, 1958, p. 151, pl. 10, fig. 32, 38, 39.
Roundya n. sp. A. Huckriede, 1958, p. 161, pl. 10, fig. 20.
Gen. et spec. indet. A. Huckriede, 1958, p. 163, pl. 10, fig. 28.
Ellisonia gradata Sweet, 1970, p. 8, pl. 1, 5, 6, 9.

Diagnosis: A species of *Ellisonia* with a skeletal apparatus that includes U-elements with a straight, denticulated posterior process; LA- and LC-elements with two anterolateral processes; LB-elements with short posterior and long anterior processes, some of which are bifid; and "white matter" that is distributed in irregular "clouds" so that all elements have a "spotted" appearance.

Material: 911 discrete elements representing the five morphologic types whose distribution is shown in Table 3.

Description: The elements on which we base our concept of the skeletal apparatus of *Ellisonia gradata* form an intergradational series of structural units that can be divided into five more or less distinct morphologic types. U-elements are bilaterally symmetrical, unpaired units with a laterally compressed, reclined cusp of triangular cross section, a straight posterior process, and anterolateral processes that form an anterior arch in the plane of the cusp's anterior face. The anterolateral processes, which are continuous with sharp-edged anterolateral costae on the cusp, project laterally, are directed downward distally, and enclose an angle of 65-70°. Each anterolateral process bears 5 or 6 discrete, anteroposteriorly compressed, erect to slightly recurved denticles, the apexes of which are directed away from the cusp. The bladelike posterior process projects straight back from the sharp posterior margin of the cusp, and is surmounted by five to eight short, broad, reclined denticles similar to those of the anterolateral processes. During growth, the anterior 2 or 3 denticles of the posterior series tend to be incorporated in the posterior margin of the cusp. The undersurface of the posterior and anterolateral processes is transversely flat or faintly channeled, and there is a slight pitlike indentation beneath the cusp where the undersurfaces of the three processes join.

LA-elements are paired, individually asymmetric units, with a long recurved cusp of subtriangular cross section, the sharp lateral edges of which are continuous with anterolateral processes of unequal length. The longer anterolateral process projects laterally from the cusp, curves gracefully downward and posteriorly toward its distal end, and bears 5 or 6 discrete, compressed, recurved denticles on its upper edge. The shorter anterolateral process projects laterally from the cusp, curves downward and anteriorly, and supports 2 to 4 short denticles of subequal length on its upper edge. With growth, all but the distal denticles of the short anterolateral process tend to be incorporated in the adjacent edge of the cusp; or they are fused into a single denticle, immediately adjacent to the cusp, which may rival or exceed the cusp in length and breadth. The posterior face of the cusp flares somewhat basally to form a short liplike structure that is regarded as an undenticulated posterior process. The undersurface is flat or faintly channeled beneath the distal extremities of small LA-elements, but a distinct pit is formed beneath the cusp and the proximal parts of the anterolateral processes. In larger LA-elements, the undersurface is flat or inverted, and only a faint indication of a basal pit is retained beneath the cusp.

LB-elements are paired, individually asymmetric structures with an erect cusp, a short, straight, posterior process, and an anterior process that projects anteriorly from the cusp and curves downward and laterally. In the simplest LB-elements, here described as LB1-elements, the anterior process bears a series of 9 to 11 discrete, sharp-pointed denticles that curve posteriorly and increase gradually in length from the distal extremity of the process toward the cusp. In more complex LB-units, here designated LB2-elements, the anterior process becomes bifid distally through development of a short secondary-lateral process that bears one or two short denticles and projects laterally from the fifth or sixth denticle of the anterior series on the convex side of the anterior process. The posterior process of both LB1 and LB2-elements is short, bladelike, and surmounted by only one or two denticles like those of the anterior process. Early-stage posterior denticles adjacent to the cusp tend to be overgrown by the cusp during ontogeny, and their former position is marked by axial strips of "white matter" in the basal part of the posterior edge of the cusp. The undersurface of both processes is a thin, flat, or convexly rounded edge, but a minute pit is recognizable beneath the cusp in all but the largest specimens at hand.

LC-elements are generally similar to LA-units in mode of denticulation and conformation of undersurfaces, but in these structures the cusp margin and the shorter anterolateral process are deflected abruptly to one side so that the units are L-shaped in superior or inferior view. Elements of this type have been included in the form genera *Apatognathus* or, more recently, *Enantiognathus,* by previous students of Triassic conodonts.

In all elements of *Ellisonia gradata* except those representing the very earliest growth stages, the cusp and process denticles are "milky" or opaque and white. However, even in large specimens, the "white matter" responsible for this feature is not concentrated in definite axial strips of uniform density, but occurs in distinct "clouds" aligned irregularly along denticle axes. The variable density of the clouds of "white matter" in these elements produces a spotted appearance that is quite distinctive and shows clearly in most of the syntypes illustrated in figures 1-8, on Plate 4.

Remarks: The fact that the element types we include in *Ellisonia gradata* form an almost continuously intergradational morphologic sequence is indicated in the trivial name and suggests that the units were all parts of the skeletal apparatus of a single conodont species. To evaluate the suggestion just noted, we applied some of the procedures used by Fager (1957) and Kohut (1969) in forming and evaluating recurrent groups. The elements under consideration form a group at the 0.7-level, in which the average index of affinity (IA) is 0.77, and the coefficient of rank concordance (W) for the 24 samples that contain all these elements is 0.41, a value that is significant at the 0.05 level. Because there seems to be significant concordance in rank and a high level of mutual co-occurrence among the elements considered, we believe it is more likely that they represent the skeletal components of a single conodont species than that they are only faunal associates. Further, the analysis of rank indicates that elements occurred in the skeleton of *Ellisonia gradata* in such a manner that LA>LB>LC>U, and the totals of Table 4 indicate a ratio of about 3:2.6:1.9:1 among these elements arranged in the same order. These calculations suggest that the skeleton of an individual representative of *Ellisonia gradata* contained no fewer than 18 elements: 2 U-, 4 LC-, 4 LB1-, 2 LB2-, and 6 LA-elements, although the actual number may have been greater than this.

Ellisonia gradata is closely related in skeletal architecture and element morphology to *E. delicatula* and *E. clarki,* and the features by which representatives of these species can be distinguished are noted in discussions of *E. clarki* and *E. delicatula*. We have examined the Salt Range specimens that Huckriede (1958) described as *Lonchodina latidentata* (Tatge), *Roundya,* n. sp. A, and Gen. et spec. indet. A, and they are clearly LB2-, U-, and LA-elements of *E. gradata,* which ranges to the top of the Lower Triassic of West

Pakistan and may be represented in even younger rocks elsewhere. We have not seen most of the younger material, however, so we have not included references to it in our synonymy.

Occurrence: Dolomite unit of Kathwai Member at Kingriali, Narmia, Zaluch, Chhidru A, and Kathwai A; Limestone unit of Kathwai Member at Kingriali, Chhidru A, Chhidru B, Khan Zaman Nala, and Kathwai B; Mittiwali Member of Mianwali Formation at Zaluch, Nammal, Ganjaroh, Chhidru A, Chhidru B, Wargal, and Kathwai A; Narmia Member of Mianwali Formation at Landa, Narmia, Zaluch, and Nammal; Lower Triassic of Nevada. A few representatives of *Ellisonia gradata* occur in samples from the "*Meekoceras* beds" of Nevada now under study by Professor James W. Collinson and Mr. Walter Hasenmueller, of The Ohio State University. The distribution of this species in West Pakistan is summarized in Appendix A and Table 3.

Types: Syntypes, OSU 28027-28033, inclusive.

Ellisonia robusta Sweet, 1970

Pl. 5, fig. 6-8, 10, 12, 16

Ellisonia robusta Sweet, 1970, p. 8, pl. 1, fig. 11, 14-16.

Diagnosis: A species of *Ellisonia* with a skeletal apparatus composed of elements with peglike denticles; U-elements with long, straight posterior processes; LA- and LC-elements with two anterolateral processes; LB-elements with long posterior processes and long anterior processes, some of which are bifid; and "white matter" that is uniformly and densely developed throughout all elements.

Material: 110 discrete elements, representing the five morphologic types whose distribution is shown in Table 4.

Description: U-elements have a long, subcylindrical cusp; a long, straight posterior process; and a pair of anterolateral processes that form an anterior arch that is essentially perpendicular to the plane defined by the cusp and posterior process. The lateral processes project laterally and downward from a point slightly anterior to the axis of the cusp. Distally, each process curves slightly toward the anterior, and each bears a series of 4 to 6 erect, peglike denticles, the apexes of which are directed away from the cusp. The posterior process, which is at least as long as the anterolateral processes, is subcircular in transverse section and bears a series of at least two discrete, peglike denticles that are inclined posteriorly at an angle of about 45°. The most anterior of these denticles is separated from the cusp by a space equivalent to its width in the large syntype (Pl. 5, figs. 8, 10), but in smaller specimens this space is occupied by several smaller denticles. This indicates that late-stage denticles incorporate several early-stage denticles. The undersurface of each process bears a very narrow groove, and a faint pit is distinguishable beneath the cusp. Marginal to the grooves, however, especially in larger specimens, there is a longitudinally striated, beveled margin, which faces partly laterally and indicates that, at least in late stages of growth, the attachment surface beneath the processes was "inverted."

LA-elements are paired, individually asymmetric units, consisting of a

TABLE 4. Distribution and Frequency of Elements of *Ellisonia robusta*, n. sp.

Sample number	Frequency of element types*				
	U	LA	LB1	LB2	LC
K4-UCS	–	2	3	1	1
K3-15	–	–	–	1	–
K3-14	2	6	13	2	9
K1-17	–	–	1	–	–
K3-12	1	2	5	9	1
K10-6	–	–	1	1	–
K3-10	4	5	10	5	8
K4-CM+LCM	–	1	8	5	3
Totals	7	16	41	24	22

* Symbols for element types are explained in the text.

stout, erect to distally recurved cusp and a pair of anterolateral processes of markedly unequal length and curvature. The longer anterolateral process projects laterally and downward from the cusp, curves slightly to the posterior distally, and supports a series of 4 or more discrete, compressed, or peglike denticles on its upper surface. The shorter anterolateral process projects directly laterally from the cusp, curves slightly to the anterior at its distal end, and bears 4 compressed, fused denticles on its upper surface. The distal 3 of these denticles are short and nodelike at all stages of growth recognized; but the one immediately adjacent to the cusp is fused to the side of the cusp and nearly two-thirds its length. The posterior side of the cusp bears a distinct costa, which is produced basally to form a short, undenticulated posterior process. On the undersurface of LA-elements there is a narrow groove beneath the processes, a shallow but capacious pit beneath the cusp, and a brim of "inverted" attachment surface along each side of the longitudinal basal grooves in specimens representing late growth stages.

LB1-elements have a reclined, subcylindrical cusp; a long, straight posterior process bearing 4 discrete, inclined, peglike denticles; and an anterior process that projects anteriorly, curves laterally downward, and bears 4 or 5 discrete, erect or recurved, peglike denticles on its upper margin. LB2-elements are identical to LB1-elements, but they have a short, undenticulated secondary process that projects laterally on the convex side of the anterior process from the side of the second denticle of the anterior series. In early-stage LB-elements, the undersurface is channeled beneath both processes and there is a capacious basal pit beneath the cusp. In larger specimens, representing later growth stages, the attachment surface is inverted marginally and the longitudinal basal grooves are rimmed by a flat surface of varying width and inclination.

LC-elements are similar to LA-units, but the shorter anterolateral process and the adjacent cusp margin curve abruptly backward so that the unit is L-shaped in superior or inferior view.

Remarks: Nearly all the specimens on which we base our concept of *Ellisonia robusta* represent late stages of growth, hence it is difficult to compare them closely with the elements of other species, which are mostly smaller. However, it is clear that *E. robusta* is most closely related in morphology to *E. gradata,* with which it occurs throughout its range. Elements of the two species are readily distinguishable by the fact that even the smallest skeletal units of *E. robusta* lack the cloudy distribution of "white matter" characteristic of *E. gradata,* and large elements are uniformly and densely white. In addition, LB-elements of *E. robusta* have a well-developed posterior process that is much longer and much better developed than is the corresponding structure in LB-elements of *E. gradata.*

Occurrence: Mittiwali Member of Mianwali Formation at Narmia, Chhidru A, Chhidru B, and Kathwai A. Distribution in West Pakistan is summarized in Appendix A and Table 4.

Types: Syntypes, OSU 28034-28038, inclusive.

Ellisonia teicherti Sweet, 1970
Pl. 4, fig. 20-28

?*Lonchodina* n. sp. A. Huckriede, 1958 (partim), p. 153, pl. 10, fig. 3 (not pl. 10, fig. 2).
?*Hindeodella* sp. a Bender & Stoppel, 1965, p. 344-345, pl. 15, fig. 6.
Ellisonia teicherti Sweet, 1970, p. 8, pl. 1, fig. 3, 4, 7, 8, 12.

Diagnosis: A species of *Ellisonia* with a skeletal apparatus that consisted of U-elements with no posterior process; and LA-, LB-, LD-, and LE-units with short anterior and long posterior processes, needlelike denticles, and an escutcheonlike attachment surface on the inner side of the element. All elements are opaque and almost uni-

TABLE 5. Distribution and Frequency of Elements of *Ellisonia teicherti*, n. sp.

Sample number	Frequency of element types*				
	U	LA	LB	LD	LE
T64-17	–	2	4	2	1
T64-15	–	–	2	–	–
K1-12	4	3	9	3	5
T62-193	–	–	–	–	1
K14-3	1	3	14	2	–
T62-192	–	1	8	9	5
T63-130	1	5	23	7	–
K3-5	–	–	–	–	1
T63-128	2	1	10	4	1
T64-14	10	33	52	48	13
T62-191	–	3	4	1	2
T64-13	1	5	8	10	–
T63-124	7	44	72	33	9
T63-123	1	4	10	6	3
T63-122	3	17	35	15	26
K4-2	–	–	2	1	–
T63-163	1	1	1	–	2
T63-165	–	–	–	1	1
T63-162	2	4	6	5	5
T63-19	3	6	11	–	4
T63-22	–	–	1	–	1
Totals	36	132	272	147	80

* Symbols for element types are explained in the text.

formly white in even earliest stages of growth.

Material: 667 discrete elements, representing the five morphologic types whose distribution is shown in Table 5.

Description: U-elements are almost bilaterally symmetrical units with an erect, compressed, subcentral cusp and two processes of equal or slightly unequal length that are produced laterally and curve slightly downward and posteriorly from faint lateral costae on the cusp. The upper edge of each lateral process bears a series of 8 to 10 small, short, needlelike denticles, and the attachment surface is an escutcheonlike area on the posterobasal side of the cusp and lateral processes.

LA-elements are similar to U-elements, but lateral processes are of distinctly unequal length. The longer process curves posteriorly in its distal portion; the shorter process is straight or curves slightly anteriorly at its distal end; and the posterior face of the cusp is twisted so that it faces the longer lateral process.

LB-elements represent a further modification of U- and LA-elements. These structures have an erect to slightly inclined cusp and are twisted so that it is more consistent with convention to describe the long process as posterior and the shorter one as anterior. The posterior process, which is straight in lateral view but irregularly sinuous in superior view, is set in small specimens with at least 15 needlelike denticles arranged in a crudely hindeodelloid series, which tends in larger specimens to be incorporated in the process bar and a few larger denticles. The short anterior process, which curves abruptly to one side, bears 3 or 4 needlelike denticles proximally and a cusp-sized denticle terminally. During growth, the cusp and the large anterior denticle overgrow the smaller denticles between them so that, in our largest specimens, there is only one recognizable denticle on the anterior process. The attachment surface is an escutcheonlike scar, which is broadest beneath the cusp and situated on the side of the element toward which the anterior process is deflected.

LD-elements, which are J-shaped in superior view, are modified LB-units in which the anterior process is strongly curved to one side, the terminal anterior denticle is greatly enlarged, and the cusp is indistinguishable from the short, needlelike denticles of the processes. As in other elements of *Ellisonia teicherti*, the attachment surface is an escutcheonlike scar on the side of the element toward which the anterior process curves.

LE-elements have a reclined cusp and anterior and posterior processes of essentially the same length. The posterior process is similar in conformation and denticulation to that of LB- and LD-elements, but the anterior process projects laterally and upward and terminates in an erect denticle that equals the cusp in size. The attachment surface is identical to that of the other elements of *Ellisonia teicherti.*

Remarks: Even the smallest elements of *Ellisonia teicherti* are opaque and larger forms are completely, uniformly, and densely white. This feature, combined with the needlelike denticulation and escutcheonlike attachment surface common to all elements, serves to distinguish elements of *E. teicherti* from all the others with which they occur and from other species of *Ellisonia.* The two Middle Permian specimens from Sicily that Bender and Stoppel (1965) describe as *Hindeodella* sp. a very strongly resemble the LB-elements of *E. teicherti* anteriorly and basally, but posterior denticulation is somewhat different from that typical of the elements of *E. teicherti.* These Sicilian specimens may indicate that *E. teicherti* had a long range and a wide geographic distribution in the Permian and earliest Triassic, but because we cannot be certain of their identity with *E. teicherti,* we attach no particular stratigraphic significance to them at this time.

Occurrence: ?M. Permian, Sicily; uppermost highly fossiliferous unit of Chhidru Formation at Chhidru B; Dolomite unit of Kathwai Member at Kingriali, Narmia, Zaluch, and Chhidru A; Limestone unit of Kathwai Member at Kingriali. Distribution in West Pakistan is summarized in Appendix A and Table 5.

Types: Syntypes, OSU 28039-28047, inclusive.

Ellisonia torta Sweet, n. sp.
Pl. 4, fig. 1-4

Diagnosis: A species of *Ellisonia* with a skeletal apparatus consisting of U-elements with a long, straight posterior process; LA- and LC-elements with two sinuous anterolateral processes; and LB-elements with short anterior and longer, sinuous, posterior processes. "White matter" is of irregular density at denticle bases, but is uniformly distributed in cusp and denticles.

Material: 49 discrete specimens, representing the four morphologic types whose distribution is shown in Table 6.

TABLE 6. Distribution and Frequency of Elements of *Ellisonia torta,* n. sp.

Sample number	Frequency of element types*			
	U	LA	LB	LC
K12-9	3	7	37	2

* Symbols for element types are explained in the text.

Description: U-elements have a stout recurved cusp at the junction of a long posterior process that bears 10 to 12 reclined denticles, the most posterior 3 of which are the largest; and two anterolateral processes, which project laterally and slightly anteriorly and downward, and bear 3 or more erect denticles. The undersurface bears a distinct pit beneath the cusp and well-developed channels beneath the processes.

In LA-elements, a stout subcentral cusp of subtriangular cross section is flanked by anterolateral processes of different length and curvature. The cusp has a flat anterior face, sharp anterolateral edges, and a rounded costa in the midportion of the posterior face, which expands slightly basally to form an undenticulated posterior process. The longer anterolateral process is sinuous, bears 9 compressed denticles, and projects laterally downward; in its midportion it is bowed posteriorly, but distally it is bowed anteriorly. The shorter anterolateral process, which bears 2 or 3 denticles, projects straight laterally from the cusp and curves slightly anteriorly at its distal end. The under edge of both processes is channeled and there is a faint pit beneath the cusp and short posterior process.

LB-elements have a stout reclined

cusp that is peglike in large specimens and situated at the junction between a long, slightly sinuous posterior process and a shorter anterolateral process. The posterior process is surmounted by 8 or 9 reclined denticles, the most posterior 3 of which are the largest and most strongly reclined; smaller denticles between these larger ones and the cusp tend with growth to be incorporated in the process bar, so that in large specimens only the cusp and most posterior denticles are distinguishable apically. The anterolateral process projects anteriorly, curves laterally and downward at its distal end, and bears 5 or 6 erect, discrete denticles on its upper surface. The underedge of both processes is channeled and there is a faint pit beneath the cusp.

LC-elements are like LA-units, but in these structures the shorter anterolateral process and the sharp cusp margin from which it projects are deflected sharply to the posterior so that the elements are L-shaped in superior or inferior view. Elements of this type have been included by previous students of Triassic conodonts in *Apatognathus* or, more recently, *Enantiognathus*.

Remarks: All the elements on which we base our concept of *Ellisonia torta* are from the same sample, and nearly all are of the same size. Consequently, we can make no very meaningful remarks about ontogeny. Morphologically, the elements of *E. torta* are most closely related to those on which we base *E. gradata* and *E. robusta*. "White matter" in elements of *E. torta* is somewhat similar in distribution to that of *E. gradata*-elements, but LA-, LB-, and LC-elements of the two species have distinctly different morphology and none of the LB-elements of *E. torta* is bifid anteriorly. Elements of *E. torta* are readily separated from those of *E. robusta* by differences in the style of denticulation and by differences in the morphology of the LB-elements of the two species.

Occurrence: Narmia Member of Mianwali Formation at Landa. Distribution in West Pakistan is summarized in Appendix A and Table 6.

Types: Syntypes, OSU 28048-28051, inclusive.

Ellisonia triassica Müller, 1956

Pl. 5, fig. 9, 13-15, 17, 18, 20-22

Ellisonia triassica Müller, 1956, p. 822, pl. 96, fig. 12-14.
Hibbardella subsymmetrica Müller, 1956, p. 825-826, pl. 96, fig. 11.
Lonchodina triassica Müller, 1956, p. 828, pl. 96, fig. 10.
Hindeodella nevadensis Müller, 1956, p. 826, pl. 96, fig. 2, 3.
Hindeodella triassica Müller, 1956, p. 826, pl. 96, fig. 4, 5.
Hindeodella raridenticulata Müller, 1956, p. 826, pl. 96, fig. 1.
Neoprioniodus unicornis Müller, 1956, p. 829, pl. 95, fig. 18.
?Ozarkodina? sp. Müller, 1956, p. 820, pl. 95, fig. 23; pl. 96, fig. 18.
Lonchodina mülleri Tatge, Huckriede, 1958, p. 151-152, pl. 10, fig. 9, 16, 17.
Lonchodina discreta Ulrich & Bassler, Huckriede, 1958, p. 150, pl. 10, fig. 21-25.
?Lonchodina sp. Huckriede, 1958, p. 153, pl. 10, fig. 4.

Diagnosis: A species of *Ellisonia* with a skeletal apparatus composed of U-elements with a long, straight posterior process; intergradational LA- and LF-elements, and LB-elements with a long posterior and a short anterior process. "White matter" is uniformly distributed in cusp and denticles and the attachment surface is completely "inverted" beneath all elements representing late stages of growth.

Material: 1,488 discrete elements, representing the four major morphologic types whose distribution is shown in Table 7.

Description: U-elements, described as the form species *Ellisonia triassica* by Müller (1956), have a recurved cusp that is subcircular in cross section; a long, high, posterior process surmounted by 5 or 6 discrete, reclined denticles, the posteriormost 3 or 4 of which are the longest and broadest; and a pair of symmetrically disposed

TABLE 7. Distribution and Frequency of Elements of *Ellisonia triassica* Müller.

Sample number	U	LA	LB	LF	Sample number	U	LA	LB	LF
K6-37	2	–	7	3	K14-6	–	1	–	–
K6-36	–	–	–	1	K3-8	–	–	1	–
K12-9	–	2	2	6	T63-136	2	2	12	–
K12-5	1	3	4	11	K3-7	3	4	15	1
K1-37	–	–	1	–	T64-21	4	4	34	2
K1-36	1	2	1	4	T64-19	2	1	4	–
K6-27A	9	11	48	16	T64-18	11	11	66	4
K13-11	–	–	3	–	T63-132	6	5	26	2
K12-21	–	1	1	2	K2-4	1	2	3	–
K4-UCL	–	1	8	4	K6-5	3	6	14	–
K4-UCS	2	–	15	3	T64-17	2	–	5	1
K13-9	7	2	51	2	K3-6	–	–	1	–
K3-15	3	13	4	4	T63-131	1	3	23	2
K3-14	10	23	26	17	T64-15	1	–	2	–
K1-17	2	8	10	4	K14-3	2	1	8	–
K3-12	3	25	28	11	T63-130	1	2	19	–
K10-6	1	2	2	2	T63-128	–	2	–	–
K3-10	11	24	37	19	T64-14	–	2	4	3
K4-CM+LCM	3	26	48	35	T62-191	–	–	2	–
T63-158	–	9	26	4	T64-13	2	–	3	–
K6-A	–	–	4	1	T63-124	2	9	13	1
K11-CL±LCL	2	2	4	3	T63-123	3	1	10	–
T63-157	21	33	166	36	T63-122	3	4	15	3
K10-5	6	–	13	1	K11-3	–	1	–	–
K2-5	–	–	2	–	T63-163	–	–	3	–
T63-155	–	–	7	6	T63-165	–	–	2	–
T63-153	–	9	7	1	T62-300	1	–	3	–
T63-145	1	1	8	1	K1-12	1	–	5	2
T63-144	1	2	1	–	T63-162	–	2	8	2
T63-139	1	–	2	2	T63-19	–	2	7	–
K10-3	2	2	6	–	Totals	140	266	860	222

* Symbols for element types are explained in the text.

anterolateral processes that project slightly anteriorly, curve gracefully laterally and downward, and bear 3 or 4 erect or slightly recurved denticles, which are anteroposteriorly compressed in small specimens but stout and peg-like in larger ones. The basal edge of the processes is transversely wedge-shaped. In small specimens and ones of intermediate size, the basal edge bears a very narrow longitudinal groove, but most of the attachment surface is "inverted" and is represented by narrow, flat, longitudinally striated faces marginal to the longitudinal grooves. There is a shallow basal pit beneath the cusp. In very large specimens, like the holotype of *Ellisonia triassica,* the attachment surface is completely "inverted"; there is no trace of grooves beneath the processes or of a pit beneath the cusp; and the marginal areas of attachment surface are half the height of the processes.

LA-elements, described by Müller (1956) as *Hibbardella subsymmetrica* and *Neoprioniodus unicornis,* have a stout recurved cusp with a broadly convex anterior face, sharp-edged anterolateral costae, and a distinct midposterior carina, the basal part of which projects posteriorly to form a short, distinct, undenticulated posterior process that is well defined in specimens of small and intermediate size, but indistinct in massive specimens representing late stages of growth. The cusp is flanked anterolaterally by downwardly-directed, posteriorly recurved processes of subequal or markedly unequal length, which are surmounted by 3 to 5 discrete, posteriorly recurved denticles that decline in length toward the extremities of the processes. In specimens of small and intermediate size, the attachment surface is marked by a narrow longitudinal channel beneath the anterolateral processes and a shallow basal pit that faces posteriorly beneath the cusp and short posterior process. Marginal to these grooves and the basal pit are strips of "inverted" attachment surface. In large specimens, like the holotype of *H. subsymmetrica,* and in very large specimens, like the holotype of *Neoprioniodus unicornis,* the attachment surface is almost completely "inverted" and the grooves and basal pit of smaller forms are barely distinguishable. Furthermore, "inverted" segments of the attachment surface develop laterally on one process (typically the shorter one), which thus has a transversely wedge-shaped lower margin, but develop essentially at right angles to the plane of the other process (typically the longer one), which thus has its greatest width basally.

LF-units, which intergrade morphologically with LA-elements, have a stout reclined cusp, the carinate side of which faces posterolaterally. The cusp is flanked by two curved processes that are anterior and posterior in conventional orientation, but clearly the same as the anterolateral processes of LA-elements. The shorter anterior process of LF-units curves anteriorly, laterally, and downward, bears 2 or 3 erect or recurved denticles, and is wedge-shaped basally. The longer process projects posteriorly from the cusp, curves later-ally distally, bears 3 or 4 discrete, reclined denticles, and has a transversely flat basal surface like that of the longer process of large LA-elements.

LF-elements grade morphologically, through forms like the one shown in fig. 22, on Plate 5, into LB-elements, which were described by Müller (1956) as *Lonchodina triassica, Hindeodella nevadensis, H. triassica,* and *H. raridenticulata.* LB-elements are somewhat variable structures, but they all have a large recurved cusp; a long posterior process with denticles like the posterior series of U-elements; and a short anterior process that curves laterally and bears a crest of 3 or 4 compressed denticles in large specimens. The attachment surface is longitudinally grooved in specimens of small and intermediate size, but almost completely "inverted" and transversely wedge-shaped in larger forms.

Remarks: The elements here included in *Ellisonia triassica* represent a morphologically intergradational series that forms a well-knit recurrent group (Fager, 1957; Kohut, 1969) with an average index of affinity of 0.71. The coefficient of rank concordance (W) among the four major element types of *Ellisonia triassica,* based on frequency data from the 21 samples that contain all four types, is 0.53, a value that is significant at the 1 percent level and suggests that the rank order of elements in the skeleton of *E. triassica* was $U<LA<LF<LB$. If, on the other hand, one regards the absolute abundance totals of Table 7 as a better measure of rank order, the position of LA- and LF-elements just given is reversed. We do not consider this a major problem in the interpretation of *Ellisonia triassica,* however, for LA- and LF-elements are completely gradational morphologically and the assignment of individual specimens to one category or another is arbitrary in many samples.

Conodont elements from Triassic rocks of widely different age and from many geographic provinces have been identified by various authors with one or more of the form species herein united as *Ellisonia triassica*. We have not included references to all these assignments, for we have not seen all the specimens involved. It may be noted, however, that the morphologic latitude represented by the sum of all these identifications is very great and we suspect that more than one multielement species is involved. Müller's type series, and collections from the Lower Triassic of Nevada assembled by Professor J. W. Collinson and Mr. Walter Hasenmueller, of The Ohio State University, indicate that all the elements of *Ellisonia triassica* are present in Nevada, that they form a distinct group with a high index of affinity, and that rank order among elements is like that suggested by the collections from West Pakistan, herein described. LF-elements in the collections from Nevada are not exactly like the ones from Pakistan, and there are minor differences in the morphology of other units of the apparatus, too. These features may ultimately be useful in defining subspecies of *Ellisonia triassica*, but, at this time, they are accorded less significance than the many obvious similarities.

Occurrence: Uppermost highly fossiliferous unit of Chhidru Formation at Wargal; Dolomite unit of Kathwai Member at Kingriali, Narmia, Zaluch, Nammal, Ganjaroh, Chhidru A, Kathwai A, and Kathwai B; Limestone unit of Kathwai Member at Kingriali, Khan Zaman Nala, and Kathwai B; Mittiwali Member of Mianwali Formation at Landa, Zaluch, Nammal, Ganjaroh, Chhidru A, Chhidru B, Wargal, and Kathwai A; Narmia Member of Mianwali Formation at Landa and Nammal; Lower Triassic of Nevada. Distribution in West Pakistan is summarized in Appendix A and Table 7. The geographic and possibly the geologic, range is probably greater than indicated, but the limits of both will depend on a study of the many elements that have been referred to the various elements of *Ellisonia triassica* in the context of the samples from which they were derived. This work has been beyond the scope of the present study.

Types: Figured hypotypes, OSU 28052-28058, inclusive.

Ellisonia? sp.
Pl. 5, fig. 5

Our collections from West Pakistan contain 22 discrete elements that resemble LB-elements of *Ellisonia triassica* in general outline and distribution of "white matter," but are readily distinguished from those elements in structure, mode of denticulation, and stratigraphic distribution. These elements are all alike in that they are laterally compressed, bladelike units with an erect to slightly recurved cusp, the anterior and posterior edges of which grade into denticulated processes. The anterior process, which curves down and laterally, bears a conspicuous terminal denticle and one or two nodelike intermediate denticles, the distal pair of which are the longest and most conspicuously reclined. The undersurface is transversely flat beneath processes and slightly indented beneath the cusp.

All the elements we describe as *Ellisonia?* sp. are about the same size, but the distribution of "white matter" in them makes it clear that the cusp, the terminal anterior denticle, and all the posterior denticles were enlarged by incorporation of adjacent early-stage denticles. This mode of development was common in *Ellisonia,* but the conformation that resulted in the large elements described here is distinctly different from that achieved in comparable elements of other species of *Ellisonia*. We question the generic reference, however, because all the other species we include in *Ellisonia* seem to have had skeletal apparatuses that included U-elements and several types of L-elements, and no such structures are associated with the few specimens to which these remarks refer. Such specimens may ultimately be found in larger collections than the ones at hand, and the affinities of the elements here described can then be determined.

Occurrence: Narmia Member of Mianwali Formation at Landa and Nammal. Distribution in West Pakistan is summarized in Appendix A.

Type: Figured specimen, OSU 28059.

Genus NEOGONDOLELLA Bender and Stoppel, 1965

Diagnosis: Neogondolella includes conodont species in which the skeletal apparatus included elements of a single morphologic type. These elements, which are elongate, paired, and individually asymmetric, have a terminal or subterminal posterior cusp; a median nodose or denticulate carina; and finely to coarsely pitted, largely unornamented, platformlike lateral extensions, which are joined posteriorly in most species by a more or less well-developed brim that encloses the posterior end of the carina. The underside of elements is marked by a longitiudinally grooved keel that widens posteriorly to enclose a pit beneath the cusp.

Type species: Gondolella mombergensis Tatge, 1956. [Bender (1967?, p. 516) cites his newly established *N. aegaea* as type of *Neogondolella*. However, *Neogondolella* was first used, without diagnosis or explanation, in connection with a well-known species (*Gondolella mombergensis* Tatge) by Bender and Stoppel in 1965. Hence, *N. mombergensis* (Tatge), not *N. aegaea* Bender, is type of *Neodondolella* by monotypy.]

Remarks: Neogondolella was inadvertently, but validly, established by Bender and Stoppel (1965, p. 343) for conodont form species from the European Triassic referred by previous authors to *Gondolella* Stauffer and Plummer, 1932. Species of the sort to which they referred are not restricted in occurrence to the European Triassic, of course, but they are represented in Lower and Middle Triassic rocks of many widely separated geographic provinces. Specimens referable to three species occur in the uppermost Permian and Lower Triassic rocks of West Pakistan, and two of these are also represented in western North America (Nevada and Idaho).

Representatives of *Neogondolella* differ markedly from those of Pennsylvanian *Gondolella*. That is, the surface of elements representing species of *Gondolella* is smooth and glassy, whereas that of elements included in *Neogondolella* is finely to coarsely pitted; the platform of *Neogondolello* elements is continued around the posterior side of the cusp as a more or less pronounced brim, but no such brim is recognizable in described elements of *Gondolella;* and the growth lamellae of *Gondolella* elements are thick, at least anteriorly, where those of *Neogondolella* elements are thin. Clark and Mosher (1966) have also shown that conspicuous differences in undersurface morphology separate *Gondolella* and *Neogondolella* elements, although they included both in *Gondolella*. Finally, it should be noted that Rhodes (1952) has shown convincingly that typical *Gondolella* elements occur with others of different morphologic type in a multielement apparatus he named *Illinella typica*. Studies of gondolella-like Triassic specimens indicates no consistent or significant association with elements of any other morphologic type, hence we conclude that the skeletal apparatus of *Neogondolella* contained elements of a single type. If this is so, *Neogondolella* was fundamentally different from *Gondolella* (or *Illinella*).

As we note in our discussion of *Neospathodus* Mosher, *Neogondolella* seems to have been very closely related in form and mode of development to both *Neospathodus* and *Xaniognathus,* although there is no evidence that elements of these three types ever occurred in the same skeletal apparatus. *Neogondolella* is distinguished primarily by the fact that its elements form a distinct platform in even the earliest ontogenetic stages, and by the fact that its elements lack a primary posterior process (although one may develop secondarily). Elements of some species of both *Neospathodus* and *Xaniognathus* develop platforms in late growth stages, and the lack of a primary posterior process is also a feature of the elements of *Neospathodus*.

Neogondolella is not known from the Lower Triassic of Europe, but it was apparently widespread and quite diversified specifically during that interval of time in West Pakistan, Timor, Kashmir, western North America, Greenland, and Spitzbergen. The type species, *N. mombergensis*, is from the German Muschelkalk, and related Middle Triassic species were apparently cosmopolitan in distribution. Mosher (1967) includes Upper Triassic gondolellas in a new genus, derived from *Neospathodus*, and states that *Neogondolella* became extinct in the late Middle Triassic.

Neogondolella carinata (Clark)

Pl. 3, fig. 1-17, 24, 26, 27

Gondolella carinata Clark, 1959, p. 308, pl. 44, fig. 15-19.——Clark, 1960, p. 125.——Müller, 1964, p. 750.——Clark and Mosher, 1966, p. 390, pl. 47, fig. 21-23.

Gondolella mombergensis Huckriede, 1958, p. 147, pl. 10, fig. 26-30 [not pl. 10, fig. 31, 33-37, 42, 43, 45, 46, 49, which =*Neogondolella mombergensis* (Tatge)].

?*Gondolella nevadensis* Clark, 1959, p. 308, pl. 44, fig. 11-14.——Clark and Mosher, 1966, p. 391-392, pl. 47, fig. 28, 29.

?*Gondolella planata* Clark, 1959, p. 309, pl. 44, fig. 8-10.——Clark, 1960, p. 124, 125. Mosher and Clark, 1965, p. 557.——Clark and Mosher, 1966, p. 392, pl. 47, fig. 26, 27.

?*Gondolella* sp. aff. *G. nevadensis* Clark, Srivastava and Mandwal, 1966, p. 621, 622, fig. 7.

Material: 4,745 discrete elements. The smallest element is 0.32 mm long; the largest is 1.13 mm in length.

Description: Most of the elements here referred to *Neogondolella carinata* (Clark) are faintly to conspicuously bowed laterally, but a few are straight and bilaterally symmetrical. The major features of these elements are displayed on Plate 3, which illustrates seven specimens chosen to exhibit the morphologic changes that take place in the growth of the skeletal elements of *Neogondolella carinata*.

Specimens representing the earliest growth stages recognized (Plate 3, fig. 16, 17) are arched, bladelike elements, with 9 or 10 short, reclined denticles, the posterior 7 or 8 of which are enclosed by upcurved lateral platforms that are not continuous posteriorly around the cusp. The underside of these small elements is marked by a narrow, conspicuous longitudinal keel, which is grooved and expands slightly to form a basal pit beneath the cusp.

In elements representing successively later growth stages (Plate 3, fig. 13-15, 10-12, 7-9, 1-3) the ratio of width to length varies considerably, the number of denticles increases to 15 or more, and lateral elements of the platform are joined by progressive development of a short brim around the posterior side of the cusp. In symmetrical elements this posterior brim develops laterally and posteriorly at the same rate. In asymmetric elements, however (e.g., Plate 3, fig. 7-9) the brim develops more rapidly on the outer and posterior sides of the cusp than on the inner side, and a conspicuous notch interrupts the inner-posterior margin of the platform until very late stages of growth when the notch is filled in.

With growth, the attachment surface and its looplike extension around the basal pit broaden and the keel-like ridge of early growth stages diminishes conspicuously in the degree to which it extends below the underside of the element. In very large specimens, like the one shown in figures 24, 26, and 27, on Plate 3, the platform is thick, downcurved laterally, and completely continuous around the cusp. The keel has almost completely disappeared from the underside, and the attachment surface has expanded to a third the width of the element. We have also noted in such large specimens that short rows of nodelike secondary denticles tend to form posterolaterally, and denticles of the carina tend to become overgrown by the platform or by adjacent denticles.

Remarks: We are inclined to suspect that the types of *Gondolella carinata, G. nevadensis,* and *G. planata* collectively represent a single species, for elements closely similar to all of them occur in our large collections from West Pakistan and are connected with one another by almost complete morphologic and/or ontogenetic transitions. Furthermore, the types of all three species came from the same sample (Clark, 1959) and they represent species that are said to have only slightly different ranges in the pre-*Meekoceras* beds of Nevada (Clark and Mosher, 1966). We have not seen the types of these species in the context of the samples from which they were drawn, however, and as isolated specimen they are readily distinguishable morphologically.

Neogondolella carinata is distinguished from *N. mombergensis* (Tatge) by elements that are broader relative to their length and by the lack of secondary posterior denticles except in erratic specimens representing very late stages of growth. In *N. mombergensis,* a conspicuous denticle or pair of denticles, develops posterior to the cusp in specimens of intermediate size, and these denticles and the cusp form a distinctive posterior cockscomb that is well shown in Tatge's (1956) illustration of the type specimens. The differences in relative dimensions between elements of *N. carinata* and those of *N. jubata,* n. sp., are shown in Figure 6, and morphologic distinctions between these two species are discussed in connection with our description of *N. jubata.*

Occurrence: Uppermost highly fossiliferous unit of Chhidru Formation at Chhidru B; White Sandstone Unit of Chhidru Formation at Chhidru A; Dolomite unit of Kathwai Member at Kingriali, Zaluch, Nammal, Ganjaroh, Chhidru A, Wargal, and Kathwai A; Limestone unit of Kathwai Member at Kingriali, Nammal, Chhidru A, Chhidru B, Khan Zaman Nala, Wargal, and Kathwai B; Mittiwali Member of Mianwali Formation at Ganjaroh, Chhidru A, Chhidru B, and Kathwai A. Also known from western United States, and possibly in the Anantnag District of the Kashmir Himalayas. Distribution of this species in West Pakistan is summarized in Appendix A.

Types: Figured hypotypes, OSU 28060-28066, inclusive.

Neogondolella elongata Sweet, n. sp.

Pl. 2, fig. 4, 5, 6-8; Pl. 3, fig. 18, 23, 25

Diagnosis: A species of *Neogondolella* characterized by skeletal elements with an average height to width to length ratio of 1:1.5:4, a platform typically confined to the posterior half of the element, and a series of discrete denticles that increase in length and width to the anterior end of the free blade.

Material: 95 discrete elements. The smallest is 0.33 mm long; the largest is 0.96 mm in length.

Description: Skeletal elements of *Neogondolella elongata* are paired, individually asymmetric, bladelike structures, with a prominent platformlike brim around the posterior half of the unit and a long free blade anteriorly. Although the ratio of height to width to length varies considerably (Figure 6), average values are 1:1.5:4. The cusp, which is the posterior denticle at all stages of growth, is short, reclined, and subcircular in transverse section. It is succeeded anteriorly by a series of 4 to 13 erect or slightly reclined denticles, which are compressed, discrete, and increase in length and width toward the anterior end of the free blade.

The platform, which is in the pos-

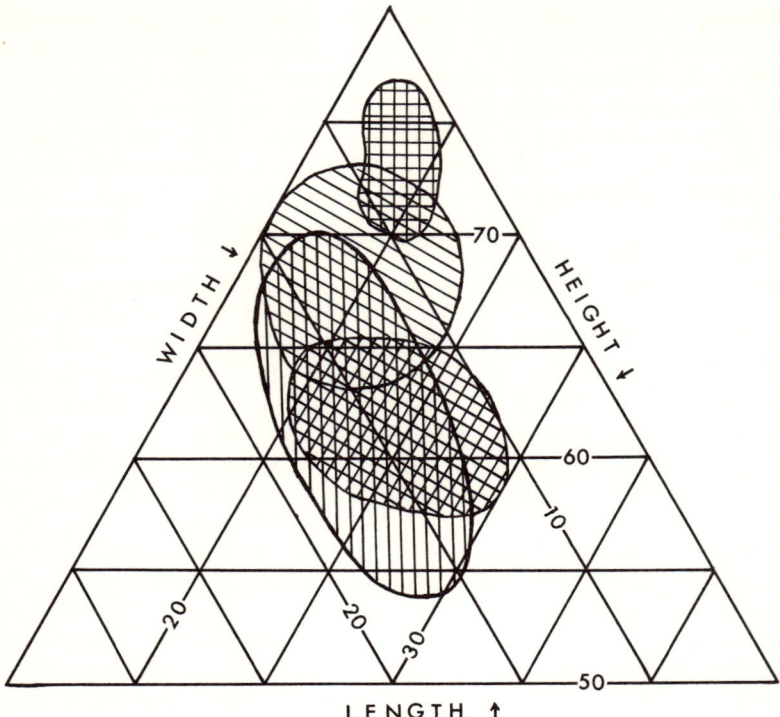

Fig. 6. Triangular diagram of width-length-height for four species of *Neogondolella*. Each field includes all measured specimens, so its size and shape are measures of the variability in proportions permitted for each species. Vertical single-lined field=*Neogondolella elongata*; oblique single-lined field=*N. jubata*; obliquely-gridded field=*N. carinata*; normally-gridded field=*N. mombergensis* (from Tatge, 1956). Dimensions are percentages; L+W+H=100 percent for each specimen.

terior half of the element, is upcurved marginally, smooth or faintly pitted surficially, and produced laterally from a prominent midlateral rib. In the smallest specimens available, the platform consists of a pair of lateral elements situated slightly anterior of the cusp. In larger specimens, the posterior margins of the lateral elements coalesce to produce a short upcurved brim around the posterior side of the cusp. Anteriorly, the platform grades abruptly into a midlateral rib on the free blade, which continues to the anterior extremity of the element.

In all but the largest specimens in our collections, the underside of the element is marked by a prominent longitudinal keel, marginal to which there may be one or two successively lower ribs representing progressive "inversion" of the attachment surface. The keel is longitudinally grooved and there is a deep pit beneath the cusp base. In the largest specimen at hand, the platform is very thick and no keel is recognizable beneath the posterior half of the element.

Remarks: Although it is clear from Figure 6 that many of the elements we include in *Neogondolella elongata* are similar in their proportions to those of both *N. carinata* and *N. jubata*, they are readily distinguished from elements of the latter two species by their short, posteriorly confined platforms, and by their relatively few denticles, which are discrete and conspicuously larger and longer anteriorly than posteriorly. To the best of our knowledge, *Neogondolella elongata* has not been reported from any area other than the one under consideration here.

Occurrence: Mittiwali Member of Mianwali Formation at Landa and Nammal; Narmia Member of Mianwali Formation at Landa and Narmia. Distribution in West Pakistan is summarized in Appendix A.

Types: Holotype, OSU 28070; paratypes, OSU 28067-28069, inclusive.

Neogondolella jubata Sweet, n. sp.
Pl. 2, fig. 1-3, 9-14, 16

Diagnosis: A species of *Neogondolella* characterized by symmetrical and asymmetrical bladelike skeletal elements with a height to width to length ratio of about 1:1.5:5; a high, even-crested carina; and a narrow, finely pitted platform that surrounds all but the anterior tenth of the unit.

Material: 236 discrete elements. The smallest is 0.15 mm long; the largest is 0.95 mm in length.

Description: Skeletal elements of *Neogondolella jubata* are either bilaterally symmetrical or distinctly asymmetrical bladelike structures with a high, conspicuous, even-crested carina and prominent platformlike brims, which are finely pitted but otherwise unornamented and surround all but the anterior tenth of the unit. As indicated in Figure 6 the ratio of height to width to length varies from about 1:1:3 to 1:2:8. However, the mean value is 1:1.5:5 and most of the elements measured have proportions that are very close to these ratios. The ratio of symmetrical to asymmetrical elements is 1:8.

The cusp, which is short, conical, reclined, and the posterior denticle at all stages of growth, is succeeded anteriorly by a series of 9 to 18 slightly inclined, laterally compressed denticles, which are confluent with adjacent denticles except apically, where they are discrete and sharply pointed. In lateral view, the upper margin is broadly arcuate and the crest of the carina is even and visible above the platform margins in elements representing all stages of growth.

In the smallest specimens at hand (e.g., Pl. 2, fig. 11-13), the platform consists of a pair of lateral structures, which project outward and curve upward from a prominent midlateral rib, but are not united posteriorly around the cusp. In specimens representing successively later stages of growth, lateral elements of the platform grow both anteriorly and posteriorly, ultimately forming a narrow brim posterior to the cusp. Development of the posterior brim is uniform laterally in elements that are bilaterally symmetrical, but in the majority that are not, the outerlateral platform element grows more rapidly posteriorly than does the innerlateral element so that there is a distinct posterolateral notch in the inner platform margin until late stages of growth, when the notch is filled in.

The underside of elements representing early growth stages is marked by a prominent longitudinal keel that is grooved and expands posteriorly into a pit beneath the cusp base. In later stages of growth, the attachment surface is progressively "inverted," and in specimens of intermediate size the keel is surrounded by two or three distinct, subparallel ridges at slightly lower elevations. In the largest specimens at hand, the platform is thick, the "inverted" portion of the attachment surface is wide, and the keel is a recognizable structure at only the anterior and posterior ends of the element.

Remarks: Although there is some overlap, most elements of *Neogondolella jubata* are readily distinguished from those of *N. mombergensis* (Tatge), *N. elongata*, n. sp., and *N. carinata* (Clark) by their proportions (Figure 6). In addition, elements of *N. jubata* are distinguished from those of the three species just cited by their high, even-crested carina and numerous denticles, and the platform of *N. jubata* elements is relatively much longer and narrower than that of elements of *N. elongata*. Finally, it should be noted that the cusp of *N. jubata* elements is terminal at all stages of growth, whereas that of *N. mombergensis* is terminal only in early growth stages and becomes the second or third denticle from the posterior end of the series in elements representing successively later stages of development.

Occurrence: Narmia Member of Mianwali Formation at Landa, Narmia, and Zaluch. Distribution in West Pakistan is summarized in Appendix A. Photographs of some of the conodont elements collected by Youngquist (1952) were sent the author in 1955 by the late Professor A. K. Miller, of the University of Iowa. These unpublished figures indicate that *Neogondolella jubata* is represented in a part of the Thaynes Formation of southeastern Idaho that also yields *Columbites-* and *Sibirites*-like ammanoids. We have no other record of the occurrence of *N. jubata*, but its presence in both West Pakistan and Idaho indicates that the species may have had a wide geographic distribution in the Lower Triassic.

Types: Holotype, OSU 28071; paratypes, OSU 28072-28074, inclusive.

Genus NEOSPATHODUS Mosher, 1968

Diagnosis: In *Neospathodus* we include a group of obviously related species, the only skeletal components of which were apparently blade-shaped elements with a well-developed anterior process; a posterior process that is short, secondary, vestigial, or absent; midlateral ribs of varying prominence that are produced laterally in late developmental stages of some species to form platforms; and a basal surface that includes a basal pit surrounded laterally and posteriorly by a looplike ridge and a laterally flaring brim, which may be wholy or partially "inverted."

Type species: Spathognathodus cristagalli Huckriede, 1958.

Remarks: The oldest representatives of *Neospathodus* are probably the specimens from the German Zechstein I described as *Spathognathodus divergens* by Bender and Stoppel (1965). Bender and Stoppel's specimens are similar in many respects to those we include in *N. dieneri*, n. sp., which is a generalized Lower Triassic form with the longest range of any of the several species we recognize. Middle and Upper Triassic species are described by Mosher (1968), who regards *Neospathodus* as the stock that gave rise to all Upper Triassic species with platform-bearing skeletal elements, but who includes in *Neospathodus* only species whose elements lack platforms.

In addition to *Neospathodus divergens* (Bender and Stoppel), which is known only from the German Zechstein, and the nine Lower Triassic species described in this report, we include in *Neospathodus* the Lower Triassic conodonts from Nevada described by Müller (1956) as *Neoprioniodus bransoni* and *N. bicuspidatus*. Elements of Müller's *Ctenognathus conservativa* and *C. discreta* also resemble those included in *Neospathodus*, but they have a well-developed posterior process in even the smallest specimens known. Such a structure appears to be a primary feature of the skeletal elements of *C. conservativa* and *C. discreta*, but not of the elements of *Neospathodus cristagalli*, type-species of *Neospathodus*, or its associates in the Salt Range Triassic. Later studies may indicate that the diagnosis of *Neospathodus* should be broadened to include species whose elements have a primary posterior process. We do not include *Ctenognathus conservativa* and *C. discreta* in *Neospathodus*, however, for only a few specimens have ever been collected and the morphology of elements representing early growth stages is unknown.

Evidence at hand is inconclusive, but it suggests that *Neospathodus* developed, perhaps iteratively, from the stock typified in our collections by the species we name *Xaniognathus curvatus* and *X. deflectens*. These two species represent a group that has been most frequently referred to *Ozarkodina* and is distinguished by elements with a long anterior process that bears a prominent longitudinal rib at midheight and a very short posterior process that is grooved basally but lacks a midlateral rib. Reduction of the short posterior process (but retention of the capacity to partially duplicate it secondarily) and concurrent broadening of the posterior half of the basal surface would produce elements like those of *Neospathodus*. A similar type of development from *Xaniognathus*, but producing elements with distinct platforms in even the earliest ontogenetic stages, would give rise to the units we regard as typical of *Neogondolella*. In short, *Xaniognathus* may be the stock from which both *Neospathodus* and *Neogondolella* were derived, and the presence of many homologous structures in the elements secreted by species of all three genera indicates a closely similar pattern of tissues in representatives of all members of this group.

Neospathodus homeri (Bender)
Pl. 1, fig. 2, 3, 9, 10

Spathognathodus homeri Bender, 1967?, p. 528, pl. 5, fig. 16, 18.
Spathognathodus homeris Bender, 1962, p. 69, 71, 73, 74 *(nomen nudum)*.
Spathognathodus homeri Bender, *in* Bender and Kockel, 1963, p. 438, 439, 441, pl. LIV (I) *(nomen nudum)*.
Spathognathodus n. sp. A Staesche, 1964, p. 289, pl. 31, fig. 1.
Neospathodus cristagalli (Huckriede) Mosher, 1968 *(partim)*, p. 930, pl. 115, fig. 1 [not pl. 115, fig. 2, which=*N. triangularis* (Bender)].

Diagnosis: A species of *Neospathodus* characterized by bladelike skeletal elements with 6 to 16 denticles; a width to height to length ratio of 1:2:3 in early growth stages and 1:2:4 in later ones; a basal margin that is straight anteriorly, but downcurved posteriorly; and a short, thin, laterally deflected secondary posterior process that bears as many as five denticles.

Material: 310 discrete elements. 142 are dextral and 168 are sinistral forms. Smallest specimen is 0.25 mm long; largest one is 0.67 mm in length.

Description: Skeletal elements of *Neospathodus homeri* are bowed, bladelike units, with a distinct midlateral rib; a width to height to length ratio that varies from about 1:2:3 in small specimens to about 1:2:4 in larger ones; and a basal margin that is straight anteriorly, but downcurved beneath the short, laterally deflected secondary posterior process. The anterior process bears 5 to 11 reclined denticles that decline in length toward the distal end of the process and are laterally confluent but apically discrete. In the smallest specimens at hand, the cusp is the terminal posterior denticle; in somewhat larger forms, however, a node on the sharp posterior edge of the cusp develops into a needlelike denticle and further growth in this manner produces a thin, laterally deflected process composed of as many as 5 denticles posterior to the cusp. In the largest specimens at hand (including the holotype) denticles formed in early growth stages tend to be incorporated into a few large denticles, and the apical profile of the element is distinctly ragged.

The undersurface, which is narrow and longitudinally grooved beneath the anterior process, widens into a broad, shallow, basal pit beneath the posterior third of the unit. Posterior to the apex of the pit, which is at the cusp base, the attachment surface is directed downward and laterally, is essentially flat, and is spatulate in outline. Even in the largest specimens at hand, no part of the attachment surface is "inverted."

"White matter" is concentrated in broad axial bands in the cusp and denticles. In specimens representing all stages of growth, the lower boundary of white matter is sharp and, in lateral view, defines a line that arches upward from the apex of the basal pit toward the distal end of the anterior process.

Remarks: The skeletal elements of *Neospathodus homeri* are morphologically most similar to the ones we include in *N. triangularis*, from which they are distinguished by the thin, laterally deflected posterior process and the spatulate shape of the posterior part of the basal margin. The elements we include in *N. homeri* almost certainly represent the species for which Bender (1962; in Bender and Kockel, 1963) has used *Spathognathodus homeris* (or *homeri*), a name that was not stabilized through description and illustration of type material for several years after it was first used. Direct comparisons between our specimens of *N. homeri* and the elements described as *Spathognathodus* n. sp. A by Staesche (1964) indicate that elements in the two collections represent the same species. Specimens of *N. homeri* also occur in undescribed collections assembled by Youngquist (1952) from the Thaynes Formation of southeastern Idaho.

Occurrence: Narmia Member of Mianwali Formation at Landa, Narmia, and Nammal; also known from Lower Triassic strata of Greece, northern Italy, and Idaho. Distribution in West Pakistan is summarized in Appendix A.

Figured hypotypes: OSU 28075; OSU 28076.

Neospathodus cristagalli (Huckriede, 1958)

Pl. 1, fig. 14, 15

Spathognathodus cristagalli Huckriede, 1958 (*partim*), p. 161-162, pl. 10, fig. 14, 15 [*non* pl. 10, fig. 1-13, 18a, 18b, which =*Neospathodus dieneri* Sweet, n. sp.].
Neospathodus cristagalli (Huckriede), Sweet, 1970, p. 9, pl. 1, fig. 18, 21.
non "*Spathognathodus*" *cristagalli* Huckriede, Lindström, 1964, p. 64 [=*Neospathodus dieneri* Sweet, n. sp.].
non Spathognathodus cristagalli Huckriede, Mosher and Clark, 1965, p. 556, 564, 565, pl. 66, fig. 8.
non Neospathodus cristagalli (Huckriede) Mosher, 1968, p. 930, pl. 115, fig. 1, 2 [pl. 115, fig. 1=*Neospathodus homeri* Bender; pl. 115, fig. 2=a juvenile specimen of *N. triangularis*, n. sp.].

Diagnosis: A species of *Neospathodus* characterized by bladelike skeletal elements with 5 to 13 denticles, a width to height to length ratio of about 1:3:4 at all stages of growth, a greatest height slightly posterior to unit midlength, a short terminal posterior cusp of broadly subtriangular lateral

---

PLATE 1

Figures are unretouched photographs of uncoated specimens, ×67.

FIGURES

1, 4.   *Neospathodus dieneri* Sweet. Lateral views of the holotype, sample K4-LCL, OSU OSU 28078.

2, 3, 9, 10. *Neospathodus homeri* (Bender). Lateral views of two hypotypes, samples K12-5, K12-9, OSU 28075, 28076.

5, 6.   *Neospathodus spathi* Sweet, n. sp. 5, Lateral view of the holotype with faint lateral platforms; 6, lateral view of a paratype, with well-developed platform, sample K1-37, OSU 28087, 28088.

7, 8.   *Neospathodus triangularis* (Bender). Lateral views, sample K1-37, OSU 28083.

11, 12.   *Neospathodus waageni* Sweet n. sp. Lateral views of the holotype, sample K4-UCS, OSU 28089.

13, 20.   *Anchignathodus typicalis* Sweet. Lateral views of, 13, paratype, 20, holotype, sample T63-122, OSU 28016, 28017.

14, 15.   *Neospathodus cristagalli* (Huckriede). Lateral views of a hypotype, sample K3-10, OSU 28077.

16, 17.   *Neospathodus pakistanensis* Sweet, n. sp. Lateral views of the holotype, sample K3-15, OSU 28084.

18, 19.   *Anchignathodus isarcicus* (Huckriede). Lateral and superior views of a hypotype, sample T62-192, OSU 28015.

PLATE 1

profile, and a basal margin that turns conspicuously upward beneath the posterior third of the unit.

Material: 1,268 discrete elements, about equally divided between dextral and sinistral forms. The smallest specimen is 0.23 mm long; the largest is 0.89 mm long.

Description: Skeletal elements of *Neospathodus cristagalli* are laterally compressed, blade-shaped units in which the ratio of width to height to length is about 1:3:4 in all stages of growth. The upper margin of the unit bears a series of 5 to 13 recurved or reclined denticles, which are discrete for most of their length in specimens representing early growth stages, but only apically in the largest specimens collected. The basal margin is straight anteriorly, but turns upward conspicuously beneath the posterior third of the unit.

The asymmetrically subpyramidal cusp is at the posterior end of the denticle series and is subtriangular in lateral view. This terminal denticle increases in length only slightly between the smallest and largest stages represented in our collections, but it increases conspicuously in width by developing a flangelike posterior margin. The outer face of the cusp is flat, but its inner side is transversely convex, and this feature is the primary criterion for distinguishing dextral from sinistral specimens.

The 4 to 12 denticles anterior to the cusp are similar to one another in lateral profile, but differ in length. New denticles are added at the anterior end of the series, and growth proceeds in such a fashion that the denticle (or denticles) at, or just behind, unit midlength are the longest. As a consequence of this pattern of growth, the lateral profile remains much the same at all stages of growth.

Elements of *Neospathodus cristagalli* are subtriangular in transverse section, but even in the smallest specimen available the otherwise planar sides are marked at midheight by a rounded longitudinal rib, the position of which on the posterior margin of the cusp is marked by a short, posteriorly convex segment in the otherwise concave profile of the cusp margin. In specimens representing successively later growth stages, this lateral rib becomes progressively more prominent, and in the largest specimens at hand, it forms a swollen, almost platformlike brim that completely surrounds the unit. In a few large specimens, there is a nodelike denticle at the level of the midlateral rib on the posterior margin of the cusp. This structure, like the brimlike rib itself, is reminiscent of comparable features in elements of some species of *Neogondolella,* and suggests close relationship.

The attachment or undersurface of *Neospathodus cristagalli* elements is morphologically complex and conspicuously different from that of the *Spathognathodus* elements they were originally thought to be. In the smallest specimens in our collections, the anterior three-quarters of the underside is a shallow, sheathed, narrowly lachrymiform, basal pit, the apex of which is at the lower end of the cusp axis. The posterior quarter of the underside, beneath the flaring, apronlike basal margin of the cusp, is convex downward in transverse profile and downwardly sigmoidal in longitudinal profile. These features suggest that the undersurface was initially a simple, completely sheathed, basal pit, but that segments of succeeding lamellae posterior to the cusp axis terminated basally along a posteriorly-directed surface of convex or sigmoidal profile, rather than growing progressively farther downward to form a gradually widening and deepening basal pit. In successively larger specimens, it can be seen that the basal pit broadens and deepens anteriorly; in later-stage elements, however, the basal pit is surrounded by a progressively broader, flat to slightly concave rim, which suggests that as the midlateral rib began to develop most conspicuously, there was decelerated growth basally. In elements representing the larger growth stages of *N. cristagalli*, the basal pit is rimmed by a looplike ridge that is quite similar in appearance and origin to the one formed basally in elements of both *Gondolella* and *Neogondolella*.

Remarks: When he established *Neospathodus cristagalli*, Huckriede (1958) included in it some Salt Range specimens that we would refer to our new species, *N. dieneri*. Although elements of *N. dieneri* and *N. cristagalli* are similar in many respects, and occur together in several of our samples, they are readily distinguished morphologically and their stratigraphic distribution in West Pakistan is not the same. That is, *N. dieneri* is first represented in the Zone of *N. kummeli* and ranges upward as far as the Zone of *N. waageni*, whereas *N. cristagalli* is restricted in occurrence to the zone that bears its name. The earliest growth stages of *N. cristagalli* represented in our collections are similar to larger elements of *N. dieneri*, but they are easily distinguished from the latter. This broad similarity between small representatives of *N. cristagalli* and larger, more typical specimens of *N. dieneri*, and the fact that *N. cristagalli* appears later than *N. dieneri*, suggests either that *N. cristagalli* developed from *N. dieneri*, or that both developed from a common ancestor such as *Xaniognathus deflectens* Sweet, n. sp.

Occurrence: Lower Triassic, West Pakistan. *Neospathodus cristagalli* is based on material collected from the section at Chhidru, West Pakistan, by Schindewolf, and its distribution in that section and in the others described in this report, defines a distinctive zone in the uppermost part of the "Lower Ceratite Limestone" and the lowermost part of the superjacent "Ceratite Marl." The species occurs in the lower part of the Mittiwali Member of the Mianwali Formation at Narmia, Ganjaroh, Chhidru A, Chhidru B, Wargal, Kathwai A, and Kathwai B. The elements from the Humboldt Range, Nevada, which were collected from the lower part of the Tobin Formation and the lower member of the Prida Formation and assigned to *N. cristagalli* by Mosher (1968), came from rocks in the *Subcolumbites* Zone and are clearly representatives of the species herein named *N. homeri* and *N. triangularis*, not *N. cristagalli*. The two Middle Triassic elements, from the Prida Formation of Nevada, identified as *N. cristagalli* by Mosher and Clark (1965) lack the distinctive posterior structures and the characteristic basal conformation of typical *N. cristagalli* elements, and are not regarded here as representatives of the species, which is thus known only from the type strata in West Pakistan. Details on the distribution and frequency of occurrence of elements of *Neospathodus cristagalli* in West Pakistan are given in Appendix A.

Type: Figured hypotype, OSU 28077.

Neospathodus dieneri Sweet, 1970

Pl. 1, fig. 1, 4

Spathognathodus cristagalli Huckriede, 1958 (*partim*), p. 161-162, pl. 10, fig. 10-13, 18a, 18b [*non* pl. 10, fig. 14, 15, which=*Neospathodus cristagalli* (Huckriede)].
"*Spathognathodus*" *cristagalli* Lindström, 1964, p. 64.
Neospathodus dieneri Sweet, 1970, p. 9, pl. 1, fig. 17.

Diagnosis: A species of *Neospathodus* characterized by blade-shaped skeletal elements with 4 to 13 denticles, a width to height to length ratio of about 1:2:2.3 at all stages of growth, a greatest height at or just anterior to the posterior end of the unit, a long terminal posterior cusp, and a basal margin that turns upward prominently beneath the posterior third to half of the unit.

Material: 7,644 discrete elements, about equally dextral and sinistral forms. Smallest specimen is 0.22 mm long; largest is 0.67 mm in length.

Description: Skeletal elements of *Neospathodus dieneri* are laterally compressed, blade-shaped units in which the ratio of width to height to length is about 1:2:2.3 at all stages of growth. The upper margin of the element bears a series of 4 to 13 erect to slightly reclined or recurved denticles of subequal width, which are sharp-pointed apically, discrete for half their length, and decrease regularly in length from the posterior to the anterior end of the unit. The basal margin is straight anteriorly, but turns upward beneath the posterior half to third of the element to form an angle of 19 to 39 degrees with the horizontal.

The cusp is the posterior denticle of the marginal series at all stages of growth. In elements like the holotype (Pl. 1, fig. 1, 4), in which the angle of postero-basal upturn is low, the cusp tends to be somewhat broader and distinctly shorter than the denticle immediately anterior to it. However, in elements in which the angle of postero-basal upturn is greater than in the holotype, the cusp tends to be either the longest denticle of the series or equal in length to the one adjacent to it anteriorly. The outer face of the cusp is flat or broadly convex; the inner face is distinctly convex transversely.

The 3 to 12 denticles anterior to the cusp are similar to the cusp and to one another in width, but they decrease regularly in length to the anterior end of the element. The greatest height of the element is at its posterior end, or just anterior to this point, at all stages of growth. Principal growth of the element is upward and anteriorly.

Elements of *Neospathodus dieneri* are subtriangular in transverse section at all stages of growth, but their sides are marked at midheight by a broad, indistinct, longitudinal rib that is somewhat more prominent in larger than in smaller elements. The under-surface is narrow and faintly grooved beneath the anterior half of the element, but broadens and turns upward beneath the posterior half to third of the unit. The basal pit is narrowly lachrymiform in outline, continuous anteriorly with the anterior groove, and surrounded laterally and posteriorly by a faintly swollen looplike ridge, beyond which the brim flares broadly outward and slightly upward.

Remarks: The elements on which we are basing *Neospathodus dieneri* were included in *N. cristagalli* by Huckriede (1958), who had only a few Salt Range samples available for study. In the large collections at hand, however, it is clear that units like those on which we base *N. dieneri* have a much longer stratigraphic range than do those typical of *N. cristagalli,* and elements of the two species are readily distinguishable in lateral profile and on comparison of their ratios of width to height to length (1:2:2.3 in *N. dieneri*; 1:3:4 in *N. cristagalli*).

Among described species that may be referable to *Neospathodus, N. dieneri* elements are grossly most similar to those from the Zechstein of Germany on which Bender and Stoppel (1965) based *Spathognathodus divergens*. Elements of *S. divergens,* however, are relatively wider than those of *N. dieneri,* and the posteriorly expanded segment of the base is longer and of a different shape than the comparable portion of *N. dieneri* elements. Elements of *N. dieneri* are also somewhat similar in proportions to the ones on which we base *N. waageni* Sweet, n. sp., but in elements of the latter species the cusp is not terminal and the profile of the upper margin is distinctly different.

Elements like the holotype of *Neospathodus dieneri* in lateral aspect, but narrower basally and with a basally widened and posteriorly sharpened cusp, grade toward those typical of *N. cristagalli,* which suggests that the latter species may have developed from *N. dieneri,* or that both species developed from a common ancestor such as *Xaniognathus deflectens* Sweet, n. sp. The latter is regarded as most likely because form-gradation between elements typical of *N. dieneri* and *N. cristagalli* is not complete, whereas forms similar to the elements of both species, but with a primary posterior process, occur in collections we are assigning to *X. deflectens*.

Occurrence: Dolomite unit of Kathwai Member at Chhidru A; Mittiwali Member of Mianwali Formation at Narmia, Nammal, Ganjaroh, Chhidru A, Chhidru B, Wargal, Kathwai A, and Kathwai B. Details on the distribution and frequency of occurrence of elements of *Neospathodus dieneri* in West Pakistan are given in Appendix A.

Type: Holotype, OSU 28078.

Neospathodus kummeli Sweet, n. sp.

Pl. 2, fig. 17-21

Diagnosis: A species of *Neospathodus* characterized by comblike skeletal elements twice as long as high; with 5 to more than 16 subequal denticles; a straight or downwardly convex basal margin; and a prominent midlateral rib that may be produced laterally into a platformlike brim in specimens representing intermediate and late growth stages.

Material: 483 discrete elements. The smallest specimen is 0.27 mm long; the largest one is 0.82 mm in length.

Description: Skeletal elements of *Neospathodus kummeli* are straight to greatly bowed, subsymmetrical or distinctly asymmetrical, blade-shaped units that are surmounted by 5 to more than 16 erect, sharp-pointed denticles of subequal size and length. These elements tend to be about twice as long as high at all stages of growth, but they vary so greatly in width as to make an average width-to-height or width-to-length ratio meaningless. That is, a midlateral rib is present at the base of the denticle series in nearly all of our specimens, but it varies greatly in prominence from one specimen to another. No such structure is discernible in a few very small specimens, but in a tenth of the larger forms, parts of the midlateral rib are produced into platformlike brims that mimic those of *Neogondolella* in shape, distribution, and development.

Small elements of *Neospathodus kummeli* (e.g., Pl. 2, fig. 17) are comb-shaped units with a terminal cusp that is laterally deflected, sharp-edged, and succeeded anteriorly by a series of 4 to 8 denticles that are nearly as long as the cusp but not as broad. Laterally there is a distinguishable midlateral rib in most small specimens, but some lack this structure. In lateral view, the basal margin is straight. The underside is narrow and longitudinally channeled beneath the anterior process. Beneath the cusp is a pyramidal basal pit, the sheath of which is more expanded on its outer than on its inner side.

In elements that are successively larger than the ones just described (e.g., Pl. 2, fig. 18), 1 or 2 denticles may develop from the sharp posterior edge of the cusp to form a thin, laterally deflected, posterior process, the basal margin of which is directed upward. In a few larger elements (e.g., Pl. 2, fig. 21), an accessory denticle also develops atop the outwardly flaring sheath of the basal pit. Elements like this have a pair of secondary processes posteriorly, and are Y-shaped in plan view.

In elements representing intermediate and late growth stages, the basal margin tends to become downwardly convex in lateral view as a result of greater lateral spreading, or "inversion," of the attachment surface in the anterior and posterior quarters of the element. The longitudinal groove beneath the anterior process of such elements tends to be bordered by flat or upwardly inclined ("inverted") brims of striated attachment surface, and the

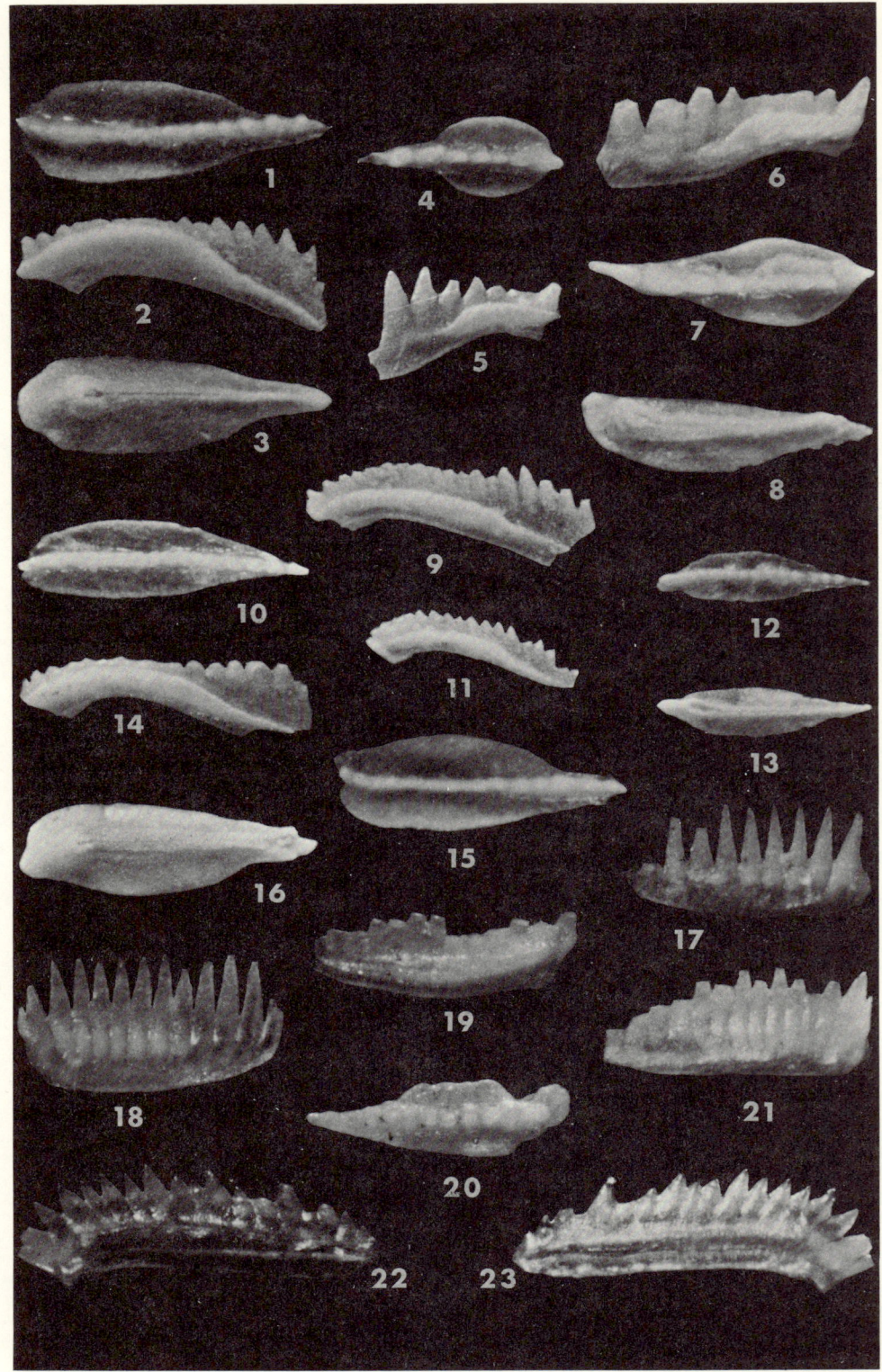

PLATE 2

basal pit of such elements is bordered by a looplike ridge or ridges of flattened or "inverted" attachment surface. There is close correspondence between width and style of development of the base and degree of development of platformlike brims on the midlateral rib. That is, in those forms with midlateral ribs and no platforms, the attachment surface is broad, flat, and essentially transverse to the blade. In specimens that have lateral platforms, the attachment surface is "inverted" and tends to be inclined upward and to face outward.

Remarks: The elements on which we base *Neospathodus kummeli* are of the same general architecture as those referred to other species of *Neospathodus,* but they are readily distinguished from them by their subequal denticles, the short, incurved, secondary posterior process, and the downwardly convex, basal margin. Specimens with platformlike brims are distinctive and would probably have been referred to a new species of *Neogondolella* if our collections had been smaller and intermediate forms were lacking.

Occurrence: Dolomite unit of Kathwai Member at Chhidru A; Mittiwali Member of Mianwali Formation at Narmia, Ganjaroh, and Chhidru A. Distribution and frequency of occurrence in West Pakistan is summarized in Appendix A.

Types: Holotype, OSU 28020; paratypes, OSU 28079, 28081, 28082.

Neospathodus triangularis (Bender)
Pl. 1, fig. 7, 8

Spathognathodus triangularis Bender, 1967?, p. 530, pl. 5, fig. 22a,b, 23.——Bender, *in* Bender and Kockel, 1963, p. 438, 439, pl. LIV (I) *(nomen nudum).*

Neospathodus cristagalli (Huckriede Mosher, 1968 *(partim),* p. 930, pl. 115, fig. 2 [not pl. 115, fig. 1, which=*N. homeri* (Bender)].

Diagnosis: A species of *Neospathodus* distinguished by blade-shaped skeletal elements with 4 to 11 denticles; a width to height to length ratio of about 1:1.5:3; a cusp that is stout and terminal, or succeeded posteriorly by only 1 or 2 short denticles; and a base that is widest, and of heart-shaped outline, in its posterior third.

Material: 115 discrete elements. Smallest specimen is 0.22 mm long; largest specimen is 0.91 mm in length.

Description: Skeletal elements of *Neospathodus triangularis* are straight bladelike units with a well-developed midlateral rib, a width to height to length ratio of about 1:1.5:3, and a basal margin that is straight or slightly downcurved posterior to the cusp. The anterior process bears 4 to 8 compressed, laterally confluent, denticles that decline slightly in length toward the anterior end of the unit. The cusp, which is the terminal posterior denticle

---

PLATE 2

All figures are unretouched photographs, ×67. Specimens shown in figures 1-16 were coated lightly with ammonium chloride before they were photographed. Other figures are of uncoated specimens.

FIGURES

1-3; 9; 10, 14,16; 11-13. *Neogondolella jubata* Sweet, n. sp. Sample K12-5.——1-3, Upper, lateral, under views of the holotype, OSU 28071.——9, Lateral view of a paratype, OSU 28074.——10, 14, 16, Upper, lateral, under views of a paratype, OSU 28072.——11-13, Lateral, upper, under views of a paratype, OSU 28073.

4, 5, 6-8. *Neogondolella elongata* Sweet, n. sp. Sample K6-31B.——4, 5, Upper, lateral views of a small paratype, OSU 28069.——6-8, Lateral, upper, under views of the holotype, OSU 28070.

17-21. *Neospathodus kummeli* Sweet, n. sp. Sample T63-155.——17, Lateral view of a paratype, OSU 28079.——18, Lateral view of holotype, OSU 28080.——19, 20, Lateral, superior views of a paratype with a well-developed platform, OSU 28081.——21, Lateral view of a paratype with a bifid posterior process, OSU 28082.

22, 23. *Neospathodus timorensis* (Nogami). Lateral views, sample T63-167, OSU 28086.

in about half of our specimens, is a stout, reclined, sharp-pointed denticle of superior size. Its lateral faces are convex, but its posterior side is flat or slightly concave and marked by a sharp-edged median costa of varying prominence. One or 2 short denticles, developed from the base of the median cusp costa, form a thin, straight, secondary posterior process in about half of the specimens at hand, and in a few very large elements the sides of the cusp develop sharp1edged longitudinal costae that continue across the upper surface of the greatly expanded base.

Beneath the anterior process, the undersurface is narrow and longitudinally grooved. Under the posterior third to half of the unit, however, is a broad, shallow basal pit, the nearly flat sheathes of which are produced about equally on either side of the blade. In plan, the lateral elements of the basal sheath have broadly rounded outlines, but posteriorly the basal margin is transversely straight or concave so that its over-all outline is triangular or heart-shaped.

The distribution of "white matter" in elements of *Neospathodus triangularis* is similar to that in elements of *N. homeri*. That is, "white matter" is concentrated in the posterior process and in broad axial strips in denticles of the anterior process. In lateral view, the lower boundary of "white matter" defines a line that arches upward from the apex of the basal pit to the base of the anteriormost denticle of the anterior process.

Remarks: Elements of *Neospathodus triangularis* are similar in many respects, including the distribution of "white matter," in those referred to as *N. homeri,* and the two types of elements occur together in Greece and in nine of our samples from the Lower Triassic of West Pakistan. The elements of *N. triangularis* are readily distinguished from those of *N. homeri,* however, by the fact that they are straight (rather than bowed), expand equally on both sides of the blade, and have a base that is heart-shaped posteriorly in plan view.

Occurrence: Narmia Member of Mianwali Formation at Landa, Narmia, and Nammal. Distribution in West Pakistan is summarized in Appendix A.

Figured hypotype: Holotype, OSU 28083.

Neospathodus pakistanensis Sweet, n. sp.

Pl. 1, fig. 16, 17

Diagnosis: A species of *Neospathodus* characterized by bladelike skeletal elements with a width-to-height-to-length ratio of 1:2:3; a short, secondary posterior process with 1 or 2 denticles; an upper edge that is arched; and a lower edge that is straight anteriorly but downcurved posteriorly.

Material: 562 discrete elements, 261 of which are dextral and 301 are sinistral forms. The smallest specimen is 0.30 mm long; the largest is 0.87 mm in length.

Description: Small elements, representing early growth stages of *Neospathodus pakistanensis,* are unarched, straight or slightly bowed blades, with a width to height to length ratio of about 1:2:3. The short, sharp-edged cusp is subtriangular in profile and is situated at the posterior end of a marginal series of 5 or 6 straight, reclined, sharp-pointed denticles, the second of which anterior to the cusp is the longest and marks the highest part of the unit. The sides of these small elements are faintly swollen at the latitude of the denticle bases. The undersurface is narrow and grooved beneath the anterior two-thirds of the unit. Beneath the posterior third, there is a broad, shallow basal pit, the posterior side of which is flat or only slightly inclined downward.

In specimens that are successively larger than the small ones just described, a width to height to length proportion of about 1:2:3 is maintained, and the second denticle anterior to the cusp continues to be the longest. At the growth stages marked by 6 to 7 denticles anterior to the cusp, however, a posteriorly-directed node appears on the sharp-edged posterior margin of the cusp at the latitude of the midlateral rib. In later growth stages, distinguished by 7 to 13 denticles, this node develops into a short, needlelike denticle posterior to the cusp and, in a very few of our specimens, an inconspicuous second posterior denticle develops from the posterior side of the first one, which tends in such specimens to be overgrown by, and incorporated in, the posterior margin of the cusp.

Concurrent with development of denticles posterior to the cusp, the midlateral rib becomes progressively more swollen and brimlike; the basal surface widens by increased outward deflection of the laminae sheathing the anterior groove and posterior pit, rather than by "inversion" or restriction of the attachment surface; the segment of the sheath posterior to the apex of the basal pit develops downward and posteriorly; and the basal margin, which is straight anterior to the cusp base, develops a characteristically downwardly deflected segment posterior to the cusp base.

The latest stages of growth represented in our collections are characterized by massive elements with blunt, laterally confluent denticles; a thick, prominent midlateral rib; and a broad basal surface marked posteriorly by a broad upturned brim of "inverted" attachment surface around the outer margin of the basal pit.

In elements of *Neospathodus pakistanensis,* "white matter" tends to be concentrated in broad columns along the axes of denticles. These columns are visible in the posterior 4 or 5 denticles of most specimens down to, or below, the level of the midlateral rib, but in more anterior denticles, the lower boundary of "white matter" is successively higher and, in lateral view, defines a line that arches slightly upward toward the anterior end of the element.

Remarks: Elements of *Neospathodus pakistanensis* are most like those of *N. cristagalli,* from which they are distinguished by being lower relative to their length, and by having a basal margin that is straight anteriorly and distinctly downcurved posteriorly. The species is not known from described material outside West Pakistan, but it is represented in undescribed collections from the Lower Triassic of Spitzbergen that are being studied by Mr. György Hamar, of Oslo University, Norway.

Occurrence: Mittiwali Member of Mianwali Formation at Narmia, Nammal, and Chhidru A. The species is also known from the Lower Triassic of Spitzbergen. Distribution and frequency of occurrence in West Pakistan are summarized in Appendix A.

Types: Holotype, OSU 28084.

Neospathodus peculiaris Sweet, n. sp.
Pl. 5, fig. 19

Diagnosis: A species of *Neospathodus* characterized by bladelike skeletal elements with 3 to 7 discrete denticles, a width to height to length ratio of about 1:2.5:3.5, a cusp that is twice the length of any other denticle, a straight basal margin, and a broadly concave under surface.

Material: 12 discrete elements. The smallest specimen is 0.40 mm long; the largest is 0.65 mm in length.

Description: Skeletal elements of *Neospathodus peculiaris* are short bladelike units in which the ratio of width to height to length is about 1:2.5:3.5. The upper margin of the unit bears a series of 3 to 7 erect or reclined denticles, which are sharply pointed, subcircular in cross section, and discrete nearly to their junction with the base. The cusp, which is terminal or the second or third denticle from the posterior end of the series, is reclined at an angle of 45° or more and is at least twice the length and width of any of the other denticles. Anterior denticles are straight, erect, or slightly reclined, and decline in length toward the process extremity. Most of our specimens have a single discrete, strongly reclined to recumbent denticle posterior to the cusp, but in 2 specimens there are no posterior denticles and in 2 others there are 2. From a study of all specimens, it is clear that denticles posterior to the cusp develop from nodes in the flaring posterobasal portion of the cusp, and that the development of these denticles is delayed until intermediate and late stages of growth. Thus the posterior denticle or denticles define a secondary posterior process that is not reflected in the structure of the basal surface.

In lateral view, the basal margin is straight or slightly uparched. In plan view, the base is lachrymiform in outline. The undersurface of most specimens is broadly concave and its deepest portion is an asymmetrical subconical pit beneath the cusp base. In the holotype and the larger specimens of our collection, the attachment surface is "inverted" posterior to the cusp base.

Remarks: Neospathodus peculiaris is represented in our collections by only a few elements, but these are readily distinguished from those of other species of *Neospathodus* by their proportions, their mode of denticulation, and by the broad concavity of their undersurfaces. A few elements that are morphologically reminiscent of those of *N. peculiaris,* but which have posterior processes that are clearly primary, are included in the rather variable samples on which we base *Xaniognathus deflectens,* n. sp. The presence of elements of this type in *X. deflectens* suggests that *N. peculiaris* may have developed from that species, which may also have been ancestral to *N. dieneri* and *N. cristagalli.*

Occurrence: Mittiwali Member of Mianwali Formation at Narmia, Chhidru A, Chhidru B, and Kathwai A. Distribution of the elements of *Neospathodus peculiaris* in West Pakistan is summarized in Appendix A.

Type: Holotype, OSU 28085.

Neospathodus timorensis (Nogami, 1968)

Pl. 2, fig. 22, 23

Gondolella timorensis Nogami, 1968, p. 127, pl. 10, fig. 17-21.

Diagnosis: A species of *Neospathodus* distinguished by bladelike skeletal elements with 6 to 17 denticles of essentially the same length; a width to height to length ratio of 1:2.8:5.7 in early stages and 1:2.3:7.5 in later stages; a laterally deflected, secondary posterior process bearing 1 or 2 denticles; and a basal margin that is deflected distinctly downward posterior to the cusp.

Material: 123 specimens; 54 are dextral and 69 are sinistral elements. The smallest specimen is 0.22 mm long; the largest is 0.82 mm in length.

Description: Skeletal elements of *Neospathodus timorensis* are laterally compressed, blade-shaped units with a crest of 6 to 17 erect to slightly reclined, sharp-pointed, apically discrete denticles of essentially the same width and length, and a basal margin that is straight anterior to the apex of the basal pit, but deflected downward and somewhat laterally posterior of that point. In the smallest specimens at hand the ratio of width to height to length is 1:2.8:5.7; in the largest ones available, this ratio is 1:2.3:7.5.

In the shortest specimen in our collection, the cusp is the longest denticle, is situated at the posterior end of the denticle series, and bears a short, clear node on its posterior margin. In successively longer specimens, the posterior node develops into a distinct denticle, which itself develops a nodelike structure on its posterior margin.

In the largest of our specimens, the cusp is the third denticle in the series, counted from the posterior end of the element. The growth series indicated by these elements makes it clear that larger elements of *Neospathodus timorensis* have a short posterior process, but that that process is a secondary development from the posterior cusp margin and not a primary structure of the element.

The 5 to 14 denticles anterior to the cusp are similar to one another, and the essentially straight line joining their apices is approximately parallel to the segment of the basal margin anterior to the apex of the basal pit. The sides of the element are marked at midheight by a lateral rib, the sharp edge of which is parallel to the basal margin for the full length of the unit. The rib is a conspicuous feature of elements representing all stages of growth, and in a few of the larger specimens at hand it is almost platformlike in development. Nodes that develop into denticles of the secondary posterior process thise at the level of the midlateral rib in a manner closely similar to that in which similar structures develop in the elements of some species of *Neogondolella* (e.g., *N. mombergensis* (Tatge).

The undersurface of the elements we identify as *Neospathodus timorensis* is similar in morphology and development to that of *N. cristagalli*, except that only very narrow brims are formed laterally in even the largest elements of *N. timorensis*, and the "inverted" segment of the attachment surface posterior to the apex of the basal pit is of limited length and is conspicuously developed in only the larger elements of our collections.

Remarks: *Neospathodus timorensis* is based on material from Portuguese Timor (Nogami, 1968) thought to be uppermost Scythian or lowermost Anisian in age. The species was referred to *Gondolella* by Nogami, but we believe it is best included in *Neospathodus*, which is very closely related to *Neogondolella* and *Xaniognathus*. All the specimens at hand come from the pisolite bed at the top of the Mianwali Formation at a single locality in the Surghar Range, West Pakistan. Elements of this species are readily distinguished from those of other species of *Neospathodus* by their even-crested lateral profile and by their conspicuous, sharp-edged, midlateral rib.

Occurrence: Narmia Member of Mianwali Formation at Narmia. Distribution and frequency of occurrence in West Pakistan is shown in Appendix A.

Type: Holotype, OSU 28086.

Neospathodus spathi Sweet, n. sp.
Pl. 1, fig. 5, 6

?*Gondolella* sp. Mosher and Clark, 1965, p. 561, pl. 66, fig. 7.

Diagnosis: A species of *Neospathodus* characterized by blade-shaped skeletal elements with 3 to 7 denticles; a width to height to length ratio that varies from 1:2.6:4 in early and intermediate growth stages to 1:2:3.6 in later ones; a terminal posterior cusp; and a prominent midlateral ridge that is produced into distinct platformlike brims in elements representing late growth stages.

Material: 67 discrete specimens; 35 are dextral and 32 are sinistral forms. Of these, 38 have platformlike brims developed from the midlateral rib. The smallest specimen is 0.27 mm long; the largest one is 0.52 mm in length.

Description: Skeletal elements of *Neospathodus spathi* are small, laterally bowed, blade-shaped units, with a terminal posterior cusp and a conspicuous midlateral rib from which platformlike brims develop in later growth stages. The ratio of width to height to length varies from 1:2.6:4 in forms without lateral platforms, to 1:2:3.6 in forms with lateral platforms. The upper margin of the element bears 3 to 7 reclined denticles, which are broad and laterally confluent for most of their length. In smaller specimens, denticle length declines from the posterior to the anterior end of the element, but in larger forms with lateral platforms, denticles anterior to the cusp tend to be of essentially equal length, or the anteriormost denticle is the longest. The basal margin is straight or faintly uparched in lateral view.

In specimens like the holotype (Pl. 1, fig. 5), a narrow midlateral rib encircles the unit, but is not a prominent structural component of the element. In larger specimens (e.g., Pl. 1, fig. 6), the midlateral rib is produced laterally to form narrow platformlike brims, which may be unequally developed laterally and unjoined posteriorly, or strongly produced laterally and completely confluent posteriorly around the cusp. Elements representing these later stages of development are, of course, very similar to forms typical of *Neogondolella*.

The underside of elements representing early growth stages is channeled and the same width as the superjacent blade for most of its length. Posteriorly, however, the underside expands to form a cup-shaped structure beneath the cusp. In elements of successively later stages of growth, the undersurface is distinctly keel-like, but becomes somewhat broader marginal to the longitudinal channel and, posterior to the cusp, accretionary lamellae tend to be progressively shorter basally as a platform is developed at midheight, so the attachment surface is gradually "inverted."

Remarks: Our collections contain elements representing a complete series of morphologic transitions between platformless units like the holotype of *Neospathodus spathi* and specimens like the paratype, which have a well-developed platform continuous around the posterior side of the cusp. Consequently, we are confident that all these elements represent the same species, and that during ontogeny platforms tended to develop on some or all of the fundamentally bladelike skeletal elements.

The five small specimens from the Middle Triassic Prida Formation of Nevada that Mosher and Clark (1965) referred to *Gondolella* sp. are similar in size, denticulation, and lateral platform development to the ones on which we base *Neospathodus spathi*. The Nevada and West Pakistan specimens may well represent the same species, but we have not had the opportunity to make direct comparisons or study the Nevada specimens in the context of their type collections. Hence we include them only with question in *Neospathodus spathi*.

Occurrence: Narmia Member of Mianwali Formation at Landa, Narmia, and Zaluch. The species may also

---

PLATE 3

All figures are unretouched photographs, ×67. Specimens shown in figures 1-18 and 23-27 were lightly coated with ammonium chloride before they were photographed. Other figures are of uncoated specimens.

FIGURES

1-3; 4-6, 7-9; 10-12; 13-15; 16, 17; 24, 26, 27. *Neogondolella carinata* (Clark). Upper, lateral under views of hypotypes representing various stages of growth and development, samples K4-11 and K3-6, OSU 28060-28066, inclusive.

18; 23, 25. *Neogondolella elongata* Sweet, n. sp.——18, Lateral view of a paratype.——23, 25, upper and under views of a paratype, sample K6-B, OSU 28067, 28068.

19. *Xaniognathus elongatus* Sweet, n. sp. Lateral view of holotype, sample T63-167, OSU 28093.

20. *Xaniognathus deflectens* Sweet, n. sp. Lateral views of holotype, sample T63-157, OSU 28092.

21. *Xaniognathus?* sp. Lateral view, sample T63-167, OSU 28094.

22. *Xaniognathus curvatus* Sweet, n. sp. Lateral view of holotype, sample T63-131, OSU 28091.

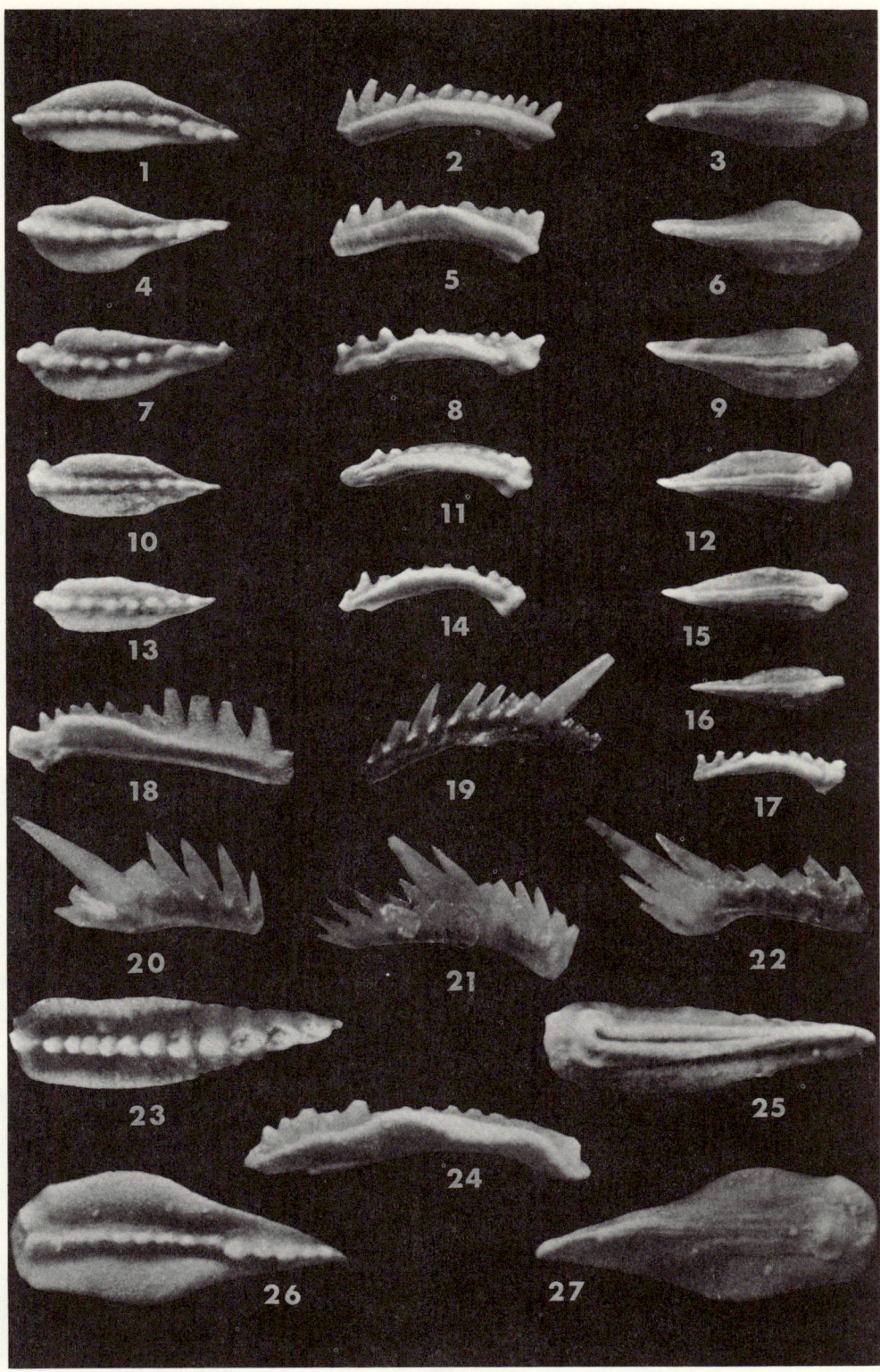

PLATE 3

occur in Middle Triassic strata of Nevada. Distribution in West Pakistan is summarized in Appendix A.

Types: Holotype, OSU 28087; paratype, OSU 28088.

Neospathodus waageni Sweet, n. sp.
Pl. 1, fig. 11, 12

Diagnosis: A species of *Neospathodus* with bladelike skeletal elements in which the height to length ratio is about 1:1 in all stages of growth, but in which the ratio of width to length (or height) changes from 1:3 in early stages to 1:2 in late stages. Denticulate margin arcuate in lateral profile, with greatest height in posterior half of element. Basal margin straight anteriorly, but deflected conspicuously upward beneath posterior half of element.

Material: 87 discrete elements; 38 are dextral and 49 are sinistral forms. Smallest element is 0.22 mm long; largest one is 0.58 mm in length.

Description: Skeletal elements of *Neospathodus waageni* are laterally compressed, blade-shaped units in which the ratio of height to length is about 1:1 in elements representing all stages of growth, but in which the ratio of width to length (or height) varies from about 1:3 in small specimens to 1:2 in the largest specimens at hand. Upper edge of element bears 5 to 12 sharp-pointed, slightly reclined denticles of subequal width that are discrete for a third to half their length. These denticles are clear along their anterior and posterior margins, but their axial portions are distinctly outlined by columns of "white matter" about a third the width of the denticles. Shortest denticles are at the anterior and posterior ends of the unit; longest one or ones just posterior of unit midlength. Basal margin straight anterior to apex of basal pit, but deflected upward in posterior half or third of element at an angle of 15° to as much as 40° from the horizontal.

In the smallest specimen at hand, the cusp is the most posterior of the marginal denticles and is only three-fourths as long as the one immediately adjacent to it. Outer cusp face broadly convex; inner face nearly flat, posterior margin sharp, concave in overall profile, but marked in lower half by short convex segment. In larger specimens, representing later growth stages, the cusp and adjacent anterior denticles retain the same relative proportions as in earliest known stage, and a single denticle develops from the convex segment of the posterior cusp margin. In the largest specimens at hand, this denticle is nearly as long and wide as the cusp, and it may be regarded as a short secondary posterior process.

In transverse section, elements of *Neospathodus waageni* are subtriangular at all stages of growth, although the sides are distinctly constricted below a faint longitudinal rib at element midheight. The undersurface is narrow anteriorly, but broadly expanded and deflected upward posteriorly. A shallow basal pit is extended anteriorly as a narrow, sheathed groove, and surrounded posteriorly and laterally by a faint looplike ridge. Beyond the looplike ridge, basal segments of the element flare broadly outward and slightly upward.

Remarks: Elements of *Neospathodus waageni* are distinguished from those of *N. cristagalli* and other known species of *Neospathodus* by their subquadrate form and by the conspicuously upturned and broadly expanded posterior segment of the basal surface. Although the specimens described in this report are the first to be illustrated, elements typical of *N. waageni* also occur in an undescribed collection that we have assembled from Lower Triassic rocks at Kotal-e-tera, Afghanistan, and in another collection from the Lower Triassic of Nevada now under study by Mr. Walter Hasenmueller and Professor James W. Collinson, of The Ohio State University. In both these collections, representatives of *N. waageni* are associated with elements of *"Gondolella" milleri* Müller, a species based on specimens from the *Meekoceras* beds in the Dinner Springs Canyon (or Crittenden Ranch) section of Elko County, Nevada. Dr. Marinus van den Boogaard, of the University of Amsterdam, has also collected several elements of *N. waageni* from the matrix surrounding a specimen of *Flemingites compressus* Waagen that was collected at Toebue Lopo, Timor, in 1916.

Occurrence: Mittiwali Member of Mianwali Formation at Zaluch and Chhidru B; Narmia Member of Mianwali Formation at Landa. The species also occurs in Lower Triassic strata of Afghanistan, Timor, and Nevada. The distribution and frequency of occurrence in West Pakistan are summarized in Appendix A.

Type: Holotype, OSU 28089.

Genus PRIONIODELLA Bassler, 1925

"Prioniodella" prioniodellides (Tatge)

Pl. 5, fig. 11

Angulodus? prioniodellides Tatge, 1956, p. 130, pl. 5, fig. 6.
Prioniodella prioniodellides (Tatge) Huckriede, 1958, p. 159, pl. 10, fig. 19; pl. 11, fig. 5, 45.

Material: 29 discrete specimens, nearly all fragmentary.

Description: Straight or slightly bowed blades that are markedly swollen at midheight and have a relatively short posterior process surmounted by 2 nearly recumbent denticles; a laterally compressed, reclined cusp; and a relatively long anterior process that bears 8 to 10 denticles that decline in length and decrease in width toward the distal end of process. Proximal denticles of both processes are nearly as long and wide as the cusp, and all denticles are discrete for at least the upper half of their length.

The basal margin is broadly arcuate: its crest is just beneath the base of the cusp. The underedge is narrow, with very shallow channels extending to the process extremities and a slightly expanded basal pit beneath the cusp.

Remarks: The preceding description is included largely for the sake of completeness and the fragmentary specimens to which it pertains are identified with *Prioniodella prioniodellides* (Tatge) because they seem to fit the rather broad diagnoses of that species given by Tatge (1956) and Huckriede (1958). However, we have not seen any specimens of Devonian *Prioniodella normalis* Ulrich and Bassler, the type form species of *Prioniodella,* so we cannot evaluate Hass's (1962) conclusion that *Prioniodella* is a synonym of *Prioniodina,* nor can we comment from experience about the elements of *Lochriea montanaensis* that Scott (1942) identified as prioniodellas. We suspect, nevertheless, that a variety of different things has been assigned to *Prioniodella,* and we express our perplexity about the utility of this generic name for Triassic species by including it in quotation marks.

The specimens at hand are somewhat similar in form to the elements of species we refer to *Xaniognathus.* However, the elements in question have a posterior process that is stouter and bears longer denticles than is typical of the elements of even extreme variants of *Xaniognathus.* Thus, for the present at least, it seems best to describe these specimens rather informally pending a study of similar forms in larger collections.

Occurrence: Mittiwali Member of Mianwali Formation at Chhidru A and Kathwai A; Narmia Member of Mianwali Formation at Nammal. The distribution of forms from West Pakisan identified as *"Prioniodella" prioniodellides* (Tatge) is given in Appendix A.

Type: Figured hypotype, OSU 28090.

Genus XANIOGNATHUS Sweet, n. gen.

Diagnosis: Xaniognathus is

erected to include conodont species with skeletal elements of a single morphologic type. These elements are blade-shaped units with a long, denticulate anterior process, which is longitudinally ribbed at midheight, and a very short posterior process that bears no midlateral rib. The undersurface of both processes is grooved, at least in early ontogenetic stages, and these grooves join in a prominent basal pit beneath the cusp.

Type species: Xaniognathus curvatus Sweet, n. sp.

Remarks: The bladelike skeletal units on which we base our concept of *Xaniognathus* are not closely associated in range or frequency of occurrence with any other elements in our collection, from which we conclude that they were the only components of the skeletal apparatus of the conodonts that secreted them. Elements like *Xaniognathus* have been included in the form genus *Ozarkodina* Branson and Mehl by most previous students of Upper Paleozoic and Triassic conodonts. However, the type form species of *Ozarkodina*, which is Middle Silurian in age, is regarded by Walliser (1964) as a component of a multielement skeletal apparatus he terms "Conodonten-Apparat J." Thus, even though elements of *Xaniognathus* have the same general form as *Ozarkodina typica* Branson and Mehl, *Ozarkodina* may ultimately have to be used for species with a much more elaborate and quite different skeletal plan than *Xaniognathus*, the skeleton of which seems to have included only modified ozarkodina-like elements.

Many post-Silurian form species have been assigned to *Ozarkodina*. Some may have been components of multielement skeletons like "Conodonten-Apparat J"; others [e.g., *Ozarkodina delicatula* (Stauffer and Plummer)] may have been parts of skeletons like the ones named *Scottella typica* by Rhodes (1952); and an additional group (like the species herein referred to *Xaniognathus*) may have been the only components in the skeletons of the animals that secreted them. We do not know which, if any, of the form species now assigned to *Ozarkodina* should be removed to *Xaniognathus*, for the relationships of these elements can be determined only by quantitative studies made in the context of the faunas from which their types were derived. Such studies are beyond the scope of this investigation, but at least some of the Permian specimens from the western United States that Clark and Ethington (1962) described as *Subbryantodus abstractus* may represent a species of *Xaniognathus*, and this is probably also true of the seemingly ubiquitous Triassic specimens included by many authors in *Ozarkodina tortilis* Tatge and related form species.

As we note in our discussion of *Neospathodus*, *Xaniognathus* may include the stock from which both *Neogondolella* and *Neospathodus* developed. That is, *Xaniognathus*, *Neogondolella*, and *Neospathodus* seem all to have included species whose skeletal apparatus consisted entirely of elements of a single type. Furthermore, in the elements of species in all three genera, the basal surface is similar and there is a tendency for the midlateral rib of the anterior process to be produced into a platformlike brim. Of these three genera, *Xaniognathus* is the logical ancestral group because it probably had the longest and widest range and because it includes elements of the most generalized form.

Xaniognathus curvatus Sweet, n. sp.

Pl. 3, fig. 22

Diagnosis: A species of *Xaniognathus* characterized by bladelike skeletal elements with a sinuous basal margin; steeply reclined cusp and denticles; a long anterior process; and a short posterior process, the denticles of which tend to be incorporated into the posterior margin of the cusp in late stages of growth.

Material: 293 discrete elements, about equally dextral and sinistral forms.

Description: Typical skeletal elements of *Xaniognathus curvatus* are denticulated blades that are both arched and bowed. These elements have a compressed, steeply reclined, apically pointed cusp; a short, inwardly curved posterior process surmounted by 1 to 3 reclined, sharp-pointed denticles; and a much longer anterior process that is either straight or bowed inwardly at its distal end and bears 4 to 8 reclined, compressed, sharp-pointed denticles, which decline in length toward the anterior end of the process. A markedly swollen longitudinal rib that occurs at midheight on the anterior process terminates at the base of the cusp. No such rib occurs on the short posterior process, which is thin and fragile at all stages of growth in which the process is recognizable. In many large specimens, it is evident that denticles of the posterior process have been incorporated in the posterior margin of the cusp, which is consequently much broader than in specimens representing early or intermediate growth stages.

In lateral view, the basal margin of typical elements of *Xaniognathus curvatus* is straight anteriorly, concave downward beneath the cusp, and convex downward beneath the short posterior process. The undersurface is grooved beneath the anterior and posterior processes, and there is a shallow basal pit beneath the cusp. In elements representing late growth stages, basal grooves and pit are bordered by a flat, longitudinally striated brim, and in some large specimens the attachment surface is "inverted" beneath the posterior process or that part of the cusp into which the posterior denticles have been incorporated.

Remarks: Elements of *Xaniognathus curvatus* are distinguished from those of closely related *X. deflectens* by their steeply reclined denticles and sinuous basal margin. In addition, elements of *X. curvatus* tend to be more distinctly compressed than those of *X. deflectens,* and the undersurface is narrower and less deeply excavated. However, these differences are largely ones of degree, and forms transitional between *X. curvatus-* and *X. deflectens-*like elements occur in collections we identify entirely with *X. curvatus.* Because the characters that distinguish *X. deflectens* exist in some elements of populations we identify as *X. curvatus* and come gradually to dominate in later populations, we conclude that *X. deflectens* developed from, and gradually replaced, *X. curvatus.*

It should be noted that in a few large specimens of *Xaniognathus curvatus,* the midlateral rib of the anterior process is produced laterally into narrow, but distinct platforms, the posterior ends of which curve slightly upward along the lateral faces of the cusp. Elements like these mimic those of *Neogondolella carinata* (Clark) in platform development, but they are readily distinguished from them by retention of a short posterior process. Such elements suggest that *Xaniognathus* and *Neogondolella* are closely related.

Occurrence: Dolomite unit of Kathwai Member at Kingriali, Narmia, Zaluch, Nammal, Ganjaroh, Chhidru A, and Kathwai A; Limestone unit of Kathwai Member at Kingriali, Chhidru A, Chhidru B, and Khan Zaman Nala; Mittiwali Member of Mianwali Formation at Chhidru A and Kathwai A. Distribution of elements of *Xaniognathus curvatus* in West Pakistan is summarized in Appendix A.

Type: Holotype, OSU 28091.

Xaniognathus deflectens Sweet, n. sp.
Pl. 3, fig. 20

Diagnosis: A species of *Xaniognathus* with bladelike skeletal elements that have an essentially straight basal margin and are characterized by discrete denticles and a broad base marked by a conspicuous basal pit and a

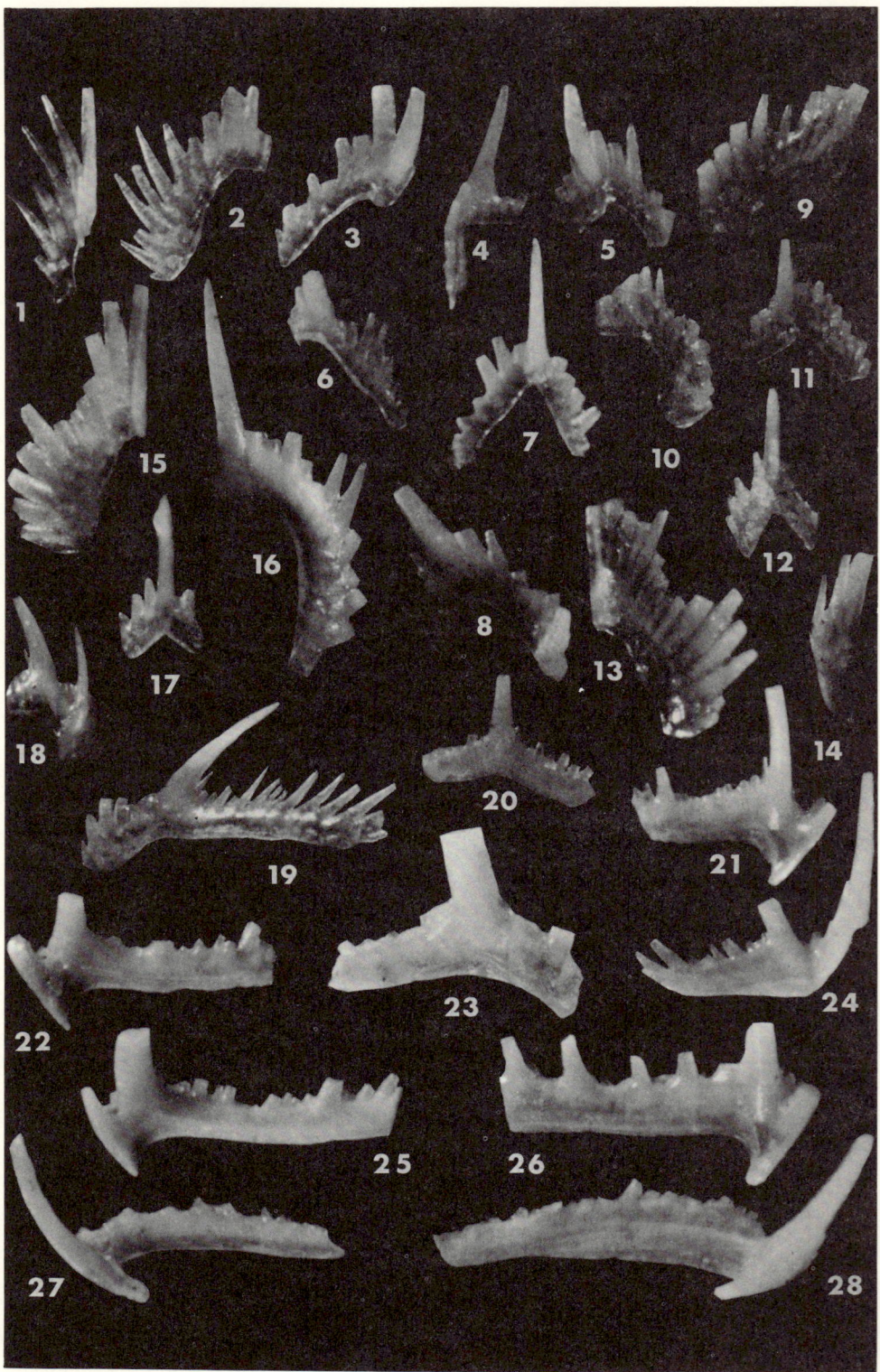

PLATE 4

broadly sheathed attachment surface beneath the processes.

Material: 703 discrete elements, which are about equally dextral and sinistral forms.

Description: The skeletal elements of *Xaniognathus deflectens* are bowed blades with a sharp-pointed, reclined cusp of subelliptical cross section; a long, stout anterior process, which is bowed inwardly and bears 3 to 6 straight, discrete, reclined, sharp-pointed denticles that decline in length toward the distal end of the process; and a very short posterior process that is thin, fragile, deflected inwardly and upwardly, and is surmounted by 1 or 2 compressed, discrete denticles that are less than half the length of the cusp. The anterior and posterior processes are compressed in elements representing early stages of development. In elements representing successively later stages, the sides of the anterior process become markedly swollen at midheight, but the posterior process, although increased slightly in length and height, remains thin and fragile, or is partially or completely incorporated into the posterior margin of the cusp.

In lateral view, the elements on which we base *Xaniognathus deflectens* display a basal margin that is essentially straight, but arches broadly upward beneath the cusp and anterior process. On the underside there is a conspicuous pit beneath the cusp and, in specimens representing early and intermediate growth stages, the attachment surface beneath the two processes is in a deep groove with thin lateral sheaths that flare slightly along the basal margin. In large specimens, representing late growth stages, the sheaths thicken appreciably, but are not extended downward; beneath the posterior process of such large elements, late lamellae are shorter basally than those preceding them and the attachment surface is partly "inverted."

Remarks: As noted on a previous page, *Xaniognathus deflectens* probably developed from *X. curvatus,* and it is appropriate to note the suggestion that *X. deflectens* was almost certainly ancestral to *Neospathodus dieneri, N. cristagalli,* and *N. peculiaris,* and may also (but far less certainly) have given rise to *N. kummeli,* as well. We suggest the development of several species of *Neospathodus* from *X. deflectens* because there was clearly a marked tendency toward suppression of the posterior process in elements of *X. deflectens,* and the rather variable collections of specimens on which we base our concept of this species include many

---

PLATE 4

All figures are unretouched photographs of uncoated specimens, ×67.

FIGURES

1-8. *Ellisonia gradata* Sweet. Syntypes. Sample K3-12.——1, Posterior view of LC-element, OSU 28027.——2, 6, Lateral views of LB-elements, OSU 28028, 28032.——3, 5, Posterior views of LA-elements, OSU 28029, 28031.——4, 7, Lateral and posterior views of U-element, OSU 28030.——8, Lateral view of LB2-element, OSU 28033.

9-14. *Ellisonia delicatula* Sweet, n. sp. Syntypes. Sample K12-9.——9, 10, Lateral views of LB-elements, OSU 28022, 28023.——11, Posterior view of LA-element, OSU 28024.——12, 14, Posterior and lateral views of U-element, OSU 28025.——13, Posterior view of LC-element, OSU 28026.

15-19. *Ellisonia clarki* Sweet, n. sp. Syntypes. Sample T63-167.——15, Posterior view of LC-element, OSU 28018.——16, Posterior view of LA-element, OSU 28019.——17, 18, Posterior and lateral views of U-element, OSU 28020.——19, Lateral view of LB-element, OSU 28021.

20-28. *Ellisonia teicherti* Sweet. Syntypes. Sample T63-122.——20, Posterior view of slightly asymmetric U-element, OSU 28039.——21, 22, 25, 26, Lateral views of four similar LB-elements, OSU 28040, 28041, 28044, 28045.——23, Posterior view of LA-element, OSU 28042.——24, Lateral view of LE-element, OSU 28043.——27, 28, Lateral views of LD-elements, OSU 28046, 28047.

elements that are morphologically intermediate between those typical of *X. deflectens* and the ones characteristic of the several species of *Neospathodus* cited. All these elements of intermediate form are referred to *X. deflectens* because the posterior process, even though short, is apparently primary and denticles posterior to the cusp of *Neospathodus*-elements seem invariably to be secondary—that is, they are delayed in their appearance until intermediate or late stages of growth.

Occurrence: Dolomite unit of Kathwai Member at Chhidru A; Mittiwali Member of Mianwali Formation at Narmia, Nammal, Ganjaroh, Chhidru A, Chhidru B, Wargal, Kathwai A, and Kathwai B. Distribution in West Pakistan is summarized in Appendix A.

Type: Holotype, OSU 28092.

Xaniognahtus elongatus Sweet, n. sp.

Pl. 3, fig. 19

?*Ozarkodina tortilis* (Tatge) Mosher and Clark, 1965, p. 563, pl. 66, fig. 11.

Diagnosis: A species of *Xaniognathus* with bladelike skeletal elements distinguished by a long anterior and a short posterior process, a broadly arcuate basal margin, and an attachment surface that is very narrow and shallowly grooved beneath the processes.

Material: 25 discrete elements, 10 of which are dextral and 15 of which are sinistral forms.

Description: Elements of *Xaniognathus elongatus* are laterally compressed blades that are both arched and bowed. The cusp is compressed, steeply reclined, sharp pointed, and twice the length of any of the process denticles. Posterior to the cusp is a short posterior process that is thin and fragile, deflected upward and inward, and surmounted by 2 or 3 short, discrete denticles. Anterior to the cusp is an anterior process that is straight and bears as many as 10 reclined, discrete denticles, which decline in length toward the distal end of the process. In specimens representing intermediate and late stages of growth, the long anterior process is marked at midheight by a distinct longitudinal rib at the base of the denticles. No such rib develops on the short posterior process, the proximal denticles of which tend to be incorporated into the posterior edge of the cusp in very large specimens.

In lateral view, the basal margin is broadly arcuate, with the apex of the arch just anterior to the cusp base. On the narrow underside, there is a deep pit beneath the cusp and grooves beneath the processes. The latter are shallow, but well defined, beneath the proximal ends of the processes, but grade distally into faintly incised lines. In the larger specimens in our collec-

---

PLATE 5

All figures are unretouched photographs of uncoated specimens, ×67.

FIGURES

1-4.   *Ellisonia torta* Sweet, n. sp. Syntypes. Sample K12-9.——1, Lateral view of LB-element, OSU 28048.——2, Posterior view of LA-element, OSU 28049.——3, Lateral view of U-element, OSU 28050.——4, Lateral view of LC-element, OSU 28051.

5.   *Ellisonia?* sp. Lateral view, sample K12-9, OSU 28059.

6-8, 10, 12, 16.   *Ellisonia robusta* Sweet. Syntypes. Sample K3-14.——6, Lateral view of LB2-element, OSU 28034.——7, Posterior view of LC-element, OSU 28035.——8, 10, Posterior, lateral views of U-element, OSU 28036.——12, Posterior view of LA-element, OSU 28037.——16, Lateral view of LB1-element, OSU 28038.

9, 13, 14, 15, 17, 18, 20-22.   *Ellisonia triassica* Müller. Hypotypes.——9, 21, 22, LB-elements, samples K4-LCM, K3-14, K6-27A, OSU 28052, 28057, 28058.——13, 14, 17, 18, U-elements, samples K6-27A, K3-12, OSU 28053, 28055.——15, LA-element, sample K3-14, OSU 28054.——20, LF-element, sample K4-CM, OSU 28056.

11.   "*Prioniodella*" *prioniodellides* (Tatge). Lateral view, sample K3-15, OSU 28090.

19.   *Neospathodus peculiaris* Sweet, n. sp. Lateral view of the holotype, OSU 28085.

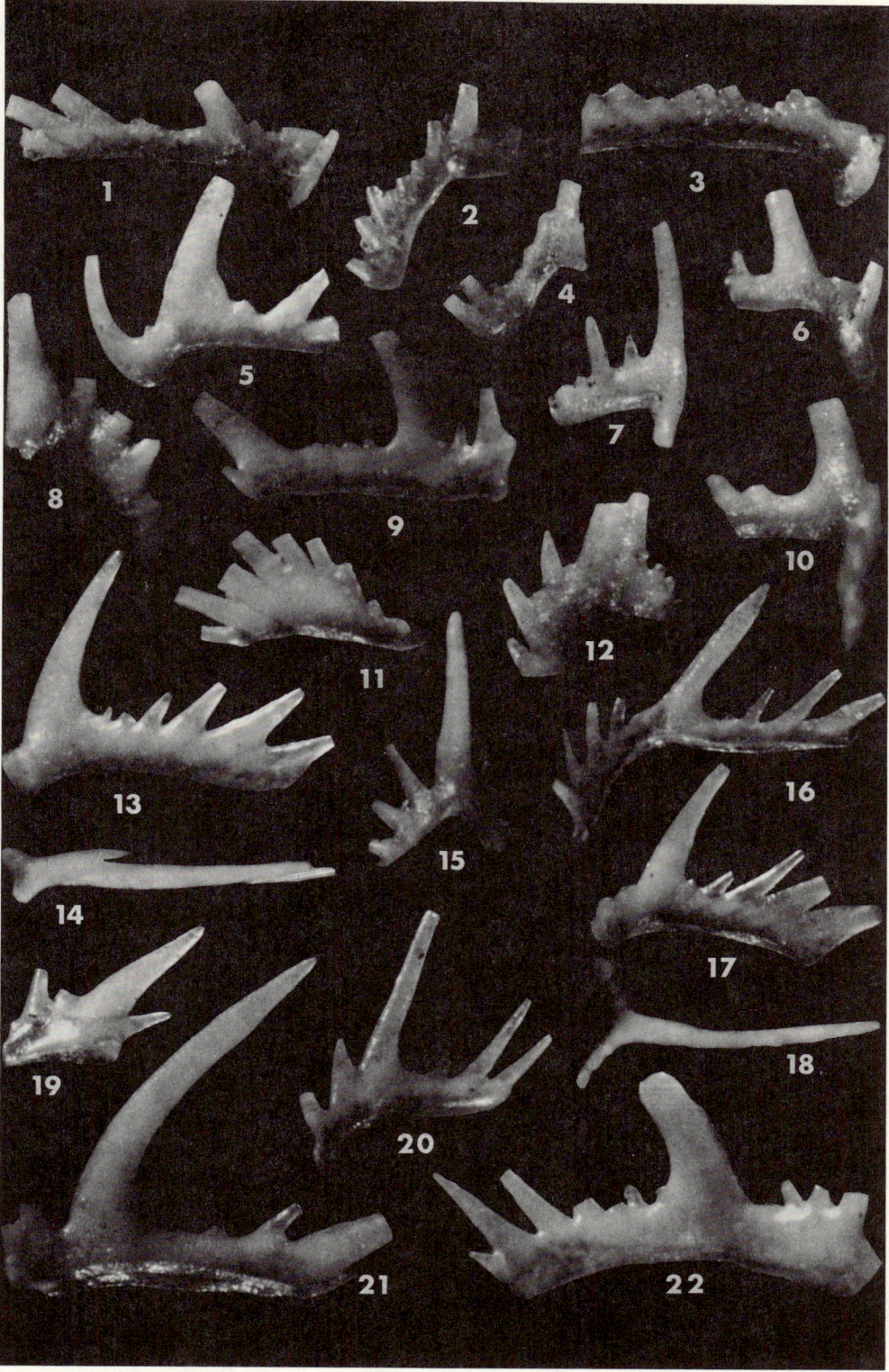

Plate 5

tions, the attachment surface is largely "inverted" posterior to the cusp base.

Remarks: Elements of *Xaniognathus elongatus* are similar to those of both *X. deflectens* and *X. curvatus,* from which they are distinguished by the relative length of the long anterior process, their shallowly grooved base, their short numerous denticles, and their broadly arcuate basal profile. The illustrated representative of the Anisian specimens from northwestern Nevada identified with *Ozarkodina tortilis* Tatge by Mosher and Clark (1965) seems to us to be more like the elements on which we base *X. elongatus* than those on which Tatge (1956) based *O. tortilis.* We suspect that Tatge (and others) have included a variety of unrelated elements in *O. tortilis.* However, the holotype of that form species has a twisted posterior process that is downwardly directed and almost as long as the anterior process, whereas in both the Nevada specimens of Mosher and Clark and the type material of *X. elongatus,* the posterior process is very short and distinctly upcurved distally.

Occurrence: Narmia Member of Mittiwali Formation at Landa, Narmia, and Nammal in West Pakistan. This species is also known from the Middle Triassic of Nevada. Distribution of this species in West Pakistan is summarized in Appendix A.

Type: Holotype, OSU 28093.

Xaniognathus? sp.

Pl. 3, fig. 21

Material: 30 discrete specimens.
Description: Compressed, slightly arched, conspicuously bowed blades, with a prominent subcentral cusp and denticulated anterior and posterior processes of essentially the same length. The cusp is reclined, sharply pointed, and is bowed inwardly distally. The anterior process, which is thicker in its basal half than in its upper portion, bears 5 to 7 compressed, discrete denticles that decline in length toward the process extremity. The posterior process, which is about as long as the anterior process, is thin and fragile and bears 4 to 6 discrete denticles that are about half the height of those on the anterior process. The posterior process is twisted in such a fashion that the tips of the denticles are directed slightly outward.

The basal margin is broadly arcuate. The underedge is narrow, faintly grooved beneath the two processes, and bears a shallow, navel-like pit beneath the cusp.

Remarks: The specimens to which the preceding description pertains may represent a species of *Xaniognathus,* the elements of which had an unusually long posterior process. However, because nearly all our specimens are thin, fragile, and incomplete, we are uncertain of their development and relationships and include this brief discussion of them largely for the sake of completeness.

Occurrence: Mittiwali Member of Mianwali Formation at Nammal; Narmia Member of Mianwali Formation at Narmia. Distribution in West Pakistan is summarized in Appendix A.

Type: Figured specimen, OSU 28094.

REFERENCES

Bassler, R. S., 1925, Classification and stratigraphic use of the conodonts: Geol. Soc. America, Bull., v. 36, p. 218-220.

Bender, Hans, 1962, Tieftriadische Hallstätter Kalke und Tuffe in Nordattika: Ges. zur Beförderung der gesamt. Naturwiss. Marburg, Sitzungsber., Bd. 84, H. 1, p. 65-79.

———, 1967?, Zur Gliederung der Mediterranen Trias II. Die Conodontenchronologie der Mediterranen Trias: Ann Géol. Pays Helléniques, v. 19, p. 465-540, 5 pl, 9 fig. [Despite the imprint date on privately circulated copies, no part of v. 19 of the Ann. Géol. Pays Helléniques had been received by an American library by October,

1969, and, although they may have been available to the author for some time, no separates seem to have been distributed by him before April or May, 1969. Thus, interested students were unaware of the existence of this paper prior to early 1969 and the effective date of publication, as defined in the International Rules of Zoological Nomenclature is still in doubt. Citation of Bender's paper herein should not be taken as establishing its publication date, but merely as recognition of the fact that his paper was published sometime before this one.]

———, and Kockel, C. W., 1963, Die Conodonten der griechischen Trias: Ann. Géol. Pays Helléniques, ser. 1, v. 14, p. 436-445, pl. LIV (I), 8 fig.

———, and Stoppel, Dieter, 1965, Perm-Conodonten: Geol. Jahrb., Bd. 82, p. 331-364, 1 fig., pl. 14-16.

Branson, E. B., and Mehl, M. G., 1938, Conodonts from the Lower Mississippian of Missouri, in Branson, E. B., et al., Stratigraphy and paleontology of the Lower Mississippian of Missouri, Part 2: Missouri Univ. Studies, v. 13, no. 4, p. 128-148, pl. 33-34.

Clark, D. L., 1959, Conodonts from the Triassic of Nevada and Utah: Jour. Paleontology, v. 33, no. 2, p. 305-312, pl. 44-45, 1 fig.

———, 1960, Triassic biostratigraphy of eastern Nevada: Intermountain Assoc. Petrol. Geol., Guidebook to 11th Ann. Field Conf., p. 122-125.

———, and Ethington, R. L., 1962, Survey of Permian conodonts in western North America: Brigham Young Univ. Geol. Studies, v. 9, pt. 2, p. 102-114, pl. 1, 2.

———, and Mosher, L. C., 1966, Stratigraphic, geographic, and evolutionary development of the conodont genus *Gondolella*: Jour. Paleontology, v. 40, no. 2, p. 376-394, pl. 45-47, 4 fig.

Clarke, W. J., 1960, Scottish Carboniferous conodonts: Edinburg Geol. Soc., Trans., v. 18, p. 1-31, pl. 1-5.

Ellison, Samuel, 1941, Revision of the Pennsylvanian conodonts: Jour. Paleontology, v. 15, no. 2, p. 107-143, pl. 20-23, 4 fig.

Fager, E. W., 1957, Determination and analysis of recurrent groups: Ecology, v. 38, p. 586-595.

Grant, R. E., 1968, Structural adaptation in two Permian brachiopod genera, Salt Range, West Pakistan: Jour. Paleontology, v. 42, no. 1, p. 1-32, pl. 1-9, 21 fig.

Hass, W. H., 1959, Conodonts from the Chappel Limestone of Texas: U.S. Geol. Survey, Prof. Paper 294-J, p. 365-399, pl. 46-50, 1 tab., 1 chart, 1 fig.

———, 1962, Conodonts, in Part W, Miscellanea, of Moore, R. C., ed., Treatise on Invertebrate Paleontology: Geol. Soc. America and Univ. Kansas Press, p. W3-W69, fig. 1-41.

Hinde, G. J., 1900, Notes and descriptions of new species of Scotch Carboniferous conodonts: Nat. Hist. Soc. Glasgow, Trans., v. 5, n. ser., pt. 3, p. 338-346, pl. 9-10.

Huckriede, Reinhold, 1958, Die Conodonten der mediterranen Trias und ihr stratigraphischer Wert: Paläont. Zeitschr., Bd. 32, p. 141-175, pl. 10-14.

Kohut, J. J., 1969, Determination, statistical analysis, and interpretation of recurrent conodont groups in the Upper Ordovician of the Cincinnati Region (Ohio, Kentucky, and Indiana): Jour. Paleontology, v. 43, no. 2, p. 392-412, 6 fig.

Kummel, Bernhard, 1954, Triassic stratigraphy of southeastern Idaho and adjacent areas: U.S. Geol. Survey, Prof. Paper 254H, p. 165-194.

———, 1966, The Lower Triassic formations of the Salt Range and Trans-Indus ranges, West Pakistan: Mus. Comp. Zool., Harvard Univ., Bull., v. 134, no. 10, p. 361-429, pl. 1-4.

———, and Teichert, Curt, 1966, Relations between the Permian and Triassic formations in the Salt Range and Trans-Indus ranges, West Pakistan: N. Jahrb. Geol. Paläont., Abh., Bd. 125, p. 297-333.

———, and ———, 1970, Stratigraphy and paleontology of the Permian-Triassic boundary beds, Salt Range and Trans-Indus ranges, West Pakistan: in Kummel, B., and Teichert, C., eds., Stratigraphic Boundary Problems: Permian and Triassic of West Pakistan: Univ. Press Kansas, Dept. Geology, Univ. Kansas Spec. Publ., 4, p. 1-110.

Lindström, Maurits, 1964, Conodonts: Elsevier Publishing Co., Amsterdam, 196 p., 64 fig.

Mojsisovics, E. V., Waagen, W., and Diener, Carl, 1895, Entwurf einer Gliederung der pelagischen Sedimente des Trias-Systems: Akad. Wiss. Wien, Sitzungsber., Bd. 104, H. 1, p. 1271-1302.

Mosher, L. C., 1967, Evolution of Triassic platform conodonts: Geol. Soc. America, Prog. 1967 Ann. Meetings, p. 156-157 (abs.).

———, 1968, Triassic conodonts from western North America and Europe and their correlation: Jour. Paleontology, v. 42, no. 4, p. 895-946, pl. 113-118.

———, and Clark, D. L., 1965, Middle Triassic conodonts from the Prida Formation of northwestern Nevada: Jour. Paleontology, v. 39, no. 4, p. 551-565, pl. 65, 66, 1 tab., 2 fig.

Müller, K. J., 1956, Triassic conodonts from Nevada: Jour. Paleontology, v. 30, no. 4, p. 818-830, pl. 95, 96.

———, and Müller, E. M., 1957, Early Upper Devonian (Independence) conodonts from Iowa, Part I: Jour. Paleontology, v. 31, no. 6, p. 1069-1108, pl. 135-142, 8 fig.

Muller, Werner, 1964, Conodonten aus der mittleren Trias der Tessiner Kalkalpen: Ecologae geol. Helvetiae, v. 57, no. 2, p. 747-753, fig. 1.

Murray, F. N., and Chronic, John, 1965, Pennsylvanian conodonts and other fossils from insoluble residues of the Minturn Formation (Desmoinesian), Colorado: Jour. Paleontology, v. 39, no. 4, p. 594-610, pl. 71-73, 2 fig.

Nogami, Yasuo, 1968, Trias-Conodonten von Timor, Malaysien und Japan (Paleontological Study of Portuguese Timor, 5): Kyoto Univ., Mem. Fac. Sci., Geol. and Mineral. ser., v. 34, no. 2, p. 115-136, pl. 8-11.

Pomesano Cherchi, Antonietta, 1967, I conodonti del Muschelkalk della Nurra (Sardegna nord-occidentale): Riv. Ital. Paleont., v. 73, no. 1, p. 205-272, pl. 12-25.

Renz, Carl, and Renz, Otto, 1948, Eine untertriadische Ammonitenfauna von der griechischen Insel Chios: Schweiz. Paläont. Abh., Bd. 66, p. 1-98, pl. 1-16.

Rexroad, C. B., and Collinson, Charles, 1961, Preliminary range chart of conodonts from the Chester Series (Mississippian) in the Illinois Basin: Illinois Geol. Survey, Circ. 319, 11 p., 1 pl.

———, and ———, 1965, Conodonts from the Keokuk, Warsaw, and Salem formations (Mississippian) of Illinois: Illinois Geol. Survey, Circ. 388, 26 p., 1 pl., 1 tab., 1 distrib. chart.

———, and Scott, A. J., 1964, Conodont zones in the Rockford Limestone and the lower part of the New Providence Shale (Mississippian) in Indiana: Indiana Geol. Survey, Bull. 30, 54 p., pl. 1-3.

Rhodes, F. H. T., 1952, A classification of Pennsylvanian conodont assemblages: Jour. Paleontology, v. 26, p. 886-901, pl. 126-129.

Schindewolf, O. H., 1954, Über die Faunenwende vom Paläozoikum zum Mesozoikum: Deutsch. Geol. Ges., Zeitschr., Bd. 105, p. 154-183.

Schmidt, Hermann, 1934, Conodonten-Funde in ursprünglichem Zusammenhang: Paläont. Zeitschr., Bd. 16, H. 1-2, p. 76-85, pl. 6.

———, and Müller, K. J., 1964, Weitere Funde von Conodonten-Gruppen aus dem oberen Karbon des Sauerlandes: Paläont. Zeitschr., Bd. 38, H. 3/4, p. 105-135, 11 fig.

Scott, H. W., 1942, Conodont assemblages from the Heath Formation, Montana: Jour. Paleontology, v. 16, p. 293-300, pl. 37-40.

Shaw, A. B., 1964, Time in Stratigraphy: McGraw-Hill Book Co., New York, 365 p.

Srivastava, J. P., and Mandwal, N. K., 1966, Record of conodonts from India: Current Science [India], v. 35, no. 24, p. 621-622, fig. 1-7.

Staesche, Ulrich, 1964, Conodonten aus dem Skyth von Südtirol: Neues Jahrb. Geologie Paläontologie, Abh., H. 119, p. 247-306, pl. 28-32, fig. 3-73.

Stauffer, C. R., and Plummer, H. J., 1932, Texas Pennsylvanian conodonts and their stratigraphic relations: Univ. Texas Bull. 3201, Contr. to Geol., pt. 1, p. 13-50, pl. 1-4, tab. 1-2.

Sweet, W. C., 1967, Sequence of Lower Triassic (Scythian) conodonts in West Pakistan: Geol. Soc. America, Progr. 1967 Ann. Meetings, p. 218 (abs.).

———, 1970, Permian and Triassic conodonts from Guryul Ravine, Vihi District, Kashmir: Univ. Kansas Paleont. Contributions, Paper 49, p. 1-10, 1 pl.

Tatge, Ursula, 1956, Conodonten aus dem Germanischen Muschelkalk: Paläont. Zeitschr., Bd. 30, p. 108-127, 129-147, pl. 5, 6.

Teichert, Curt, 1966, Nomenclature and correlation of the Permian "Productus Limestone," Salt Range, West Pakistan: Geol. Survey Pakistan, Rec., v. 15, pt .1, 20 p., 7 fig.

Ulrich, E. O., and Bassler, R. S., 1926, A classification of the toothlike fossils, conodonts, with descriptions of American Devonian and Mississippian species: U.S. Nat. Mus., Proc., v. 68, art. 12, p. 1-63, pl. 1-11.

Walliser, O. H., 1964, Conodonten des Silurs: Hess. Landesamt. Bodenforsch., Abh., H. 41, 106 p., 10 fig., 2 tab., 32 pl.

Youngquist, Walter, 1952, Triassic conodonts from southeastern Idaho: Jour. Paleontology, v. 26, no. 4, p. 650-655.

———, and Miller, A. K., 1949, Conodonts from the Late Mississippian Pella beds of south-central Iowa: Jour. Paleontology, v. 23, p. 617-622, pl. 101.

APPENDIX A

In the following Table 1, section names and field numbers of conodont-bearing samples are listed in the two left-hand columns. Within each section, samples are listed in ascending order, with the oldest listed first (e.g., K1-12 is the oldest sample in the Narmia section; K1-50 + T63-167 are the youngest). The position of samples in the sections from which they were derived is shown in Figures 2 and 3 of this paper. For example, the following table shows that sample K1-17 is the third sample above the base of that part of the Narmia section shown in Figure 2, or the second sample above the base of the part of the same section shown in Figure 3. Exact occurrence data are found in Kummel and Teichert (1970, Appendix B).

In the following table, numbers in the matrix are the numbers of representatives of species collected from the samples indicated. Species numbers across the top of the matrix refer to the following list:

1. *Anchignathodus isarcicus* (Huckriede)
2. *Anchignathodus typicalis* Sweet
3. *Ellisonia clarki* Sweet, n. sp.
4. *E. delicatula* Sweet, n. sp.
5. *E. gradata* Sweet
6. *E. robusta* Sweet
7. *E. teicherti* Sweet
8. *E. torta* Sweet, n. sp.
9. *E. triassica* Müller
10. *E.?* sp.
11. *Neogondolella carinata* (Clark)
12. *N. elongata* Sweet, n. sp.
13. *N. jubata* Sweet, n. sp.
14. *Neospathodus homeri* (Bender)
15. *Neospathodus cristagalli* (Huckriede)
16. *N. dieneri* Sweet
17. *N. kummeli* Sweet, n. sp.
18. *N. triangularis* (Bender)
19. *N. pakistanensis* Sweet, n. sp.
20. *N. peculiaris* Sweet, n. sp.
21. *N. timorensis* (Nogami)
22. *N. spathi* Sweet, n. sp.
23. *N. waageni* Sweet, n. sp.
24. *"Prioniodella" prioniodellides* (Tatge)
25. *Xaniognathus curvatus* Sweet, n. sp.
26. *X. deflectens* Sweet, n. sp.
27. *X. elongatus* Sweet, n. sp.
28. *X.?* sp.

TABLE 1. Localities and Field Numbers of All Conodont-Bearing Samples Described in This Report

Loc.	Sample Number	Fig.	Conodont species																											
			1	2	3	4	5	6	7	8	9	10	11	12	13	14	15	16	17	18	19	20	21	22	23	24	25	26	27	28
Narmia	T63-163	2		11					5		3																			
	T63-165	2		4					2		2																			
	T62-300	2			5						4																			
	T63-162	2		9					22		12															1			
	T63-19	2		10				1	24		9																			
	T63-22	2							2																					
	K1-12	2		15					24																	2			
	K1-16	2,3					2											4	115											
	K1-17	2,3					29	1			24						365	1536				2						131		
	K1-21	3																			6								1		
	K1-36, K1-37	3				5	34				9		54	108		19				30				57					3	7
	K1-50+T63-167	3			76		74						1			46				21			123					12		9
Ganjaroh	K24	2		1						6			91														9			
	K2-5	2						5		2			53					4	25								2	15		
Chhidru A (West)	K3-2A	2		2									2																	
	T63-122	2	1	121					96		25		1																	
	T63-123	2		73					24		14		2																	
	T63-124	2		337					165		25		43														1			
	T63-128	2	1	33					18		2		125														1			
	K3-5	2		4					1				3														1			
	T63-130	2	1	53			3		36		22		503														28			
	K3-6+T63-131	2		7			11				30		289														20			
	T63-132	2		5			14		39		39		458														38			
	K3-7	2					11				23		18														4			
	T63-136	2					1				16		175														4			
	K3-8	2									1		44															3		

TABLE 1 (continued).

Loc.	Sample Number	Fig.	\multicolumn{28}{c}{Conodont species}
			1	2	3	4	5	6	7	8	9	10	11	12	13	14	15	16	17	18	19	20	21	22	23	24	25	26	27	28
Chhidru A (West)	T63-153	2					32				17		310														33	3		
	K3-3+T63-155	2					19				13		17						6	343							1	12		
	T63-157	2					53				256		38					2591										169		
	T63-158	2					20				39						87	768										58		
	K3-10	2,3					72	32			91						177	457				3				3		47		
	K3-12	2,3					64	18			67						260	475				1						56		
	K3-14	2,3					135	32			76						100	256				2				6		33		
	K3-15	3					24	1			24							109			371					8		35		
Chhidru B (East)	K4-2	2		4					3				36																	
	K4-11	2					2						36														2			
	K4-14	2											39															2		
	K4-CM+K4-LCM	2					37	17			112						187	425				3						44		
	K4-UCS	3				15	10	7			20							3							66					
	K4-UCL	3					1				13												1							
Nammal	K6-5	2		1							23		347														10			
	K6-6	2											73																	
	K6-8A	2,3																11										1		
	K6-A	2,3					23				5							148										14		
	K6-8B	3																2												
	K6-8C	3																15			9									
	K6-B	3					2						10								162				2		6		

274

TABLE 1 (continued).

Loc.	Sample Number	Fig.	\multicolumn{28}{c}{Conodont species}																										
			1	2	3	4	5	6	7	8	9	10	11	12	13	14	15	16	17	18	19	20	21	22	23	24	25	26	27	28
Nammal	K6-8D	3																			2									
	K6-8F	3																			7								3		
	K6-8G	3																3			5										
	K6-27A	3				1	24				84		13																		14
	K6-30	3													52				14											4	
	K6-31B	3					3						12		30	12			2											3	
	K6-36	3			2		4				1	4			3	4			4							5				
	K6-37	3			5		1				12	2			2	3															
Kathwai A	K10-2	2										15																			
	K10-3	2					7				10	207															7			
	K10-5	2,3					6				20	3						82									2	8		
	K10-6	2,3					47	2			7						78	323				1				5		38		
Wargal	K11-3	2									1		20																		
	K11-9	2																													
	K11-10	2									3																				
	K11-CL, K11-LCL	2					10				11							351											21		
	K11-15	2															1	1													
Landa	K12-21	3									4			1											4						
	K12-1	3				2	1							1	20					1				5							
	K12-2	3																		1											
	K12-5	3			12		49				19	3			18	41			33									3		

Stratigraphic Boundary Problems

TABLE 1 (continued).

Loc.	Sample Number	Fig.	1	2	3	4	5	6	7	8	9	10	11	12	13	14	15	16	17	18	19	20	21	22	23	24	25	26	27	28
Landa	K12-7	3				2										7				2										
Landa	K12-9	3			22		6		49	10	12					159				13										
Landa	K12-10	3														8				2										
Landa	K12-12	3									1																			
Zaluch	T62-191	2	14			1		10		2																				
Zaluch	T62-192	2	1	64				23				11															1			
Zaluch	T62-193	2	1					1																						
Zaluch	K13-9	3		23		16				62														14						
Zaluch	K13-11	3		3						3														2						
Zaluch	K13-13	3		11		1									3					1				5						
Kathwai B	K14-3	2	25					20		11																				
Kathwai B	K14-6	2				1				1		7																		
Kathwai B	K14-9	2															13	24											3		
Kingriali	T64-13	2	62					24		5																				
Kingriali	T64-14	2	2	234		1		156		9		316															16			
Kingriali	T64-15	2	6					2		3		110															3			
Kingriali	T64-17	2	9			10		9		8		321															35			
Kingriali	T64-18	2				20				92		334															49			
Kingriali	T64-19	2				5				7		96															6			
Kingriali	T64-21	2				15				44		231															13			
*	T63-145					2				11		177															8			
*	T63-144									4		77															3			

* Khan Zaman Nala.

Acritarchs and Tasmanitids from the Chhidru Formation, Uppermost Permian of West Pakistan

WILLIAM A. S. SARJEANT
Department of Geology, The University, Nottingham, England

ABSTRACT

In the assemblages here described, 17 species of acritarchs and one species of Prasinophycean alga are present: their dimensions are uniformly small. Two species are new *(Micrhystridium karamurzae* and *M. pakistanense)*. The diagnosis of an existing species, *Veryhachium valensii*, is emended and a new combination, *Deunffia unispinosa* (Schön), is proposed. The paleoecological relationships and stratigraphic value of the assemblages are considered.

INTRODUCTION

Records of acritarchs from the Permian are extremely sparse. The first mention was by Wilson (1960), who recorded an unnamed spinose "hystrichosphere" from the Flowerpot Formation (Middle Permian) of Oklahoma, USA. In 1961 Medvedeva and Chepilova recorded leiospheres in petroleum, considered to be of Permian age, from the Volga-Ural region of the USSR: and, in the same year, Jékhowsky described polygonomorphid acritarchs from unnamed Permo-Triassic formations of Yugoslavia, Libya, Tunisia, and Madagascar. In 1962 Jansonius included descriptions of acritarchs in an account of the palynology of the Belloy Formation (Middle Permian) of western Canada. Tasch (1962, 1963a, 1963b) recorded acritarchs from the Wellington Formation (Lower Permian) of Kansas, USA. In 1963 Wall and Downie described acanthomorphid and polygonomorphid acritarchs from English Lower Permian marls; and Schaarschmidt described assemblages from the Zechstein of Germany. Most recently, Balme and Playford (1967) recorded what they considered to be a nonmarine acritarch from the Late Permian of the Prince Charles Mountains, Antarctica.

Tasmanitids were first described from the Permian "white coal" (Tasmanite) of the Mersey River area in Tasmania, by Newton (1875). A long series of subsequent papers have contained descriptions, or incidental mentions, of specimens from these deposits, but descriptions from elsewhere in the world have been extremely few: Segroves (1967) noted their presence in the Western Australia Permian and Efremova (1967) recorded tasmanitids from two areas in the USSR.

In the course of palynological studies of samples from various Permo-Triassic sequences in the Salt Range and Surghar Range of Pakistan, Dr. Basil E. Balme (University of Western Australia) encountered acritarchs at a number of horizons. I was invited to prepare an account of these and a batch of samples was sent to me: unforunately, these were lost in the mail. A second batch of samples was then forwarded: these suffered considerable delays en route and I was left with insufficient time to make a complete study of all horizons containing acritarchs. I considered it better to make a detailed examination of a restricted number of assemblages, rather than to make a superficial study of a large number.

The sequence selected contained samples from the topmost Chhidru Formation of Munta Nala near Wargal, in the Salt Range (Kummel and Teichert, 1970, loc. 7, unit 5), considered to represent the topmost Permian in this area. Three assemblages were available, all taken from the dark gray shale at the top of the white sandstone unit. The concentrate from the topmost sample (1 foot below the top of the formation) contained much fine debris and minute mineral crystals: a few poorly preserved, long-spined *Micrhystridium* were present, but the numbers and quality were insufficient to permit a full study. In contrast, two lower horizons (respectively 2 feet and 3 feet below the top) contained rich and varied acritarch assemblages, though the overall size of the species represented was uniformly small (shell diameter less than 30μ). The preservation of specimens in the 3-foot horizon was markedly better: attention was therefore concentrated on this horizon, since the species content, and their relative proportions, in the two horizons is virtually identical. Three hundred specimens were initially counted and (where orientation and preservation permitted) measured: further traversing was continued to find good specimens of less common species, a total of 374 specimens being identified and allotted numbers (the total considered was approximately 1200). Of these, 372 were acritarchs. Two tasmanitids were also noted, but dinoflagellate cysts were not encountered.

All figured specimens are lodged in the collections of the University of Kansas Museum of Invertebrate Paleontology, Lawrence, Kansas. The Museum numbers are given in the species descriptions. Representative slides of the assemblages studied are lodged in the collections of the British Museum (Natural History), London.

SYSTEMATIC SECTION

INCERTAE SEDIS

Group ACRITARCHA Evitt

Four subgroups are represented in the horizons studied; their percentage distribution is shown in Table 1. The assemblage is dominated by Acanthomorphitae and Polygonomorphitae, the latter considerably outnumbering the former and represented only by the genus *Veryhachium,* the single species *Veryhachium valensii* (Valensi) alone forming 61.66 percent of the acritarch assemblage. The Acanthomorphitae are represented exclusively by the genus *Micrhystridium,* the most abundant species being *Micrhystridium inconspicuum* (Defl.), which forms 11.66 percent of the assemblage.

Two other subgroups are present in low numbers. The Netromorphitae are represented by two species, both

TABLE 1. Percentage Distribution of Acritarch Subgroups in the 3-foot Horizon, Chhidru Formation, Pakistan.

Subgroup	Total in first 300 specimens	Total examined	Percentage based on first 300
Acanthomorphitae	87	129	29.00
Polygonomorphitae	209	231	69.66
Netromorphitae	4	10	1.33
Prismatomorphitae	..	2	..
Total	300	372	100.00

TABLE 2. Percentage Distribution of Species in the 3-Foot Horizon, Chhidru Formation (Topmost Permian), Pakistan, Compared with Their Previously Known Distribution.

Species	Number of specimens	Percent (in first 300)	Known stratigraphic range
Micrhystridium inconspicuum (Defl.) Deflandre	42 (35)	11.66	Permian - Upper Cretaceous
M. breve Jansonius	20 (7)	2.33	Permian - Middle Jurassic
M. karamurzae Sarjeant sp. nov.	17 (11)	3.66	Triassic
M. densispinum Valensi	4 (3)	1.00	Middle Jurassic
M. setasessitante Jansonius	2 (1)	0.33	Permian
M. pakistanense Sarjeant, sp. nov.	14 (12)	4.00	
M. circulum Schön	9 (6)	2.00	Triassic
M. aff. *keratoides* Spode	11 (6)	2.00	Permian - Triassic
M. microspinosum Schaarschmidt	10 (6)	2.00	Permian - Triassic
Veryhachium valensii (Valensi) Downie and Sarjeant emend.	188 (185)	61.66	Permian - Jurassic
V. irregulare forma *subtetraedron* Jekhowsky	7 (5)	1.66	Permian - Triassic
V.? riburgense Brosius and Bitterli	26 (12)	4.00	Permian - Triassic
Wilsonastrum colonicum Jansonius	10 (7)	2.33	Permian - Triassic
Polyedryxium sp.	2 (..)	--	
Leiofusa stassfurtensis Schön	7 (3)	1.00	Triassic
Deunffia unispinosa (Schön) comb. nov.	3 (1)	0.33	?Permian - Triassic
Tasmanites sp.	2		
TOTAL 17 species	372 (300)+2	100.00	

present in small numbers: and the Prismatomorphitae are recorded for the first time from the Permian, being represented by two specimens of an undescribed species. It is noteworthy, and surprising, that the Sphaeromorphitae are unrepresented: a single specimen encountered may have been a representative of the genus *Leiosphaeridia*, but preservation was too poor for this to be certain.

The distribution numbers of individual species are shown in Table 2, their previously known range being summarized.

Subgroup ACANTHOMORPHITAE
Downie, Evitt, and Sarjeant, 1963

Genus MICRHYSTRIDIUM
Deflandre, 1937, emend. Sarjeant, 1967

Micrhystridium inconspicuum
(Deflandre, 1935) Deflandre, 1937

Fig. 1a-d, 2; Plate 1, fig. 3, Plate 2, fig. 17

?*Echinum micraster* Meunier, 1910, p. 71, pl. 4, fig. 34.
Hystrichosphaera inconspicua Deflandre, 1935, p. 233, pl. 9, fig. 11-12.
Micrhystridium inconspicuum (Defl.). Deflandre, 1937, p. 32, pl. 12, fig. 11-13.
M. inconspicuum (Defl.), Valensi, 1947, p. 817, fig. 8.

M. inconspicuum (Defl.), Deflandre, 1947, p. 6, fig. 7-12.
M. inconspicuum (Defl.), Deflandre, 1952, fig. 26-29.
M. inconspicuum (Defl.), Valensi, 1953, p. 53-55, pl. 2, fig. 10; pl. 7, fig. 14, 15, 20, 23, 36; pl. 8, fig. 1-5, 7-10, 14, 15, 17-20, 23-29; pl. 14, fig. 8-11.
?Hystrichosphere. Kara-Murza, 1957, pl. 1, fig. 24.
M. inconspicuum (Defl.), Sarjeant, 1959, p. 340, fig. 7b.
M. inconspicuum (Defl.), Sarjeant, 1960, p. 398, pl. 14, fig. 18, fig. 1f, g, tab. II.
M. cf. *inconspicuum* (Defl.), Sarjeant, 1960, p. 398, fig. 1k.
M. inconspicuum (Defl.), Brosius and Bitterli, 1961, p. 40, pl. 2, fig. 17-18, fig. 8.
M. inconspicuum (Defl.), Sarjeant, 1961, p. 105, pl. 13, fig. 5, fig. 8f, 15.
M. inconspicuum (Defl.), Sarjeant, 1962a, pl. 2, fig. 13, fig. 8b, tab. 4.
M. inconspicuum (Defl.), Sarjeant, 1962b, tab. 2, 3.
Micrhystridium cf. *M. inconspicuum* (Defl.), Jansonius, 1962, p. 85, pl. 16, fig. 60-61, fig. 3i-j.
?*Archaeohystrichosphaeridium aculeatum* Isagulova, 1963, p. 1157, fig. 148.
M. inconspicuum (Defl.), Chornaya, 1963, p. 284, pl. 7, fig. 13.
?*Micrhystridium* sp. C. Schaarschmidt, 1963, p. 67, pl. 19, fig. 16.
M. inconspicuum (Defl.), Downie and Sarjeant, 1963, p. 92.
M. inconspicuum (Defl.), Sarjeant, 1964, tab. 4.
?*M. inconspicuum* (Defl.), Spode, 1964, p. 365, pl. 38, fig. j, fig. 2j.
M. inconspicuum (Defl.), Gocht, 1964, p. 124, fig. 44.
M. inconspicuum (Defl.), Downie and Sarjeant, 1964, p. 131.
M. inconspicuum (Defl.), Deflandre and Deflandre, 1965a, p. 2.
M. inconspicuum (Defl.), Sarjeant, 1965, p. 177, pl. 1, fig. 4, tab. I.
M. inconspicuum (Defl.), Deflandre and Deflandre, 1965b, fiches 2251-68.
M. cf. *inconspicuum* (Defl.) =*Archaeohystrichosphaeridium aculeatum* Isagulova. Deflandre and Deflandre, 1965b, fiche 2271.
Micrhystridium sp. plur. (pars), Deflandre and Deflandre, 1965b, fiche 2448 (in a reallocation of some of Kara-Murza's forms).
M. inconspicuum (Defl.), Norris and Sarjeant, 1965, p. 40.
?*M. lymensis* var. *gliscum* (pars), Wall, 1965, p. 158, pl. 2, fig. 13.
M. inconspicuum (Defl.), Staplin, Jansonius and Pocock, 1965, p. 180.
M. inconspicuum (Defl.), Sarjeant, 1967, p. 202-3, pl. 1, fig. 2, fig. 10.
M. inconspicuum (Defl.), Sarjeant, 1968, pl. 2, fig. 12, tab. 2A.
M. inconspicuum (Defl.), Gitmez, 1969, p. 3/ 7-8, pl. 1, fig. 8, pl. 8, fig. 11, pl. 11, fig. 6, tab. 4B.

Figured Specimens: i. 1,800,337 (Pl. 2, fig. 17). ii. 1,800,338 (Pl. 1, fig. 3; Fig. 1a). iii. 1,800,339 (Fig. 1b). iv. 1,800,340 (Fig. 1c). v. 1,800,341 (Fig. 1d).

Dimensions: i. Shell 10x9μ, spines 7.5μ. ii. Shell 9x8μ, spines c.5μ. iii. Shell 9x8μ, spines c.4.5μ. iv. Shell 8x7μ, spines c.6μ. v. Shell 12x9.5μ, spines c.6μ. Overall range: shell 8-12.5μ (mean 10.02μ) X7-10.5μ (mean 9.02μ), spines c.3-7.5μ (mean 4.45μ). Measurements, as taken, accurate to within 0.5μ. 21 specimens measured.

Diagnosis: "The shell, of small size and globular or slightly polyhedral shape, bears, distributed over its entire surface, spines whose length varies from a third to two-thirds of its diameter; these spines are hollow and their cavity communicates with that of the shell; generally pointed, they are straight and rigid, slightly curved, recurved or sometimes flexuous; their number exceeds fifteen and may reach forty. The color of the shell is generally deep brown." (Transl.)

Discussion: This is a species that is characterized by its small dimensions. Within its considerable stratigraphic range (certainly Permian to Upper Cretaceous), there is some variation in its morphology and it is probable that this is a "sack" species containing a range of similar, but not identical, types. Deflandre's type specimens, from the Upper Cretaceous, appear to have broader-based spines, fewer in number, when compared with the amply-illustrated Middle and Upper Jurassic assemblages from France and England (Valensi, 1953; Sarjeant, 1959, 1960, 1961, 1965, 1967, 1968; Gitmez, 1969). Valensi distinguished two forms in his searching study: forma *bullosa,* with 20-30 spines always about 2μ in length, and forma *helios,* with spines regularly spaced and of length about equal to one-third the shell diameter (1953, p. 54-55). His illustrations show a great diversity of types, encompassing all four variations here illustrated and including some that are very similar to the type material. Establishment of whether this is a continuously varying morphological plexus, or whether a group of species of related but distinguishable morphology are being lumped together, would require, firstly, a searching restudy of the type material and of additional Upper Cretaceous assemblages; and secondly, careful measurement and spine-counts for a large number of specimens. In view of the slight stratigraphic interest it affords, such a prolonged study of this species seems scarcely justifiable.

The forms from the Pennsylvanian (Westphalian) assigned to this species by Spode (1964) have somewhat

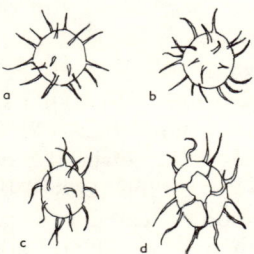

Fig. 1. *Micrhystridium inconspicuum* (Deflandre) Deflandre. Four specimens, showing variation in size and in the length and attitude of spines. An opening, of irregular shape, is shown in (d).

stronger and broader-based spines. I believe that Spode's contention, that *M. inconspicuum* is conspecific with *M. westphalienum* Stockmans and Willière (1962c), is incorrect and that his forms should be reassigned to the latter species.

The absence of any mention of *M. inconspicuum* in the Lower Jurassic assemblages described by Wall (1965) is remarkable. His species *M. lymense*, in its typical form (var. *lymensis*), is certainly distinct from *M. inconspicuum*. However, it is possible that some of the forms assigned to *M. lymense* var. *glisca* might be attributable to *M. inconspicuum*, since Wall (p. 157) defines the species to comprise thin-walled forms as well as forms with thick, double walls. The content of *M. lymense* merits reconsideration.

M. inconspicuum exhibits a striking degree of morphological similarity to a problematical species, *Echinum micraster*, recorded from present day Arctic marine plankton by Meunier (1910). A reexamination of Meunier's results is overdue; the nature and affinities of the forms he records are far from clear and would be of great interest in view of the uncertain affinities of the Micrhystridia.

Micrhystridium breve Jansonius, 1962

Fig. 2, 3a, b; Pl. 2, fig. 5, 9

Micrhystridium recurvatum forma *brevispinosa* Valensi, 1953, p. 44, pl. 6, fig. 9-10.
Micrhystridium aff. *fragile* Defl. (pars), Brosius and Bitterli, 1961, p. 41, pl. II, fig. 15-16; Fig. 6b-d.
Micrhystridium breve Jansonius, 1962, p. 85, pl. 16, fig. 62-3, 66; Fig. 3f, l, m.
M. recurvatum forma *brevispinosa* Val., Downie and Sarjeant, 1964, p. 133.
M. breve Jan., Downie and Sarjeant, 1964, p. 130.
M. recurvatum forma *brevispinosa* Val., Deflandre and Deflandre, 1965b, fiche 2354.
M. breve Jan., Deflandre and Deflandre, 1965b, fiche 2192.

Figured Specimens: i. 1,800,342 (Pl. 2, fig. 9; Fig. 3a). ii. 1,800,343 (Pl. 2, fig. 5; Fig. 3b).

Dimensions: i. Shell 15x13μ, spines 5μ. ii. Shell 17x16μ, spines c.5μ. Overall range; shell 11-19.5μ (mean 15.38μ) X10-18μ (mean 13.68μ), spines 3.5-6.5μ (mean 4.28μ). Measurements, as taken, accurate to within 0.5μ. 16 specimens measured.

Diagnosis: "Vesicle roughly circular to polyangular in outline; wall firm, yellow to light brown; spines numerous (30-50), 50 in holotype, relatively short, tapering to sharp point, hollow, at basis not inflated or thickened; diameter vesicle 15 (21)25μ; spines approx. 2μ at basis, 3-5μ in length."

Discussion: In view of the high degree of morphological similarity, in both description and figures, I consider that Jansonius' Permian species *M. breve* and Valensi's Middle Jurassic "forma *brevispinosa*" are one and the same species. The diagnosis for the latter is as follows: ". . . forma *brevispinosa* possesses processes that are always incurved, with sometimes a small inflation at their base, short (from 1/6 to 1/4 of the shell diameter) and 20-27 in number; the shell diameter lies between 12 and 14μ and the total span goes from 17 to 22μ" (Transl.). The spine numbers quoted by Valensi are certainly incorrect, since his two figures respectively illustrate forms with 25 and 35 spines visible *on the upper surface alone,* indicating a probable *total* number of 32+ and 42+ respectively. The ranges of dimensions overlap. The "International Code of Botanical Nomenclature," Art. 60, states: "When the rank of a genus or infrageneric taxon is changed, the correct name or epithet is the earliest legitimate one available in the new rank. In no case does a name or an epithet have priority outside its own rank." Thus Jansonius' name, as the first species name proposed for this form, is valid.

Three species in the assemblages here described share the characteristics of ovoid shape and simple spines of length varying around one-third of the longest diameter; *M. inconspicuum, M. breve,* and *M. karamurzae.* In order to assess whether they could be differentiated on dimensions alone, the shell measurements were plotted graphically. Since the mean spine lengths for the three species are virtually identical, the scatter on a ternary diagram was confused; in the accompanying graph (Fig. 2), the shell dimensions only are plotted. It may be seen that *M. inconspicuum* is readily differentiated by its smaller dimensions, but that *M. breve* and *M. karamurzae* cannot be separated on this character. They are, however, differentiated by spine number, *M. karamurzae* having consistently fewer spines (c.15-25 as against c.30-50) and spines that are stouter and less acute. All other species of *Micrhystridium* in this assemblage have more nearly spheroidal shells.

Micrhystridium karamurzae sp. nov.

Fig. 2, 3h; Pl. 1, fig. 17; Pl. 2, fig. 6

?Hystrichosphaerid, Kara-Murza, 1957, pl. 1, fig. 22.
?*Micrhystridium* sp. plur. (pars), Deflandre and Deflandre, 1965b, fiche 2448, fig. 22 (in a reallocation of some of Kara-Murza's forms).

Derivation of Name: Named after E. N. Kara-Murza, pioneer worker on Triassic and Cretaceous acritarchs from the Soviet Arctic.

Holotype: 1,800,344 (Pl. 2, fig. 6; Fig. 3h), Chhidru Formation 3 feet below top (uppermost Permian), Munta Nala, near Wargal, Salt Range, Pakistan.

Paratype: 1,800,345, same locality and horizon.

Dimensions: Holotype: shell 19.5x 18μ, spines c.4.5μ. Paratype: shell 18x 14μ, spines c.5μ. Overall range: shell 10-19.5μ (mean 15.14μ) x10-18μ (mean 13.27μ): spines c.3-5μ (mean 4.18μ). Measurements, as taken, accurate to within 0.5μ. 11 specimens measured.

Diagnosis: A species of *Micrhystridium* having an ovoid shell bearing c.15-25 spines. The spines vary in length from about one-quarter to one-third of the shortest shell diameter; they are stout, without marked basal widening, and somewhat bluntly pointed. The shell opens by the loss of an irregular section, typically by loss of one pole. Shell wall thin, probably but not certainly composed of a single wall layer.

Discussion: Micrhystridium karamurzae sp. nov., despite its simple morphology, may be readily differentiated from all described species on the basis of spine number, length and character. The most closely comparable species is *Micrhystridium angustum* Staplin 1961,

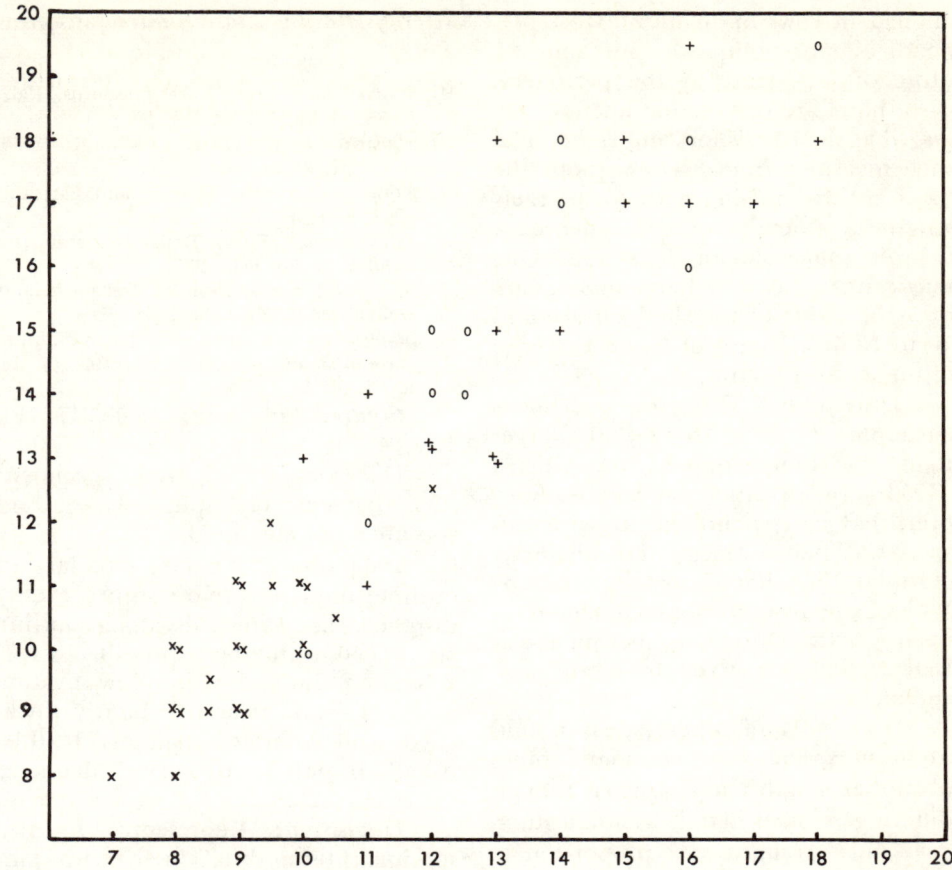

Fig. 2. Comparison of shell length (vertical) and breadth in three species of *Micrhystridium* [×, *M. inconspicuum* (Deflandre) Deflandre; +, *M. breve* Jansonius; O, *M. karamurzae* sp. nov.].

from the Devonian of Canada, which has spines similar in length and number, but much more acute; its shell is spheroidal, not ovoidal. *M. circulum* Schön, 1967, has a markedly spheroidal shell and longer, more pointed processes. (Comparisons with *M. breve* are made in the preceding section: and the dimensions of the 11 forms measured are plotted in Fig. 2.)

Kara-Murza's "hystrichosphere" from the Triassic (Carnian) of a borehole at Zhurgung-Tumus, Soviet Arctic, appears from her figure to be probably referable to this new species.

Micrhystridium densispinum Valensi
Fig. 3c; Pl. 1, fig. 7

Micrhystridium densispinum Valensi, 1953, p. 52-3, pl. 7, fig. 6-9, 13, 17-18, pl. 14, fig. 4.

M. densispinum Val., Valensi, 1955, p. 589, pl. 3, fig. 10.

M. densispinum Val., Downie and Sarjeant, 1963, p. 93.

?*M.* cf. *densispinum* Val., Gocht, 1964, p. 124, fig. 45.

M. densispinum Val., Downie and Sarjeant, 1964, p. 131.

M. densispinum Val., Deflandre and Deflandre, 1965b, fiches 2208-10.

?*M.* cf. *densispinum* (of Gocht), Deflandre and Deflandre, 1965b, fiche 2211.

Figured Specimen: 1,800,346 (Pl. 1, fig. 7: Fig. 3c).

Dimensions: Figured specimen: shell 13×12.5μ, spines c.2μ. Only one specimen measured.

Diagnosis: "The shell, of small dimension, is spherical or ellipsoidal and covered by a large number of hairs aligned in rows or in files almost perpendicular or oblique to one another. More or less dense at the periphery, these hairs are of variable fineness, enlarged at the base, short and rigid. The shell measures from 5-11μ without the hairs and from 7-14μ with them; their length is generally of the order of a micron, sometimes of 2 or 2.5μ. One may count at least 20 hairs on a square of 5μ of a side of the shell surface and 10 to 12 in a length of 5μ on the edge of the shell" (Transl.).

Discussion: The four specimens encountered were somewhat larger than the range quoted by Valensi (1953). In all other particulars, however, they correspond exactly with this species. Their surfaces were distinctly granular, a character suggested by Valensi's figures but not commented on in his text. Only one specimen was well enough preserved to merit measurement.

A number of other Paleozoic and Mesozoic species have abundant spines of similar length and character, though differing in such details as spine number and arrangement, proportionate length, or character of spine tips. Micrhystridia of this general type range from Ordovician certainly into the Upper Jurassic.

Micrhystridium setasessitante Jansonius

Fig. 3d; Pl. 1, fig. 4

Micrhystridium setasessitante Jansonius, 1962, p. 85, pl. 16, fig. 50; Fig. 3d.
?*Micrhystridium* sp. C, Schaarschmidt, 1963, p. 67, pl. 19, fig. 16.
M. setasessitante Jans., Downie and Sarjeant, 1964, p. 133.
M. setasessitante Jans., Deflandre and Deflandre, 1965b, fiche 2371.
?*M.* sp. C (of Schaarschmidt), Deflandre and Deflandre, 1965b, fiche 2443.
Filisphaeridium setasessitante (Jans.) Staplin, Jansonius and Pocock, 1965, 192, pl. 18, fig. 3, fig. 2.

Figured Specimen: 1,800,347 (Pl. 1, fig. 4; Fig. 3d).

Dimensions: Figured specimen: shell diameter 14μ, spines 3μ. Second specimen not measured.

Diagnosis: "Vesicle circular in outline: numerous setose spines, evenly disposed, not crowded, along outline 30 (34) 40 in number; spines ½μ wide, 2-4μ long, implanted on low hyaline warts (1μ) and near tip slightly thickened and rounded; spines flexible, usually in part gently curved; diameter vesicle 15 (22)25μ."

Discussion: Represented by two specimens, according closely with Jansonius's description but without clear indication of his "low hyaline warts."

In 1965 Staplin, Jansonius, and Pocock utilized this species as type for

PLATE 1

Figures 1-8 are phase-contrast microphotographs; Figures 9-19 were photographed by reflected light. Magnification ×1000.

FIGURES
1. *Veryhachium valensii* (Valensi) Downie and Sarjeant, emend.
2. *V. irregulare* forma *subtetraedron* Jekhowsky.
3. *Micrhystridium inconspicuum* (Deflandre) Deflandre.
4. *M. setasessitante* Jansonius.
5. *Veryhachium valensii* (Valensi) Downie and Sargeant, emend.
6. *Deunffia unispinosa* (Schön) comb. nov.
7. *Micrhystridium densispinum* Valensi.
8. *Tasmanites* sp.; detail of surface, showing micropores.
9. *Leiofusa stassfurtensis* Schön.
10-11. *Veryhachium? riburgense* Brosius and Bitterli.
12. *Tasmanites* sp.
13. *Veryhachium valensii* (Valensi) Downie and Sarjeant, emend.
14-15. *V.? riburgense* Brosius and Bitterli.
16. *Deunffia unispinosa* (Schön) comb. nov.
17. *Micrhystridium karamurzae* sp. nov.
18-19. *Veryhachium? riburgense* Brosius and Bitterli.

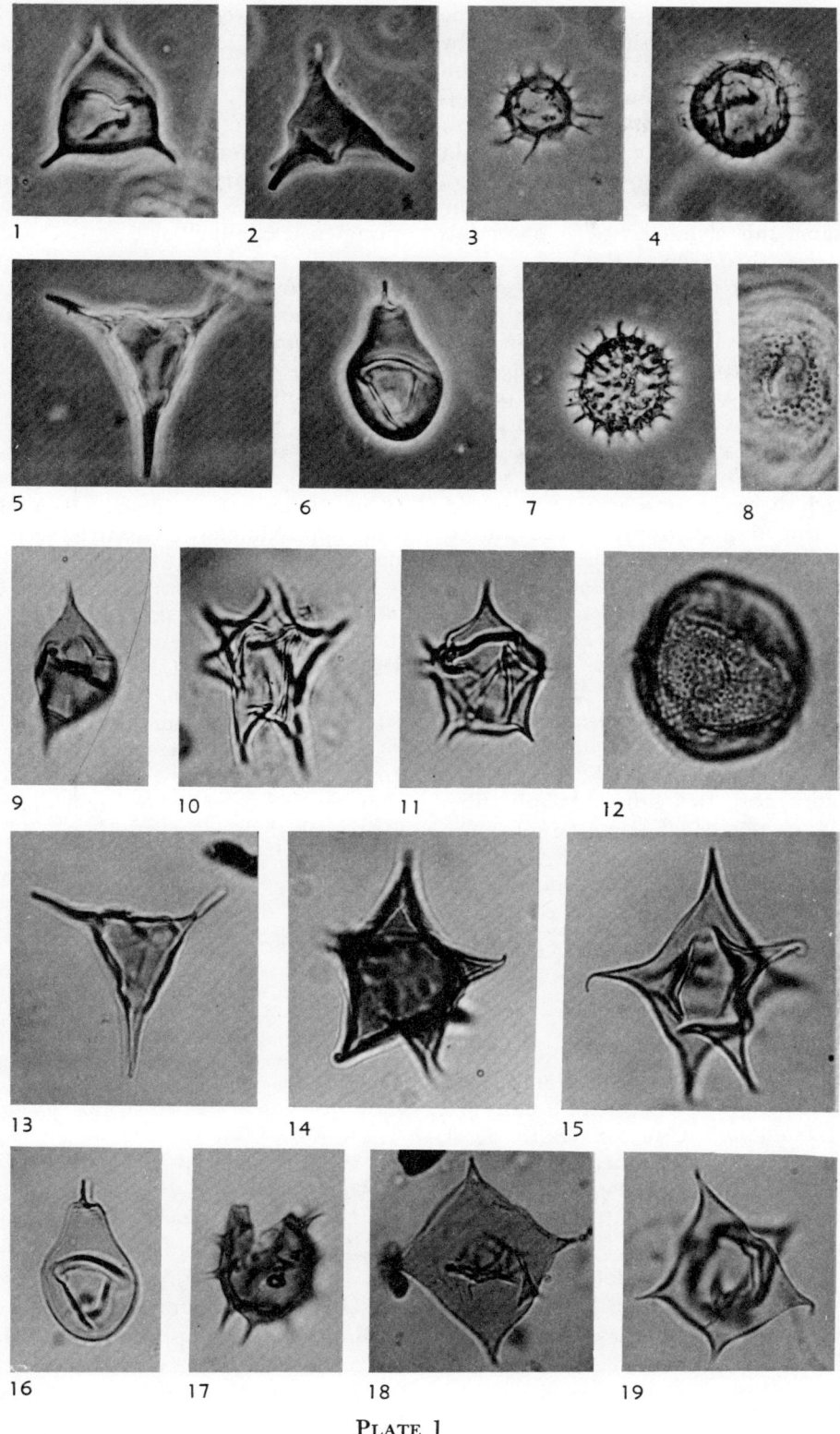

Plate 1

their new genus *Filisphaeridum*. The desirability of the single criterion by which this proposed genus is distinguished from *Micrhystridium*—its possession of solid, rather than hollow spines—is considered by the present author to be highly questionable, in view of the extreme difficulty of confirming this characteristic in species as small as this, even at the highest magnifications available under the optical microscope. (Indeed, the spines of the Pakistan specimens of *M. setasessitante*, although extremely slender, appear to be hollow, at least in their proximal portion.) It is therefore considered preferable to retain this species in the genus *Micrhystridium*, a genus within which there are considered to be species showing all gradations between entirely solid and entirely hollow spines, comparable with the variation considered to occur in the genus *Solisphaeridium* Staplin, Jansonius, and Pocock (1965, p. 183-184, Fig. 10).

Two species in the Middle Jurassic, *Micrhystridium castaninum* Valensi 1953 and *M. bigoti* Deflandre, 1947 (transferred to *Tenuà* by Deflandre and Deflandre, 1965b), differ only in details of spine number, arrangement, and form. A Devonian species, *Micrhystridium alloiteaui* Deunff 1955, is similar, but has brief bifurcations at the spine tips.

Micrhystridium pakistanense sp. nov.

Fig. 3e; Pl. 2, fig. 3

Derivation of Name: Named from its country of first recorded occurrence.

Holotype: 1,800,348 (Pl. 2, fig. 3; Fig. 3e), Chhidru Formation 3 feet below top (uppermost Permian), Munta Nala near Wargal, Salt Range, Pakistan.

Dimensions: Holotype: shell diameter 21μ, spines 2.5μ. Overall range: shell diameter $18-21.5\mu$ (mean 19.66μ), spines $2.5-4.5\mu$ (mean 3.58μ). Measurements, as taken, accurate to within 0.5μ. Six specimens measured (all others seen were either severely damaged or obscured).

Diagnosis: A species of *Micrhystridium* having a spheroidal shell bearing a large number (c.55-75) of short, stout spines, generally blunt-ended, rarely briefly bifurcate or more deeply bifurcate in Y-fashion. (In all specimens examined, at least one spine was bifurcate; in no case were more than 3 spines recognized to be bifurcate). The spines are hollow, their cavity communicating directly with the shell interior: their length ranges from about one-ninth to one-fifth of the shell diameter. Shell wall relatively thin, composed of a single layer or two closely appressed layers. The shell opens by loss of an irregular (sometimes approximately crescentric) section of its wall.

Discussion: Micrhystridium pakistanense differs, in the number and character of its spines, from all other described species. *Micrhystridium intromittum* Wall, 1965, shares the character of having dominantly simple, rarely bifurcate spines; but it has fewer spines (8-20) and they are proportionately much longer (about one-quarter to one-half the shell diameter).

Micrhystridium circulum Schön, 1967

Fig. 3f,g; Pl. 2, fig. 10, 14

?*Micrhystridium* cf. *albertensis* Staplin, Schaarschmidt, 1963, p. 66, pl. 18,, fig. 11-13; Fig. 22.

?*Micrhystridium* cf. *albertensis* Staplin (of Schaarschmidt), Deflandre and Deflandre, 1965b, fiche 2177.

M. cf. *fragile* Deflandre, Medd, 1966, p. 352-3, pl. 59, fig. 1, 2a-b.

Micrhystridium circulum Schön, 1967, p. 528, pl. 1, fig. 1-5, fig. 1, no. 8.

Figured Specimens: i. 1,800,349 (Pl. 2, fig. 10, Fig. 3f). ii. 1,800,350 (Pl. 2, fig. 14; Fig. 3g).

Dimensions: Figured specimens: i. Shell diameter 18μ, spines $c.8\mu$. ii. Shell diameter 20μ, spines $c.6\mu$. Overall range: shell diameter $14-20\mu$ (mean 16.33μ), spines $4.5-9\mu$ (mean 7.00μ).

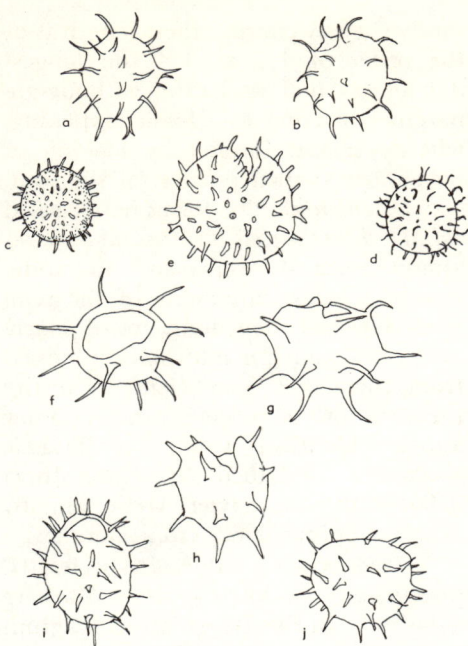

FIG. 3. a-b, *Micrhystridium breve* Jansonius; c, *M. densispinum* Valensi; d, *M. setasessitante* Jansonius; e, *M. pakistanense* sp. nov.; f-g, *M. circulum* Schön; h, *M. karamurzae* sp. nov.; i-j, *M.* aff. *keratoides* Spode.

Measurements, as taken, accurate to within 0.5μ. Six specimens measured.

Diagnosis: "A characteristic species with round central body and 6-15 hollow processes. The diameter amounts on average to 20μ. The processes are shorter than the radius and at the base are not, or only very slightly, widened" (Transl.).

Discussion: The morphology of the forms from Pakistan accords exactly with that of Schön's forms from the Lower Triassic of Germany and Medd's forms from the Lower Triassic of England. The electron micrographs prepared by Medd show the shell surface to be covered by a regular arrangement of small granules (diameter c.500Å). The Pakistan specimens are typically opened by loss of an irregular to subpolygonal portion of the shell surface. Confirmation of the occurrence of these characters in the type material is needed before an emendation of the diagnosis can be justified; and, indeed, though this species is objectively a valid one, it is probably a subjective synonym of the well-known Jurassic species *Micrhystridium fragile* Deflandre, 1947. The diagnosis of the latter species merits quotation: "The more or less regularly spheroidal shell, of medium size ($10\text{-}15\mu$ in diameter) bears a small number of processes (8 to 15), exceptionally as many as 30; thin and pointed, these processes are generally flexuous, rarely straight, rigid and strong or curved. Typically their length attains or exceeds the shell diameter, but it may descend sometimes to a third of this diameter" (Transl.). In a more extended study, Valensi (1953, p. 42) noted that the shell diameter of Middle Jurassic forms varied from $7\text{-}20\mu$ and the total span (shell and processes) from 15 to 40μ. In a study of variation in Upper Jurassic forms of *M. fragile* from England, I found a range of shell diameters from 8 to 20μ, a variation in shell shape from spheroidal to polygonal and a range of relative spine lengths from one-third to slightly greater than the shell diameter (Sarjeant, 1960, p. 399). In a subsequent study of French Upper Jurassic *M. fragile*, the mean shell diameter was found to be 15.69μ and the mean overall span 27.56μ (Sarjeant, 1965, tab. 2).

Thus, the separate identity of *M. circulum* must be considered suspect; the characters differentiating it from *M. fragile* are the slightly larger average diameter of the type material and the fact that the average spine length is less great.

Micrhystridia of this general type (spherical, with a low number of quite long processes without basal inflations) have a considerable geologic range. They include *Micrhystridium shinetonese* Downie, 1958, from the Ordovician, *M. nanum* (Deflandre, 1945) Deflandre and Deflandre, 1965, and *M. coronatum* Stockmans and Willière, 1963, from the Silurian; *M. lejeunei* Stockmans and Willière,

1962b, *M. pascheri* Stockmans and Willière, 1962a, *M. albertense* Staplin, 1961 and *M. spinoglobosum* Staplin, 1961, from the Devonian; *M. parvidumeti* Stockmans and Willière, 1962c, from the Upper Carboniferous; *M.* cf. *M. fragile* Defl. in Jansonius, 1962, from the Permian (the diameter and the number of spines of these specimens preclude their allocation to *M. circulum*); *M. fragile* Deflandre, 1947, from the Jurassic; *M. singulare* Firtion, 1952, from the Cretaceous; and forms assigned to *M. fragile* Deflandre by Takahashi, 1964, from the Oligocene.

Micrhystridium aff. *M. keratoides*
Spode, 1964

Fig. 3i, j; Pl. 2, fig. 4, 8

Hystrichosphaerids, Kara-Murza, 1957, pl. 1, fig. 20, 23.
aff. *Micrhystridium pelagicum*, Stockmans and Willière, 1962a, p. 71, pl. 2, fig. 4, fig. 21.
Micrhystridium cf. *stellatum* Defl., Schaarschmidt, 1963, p. 66, pl. 19, fig. 1.
aff. *M. pelagicum* Stock. and Will., Downie and Sarjeant, 1964a, p. 132.
aff. *Micrhystridium keratoides*, Spode, 1964, p. 365, pl. 38, fig. k, fig. 2k.
aff. *M. pelagicum* Stock. and Will., Deflandre and Deflandre, 1965b, fiche 2324.
M. cf. *stellatum* Defl. (of Schaarschmidt), Deflandre and Deflandre, 1965b, fiche 2049.
aff. *M. keratoides* Spode, Deflandre and Deflandre, 1965b, fiche 2274.
Micrhystridium sp. Deflandre and Deflandre, 1965b, fiche 2447, 2453 (in a reallocation of Kara-Murza's forms).

Figured Specimens: i. 1,800,351 (Pl. 2, fig. 8; Fig. 3i—the latter showing the undersurface). ii. 1,800,352 (Pl. 2, fig. 4; Fig. 3j).

Dimensions. i. Shell 20x16μ, spines c.4μ. ii. Max. shell diameter 17.5μ, spines 3.5μ. Overall range: max. shell diameter 15-20μ (mean 17-30μ), spines 1-4μ (mean 2.70μ). Measurements, as taken, accurate to within 0.5μ. Five specimens measured.

Description: Shell spherical to ovoidal, bearing 35-50 quite short slightly tapering spines, hollow but with closed, pointed tips. The spines appear stiff; they are generally straight, rarely slightly curved: their length is of the order of 1/5 to 1/4 the longest diameter. Shell wall thin, without apparent ornament. Shell opening, where present, formed by loss of an irregularly shaped portion of the wall.

Discussion: The forms here classed as *Micrhystridium* aff. *M. keratoides* almost certainly represent an undescribed species; but many of the relatively few specimens found were poorly preserved, and the information gained from them was insufficient to justify the erection of a new species. They appear identical with the Triassic specimens described by Kara-Murza (1961) from the Soviet Arctic, and by Schaarschmidt (1962) from Germany.

Two existing Paleozoic species, *M. pelagicum* Stockmans and Willière (1962a) from the Devonian of Belgium and *M. keratoides* Spode (1964) from the Carboniferous of England, resemble these Permian specimens in shape and in the possession of short spines in similar number. The former species (apparently based on a single specimen) is smaller and its processes appear more flexible; the latter has spines that appear more rigid, but which are again generally curved and are more broadly conical. Two other species (*M. parinconspicuum* Deflandre, 1942, Silurian; and *M. recurvatum* Valensi, 1953, Jurassic) are also of this general type but differ in spine form and number.

Micrhystridium microspinosum
Schaarschmidt, 1962

Fig. 4; Pl. 2, fig. 7, 11

?*Micrhystridium nannacanthum* Defl., Valensi, 1953, p. 51, pl. 2, fig. 7.
Hystrichosphaerids, Kara-Murza, 1961, pl. 1, fig. 7-8, ?9.
?*Micrhystridium parvispinum* Defl., Brosius and Bitterli, 1957, p. 40, pl. 2, fig. 13, 14; Fig. 7a-b.
Micrhystridium microspinosum Schaarschmidt, 1963, p. 67, pl. 19, fig. 8-10; Fig. 23.
?*M. parvispinum* Defl. (of Brosius and Bitterli), Deflandre and Deflandre, 1965, fiche 2320.

M. microspinosum Schaar., Deflandre and Deflandre, 1965b, fiche 2287.

Micrhystridium sp., Deflandre and Deflandre, 1965b, fiches 2444-5 (in a reallocation of Kara-Murza's forms).

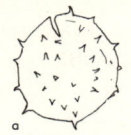 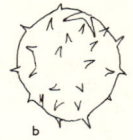

FIG. 4. *Micrhystridium microspinosum* Schaarschmidt. Extremes in the range of morphologic variation.

Figured Specimens: i. 1,800,353 (Pl. 2, fig. 7; Fig. 4a). ii. 1,800,354 (Pl. 2, fig. 11; Fig. 4b).

Dimensions: i. Shell diameter 21μ, spines c.1μ. ii. Shell diameter $19 \times 21\mu$, spines 2μ. Overall range; shell diameter $17\text{-}22\mu$ (mean 19.12μ), spines $1\text{-}2\mu$ (mean 1.86μ). Measurements, as taken, accurate to within 0.5μ. Eight specimens measured.

Diagnosis: "Shell $14\text{-}16\mu$ large, mostly broadly oval to circular, very delicate, with very many pointed, conical processes, scarcely 1μ long, equally distributed over the whole shell surface" (Transl.).

Discussion: The Pakistan specimens compare closely with Schaarschmidt's species and are confidently attributed to it, despite their larger mean size. They exhibit openings in the form of splits, generally crescentic. The figured specimens represent extremes in morphology—the morphology is, indeed, extremely constant.

Some forms figured by Kara-Murza (1957) from the Triassic (Carnian) of Arctic USSR, appear identical with this species: and a high degree of similarity is evident with Middle Jurassic forms allocated to the Silurian species *Micrhystridium nannacanthum* by Valensi (1953), and with Triassic forms allocated to the Carboniferous species *M. parvispinum* by Brosius and Bitterli (1961).

Micrhystridia of this general morphological type (spheroidal to broadly ovate, with a widely spaced scatter of very short but strong spines) have a long stratigraphic range. The earliest forms, *Micrhystridium antarcticum* (Timofeev, 1959) Deflandre and Deflandre, 1965b, and *Micrhystridium resistens* (Timofeev, 1959) Deflandre and Deflandre, 1965b, from the Lower Cambrian, may well be the descendents of Late Precambrian sphaeromorphid acritarchs. Other Paleozoic representatives include *M. nannacanthum* Deflandre, 1942 (Silurian to Carboniferous) and *M. parvispinum* Deflandre, 1946 (Carboniferous). Related Mesozoic forms are the Triassic species *M. triassicum* Jansonius, 1962; the Lower Jurassic *M. minutispinum* Wall, 1965 (differing from *M. microspinosum* only in its lower number of spines); and the Lower to Upper Jurassic *M. rarispinum* Sarjeant, 1960. Two thicker-walled forms, ?*M. crassimuratum* Sarjeant, 1968 (Upper Jurassic), and *M. pachydermum* Deflandre and Cookson, 1955 (Upper Cretaceous–Lower Eocene), exhibit similarity in all other respects to this group.

Subgroup POLYGONOMORPHITAE
Downie, Evitt, and Sarjeant, 1963

Genus VERYHACHIUM Deunff 1958 emend. Downie and Sarjeant, 1963

Discussion: The name *Veryhachium*, accompanied by a brief diagnosis, was first introduced by Deunff in 1954; but the type species selected, *Veryhachium trisulcum*, was not validly published until 1958, since the original illustration (Deunff, 1951; as *Hystrichosphaeridium trisulcum*) was unaccompanied by a diagnosis. It is therefore not altogether clear at which date this generic name is to be considered validly published, but Articles 37 and 45 of the "International Code of Botanical Nomenclature" are here interpreted as indicating that the later date (1958) is the date of valid publication. On this basis, all species published as species of this genus, and all transfers to it only became valid in 1958 (cf. Art. 43).

The circumscription of this genus has for long been recognized as presenting problems. Downie and Sarjeant (1963, p. 84) note that: "It is evident from recent publications . . . that morphological transitions exist between *Veryhachium* and both *Micrhystridium* and *Baltisphaeridium*"; but they considered that the name was a useful one and proposed that it be applied to ". . . forms having a low number (generally 3-8) of hollow spines arising from a polygonal or subpolygonal test" (p. 85). Their emendation of the generic diagnosis (1963, p. 93) was accepted by Deunff (e.g., 1966, p. 43).

The most useful document on *Veryhachium* is an unpublished report, "Preliminary Report on the Genus *Veryhachium*," prepared by Downie in 1964 and circulated among members of Groupe 9, Acritarcha, of the Commission International sur le Microflore du Paléozoique. I have employed this extensively, as a source of data and of ideas, in preparing the section that follows. At the time it was written, pylomes were not known to be developed by members of this genus. The development of slit-like pylomes, parallel to a side in forms with a triangular test, was first noted by Calandra in 1964 (see Sarjeant, 1967, p. 203). More recently, Deunff (1968) has recorded the development of circular to quadrangular openings in a species with a subquadratic to lozenge-shaped shell, *V. miloni*. It remains true, however, that regularly formed openings have not been reported from the majority of species of this genus.

Veryhachium valensii (Valensi, 1953) Downie and Sarjeant, emend. 1964

Fig. 5, 6; Pl. 1, fig. 1, 5, 13; Pl. 2, fig. 12, 13

Micrhystridium polyedricum forma *reducta* Valensi, 1953, p. 66-71, pl. 10, fig. 28.
Cysts of Pyrrhophyta? Kara-Murza, 1957, pl. 1, fig. 25, 26, 27, 33-4, ?36.
Veryhachium reductum (Deunff 1958), Jekhowsky, 1961, p. 210, 212, pl. 2.
V. reductum (Deunff) Jekh., Brosius and Bitterli, 1961, p. 36, pl. 1, fig. 3-6; Fig. 1a-e.
?*Simsangia trispinosa* Baksi, 1962, p. 18, pl. 3, fig. 34.
V. reductum (Deunff) Jekh., Wall and Downie, 1963, p. 780, pl. 112, fig. 7-9.
Veryhachium valensii (Valensi, 1953), Downie and Sarjeant, 1963, p. 94, *nom. nud.* (proposed as *nom. nov.* pro *M. polyedricum* f. *reducta*; invalid under Art. 33, para. 4, I.C.B.N.).
V. valensii (Val.) Downie and Sarjeant, Sarjeant, 1964, tab. 3, *nom. nud.*
V. reductum (Deunff) Jekh., Spode, 1964, p. 362-3, pl. 38 a-d=fig. 2a-d.
Veryhachium valensii (Valensi, 1953), Downie and Sarjeant, 1964, p. 153 (nom. nov. pro *M. polyedricum* f. *reducta*, here validly published).
V. reductum (Deunff) Jekh., Downie and Sarjeant, 1964, p. 152.
?*Simsangia trispinosa* Baksi, Downie and Sarjeant, 1964, p. 145.
V. reductum (Deunff), Jekh., Wall, 1965, p. 160, pl. 4, fig. 10-11.
?*S. trispinosa* Baksi, Norris and Sarjeant, 1965, p. 55.
V. reductum (Deunff) Jekh., Sarjeant, 1965, p. 180, pl. 1, fig. 8, tab. I.
V. valensii (Val.) Downie and Sarj., Sarjeant, 1965, p. 180, pl. 1, fig. 9.
V. valensii (Val.) Downie and Sarj., Deflandre and Deflandre, 1965b, fiche 2507.
V. reductum (Deunff) Jekh., Schön, 1967, p. 527, fig. 1.

Figured Specimens: i. 1,800,355 (Pl. 1, fig. 1; Fig. 5a). ii. 1,800,356 (Pl. 1, fig. 5; Fig. 5c). iii. 1,800,357 (Pl. 2, fig. 12; Fig. 5b). iv. 1,800,358 (Pl. 2, fig. 13; Fig. 5e). v. 1,800,359 (Fig. 5d). vi. 1,800,360 (Fig. 5f).

Dimensions: i. Shell (longest side) 15μ, spines c.4μ. ii. Shell (longest side) 12μ, spines 8μ. iii. Shell (longest side) 16μ, spines c.3μ. iv. Shell (longest side) 16μ, spines 4.5μ. v. Shell (longest side) 14μ, spines c.5μ. Overall range: shell (longest side) 11-18μ (mean 15.15μ), spines 2.5-8.5μ (mean 5.63μ). Measurements, as taken, accurate to within 0.5μ. 34 specimens measured.

Emended Diagnosis: A species of *Veryhachium* of small size, compressed and with the outline of an equilateral or isosceles triangle, from whose corners arise three spines of variable length. Length of sides typically less than 20μ; length of spines also typically less than 20μ. Sides convex, flat or slightly concave. Character of shell opening not elucidated.

Holotype: BS 36, chert, Lessart, Poitou, France (Middle Jurassic; Bathonian). Lodged in the collections of the Laboratoire de Micropaléontologie, Ecole Pratique des Hautes Etudes, Paris.

Discussion: In his account of Permo-Triassic acritarchs, Jekhowsky elevated *Veryhachium trisculum* var. *reductum* Deunff, 1958, to specific status and allocated his forms to this species. The desirability of this procedure seems questionable, on two counts; firstly, because Deunff clearly demonstrates (e.g., 1958, pl. 1) that his forms are extremes in a continuously varying morphological plexus whose most typical members (*V. reductum sensu stricto*) have extremely long processes, so that a varietal status for the short-spined end-members seems entirely appropriate: secondly, because not only are Jekhowsky's Permian and Triassic assemblages much less variable in morphology, but also the size range, throughout the many localities studied (Yugoslavia, Tunisia, Libya, Madagascar), is more restricted. Length of sides of shell 10-25μ, as against a shell height, from a side to an angle, of 20-40μ; the difference in size ranges is thus even greater than the figures suggest at first sight. [Note: This small size is consistent in all described Permo-Triassic and Jurassic assemblages. Brosius and Bitterli quote shell "diameter" as 10-15μ: Schön quotes no dimensions, but the small size of his specimens is suggested by the scale of his figure of the species; Valensi's holotype has a side-length of about 10μ; the two specimens illustrated by Wall have side-lengths of about 13μ and 16.5μ respectively; and the French Upper Jurassic specimens allocated to *V. valensii* have shell dimensions of 11-14μ x 13-20μ (Sarjeant, 1965: the dimensions of the specimens placed in *V. reductum* are quoted only in terms of overall span, but they would be of comparable shell size).]

It is thus my opinion that the Permian to Jurassic species here discussed should be distinguished by a separate species name, as *Veryhachium valensii*; and that the name *reductum* should be retained as a varietal name for extreme members of the *Veryhachium trisulcum* plexus (Devonian), as originally proposed by Deunff (1958).

The original diagnosis of *Micrhystridium polyedricum* forma *reducta* by Valensi was extremely brief (1953, p. 61): ". . . the shell, of triangular and probably tetrahedral outline, shows only three processes which prolong its angles" (Transl.).

Through the courtesy of Professor Georges Deflandre, I was able to examine the type material during a visit to Paris in 1961: there is no indication of a "tetrahedral" outline. The diagnosis is here emended, since, as it stands, it is much too widely drawn and would apply equally to the representatives of at least 12 other species of *Veryhachium*.

This is the most abundant species in the assemblages here described, over 500 specimens having been encountered; but many specimens were distorted, set obliquely, or obscured. Only

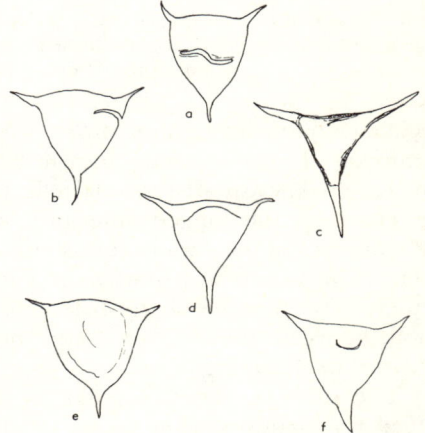

Fig. 5. *Veryhachium valensii* (Valensi) Downie and Sarjeant, emend. Specimens illustrating the range of morphologic variation (a, b, d and f show different types of shell openings).

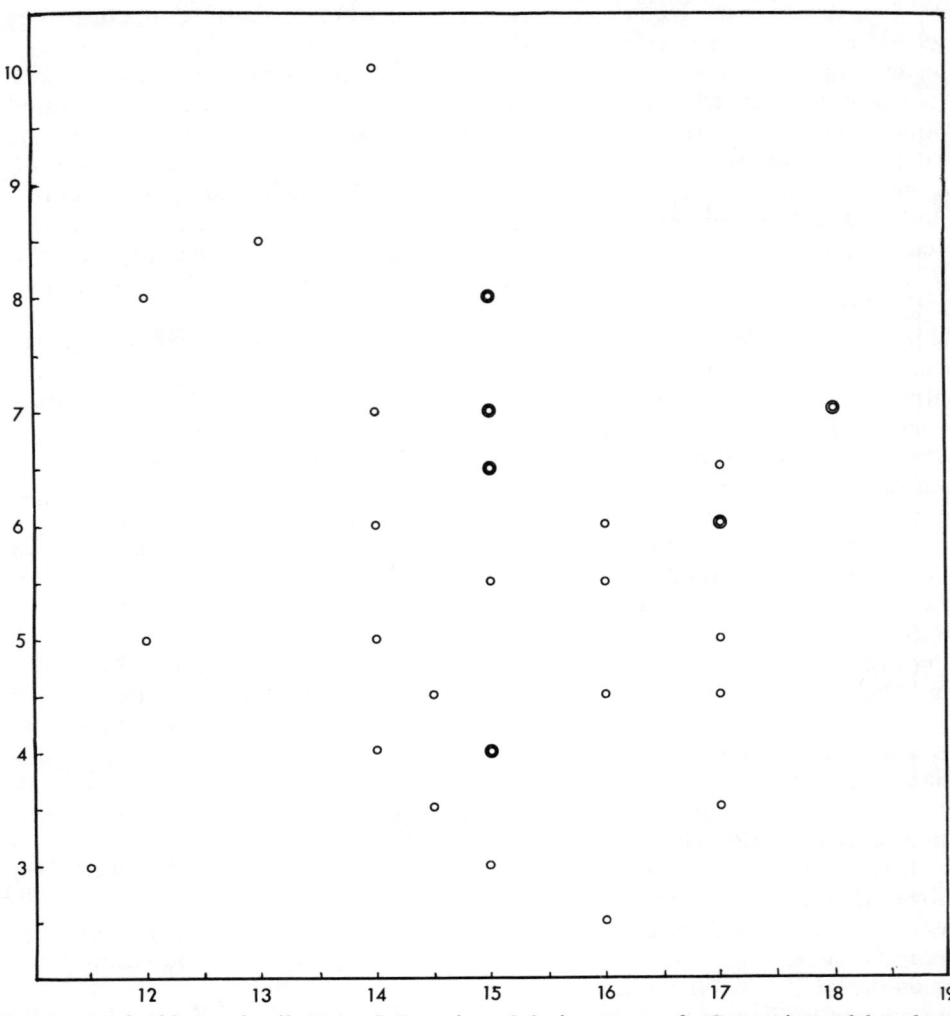

FIG. 6. *Veryhachium valensii* (Valensi) Downie and Sarjeant, emend. Comparison of length of longest measurable side (straight-line measurement: horizontal) with length of longest measurable spine. Double rings indicate 2 specimens.

specimens positioned more or less in a horizontal plane were measured: and, to obviate any distortional effect, the length of the longest side and of the longest spine was consistently taken. The results are plotted in Fig. 6; there is seen to be no consistent relation between these factors and no tendency to clustering.

The sides of the triangular shell ranged from convex (Fig. 5a, e) to almost flat (Fig. 5d) or even slightly concave (Fig. 5c): there seems to be an inverse relation between spine length and shell convexity, the forms with concave spines having the longest spines. Convexity and concavity of shell sides has been widely used as a means of distinguishing species (e.g., by Stockmans and Willière, 1963); but this procedure must be considered suspect, unless the absence of intermediates can be convincingly demonstrated. The Pakistan assemblage includes forms whose overall shape is almost an equilateral triangle (e.g., Fig. 5b) and others in the form of an isosceles triangle with a distinct longer axis (e.g., Fig. 5a); this variation seems independent of spine length.

Again, a similar variation has been used as a means of distinguishing species (e.g., by Stockmans and Willière, 1963); again, I feel it should be used with extreme caution.

The manner of shell opening was not determined with any certainty. In several specimens, a slit is present, at right angles to an axis of the triangular shell. In some instances, this is of the "letter-box flap" type originally recorded in *Veryhachium* by Calandra (see Sarjeant, 1967, p. 203); this is illustrated in Fig. 5d,e,f. In other cases, a more irregular slit is present in transverse position (Pl. 1, fig. 1; Fig. 5a); and, in at least one instance, the slit is only visible marginally (Fig. 5b). In the majority of specimens, no opening is visible: this might be the result of the complete closing of a flap-like aperture, but the present evidence does not permit confidence.

The stratigraphic range of *V. valensii* certainly extends from Permian to Upper Jurassic (Callovian). Its presence in the Upper Carboniferous also appears highly probable, since Spode (1964, p. 362), although he does not quote the dimensions of his forms, comments that they are "morphologically identical" with Jekhowsky's forms. An isolated record from Tertiary sediments of Assam (Baksi, 1962) of forms closely similar in morphology and dimensions (long axis 22μ, spines 6-8μ) to *Veryhachium valensii*, must at present be viewed with doubt. Baksi's genus *Simsangia* is a subjective junior synonym of *Veryhachium,* whose type-species, *V. trisulcum* Deunff, 1958, has a subtriangular shell bearing three spines.

Forms of *Veryhachium* having a roughly triangular shell, with three simple spines, are a common element in Paleozoic assemblages. The earliest forms appear in the Ordovician: *V. cucruse* Timofeev, 1962, *V. trisulcum* Deunff (1951), 1958, and *V. piliferum* Martin, 1966. *V. delmeri* Stockmans and Willière, 1963, *V. exile* Timofeev, 1962, *V. geometricum* (Deflandre 1942) Stockmans and Willière, 1963, *V. limaciforme* Stockmans and Willière 1963, *V. trispinosum* (Eisenack, 1938) Deunff, 1958, and *V. wenlockium* (Downie 1959) Stockmans and Willière, 1963, are Silurian species, whilst *V. cochinum* Cramer, 1964b, *V. downiei* Stockmans and Willière, 1963, *V. libratum* Deunff (1957), 1958, and *V. helenae* Cramer, 1964b, range from Silurian to Devonian. In the Devonian, five further species are present: *V. asymmetricum* Deunff (1954), 1958, *V. brevitrispinosum* Stapin, 1961, *V. centrigerum* Deunff (1957), 1958, *V. aquila* Deunff, 1966, and *V. trispininflatum* Cramer, 1964b.

Veryhachium irregulare Jekhowsky, 1961

Veryhachium irregulare Jekh. forma *subtetraedron* Jekhowsky, 1961

Fig. 7b; Pl. 2, fig. 15

Veryhachium irregulare forma *subtetraedron* Jekhowsky, 1961, p. 208, pl. 1, fig. 4-9.
V. irregulare forma *subtetraedron* Jekh., Downie and Sarjeant, 1963, p. 94.
Veryhachium hyalodermum (Cookson) Schaarschmidt, 1963, p. 62-3, pl. 17, fig. 1-7; Fig. 12.
V. irregulare forma *subtetraedron* Jekh., Wall and Downie, 1963, p. 781, pl. 113, fig. 7; Fig. 1.
V. irregulare forma *subtetraedron* Jekh., Downie and Sarjeant, 1964, p. 151.
V. irregulare forma *subtetraedron* Jekh., Wall, 1965, p. 160, pl. 3, fig. 21.
Veryhachium reductum Deunff (pars). Medd, 1966, p. 353, fig. 1a, ?1b-c.
V. hyalodermum (Cookson) Schaarschm., Schön, 1967, p. 527, fig. 24.

Figured Specimen: 1,800,361 (Pl. 2, fig. 15; Fig. 7b).

Dimensions: Figured specimens:

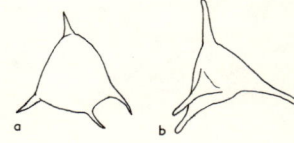

FIG. 7. a, *Wilsonastrum colonicum* Jansonius; b, *Veryhachium irregulare* forma *subtetraedron* Jekhowsky.

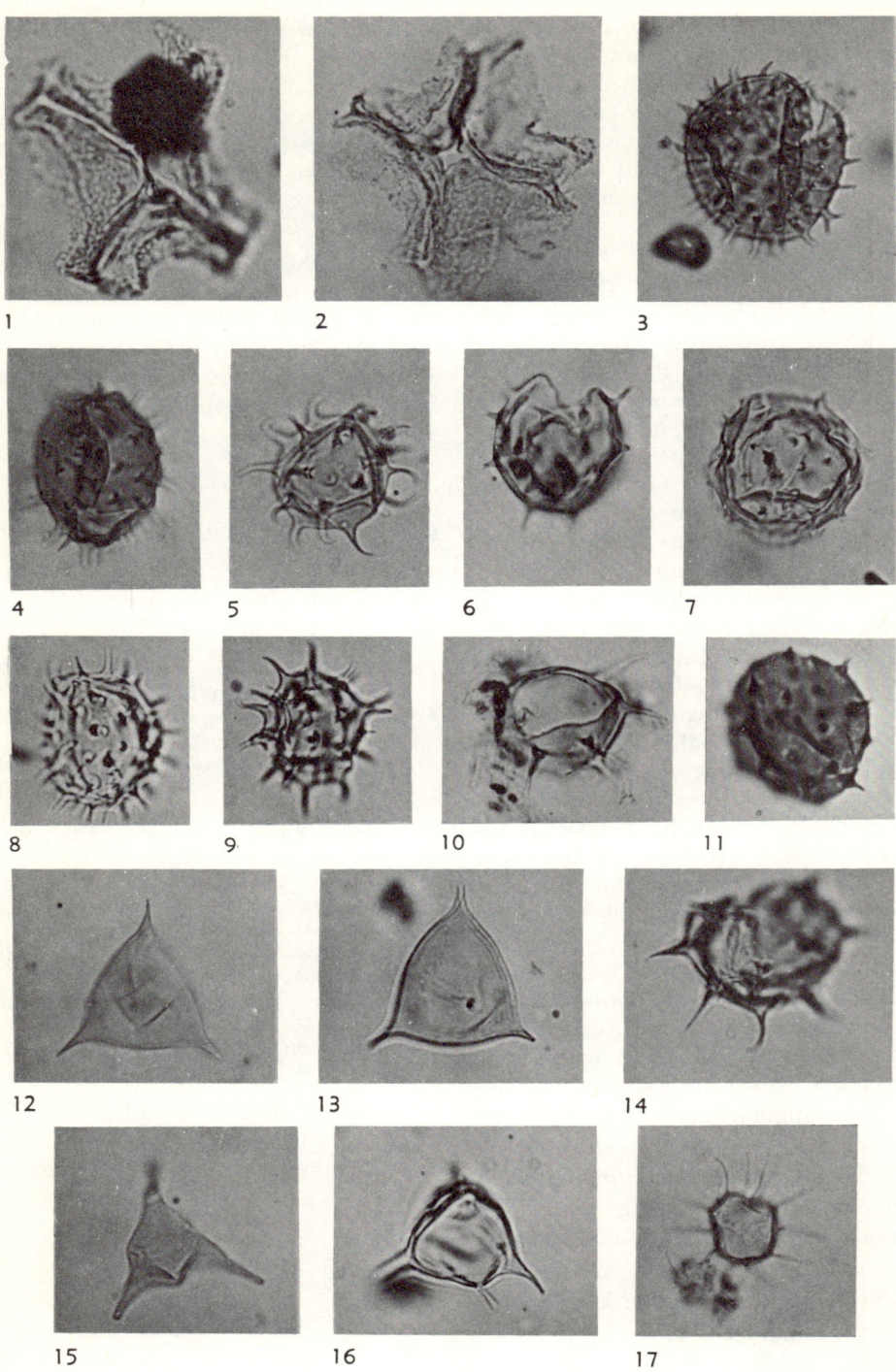

Plate 2

length of sides of shell 11μ, spines c.7μ. Second specimen; length of sides of shell 16μ, spines 10.5μ. Two specimens measured.

Diagnosis: "Central body subtetrahedral, convex, four appendages (one at each 'summit')" (Transl.). [Note: The species diagnosis stresses the small size ('diameter' between 10 and 20μ, spines 5 to 15μ long) and the hollow nature of the spines.]

Discussion: The species *V. irregulare* was recognized by Jekhowsky to be "very polymorphic" and he distinguished four principal "forms." Only one of these forms is present in the Pakistan Permian assemblage studied, represented by very few specimens. The spines in some instances have acute extremities, but the specimen illustrated (Pl. 1, fig. 15; Fig. 7b) has spines with blunt tips. Medd (1966, fig. 1) figures three types of small *Veryhachium*, all of which he places in *V. reductum* (=*V. valensii* as redefined herein). On morphological grounds, the first of these (1a) is definitely *V. irregulare* f. *subtetraedron*; the second and third only possibly so (his specimen 1b is probably *V. europaeum* Stockmans and Willière, 1960). Whether or not such forms are within the morphological plexus centered on *V. valensii* is arguable (see discussion under *Wilsonastrum colonicum*).

The known stratigraphic range for this form is Permian to Lower Jurassic. A closely related species (possibly synonymous), *Veryhachium tetraxis* (Sarjeant, 1960) Downie and Sarjeant, 1964, is present in the Upper Jurassic. *Veryhachium hyalodermum* (Cookson, 1956) Schaarschmidt, 1963, has much longer spines in proportion to shell size. It was described from the Cretaceous, but has since been recorded from the Upper Jurassic (Gitmez, 1969). Alberti (1961, p. 32-33) attributes both Cretaceous and Middle Jurassic specimens to this species; the latter are not separately described or illustrated and may well be specimens either of *V. tetraxis* or of *V. irregulare* f. *subtetraedron*. Wall (1965) has also recorded the species *V. tetraedron* Stockmans and Willière, 1960, in the Lower Jurassic.

In the Paleozoic, species of *Veryhachium* with tetrahedral shells bearing 4 simple spines appear first in the Upper Silurian, with the species *V. scabratum* Cramer, 1964b (ranging into the Lower Devonian). Four other species are also represented in the Devonian: *V. europaeum* Stockmans and Willière, 1960, *V. inflatissimum* Cramer, 1964, *V. legrandi* Stockmans and Willière, 1962a, and *V. tetraedron* Deunff, 1954.

PLATE 2

All photographs taken by reflected light. Magnification ×1000.

FIGURES

1-2. *Polyedryxium* sp.
3. *Micrhystridium pakistanense* sp. nov.
4. *M.* aff. *keratoides* Spode.
5. *M. breve* Jansonius.
6. *M. karamurzae* sp. nov.
7. *M. microspinosum* Schaarschmidt.
8. *M.* aff. *keratoides* Spode.
9. *M. breve* Jansonius.
10. *M. circulum* Schön.
11. *M. microspinosum* Schaarschmidt.
12-13. *Veryhachium valensii* (Valensi) Downie and Sarjeant, emend.
14. *Micrhystridium circulum* Schön.
15. *Veryhachium irregulare* forma *subtetraedron* Jekhowsky.
16. *Wilsonastrum colonicum* Jansonius.
17. *Micrhystridium inconspicuum* (Deflandre) Deflandre.

Veryhachium? riburgense Brosius and Bitterli, 1961

Fig. 8; Pl. 1, fig. 10, 11, 14, 15, 18, 19

Veryhachium? riburgense Brosius and Bitterli, 1961, p. 39, pl. 2, fig. 7-12; Fig. 4a-d, 5a-d.
V. riburgense Bros. and Bitt. (pars) [sic], Downie and Sarjeant, 1963, p. 94.
Veryhachium cf. *nasicum* (Stockmans and Willière), Schaarschmidt, 1963, p. 63, pl. 17, fig. 11-14; Fig. 14.
Polyedryxium kraeuselianum, Schaarschmidt, 1963, p. 65, pl. 17, fig. 16, 17, pl. 18, fig. 1-3; Fig. 19.
Veryhachium sedecimspinosum Staplin, Schaarschmidt, 1963, p. 63-64, pl. 19, fig. 2, 4, 5; Fig. 15.
?*V. riburgense* Bros. and Bitt., Downie and Sarjeant, 1964, p. 152.
V. nasicum (Stock and Will.), Schaarschm., Schön, 1967, p. 527, fig. 1.
V. cf. *nasicum* (Stock and Will.) Schaarschm., Schön, 1967, p. 527, fig. 1.

Figured Specimens: i. 1,800,362 (Pl. 1, fig. 10; Fig. 8c). ii. 1,800,363 (Pl. 1, fig. 11; Fig. 8g). iii. 1,800,364 (Pl. 1, fig. 14; Fig. 8e). iv. 1,800,365 (Pl. 1, fig. 15; Fig. 8d). v. 1,800,366 (Pl. 1, fig. 18; Fig. 8a, b). vi. 1,800,367 (Pl. 1, fig. 19; Fig. 8f).

Dimensions: Figured specimens: i. Shell cross-measurement c.13μ, spines up to 12μ. ii. Shell c.$16 \times 15\mu$ in cross-measurement, spines up to 5μ. iii. Shell cross-meaurement c.14μ, spines up to 11μ. iv. Shell cross-measurement c.20μ, spines up to 8μ. v. Shell c.$19 \times 16\mu$ in cross-measurement, spines up to 9μ. Overall range: shell cross-measurement $13\text{-}22\mu$ (mean 17.5μ), max. spine length $2\text{-}11\mu$ (mean 7μ). Measurements, as taken, accurate to within 0.5μ. Ten specimens measured.

Diagnosis: "The test is polygonal, often folded, with 6 to 12 (normally 9) acute-angled processes merging with a widened base into the test." The authors distinguish two variants:

i. forma *regulare*. "The polygonal test bears broad-based spines of which three together always seem to form an equilateral triangle."
ii. forma *irregulare*. "The test is pillow-shaped, polygonal. The processes are arranged irregularly and are normally shorter and wider at their base than those of the forma *regulare*."

They note the existence of transitional stages between these forms.

Discussion: In considering forms showing a degree of variation such as is exhibited by the representatives of this species, two approaches can be taken; to distinguish a number of types, on the basis of such features as spine number or shell shape or proportions, or to recognize that one is dealing with a morphological complex, which cannot be meaningfully subdivided at any level higher than that of subspecies. These contrasting approaches have been adopted by Schaarschmidt (1963) and by Brosius and Bitterli (1961). The former author attempted to subdivide the complex, placing its components into three species classed into two different genera; his allocation of one species to *Polyedryxium* is unacceptable, for it is morphologically wholly dissimilar from the species with prismatic shells which characterize the genus and what he took to be "swellings" *(Wülste)*

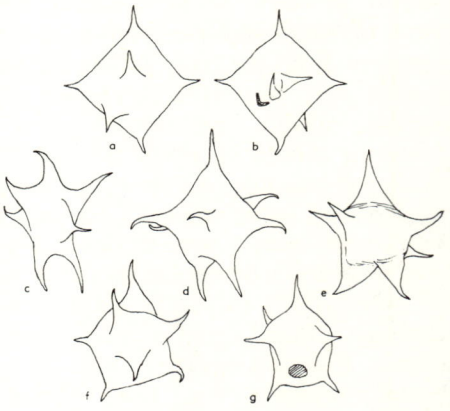

FIG. 8. The range of morphology in *Veryhachium? riburgense* Brosius and Bitterli. Shell openings are shaded. (In specimens f and g, one spine is completely concealed on the lower surface.)

linking the processes are surely only folds. Brosius and Bitterli, in contrast, found themselves unable to make meaningful subdivisions and placed all their specimens into a single species, whilst recognizing that there was considerable variation within it. They were, however, concerned regarding the matter of generic allocation, noting that, whereas some variants had distinctly angular shells and could be unhesitatingly allocated to *Veryhachium*, others had rounded shells and might be assigned to *Micrhystridium* or *Baltisphaeridium*.

The Pakistan forms fall within the *?V. riburgense* plexus. Apparent shape is clearly controlled by three factors: aspect, process number and relative development, and diagenetic history. Extreme members (e.g., Fig. 8a,b and 8e) would, in isolation, be considered to represent wholly different species; but the assemblage, considered as a whole, defies such subdivisions except on arbitrary criteria. Spine numbers might be used, forms with 7, 8 and 9 spines being present: but there is a variation in spine development on individual specimens, with particular spines shorter or less massive than the others (e.g., Fig. 8b,c), so that this appears unsatisfactory. It is frequently difficult to count spine numbers, because of unfavorable aspect. I have therefore followed the procedure of Brosius and Bitterli and classed all variants within the single species: though it should be noted that, of 20 specimens capable of close examination, 12 had seven spines, 7 had eight spines and only 1 was recognized to have 9 spines, whereas the latter number was typical for their Triassic assemblage. Both variants were represented (forma *regulare*, Fig. 8d-g; forma *irregulare*, Fig. 8a-c), though it was not always found easy to distinguish them.

Comparable Paleozoic species include *Veryhachium balticum* (Eisenack, 1951) Eisenack, 1959, from the Ordovician of the Baltic region, and two species from the Upper Devonian of Belgium, *V. belgicum* (Stockmans and Willière, 1960) Stockmans and Willière, 1960a, and *V. vandenbergheni* Stockmans and Willière, 1962b. Some of the species attributed to *M. stellatum* Deflandre, 1942, by Cramer (1964b), from the Spanish Upper Silurian and Devonian, and by Wall and Downie (1963) and Wall (1965), respectively from the Permian and the Lower Jurassic of England, are of subpolygonal shell shape and resemble *?V. riburgense*. The Permian forms attributed to the Silurian species *V. rhomboidium* Downie, 1959, by Wall and Downie, 1963, are likewise closely comparable to *V. ?riburgense*, differing only in greater spine length. (The group of forms which these authors designate "the *V. ?irregulare* complex" is surely ancestral to *V. ?riburgense*.) The species *Veryhachium aster* Sarjeant, 1967, from the Middle to Upper Jurassic of France and England, is likewise similar in morphology.

Genus WILSONASTRUM
Jansonius, 1962

Wilsonastrum colonicum Jansonius, 1962

Fig. 7a; Pl. 2, fig. 16

Wilsonastrum colonicum Jansonius, 1962, p. 89, pl. 16, fig. 42-49, 58; Fig. 3g.
W. colonicum Jansonius, Downie and Sarjeant, 1963, p. 155.
Veryhachium sp. Wall and Downie, 1963, p. 773, fig. 2; Pl. 114, fig. 12.
W. colonicum Jans., Norris and Sarjeant, 1965, p. 63.
W. colonicum Jans., Staplin, Jansonius, and Pocock, 1965, p. 184-5.
W. colonicum Jans., Medd, 1966, p. 353-354.

Figured Specimen: 1,800,368 (Pl. 2, fig. 16; Fig. 7a).

Dimensions: Figured specimen: length of sides of shell c.12μ, spines at angles 6μ, adventitious spine 3.5μ. Overall range: length of sides of shell 12-15μ (mean 13μ), spines at angles 4.5-6μ (mean 5.5μ), adventitious spine

3.5-4μ (mean 3.75μ). Measurements, as taken, accurate to within 0.5μ. Five specimens measured.

Diagnosis: "Vesicle triangular sublenticular to tetrahedral; wall laevigate, moderately thin, but secondary folding not common; corners drawn out into tapering arms or spines, usually three, occasionally four and rarely five in number; spines may be pointed but usually are gently tapering at the base, grading to cylindrical and rounded at the tip; approximately near the center of the face a local thickening or small fold, marking the position of a thin bristle which may be present (in our material in 20 percent of the individuals): the bristle is very diaphanous and oil immersion is normally needed for observation: size of vesicle 16 (25)28μ, of spines 3 (11)16μ x1 (1½)2μ, usually all of same size in each individual: bristle 5x½μ."

Discussion: This genus, with its single species, is characterized by the possession of the "bristle"; in all other respects it is indistinguishable from *Veryhachium*. The author (p. 88-89) notes the absence of the bristle in 80 percent of specimens; distinction is then to be made, presumably, on the presence of the "local thickening or small fold"—a most difficult feature to confirm with any certainty.

On the basis of study of the Pakistan specimens and of Jansonius' figures, I believe that what he terms a "flagellum or bristle" (p. 88) is, in fact, a small spine, hollow at the base but possibly solid towards its tip, markedly less long than the spines at the angles and developed without modification of the shell shape. Without this additional spine, which I here term "adventitious," the specimens examined (all of which were of triangular shape and had spines at the angles) would be indistinguishable from *Veryhachium valensii* (Valensi) Downie and Sarjeant, 1964. I believe it probable that the 80 percent of Canadian and Australian Permian specimens, which lacked this special feature, would be equally capable of assignation to the latter species.

It is arguable that *"W. colonicum"* represents an intermediate stage between three-spined, triangular forms *(V. valensii)* and four-spined tetrahedral forms *(V. irregulare* f. *subtetraedron* Jekhowsky, 1961), in which the fourth spine has appeared but is insufficiently developed to produce any modification of shell shape. (The variable degree of development of the different spines in another species, *?V. riburgense* Brosius and Bitterli, 1961, has already been noted.) If this is indeed the case, then a separate generic status for *colonicum* is scarcely merited.

Subgroup **PRISMATOMORPHITAE**
Downie, Evitt, and Sarjeant, 1963

Genus **POLYEDRYXIUM** Deunff, 1954

Polyedryxium sp.

Fig. 9; Pl. 2, fig. 1-2

Figured Specimens: i. 1,800,369 (Pl. 2, fig. 1; Fig. 9). ii. 1,800,370 (Pl. 2, fig. 2).

Dimensions: Figured specimens: i. Length (measured parallel to the central axis of the shell) 35μ: breadth (at right angles to this direction) 28μ. ii. Length 30μ, breadth 30μ. Only two specimens encountered.

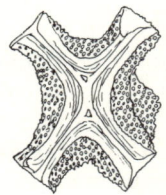

Fig. 9. *Polyedryxium* sp.

Description: A species of *Polyedryxium* having a cruciform shell (+ or ×), with or without an opening in median position, the arms of the cross being linked by a perforate, aliform membrane.

The two specimens found were dissimilar in shape. Specimen i. had the apperance of a St. Andrew's Cross with expanded tips; the surrounding mesh was slightly torn and an opening (probably also a tear) was present at the tip of one arm (Fig. 9). There are also two small openings in the central shell axis. Specimen ii., in contrast, had the form of a regular St. George's cross, with a quadrate central opening (pylome) clearly developed. It is possible that this is the true shape and that the X-shape results from distortion, which has also produced a closing-up of the central part of the opening, leaving only two smaller openings.

The morphology is difficult to describe with precision, for want of any established terminology. Each arm of the cross has a central raised axis, widening, almost like a Maltese cross, at the tip. The sides of the arms have a concave slope, down to the position of the "wing"; this is two-layered, formed by the folding-over of a single membrane, and is densely perforate.

Discussion: Although certainly assignable to the genus *Polyedryxium*, the specimens here described do not accord with any existing species. The closest analogy is with *P. deflandrei* Deunff, 1954, from the Devonian; but this has arms branching at the tips and a surrounding "wing" with median outgrowths. We are clearly here concerned with a new species; but two specimens, of such widely different character, are certainly insufficient to serve as basis for a new taxon; further specimens must be awaited.

Subgroup NETROMORPHITAE
Downie, Evitt, and Sarjeant 1963

Genus LEIOFUSA Eisenack, 1938, emend.
Combaz, Lange, and Pansart, 1967

Leiofusa stassfurtensis Schön, 1967
Fig. 10b; Pl. 1, fig. 9

Cyst of Pyrrhophyta, Kara-Murza, 1957, pl. 1, fig. 28.
Leiofusa stassfurtensis, Schön, 1967, p. 531, pl. 1, fig. 8-14; Fig. 1, nos. 10-12; Fig. 2.

Figured Specimen: 1,800,371 (Pl. 1, fig. 9; Fig. 10b).

Dimensions: Figured specimen: Total length 28μ, shell length 16μ, breadth 11μ, processes c.6μ. Overall range: total length 26-32μ (mean 28.70μ), shell length 15-16μ (mean 15.75μ), breadth 8-11μ (mean 9.50μ), processes 6-8μ (mean 6.5μ). Dimensions, as measured, accurate to 0.5μ. 4 specimens measured.

Diagnosis: "A short and stout species of *Leiofusa,* with two mutually opposite thorn-like processes. The relationship of length to breadth amounts on average to 1-3, the long axis measures on average 21μ and the processes 3μ" (Transl.).

Discussion: The length of the processes of the Pakistan specimens is consistently at the upper end of the range of 1-6μ quoted by Schön (p. 531): in all other respects, they accord exactly with his description. There is a suggestion of a flap-like opening, like that of some *Veryhachium* species (seen midway between the median point and lower end of the shell in Fig. 10b). A circular central pylome, such as appears typical for the genus (see Combaz, Lange, and Pansart, 1967), was not seen in this material and was not recorded by Schön.

FIG. 10. a, *Deunffia unispinosa* (Schön), comb. nov.; b, *Leiofusa stassfurtensis* Schön.

A specimen illustrated from the Russian Triassic by Kara-Murza (1957) appears attributable to this species.

Five Paleozoic species share with *L. stassfurtensis* the character of a well-marked central body and two axial processes: these are *L. tumida* Downie, 1959, from the English Silurian; the Devonian species *L. pumilia* Deunff, 1966, from Tunisia; *L. cantabrica*

Cramer, 1964a, *L. bernesgra* Cramer, 1964a, and *L. banderilla* Cramer, 1964a, from Spain. However, stronger morphological analogies exist with two Mesozoic species: *Leiofusa jurassica* Cookson and Eisenack, 1958 (recorded also from the Triassic by Wall and Downie, 1963) and *L. lidiae* Górka, 1963, from the Upper Cretaceous of Poland.

Genus DEUNFFIA Downie, 1960

Deunffia unispinosa (Schön, 1967) comb. nov.

Fig. 10a; Pl. 1, fig. 6, 16

Hystrichosphaerid?, Jansonius, 1962, p. 87, pl. 16, fig. 68, ?67.
Leiofusa unispinosa, Schön, 1967, pl. 1, fig. 6-7; Fig. 1, no. 9.

Figured Specimen: 1,800,372 (Pl. 1, fig. 6, 16; Fig. 10a).

Dimensions: Figured specimen: Total length 23μ, shell alone 20μ, max. breadth 18.5μ, spine 3μ. Other two specimens set obliquely: approx. dimensions: overall length 20μ and 20μ, shell length 14μ and 15μ; max. breadth 12μ and 12μ: spine 6μ and 5μ. Three specimens seen.

Diagnosis: "A species . . . with round or oval body-outline and only one short process. The relationship of length and breadth amounts on average to 1 to 2" (Transl.).

Discussion: In view of its possession of only a single process, this species is here reassigned to the genus *Deunffia*, since the emended diagnosis of *Leiofusa* specifies that an appendage must be present at each of the two shell extremities (Combaz, Lange, and Pansart, 1967, p. 297).

Of two forms figured by Jansonius, at least one (1962, pl. 16, fig. 68) appears attributable to this species: the attribution of the second (*ibid.*, pl. 16, fig. 67) to this species is less clear. (The age of these forms, i.e., whether Permian or Triassic, is not made clear.)

The best displayed of the three Pakistan specimens, here figured, is strikingly pear-like in appearance, this apparently resulting from folding of the wall (cf. Pl. 1, fig. 6). The others, set obliquely, were more exactly typical.

The only closely similar described species is *D. brevispinosa* Downie, 1960, from the English Silurian.

Class PRASINOPHYCEAE

Family TASMANACEAE Sommer, 1956

Genus TASMANITES Newton, 1875

Tasmanites sp.

Pl. 1, fig. 8, 12

Figured Specimen: 1,800,373 (Pl. 1, fig. 8, 12).

Dimensions: Figured specimen: Maximum diameter (specimen folded) 26μ. Second specimen; diameter 22.5μ.

Description: A species of *Tasmanites* of small size, with a pattern of regularly spaced micropores (the presence of ultrapores, detectable only by electron microscopy, could not be confirmed). Wall of moderate thickness, folded. No opening detected.

Discussion: The small size of these forms is remarkable, for tasmanitids frequently measure several hundred microns in diameter. (*T. minutus* Eisenack, 1965, although distinguishable largely by its small size, is nonetheless several times larger than these forms—50 to 70μ). They may well represent an undescribed species, but more abundant material is necessary before this can be established.

CONCLUSIONS

Two striking features emerge from the examination of this uppermost Permian acritarch assemblage. First of all, the size of the species represented in the 3-foot and 2-foot horizons is uniformly quite small and their morphology is generally quite simple. In his study of Lower Jurassic assemblages, Wall (1965) suggested that members of the Acanthomorphitae (*Micrhystridium* in particular) favored an inshore, partly enclosed environment, whereas the Polygonomorphitae *(Veryhachium)* and Netromorphitae *(Leiofusa* etc.) favored an open-sea environment. On this basis, the assemblage here described would suggest an intermediate environment—not far offshore, but not, apparently, enclosed. It is therefore interesting to note that Kummel and Teichert (1966, p. 324) consider the Chhidru Formation to be "a deposit of a type usually associated with relatively shallow-water shelf environments where conditions favorable to moderate carbonate deposition existed and varying amounts of detrital quartz were added from external sources." The topmost assemblage examined (1-foot horizon) was virtually barren of microplankton: at this level, Kummel and Teichert (*ibid.*, p. 325) note a sudden change in conditions, producing a flood of detrital quartz and the extinction of many lifeforms.

The second striking feature is that this is very definitely an intermediate assemblage, with characteristics both of Permian and of Triassic. Nine species were previously recorded from both systems: three were hitherto known only from Mesozoic deposits and one only from the Permian (a third, *Polyedryxium* sp., is strikingly Paleozoic in character). The remaining three species are new. The most similar described assemblages are those described from Canada by Jansonius (1962), dated as Middle Permian (Leonardian) to Lower Triassic (Anisian); the assemblage described by Schaarschmidt (1963) from the Zechstein (Upper Permian) of Germany; and those from the Bunter (Lower Triassic, Scythian) of Germany, described by Schön (1967), from the Triassic of Arctic Russia by Kara-Murza (1957) and from the Middle Triassic of Switzerland by Brosius and Bitterli (1961). None of these assemblages affords an exact parallel in age, however: all are slightly, or considerably, older or younger. In each case the overlap in species content is only partial: they all contain numerous elements not present in the Pakistan assemblages. The pronounced dissimilarity of the English assemblage of early Late Permian (Lower Zechstein) age assemblage, described by Wall and Downie (1963), may be remarked: and there were no elements in common with Tasch's (1963a) Middle Permian assemblage, which may well represent an entirely different environment.

The assemblage thus affords unexpected promise that it might be possible to differentiate broad stratigraphic zones in the Permian and Triassic, from study of the acritarchs. The degree of environmental control, however, remains to be assessed.

ACKNOWLEDGMENTS

The author would like to thank Dr. Curt Teichert (University of Kansas) and Dr. Basil E. Balme (University of Western Australia) for making these assemblages available for study. He would also like to acknowledge helpful discussions of data with Dr. Peter Harvey and Dr. Ian D. Sutton (University of Nottingham) and the critical reading of the manuscript by Dr. Charles Downie (University of Sheffield).

REFERENCES

Alberti, G., 1961, Zur Kenntnis mesozoischer und alttertiärer Dinoflagellaten und Hystrichosphaerideen von Nord—und Mitteldeutschland sowie einigen anderen europäischen Gebieten: Palaeontographica, Ser. A, v. 116, p. 1-58, pl. 1-12.

Baksi, S. K., 1962, Palynological investigation of Simsang River Tertiaries, South Shillong Front, Assam: Bull. Geol. Mining Metall. Soc. India, no. 26, p. 1-22, pl. 1-5.

Balme, B. E., and Playford, G., 1967, Late Permian plant microfossils from the Prince Charles Mountains, Antarctica: Rev. Micropaléontologie, v. 10, p. 179-192, pl. 1-2.

Brosius, M., and Bitterli, P., 1961, Middle Triassic hystrichosphaerids from Saltwells Riburg -15 and -17: Bull. Ver. Schweizer. Petrol.-Geol. Ing., v. 28, p. 33-49, pl. 1-2.

Chornaya, O., 1963, Spory, pyltsa i mikroplankton iz Mezozoya Zapadnykh Karpat (Spores, pollen and microplankton from the Mesozoic of western Carpathians): Geol. Sbornik, v. 14, p. 283-286, pl. 5-8.

Combaz, A., Lange, F. W., and Pansart, J., 1967, Les "Leiofusidae" Eisenack, 1938: Rev. Palébot. Palynologie, v. 1, p. 291-307, pl. 1-2.

Cookson, I. C., 1956, Additional microplankton from Australian late Mesozoic and Tertiary sediments: Austral. Jour. Mar. Freshw. Research, v. 7, p. 183-191, pl. 1-2.

——, and Eisenack, A., 1958, Microplankton from Australian and New Guinea Upper Mesozoic sediments: Proc. Roy. Soc. Victoria, v. 70, p. 19-79, pl. 1-2.

Cramer, F. H., 1964a, Some acritarchs from the San Pedro Formation of the Cantabric Mountains in Spain: Soc. Géologie Hydrologie Belge, Bull., v. 73, p. 33-38.

——, 1964b, Microplankton from three Palaeozoic formations in the province of Léon (N.W. Spain): Leidse Geol. Meded., v. 30, p. 253-361, pl. 1-24.

Deflandre, G., 1935, Considerations biologiques sur les micro-organismes d'origine planctonique conservés dans les silex de la craie: Bull. Biol. France Belgique, v. 69, p. 213-244, pl. 5-9.

——, 1937, Microfossiles des silex crétacés II. Flagellés incertae sedis. Hystrichosphaeridées. Sarcodinés. Organismes divers: Ann. Paléontologie, v. 26, p. 51-103, pl. 8-18.

——, 1942, Sur les hystrichosphères des calcaires siluriens de la Montagne Noire: C. R. Acad. Sci. [Paris], v. 215, p. 475-476.

——, 1945, Microfossiles des calcaires siluriens de la Montagne Noire: Ann. Paléontologie, v. 31, p. 41-76, pl. 1-3.

——, 1946, Radiolaires et Hystrichosphaeridés du Carbonifère de la Montagne Noire: C. R. Acad. Sci. [Paris], v. 233, p. 515-517.

——, 1947, Sur quelques micro-organismes planctoniques des silex jurassiques: Bull. Inst. Oceanographie Monaco., no. 921, p. 1-10.

——, 1952, Protistes. Généralités. Sousembranchement des Flagellés. Groupes incertae sedis: in Piveteau, J., ed., Traité de Paléontologie, I. Masson, Paris, p. 89-95, 99-130.

Deflandre, G., and Cookson, I. C., 1955, Fossil microplankton from Australian late Mesozoic and Tertiary sediments: Austral. Jour. Mar. and Freshw. Research, v. 6, p. 242-313, pl. 1-9.

——, and Deflandre, M., 1965a, Remarques critiques sur le genre *Micrhystridium* Defl. Paris: Multicop. Lab. Micropaléont. E. P. H. E., 9 p.

——, and ——, 1965b, Fichier micropaléontologique général. Série 13, Acritarches II. Acanthomorphitae I. Genre *Micrhystridium* Deflandre sens. lat.: Paris, Centre Natl. Rech. Scient., p. I-V: fiches 2176-2521.

Deunff, J., 1951, Sur la présence de microorganismes (Hystrichosphères) dans les schistes ordoviciens du Finistère: C. R. Acad. Sci. [Paris], v. 233, p. 321-3.

——, 1954, *Veryhachium,* genre nouveau d'Hystrichosphères du Primaire: C. R. Somm. Soc. Géol. France, no. 13, p. 305-306.

——, 1955, Un microplankton fossile dévonien à Hystrichosphères du Continent Nord-Americain: Bull. Microscopie Appliqué, sér. 2, v. 5, p. 138-147, pl. 1-4.

——, 1957, Microorganismes nouveaux (Hystrichospères) du Dévonien de l'Amerique du Nord: Soc. Géol. Mineral. Bretagne, Bull., v. 2, p. 5-14.

——, 1958, Micro-organismes planctoniques du Primaire armoricain I. Ordovicien du Veryhac'h (Presqu' Ile de Crozon): Bull. Soc. Géol. Minéral. Bretagne, new ser., v. 2, p. 1-141, pl. 1-12.

——, 1966, Recherches sur les microplanctons du Dévonien (Acritarches et Dinophyceae): Rennes, The Author, 168 p., 14 pl.

——, 1968, Sur une forme nouvelle d'Acritarche possédent une ouverture polaire (*Veryhachium miloni* n. sp.) et sur la presence d'une colonie de *Veryhachium* dans le Trémadocien marocain: C. R. Acad. Sci. [Paris], sér. D, v. 267, p. 46-49.

Downie, C., 1958, An assemblage of microplankton from the Shineton Shales (Tremadocian): Yorkshire Geol. Soc., Proc., v. 31, p. 331-349, pl. 16-17.

——, 1959, Hystrichospheres from the Silurian Wenlock Shale of England: Palaeontology, v. 2, p. 36-71, pl. 10-12.

——, 1960, *Deunffia* and *Domasia,* new genera of Hystrichospheres: Micropalaeontology, 6, p. 197-202, pl. 1.

——, Evitt, W. R., and Sarjeant, W. A. S., 1963, Dinoflagellates, hystrichospheres and the classification of the acritarchs: Stanford Univ. Publ., Geol. Sci., v. 7, p. 1-16.

———, and Sarjeant, W. A. S., 1963, On the interpretation and status of some hystrichosphere genera: Palaeontology, 6, p. 83-96.

———, and ———, 1964, Bibliography and index of fossil dinoflagellates and acritarchs: Geol. Soc. Amer. Mem., no. 94, p. 1-180.

Efremova, G. D., 1967, K voprosu o sistematicheskom polozhenii nekotorykh mikrofossiliy Permskikh otlozheniy (Concerning the question of the systematic positioning of certain Permian microfossil deposits): in T. F. Vozzhennikova et al., eds., Iskopayemoye Vodorosli SSSR, Moscow, Akad. Nauk SSSR, Sibirskoe Otdel., Inst. Geol. Geofiz., 148 p., 30 pl. (p. 108-11, pl. 23-4).

Eisenack, A., 1938, Hystrichosphaerideen und verwandte Formen im baltischen Silur: Zeitsch. Geschiebeforschung, v. 14, p. 1-30, pl. 1-4.

———, 1951, Uber Hystrichosphaerideen und andere Kleinformen aus baltischem Silur und Kambrium: Senckenbergiana, v. 32, pl. 187-204, pl. 1-4.

———, 1959, Neotypen baltischer Silur-Hystrichosphären und neue Arten: Palaeontographica, ser. A, v. 112, p. 193-211, pl. 15-17.

———, 1965, Die Mikrofauna der Ostseekalke. I. Chitinozoen, Hystrichosphären: Neues Jahrb. Geologie Paläontologie, pl. 9-13.

Firtion, F., 1952, Le Cenomanien inférieur du Nouvion-en-Thiérache: examen micropaléontologique: Soc. Géol. Nord., Ann., v. 72, p. 150, 164, pl. 8-10.

Gitmez, G. U., 1969, Dinoflagellate cysts and acritarchs from the basal Kimmeridgian (Upper Jurassic) of England, Scotland and France: Bull. Brit. Mus. (Nat. Hist.), Geology, v. 18, p. 231-330, pl. 1-14.

Gocht, H., 1964, Planktonische Kleinformen aus dem Lias/Dogger Grenzbereich Nord- und Süddeutschlands: Neues Jahrb. Geologie Paläontologie, Abh., v. 119, p. 113-133.

Górka, H., 1963, Coccolithophoridées, dinoflagellés, hystrichosphaeridés et microfossiles incertae sedis du Crétacé supérieur de Pologne: Acta Palaeont. Polon., v. 8, p. 33-90, pl. 1-11.

Isagulova, E., 1963, Gistrikhosfery v yurskikh otlozheniyakh Lvovsko-Volynskogo Kamennougol'nogo Basseyna (Hystrichosphaerids in Jurassic deposits of the Lvov-Volyn coal-bearing basin): Akad. Nauk SSSR, Dokl., v. 148, p. 1156-1158.

Jansonius, J., 1962, Palynology of Permian and Triassic sediments, Peace River area, Western Canada: Palaeontographica, ser. B, v. 110, p. 35-98, pl. 11-16.

Jekhowsky, B. de, 1961, Sur quelques hystrichosphères permo-triasiques d'Europe et d'Afrique: Rev. Micropaléontologie, v. 3, p. 207-212, pl. 1-2.

Kara-Murza, E. N., 1957, Verkhnemelovye i triasovye Hystrichosphaeridae Sovetskoi Arktiki (Upper Cretaceous and Triassic Hystrichosphaeridae of the Soviet Arctic): Nauch.-Issled. Inst. Geol. Arktiki, Sbornik Statei Paleont. i Biostratig., v. 4, p. 64-68.

Kummel, B., and Teichert, C., 1966, Relations between the Permian and Triassic formations in the Salt Range and Trans-Indus ranges, West Pakistan: Neues Jahrb. Geologie Paläontologie, Abh., v. 125 (Festband Schindewolf), p. 297-333, pl. 27, 28.

——— and ———, 1970, Stratigraphy and paleontology of the Permian-Triassic boundary beds, Salt Range and Trans-Indus ranges, West Pakistan: in Kummel, B., and Teichert, C., eds., Stratigraphic Boundary Problems: Permian and Triassic of West Pakistan: Univ. Press Kansas, Dept. Geology, Univ. Kansas Spec. Publ. 4, p. 1-110.

Martin, F., 1966, Les acritarches du sondage de la brasserie Lust, à Kortrijk (Courtrai), (Silurien belge): Soc. Belge Géologie, Bull., v. 74, p. 1-47, pl. 1.

Medd, A., 1966, The fine structure of some Lower Triassic acritarchs: Palaeontology, v. 9, p. 351-354, pl. 59.

Medvedeva, A. M., and Chepilova, I. K., 1961, Protoleiosphaeridium sorediforme Tim. and Pr. conglutinatum Tim. nefty i porod Volgo-Ural-skoy Oblasty (Protoleiosphaeridium sorediforme Tim. and Pr. conglutinatum Tim. from the petroleum and rocks of the Volga-Ural region): Akad. Nauk SSSR, Dokl., v. 139, no. 2, p. 461-462 (Engl. transl.: Doklady Acad. Sci. USSR, Earth Sci. Sect., v. 139, nos. 1-6, p. 847-848, 1963).

Meunier, A., 1910, Microplankton des mers de Barents et de Kara: Brussels, Campagne Arctique de 1907, pt. 5, xviii + 355 pp., 37 pl., 2 maps.

Newton, E. T., 1875, On "Tasmanite" and Australian "White Coal": Geol. Mag., ser. 2, v. 2, p. 337-342, pl. 10.

Norris, G., and Sarjeant, W. A. S., 1965, A descriptive index of genera of fossil Dinophyceae and Acritarcha: N. Zealand Geol. Survey, Paleont. Bull. no. 40, 72 p.

Sarjeant, W. A. S., 1959, Microplankton from the Cornbrash of Yorkshire: Geol. Mag., v. 96, p. 329-346, pl. 13.

———, 1960, Microplankton from the Corallian rocks of Yorkshire: Yorkshire Geol. Soc., Proc., v. 32, p. 389-408, pl. 12-14.

———, 1961, Microplankton from the Kellaways Rock and Oxford Clay of Yorkshire: Palaeontology, v. 4, p. 90-118, pl. 13-15.

———, 1962a, Upper Jurassic microplankton from Dorset, England: Micropaleontology, v. 8, p. 255-268, pl. 1-2.

———, 1962b, Microplankton from the Ampthill Clay of Melton, south Yorkshire: Palaeontology, v. 5, p. 478-497, pl. 69-70.

———, 1964, The stratigraphic application of fossil microplankton (dinoflagellates and hystrichospheres) in the Jurassic: Colloque de Jurassique, Luxembourg 1962, Vol. C. R. et Mém., p. 441-448.

———, 1965, Microplankton from the Callovian (*S. calloviense* Zone) of Normandy: Rev. Micropaléontologie, v. 8, p. 175-184, pl. 1.

———, 1967, Observations on the acritarch genus *Micrhystridium* Deflandre: Same, v. 9, p. 201-208, pl. 1.

———, 1968, Microplankton from the Upper Callovian and Lower Oxfordian of Normandy: Same, v. 10, p. 221-242, pl. 1-3.

Schaarschmidt, F., 1963, Sporen und Hystrichosphaerideen aus dem Zechstein von Büdingen in der Wetterau: Palaeontographica, ser. B, v. 113, p. 39-91, pl. 11-20.

Schön, M., 1967, Hystrichosphaerideen aus dem mittleren Buntsandstein von Thüringen: Monatsber. Deutsch. Akad. Wiss., v. 9, p. 527-535, pl. 1.

Segroves, K. L., 1967, Cutinized microfossils of probable nonvascular origin from the Permian of Western Australia: Micropaleontology, v. 13, p. 289-305, pl. 1-3.

Sommer, F. W., 1966, South American Paleozoic sporomorphae without haptotypic structures: Micropaleontology, v. 2, p. 175-181, pl. 1-2.

Spode, F., 1964, A new record of hystrichospheres from the Mansfield Marine Band, Westphalian: Yorkshire Geol. Soc. Proc., v. 34, p. 357-370, pl. 38.

Staplin, F. L., 1961, Reef-controlled distribution of Devonian microplankton in Alberta: Palaeontology, v. 4, p. 392-424, pl. 48-51.

———, Jansonius, J., and Pocock, S. A. J., 1965, Evaluation of some acritarchous hystrichosphere genera: Neues Jahrb. Geologie Paläontologie, Abh., v. 123, p. 167-201, pl. 18-20.

Stockmans, F., and Willière, Y., 1960, Hystrichosphères du Dévonien belge (Sondage de l'Asile d'aliénés à Tournai): Senckenbergiana Lethaea, v. 4, p. 1-11, pl. 1-2.

———, and ———, 1962a, Hystrichosphères du Dévonien belge (Sondage de l'Asile d'aliénés à Tournai): Soc. Belge Géologie, Bull., v. 71, p. 41-77, pl. 1-2.

———, and ———, 1962b, Hystrichosphères du Dévonien belge (Sondage de Wépion): Soc. Belg. Géol., Bull., v. 71, p. 83-99, pl. 1-2.

———, and ———, 1962c, Description de trois Hystrichosphères: in W. P. Van Leckwijk and C. H. Chesaux: Etude de l'horizon marin de Petit Buisson dans la partie occidentale du Massif du Borinage: Centre Nat. Géologie Houillière, Publ. no. 5, p. 11-30, pl. 1.

———, and ———, 1963, Les Hystrichosphères ou mieux les Acritarches du Silurien belge. Sondage de la Brasserie Lust à Courtrai. Soc. Belge Géologie, Bull., v. 71, p. 450-481, pl. 1-3.

Tasch, P., 1962, Paleoecologic significance of newly discovered Hystrichosphaerids from the Kansas Permian (Artinskian) (abs.): Int. Conf. on Palynology, Tucson, 1962, Pollen et Spores, v. 4, p. 382.

———, 1963a, Hystrichosphaerids and dinoflagellates from the Permian of Kansas: Micropaleontology, v. 9, p. 332-6, pl. 1.

———, 1963b, Fossil content of salt and associated evaporites: Ohio Geol. Soc. Symposium on Salt, 1963, p. 96-102.

Takahashi, K., 1964, Microplankton from the Asagai Formation in the Joban Coal-field: Palaeont. Soc. Japan, Trans. Proc., new ser., no. 54, p. 201-214, pl. 30-33.

Timofeev, B. V., 1959, Drevneyshaya flora pribaltiki i ee stratifraficheskoe znachenie (The ancient flora of the Baltic region and its stratigraphic significance): Leningrad, V.N.I.G.R.I., Mem., no. 129, 350 p., 25 pl.

———, 1962, Teodolitiyye paleontologicheskiy stolik (novyee metod issledovaniya iskopayemogo mikroplanktona) [A small palaeontological theodolite table (a new method of research into fossil microplankton)]: Trudy V.N.I.G.R.I., no. 196, no. 3, p. 601-647, pl. 1-20.

Valensi, L., 1947, Note préliminaire à une étude des microfossiles des silex jurassiques de la région de Poitiers: C. R. Acad. Sci. [Paris], v. 225, p. 816-818.

———, 1953, Microfossiles des silex du Jurassique moyen. Remarques pétrographiques: Soc. Géol. France, Mém., no. 68, p. 1-100, pl. 1-16.

———, 1955, Etude micropaléontologique des silex du Magdalénien de Saint-Amand (Cher.): Soc. Préhist. France, Bull., v. 52, p. 584-596, pl. 1-5.

Wall., D., 1965, Microplankton, pollen and spores from the Lower Jurassic in Britain: Micropaleontology, v. 11, p. 151-190, pl. 1-9.

———, and Downie, C., 1963, Permian hystrichospheres from Britain: Palaeontology, v. 5, p. 770-784, pl. 112-14.

Wilson, L. R., 1960, A Permian hystrichosphaerid from Oklahoma: Oklahoma Geology Notes., v. 20, p. 170.

Palynology of Permian and Triassic Strata in the Salt Range and Surghar Range, West Pakistan

B. E. BALME

Department of Geology, University of Western Australia

CONTENTS

Abstract   306
Acknowledgments   306
Introduction   306
Previous Palynological Work   309
Samples Examined   311
Techniques   315
Storage of Material   315
Preservation of Microfossils   316
Notes on Systematics   316
Terminology   319
  Morphological Terms   319
  Expression of Frequencies   320
Systematic Palynology   320
Assemblage Lists   417
  Amb Formation   417
  Wargal Limestone   418
  Chhidru Formation   419
  Mianwali Formation   421
    Kathwai Member   421
    Mittiwali Member   421
    Narmia Member   422
  Tredian Formation   423
Succession of Spore Assemblages   424
  Assemblages of the Permian   424
    Features Common to the Permian Assemblages   425
    Changes in Spore Assemblages within the Permian Succession   427
    Comparison with Other Upper Permian Spore Assemblages   431
    Summary   439
  Assemblages of the Triassic   439
    Early Scythian Assemblages   439
    Significance of the Older Scythian Assemblages   440
    Late Scythian and Middle Triassic Assemblages   442
    Significance of Late Scythian and Middle Triassic Assemblages   444
References   446

ABSTRACT

This paper deals with the palynological examination of 57 samples of sediments, six of which came from a single locality in the Surghar Range, and the remainder from eight localities in the Salt Range, West Pakistan. The material ranges in age from Lower Permian (Artinskian) to Middle Triassic, but most of the samples are from the uppermost Permian and Lower Triassic.

Seventy-nine species of spores and pollen grains, 23 of which are new, have been described and illustrated. *Simeonospora* (type species: *S. khlonovae* sp. nov.) and *Paravittatina* (type species: *Decussatisporites lucifer* Bharadwaj and Salujha) have been instituted as new form genera.

Implications of the palynological data are discussed in relation to biostratigraphy and the plant geography of late Paleozoic and early Mesozoic times. They provide evidence for sweeping floral changes in the Salt Range area, coincident with the Permian-Triassic boundary, and for further extensive modifications during the late Scythian. Spore and pollen assemblages from the Late Permian Chhidru Formation appear to have derived from a transitional flora, but to have much in common with those from strata of similar age in Madagascar.

Triassic assemblages both in their composition and succession closely resemble those from the Perth Basin, Western Australia.

ACKNOWLEDGMENTS

I am initially deeply indebted to Dr. Curt Teichert, University of Kansas, and Dr. Bernard Kummel, Harvard University, both for suggesting this study and for providing the materials upon which it was based. In addition they have supplied me with unpublished information on the stratigraphy of the Salt Range and commented on various aspects of the investigation and the manuscript.

Many palynologists have influenced me in conversation and correspondence, but I am especially grateful to the following investigators: Dr. W. Klaus, Vienna, for his help in the identification of Permian disaccate pollen species and for supplying comparative material from the Austrian Permian; Dr. J. Jansonius and Dr. S. A. J. Pocock, both of Calgary, and Dr. L. R. Wilson, Oklahoma, who generously provided representative slides of spore-pollen assemblages from North American Permian and Triassic strata; Dr. B. Owens, Geological Survey of Great Britain, Leeds, for sending me strew mounts of assemblages from the Upper Permian of Britain and Dr. G. F. Hart, now of Louisiana State University, for making available to me several of his translations of Russian papers. Dr. Geoffrey Playford, University of Queensland, commented most helpfully on many of the Triassic species. Dr. R. M. Kosanke, United States Geological Survey, Denver, read the draft manuscript and offered valuable criticisms.

Within the Department of Geology, University of Western Australia, Dr. P. J. Coleman has provided stimulating discussion on a variety of problems concerning Permian stratigraphy and paleogeography. Mr. K. C. Hughes has been responsible for all the photomicrography, and Mrs. J. White has been a most patient typist during the preparation of the manuscript.

Finance for the investigation was provided by a Research Grant from the University of Western Australia.

INTRODUCTION

Permian and Triassic strata in the Salt Range and Trans-Indus ranges of West Pakistan form one of the most celebrated geological successions in the World. They overlie sediments of Cambrian age and make up a compressed section of predominantly stable shelf marine deposits, ranging in age from

Lower Permian (Sakmarian) to early Middle Triassic. The Artinskian, Upper Permian, and Scythian parts of the section contain rich invertebrate faunas which serve as the basis of a mondial biostratigraphic standard for the uppermost Permian and Lower Triassic. According to Kummel and Teichert (1966) the Permian-Triassic contact is a paraconformity, although they have not committed themselves as to the possible duration of the break in deposition separating the two Systems. In the geological sense the hiatus cannot be a long one, and certain aspects of the transition from the Paleozoic to Mesozoic Era may be studied in the Salt Range, with a basic paleontological control available in few other places in the World. The area of the present Salt Range is also a key one to the understanding of Late Paleozoic phytogeography, occupying, as it did, a peripheral position relative to the "Gondwanaland" continents in the south and the Tethys Sea to the north. Some writers (e.g., Wadia, 1934; Sahni, 1936) have suggested the existence of an archipelagic migration route in the Punjab-Kashmir area, along which southern elements moved into Angaraland during the Artinskian and Late Permian. Convincing phytogeographical interpretations are, however, scarcely possible solely on the basis of plant megafossils, for remains of the larger plant organs are, as might be expected, not commonly found in the dominantly marine, and often calcareous, sediments of the Salt Range.

Plant megafossils of Permian and Triassic age have nonetheless been reported from a few localities. *Glossopteris* and *Gangamopteris* were recorded from a band of carbonaceous shale in the upper part of the "Lower Productus Limestone" in Warchha Valley (Jarhanwala Nala) by Reed, Cotter, and Lahiri (1930), and a more varied assemblage of Permian plants was collected by Dr. E. R. Gee from the Speckled sandstone (Warchha Formation) at Kathwai in the central Salt Range. This came from a horizon lying 20-25 feet above the "Talchir" Boulder bed (Tobra Formation) and included *Glossopteris, Gangamopteris, Samaropsis,* and *Ottokaria* (Virkki, 1939; Sahni in Virkki, 1946). Both these plant-bearing horizons are of Artinskian age, and according to Gee (quoted by Sahni in Virkki, 1946) the Kathwai deposit is as old or older than any other occurrence of elements of the *Glossopteris* flora in India (or Pakistan). It is probably more or less synchronous with the *Gangamopteris* bed of Kashmir (Seward and Smith-Woodward, 1905). Insofar as inferences are justified from this scanty megafossil evidence, the Lower Permian floras of the Salt Range are allied to those of the southern continents and peninsular India. The situation is obscure in the Late Permian and early Triassic; identifiable plant megafossils are unknown from the Upper Permian Wargal Limestone and Chhidru Formation, and the Scythian Mianwali Formation contains little obvious vegetable material. Fragmentary plant remains have been recovered from the Landa Member of the Tredian Formation at a locality near Sarai village in the Salt Range. These were first described by Sitholey (1943) who identified *Equisetites, Cladophlebis,* and a pinnate microsporophyll which he assigned to the new genus *Indotheca* Sitholey. This last-mentioned organ may, according to Pant (1949), who has re-examined the material, belong to the Corystospermaceae.

Apart from the remains of larger organs, plant microfossils, including spores, pollen grains, cuticles, and other finely comminuted tissues, have been known for many years to be common in some sediments of the Salt Range. They were recognized first by Professor Sahni and his students, and following the pioneering studies of Virkki (1937, 1939, 1946) it was clear that plant microfossils were likely to be most important in Salt Range stratigraphy.

With this in mind, suites of sam-

TABLE 1. Permian and Triassic Stratigraphic Units in the Salt Range and Trans-Indus Ranges.

System	Series	Stage	Formation	Rock unit	Pre-existing informal names
TRIASSIC	MIDDLE	Tredian Formation		Khatkiara Member	Kingriali sandstones
TRIASSIC	MIDDLE	Tredian Formation		Landa Member	Kingriali sandstones
TRIASSIC	LOWER	Scythian	Mianwali Formation	Narmia Member	Topmost limestone Dolomite beds
TRIASSIC	LOWER	Scythian	Mianwali Formation	Mittiwali Member	Bivalve limestone
TRIASSIC	LOWER	Scythian	Mianwali Formation	Mittiwali Member	Ceratite beds
TRIASSIC	LOWER	Scythian	Mianwali Formation	Kathwai Member	Upper Productus limestone
PERMIAN	UPPER	Chhidruan	Zaluch Group	Chhidru Formation	Upper Productus limestone
PERMIAN	LOWER	Guada-lupian	Zaluch Group	Wargal Limestone	Middle Productus limestone
PERMIAN	LOWER	Artinskian	Zaluch Group	Amb Formation	Lower Productus limestone

ples for palynological examination were collected at several localities by Curt Teichert and Bernhard Kummel during their field investigations in the Salt Range and Trans-Indus ranges in the years 1961-1963. Teichert and Kummel were concerned jointly with problems associated with the Permian-Triassic boundary, and sampling was therefore most detailed within the uppermost Permian and basal Triassic units. The complete stratigraphic spread of the material collected was nevertheless wide and covered the whole of the Zaluch Group of Teichert (1966) as well as all the units of the Triassic succession (Kummel, 1966). Table 1 presents, in summary, the stratigraphic succession recognized in the Salt Range and discussed in detail in the papers of Teichert and Kummel already cited.

At the collectors' request, I examined two of these samples, one from the Chhidru Formation and the other from the Mittiwali Member of the Mianwali Formation, towards the end of 1964. Abundant plant microfossils were obtained from both, and the Permian material was especially diverse and well preserved. With this encouragement, a further 82 lithologically promising samples were selected from their collections by Dr. Kummel and Dr. Teichert. These were prepared during the early months of 1965, and acid insoluble microfossils were recovered from 57 of them, although the preservation of many assemblages was poor. This applied particularly to those from the Kathwai and Mittiwali Members of the Mianwali Formation, in which the spores and pollen grains (but less commonly the acritarchs) almost invariably showed signs of pyritic corrosion. Assemblages from the Chhidru Formation were, by contrast, usually outstandingly well preserved. The most detailed stratigraphic collections examined came from Nammal Gorge (loc. 4) and Wargal (loc. 7), two localities in the central part of the Salt Range. A further sequence of closely spaced samples taken at the Chhidru West locality (loc. 6A) yielded virtually no useful material, and almost all the sediments from this section were completely devoid of insoluble organic remains. Presumably this absence of plant microfossils is due to the more extensive weathering of the Chhidru section, although this is not obvious from a visual inspection of the lithologies of the samples.

Work on the project described in this account began in April, 1965, initially with the object of describing, illustrating, and classifying the spores and pollen grains occurring commonly in Permian and Triassic sediments of the Salt Range. Subsequently, the distributions of the various forms were studied, particularly in relation to the Permian-Triassic boundary, in an endeavor to determine whether this was coincident with extensive floral modifications. Finally, an attempt has been made to analyze the palynological data to see whether they throw any light on phytogeographic problems of the late Paleozoic and early Mesozoic.

PREVIOUS PALYNOLOGICAL WORK

Although previous publications on plant microfossils from Salt Range sediments contain little biostratigraphic emphasis, they include several important contributions, and the area has a notable place in palynological history. It was from Lower Permian shales of Kathwai that Dr. Chinna Virkki first recovered the taeniate and monosaccate pollen grains which have proved ubiquitous and abundant in Permian strata of "Gondwanaland." During the 1940's the Salt Range also became the center of a celebrated palynological controversy, that concerning the age of the Saline Series, the basal stratigraphic unit of the Salt Range succession.

Dr. Virkki's position as a pioneer of stratigraphic palynology is assured, although she was deprived of some due credit by the Second World War which delayed publication of her full results for six years. Her two earlier papers dealt with disaccate Striatiti and monosaccate pollen grains, which have been subsequently referred to *Nuskoisporites* and other form genera, isolated from samples collected near the base of the Speckled sandstone at Kathwai. She was able to demonstrate that closely similar pollen grains were present in Australian Permian sediments, and hence were potentially of considerable biostratigraphic significance. Full details of Virkki's investigations were published in 1946, in a paper which is still an essential reference for workers on Permian palynology. In this, assemblages recovered from the Speckled sandstone were described and illustrated, and in addition rich microfloras were reported from the upper part of the "Lower Productus limestone" at Warchha and Jhallewali. These last two were stratigraphically the highest samples examined by Virkki from the Salt Range, although her paper also included discussion of Upper Permian microfloras from peninsular India and Australia.

Prior to Virkki's paper, Sitholey (1943) had noted microfossils associated with the remains of larger plant organs in Triassic sediments exposed on Sakesar Ridge, about 0.75 mile east of Sarai village. Among the plant microfossils identified by Sitholey were siliceous casts of trilete megaspores and microspores, and he noted also the presence of small siliceous discs and oval bodies but declined to speculate on their origins. Subsequently, Pant (1949) suggested that these were internal casts of monosaccate and disaccate pollen grains. Sitholey (1951) has accepted this interpretation and pointed out that the disaccate forms are probably attributable to *Alisporites* Daugherty.

The studies of Sitholey and Virkki were overshadowed at the time of their publication by a revival of the recurrent controversy concerning the age of the Saline Series. This is overlain at some localities by Cambrian strata, with what most early field geologists accepted as a normal sedimentary contact. In 1943, Sahni and Trivedi announced that they had found angiospermous and coniferalean wood fragments, together with an articulated insect skeleton, in dolomitic laminae occurring within the Saline Series. This evidence, if accepted, clearly demonstrated that the salt-bearing strata were of late Cretaceous or Cenozoic age and that their stratigraphic position was the result of overthrusting. For some years a lively debate ensued, primarily between paleobotanists, who were in the main inclined to accept Sahni's conclusions, and regional stratigraphers, who tended to discount them. Even today the question is apparently not resolved to everyone's satisfaction (*vide* Krishnan, 1963), despite the overwhelming evidence now available that the Saline Series is no younger than Middle Cambrian. This evidence has been summarized by Schindewolf and Seilacher (1955) and Teichert (1964), who emphasize that detailed field studies have incontestably confirmed the view of Gee and other previous investigators that the contact between the Saline Series and the overlying Purple sandstone is conformable. In addition, several experienced palynologists who attempted to reproduce Sahni's results were unable to recover microfossils suggesting a post-Cambrian age for any part of the Saline Series. The only objects of possible organic origin encountered by these investigators were opaque spheres ranging between 40-150μ in diameter and small pyritic spines about 0.3 mm long. The latter may well be of organic origin (Schopf quoted by Teichert, 1964) but have no chronological significance.

As one would expect, Sahni was highly conscious of the possibility of contamination and took precautions against it. Even so, his results are explicable in no other terms, and they provide still an important lesson in the dangers of basing sweeping biostratigraphic conclusions on small numbers of plant microfossils obtained from rocks which have been exposed to the atmosphere for even a short time. Field, laboratory, or museum contamination must also be invoked to explain the remains of advanced plants, including Permian pollen grains, reported from the Salt pseudomorph beds, Magnesian sandstone, Neobolus shales and Purple standstone (Ghosh and Bose, 1947; Ghosh, Sen, and Bose, 1951; Jacob, Jacob, and Srivastava, 1953). This interpretation also seems the only one to apply to the essentially late

Paleozoic spore assemblages reputedly occurring in the Cambrian of Kashmir (Jacob, Jacob, and Srivastava, 1953; Ghosh and Bose, 1952).

Several brief accounts of palynological examinations of post-Cambrian sediments have appeared since Virkki's paper. Ghosh and Bose (1951) investigated samples from the "Productus limestone" and the "Ceratite beds," but failed to recover spores or pollen grains. A Jurassic microflora from a sample collected by Dr. Sitholey from Nammal Gorge was described and illustrated by Sah (1955). This was well preserved and contained characteristic Mesozoic elements such as *Matonisporites, Classopollis,* and *Applanopsis.*

Pant and Srivastava (1964) have published the most recent contribution to Salt Range palynology. This deals with the results of a re-examination of the Triassic material previously studied by Sitholey (1943, 1951) and Pant (1949). Although the preservation of this material is in general unsatisfactory, it contains a varied assemblage of plant microfossils, including megaspores, microspores, and pollen grains. A number of these were described and illustrated as new form species by Pant and Srivastava.

SAMPLES EXAMINED

Details of the 57 samples from which plant microfossils were recovered are set out below. Sampling localities are shown on Figure 1 and the approximate stratigraphic position of most of the samples listed is indicated on Figures 2 and 3. Sample numbers prefixed in the lists by U.W.A. refer to the catalog of the collection of the Department of Geology, University of Western Australia, where bulk samples of the sediments examined are retained. Field numbers are those of Curt Teichert (prefixed T) and Bernhard Kummel (prefixed K). Complete information on these localities may be found in Kummel and Teichert (1970, Appendix B).

Landa Pusha, Surghar Range (loc. 1)

U.W.A. 54640, dark gray shale. Field no. K12-18A. Landa Member, Tredian Formation, 5 feet above base. Middle Triassic.

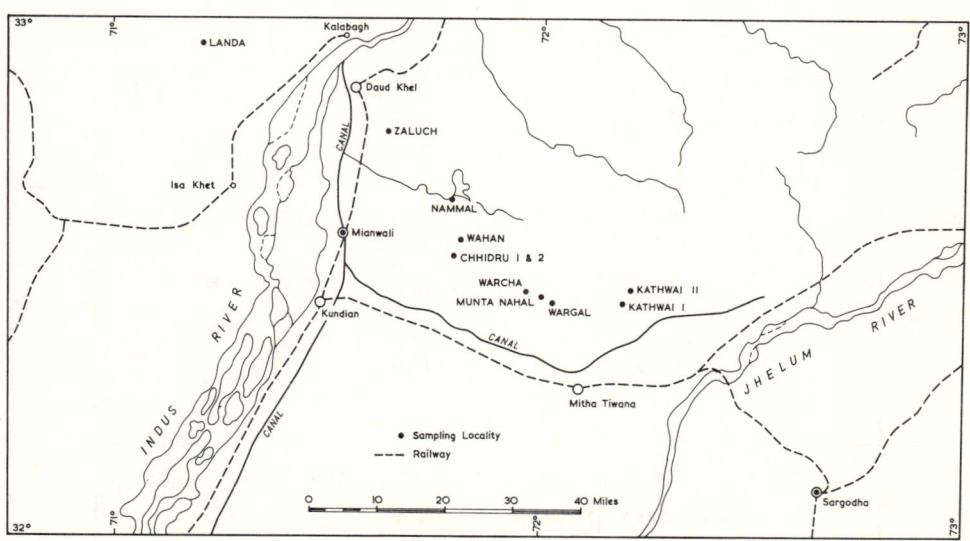

FIG. 1. Map of part of the Salt Range and Trans-Indus ranges of West Pakistan showing localities of palynological samples.

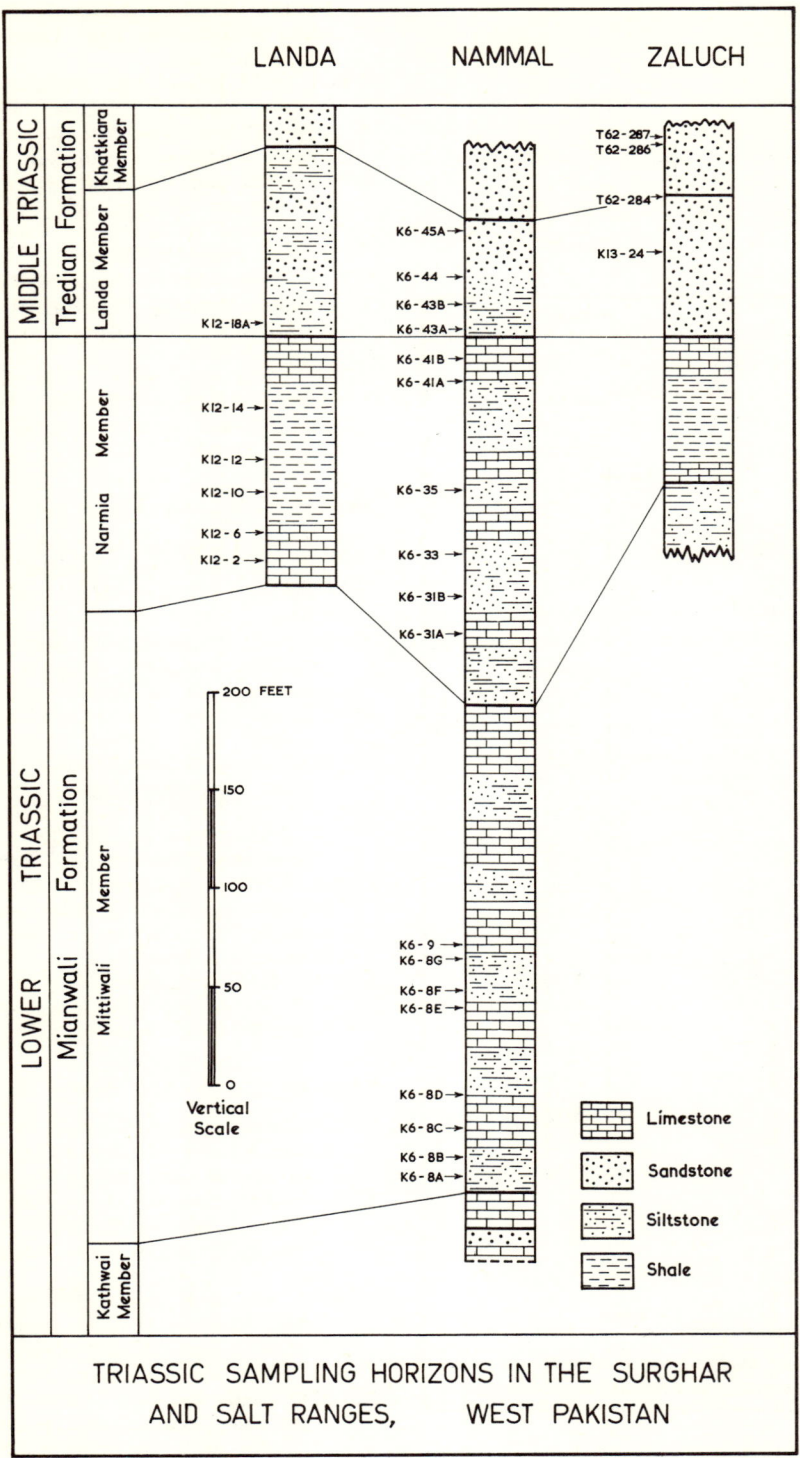

FIG. 2. Stratigraphic position of Triassic samples (indicated by field numbers) in sections in the Surghar Range and Salt Range.

U.W.A. 54641, gray shale. Field no. K12-14. Narmia Member, Mianwali Formation, about 90 feet above base. Lower Triassic (upper Scythian).

U.W.A. 54652, brown shale. Field no. K12-12. Narmia Member, Mianwali Formation, about 65 feet above base. Lower Triassic (upper Scythian).

U.W.A. 54643, pale gray shale. Field no. K12-10. Narmia Member, Mianwali Formation, about 50 feet above base. Lower Triassic (upper Scythian).

U.W.A. 54644, black shale. Field no. K12-6. Narmia Member, Mianwali Formation, about 30 feet above base. Lower Triassic (upper Scythian).

U.W.A. 54645, gray shale. Field no. K12-2. Narmia Member, Mianwali Formation, about 12 feet above base. Lower Triassic (upper Scythian).

Zaluch Nala, Salt Range (loc. 3)

U.W.A. 54633, brownish silty shale. Field no. T62-287. Khatkiara Member, Tredian Formation, 55 feet above base. Middle Triassic.

U.W.A. 54634, brownish silty shale. Field no. T62-286. Khatkiara Member, Tredian Formation, 50 feet above base. Middle Triassic.

U.W.A. 54635, gray siltstone. Field no. T62-284. Landa Member, Tredian Formation, 73 feet above base. Middle Triassic.

U.W.A. 54636, brownish shale. Field no. K13-24. Landa Member, Tredian Formation, 44 feet above base. Middle Triassic.

U.W.A. 54637, brownish silty shale. Field no. T62-173. Chhidru Formation, basal shale member, 2-3 feet above base. Upper Permian (Chhidruan).

U.W.A. 54638, gray shale. Field no. T62-53. Amb Formation, 127 feet above base. Lower Permian (Artinskian).

Nammal Gorge, Salt Range (loc. 4)

U.W.A. 54610, gray shale. Field no. K6-45A. Landa Member, Tredian Formation, topmost part of unit. Middle Triassic.

U.W.A. 54611, brownish shale. Field no. K6-44. Landa Member, Tredian Formation, about 32 feet above base. Middle Triassic.

U.W.A. 54612, gray shale. Field no. K6-43B. Landa Member, Tredian Formation, 14 feet above base. Middle Triassic.

U.W.A. 54613, brownish shale. Field no. K6-43A. Landa Member, Tredian Formation, 3 feet above base. Middle Triassic.

U.W.A. 54614, dark gray shale. Field no. K6-41B. Narmia Member, Mianwali Formation, 182 feet above base. Lower Triassic (upper Scythian).

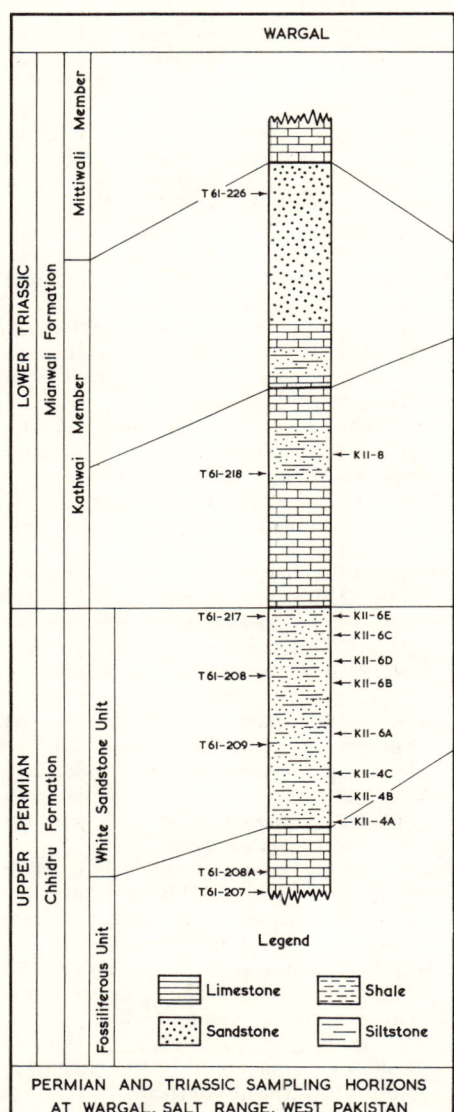

Fig. 3. Stratigraphic position of Permian and Triassic samples (indicated by field numbers) at Wargal, Salt Range.

U.W.A. 54615, gray shale. Field no. K6-41A. Narmia Member, Mianwali Formation, 167 feet above base. Lower Triassic (upper Scythian).

U.W.A. 54616, brownish shale. Field no. K6-35. Narmia Member, Mianwali Formation, about 110 feet above base. Lower Triassic (upper Scythian).

U.W.A. 54617, gray shale. Field no. K6-33. Narmia Member, Mianwali Formation, about 77 feet above base. Lower Triassic (upper Scythian).

U.W.A. 54618, gray shale. Field no. K6-31B. Narmia Member, Mianwali Formation, 55 feet above base. Lower Triassic (upper Scythian).

U.W.A. 54619, greenish-brown shale. Field no. K6-31A. Narmia Member, Mianwali Formation, 34 feet above base. Lower Triassic (upper Scythian).

U.W.A. 54621, gray and green siltstone. Field no. K6-9. Mittiwali Member, Mianwali Formation, about 123 feet above base. Lower Triassic (mid-Scythian).

U.W.A. 54622, dark gray shale. Field no. K6-8G. Mittiwali Member, Mianwali Formation, 112 feet above base. Lower Triassic (mid-Scythian).

U.W.A. 54623, green shale. Field no. K6-8F. Mittiwali Member, Mianwali Formation, 99 feet above base. Lower Triassic (mid-Scythian).

U.W.A. 54624, gray-green shale. Field no. K6-8E. Mittiwali Member, Mianwali Formation, 84 feet above base. Lower Triassic (mid-Scythian).

U.W.A. 54625, green shale. Field no. K6-8D. Mittiwali Member, Mianwali Formation, 54 feet above base. Lower Triassic (mid-Scythian).

U.W.A. 54628, green shale. Field no. K6-8A. Mittiwali Member, Mianwali Formation, 8 feet above base, Lower Triassic (Scythian).

U.A.W. 54629, yellowish-gray sandstone. Field no. T62-87. Chhidru Formation, 148 feet above base. Upper Permian (Chhidruan).

U.W.A. 54630, brownish shale. Field no. T62-73. Chhidru Formation, basal shale member, 2-3 feet above base. Upper Permian (Chhidruan).

U.W.A. 54631, black laminated shale and siltstone. Field no. T61-81. Wargal Limestone, 130 feet above base. Upper Permian (Guadalupian).

U.W.A. 54632, black siltstone. Field no. T61-64. Wargal Limestone, base. Upper Permian (Guadalupian).

Chhidru, Salt Range (loc. 6A)

U.W.A. 54671, pale gray shale. Field no. T61-61. Chhidru Formation, just below top. Upper Permian (Chhidruan).

Dhodha Wahan, Salt Range

U.W.A. 54672, fine-grained gray sandstone. Field no. T62-320. Amb Formation, 67-70 feet above base. Lower Permian (Artinskian).

Warchha Water Tank, Salt Range

U.W.A. 54673, dark gray shale. Field no. T62-350. Amb Formation, 160 feet above base. Lower Permian (Artinskian).

Wargal, Salt Range (loc. 7)

U.W.A. 58127, gray shale (slides only retained). Field no. T61-206. Chhidru Formation, 3.5 feet below top. Upper Permian (Chhidruan).

U.W.A. 54646, yellow-gray shale. Field no. T61-226. Kathwai Member, Mianwali Formation, 14 feet above base. Lower Triassic (lower Scythian).

U.W.A. 54647, yellow-gray shale. Field no. K11-8. Kathwai Member, Mianwali Formation, about 7 feet above base. Lower Triassic (lower Scythian).

U.W.A. 54648, gray shale. Field no. T61-218. Kathwai Member, Mianwali Formation. 7.5 feet above base. Lower Triassic (lower Scythian).

U.W.A. 54671, pale gray shale. Field no. T61-217. Chhidru Formation, 0.5 feet from top. Upper Permian (Chhidruan).

U.W.A. 54649, gray shale. Field no. K11-6E. Chhidru Formation, upper 1 foot of unit. Upper Permian (Chhidruan).

U.W.A. 54651, black shale. Field no. T61-207. Chhidru Formation, about 15 feet from top. Upper Permian (Chhidruan).

U.W.A. 54650, pale gray shale. Field no. K11-6C. Chhidru Formation, 2 feet from top. Upper Permian (Chhidruan).

U.W.A. 54639, yellow-gray shale. Field no. T61-208. Chhidru Formation, 3.5 feet from top. Upper Permian (Chhidruan).

U.W.A. 54652, gray shale. Field no. K11-6D. Chhidru Formation, 3 feet from top. Upper Permian (Chhidruan).

U.W.A. 54654, greenish and yellow shale. Field no. K11-6B. Chhidru Formation, from middle of white sandstone unit. Upper Permian (Chhidruan).

U.W.A. 54655, gray shale. Field no. K11-6A. Chhidru Formation, 5.5 feet from top. Upper Permian (Chhidruan).

U.W.A. 54656, gray shale. Field no. T61-209. Chhidru Formation, 5.5 feet from top. Upper Permian (Chhidruan).

U.W.A. 54657, gray shale. Field no. K11-4C. Chhidru Formation, 7 feet from top. Upper Permian (Chhidruan).

U.W.A. 54658, gray-green shale. Field no. K11-4B. Chhidru Formation, 8 feet from top. Upper Permian (Chhidruan).

U.W.A. 54660, gray-green siltstone. Field no. K11-4A. Chhidru Formation, from bottom of unit. Upper Permian (Chhidruan).

U.W.A. 54663, greenish-gray shale. Field no. T61-208A. Chhidru Formation, 12 feet from top. Upper Permian (Chhidruan).

U.W.A. 54667, gray shale. Field no. K11-2D. Chhidru Formation, 14 feet from top. Upper Permian (Chhidruan).

U.W.A. 54668, greenish shale. Field no. K11-2C. Chhidru Formation, 15 feet from top. Upper Permian (Chhidruan).

U.W.A. 54666, yellow and gray silty shale. Field no. K11-2F. Chhidru Formation, 18 feet from top. Upper Permian (Chhidruan).

Kathwai, Salt Range (loc. 9)

U.W.A. 54669, dark gray fissile shale. Field no. T61-178. Chhidru Formation, about 80 feet below top. Upper Permian (Chhidruan).

TECHNIQUES

Samples were prepared by crushing 50g or so of the sediment to about 70B.S. size. The crushed samples were reduced by quartering to 8-10 gm and macerated, using the modified Schultze technique described by Balme and Hassell (1962). This method gave good results for most of the Permian sediments and also for those from the upper part of the Triassic succession. It was less successful when dealing with samples from the Mianwali Formation, most of which proved impossible to free of finely disseminated inorganic material by chemical means alone. A combination of Schultze solution followed by 5 percent NaOH was also clearly too drastic for some of the Lower Triassic assemblages, in which the spores and pollen grains, although not the acritarchs, were swollen and decolorized. Some improvement in the state of preservation of the microfossils was achieved by using "Chlorox" type oxidants instead of Schultze solution, but the problem of obtaining unattacked specimens from the slightly weathered basal Triassic sediments was not satisfactorily solved.

Gravity separation using $ZnCl_2$ (S.G. 1.95), following the method described by Funkhouser and Evitt (1959), and slightly modified by Edgell (1965), was used to remove inorganic material from samples with a high quartz and heavy mineral content. This resulted in separations almost completely free of mineral matter, but in which the bulk of organic material had also been reduced. Examination of the rejected "sinks" showed that they contained numbers of spores and pollen grains which had presumably adhered to the inorganic particles. There is, however, no evidence that this removal of plant microfossils was in any way selective. The panning technique, also described by Funkhouser and Evitt (1959), was used with success in cleaning some of the highly mineralized samples.

Specimens designated as types or illustrated in the plates are, with few exceptions, preserved as single grain mounts. In these the specimen is embedded in a spot of glycerine jelly about 1 mm in diameter which is surrounded, below the cover slip, by beeswax. Margins of the cover slip are sealed with gold-size. A Leitz "Ortholux" microscope (No. 540142) was used throughout the investigation, and the majority of photomicrographs were taken with a Pl Apo Oel 100/1.32 objective and a Leica M1 camera body.

STORAGE OF MATERIAL

Specimens designated as types or used for purposes of illustration are lodged in the collections of the Department of Geology, University of Western Australia. Strew mounts of each of the assemblages examined, as well as bulk samples of the rock samples processed, are stored in the same repository. Sample numbers used in the text and prefixed by the letters U.W.A. refer to the catalog of the Department of Geology.

PRESERVATION OF MICROFOSSILS

Generally speaking, the best preserved assemblages were obtained from the Chhidru Formation. Except for those collected at Chhidru itself, samples from this unit almost always yielded rich and diverse microfloras. Spores and pollen were also abundant in the few samples examined from the Wargal Limestone, although the quality of preservation was poor and many of the specimens were impossible to identify with confidence. Inferior preservation, too, characterized assemblages from the Amb Formation, with the spectacular exception of the single sample from Warchha Water Tank which provided some of the finest material examined.

Acid insoluble microfossils, especially acritarchs, were abundant in most samples from the Triassic strata, although pollen grains and spores from the Mianwali Formation were usually corroded, especially in sediments from the Kathwai and Mittiwali Members. There was a progressive improvement in preservation of the plant microfossils towards the top of the Triassic section, and excellent assemblages were obtained from the Khatkiara Member of the Tredian Formation. Variations in the quality of preservation of the Triassic assemblages seem to be attributable solely to differences in the original conditions of sedimentary deposition. There is no evidence that the microfossils have been affected by the strong tectonism which has dislocated the Salt Range area.

NOTES ON SYSTEMATICS

This investigation has two main objects: first, to find whether there is any palynological evidence for major floral changes in the area of the present Salt Range in Late Permian and Early Triassic times; second, to provide an admittedly embryonic basis for paleofloristic comparisons between the Salt Range and other parts of the world. The systematic practices adopted are, in my judgment, those which allow the most useful generalizations within the scope of these two aims. Detailed discussions of certain classificatory and nomenclatural procedures are given in the section dealing with systematic palynology, but a brief preliminary justification of the taxonomic approach is also warranted.

Attempts to establish coherent nomenclatural procedures for fossil dispersed spores and pollen grains were initiated by Schopf, Wilson, and Bentall (1944), but the great bulk of publications dealing with systematics has appeared since 1950. After this relatively short time palynological nomenclature is approaching the situation which some micropaleontologists believe has overtaken the nomenclature of Foraminifera, in which, it is said, a superfluity of names chokes the literature and inhibits progress in a whole field of study. No doubt such a trend was inevitable, given the rapid expansion of palynological studies during the last decade, together with their strong utilitarian emphasis. Equally certainly the multiplication of species which are inadequately typified and cannot be used with confidence by other workers is inimical to the development of palynology as a branch of biological science concerned with the elucidation of aspects of floral history.

Efforts to redeem the situation, at

least as it affects Paleozoic spores, have been made by various subcommittees of the *Commission Internationale de Microflore du Paléozöique* in a series of invaluable reports. The volume of publication on palynological topics is nowadays so great, however, that the task of similar subcommittees in the future is likely to become sisyphean. It seems inevitable that eventually additional controls relating to palynological nomenclature will need to be incorporated within the framework of the International Code of Botanical Nomenclature to prevent the proliferation of unsatisfactorily defined taxa. Some taxonomic discussion along these lines has been initiated by Hughes (1964). His suggestions are controversial but serve to emphasize that the authority of the present International Code is not always an unqualified blessing to palynological systematists who, under the rules of priority, are forced to consider poorly illustrated taxa lacking adequate diagnoses and created without regard to their actual or potential usefulness. These difficulties are, of course, common to the systematics of all fossil organisms: they are accentuated in palynology by the great abundance of plant microfossils, and by the fact that material on which many of the early taxa were based has seldom been preserved.

Dispersed spores and pollen grains are not easy to classify effectively. They have few morphological characters from which classificatory criteria may be selected, and their profusion and variety in many sedimentary rocks create problems that specialists in most other fossil groups are able to avoid. Because of the frequent absence of clear-cut morphological characters in fossil spores, there has been a tendency to stress minor variations of dubious significance in an effort to obtain a more refined classification. The futility of continued splitting of paleontological taxa on the basis of morphology, in the hope of producing a classification which more closely accords with reality, has been pointed out by Arkell (1956, p. 97) and other taxonomists. When dealing with categories of dispersed spores, indiscriminate splitting may not only be meaningless but sufficiently misleading to obscure implications of considerable paleobotanical significance. Marked variations within the spores or pollen of single fructifications, both living and fossil, have been so frequently described (cf. *inter alia* Potonié and Schweitzer, 1960; Cranwell, 1961; Wilson, 1963) that no further argument is necessary to support the contention that morphology alone can seldom serve as a satisfactory basis for the classification of fossil dispersed spores and pollen grains. Obviously taxa can be discriminated only on the basis of morphology, but whether any purpose is served by distinguishing them is judged mainly on botanical analogies and on paleoecological, paleogeographical, or stratigraphical considerations. These comments are truistic and are reiterated solely because of the faith, implicit in some palynological papers and explicit in others, that increased refinement of the morphological criteria for discriminating form categories will somehow lead eventually to a phylogenetic classification of fossil spores.

It is possible to overemphasize the artificiality of taxa of dispersed spores and pollen grains. Certainly the structure of palynological classification reflects phylogeny only in the crudest way, and most people accept that it will always be imperfect. There is, nevertheless, no point in accepting the rigors of Linnéan classification unless it is considered to serve a biological purpose, and at any time the classification of dispersed spores should not conflict with what is known, or can be reasonably presumed, about their natural affinities. Relationships of the majority of fossil spores can, of course, only be inferred in the broadest terms, although there have been considerable additions to our knowledge of the spores of fossil fructifications during the past ten years. These data, together with earlier information, have been collated by Potonié and form the basis of his invaluable summary of the known morphology of *in situ* fossil spores (Potonié, 1962a). A number of able paleobotanists are nowadays concentrating their attention on fossil fructifications, and the possibility that we may eventually be able to make reasonable inferences concerning the natural affinities of at least the more common fossil spore and pollen types seems far less remote than it did a few years ago. Improvement in the optical qualities of microscopes, the more widespread use of microtome sections, and the application of sophisticated techniques such as electron microscopy (*vide* Pettitt and Chaloner, 1964) to the study of fossil spores may all reduce the artificiality of palynological systematics. It is, therefore, important that any "scheme" of classification should be sufficiently flexible to incorporate changing concepts with a minimum of disruption. I am not convinced that the multiplicity of suprageneric taxa erected in recent years has advanced this objective. If, as Potonié and Kremp (1954) originally proposed, the framework of higher categories is intended to serve as a key, it fulfills a useful purpose, although it is difficult to see how its utility is increased by affixing a polysyllabic name to every minor morphological category. Scientifically, the only justification for the adoption of suprageneric categories is that these comprise groups of lower taxa which may be reasonably inferred to have natural affinities, however broad. Many of the higher taxa in current use are of this type: the primary subdivision of Potonié and Kremp into Sporites and Pollenites represents a crude natural grouping and, at a lower systematic level, most palynologists would accept that such an infraturma as the disaccate Striatiti has phylogenetic significance, even though its affinities are unknown. Given the present lack of understanding of the evolutionary significance, if any, of most features of spore and pollen morphology, the natural basis of any existing classification can be nothing but rudimentary. Progress towards lesser imperfection will be slow, but it is important in the meantime to avoid any suggestion of taxonomic inflexibility by entrenching too firmly any particular scheme of classification.

Dettmann (1963) has criticised certain aspects of the suprageneric classification of Potonié and Kremp and proposed a modified system, in an attempt to eliminate inconsistencies which permitted spores of similar morphology to be grouped under two or more suprageneric categories of equal rank. The hope of achieving complete consistency within the bounds of any morphological classification is illusory, and Dettmann's scheme does not, for example, avoid the difficulty of classifying form genera such as *Densosporites* Berry and *Geminospora* Balme, in which the intexine is sometimes detached from the exoexine and sometimes not (cf. Allen, 1965). It does, however, because of its emphasis on exine stratification, ameliorate one important difficulty of the classification of Potonié and Kremp in which trilete spores with a cavate exine (*sensu* Dettmann, 1963) were assigned to the Anteturma Pollenites. In this account the broad outlines of Dettmann's classification have been accepted, including her emendation of Perinotrilites (Erdtman), although there may be force in the contention of Richardson (1965) that forms with a membranous "perispore"-like outer exine layer (i.e., the Perinotriliti of Potonié, 1956) warrant recognition as a distinct group. At present, however, few form genera assignable to the Perinotriliti (*sensu* Potonié) have been validly published, and only one, *Perotrilites* Couper, has gained wide acceptance. They can, therefore, be adequately separated at the generic level until more is known of their distribution, variation, and possible affinities.

In selecting species for description

and illustration, I have, in general, only concerned myself with forms occurring commonly in one or more of the assemblages. Descriptions of species are based on the examination of 30 or more specimens, except for a few morphologically distinctive species which are rare in the Salt Range sediments but have been previously described from other parts of the world.

Synonymy lists are intended to convey, as far as possible, my concept of the scope of the various form species and to outline their nomenclatural history. No attempt has been made to incorporate the majority of species of disaccate pollen grains described in the Russian literature into the synonymy lists. Some of these are undoubtedly similar to types described in the present account, but the general lack of photomicrographs in the published Russian work makes detailed comparisons impracticable. Disaccate pollen grains are notoriously difficult to classify, and over a hundred names of form genera have been proposed, validly and invalidly, to accommodate them. The majority of these genera are obviously unnecessary, but there are some which have been carefully defined and well-illustrated; *Paravesicaspora* Klaus and *Rhizomospora* Wilson are two examples, the usefulness of which is still *sub judice*. Subsequent work may show that these merit recognition, but in the meantime I have used as few form genera of disaccate pollen as seemed possible, on the argument that it is easier for future workers to subdivide broad categories than combine narrow ones. The elucidation of the disaccate Striatiti by Hart (1964) has been accepted insofar as it deals with genera. However, most of the species of disaccate Striatiti recognized by Hart are inadequately defined on the basis of his descriptions and differential diagnoses.

As far as possible a conservative approach to nomenclature has been adopted, and I have avoided taxonomic revision unless it appeared essential.

Comments appearing under the heading *Possible affinities* are intended to draw attention to morphological similarities between the form species under discussion and the spores or pollen of fossil fructifications of known affinities. They are not to be construed as clear evidence for the presence of the plant groups mentioned in the floras of the Salt Range.

TERMINOLOGY

MORPHOLOGICAL TERMS

Almost all the morphological terms used in the systematic descriptions are defined in the comprehensive glossary recently prepared by Kremp (1965). Some of these terms have been used in conflicting senses, and when doubt exists the usage of Dettmann (1963, p. 16-19) has been followed. The mode of citing dimensions of disaccate pollen grains also follows Dettmann (1963, fig. 3e), except that I have used the terms *total breadth* instead of *overall breadth* and *corpus height* in place of *corpus length*. Saccate pollen grains are described using the terminology of Erdtman (1957, p. 3-4) which has been subsequently adopted for fossil pollen by Townrow (1962), Harris (1964), and other authors.

Morphological terms which do not appear in the glossaries of Kremp or Dettmann or which are used here in any equivocal sense are discussed individually below:

Cavate—Dettmann's application of the term *cavate* to spores *(sensu stricto)* has been criticized on the grounds that it was first applied to gymnosperm pollen grains and, according to Richardson (1965), was originally synonymous with *saccate*. Although the original authors certainly regarded the *saccate* (or *vesiculate*) condition as a special case of cavate pollen grain, they almost certainly intended the term to have wider connotation. In their revised definition (Faegri and Iversen, 1964, p. 219) the meaning of *cavate* is given simply as "with ektexine loosened from intexine." By analogy with many other palynological terms the extension of *cavate* to include both spores and pollen grains with distinctly separated exine layers seems justified.

Clefts—see *taeniae*.

Spore assemblage—As used in the text the term *spore assemblage* embraces all those entities which may be reasonably accepted as providing the basic link between the diploid and haploid generations. It thus includes megaspores, microspores, the spores of homosporous plants, and pollen grains. In addition, it incorporates bodies, commonly referred to the Acritarcha, the origin and function of which cannot be assessed with any confidence.

Taeniae— (cf. Jansonius, 1962). Strips of exoexine usually running transversely across the cappa of sáccate or subsaccate pollen grains and separated from one another by clefts in which the intexine is exposed. Less frequently taeniae also cross the cappula, usually at right angles to those on the cappa (e.g., in *Hamiapollenites* Wilson).

EXPRESSION OF FREQUENCIES

Relative abundance of the species in the assemblages in which they occurred has been estimated by counting 150-250 specimens in each slide. Frequencies are expressed in descriptions in the following terms:

Very rare—Not appearing in counts, but specimens observed in general scanning; rare—1 percent or less; common—>1 percent <10 percent; abundant—>10 percent <25 percent; dominant—>25 percent.

Dimensions: Unless otherwise indicated, measurements of dimensions are based on specimens preserved in polar view. They are expressed as ranges with the arithmetic mean of the values given in parentheses.

SYSTEMATIC PALYNOLOGY

Anteturma SPORITES H. Potonié, 1893
Turma TRILETES Reinsch emend. Dettmann, 1963
Suprasubturma ACAVATITRILETES Dettmann, 1963
Subturma AZONOTRILETES Luber emend. Dettmann, 1963

Genus PUNCTATISPORITES
Ibrahim emend. Potonié and Kremp, 1954

Type species (original designation): *Punctatisporites punctatus* Ibrahim.

Punctatisporites fungosus Balme

Plate 2, figure 9

Punctatisporites fungosus Balme, 1963, p. 16, pl. 4, fig. 10-11.
?Punctatisporites punjabensis Pant and Srivastava, 1964, p. 86-87, pl. 17, fig. 20-21, fig. 2E-F.

Occurrence: Mianwali Formation: rare to abundant.

Representative Specimens: U.W.A. 57738, 57739, 57740. Field no. T61-226, Wargal, Salt Range; Kathwai Member, Mianwali Formation.

Description: Amb circular, exine rarely folded, often ruptured in continuity with laesurae. Trilete, tetrad scar distinct, laesurae extending one-third to two-thirds radius, often of unequal length. Exine 3-5μ thick, psilate or faintly punctate, many specimens with fine pits or caniculae developed in the vicinity of the proximal pole. These are thought to arise from post-depositional corrosion.

Dimensions (25 specimens): Diameter: 65 (81) 110μ.

Remarks: Except for their generally somewhat thinner exines, specimens from the Salt Range agree closely with those from the Scythian Kockatea Shale in Western Australia (Balme, 1963). Clearly there are difficulties in discriminating species within a form genus with so few clear-cut characters as *Punctatisporites*. The distinguishing features of *P. fungosus* are taken to be its laesurae of unequal length and fairly thick, faintly structured exine. *P. triassicus* Schulz from the middle Bunter of Halle (Schulz, 1964) may be synonymous with *P. fungosus,* but is described as having a thinner exine and more pronounced structure.

From the description of Pant and Srivastava (1964) their species *Punctatisporites punjabensis* is similar to *P. fungosus,* although somewhat larger.

Genus CALAMOSPORA Schopf, Wilson, and Bentall, 1944

Type species (original designation): *Calamospora hartungiana* Schopf.

Calamospora landiana Balme, n. sp.

Plate 1, figures 16-17

Occurrence: Mianwali Formation: very rare to common. Tredian Formation: very rare to common. Chhidru Formation: very rare.

Representative Specimens: Holotype: U.W.A. 57741. Others: U.W.A. 57742, 57743, 57744. Field no. K12-6, Landa Pusha, Surghar Range, Narmia Member, Mianwali Formation.

Description: Amb circular, distorted in most specimens by compressional folding. Trilete, tetrad scar indistinct, laesurae extending one-third to one-half radius, commissure often open. Exine less than 1μ thick, closely and finely granulate, with frequent concentric and random folds giving it a wrinkled appearance. Contact areas darker in color and somewhat thicker than remainder of exine. Most specimens are somewhat corroded.

Dimensions (30 specimens): Diameter: 60 (70) 89μ.

Derivation of name: From Landa Pusha (=Creek) in the Surghar Range.

Remarks: Although described initially from Carboniferous sediments the genus *Calamospora* has been frequently used for Permian and early Mesozoic spores. Many of the dispersed spores assigned to *Calamospora* are undoubtedly of calamarian origin and *C. landiana* is similar to the spores of *Paracalamostachys heterospora* Remy and Remy from the Permian of Thuringia (Remy and Remy, 1958). The characters of the form genus are not, however, especially distinctive, and there are no grounds for inferring that all dispersed spores assignable to *Calamospora* derived from calamarian or even equisetalean plants.

Calamospora landiana appears distinct from previously described members of the form genus which, like *Punctatisporites,* is difficult to subdivide. It is larger than any previously described Mesozoic form assigned to the genus (cf. discussion by Playford, 1965, and Bharadwaj and Singh, 1964), and its pronounced darkened contact area is also distinctive.

Possible affinities: Equisetales.

Genus LEIOTRILETES Naumova ex Potonié and Kremp, 1954

Type species (designated Potonié and Kremp, 1954): *Leiotriletes sphaerotriangulus* (Loose).

Leiotriletes sp. cf. *L. adnatus* (Kosanke)

Plate 1, figures 1-3

Granulatisporites adnatus Kosanke, 1950, p. 20, pl. 3, fig. 1.
Leiotriletes adnatus (Kosanke), Potonié and Kremp, 1955, p. 39, pl. 11, fig. 11.

Occurrence: Amb Formation: rare. Wargal Limestone: rare. Chhidru Formation: rare to common.

Representative Specimens: U.W.A. 57745, 57746, 57747. Field no. K11-6B, Wargal, Salt Range, Chhidru Formation.

Description: Amb triangular, angles broadly rounded, sides slightly to markedly concave. Trilete, tetrad scar distinct, length of laesurae variable, but never reaching the equatorial margin and usually about two-thirds radius. Exine thickness 1μ or less, rigid and seldom folded, slightly darkened and thicker along the margins of the laesurae. Exine laevigate but some specimens show faint intragranulate structure in the vicinity of the proximal pole.

Dimensions (30 specimens): Median diameter: 29 (33) 36μ.

Remarks: Simple, trilete, spores conforming to the general morphology of *Leiotriletes adnatus* have been described under a variety of specific names and assigned to several different form genera. Size, fine structure of the exine, and the prominence or otherwise of a darkened contact area have been used as bases for discrimination. Apart, possibly, from large size differences, such criteria are unlikely to allow confident differentiation of form species, especially in indifferently preserved material. Most of the specimens assigned here to *Leiotriletes* sp. are entirely laevigate and are not, therefore, placed in direct synonymy with *L. adnatus* which is said to be faintly punctate near the scar.

Trilete spores with the general habit of *Leiotriletes* sp. are known from Lower Carboniferous sediments (*vide* Staplin, 1960; Sullivan, 1964) and are common in Late Paleozoic, Mesozoic, and Cenozoic assemblages. Post-Paleozoic specimens have been frequently assigned to *Cyathidites* Couper or *Deltoidospora* Miner.

Possible affinities: A variety of Paleozoic and Mesozoic filicalean fructifications are known to have produced spores resembling *Leiotriletes adnatus*. These include species of *Coniopteris* (Couper, 1958; Doludenko, 1960), *Oligocarpia* (Remy and Remy, 1957), *Myriotheca* (Sellards, 1902), and *Renaultia* (Kidston, 1923).

Genus LOPHOTRILETES Naumova ex Potonié and Kremp, 1954

Type species (designated Potonié and Kremp, 1954): *Lophotriletes gibbosus* (Ibrahim, 1933).

Lophotriletes novicus Singh

Plate 1, figures 6-9

Lophotriletes novicus Singh, 1964, p. 247, pl. 44, fig. 24-25.

Occurrence: Amb Formation: very rare. Wargal Limestone: very rare. Chhidru Formation: very rare to common.

Representative Specimens: U.W.A. 57751, 57752, 57753, 57754. Field no. K11-2D, Wargal, Salt Range, Chhidru Formation.

Description: Amb triangular, angles broadly rounded, sides straight to slightly concave. Trilete, tetrad scar

PLATE 1

Permian and Triassic Azonotriletes. All ×600.

FIGURES

1-3. *Leiotriletes* sp. cf. *L. adnatus* (Kosanke). Lower to Upper Permian.—1. Unnumbered specimen from sample 54654.—2. Specimen 57745.—3. Specimen 57746.

4-5. *Acanthotriletes tereteangulatus* Balme and Hennelly. Lower to Upper Permian.—4. Specimen 57756.—5. Specimen 57755.

6-9. *Lophotriletes novicus* Singh. Lower to Upper Permian.—6. Specimen 57752.—7. Specimen 57753.—8. Specimen 57754.—9. Unnumbered specimen from sample 54667.

10-12. *Cyclogranisporites arenosus* Mädler. Upper Lower Triassic.—10. Specimen 57758.—11. Specimen 57757.—12. Specimen 57759.

13-15. *Verrucosisporites narmianus* Balme, n. sp. Upper Lower to Middle Triassic.—13. Specimen 57764.—14. Holotype, specimen 57761.—15. Specimen 57763.

16-17. *Calamospora landiana* Balme, n. sp. Upper Permian to Middle Triassic.—16. Specimen 57742.—17. Holotype, specimen 57741.

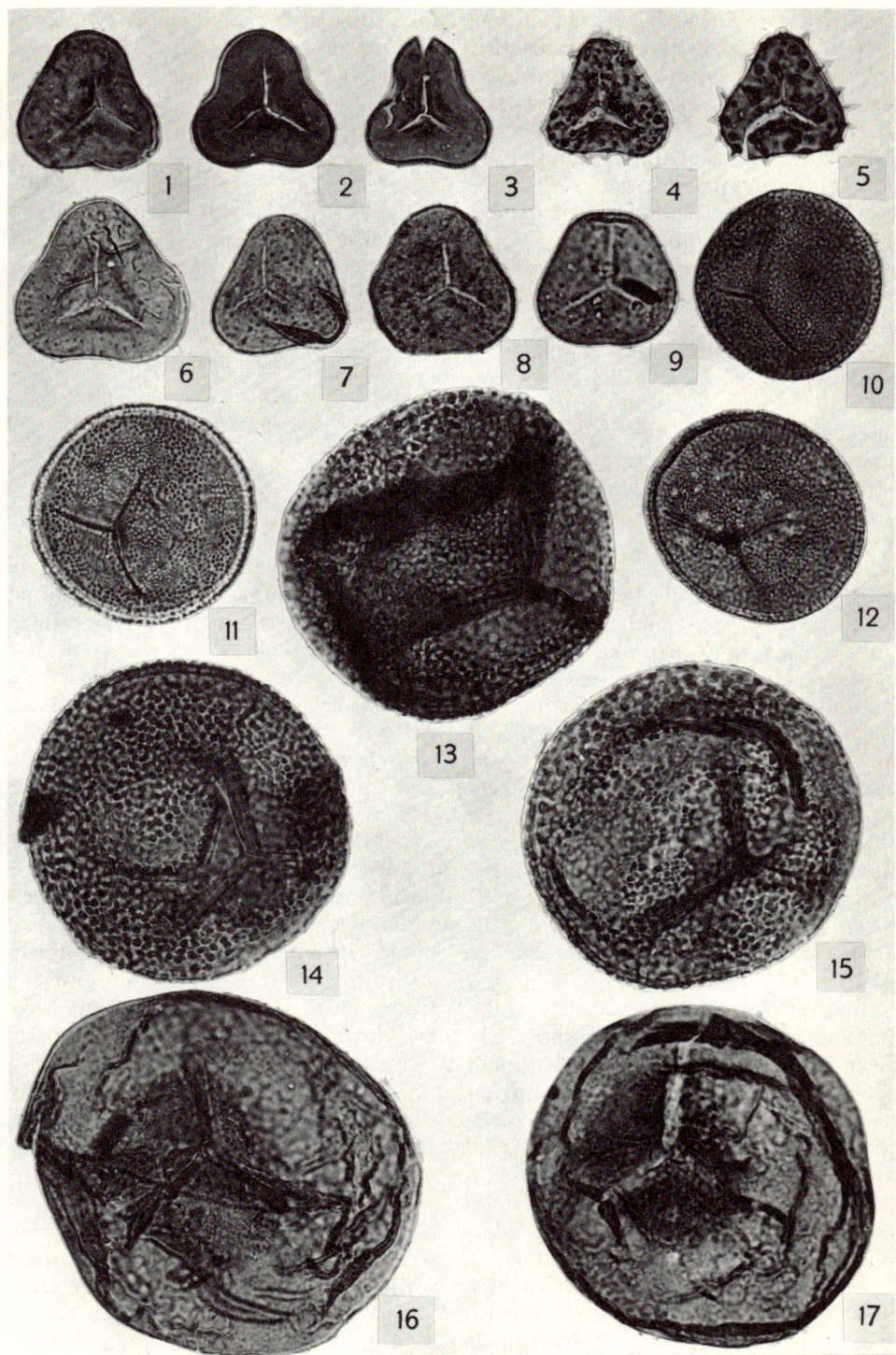

Plate 1

distinct, laesurae extending about two-thirds radius, commissure usually slightly open. Exine less than 1μ thick, sculptured proximally and distally with small spinulae about 1μ in basal diameter, up to 2μ high and 1-4μ apart. Interradial areas usually darker in color and slightly thickened in the vicinity of the proximal pole.

Dimensions (30 specimens): Median diameter: 27 (32) 37μ.

Remarks: Lophotriletes novicus belongs to a group of morphologically analogous, mainly Upper Paleozoic, trilete spores with rounded angles, a tendency to concave sides, and frequently arcuate areas of weak exinal thickening in the polar interradial areas. Within the group one finds a variety of sculptural elements including punctae, grana, spinulae, spinae, and bacula. Subdivision of the group is possible only on the basis of sculpture; but many forms bear more than one type of process, and there appears to be a constant gradational series from laevigate forms at one extreme to coarsely baculate at the other. Bharadwaj and Singh (1964) suggest, without qualification, that the presence or absence of interradial exinal thickening is a useful systematic character. *Prima facie* evidence does not support their contention, for, as noted previously, the spores of single species of *Coniopteris* may or may not show this feature. It is true that the various species of *Lophotriletes* illustrated from the Westphalian of Germany by Potonié and Kremp seldom show any clear interradial thickenings, although they may be present in the illustration of *L. gibbosus* (Ibrahim) the type species of *Lophotriletes* (see Potonié and Kremp, 1955, pl. 14, fig. 221). Undoubtedly, they occur in *Lophotriletes tribulosus* Sullivan from the early Westphalian of England (Sullivan, 1964).

Lophotriletes novicus Singh was described from the Permian of Iraq and is conceivably a junior synonym of *L. rarus* Bharadwaj and Salujha from the Raniganj Coal Measures. The latter species is not clearly illustrated, and *L. novicus* may be differentiated for the present by its more abundant sculptural elements.

Genus ACANTHOTRILETES
Naumova ex Potonié and Kremp, 1954

Type species (designated Potonié and Kremp, 1954): *Acanthotriletes ciliatus* (Knox).

Acanthotriletes tereteangulatus Balme and Hennelly
Plate 1, figures 4-5

Raistrickia sp., Rilett, 1954, p. 35, pl. 5, fig. 1.
Acanthotriletes tereteangulatus Balme and Hennelly, 1956, p. 247-48, pl. 2, fig. 27-29.
Acanthotriletes Bharadwaj, 1962, p. 79-80, pl. 1, fig. 23-24.
Lophotriletes sparsus Singh, 1964, p. 247, pl. 44, fig. 23.

Occurrence: Amb Formation: very rare to common. Wargal Limestone: rare. Chhidru Formation: very rare to common.

Representative Specimens: U.W.A. 57755, 57756. Field no. K11-6C, Wargal, Salt Range, Chhidru Formation.

Description: Amb triangular, angles rounded, sides straight to concave. Trilete, tetrad scar distinct, laesurae extending about two-thirds radius. Exine less than 1μ thick, slightly darker and thickened in the interradial areas adjacent to the proximal pole. Proximal and distal surfaces sculptured with coni and fine tapering spinae which are usually curved. Sculptural elements 1-3μ long, 1μ basal diameter, 1-4μ apart.

Dimensions (8 specimens): Median diameter: 23-29μ.

Remarks: Acanthotriletes tereteangulatus is a ubiquitous form in the Australian Permian in which it is usually most common in early Artinskian sediments. Closely similar or identical forms have also been recorded from Africa (Rilett, 1954; Hart, 1963, 1965a), Antarctica (Balme and Playford, 1967), and the Raniganj Coal Measures of India (Ghosh and Sen,

1948; Bharadwaj, 1962). *Azonotriletes trisulcus* Andreeva et al. from the Russian Permian is also a similar form (Andreeva et al., 1956, p. 244, pl. 47, fig. 33).

The species is assigned to *Acanthotriletes* because spinae are the dominant sculptural elements, although coni also occur. The relative proportions of spinae and coni may be affected by the quality of preservation, so the generic assignment of the species is somewhat subjective.

Possible affinities: Spores of broadly similar morphology occur in the filicalean organ *Sphyropteris* cf. *boenischi* Stur (Remy and Remy, 1957).

Genus CYCLOGRANISPORITES
Potonié and Kremp, 1954

Type species (original designation): *Cyclogranisporites leopoldi* (Kremp).

Cyclogranisporites arenosus Mädler

Pl. 1, figures 10-12

Cyclogranisporites arenosus Mädler, 1964, p. 96-97, pl. 8, fig. 11.

Occurrence: Narmia Member, Mianwali Formation: very rare to common. Tredian Formation: rare to common.

Representative Specimens: U.W.A. 57757, 57758, 57759. Field no. K6-33, Nammal Gorge, Salt Range, Narmia Member, Mianwali Formation.

Description: Amb circular, periphery finely notched, usually preserved in off-polar compressions. Trilete, tetrad scar distinct, laesurae extending one-half to two-thirds radius, commissure slightly open in most specimens. Exine 2-4μ thick, sculptured proximally and distally with closely spaced but clearly separated grana. Grana circular or rounded polygonal in plan, basal diameter 0.5-1μ, height up to 1μ, 0-2μ apart.

Dimensions (25 specimens): Diameter: 36 (45) 54μ.

Remarks: Cyclogranisporites arenosus is a distinctive form characterized by its almost perfect sphericity when uncompressed, its relatively thick exine, and the clarity and curious appearance of its sculpture. Mädler (1964, p. 97) speaks of the "lichtbrechenden" properties of the individual sculptural elements and, in high focus, each appears brightly illuminated and clearly defined.

Mädler's type material came from the lower Keuper of Bad Harzburg in Germany, and this is the only previous record of *Cyclogranisporites arenosus*.

Genus VERRUCOSISPORITES
Ibrahim emend. Smith and Butterworth, 1967

Type species (original designation): *Verrucosisporites verrucosus* Ibrahim.

Verrucosisporites narmianus Balme, n. sp.

Plate 1, figures 13-15

Occurrence: Narmia Member, Mianwali Formation: rare to common. Landa Member, Tredian Formation: very rare to common.

Representative Specimens: Holotype: U.W.A. 57761. Others: U.W.A. 57762, 57763, 57764. Field no. K12-6. Landa Pusha, Surghar Range, Narmia Member, Mianwali Formation.

Description: Amb circular usually slightly distorted by folding, periphery notched. Trilete, tetrad scar prominent, laesurae extending two-thirds radius, commissure bordered by raised, slightly sinuous labrae. Exine 2-4μ thick, uniformly sculptured with close-packed discrete verrucae of variable shape and size. Individual elements 1-4μ in basal diameter, 1-2μ high, and less than 1μ apart. Sculpture reduced in size in the vicinity of the proximal pole.

Dimensions (30 specimens): Diameter: 68 (73) 79μ.

Derivation of name: From the village of Narmia in the Surghar Range.

Remarks: Mädler (1964) created the genus *Cyclotriletes* to incorporate trilete spores with a circular amb, labrate tetrad scar and sculpture of close-packed but discrete grana, verrucae or low coni. It thus includes forms which apart from their labrate laesurae would be assigned to *Cyclogranisporites* or *Verrucosisporites*. Smith and Butterworth (1967, p. 147) do not exclude labrate forms from *Verrucosisporites,* and on purely morphological grains there is little ground for accepting *Cyclotriletes,* although Mädler makes the additional point that granulate and verrucate forms with labra are characteristically Triassic in Germany. If their stratigraphic distribution proves to be similar elsewhere the recognition of *Cyclotriletes* as a separate form genus will be justified.

Verrucosisporites narmianus shows little morphological variation and the specimens measured lie within a narrow size range. *Cyclotriletes oligogranifer* Mädler is finely granulate and infrapunctate. *C. microgranifer* Mädler and *C. pustulatus* Mädler are smaller than *V. narmianus* and are sculptured with grana, pustulae, or coni. A marked darkened contact area characterizes *C. granulatus* Mädler, and in *C. triassicus* Mädler the sculptural elements are widely separated and the exine infrapunctate.

In Germany *Cyclotriletes* is, according to Mädler, most common in the upper Buntsandstein although two species range into the lower Keuper (Mädler, 1964, p. 128).

Possible affinities: Mädler suggests that *Cyclotriletes* has osmundaceous origins, although no closely comparable forms are known from any living or fossil group.

Verrucosisporites sp. cf.
V. planiverrucatus Imgrund
Plate 2, figures 10-11

Verrucosisporites planiverrucatus Imgrund, 1960, p. 163, pl. 15, fig. 60.

Occurrence: Amb Formation: rare.
Representative Specimens: U.W.A. 57765, 57766, 57767. Field no. Y62-350, Warchha Water Tank, Salt Range, Amb Formation.

Description: Amb circular, periphery notched, specimens often ruptured in continuity with laesurae, but seldom folded. Trilete, tetrad scar often obscured by sculpture, but sometimes distinct, laesurae one-half to two-thirds radius. Exine 2-3μ thick, sculptured proximally and distally with closely packed low cones and rounded verrucae, about 100 of which are visible along the periphery. Individual elements of irregular shape and size, circular to rounded polygonal in plan, apices flattened, gently rounded, or acute. Basal diameter of sculptural elements usually between 2-5μ, height 1-2μ, and 0.5μ apart. Adjacent elements occasionally fused at their bases to form incipient cristae.

Dimensions (30 specimens): Diameter: 71 (78) 85μ.

Remarks: Although not common, *Verrucosisporites* cf. *V. planiverrucatus* is a well-characterized component of the few microfloras examined from the Amb Formation and has not been found in those from other units. It differs from most previously described species of *Verrucosisporites* in its large number of closely packed sculptural elements of variable form. *V. sinensis* Imgrund from the Lower Permian of China is similar, but the processes in this species are consistently flattened. In *V. trisecatus* Balme and Hennelly the tetrad scar is more prominent, the sculpture coarser, and the individual elements usually more or less hemispherical. The size of *Cyclobaculisporites indicus* Bharadwaj and Singh is larger, it seems more fragile than the present species, and its sculpture is said to be baculate, although this does not show clearly in the illustrations of Bharadwaj and Singh (1964, pl. 2, fig. 45-46).

Verrucosisporites planiverrucatus is a poorly illustrated species from the

Lower Permian of the Kaiping Basin which needs further investigation. As far as can be judged from Imgrund's description and illustration, its morphology agrees with that of the Salt Range form and the two are tentatively compared.

Possible affinities: Remy and Remy (1957) have described spores with the morphological characters of *Verrucosisporites* from zygopterid fructifications.

Genus CAMPTOTRILETES
Naumova ex Potonié and Kremp, 1954

Type species (designated Potonié and Kremp, 1954): *Camptotriletes corrugatus* (Ibrahim).

Camptotriletes warchianus Balme, n. sp.

Plate 3, figures 12-13

Occurrence: Amb Formation: rare. Chhidru Formation: very rare.

Representative Specimens: Holotype: U.W.A. 57768. Others: U.W.A. 57769, 57770, 57771. Field no T62-350, Warchha Water Tank, Salt Range, Amb Formation.

Description: Amb triangular, angles fairly sharply rounded, sides straight to slightly convex. Trilete, tetrad scar distinct, laesurae extending almost to the equator, commissure often open. Exine 2-4μ thick, sculptured proximally and distally with close-packed coarse elements of variable form. Low, sinuous cristate processes formed by the fusion of the bases of adjacent stout cones are the most common, but rod-like sub-bacula and discrete coni also occur. Processes 1-3μ high.

Dimensions (30 specimens): Median diameter: 63 (75) 81μ.

Derivation of name. From the village of Warchha in the Salt Range.

Remarks: Camptotriletes warchianus is not easy to assign to a morphological category because of variability in the form of its sculptural elements. Some specimens might be appropriately assigned to *Lophotriletes,* and others are close to *Converrucosisporites* Potonié and Kremp. In most specimens, however, cristae or sub-cristae predominate.

Singh (1964) has described a similar, perhaps identical, form from the Permian of Iraq under the name *Camptotriletes* sp. The single illustration of this species appears to have more densely packed sculptural elements than *C. warchianus*. Spore 27 of Virkki (Virkki, 1946, p. 130, fig. 42) resembles *C. warchianus,* although it was originally described as spinose, and the photographic illustration does not allow detailed comparisons.

Genus OSMUNDACIDITES Couper, 1953

Type species (original designation): *Osmundacidites wellmanii* Couper.

Osmundacidites senectus Balme

Plate 2, figures 7-8

Osmundacidites senectus Balme, 1963, p. 17, pl. 4, fig. 1-2.

Occurrence: Chhidru Formation: very rare to rare. Mianwali Formation: rare. Tredian Formation: common to abundant.

Representative Specimens: U.W.A. 57772, 57773, 57774. Field no. T62-287. Zaluch Nala, Salt Range, Khatkiara Member, Tredian Formation.

Description: Amb circular, usually distorted by compressional folding, periphery finely notched. Trilete, tetrad scar distinct, laesurae extending three-fourths radius, bordered by narrow, silghtly darkened labra. Exine about 1μ thick, sculptured proximally and distally by fine elements including coni, bacula, and irregular grana. Individual elements 1-2μ high and up to 2μ in basal diameter.

Dimensions (15 specimens): Diameter: 57-72μ.

Remarks: Osmundacidites senectus from the Lower Triassic of Western Australia was originally distinguished

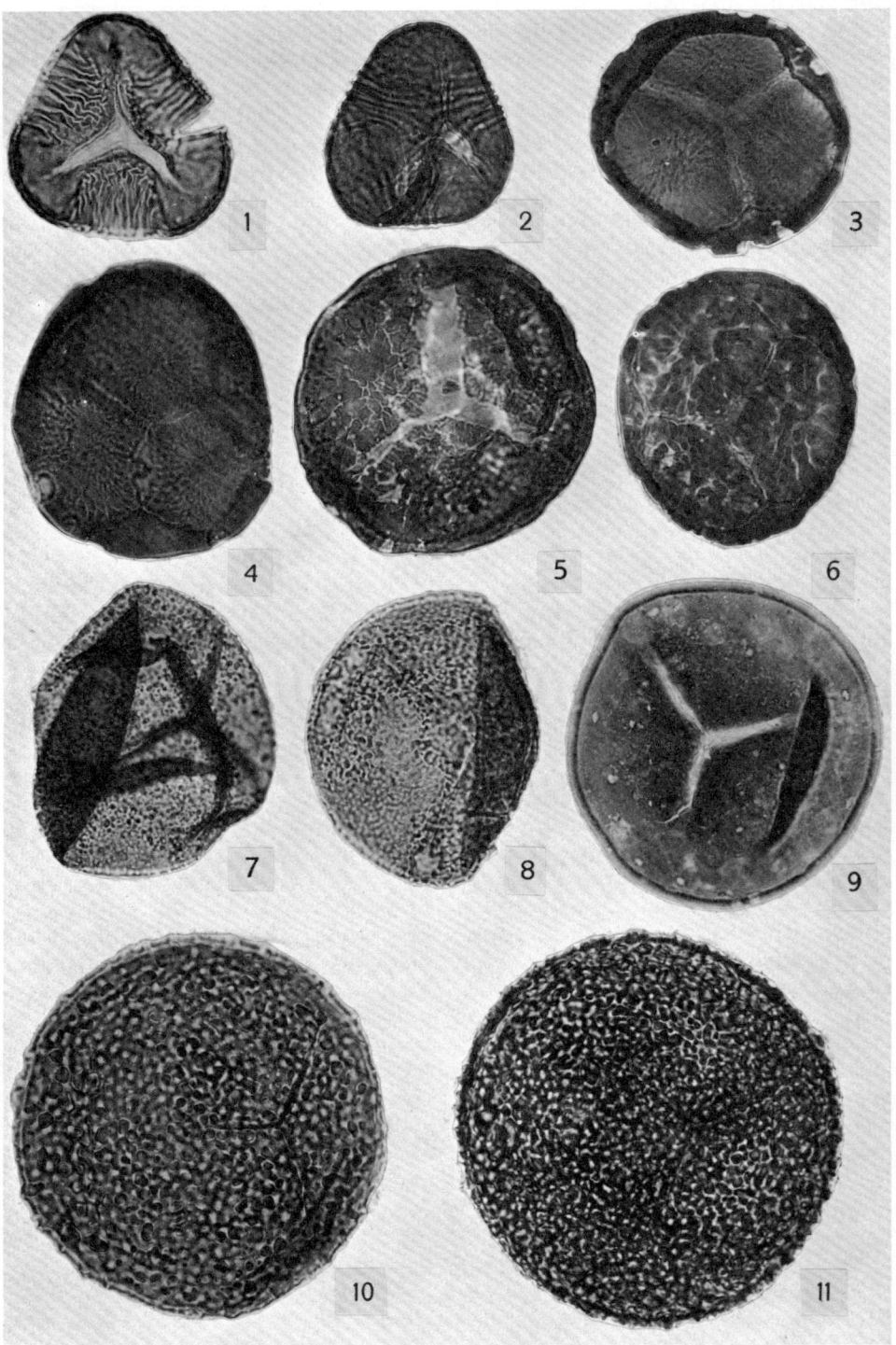

PLATE 2

from *O. wellmanii* Couper on the basis of its finer and more regular sculptural elements. Other authors have, however, expanded the concept of *O. wellmanii* to include forms closely resembling the holotype of *O. senectus* (cf. Playford, 1965). The extension of *O. wellmanii* appears to have the imprimatur of Couper himself (cf. Couper, 1953, 1958), and future work may show that nothing is achieved by distinguishing the two form species.

Possible affinities: Osmundaceae.

Genus SIMEONOSPORA Balme, n. gen.

Type species (here designated): *Simeonospora khlonovae* Balme, n. sp.

Diagnosis: The name *Simeonospora* is proposed as a form genus for dispersed spores having the following morphological characters:

Amb circular or subcircular. Trilete with strongly defined sunken contact areas. Exine fairly thick, appearing as a false equatorial rim in optical section. Contact areas sculptured with flattened rugulae and verrucae. Proximal sculptural elements with a tendency to radial alignment from the center of each contact depression. Additional sculpture may or may not be present on the distal surface. Diameter of known species less than 100µ.

Remarks: Simeonospora resembles in its gross morphology *Retusotriletes* (Naumova), which has been used to include forms with distal and occasionally proximal sculpture. In *Retusotriletes birealis* McGregor from the Devonian of Canada slight radial alignment of the proximal sculpture has been noted (McGregor, 1964). A recent emendation of *Retusotriletes* by Richardson (1965) excludes sculptured forms from that genus. Species assigned here to *Simeonospora,* in which sculpture is virtually confined to the contact faces, are considered sufficiently different from any described species of *Retusotriletes* to warrant separate generic status.

Emphanisporites McGregor, an important Devonian form, is one of the few other genera of dispersed spores in which the sculpture is predominantly proximal. In *Emphanisporites,* however, it consists of low, flattened, clearly defined ridges which, in most species, radiate from the proximal pole. *Emphanisporites erraticus* (Eisenack) has ridges which show a tendency to radial arrangement on each contact facet. In all species of *Emphanisporites* the proximal ridges form a stronger and more regular radial pattern than they do in *Simeonospora,* and the former genus does not display the sharply delimited contact areas.

Khlonova (1960, 1961) has described forms from the early Late Cretaceous of Siberia and the Ukraine which are possibly embraced by the genus *Simeonospora*. She assigned these to *Stenozonotriletes* and created three species, *Stenozonotriletes radiatus, Stenozonotriletes exuperans,* and *Stenozonotriletes stellatus,* all of which bear sculpture on the contact areas. In two of Khlonova's species, *S. radiatus* and *S. stellatus,* the pattern of proxi-

---

PLATE 2

Permian and Triassic Azonotriletes. All ×600.

FIGURES

1-2. *Tigrisporites playfordi* de Jersey and Hamilton. Upper Permian to Middle Triassic. ——1. Specimen 57780.——2. Specimen 57778.

3-6. *Simeonospora khlonovae* Balme, n. sp. Upper Lower Triassic.——3. Holotype, specimen 57828.——4. Specimen 57776.——5. Specimen 57777.——6. Specimen 57775.

7-8. *Osmundacidites senectus* Balme. Upper Permian to Middle Triassic.——7. Specimen 57774.——8. Specimen 57772.

9. *Punctatisporites fungosus* Balme. Lower Triassic. Specimen 57740.

10-11. *Verrucosisporites* sp. cf. *V. planiverrucatus* Imgrund. Lower Permian.——10. Specimen 57766.——11. Specimen 57767.

mal sculpture is radial and similar to that of *S. khlonovae*. However, additional distal sculpture of low verrucae is also present, at least in *S. radiatus*.

In the Soviet Union *S. radiatus* is taken as an index fossil for the early Late Cretaceous, although it ranges into the Senonian and occasional specimens occur in the Aptian-Albian. There are no published records from outside Russia of Cretaceous forms resembling *S. radiatus,* although a spore similar to *S. stellatus* occurs extremely rarely in Lower Cretaceous sediments in the Perth Basin, Western Australia.

Dulhuntyispora ? minuta Jansonius from the Lower Triassic of Canada may well be assignable to *Simeonospora*, although it is much smaller than *S. khlonovae* and bears annular distal thickenings. Jansonius (1962) placed the species provisionally in *Dulhuntyispora* Potonié, but this is clearly inappropriate, for it lacks the characteristic exoexinal "blisters" of the Australian Permian genus.

Simeonospora khlonovae Balme, n. sp.

Figure 4; Plate 2, figures 3-6

Occurrence: Narmia Member, Mianwali Formation: very rare to rare.

Representative Specimens: Holotype: U.W.A. 57828. Others: U.W.A. 57775, 57776, 57777. Field no. K12-6, Landa Pusha, Surghar Range, Narmia Member, Mianwali Formation.

Description: Amb circular to subcircular, periphery broadly undulate. Trilete, tetrad scar distinct, laesurae extending two-thirds radius or more, commissure usually open and with slightly ragged margins. Contact areas rounded to pentagonal, markedly depressed, bounded by low curvaturae about 0.5μ wide at their crests. Exine sculptured within contact areas with complex patterns of flattened sinuous rugulae and verrucae which tend to radiate from the center of each contact depression and terminate either at the edges of the laesurae or at the inner margin of the curvaturae. Sculptural elements 0.5-2μ wide; exine 3-5μ thick appearing in optical section at the equator as a false rim; exine outside contact areas laevigate or with low broad rugulae on the distal surface.

Dimensions (20 specimens): Diameter: 60 (65) 73μ.

Derivation of name: After Dr. A. F. Khlonova, who first described morphologically comparable spores from the Cretaceous of the USSR.

Remarks: Simeonospora khlonovae was found only in samples from the Narmia Member of the Mianwali Formation. It is rare, even at the type locality, and no clear biostratigraphic significance can be attached to its apparently restricted distribution. Quantitatively, the species is an insignificant component of the Salt Range Triassic assemblages, but it may be an important form, because of its unusual morphology and its resemblance to species that have been regarded as Cretaceous indices by some workers.

Stenozonotriletes radiatus Khlonova differs in the more precisely radial arrangement of its proximal sculpture, its distal processes and smaller size. *S. stellatus* Khlonova is smaller and has very fine sculpture. The Lower Triassic species *Dulhuntyispora ? minuta* Jansonius has a diameter of 24-28μ and bears annular exinal thickenings on the distal face.

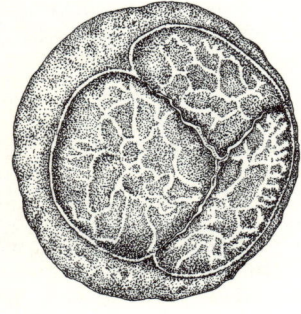

Fig. 4. *Simeonospora khlonovae* Balme, n. gen. et n. sp. Reconstruction in proximal view, $\times 600$.

Genus TIGRISPORITES Klaus, 1960

Type species (original monotypy): *Tigrisporites halleinis* Klaus.

Klaus based the genus *Tigrisporites* on specimens from the Carnian of Austria, and it has been subsequently used by Playford (1965) for spores of Triassic age from Tasmania. It is characterized mainly by the presence of radially disposed rugulae running from the proximal to the distal face in the interradial areas. Playford discussed the distinction between *Tigrisporites* and *Tripartina* Malyavkina, concluding that in the latter genus the sculptural elements were not visible as projections along the equatorial margin. *Costaspora* Staplin and Jansonius from the Mississippian of Canada (Staplin, 1960) is a genus basically similar to *Tigrisporites*, which could, as defined, include the spores assigned by Playford to *Tigrisporites* sp. *Costaspora* is monotypic and its type species, *C. radiosa* Staplin and Jansonius, bears distal sculpture of grana and rounded verrucae in addition to the interradial rugulae, which are usually confined to the proximal face. For the present, therefore, *Tigrisporites* is retained for those forms in which the rugulae clearly pass from the proximal to the distal face and which bear no other pronounced sculptural elements.

Tigrisporites playfordi de Jersey and Hamilton

Plate 2, figures 1-2

Tigrisporites sp., Playford, 1965, p. 186, pl. 8, fig. 3-4.
Tigrisporites playfordi de Jersey and Hamilton, 1967, p. 10, pl. 5, fig. 5-7.

Occurrence: Chhidru Formation: very rare to rare. Khatkiara Member, Tredian Formation: very rare.

Representative Specimens: U.W.A. 57778, 57779, 57780. Field no. K11-6D, Wargal, Salt Range, Chhidru Formation.

Description: Amb triangular, angles rounded, sides slightly concave. Trilete, tetrad scar distinct, laesurae extending almost to equator, commissure bordered by slightly thickened labra. Exine about 2μ thick, bearing in the interradial areas 6-12 low, rounded rugulae which arise in the area of the proximal pole and pass on to the distal surface. Rugulae about 0.5μ high, up to 1μ wide, and 0.5μ apart; in some specimens dichotomizing or anastomosing.

Dimensions (14 specimens): Median diameter: 39 (42) 47μ.

Remarks: Tigrisporites playfordi (as *Tigrisporites* sp.) has been previously reported from the Tiers Formation of Middle to Late Triassic age in northeastern Tasmania (Playford, 1965) and the Middle Triassic Moolayember Formation in the Bowen Basin, Queensland (de Jersey and Hamilton, 1967). Occasional specimens also occur in Middle Triassic strata in the Perth Basin, Western Australia. Playford has drawn attention to records from Russian Triassic and Jurassic deposits of species bearing some resemblance to *T. playfordi*. Such forms have not, to my knowledge, been reported from Permian deposits.

Genus INDOSPORA Bharadwaj, 1962

Type species (original designation): *Indospora clara* Bharadwaj.

Indospora clara Bharadwaj

Plate 3, figure 11

Indospora clara Bharadwaj, 1962, p. 83, pl. 3, fig. 54-55.

Occurrence: Chhidru Formation: very rare.

Representative Specimen: U.W.A. 57781. Field no. K11-6C, Wargal, Salt Range, Chhidru Formation.

Discussion: Specimens referred here to *Indospora clara* fall within the limits of the original diagnosis except for their generally larger size. Bharadwaj (1962) states that the diameter varies from 49μ to 64μ, whereas only two of the seven specimens measured from the Chhidru Formation fell within this range; the other five were larger and the biggest had a median diameter of 81μ.

Despite its rarity *Indospora clara* is an interesting component of the Upper Permian assemblages in the Salt Range. It had been previously reported in published work only from the Raniganj Coal Measures in which it appears to be fairly common in certain seams (Bharadwaj and Salujha, 1965a,b). Recently it has been found in the Wagina Sandstone, an Upper Permian unit occurring in the northern part of the Perth Basin, Western Australia.

Two additional species of *Indospora* have been described by Peppers (1964) from the Pennsylvanian of Illinois. These attributions are justifiable if Bharadwaj's genus is broadly interpreted, but both species lack a distal reticulum and neither closely resembles *I. clara*.

Subturma ZONOTRILETES Waltz, 1935

Genus TRIQUITRITES Wilson and Coe emend. Schopf, Wilson, and Bentall, 1944

Type species (original monotypy): *Triquitrites arculatus* Wilson and Coe.

The morphology and distribution of the various species of *Triquitrites* have recently been discussed by Sullivan and Neves (1964). Forms possessing the generic characters of *Triquitrites* range from the Tournaisian, but are most abundant in Westphalian deposits in which some species are stratigraphically important. Since the review by Sullivan and Neves was prepared, additional species of *Triquitrites* have been described by Ouyang (1962) from the Lungtan Series in South China. The age of these deposits is in dispute, but they have usually been regarded as Permian, even Late Permian. Ouyang draws attention to the notable similarities between his assemblages and those from the upper Westphalian of Europe, and his palynological evidence points towards a Late Carboniferous rather than a Permian age. *Triquitrites* has rarely been recorded from sediments younger than Stephanian. A large form referable to the genus occurs in the early Artinskian of the Canning Basin, Western Australia, and Imgrund (1960) described one species, the assignation of which is doubtful, from strata dated as Lower Permian from the Kaiping Basin of China.

Triquitrites proratus Balme, n. sp.

Plate 3, figures 6-8

Occurrence: Chhidru Formation: very rare to rare.

Representative Specimens: Holotype: U.W.A. 57782. Others: U.W.A. 5773, 57784, 57785. Field no. K11-6D, Wargal, Salt Range, Chhidru Formation.

Description: Auriculate, amb triangular, sides concave. Trilete, tetrad

PLATE 3

Permian and Triassic Azonotriletes, Zonotriletes, and Perinotriletes. All ×600.

FIGURES

1-5. *Nevesisporites fossulatus* Balme, n. sp. Upper Permian to ?Middle Triassic.——1. Unnumbered specimen from sample 54650.——2. Specimen 57792.——3. Specimen 57791. 4. Holotype, specimen 57789.——5. Specimen 57790.

6-8. *Triquitrites proratus* Balme, n. sp. Upper Permian.——6, Specimen 57781.——7. Holotype, specimen 57782.——8. Specimen 57783.

9-10. *Iraqispora labrata* Singh. Upper Permian.——9. Specimen 57786.——10. Specimen 57788.

11. *Indospora clara* Bharadwaj. Upper Permian. Specimen 57781.

12-13. *Camptotriletes warchianus* Balme, n. sp. Lower to Upper Permian.——12. Specimen 57771.——13. Holotype, specimen 57768.

14-15. *Perotrilites* sp. Upper Lower Triassic.——14. Specimen 57794, with perispore-like exine layer ruptured.——15. Specimen 57793.

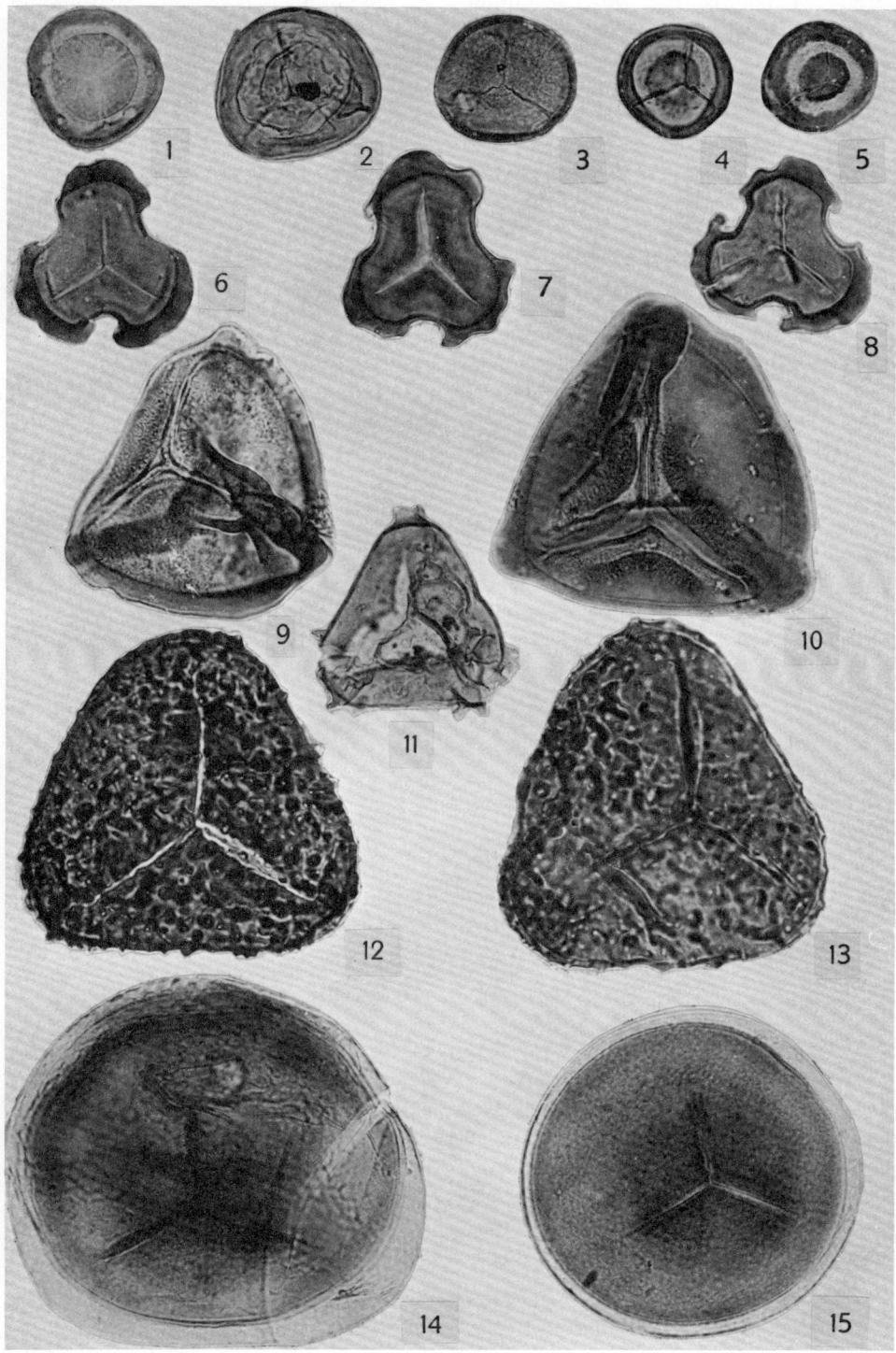

PLATE 3

scar distinct, laesurae extending about three-fourths distance to the inner margin of the auriculae, sometimes bordered by slightly thickened darker exine. Exine 1-2µ thick on proximal and distal faces, smooth or faintly maculate. Auriculae 2-5µ wide, outer margins undulate, extremities usually markedly prolonged and incurved.

Dimensions (20 specimens): Median diameter (including auriculae): 33 (39) 46µ.

Derivation of name: Greek πρωρα *(prora)* = the prow of a ship, referring to the shape of the auriculae.

Remarks: Triquitrites proratus is an uncommon but distinctive species recovered only from samples of the upper part of the Chhidru Formation. It differs from previously described species of *Triquitrites* in the shape of its hyaline auriculae and in its fairly markedly concave sides. Auriculae of this general type are found in the typically Lower Carboniferous genus *Tripartites,* but Sullivan and Neves (1964) regard distal plication of the auriculae as a definitive character of *Tripartites.*

Triquitrites iraqiensis Singh from the Permian of Iraq (Singh, 1964) may be of the same general geological age as *T. proratus,* but is smaller and its auriculae less strongly developed. *Triancoraesporites communis* Schulz from the Rhaetic of Thuringia (Schulz, 1962) has laterally prolonged angles but lacks the auriculae of *Triquitrites,* and has been subsequently referred to *Waltzispora* Stapin by Reinhardt (1964). The same species with more pronounced angular prolongations has been illustrated by McGregor (1965) from the Upper Triassic of the Canadian Arctic.

Genus IRAQISPORA Singh, 1964

Type species (original designation): *Iraqispora labrata* Singh.

Iraqispora labrata Singh
Plate 3, figures 9-10

Iraqispora labrata Singh, 1964, p. 243, pl. 44, fig. 7-8.

Occurrence: Chhidru Formation: very rare to rare.

Representative Specimens: U.W.A. 57786, 57787, 57788. Field no. K11-2D, Wargal, Salt Range, Chhidru Formation.

Description: Cingulate, amb triangular, angles sharply rounded, sides straight to convex. Trilete, tetrad scar distinct, laesurae extending almost to inner margin of cingulum, bent in vicinity of proximal pole. Commissure mordered by labra about 2µ wide and 1-2µ high. Exine with a finely spongeous texture which is accentuated in some specimens in the vicinity of the proximal pole. Proximal pole bearing arcuate thickenings in the interradial areas. Cingulum unstructured, 2-10µ wide, usually broader at the angles.

Dimensions (10 specimens): Median diameter: 57 (63) 68µ.

Remarks: All the specimens measured from the Salt Range were larger than the single example for which Singh (1964) quoted dimensions. In other respects, however, these forms agree precisely with Singh's diagnosis and are placed in synonymy with his species.

Genus NEVESISPORITES de Jersey and Paten, 1964

Type species (original designation): *Nevesisporites vallatus* de Jersey and Paten.

The genus *Nevesisporites* was based on type material from the Jurassic of Queensland (de Jersey and Paten, 1964, p. 8) and designed to include circular, trilete forms which are thickened equatorially and sculptured only on the proximal face. Playford (1965) has described a further species, *N. limulatus* from the Triassic of Tasmania, with more uniform sculptural elements and, in some specimens, a thickened area at the distal pole.

No species have previously been assigned to the genus from extra-Australian sediments.

Nevesisporites fossulatus Balme, n. sp.

Plate 3, figures 1-5

Occurrence: Chhidru Formation: very rare to common. ? Khatkiara Member, Tredian Formation: very rare.

Representative Specimens: Holotype: 57789. Others: U.W.A. 57790, 57791, 57792. Field no. K11-6C, Wargal, Salt Range, Chhidru Formation.

Description: Equatorially thickened, amb circular to strongly rounded triangular. Trilete, tetrad scar distinct, laesurae extending to inner margin of the equatorial thickening, commissure bordered by narrow, slightly sinuous, raised labra. Equatorial thickening in the form of a distal annular crassitude 2-5μ wide rather than a true cingulum, although some specimens also possess a narrow equatorial rim. Exine of proximal face thin and translucent, bearing scattered and usually sparse grana 1μ or less in diameter. Distal face smooth with a faint to distinct, subcircular, darkened, polar exinal thickening.

Dimensions (30 specimens): Diameter: 26 (30) 36μ.

Derivation of name. Latin *fossula* = a little ditch, referring to the distal annular area of thinner exine lying between the polar and equatorial thickenings.

Remarks: In the constant presence of a distal polar thickening, combined with an equatorial crassitude, *Nevesisporites fossulatus* resembles forms assigned by other authors to *Stereisporites* Pflug or the various synonyms of that genus. *Leiotriletes furcatus* Bolkovitina also bears some resemblance to *N. fossulatus* (cf. Bolkhovitina, 1956, pl. 3, fig. 22), although it lacks sculpture and, according to Bolkhovitina, the polar exinal thickening is proximal. No distal polar thickening was noted in the type species of *Nevesisporites* by de Jersey and Paten, but Playford (1965) has, by implication, extended the concept of the genus to include forms possessing this feature. In *Nevesisporites fossulatus* the distal thickening is constant and the species is smaller than either *N. vallatus* (35-47μ) or *N. limulatus* (36-47μ). It shares with both these species proximal sculpture and an equatorial or subequatorial crassitudinous thickening.

Nevesisporites fossulatus was recorded definitely only from the upper part of the Chhidru Formation, where it is sometimes common. One or two specimens of *Nevesisporites* were also found in the Khatkiara Member of the Tredian Formation, but these had more densely packed sculpture and probably belong to another species. The Permian occurrence extends considerably the stratigraphic range of *Nevesisporites*, which has been previously recorded only from Jurassic and Middle to Upper Triassic strata.

Suprasubturma PERINOTRILITES Erdtman emend. Dettmann, 1963

a. Forms without equatorial modifications.

Genus PEROTRILITES Erdtman ex Couper, 1953

Type species (designated Couper, 1953): *Perotrilites granulatus* Couper.

As circumscribed by Couper the genus *Perotrilites* includes all trilete spores with a diameter less than 200μ, whether smooth or sculptured, which possess a thin, transparent "perispore" (= sculptine *sensu* Harris, 1955). The genus is not entirely satisfactory, partly because the term "perispore" is a subjective one and partly because the type species of *Perotrilites* has an unusual morphology, which is not completely clarified either by Couper's description or his illustrations. Couper in his description of *P. granulatus* described the sculptine as finely granular, but in both his illustrations it appears spinulose. The exine enveloped by the sculptine is also said to be finely granulate distally, but bears spinules or coni in the illustration of the holotype (Couper, 1953, pl. 3, fig. 28). By analogy with other genera of trilete spores, a case exists for restricting the name *Perotrilites* to forms with sculptural elements at least on the inner exine layer. *Perotrilites* in the broad sense of Couper's diagnosis has, nonetheless, been widely used, especially for Devonian and Lower Carboniferous spores. Other form genera have been proposed to accommodate cavate trilete spores, with thin transparent sculptine, but none has a completely adequate differential diagnosis. Among such genera are *Diaphanospora* Balme and Hassell, *Proprisporites* Neves, *Hymenospora* Neves, and *Ricaspora* Bharadwaj and Salujha. The diagnosis of *Ricaspora* is virtually identical with that of *Perotrilites*, but from their comparisons it is clear that Bharadwaj and Salujha (1964, p. 190-191) intended to restrict *Perotrilites* to scuptured forms with a triangular amb. They did not, however, differentiate *Ricaspora* from any of the other previously published genera of forms with a perinoid exine layer. *Ricaspora* may prove a useful genus in the future, but its use has not been extended here because of the necessity for clarification of the status of several perinoid form genera.

Perotrilites sp.
Plate 3, figures 14-15

Occurrence: Mianwali Formation: very rare to rare.

Representative Specimens: U.W.A. 57793, 57794, 57795. Field no. K6-41B, Nammal Gorge, Salt Range, Narmia Member, Mianwali Formation.

Description: Spore cavate, perinoid, amb circular. Trilete, tetrad scar distinct, laesurae extending about two-thirds radius of the intexine, commissure bordered by lips less than 1μ wide and about 1μ high. Intexine 1-2μ thick, maculate to weakly granulate, with a darkened area in the vicinity of the proximal pole. Exoexine perinoid, translucent without clear structure or sculpture, margin usually slightly eccentric with respect to that of the intexine, apparently attached proximally and sometimes invaginated along the laesurae. In some specimens the exoexine is partly removed, and it is probable that in others it is missing completely, so that they would be recorded as *Punctatisporites* or *Cyclogranisporites*.

Dimensions (11 specimens): Total diameter: 74-95μ. Diameter of intexine: 67-83μ.

Remarks: Perotrilites sp. is a rare form which appears to be confined to the Lower Triassic in the Salt Range. Its main interest lies in its basic resemblance to *Ricaspora granulata* Bharadwaj and Salujha from the Upper Permian of the Raniganj Coal Measures. The Indian species is distinct from *Perotrilites* sp., however, for it has more prominent laesurae, lacks a darkened polar area, and its exoexine is granulate.

b. Forms with equatorial modifications.

Genus KRAEUSELISPORITES
Leschik emend. Jansonius, 1962

Type species (original designation): *Kraeuselisporites dentatus* Leschik.

Despite the emendation by Jansonius and the comments of Dettmann (1963, p. 77) the circumscription of *Kraeuselisporites* is still not firmly established. Leschik's original view, that the genus is alete, has been accepted by a number of recent authors, and it is true that no clear tetrad scar is visible on the illustration of the type species. Other specimens assigned by Leschik to *Kraeuselisporites* display what are almost certainly traces of tetrad markings (cf. Leschik, 1955, pl. 4, fig. 20-21), and on general morphological grounds it seems most unlikely that spores with the form of Leschik's various species of *Kraeuselisporites* are alete. Reinhardt and Schmitz (1965) are strongly of the opinion that Leschik's specimens are trilete, although overmacerated, and unless the correctness of the original interpretation is demonstrated by reexamination of the type material of *Kraeuselisporites* their opinion may be accepted.

A more difficult question is whether it is justifiable to assign spores which are clearly cavate to *Kraeuselisporites*. No detached intexine is visible in any of Leschik's original species, although confident inferences are scarcely possible on the basis of the illustrations. Jansonius in emending the genus made no mention of a cavate condition and has subsequently stated (Jansonius, personal communication) that the Canadian species of *Kraeuselisporites* showed no obviously detached intexine.

Zonate flanged spores with distal apiculate sculpture have been assigned to *Kraeuselisporites* by some authors (e.g., Balme, 1963; Reinhardt and Schmitz, 1965) and these attributions have been in part accepted. In the present account the genus is used in the sense of Jansonius with the proviso that it may include both cavate and acavate species. If this usage is accepted, the genus *Indotriradites* Tiwari is superfluous, although it may be adopted for cavate species if study of the type material suggests that *Kraeuselisporites* should be restricted to acavate forms.

Kraeuselisporites rallus Balme, n. sp.

Figure 5; Plate 4, figures 9-13

Occurrence: Chhidru Formation: rare to dominant.

Representative Specimens: Holotype U.W.A. 57796. Others: U.W.A. 57797, 57798, 57799. Field no. K11-6C, Wargal, Salt Range, Chhidru Formation.

Description: Cavate, zonate, amb strongly rounded triangular, periphery spinose. Trilete, tetrad scar distinct, laesurae extending to the inner margin

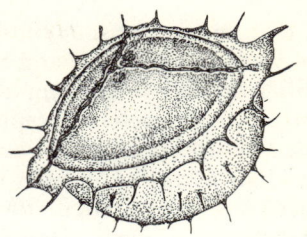

Fig. 5. *Kraeuselisporites rallus* Balme, n. sp. Reconstruction in semilateral view, ×600.

of zona sometimes traceable on to the zona itself. Commissure bordered by weak slightly sinuous lips. Exoexine less than 1μ thick, scabrate, with a thin, translucent, spinose zona. Form of zona variable, of uneven width, and in many specimens discontinuous, so that the zona is partly represented by an equatorial ring of spines, Spines projecting radially from zona and about 1μ in basal diameter and $2\text{-}5\mu$ long. Distal surface sculptured irregularly with spines similar to those on the zona, but proximal face free of sculpture or bearing sparse minute spinules. Intexine subcircular, seldom folded and almost filling the exoexinal cavity, slightly thickened and darker in color in the interradial areas immediately adjacent to the proximal pole.

Dimensions (30 specimens): Total diameter: 43 (52) 61μ. Diameter of exoexinal cavity: 32 (43) 53μ.

Derivation of name: Latin *rallus* = thin, finely textured, from the filmy, hyaline, zona.

Remarks: Kraeuselisporites rallus is a distinctive and commonly abundant form occurring in the upper part of the Chhidru Formation at several localities. It is characterized primarily by its irregular, prominently spinose zona, which in many specimens is virtually reduced to an equatorially arranged ring of spines. Only one previously described species bears any noteworthy resemblance to *K. rallus*. This is *K. altmarkensis* (Schulz), known to range in Germany from upper Buntsandstein to Lias *a* (Schulz, 1962; Reinhardt and Schmitz, 1965). Schulz originally assigned *K. altmar-*

kensis to a new genus, *Heliosporites*, but the view of Reinhardt and Schmitz that this is a junior synonym of *Kraeuselisporites* is at present accepted.

Kraeuselisporites altmarkensis is smaller than *K. rallus* and has a finely structured (? intrareticulate) zona. *Styxisporites reissingeri* Danzé-Corsin and Laveine, a species known from the Lower Jurassic of France (Briche, Danzé-Corsin, and Laveine, 1963) and Arctic Canada (McGregor, 1965, pl. 4, fig. 20), appears to be a junior synonym of *K. altmarkensis*.

Possible affinities: Lycopodiales.

Kraeuselisporites cuspidus Balme, 1963
Plate 5, figure 12

Kraeuselisporites cuspidus Balme, 1963, p. 19-20, pl. 5, fig. 9-11, fig. 2c-d.
Kraeuselisporites cuspidus Balme, Playford, 1965, p. 189, pl. 8, fig. 13.

Occurrence: Chhidru Formation: very rare. Mittiwali Member, Mianwali Formation: very rare.

Representative Specimen: U.W.A. 57800. Field no. K6-8D, Nammal Gorge, Salt Range, Mittiwali Member, Mianwali Formation.

Remarks: Specimens assigned to *Kraeuselisporites cuspidus* agreed precisely with the original diagnosis except for their smaller size. The largest example of the species measured in the Salt Range material had a diameter of 69μ, whereas the range in size of specimens from the type locality was 71-112μ.

In Western Australia *Kraeuselisporites cuspidus* is known only from Scythian strata, and Playford (1965) illustrated a single specimen from the Early Triassic of Tasmania.

Possible affinities: Lycopodiales.

Kraeuselisporites wargalensis Balme, n. sp.
Plate 4, figures 14-17

Occurrence: Chhidru Formation: rare to abundant.

Representative Specimens: Holotype: U.W.A. 57801. Others: U.W.A. 57802, 57803, 57804. Field no. K11-2D, Wargal, Salt Range, Chhidru Formation.

Description: Zonate, cavate, amb triangular, angles fairly sharply rounded. Trilete, tetrad scar faint to distinct, laesurae extending to inner margin of zona, traceable on to the zona in some specimens. Commissures bordered in most specimens by narrow raised labra. Exoexine 1-2μ thick on the distal face, thinner on the proximal with a translucent, unstructured zona 7-10μ wide. Apart from the zona the exoexine has a fine spongeous texture and is sculptured on the distal face with irregularly disposed spines or spino-mamillate processes. Sculptural elements about 2μ in basal diameter, 2-4μ long. Occasional spines occur on the inner part of the zona, but proximal face devoid of sculpture. Intexine thin, transparent, almost filling the exoexinal cavity, bearing three apical papillae and usually showing marginal crescentic folds. A faint trilete scar occurs on the proximal side of the intexine.

Dimensions (30 specimens): Total diameter: 40 (50) 58μ. Diameter of exoexinal cavity: 27 (33) 39μ.

Derivation of name: From the village of Wargal in the Salt Range.

Remarks: Kraeuselisporites wargalensis is common in most assemblages from the Chhidru Formation. It differs from any species of *Kraeuselisporites* described by Jansonius (1962) in its clearly defined detached intexine. *Cirratriradites splendens* Balme and Hennelly is often cavate, a fact not noted in the original description (cf. Balme and Hennelly, 1956b, pl. 5, fig. 59), but is considerably larger than *K. wargalensis* and has a thicker, more fleshy zona. *Kraeuselisporites chamotti* Cousminer is larger with a heavier flange and finer sculpture. *Indotriradites korbaensis* Tiwari has closely packed sculptural elements and bears a close resemblance to *Cirratriradites africanensis* Hart from which it was not differentiated by Tiwari (1964).

Leschik's forms of *Kraeuselisporites* are difficult to compare with *K. wargalensis* owing to the poor definition of their illustrations, and doubts have been expressed as to whether the species described by Leschik are in practice distinguishable from one another. The characters of *Cirratriradites australiensis* Hart are also not completely clear from its illustration (Hart, 1963, fig. 2L). It appears to be equatorially thickened and is perhaps best regarded as bearing a wedge-shaped cingulum, rather than a zona.

Possible affinities: Lycopodiales.

Genus DENSOISPORITES Weyland and Krieger emend. Dettmann, 1963

Type species (original designation): *Densoisporites velatus* Weyland and Krieger.

Dettmann (1963, p. 83) has given a full commentary on the circumscription of the genus *Densoisporites*. According to her, it is characterized by a finely spongeous, but unsculptured, exoexine, a thickened equatorial rim or cingulum, and a proximally attached papillate intexine. Trilete spores of this type occur commonly in Mesozoic sediments, especially in the Triassic, early Jurassic, and Lower Cretaceous.

Since Dettmann's account was written, Pocock (1964, p. 180) has instituted the form genus *Lygodiidites*, which is not clearly differentiated from *Densoisporites*. Morphologically, the two genera are closely similar, and *Lygodiidites balmei* Pocock appears indistinguishable from *Densoisporites velatus* Weyland and Krieger.

A lycopodiaceous origin has usually been inferred for spores of the *Densoisporites* type (Potonié, 1956; Balme, 1963), although Pocock (1964) has pointed out that they are also produced by some living species of *Lygodium*. According to Schulz (1964) the spores of *Pleuromeia rossica* Neyburg (Neuberg) from the Lower Triassic of the USSR are also assignable to *Densoisporites*.

Densoisporites playfordi (Balme, 1963)

Plate 5, figures 4-6

Lundbladispora playfordi Balme, 1963, p. 23, pl. 5, fig. 4-8, fig. 2a, b.
Densoisporites playfordi (Balme), Dettmann, 1963, p. 83.
Densoisporites playfordi (Balme), Playford, 1965, p. 191, pl. 9, fig. 6.

Occurrence: Chhidru Formation: very rare. Mianwali Formation: common to dominant.

Representative Specimens: U.W.A. 57805, 57806, 57807, 57808. Field no. K6-8D, Nammal Gorge, Salt Range, Mittiwali Member, Mianwali Formation.

Description: Cingulate, cavate, amb rounded triangular. Trilete, tetrad scar faint to distinct, laesurae extending almost to the equator. Exoexine 2-4μ thick, finely structured with a spongeous texture and thickened equatorially in a narrow rim 2-5μ wide. Intexine thin, smooth, and unstructured with three apical papillae and attached to the exoexine in the area of the proximal pole.

Dimensions (20 specimens): Total diameter: 42-91μ. Diameter of intexine: 27-56μ.

Remarks: The range in size of specimens from the Salt Range is greater than that quoted by Balme (1963) for those from the Kockatea Shale, Western Australia. Obviously forms otherwise closely similar vary widely in size, and *Densoisporites playfordi* shows a marked tendency to swell during the process of maceration. This happens in assemblages in which other species are not apparently affected, and the susceptibility of *D. playfordi* to attack by maceration reagents may be related to its finely spongeous exoexine.

Densoisporites playfordi characterizes Scythian assemblages in Western Australia and has been recorded also from Early Triassic strata in Tasmania (Playford, 1965).

Possible affinities: Lycopodiales (Selaginellaceae); cf. Lundblad (1948).

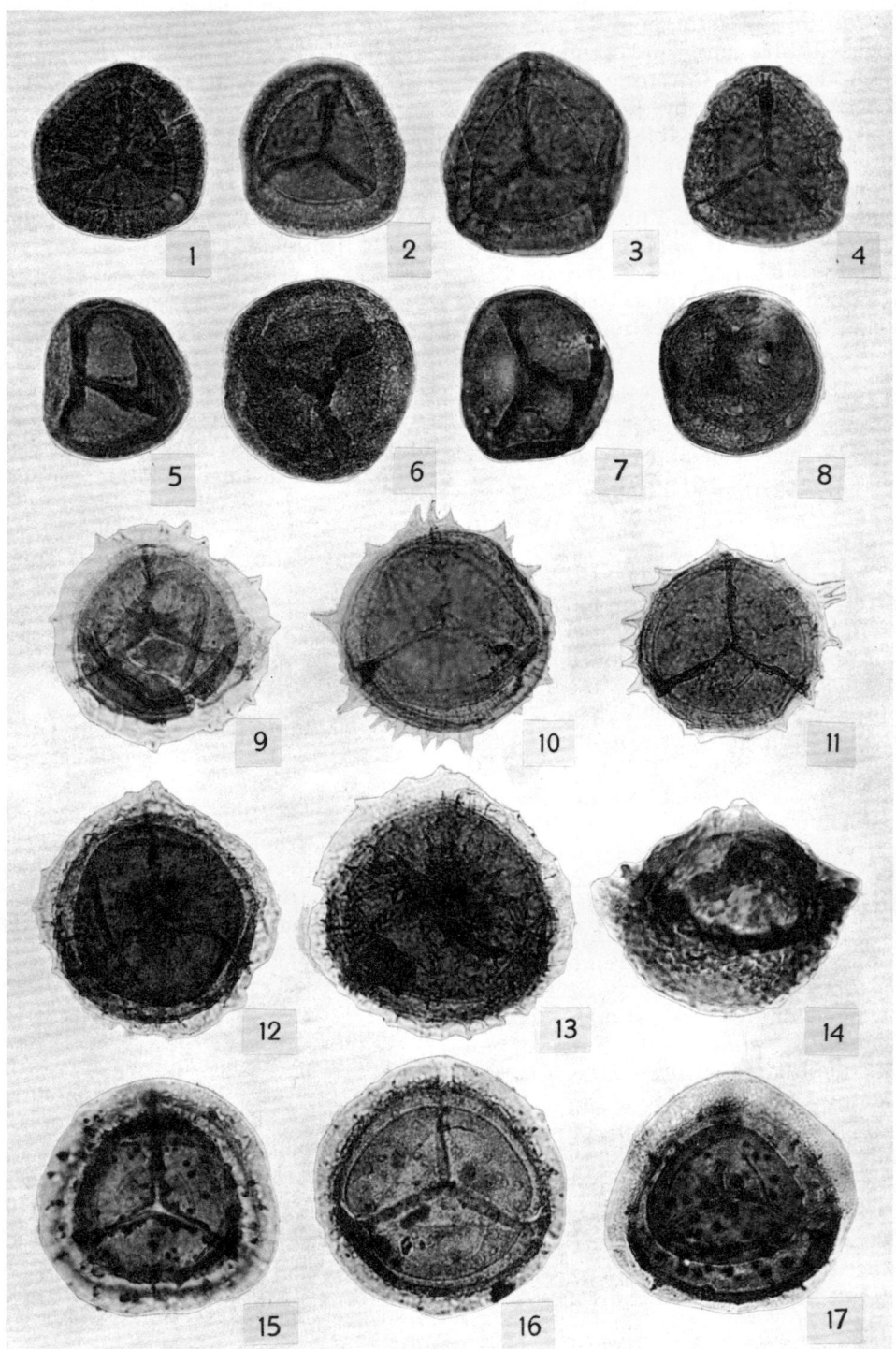

PLATE 4

Densoisporites nejburgii Schulz, comb. nov.

Plate 4, figures 5-8

Lundbladispora nejburgii Schulz, 1964, p. 604, pl. 2, fig. 8-9.
Endosporites? roeticus Reinhardt, 1964, p. 611, pl. 1, fig. 4.
Lundbladispora nejburgii Schulz, Reinhardt, and Schmitz, 1965, p. 22, pl. 1, fig. 11-12.

Occurrence: Mittiwali and Narmia Members, Mianwali Formation: rare to dominant. Landa Member, Tredian Formation: rare to dominant.

Representative Specimens: U.W.A. 57809, 57810, 57811, 57812, 57813. Field no. K12-6, Landa Pusha, Surghar Range, Narmia Member, Mianwali Formation.

Description: Cavate, amb circular to strongly rounded triangular. Trilete, tetrad scar usually distinct, laesurae extending almost to the equator, accentuated in many specimens by heavy plicate labra about 3μ high. Exoexine 2-3μ thick, with a finely spongeous structure, equatorial thickening slight or absent. Intexine indistinct, thin, subcircular, sometimes with marginal concentric folds. Margin of intexine usually excentric with respect to that of the exoexine. Apical papillae not obvious.

Dimensions (40 specimens): Diameter: 35 (40) 46μ. Diameter of intexine: 26 (30) 37μ.

Remarks: Densoisporites nejburgii was first described from the middle and upper Buntsandstein of Germany. Specimens assigned here to the species agree with the original diagnosis and resemble the illustrations given by Schulz (1964) and Reinhardt and Schmitz (1965). *Densoisporites playfordi* is generally larger than *D. nejburgii*, which has a narrow size range, has a more triangular amb, clearly defined intexine, and marked equatorial rim. The absence of an obligate equatorial thickening in *D. nejburgii*, if strictly interpreted, precludes its assignation to *Densoisporites*, although in other respects it conforms to the concept of that genus. It is excluded from *Lundbladispora*, following Playford's emendation, by its lack of distal sculpture.

Possible affinities: Schulz noted the resemblance between *Densoisporites nejburgii* and the spores of *Pleuromeia rossica* Neuberg from the Triassic of the USSR, and forms with a broadly similar morphology were illustrated as *Pleuromeia* sp. by Chalyshev and Varyukhina (1962, pl. 3, fig. 4a,b).

Possible affinities: Lycopodiales.

Densoisporites complicatus Balme, n. sp.

Plate 4, figures 1-4

Occurrence: Chhidru Formation: rare to common.

Representative Specimens: Holotype: U.W.A. 57814. Others: U.W.A. 57815, 57816, 57817. Field no. K11-6D, Wargal, Salt Range, Chhidru Formation.

PLATE 4

Permian and Triassic Perinotriletes. All $\times 600$.

FIGURES

1-4. *Densoisporites complicatus* Balme, n. sp. Upper Permian.——1. Holotype, specimen 57814.——2. Unnumbered specimen from sample 54652.——3. Specimen 57815.——4. Specimen 57816.

5-8. *Densoisporites nejburgii* Schulz, comb. nov. Lower to Middle Triassic.——5. Specimen 57812.——6. Specimen 57813.——7. Unnumbered specimen from sample 54644.——8. Specimen 57809.

9-13. *Kraeuselisporites rallus* Balme, n. sp. Upper Permian.——9. Holotype, specimen 57796.——10. Specimen 57799.——11. Specimen 57798.——12. Specimen 57797.——13. Unnumbered specimen from sample 54650.

14-17. *Kraeuselisporites wargalensis* Balme, n. sp. Upper Permian.——14. Specimen 57819, equatorial view with pyramidal proximal face uppermost.——15. Specimen 57803.——16. Specimen 57802.——17. Holotype, specimen 57801.

Description: Cingulate, cavate, amb triangular, angles sharply rounded. Trilete, tetrad scar clearly defined, laesurae extending to equatorial margin, bordered by narrow sinuous labra. Exoexine 2-3µ thick on distal face, which has a fine spongeous structure: proximal face thinner, translucent, without obvious structure, but bearing fine plications radiating from the proximal pole. Exoexine in most specimens thickened equatorially into a rim 2-3µ wide in polar view. Intexine triangular with sharply rounded angles, about 1µ thick, smooth or faintly punctate, with or without clearly defined apical papillae.

Dimensions (30 specimens): Median diameter: 33 (40) 49µ. Median diameter of intexine: 22 (29) 38µ.

Derivation of name: Latin *complicare* = to fold up, from the plicate appearance of the proximal face.

Remarks: Densoisporites complicatus is differentiated from other species of the genus by its consistently labrate laesurae, plicate proximal face, and sharply triangular intexine. It does resemble, however, certain species which have been assigned to other form genera. *Psilatriletes circumundulatus* Brenner from the Lower Cretaceous of Maryland (Brenner, 1963, p. 67-68, pl. 20, fig. 4-5) is closely similar but appears to lack fine exinal structure, has a relatively larger intexine and an undulate equatorial margin.

Anguisporites anguinus Potonié and Klaus, which was originally interpreted as an acavate cingulate form (Potonié and Klaus, 1954), bears at least a superficial resemblance to *Densoisporites complicatus* and should perhaps be reexamined. Other species assigned to *Anguisporites* (e.g., *A. intonsus* Wilson and *A. minutus* Singh) may, judging from their illustrations, also be cavate, but none is likely to be confused with *D. complicatus*.

Medvedeva (1960, pl. 14, fig. 12-14) has illustrated species of *Hymenozonotriletes* which also appear cavate, but close comparisons with *D. complicatus* are not possible.

Possible affinities: Lycopodiales.

Genus LUNDBLADISPORA Balme emend. G. Playford, 1965

Type species (original designation): *Lundbladispora willmotti* Balme.

As originally defined *Lundbladispora* included three species, one of which (*L. playfordi* Balme) Dettmann (1963) subsequently transferred to *Densoisporites* Weyland and Krieger emend. Dettmann, although she proposed no formal emendation of *Lundbladispora*. This was done by Playford (1965), who excluded from the genus forms with unsculptured exines. In the amended sense, therefore, *Lundbladispora* is similar to *Densoisporites* except that its exoexine bears a predominantly distal sculpture of spines, cones, and grana. Dispersed spores of this type have been recorded from the Triassic of Australia (Balme, 1963;

PLATE 5

Upper Permian and Triassic Perinotriletes. All ×600.

FIGURES

1-3. *Lundbladispora brevicula* Balme. Lower to Middle Triassic.——1. Specimen 58130.——2. Specimen 58129.——3. Specimen 58128.

4-6. *Densoisporites playfordi* (Balme). Upper Permian to Lower Triassic.——4. Specimen 57806.——5. Specimen 57809, showing angular corrosion pits resulting from crystallization of syngenetic pyrites.——6. Specimen 57805.

7-11. *Lundbladispora obsoleta* Balme, n. sp. Upper Permian to Lower Triassic.——7. Unnumbered specimen from sample 54625.——8. Specimen 57823, semiequatorial view.——9. Holotype, specimen 57818, with pyrites corrosion on the proximal face.——10. Unnumbered specimen from sample 54625, with prominent lips bordering the trilete commissure.——11. Specimen 57822.

12. *Kraeuselisporites cuspidus* Balme. Specimen 57800.

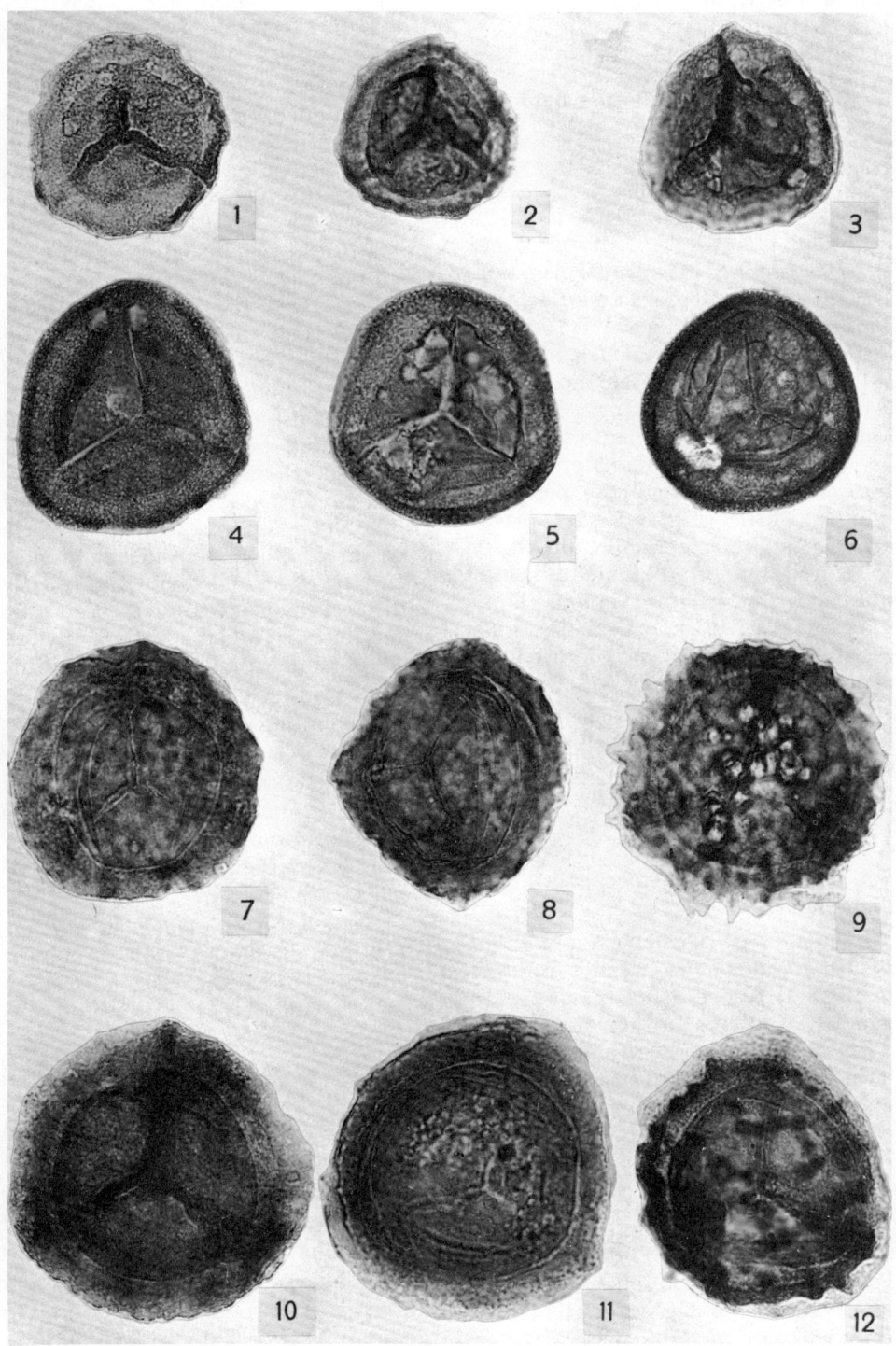

PLATE 5

Playford, 1965; Evans, 1966) and the Late Pennsylvanian of Illinois (Peppers, 1964).

Lundbladispora brevicula Balme
Plate 5, figures 1-3

Lundbladispora brevicula Balme, 1963, p. 23-24, pl. 4, fig. 8-9.
Lundbladispora brevicula Balme, Playford, 1965, p. 190, pl. 8, fig. 20-21, pl. 9, fig. 1-2.

Occurrence: Mianwali Formation: very rare to abundant. Landa Member, Tredian Formation: very rare to rare.

Representative Specimens: U.W.A. 58128, 58129, 58130. Field no. K6-8G, Nammal Gorge, Salt Range, Mittiwali Member, Mianwali Formation.

Description: Cingulate, cavate, amb rounded triangular, periphery spinose. Trilete, tetrad scar prominent, laesurae extending almost to equatorial margin, bounded in most specimens by narrow sinuous labra. Exoexine 1-2μ thick with a fine spongeous structure, thickened in a narrow equatorial rim about 2μ wide in polar view. Distal surface of exoexine bearing scattered spines about 1μ in basal diameter and 2-3μ long; spines also visible around equatorial margin. Intexine thin, smooth, and usually with concentric folds.

Dimensions (10 specimens): Median diameter: 41-51μ.

Remarks: Specimens of *Lundbladispora brevicula* are not common in the Salt Range deposits but are apparently confined to Triassic strata. They are known also from the Early Triassic of Western Australia and Tasmania.

Possible affinities: Lycopodiales.

Lundbladispora obsoleta Balme, n. sp.
Plate 5, figures 7-11

Occurrence: Chhidru Formation: very rare to rare. Mianwali Formation: very rare to abundant.

Representative Specimens: Holotype U.W.A. 57818. Others: U.W.A. 57819, 57820, 57821, 57822, 57823. Field no. K6-8D, Nammal Gorge, Salt Range, Mittiwali Member, Mianwali Formation.

Description: Cingulate, cavate, amb rounded triangular, periphery ragged. Trilete, tetrad scar indistinct, laesurae extending to equatorial margin, narrow labra present in some specimens. Exine thickened equatorially into a pronounced wedge-shaped cingulum 7-10μ wide, becoming translucent towards the periphery. Distal exoexine 5-6μ thick, proximal face thinner, exoexine with a fine to moderately coarse spongeous structure. Distal surface hemispherical, bearing conical or mamilloid processes 1-3μ high and 2-4μ basal diameter. Sculptural elements present on the cingulum and visible along equator. Intexine clearly defined, thin and translucent, almost filling the exoexinal cavity and usually with marginal, concentric, compressional folds. Intexine proximally attached with faint apical papillae visible in some specimens.

Dimensions (30 specimens): Median diameter: 58 (71) 82μ. Diameter of exoexinal cavity: 43 (53) 63μ.

Derivation of name: Latin *obsoleta* = worn out, from the persistent liability of the exoexine to post-depositional damage.

Remarks: Lundbladispora obsoleta is common in most samples from the Mianwali Formation, but most specimens are indifferently preserved. As in *Densoisporites playfordi* the spongeous exoexine of *L. obsoleta* seems prone to partial destruction, either during the initial stages of preservation or as a result of laboratory treatment.

Lundbladispora obsoleta differs from *L. willmotti* in its broader and strongly pronounced cingulum, its more sharply triangular amb, and slightly coarser spongeous structure. In *L. simoni* Peppers the distal sculpture is verrucate granulate.

Possible affinities: Lycopodiales.

Genus GUTHOERLISPORITES
Bhardwaj, 1954

Type species (original designation): *Guthoerlisporites magnificus* Bhardwaj.

Playford and Dettmann (1965) have adopted *Guthoerlisporites* as a broad form category for cavate, trilete spores with a finely and evenly reticulate exoexine. In doing so they admitted that Bhardwaj's genus needed clarification, particularly in regard to its differentiation from *Remysporites* Butterworth and Williams and *Wilsonites* Kosanke. *Guthoerlisporites* lacks completely adequate illustration, but the photomicrograph of the diplotype of *G. magnificus* given by Bhardwaj (1955, pl. 2, fig. 12) suggests that it has a fairly coarse, and somewhat disorderly, intrareticulum developed in the detached areas of the exoexine which therefore forms a true saccus, in the sense in which the term is used in this account. *Guthoerlisporites cancellosus* Playford and Dettmann has an exoexine with a markedly different structure. Here the intrareticulum is extremely fine, regular, and sharply defined, with rounded or slightly angular lumina less than 1μ in diameter. Such spores are known from the European Zechstein (Leschik, 1956; Grebe and Schweitzer, 1962; Klaus, 1963), but in Australia appear to occur only in the Triassic, in which they are long-ranging, at least in Tasmania (Playford, 1965).

Because of their distinctive morphology and probable biostratigraphic importance, *Endosporites velatus* Leschik, *Endosporites hexareticularis* Klaus, and *Guthoerlisporites cancellosus* Playford and Dettmann could justifiably be associated under a distinct form genus. Such forms are, however, too rare in Salt Range strata to allow comprehensive study, and Playford and Dettmann's nomenclature is tentatively followed.

Guthoerlisporites cancellosus Playford and Dettmann

Plate 6, figures 14-15

Guthoerlisporites cancellosus Playford and Dettmann, 1965, p. 147-148, pl. 14, fig. 33-35.
Guthoerlisporites cancellosus Playford and Dettmann, Playford, 1965, p. 192, pl. 9, fig. 18.

Occurrence: Chhidru Formation: very rare to rare. Mittiwali Member, Mianwali Formation: very rare. Landa Member, Tredian Formation: very rare.

Representative Specimens: U.W.A. 57824, 57825, 57826, 57827. Field no. T61-178, Kathwai, Salt Range, Chhidru Formation.

Description: Cavate, amb circular, usually distorted by folding. Trilete, tetrad scar faint to distinct, laesurae extending to about the margin of the intexine. Exoexine less than 1μ thick, translucent, often with occasional radial folds, finely and extremely regularly microreticulate. Lumina of reticulum less than 1μ in diameter, rounded or slightly angular; muri about 0.5μ wide, sharply defined in plane of optimum focus. Intexine circular or distorted by compression, 1-3μ thick, smooth.

Dimensions (5 specimens): Diameter of exoexine: 92-109μ. Diameter of intexine: 37-46μ.

Remarks: Guthoerlisporites cancellosus may prove difficult to separate from *Endosporites velatus* Leschik and *Endosporites hexareticularis* Klaus. *E. velatus* was described as granulate by Leschik (1956), although this interpretation does not seem to have been accepted by Potonié (1958, p. 40), who suggested that the species was referable to *Guthoerlisporites*. *E. hexareticularis* is, for the present, distinguished from *G. cancellosus* by its slightly coarser exoexine structure and hexagonal lumina.

The structure of the exoexine in *Guthoerlisporites cancellosus* approaches that of the columellate sacci of dispersed pollen grains normally accepted as gymnospermous. In the extreme regularity and clear definition of its exoexinal microreticulum, however, *G. cancellosus* has no close analogues among the pollen grains of any known living or fossil gymnosperms. It is presumed, therefore, to be a pteridophyte spore.

Turma MONOLETES Ibrahim, 1933
Suprasubturma ACAVATOMONOLETES Dettmann, 1963
Subturma AZONOMONOLETES Luber, 1935

Genus LAEVIGATOSPORITES Ibrahim, 1933

Type species (original designation): *Laevigatosporites vulgaris* Ibrahim.

Laevigatosporites callosus Balme, n. sp.

Plate 6, figures 16-18

Occurrence: Chhidru Formation: very rare to common.

Representative Specimens: Holotype: U.W.A. 57829. Others: U.W.A. 57830, 57831, 57832. Field no. K11-6A, Wargal, Salt Range, Chhidru Formation.

Description: Amb broadly oval, frequently preserved in off-polar compressions, proximal face somewhat flattened. Monolete, tetrad scar distinct, laesura extending about two-thirds length of spore, commissure set in a distinct groove which is, in some specimens, bordered by weak labra. Exine 2-4μ thick, rarely folded, smooth and structureless: in many specimens the exine is locally thickened in an area about 10μ in diameter on the distal face.

Dimensions (30 specimens): Length: 69 (77) 89μ. Breadth: 51 (62) 76μ.

Derivation of name: Latin *callosus* = hard-skinned, from the heavy exine.

Remarks: Opinion has been divided on whether unsculptured monolete spores with a relatively thick rigid exine should be assigned to *Laevigatosporites*, the type species of which is thin-walled and readily folded. A case certainly exists, on stratigraphic grounds, for separating the two form categories, as species with thick exines are characteristically post-Paleozoic, and *Monolites* Cookson ex Potonié has been adopted by some authors (e.g., Couper, 1958). In practice, however, there seems no possibility of finding any satisfactory criteria for separating a genus based on *Monolites major* Cookson from the pre-existing *Laevigatosporites*.

Laevigatosporites callosus is distinguished from other species of the genus especially by the curious callus-like thickening that frequently occurs on its distal surface. Even specimens which lack the thickened area are well characterized by their combination of large size, thick exine, and prominent tetrad scar with its pronounced groove of commissure. *Latosporites planorbis* Imgrund from the Kaiping Basin (Imgrund, 1960) is of comparable size, but has a thinner exine and more nearly circular amb. In *Latosporites intragranulosus* Singh, the exine is finely sculptured.

Although never common, *Laevigatosporites callosus* is a distinctive form which occurred in most samples from the Chhidru Formation. It was not identified in any other stratigraphic unit.

Genus POLYPODIISPORITES Potonié and Gelletich ex Potonié, 1956

Type species (designated Potonié, 1956): *Polypodiisporites favus* (Potonié).

Polypodiisporites embraces monolete spores sculptured proximally and distally with flattened verrucae or polygonal processes. The sculptural elements are closely and evenly spaced so that the surface of the exine appears negatively reticulate. Playford and Dettmann (1965, p. 150) discussed the long and rather confused history of the generic name and concluded that *Thymospora* Wilson and Venkatachala is a junior subjective synonym of *Polypodiisporites*. In the absence of an adequate differential diagnosis of *Thymospora*, this opinion has been accepted here.

Polypodiisporites mutabilis Balme, n. sp.

Plate 6, figures 7-9

Occurrence: Chhidru Formation: rare to common.

Representative Specimens: Holotype: U.W.A. 57833. Others: U.W.A. 57834, 57835, 57836. Field no. K11-2D, Wargal, Salt Range, Chhidru Formation.

Description: Amb broadly oval, periphery slightly undulate, planoconvex to biconvex in lateral view. Monolete or asymmetrically trilete, tetrad scar often obscured by sculpture, laesura extending about three-fourths length of the spore. Exine 2-3μ thick, sculptured proximally and distally with irregular verrucae or rugulae with gently rounded crests. Sculptural elements about 2μ wide, 1-2μ high and separated by sinuous channels less than 1μ wide which form a well-defined negative reticulum.

Dimensions (30 specimens): Length: 35 (41) 46μ. Breadth: 28 (35) 41μ.

Derivation of name: Latin *mutabilis* = changeable, from the variable form of the tetrad scar.

Remarks: Polypodiisporites mutabilis is distinguished from other species of *Polypodiisporites* principally on the basis of its ambivalent tetrad scar, which may be monolete or asymmetrically trilete. In this character the species resembles *Leschikisporis* Potonié, which Bharadwaj and Singh (1964) consider, correctly I believe, to be essentially monolete. Similar variations in the tetrad scar have been noted in the Carboniferous form species *Thymospora pseudothiesseni* (Kosanke), which has been studied in detail by Wilson and Venkatachala (1963). Trilete and monolete spores are also known to occur together in many species of living pteridophytes, especially among the ferns (cf. Selling, 1946). Specimens assigned here to *P. mutabilis* have a basic bilateral symmetry even though the tetrad scar may be linear, geniculate, or skewed trilete, and are most appropriately regarded as monolete for the purpose of classification.

Polypodiisporites ipsviciensis (de Jersey) is smaller than *P. mutabilis* and has a relatively coarser sculpture. *Thymospora opaqua* Singh is also small and appears to have weak labra along the scar.

Genus RETICULOIDOSPORITES Pflug, 1953

Type species (original designation): *Reticuloidosporites dentatus* Pflug.

Dettmann (1963, p. 86) has discussed the circumscription of *Reticuloidosporites* and concluded that it is distinguished from *Polypodiisporites* by the form and arrangement of its sculptural elements. In *Reticuloidosporites* they are elongate, more widely separated, and do not give rise to the negative reticulum which characterizes *Polypodiisporites*. Clearly there is a morphological convergence between the two form genera, and the distinction between them is an arbitrary one at the extremes of variation.

Reticuloidosporites warchianus Balme, n. sp.

Plate 6, figures 4-6

Occurrence: Amb Formation: rare. Chhidru Formation: very rare to rare.

Representative Specimens: Holotype: U.W.A. 57837. Others: U.W.A. 57838, 57839, 57840. Field no. T62-350, Warchha Water Tank, Salt Range, Amb Formation.

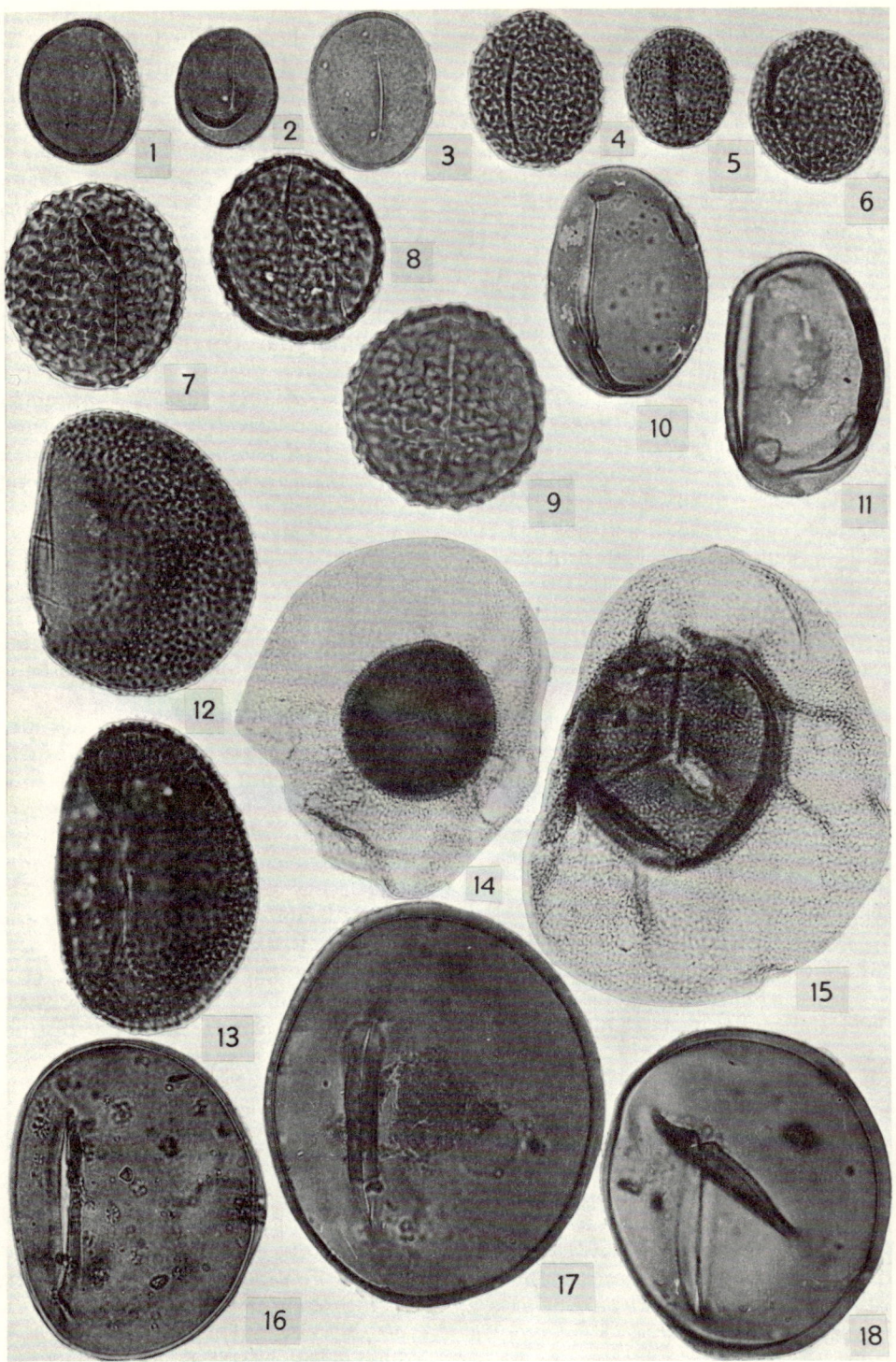

Plate 6

Description: Amb oval, outline plano-convex in lateral view, periphery undulate. Monolete, tetrad scar distinct, laesura extending almost the full length of the proximal face, commissure bordered by narrow lips less than 1μ wide. Exine 1-2μ thick, sculptured proximally and distally by processes of variable form. Consisting predominantly of low sinuous ridges about 1μ high, 1μ wide, and 1-3μ apart, but occasional coni, grana, and verrucae may also occur.

Dimensions (30 specimens): Length: 22 (24) 28μ. Breadth: 18 (20) 23μ.

Derivation of name: From the village of Warchha in the Salt Range.

Remarks: Although forms placed in *Reticuloidosporites* are considered, on the grounds of constant association and general morphological similarity, to form a homogenous population, they could, on the criterion of sculpture, be assigned to two or three distinct form genera. The majority of specimens are closest to *Reticuloidosporites*, as their sculptural elements are too irregular and widely spaced to form the well-defined negative reticulum, which is regarded as an essential character of *Polypodiisporites*.

Reticuloidosporites warchianus is smaller and has finer sculpture than *Polypodiisporites mutabilis*. Its tetrad marking is also invariably monolete.

Genus POLYPODIIDITES Ross, 1949

Type species (by original monotypy): *Polypodiidites senonicus* Ross.

Polypodiidites is used here to accommodate monolete spores with verrucate or spinose sculpture, which is absent or much reduced on the proximal face.

Polypodiidites sp.

Plate 6, figures 12-13

Occurrence: Chhidru Formation: very rare to rare.

Representative Specimens: U.W.A. 57841, 57842, 57843. Field no. K11-2C, Wargal, Salt Range, Chhidru Formation.

Description: Amb oval, periphery notched, plano-convex or concavo-convex in lateral view, proximal face strongly inflated. Monolete, tetrad scar distinct, laesura extending about three-fourths length of proximal face, commissure bordered by narrow lips in some specimens. Contact areas devoid of sculpture, weakly to markedly depressed. Exine about 2μ thick, densely sculptured on the distal face with small, discrete verrucae and coni. Individual elements 1-2μ in basal diameter, about 1μ high and 1-2μ apart.

Dimensions (10 specimens): Length: 45 (60) 68μ. Breadth: 32 (41) 44μ.

PLATE 6

Permian and Triassic Azonomonoletes and Perinotriletes. All ×600.

FIGURES

1-3. *Punctatosporites* sp. cf. *P. minutus* Ibrahim. Upper Permian.——1. Specimen 57851.——2. Specimen 57850.——3. Specimen 57849.

4-6. *Reticuloidosporites warchianus* Balme, n. sp. Lower to Upper Permian.——4. Holotype, specimen 57837.——5. Specimen 57838.——6. Specimen 57839.

7-9. *Polypodiisporites mutabilis* Balme, n. sp. Upper Permian.——7. Specimen 57835.——8. Specimen 57834.——9. Holotype, specimen 57833.

10-11. *Lunulasporites vulgaris* Wilson. Upper Permian.——10. Specimen 57845.——11. Specimen 57846.

12-13. *Polypodiidites* sp. Upper Permian.——12. Specimen 57842, lateral view with proximal side facing left.——13. Specimen 57843, proximal view showing smooth contact faces.

14-15. *Guthoerlisporites cancellosus* Playford and Dettmann.——Upper Permian to Middle Triassic.——14. Specimen 57826.——15. Specimen 57827.

16-18. *Laevigatosporites callosus* Balme, n. sp. Upper Permian.——16. Holotype, specimen 57829, showing distal exinal thickening on right hand margin of photograph.——17. Specimen 57832.——18. Specimen 57831.

Remarks: Polypodiidites sp. is a rare but distinctive form, recognized only in samples from the Chhidru Formation. It differs from any species previously assigned to *Polypodiidites* in its smaller, more widely separated, sculptural elements and its prominent smooth contact areas.

Genus LUNULASPORITES Wilson, 1962

Type species (original designation): *Lunulasporites vulgaris* Wilson.

Lunulasporites vulgaris Wilson

Plate 6, figures 10-11

Lunulasporites vulgaris Wilson, 1962, p. 13, pl. 1, fig. 8.

Occurrence: Chhidru Formation: very rare.

Representative Specimens: U.W.A. 57844, 57845, 57846, 57847. Field no. K11-6C, Wargal, Salt Range, Chhidru Formation.

Description: Amb oval, concavo-convex in lateral view. Monolete, tetrad scar distinct, laesura extending about three-fourths length of spore body. Contact areas markedly developed with slight curvaturae, which are more pronounced near the extremities of the laesura. Exine about 2μ thick, seldom folded, smooth or faintly punctate.

Dimensions (10 specimens): Length: 39 (49) 62μ. Breadth: 33 (37) 43μ.

Remarks: Wilson (1962) illustrated *Lunulasporites vulgaris* only with a single specimen, so the possible range of variation within the species is difficult to judge. Specimens from the Salt Range are thinner walled than the range quoted in the original specific diagnosis, but otherwise seem inseparable. The species is rare in the Salt Range sediments, but appears to be confined to the Chhidru Formation. Similar but perhaps not identical forms occur occasionally in Upper Permian sediments from the Perth Basin, Western Australia.

Genus PUNCTATOSPORITES Ibrahim, 1933

Type species (by original monotypy): *Punctatosporites minutus* Ibrahim.

Punctatosporites sp. cf. *P. minutus* Ibrahim

Plate 6, figures 1-3

Punctato-sporites minutus Ibrahim, 1933, p. 40, pl. 5, fig. 33.

Punctatosporites minutus Ibrahim, Potonié and Kremp, 1956, p. 143, pl. 19, fig. 439-441.

Occurrence: Wargal Limestone: rare. Chhidru Formation: rare to common.

Representative Specimens: U.W.A. 57848, 57849, 57850, 57851. Field no. K11-2C, Wargal, Salt Range, Chhidru Formation.

Description: Amb broadly oval. Monolete, tetrad scar distinct, laesura extending about two-thirds length of proximal face, commissure bordered by weak labra less than 1μ wide. Exine about 1μ thick, rigid and rarely folded, sculptured with faint maculae less than 1μ in diameter.

Dimensions (30 specimens): Length: 20 (24) 28μ. Breadth: 15 (19) 23μ.

Remarks: Many species of *Punctatosporites* have been described, all of them characterized by small size and maculate-granulate sculpture. The main bases for discrimination within the genus have been size and minor sculptural differences, and some of the proposed distinctions are of dubious value. Published species fall into two groups. One, exemplified by *Punctatosporites (Marattisporites) scabratus* Couper, with faint sculpture, and the other, represented by *Punctatosporites walkomi* de Jersey, bearing clearly defined grana. Both groups range from the Carboniferous onwards, and neither appears to have much biostratigraphic significance.

The sculptural pattern of *Punctatosporites* sp. is fainter than that of any of the specimens of *P. minutus* illustrated by Potonié and Kremp (1956), and the two forms are possibly distinguishable on this basis. Until more adequate information becomes available on the range of acceptable variation within the European Carboniferous species, they are placed in tentative synonymy.

Possible affinities: Spores closely resembling *Punctatosporites* have been recovered from a number of fossil marattiaceous fructifications (Couper, 1958; Potonié, 1962), and modern members of the Marattiaceae produce small monolete or pseudo-trilete spores with a granulate or spinulose sculpture.

Suprasubturma PERINOMONOLITES Erdtman, 1947

Genus ARATRISPORITES Leschik emend. Playford and Dettmann, 1965

Type species (original designation): *Aratrisporites parvispinosus* Leschik.

There seems no doubt of the biostratigraphic importance of the Triassic genus *Aratrisporites* which has been discussed by several authors, most fully by Playford and Dettmann (1965) and Playford (1965). It has a wide geographic distribution in Triassic strata, and records exist from western and central Europe (Leschik, 1955; Klaus, 1960; Bharadwaj and Singh, 1964), the USSR (Chalyshev and Varyukhina, 1962; Aristova, 1963), and Australia (Playford and Dettmann, 1965; Playford, 1965). As far as can be judged at present, *Aratrisporites* is mainly characteristic of Middle and Upper Triassic strata, although it has been recorded from the lower Scythian of Russia.

Spores of the *Aratrisporites*-type have been reported from four Triassic lycopsid fructifications: *Lycostrobus scotti* (Nathorst), and recently from three species of *Cylostrobus* Helby and Martin, from the Narrabeen Group in New South Wales, Australia (Helby and Martin, 1965). Such unusual morphological characters are combined in these spores that the presence of *Aratrisporites* in an assemblage may be confidently assumed to indicate a lycopsid element in its parent flora.

Aratrisporites fischeri (Klaus)

Plate 7, figures 5-7

Saturnisporites fischeri Klaus, 1960, p. 144-145, pl. 42, fig. 35.
Aratrisporites fischeri (Klaus), Playford and Dettmann, 1965, p. 152.

Occurrence: Narmia Member, Mianwali Formation: very rare to rare. Landa Member. Tredian Formation: rare.

Representative Specimens: U.W.A. 57856, 57857, 57858, 57859. Field no. K12-12, Landa Pusha, Surghar Range, Narmia Member, Mianwali Formation.

Description: Cavate, amb oval, usually distorted by compressional folding, periphery spinose. Monolete, scar distinct to prominent, laesura extending full length of proximal face, commissure sinuous, bordered and, in most specimens, obscured by raised, rather ragged lips 5-8μ high. Exoexine about 1μ thick with a scabrate surface and bearing scattered spinules 1-2μ long and less than 1μ in basal diameter. Intexine usually clearly defined, proximally attached and unfolded. Length and breadth of the intexine about one-half of the compressed exoexine.

Dimensions (15 specimens): Exoexine: length, 84 (91) 103μ; breadth, 62 (67) 74μ. Intexine: length, 38 (43) 46μ; breadth, 29 (32) 35μ.

Remarks: Specimens of *Aratrisporites fischeri* from the Salt Range agree in dimensions and morphology with Klaus's description of material from the type locality. The Austrian specimens are Carnian in age and therefore younger than those encountered in the present study. *Aratrisporites banksi* Playford from Tasmanian strata of probable Middle Triassic age bears a general resemblance to *A. fischeri* but is sculptured with rugulae and coni.

Possible affinities: Lycopodiales.

Aratrisporites paenulatus Playford and Dettmann

Plate 7, figures 1-4

Aratrisporites paenulatus Playford and Dettmann, 1965, p. 154, pl. 15, fig. 44-45.

Occurrence: Narmia Member, Mianwali Formation: rare to dominant. Tredian Formation: very rare to common.

Representative Specimens: U.W.A. 57852, 57853, 57854, 57855. Field no. K6-33, Nammal Gorge, Salt Range, Narmia Member, Mianwali Formation.

Description: Cavate, amb broadly oval, periphery spinose. Monolete, tetrad scar distinct, laesura extending full length of proximal face, commissure bordered and usually obscured by sinuous raised lips 3-7μ high. Exoexine about 1μ thick, finely structured with a scabrate surface pattern, bearing sparsely distributed spinules 1-2μ in basal diameter, 2-3μ high, and 2-5μ apart.

Dimensions (30 specimens): Exoexine: length, 31 (39) 44μ; breadth, 22 (28) 36μ. Intexine: length, 24 (30) 39μ; breadth, 19 (23) 28μ.

Remarks: About twenty species of *Aratrisporites* have now been described, distinguished from one another on the basis of size and sculptural patterns. Some of these may prove difficult to maintain as further data accumulate, for, as Helby and Martin (1965) have noted, the sculptural patterns of *Aratrisporites* appear to be easily affected by preservational differences. In many of the specimens included here in *Aratrisporites paenulatus* the spines are sparser than they are in the type population of Playford and Dettmann. Evidence of exine corrosion is, however, so frequent in the Salt Range material that density of sculptural processes is considered an unreliable criterion. *Aratrisporites paenulatus* is considerably smaller than *A. fischeri* and has a relatively larger but less clearly defined intexine.

Possible affinities: Lycopodiales.

Anteturma POLLENITES Potonié, 1931
Turma SACCITES Erdtman, 1947
Subturma MONOSACCITES Chitaley emend. Potonié and Kremp, 1954

Group A: Radiosymmetrical.

Genus DENSIPOLLENITES Bharadwaj, 1962

Type species (original designation): *Densipollenites indicus* Bharadwaj.

Densipollenites encompasses monosaccate pollen grains without tetrad markings or obvious germinal aperture, in which the intexine is detached from the exoexine except for a small area in the vicinity of one pole. In compressed specimens, therefore, the corpus is almost invariably placed excentri-

---

PLATE 7

Permian Monosaccites and Triassic Perinomonolites. All ×600.

FIGURES

1-4. *Aratrisporites paenulatus* Playford and Dettmann. Upper Lower to Middle Triassic. 1. Specimen 57853.——2. Unnumbered specimen from sample 54617.——3. Specimen 57855.——4. Specimen 57854.

5-7. *Aratrisporites fischeri* (Klaus). Upper Lower to Lower Middle Triassic.——5. Specimen 57859 with torn exoexine.——6. Specimen 57858.——7. Specimen 57856.

8. *Densipollenites indicus* Bharadwaj. Upper Permian. Specimen 57860, with strongly defined intexine.

9. *Plicatipollenites indicus* Lele. Lower to Upper Permian. Specimen 57867.

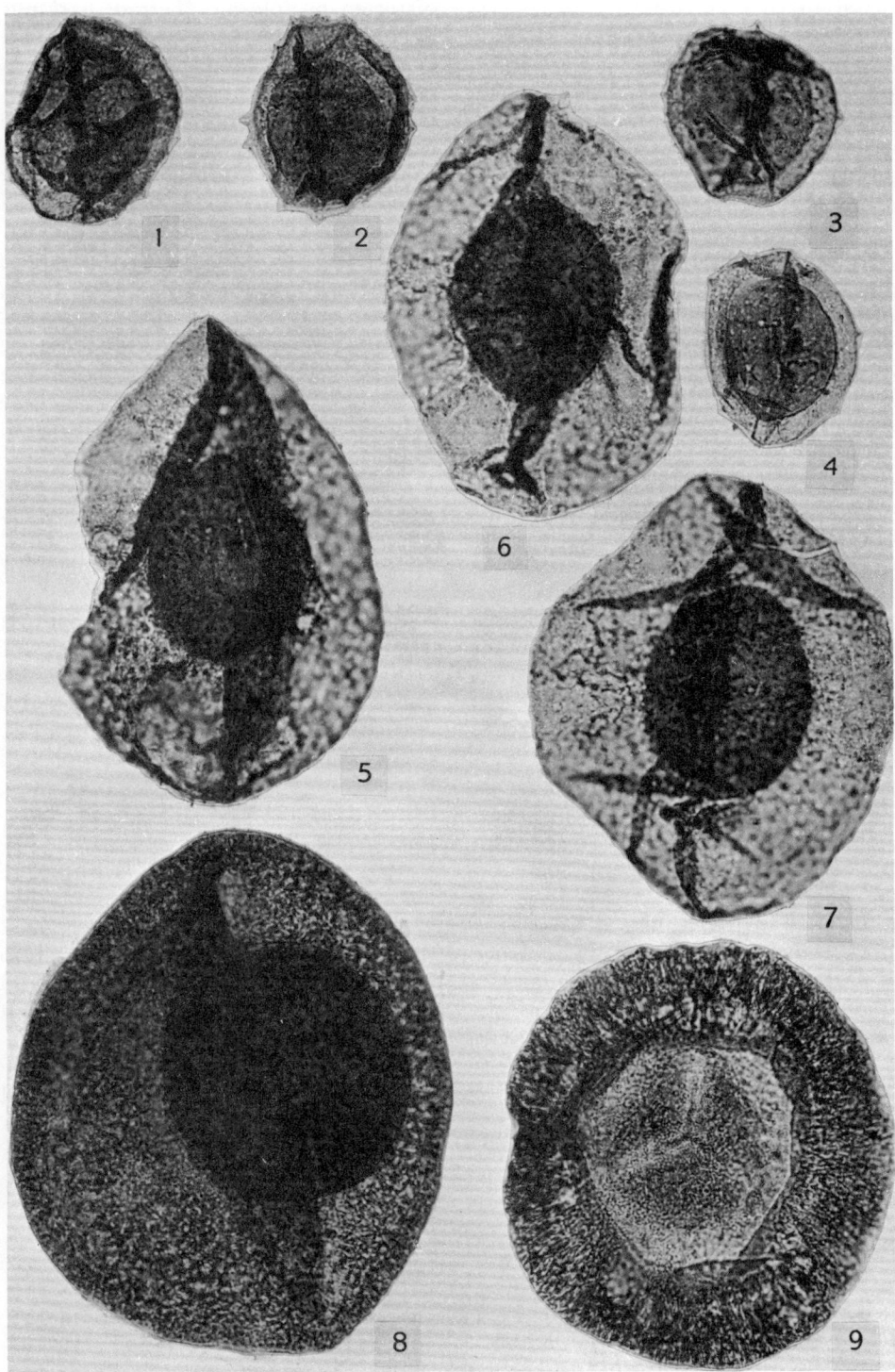

Plate 7

cally with respect to the periphery of the saccus.

Hart (1965b, p. 99) placed *Densipollenites* in synonymy with *Sehorisporites* Sukh Dev, the type species of which was said to come from the Jabalpur Series, Madhya Pradesh, and to be of Jurassic age (Sukh Dev, 1961). There is, unfortunately, some doubt about the true stratigraphic position of Sukh Dev's assemblages, which contain an admixture of Permian and Jurassic forms; and, although his new taxa conform with the requirements for valid publication, they are likely to prove biostratigraphically misleading. Apart from this, there is no certainty that *Sehorisporites* and *Densipollenites* are identical. From its single illustration *Sehorisporites indicus* Sukh Dev, the type species of *Sehorisporites* is characterized by saccus folds radiating from the corpus, a feature which does not occur in *Densipollenites*. In addition, Sukh Dev makes no mention in the diagnosis of *Sehorisporites* of any eccentricity of its corpus, which suggests that its saccus is not completely detached on either the proximal or distal faces.

For these reasons I consider *Densipollenites* to be a distinct and well-characterized form genus which may prove stratigraphically and paleobotanically important. Apart from its occurrence in the Raniganj Coal Measures, it has been previously recorded from the Upper Permian of Antarctica (Balme and Playford, 1967) and the Perth Basin, Western Australia (K. L. Segroves, unpublished data).

Densipollenites indicus Bharadwaj

Plate 7, figure 8

Densipollenites indicus Bharadwaj, 1962, p. 87, pl. 6, fig. 103-104.

Occurrence: Chhidru Formation: very rare to rare.

Representative Specimens: U.W.A. 57860, 57861, 57862. Field no. K11-4C, Wargal, Salt Range, Chhidru Formation.

Description: Monosaccate, outline circular to rounded triangular, often distorted by folding. Tetrad markings or obvious germinal opening absent. Corpus circular and usually excentrically placed with respect to the margin of the saccus; intexine of variable thickness, but typically $3-4\mu$; dark in color and with a faintly rugulose surface. Exoexine detached from intexine except for a small area on the (?) distal face and forming a columellate saccus. Saccus exoexine $5-6\mu$ thick, coarsely and densely intrareticulate, structure becoming finer towards the undetached area. Where it is unseparated from the intexine the exoexine is thinner and finely structured. Margins of exoexine limboid in compressed specimens because of the considerable thickness of the columellae.

Dimensions (20 specimens): Total diameter: 86 (104) 124μ. Diameter of corpus: 38 (50) 59μ.

Remarks: Bharadwaj and Salujha (1964) recognized three species of *Densipollenites* which were distinguished by differences in the thickness and prominence of their intexines. Observations of variations in the Salt Range assemblages suggest that these distinctions may be difficult to maintain in all areas. The specimen illustrated (pl. 7, fig. 8) falls within the circumscription of *D. indicus*, but variants occurred which possessed thinner intexines and graded into *Densipollenites invisus* Bharadwaj and Salujha. In a few specimens in which the exoexine resembled that of *Densipollenites*, no intexine at all could be discerned. These variants all occurred in single assemblages, and it was not considered expedient, for the purposes of the present study, to recognize more than one species of *Densipollenites*.

Genus CORDAITINA Samoilovich, 1953

Type species (original designation): *Cordaitina uralensis* (Luber).

No doubt exists as to the validity of *Cordaitina*, which has a proper diagnosis (Samoilovich, 1953) and adequately illustrated type species (Luber in Luber and Waltz, 1941). Broadly the diagnostic generic characters are a circular to oval amb, equatorially detached and inflated saccus and fairly thin corpus. By implication in Samoilovich's diagnosis, and explicitly in Luber's description of the type species, *Cordaitina* has no tetrad mark or defined germinal area. *Cordaitina* has been adopted by some Russian authors, although seldom in the precise sense of Samoilovich (cf. Medvedeva, 1960), but in general seems to have gained no wide acceptance, even in the USSR. Potonié (1958, p. 43) treated *Cordaitina* cursorily, and his comments imply that it is indistinguishable from *Perisaccus* Naumova emend. Potonié. *Perisaccus verrucalatus* Naumova, the type species of *Perisaccus*, came from the Upper Devonian of the USSR, and there is nothing in Naumova's original description or illustration to support Potonié's interpretation of its morphology. I have found forms which appear identical with *Perisaccus verrucalatus* to be common in certain Upper Devonian marine deposits in Libya, and in my view they are of nonvascular origin. Even if this is not so, they are in no way analogous in morphology to the pollen of any known gymnospermous plant, living or extinct. Although the matter will only be settled following recourse to Naumova's material, the emendations of *Perisaccus* by Klaus (1963) and Clarke (1965b) seem, with little doubt, to exclude the type species and are thus invalid. Similarly, the assignment of Permian monosaccate species to *Perisaccus* by Pant and Srivastava (1964, 1965) is inappropriate.

Hart (1965b) is the first non-Russian author to consider seriously the status of *Cordaitina*, which he emended to include all radially symmetrical forms with an inflated equatorially detached saccus, apart from those assignable to *Nuskoisporites* on the bases of a strong trilete scar and marginal limbus. Several recently proposed form genera overlap with Hart's concept of *Cordaitina*, among them *Plicatipollenites* Lele, *Virkkipollenites* Lele, *Barakarites* Bharadwaj and Tiwari, *Parasaccites* Bharadwaj and Tiwari, and *Heliosaccus* Mädler. Except for *Barakarites*, the morphology of which is clear cut, these monosaccate form genera appear difficult to distinguish, and their usefulness will be judged only in the light of future data. One established fact, however, is that the overwhelming majority of monosaccate pollen grains from the Permian of the southern continents possess a tetrad marking, although it may be faint. Of the 15 Russian species assigned by Hart to *Cordaitina*, on the other hand, only five have tetrad markings, and of these at least one, *C. psiloptera* (Luber), is doubtfully saccate. Admittedly none of the Russian forms has been fully illustrated, but there appears to be a case on geographic grounds for separating at generic level the radially symmetrical, monosaccate forms without tetrad marks. They are assigned here to *Cordaitina*, using that genus in the original sense of Samoilovich.

Cordaitina gunyalensis Pant and Srivastava, comb. nov.

Plate 8, figures 1-2

Perisaccus gunyalensis Pant and Srivastava, 1964, p. 87-88, pl. 17, fig. 24-25.

Occurrence: Mittiwali Member, Mianwali Formation: very rare to rare. Narmia Member, Mianwali Formation: very rare to common.

Representative Specimens: U.W.A. 57863, 57864, 57865, 57866. Field no. K6-41B. Nammal Gorge, Salt Range, Narmia Member. Mianwali Formation.

Description: Monosaccate, amb circular to oval. Tetrad markings and defined germinal area absent. Corpus circular to oval, diameter about two-thirds that of total diameter, exine of corpus thin, translucent, and faintly intrapunctate, frequently missing, so that the species is represented only by

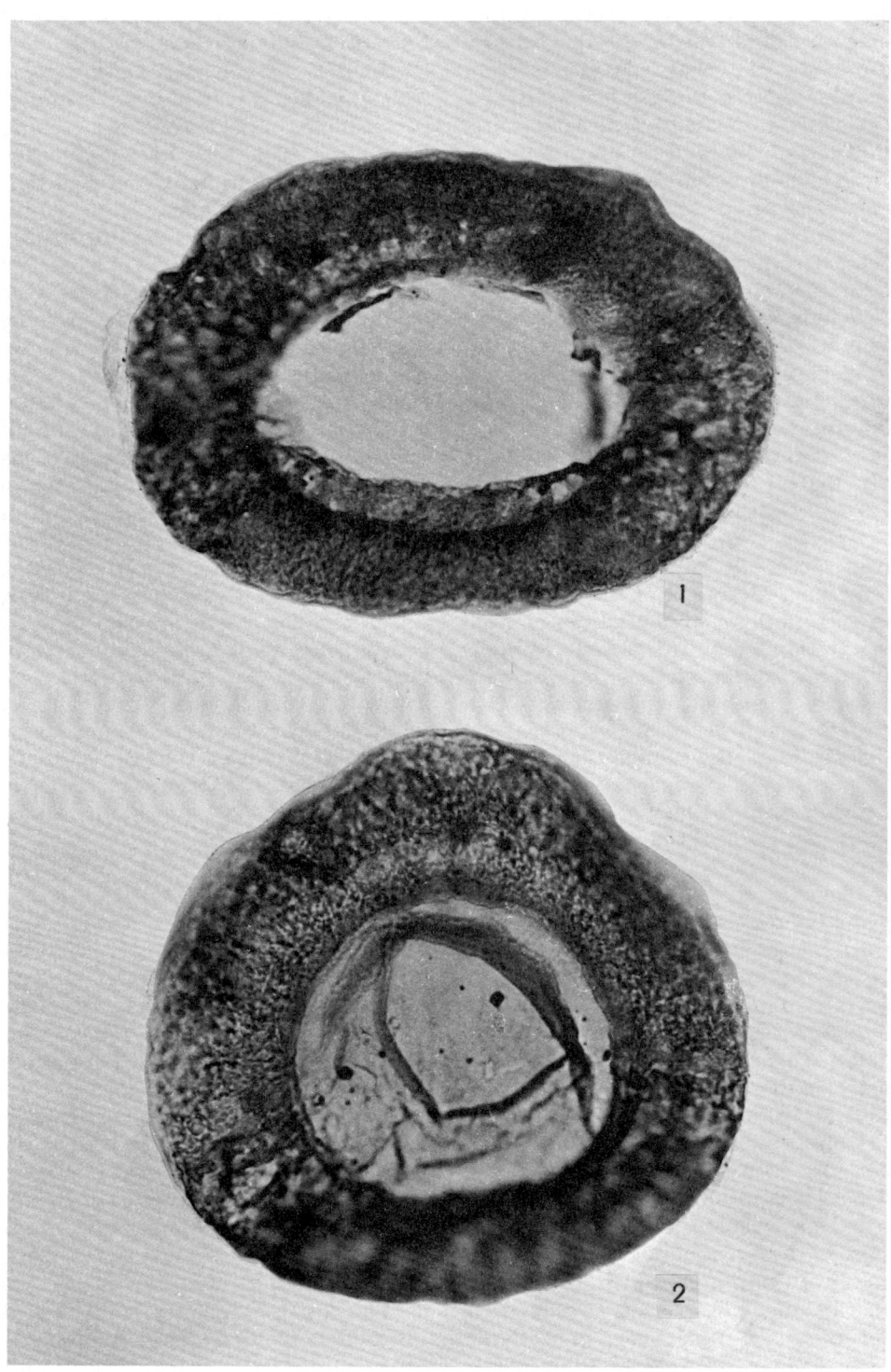

Plate 8

a hollow saccus. Corpus usually with random plications and in addition showing, in many specimens, circumpolar intexinal folds below the base of the saccus. Saccus detached equatorially with a wider overlap on one (? distal) face of the corpus than the other. Saccus exoexine about 3μ thick, intrareticulate, brochi $1\text{-}4\mu$ in diameter with a tendency to radial elongation. Marginal limbus absent.

Dimensions (20 specimens): Total diameter: 110 (142) 175μ. Corpus (intexine) diameter: 85 (108) 139μ. Maximum width of saccus: 28 (36) 51μ.

Remarks: Although Pant and Srivastava gave no precise stratigraphic horizon for the original population of *Cordaitina gunyalensis,* it probably came from the Landa Member of the Tredian Formation; that is, from a higher horizon than any of the specimens encountered during the present investigation. The species, although seldom common, appears to be confined to Triassic strata in the Salt Range. It shows a wide range in size, but is always considerably larger than any of the Russian species assigned to *Cordaitina* by Hart.

Genus PLICATIPOLLENITES Lele, 1964

Type species (original designation): *Plicatipollenites indicus* Lele.

With certain reservations, which have been briefly summarized in the discussion of *Cordaitina,* the form genus *Plicatipollenites* is used here to accommodate monosaccate, trilete, radially symmetrical forms, which develop prominent circumpolar folds in the intexine below the base of the saccus. Such forms were rarely encountered during the present investigations, although from Virkki's (1946) account they are common in the lower units of the Permian succession.

Plicatipollenites indicus Lele

Plate 7, figure 9

Plicatipollenites indicus Lele, 1964, p. 152-154, fig. 3 a-f, 12a; pl. 1, fig. 6-10.

For further synonymy see Lele (1964, p. 154).

Occurrence: Amb formation: very rare. Chhidru Formation: very rare.

Representative Specimen: U.W.A. 57867. Field no. T61-207, Wargal, Salt Range, Chhidru Formation.

Description: Monosaccate, amb circular, periphery slightly ragged. Trilete, scar distinct, laesurae extending almost to the proximal saccus root. Corpus circular, exoexine of corpus intrareticulate to intravermiculate, intexine very thin with circumpolar folds developed below the distal saccus root. Saccus exoexine $1\text{-}2\mu$ thick, densely intrareticulate, brochi with a marked tendency to radial elongation. Saccus detached equatorially with a wider overlap of intexine on the distal face than the proximal.

Dimensions (5 specimens): Total diameter: $90\text{-}108\mu$. Diameter of corpus: $65\text{-}77\mu$. Maximum width of saccus: $20\text{-}23\mu$.

Remarks: Although I agree in general with Lele's interpretation of *Plicatipollenites,* his figures give a misleading picture of the relation between intexine and exoexine. This results from his following the common tendency of regarding the saccus as an appended structure rather than a simple equatorial detachment of the outer layer of the exine. Klaus (1963, p. 256) has illustrated the essential morphology of monosaccate forms in his

---

PLATE 8

Triassic Monosaccites. Both $\times 600$.

FIGURES

1-2. *Cordaitina gunyalensis* Pant and Srivastava, comb. nov. Lower Triassic.——1. Specimen from strew mount 57863, with corpus missing.——2. Intact specimen, also from strew mount 57863.

diagrammatic reconstructions of *Nuskoisporites dulhuntyi* Potonié and Klaus, and with little modification this applies to *Plicatipollenites*.

Forms similar to *Plicatipollenites* characterize Permian strata in the southern continents and peninsular India. They have been recorded also from the Permian of Iraq (Singh, 1964), and in some of the illustrations of Medvedeva (1960) Permian species from Russia appear to possess the characteristic intexinal folds.

Group B: Bilaterally symmetrical.

Genus POTONIEISPORITES Bhardwaj, 1954

Type species (original designation): *Potonieisporites novicus* Bhardwaj.

The form genus *Potonieisporites* was based on material of Stephanian age from the Saar Basin and originally illustrated by line drawings, although photomicrographs of a "diplotype" were subsequently provided by Bhardwaj (1955). Bharadwaj (1964) has revised his interpretation of the form genus and placed in synonymy with it *Sahnites* Pant and *Vestigisporites* Balme and Hennelly. The latter form genus is, however, basically disaccate with inflated, distally inclined sacci, and should be retained in a separate category. *Sahnites* is difficult to assess, as the type species is represented only by a single photograph (Mehta, 1944). This specimen is clearly disaccate and, although it could be an extreme variant of *Potonieisporites*, it seems equally likely to belong to *Vestigisporites*. As *Sahnites* is junior to both *Potonieisporites* Bhardwaj and *Vestigisporites* Balme and Hennelly, and there seems no prospect of clarifying further the morphology of its holotype, the name may be safely discarded.

Potonieisporites novicus Bhardwaj

Plate 9, figures 1-2

Spore 34, Virkki, 1946, p. 130, fig. 43, pl. 6, fig. 78-79.
Potonieisporites novicus Bhardwaj, 1954, p. 520-521, fig. 10.
Potonieisporites novicus Bhardwaj, 1955, pl. 2, fig. 13-14.
Potonieisporites bilateralis Singh, 1964, p. 254-55, fig. 3, pl. 45, fig. 16.

Occurrence: Amb Formation: rare to common. Wargal Limestone: very rare to rare.

Representative Specimens: U.W.A. 57868, 57869, 57870. Field no. T62-350, Warchha Water Tank, Salt Range, Amb Formation.

Description: Monosaccate, bilaterally symmetrical, amb transversely oval. Corpus transversely oval, proximal face bearing a clearly defined, straight or geniculate tetrad scar usually with darkened margins. Exoexine of corpus finely intrareticulate or intravermiculate. Saccus equatorially detached with a slight distal inclination, saccus exoexine 4-5μ thick, finely and densely intrareticulate, brochi with marked radial elongation, especially near the proximal and distal sacci bases. Circumpolar intexinal folds present below the distal base of the saccus in some specimens.

Dimensions (30 specimens): Total breadth: 113 (157) 188μ; total height: 86 (115) 135μ. Corpus breadth: 70 (92) 124μ; corpus height: 60 (81) 92μ. Maximum breadth of saccus: 35-60μ.

Remarks: It is dubious whether the various published species of *Poto-*

PLATE 9

Permian Monosaccites: Bilaterally symmetrical forms. Both ×600.

FIGURES

1-2. *Potonieisporites novicus* Bhardwaj. Lower to lower Upper Permian.
——1. Specimen 57868, with poorly defined, elongate, transverse scar.——2. Specimen 57869, with short strongly defined geniculate scar.

PALYNOLOGY OF PERMIAN AND TRIASSIC 359

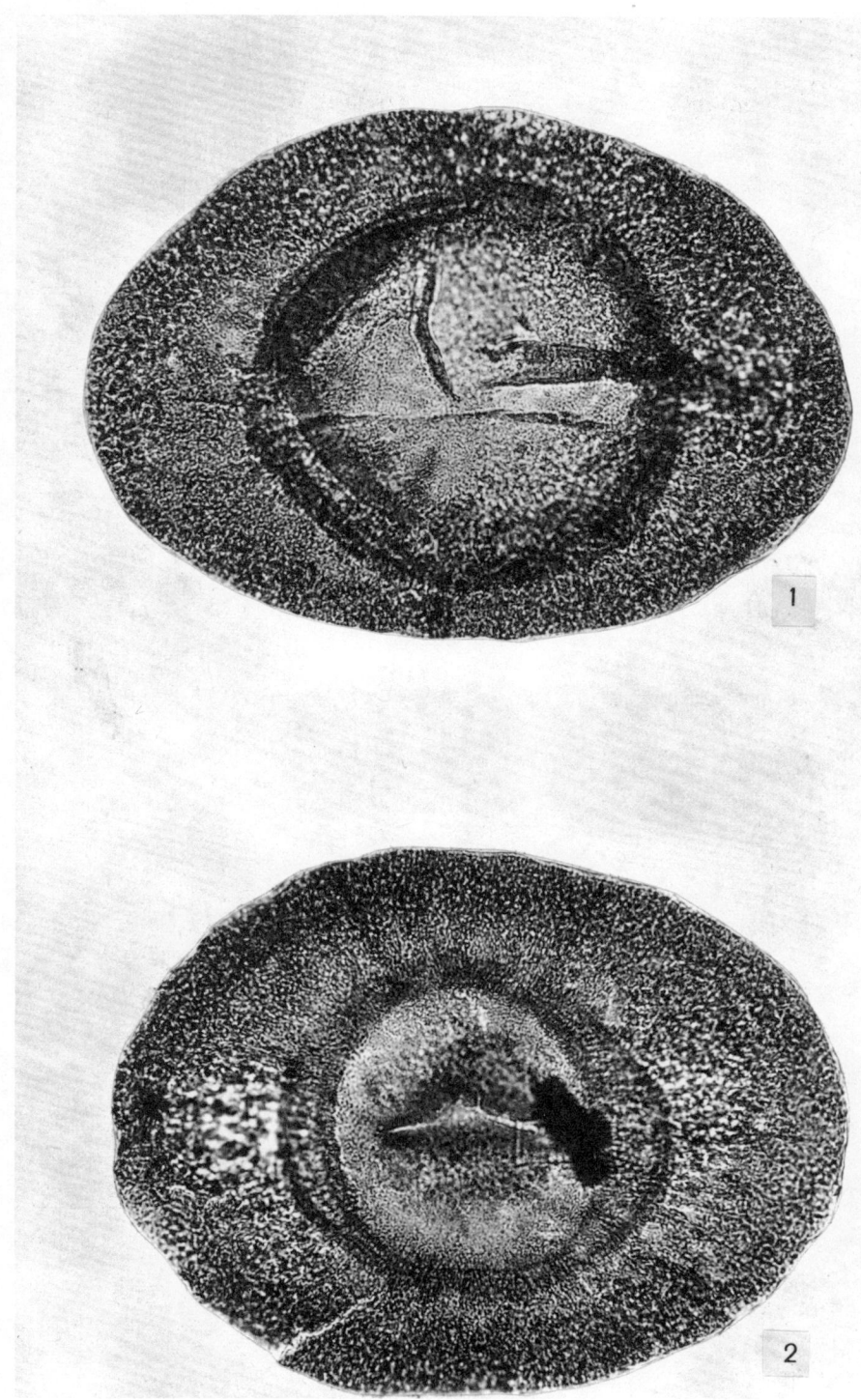

PLATE 9

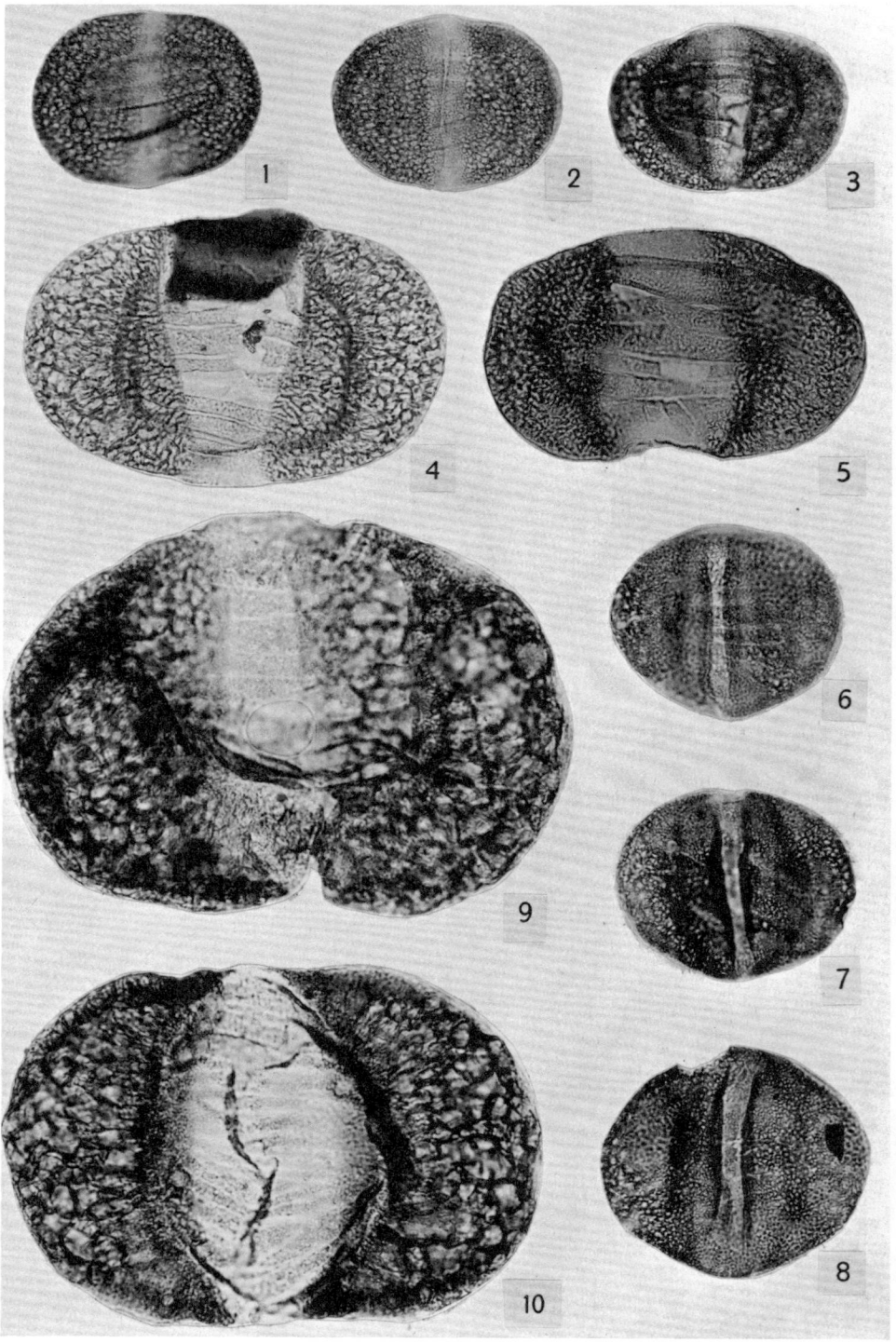

PLATE 10

nieisporites are adequately distinguished, particularly in the absence of detailed illustrations of the type species. Potonié and Lele (1961) have discussed the difficulty of separating *Potonieisporites neglectus* Potonié and Lele from *P. novicus* Bhardwaj but concluded that *P. neglectus* could be recognized by its rounded polygonal corpus and pronounced bilateral symmetry.

Potonieisporites bhardwaji Remy and Remy is said to be distinguishable from *P. novicus* by its larger size and longer monolete tetrad scar (Remy and Remy, 1961; Clarke 1965b) but these criteria are not convincing. In samples from the Amb Formation which contain many specimens of *Potonieisporites*, it is possible to find examples of most described form species of the genus. Some specimens show intexinal folds, others do not, the monolete scar may be long, short or almost nonexistent, the corpus may be distinct or faint. I can find no justification for treating these other than as a single population and have accordingly referred them to *P. novicus* Bhardwaj.

Possible affinities: Bharadwaj (1964) has re-examined pollen grains from fructifications of five species of Upper Paleozoic conifers described by Florin. These included representatives of *Lebachia, Ernestiodendron,* and *Walchianthus,* all of which produced pollen grains bearing a clear resemblance to *Potonieisporites.* It should nevertheless be remembered that bilaterally symmetrical monosaccate pollen grains are known also from Permian fructifications of presumed pteridospermic origin (Remy and Remy, 1958), so that their presence in a fossil pollen assemblage cannot be taken as clear evidence for a voltziacean component in its parent flora.

Subturma DISACCITES Cookson, 1947

a. Infraturma Striatiti Pant, 1954

Genus PROTOHAPLOXYPINUS Samoilovich emend. Hart, 1964

Type species (original designation): *Protohaploxypinus latissimus* (Luber).

Although Hart has made a convincing case for the validity and acceptability of *Protohaploxypinus* as a form genus, the type species is still inadequately known. No photographs of *P. latissimus* have, to my knowledge, ever been published, and Hart, in his restated diagnosis, gave no details of the structure of the cappa, cappula, or saccus exoexine. *Protohaploxypinus* has, following Hart's work, been fairly widely adopted by western palynologists (e.g., Clarke, 1965b; Playford, 1965; Goubin, 1965), although it seems to have gained no general acceptance in the USSR, where Zauer (1965) has assigned the type species to *Striatohaploxypinus* Zauer, a form genus of uncertain credentials.

Protohaploxypinus is used here as

PLATE 10

Permian Striatiti: Haploxylonoid forms. All ×600.

FIGURES
1-3. *Protohaploxypinus limpidus* (Balme and Hennelly). Lower to Upper Permian.——1. Specimen 57871.——2. Specimen 57873.——3. Specimen 57874.
4-5. *Protohaploxypinus varius* Bharadwaj, comb. nov. Lower to Upper Permian.——4. Specimen 57883.——5. Specimen 57885.
6-8. *Protohaploxypinus diagonalis* Balme, n. sp. Lower to Upper Permian.——6. Specimen 57880.——7. Specimen 57882.——8. Holotype, specimen 57879.
9-10. *Protohaploxypinus microcorpus* (Schaarschmidt). Upper Permian.——9. Specimen 57887. 10. Specimen 57889.

a form genus to include disaccate, haploxylonoid pollen grains with a thin intrareticulate taeniate cappa and more or less parallel-sided cappula. The taeniae are usually not strongly delineated.

Protohaploxypinus limpidus (Balme and Hennelly)

Plate 10, figures 1-3

Lueckisporites limpidus Balme and Hennelly, 1955 (May), p. 94, pl. 3, fig. 29-32.
Striatites sewardi (Virkki) partim., Pant, 1955 (Oct.), p. 762-763, pl. 19, fig. 3, *non* pl. 19, fig. 4-5.
Lunatisporites limpidus (Balme and Hennelly), Potonié, 1958, p. 53.
Taeniaesporites discurrens Leschik, 1959, p. 70-71, pl. 5, fig. 34.
Protohaploxypinus sewardi (Virkki), Hart, 1964, p. 1180, fig. 11.
Protohaploxipinus limpidus (Balme and Hennelly), Balme and Playford, 1968, p. 185, pl. 1, fig. 5.

Occurrence: Amb Formation: abundant to dominant. Wargal Limestone: common. Chhidru Formation: common to dominant.

Representative Specimens: U.W.A. 57871, 57872, 57873, 57874. Field no. K11-6C, Wargal, Salt Range, Chhidru Formation.

Description: Disaccate, haploxylonoid. Corpus circular or oval with either a slight transverse or longitudinal elongation. Cappa thin, translucent, finely intrareticulate, divided by narrow transverse clefts into 5-8 taeniae. Sacci almost hemispherical in polar view, distally inclined, saccus exoexine about 2μ thick, intrareticulate, brochi equidimensional, 1-2μ in diameter. Cappula nearly parallel-sided, breadth about one-fourth that of corpus, thin and translucent. Exine of cappula finely structured near the sacci bases, becoming structureless in the area of the distal pole.

Dimensions (30 specimens): Total breadth: 43 (53) 64μ. Corpus breadth: 27 (32) 38μ. Sacci breadth: 19 (24) 30μ. Breadth of cappula: 5 (7) 11μ. Corpus height: 27 (32) 38μ.

Remarks: Specimens of *Protohaploxypinus limpidus* from the Salt Range are generally smaller and fall within a narrower size range than those originally described by Balme and Hennelly. In other respects they are inseparable, and there is no present justification for attempting the subdivision of a form category which undoubtedly includes pollen grains from a variety of widely dispersed Permian gymnosperms.

The species occurs, usually abundantly, in all the Permian samples examined, but was not recorded from the Triassic. Similar forms are common in Permian sediments of late Sakmarian or younger age throughout Australia. They are known also from Permian strata in Africa (Hart, 1960, 1963, 1965a; Leschik, 1959), India (Virkki, 1946; Bharadwaj, 1962), and Antarctica (Balme and Playford, 1967). Russian workers have frequently illustrated pollen grains of apparently similar morphology from Upper Permian deposits, but no closely comparable types have been recorded from Europe or North America.

Protohaploxypinus goraiensis (Potonié and Lele)

Plate 11, figures 1-3

Lunatisporites goraiensis Potonié and Lele, 1961, p. 32, pl. 3, fig. 70-73.
Faunipollenites Bharadwaj, 1962, pl. 18, fig. 232.
Protohaploxypinus goraiensis (Potonié and Lele), Hart, 1964, p. 1180, fig. 13.

Occurrence: Amb Formation: rare to common. Wargal Limestone: rare.

PLATE 11
Permian Striatiti: Haploxylonoid forms. All ×600.
FIGURES
1-3. *Protohaploxypinus goraiensis* (Potonié and Lele). Lower to Upper Permian.——1. Specimen 57875.——2. Specimen 57877.——3. Specimen 57878.

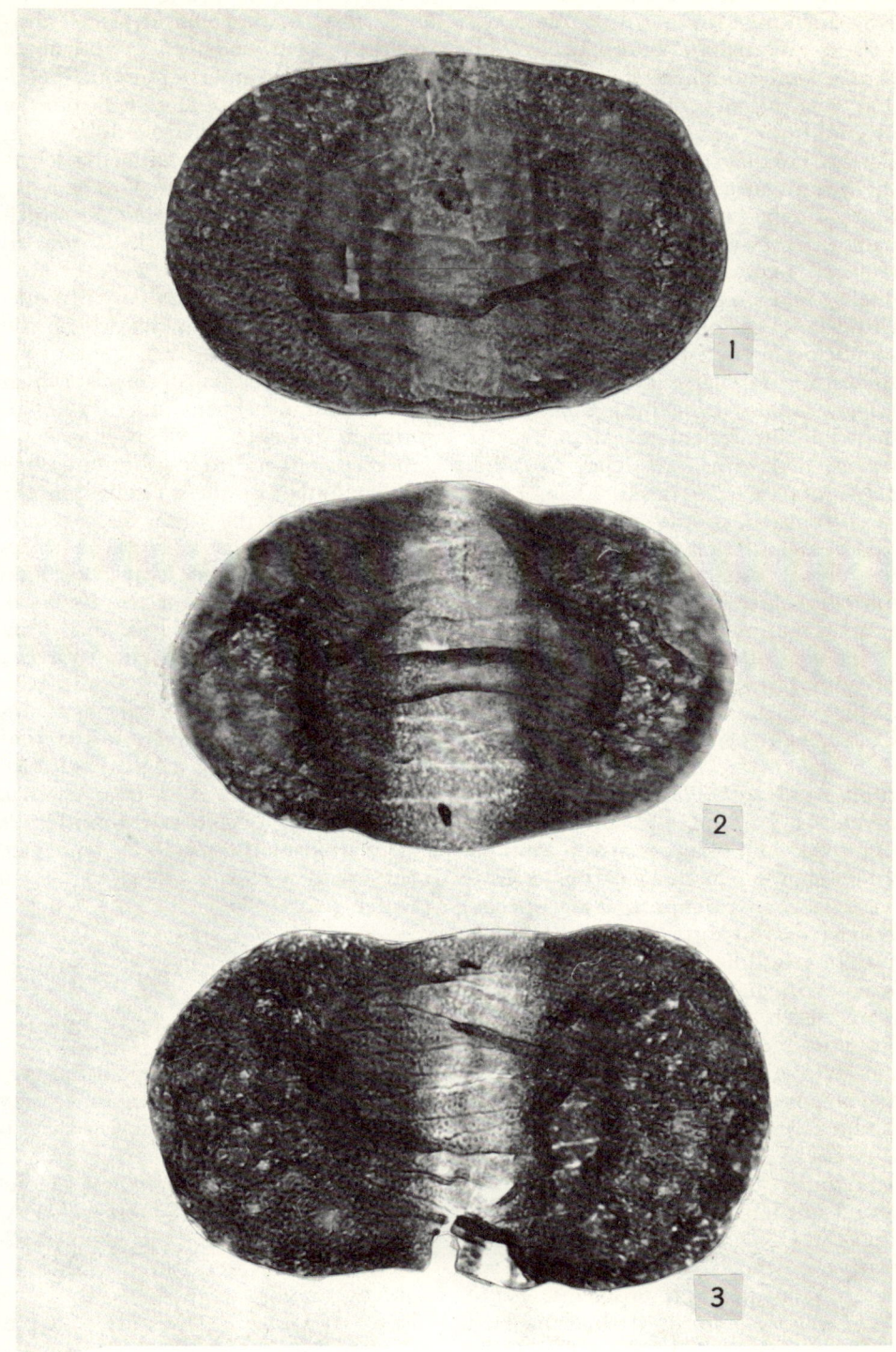

PLATE 11

Chhidru Formation: very rare to rare.

Representative Specimens: U.W.A. 57875, 57876, 57877, 57878. Field no. T62-50, Warchha Water Tank, Salt Range, Amb Formation.

Description: Disaccate, haploxylonoid or slightly diploxylonoid. Corpus circular to oval with either a slight transverse or longitudinal elongation. Cappa 1-2µ thick, divided by narrow transverse clefts into 8-12 taeniae. Exoexine of cappa distinctly, finely and regularly intrareticulate with a low marginal crest developed in some specimens. Sacci almost hemispherical in polar view, inclined distally, saccus exoexine 2-3µ thick, densely intrareticulate, diameter of brochi 1-3µ with a tendency to radial elongation near the proximal and distal sacci roots. Cappula almost parallel-sided, breadth less than one-fifth that of corpus, usually very narrow, thin and translucent with fine structure near its margins.

Dimensions (60 specimens): Total breadth: 85 (112) 155µ. Corpus breadth: 51 (69) 111µ. Breadth of cappula: 2-22µ. Height of corpus: 55 (75) 108µ.

Remarks: The distinguishing features of *Protohaploxypinus goraiensis* given by Potonié and Lele were its haploxylonoid shape, narrow cappula, and poorly defined corpus. Hart (1964), when recombining the species, restricted it to forms with 7-8 taeniae; but this is difficult to justify, for the original diagnosis stated only that there were seven or more, and none of Potonié and Lele's illustrations shows clearly the number of taeniae. Most specimens assigned here to *P. goraiensis* had ten taeniae.

Large haploxylonoid taeniate pollen grains are ubiquitous in Permian strata in the southern continents, India and the USSR. Various specific names have been used for them and there are many nomenclatural problems. *Protohaploxypinus amplus* (Balme and Hennelly), as originally conceived, included forms assigned here to *P. goraiensis*, but separation of the two species is warranted if only on the grounds that forms closely resembling the holotype of *P. amplus* were not identified during the present study. *P. amplus* needs more rigid circumscription but, pending reexamination of material from the type locality, it is taken to differ from *P. goraiensis* in the following characters:

(a) Relatively large sacci, which are slightly higher than the corpus in many specimens.

(b) A wide cappula, usually one-fourth to one-fifth as broad as the corpus.

(c) Clearly defined corpus, which is, in most specimens, slightly longitudinally elongate.

d) Tendency to develop intexine folds paralleling the margins of the cappula below the distal base of the sacci. *Protohaploxypinus pennatulus* (Andreeva) from the Upper Permian Erunakova "horizon" in the Kuznetsk Basin may be indistinguishable from *P. goraiensis*. Andreeva (in Andreeva et al., 1956) assigned both taeniate and nontaeniate forms to her species *Coniferales pennatulus*, merely stating that some specimens are slightly ribbed. The holotype shows 11 clear taeniae and appears to be excluded from Hart's emendation of the species in which he restricts it to forms possessing 14-16 taeniae (Hart, 1964).

Protohaploxypinus diagonalis Balme, n. sp.

Plate 10, figures 6-8

Occurrence: Amb Formation: rare to abundant. Wargal Limestone: very rare. Chhidru Formation: very rare to rare.

Representative Specimens: Holotype: U.W.A. 57879. Others: U.W.A. 57880. 57881, 57882. Field no. T62-350, Warchha Water Tank, Salt Range, Amb Formation.

Description: Disaccate, haploxylonoid. Corpus longitudinally oval, poorly defined. Cappa finely intrareticulate, structure sharply defined and merging without abrupt transition into that of sacci; subdivided into imperfect taeniae by 4-8 poorly defined oblique clefts, which may be simple or forked. Sacci almost hemispherical, distally inclined, saccus exoexine 1-2μ thick, finely intrareticulate, brochi 1μ or less in diameter, becoming coarser near the lateral sacci margins. Cappula breadth less than about one-fourth that of the corpus, parallel-sided or slightly constricted at the distal pole. Exine of cappula thin, faintly and indeterminately structured.

Dimensions (40 specimens): Total breadth: 40 (49) 60μ. Corpus breadth: 23 (28) 33μ. Breadth of cappula: 2-9μ. Corpus height: 32 (41) 51μ.

Derivation of name: Latin *diagonalis* = oblique, referring to direction of the proximal clefts.

Remarks: *Protohaploxypinus diagonalis* resembles *Striatites angulistriatus* Klaus in its general morphology and the oblique orientation of its proximal clefts. The Austrian species is a little smaller, however, and has a clearly defined corpus and well-marked taeniae.

Protohaploxypinus varius Bharadwaj, comb. nov.

Plate 10, figures 4-5

Lueckisporites amplus Balme and Hennelly, 1955, *partim*, pl. 3, fig. 27, non pl. 3, fig. 24-26.
Faunipollenites varius Bharadwaj, 1962, p. 95, pl. 18, fig. 230.
Faunipollenites varius Bharadwaj, Bharadwaj and Salujha, 1964, p. 210, pl. 11, fig. 150.

Occurrence: Amb Formation: very rare to common. Wargal Limestone: very rare to rare. Chhidru Formation: very rare to common.

Representative Specimens: U.W.A. 57883, 57884, 57885, 57886. Field no. K11-6A, Wargal, Salt Range, Chhidru Formation.

Description: Disaccate, haploxylonoid. Corpus poorly defined, circular to oval with a slight transverse or longitudinal elongation. Exine of cappa about 1μ thick, exoexine intrareticulate or intravermiculate with clearly defined structural elements, intexine very thin, translucent. Cappa divided into 6-9 taeniae by well-defined clefts, which are sometimes branched to give wedge-shaped taeniae. Sacci almost hemispherical in polar view, saccus exoexine 2-3μ thick, strongly and coarsely intrareticulate, brochi diameter 2-4μ with slight radial elongation near the proximal sacci bases. Cappula parallel-sided, breadth about one-half that of corpus, exine of cappula thin, translucent, usually faintly wrinkled.

Dimensions (30 specimens): Total breadth: 78 (86) 92μ. Corpus breadth: 48 (56) 65μ. Breadth of cappula: 21 (27) 35μ. Corpus height: 43 (53) 67μ.

Remarks: *Faunipollenites* was erected by Bharadwaj to include taeniate, haploxylonoid pollen grains with an ill-defined corpus and intrareticulate cappa. As defined, therefore, it appears to include the type species of *Protohaploxypinus* Samoilovich and on this ground was rejected by Hart (1964) as superfluous. Hart's view is accepted provisionally although, as pointed out elsewhere, the type species of *Protohaploxypinus* has not been fully described.

Faunipollenites varius has been previously illustrated by two specimens (Bharadwaj, 1962, pl. 18, fig. 230; Bharadwaj and Salujha, 1964, pl. 11, fig. 150), and it is difficult to determine from its diagnosis what range of variation Bharadwaj accepts. Hart placed *F. varius* in synonymy with *Protohaploxypinus limpidus* (Balme and Hennelly) but I believe these two species may be reliably distinguished. *F. varius* is larger, has a broad cappula, a corser saccus intrareticulum, and more strongly defined taeniae. In many of the specimens from the Salt Range the proximal clefts are wide (cf. pl. 10, fig. 4) and the corpus distorted. As the intexine within the clefts is invariably ruptured, the wide separation of adjacent taeniae is taken to be a secondary feature.

Certain species used by Russian authors appear close to *Protohaploxypinus varius*. An example is *Striatopodocarpites oblongatus*, illustrated but not defined by Zauer (1965, pl. 11, fig. 1).

Protohaploxypinus microcorpus (Schaarschmidt)

Plate 10, figures 9-10

Striatites jacobii auct. non Jansonius, Klaus, 1963, p. 322, pl. 17, fig. 79.
Striatites microcorpus Schaarschmidt, 1963, p. 55, pl. 14, fig. 6-7.
Protohaploxypinus microcorpus (Schaarschmidt) Clarke, 1965b, p. 338, pl. 41, fig. 3.

Occurrence: Chhidru Formation: very rare to rare.

Representative Specimens: U.W.A. 57887, 57888, 57889, 57890. Field no. K11-6A, Wargal, Salt Range, Chhidru Formation.

Description: Disaccate, haploxylonoid or slightly diploxylonoid. Corpus longitudinally elongate oval, often not clearly visible. Cappa dissected by between 12-18 poorly delineated, narrow clefts, into ill-defined taeniae, which are usually of uneven width and occasionally discontinuous and wedge-shaped. Exoexine of cappa about 1μ thick, imperfectly intrareticulate with structural elements of variable size and shape. Sacci semicircular or slightly crescentic in polar view with a pronounced distal inclination and sometimes joined equatorially. Saccus exoexine about 3μ thick, densely intrareticulate, brochi coarser near the distal saccus roots where they have a maximum diameter of about 4μ. Cappula longitudinally elongate oval or nearly parallel-sided, breadth about one-third that of the corpus, finely structured and often ruptured.

Dimensions (10 specimens): Total breadth: $114\text{-}132\mu$. Corpus breadth: $60\text{-}72\mu$. Breadth of cappula: $18\text{-}26\mu$. Height of corpus: $80\text{-}96\mu$.

Remarks: Specimens from the Salt Range referred to *Protohaploxypinus microcorpus* agree with Schaarschmidt's diagnosis except for their somewhat coarser saccus infrareticulum. However, this character is variable in specimens from the European Zechstein, which Clarke regards, with reason, to fall within the scope of Schaarschmidt's species.

P. microcorpus is never common in Salt Range sediments, but it is a distinctive form which has seldom been reported other than in the Upper Permian of Europe. A specimen illustrated by Hemer (1965) as *Protohaploxypinus jacobii* (Jansonius) Hart, from the Upper Permian of Saudi Arabia might be more appropriately referred to *P. microcorpus*.

Genus STRIATOPODOCARPITES Zoricheva and Sedova ex Sedova emend. Hart, 1964

Type species (original designation): *Striatopodocarpites tojmensis* Sedova.

In 1962 Dr. G. F. Hart drew attention to a little known paper by Sedova (1956) in which she validated *Striatopodocarpites*, a generic name previously used informally by Zoricheva and Sedova (1954). As subsequently emended by Hart (1964), the main characters of *Striatopodocarpites* are its diploxylonoid form, relatively large sacci, and circular to subcircular corpus with more than about four proximal taeniae.

Although there is no question that the form of publication of the name *Striatopodocarpites* is valid, the legitimacy of the genus has been questioned on the grounds of superfluity. Venkatachala and Kar (1964) believe that *Striatopodocarpites* encompasses the same forms as *Lunatisporites* Leschik and that the appropriate genus for pollen grains combining the characters attributed by Hart to *Striatopodocarpites* is *Strotersporites* Wilson. Such an argument is difficult to sustain. If *Lunatisporites acutus*, the type species of *Lunatisporites*, is identical with *Taeniaesporites kraeuseli*, as some authors have maintained, the characters of *Lunatisporites* clearly differ from those of *Striatopodocarpites*. If, as Potonié (1958) believes, *L. acutus* and *T. kraeuseli* are distinct, it is impossible to infer any coherent generic circumscription of *Lunatisporites* from Leschik's brief description and single illustration of the type species.

Striatites Pant, which was first validly published in 1955, has priority over *Striatopodocarpites,* and there is certainly a possibility that the two names are synonyms. Unfortunately, it is at present impossible to determine the precise morphology of the holotype of *Pityosporites sewardi* Virkki, which was cited by Pant (1955) as the type species of *Striatites* (cf. the varying interpretation of Pant, 1955, Bharadwaj, 1962, and Hart, 1964). If Virkki's original slides are found, the status of *Striatites* should certainly be reassessed, but until then there seems no alternative but to regard it as a *nomen ambiguum* and allow it to lapse.

Striatopodocarpites cancellatus (Balme and Hennelly)
Plate 12, figures 1-3

Pityosporites sewardi Virkki, 1937, *partim,* pl. 32, fig. 1c.
Pityosporites sewardi Virkki, 1946, *partim,* pl. 15, fig. 191.
Lueckisporites cancellatus Balme and Hennelly, 1955, p. 92-93, pl. 2, fig. 11-15.
Striatites cancellatus (Balme and Hennelly), Potonié, 1958, p. 51.
Striatites cancellatus (Balme and Hennelly), Hart, 1960, p. 6-7, pl. 1, fig. 10.
Striatopodocarpites cancellatus (Balme and Hennelly), Hart, 1964, p. 1182, fig. 22.

Occurrence: Amb Formation: very rare to rare. Chhidru Formation: very rare to rare.

Representative Specimens: U.W.A. 57891, 57892, 57893, 57894. Field no. K11-6C, Wargal, Salt Range, Chhidru Formation.

Description: Disaccate, diploxylonoid. Corpus circular, cappa $1-2\mu$ thick, dark in color, and without determinable exoexine structure in any of the specimens examined. Cappa divided by narrow transverse clefts into 5-8 taeniae, which are usually continuous across the whole breadth of the cappa but are sometimes wedge-shaped or constricted near the distal sacci roots. A narrow, rather ragged, marginal crest is present in some specimens. Sacci thin, somewhat flaccid, with a marked distal inclination, saccus exoexine less than 1μ thick, finely intrareticulate. Cappula nearly parallel-sided, breadth about one-fifth that of the corpus or less.

Dimensions (10 specimens): Total breadth: $39\text{-}48\mu$. Corpus diameter: $20\text{-}23\mu$. Breadth of cappula: $1\text{-}3\mu$. Height of sacci: $22\text{-}29\mu$.

Remarks: Striatopodocarpites cancellatus is widespread, although seldom common in Permian sediments from the "Gondwanaland" countries. In Australia it occurs in both Lower and Upper Permian deposits, although it is more common in the latter, especially in New South Wales and Queensland (Balme and Hennelly, 1955; Hill and Woods, 1964). Virkki (1946) recorded similar forms from the Pali Beds, Rewa, and they are known also from the Raniganj Coal Measures (Bharadwaj, 1962). In Africa *S. cancellatus* occurs in Lower Permian strata in Tanzania (Hart, 1960) and perhaps also in early Karroo deposits in the Orange Free State (Hart, 1963). Balme and Playford (1967) found small numbers of the species in Upper Permian sediments from the Amery Formation, Prince Charles Mountains, East Antarctica.

In the northern hemisphere, apart from India, *Striatopodocarpites cancellatus* appears to be absent from, or insignificant in, Permian microfloras, for no obviously similar forms have been illustrated in the now numerous publications dealing with Permian palynology.

Striatopodocarpites rarus Bharadwaj and Salujha, comb. nov.
Plate 12, figures 10-12

Lahirites rarus Bharadwaj and Salujha, 1964, p. 206-207, fig. 6D, pl. 9, fig. 128-130.

Occurrence: Amb Formation: very rare to rare. Chhidru Formation: very rare to rare.

Representative Specimens: U.W.A. 57895, 57896, 57897, 57898. Field no. K11-2C, Wargal, Salt Range, Chhidru Formation.

Description: Disaccate, diploxylonoid. Corpus circular, cappa about 2μ thick, finely and densely structured, appearing intrapunctate in surface focus. Cappa divided by narrow transverse clefts into 5-9 clearly defined taeniae, which are usually continuous over the full breadth of the cappa but occasionally wedge-shaped. Narrow marginal crest developed at the proximal margin of the cappa. Sacci large and strongly distally inclined; saccus exoexine about 1μ thick, finely to fairly coarsely intrareticulate, brochi radially elongate near the proximal sacci roots. Cappula breadth about one-third that of the corpus or less, more or less parallel-sided, usually bordered by longitudinal, crescentic, intexinal folds. Exine of cappula thin, translucent, faintly wrinkled.

Dimension (30 specimens): Total breadth: 74 (86) 99μ. Corpus diameter: 33 (38) 44μ. Breadth of cappula: 5 (10) 18μ. Height of sacci: 43 (48) 51μ.

Remarks: Bharadwaj and his co-workers have adopted a refined subdivision of the Striatiti which demands extremely well-preserved material, for it relies heavily on details of the structure of the cappa exoexine for discrimination at both generic and specific levels. Most of the specimens assigned here to *Striatopodocarpites rarus* agree well with the original diagnosis, but for quantitative purposes the species has been interpreted more widely than Bharadwaj and Salujha's classification would allow.

Pollen grains of broadly similar morphology to *Striatopodocarpites rarus* have been widely reported from India and the southern continents. They occur also in the Permian of Iraq (Singh, 1964) and the Soviet Union (Zoricheva and Sedova, 1954). *Striatopodocarpites octostriatus* Hart is close to *S. rarus*, but its present diagnosis allows no variation from eight taeniae and it may not possess a marginal crest.

Virkki (1946, pl. 2, fig. 23; pl. 5, fig. 60; pl. 13, fig. 168) has illustrated diploxylonoid Striatiti, which resemble *Striatopodocarpites rarus* in a general way, but confident comparisons are not possible.

Striatopodocarpites pantii Jansonius, comb. nov.

Plate 12, figures 7-9

Striatites samoilovitchii var. *pantii*, Jansonius, 1962, p. 68, pl. 14, fig. 14-15.
?*Taeniaesporites samoilovitchii pantii* (Jansonius), Klaus, 1963, p. 312-313, pl. 14, fig. 71-73.
Strotersporites pantii (Jansonius), Goubin, 1965, p. 1424, pl. 2, fig. 7-8.

Occurrence: Amb Formation: very rare. Chhidru Formation: very rare to common.

Representative Specimens: U.W.A. 57899, 57900, 57901, 57902. Field no. T61-226, Wargal, Salt Range, Chhidru Formation.

Description: Disaccate, slightly diploxylonoid. Corpus strongly defined, circular or oval with either a transverse or longitudinal elongation. Cappa 2-3μ thick, finely intrareticulate, divided by transverse and oblique clefts into 5-9 taeniae, which may be continuous across the full breadth of the cappa or wedge-shaped. Sacci broadly crescentic in polar view, strongly inclined distally, saccus exoexine, about 3μ thick, densely and fairly coarsely intrareticulate with radially elongate brochi, especially near the sacci bases. Cappula longitudinally elongate oval to rounded quadrilateral, breadth about one-third that of the corpus, bordered by crescentic intexinal folds lying below the sacci bases.

Dimensions (30 specimens): Total breadth: 92 (108) 124μ. Corpus breadth: 50 (59) 70μ. Breadth of cappula: 12 (17) 29μ. Height of corpus: 48 (49) 75μ. Height of sacci: 52 (69) 80μ.

Remarks: Striatites samoilovitchii var. *pantii* was first described from the Lower Triassic of Canada by Jansonius, who distinguished it from *Striatites samoilovitchii* var. *samoilovitchii* Jansonius by the form of its polar taeniae, which tended to be interrupted, wedge-shaped or oblique (cf. Jansonius, 1962, pl. 14, fig. 14-16). Klaus used the same varietal name for pollen grains from Upper Permian strata in Austria, although he recombined it under *Taeniaesporites*. There seems, however, some doubt of the identity of Klaus's specimens, for despite their general similarity to *S. samoilovitchii pantii* they possess a transverse monolete tetrad mark which is not visible in either of the illustrations of Jansonius.

Playford (1965) suggested that *Striatites samoilovitchii* var. *pantii* warranted recognition as a separate species and this had, in fact, been done by Goubin while his paper was awaiting publication. Goubin, however, accepted the view of Venkatachala and Kar that *Striatopodocarpites* is invalid and assigned the species to *Strotersporites* Wilson.

Mean dimensions of specimens of *Striatopodocarpites pantii* from the Salt Range are considerably larger than those given by Jansonius and, indeed, are greater than the extended size range allowed by Goubin. There is, however, an overlap with Goubin's figures, and the order of size of the specimens from Canada, Madagascar, and Pakistan is comparable.

The species was not encountered in Triassic strata in the Salt Range, although it ranges in Madagascar from Upper Permian to Middle Triassic and in western Canada is common in the Scythian (Jansonius, 1962). Closely similar pollen grains are not known from either Permian or Triassic strata in any other "Gondwanaland" country unless one includes Playford's record of *Protohaploxypinus samoilovitchii samoilovitchii* (Jansonius) from the Triassic of Tasmania.

Genus TAENIAESPORITES Leschik emend. Klaus, 1963

Type species (original designation): *Taeniaesporites kraeuseli* Leschik.

Although it has not been universally accepted (*vide* Grebe and Schweitzer, 1962; Schaarschmidt, 1963; Mädler, 1964), the name *Taeniaesporites*, despite the initial confusion concerning its application, has been used widely for certain Upper Permian and Triassic pollen grains from many parts of the world. There is admittedly a problem concerning the application of *Lunatisporites* Leschik. This was instituted in the same publication as *Taeniaesporites* with *Lunatisporites acutus* Leschik as type species. Several subsequent authors have stated that *L. acutus* is identical with *Taeniaesporites kraeuseli* Leschik, the type species of *Taeniaesporites*. This may be so, although it is impossible to tell from Leschik's single illustration of *L. acutus*, and the species were accepted as distinct by Potonié (1958). As there seems no possibility of clarifying the morphology of *L. acutus*, the name *Lunatisporites* should be abandoned.

Taeniaesporites has proved a difficult genus to define satisfactorily, although paradoxically its component species are among the most clearly characterized members of the Striatiti, and in practice there is little disagreement about their identification. Several authors (e.g., Jansonius, 1962; Balme, 1963; Klaus, 1963; Clarke, 1965b) have discussed the morphology and circumscription of *Taeniaesporites*, and in broad terms the interpretation of Clarke is accepted here, although I prefer not to lay down any hard and fast maximum for the number of proximal taeniae. Almost invariably there are less than five but, as Klaus points out, occasional specimens possess additional subsidiary clefts, which increases the complexity of the dissection of the cappa. The most obvious morphological feature of the haploxylonoid

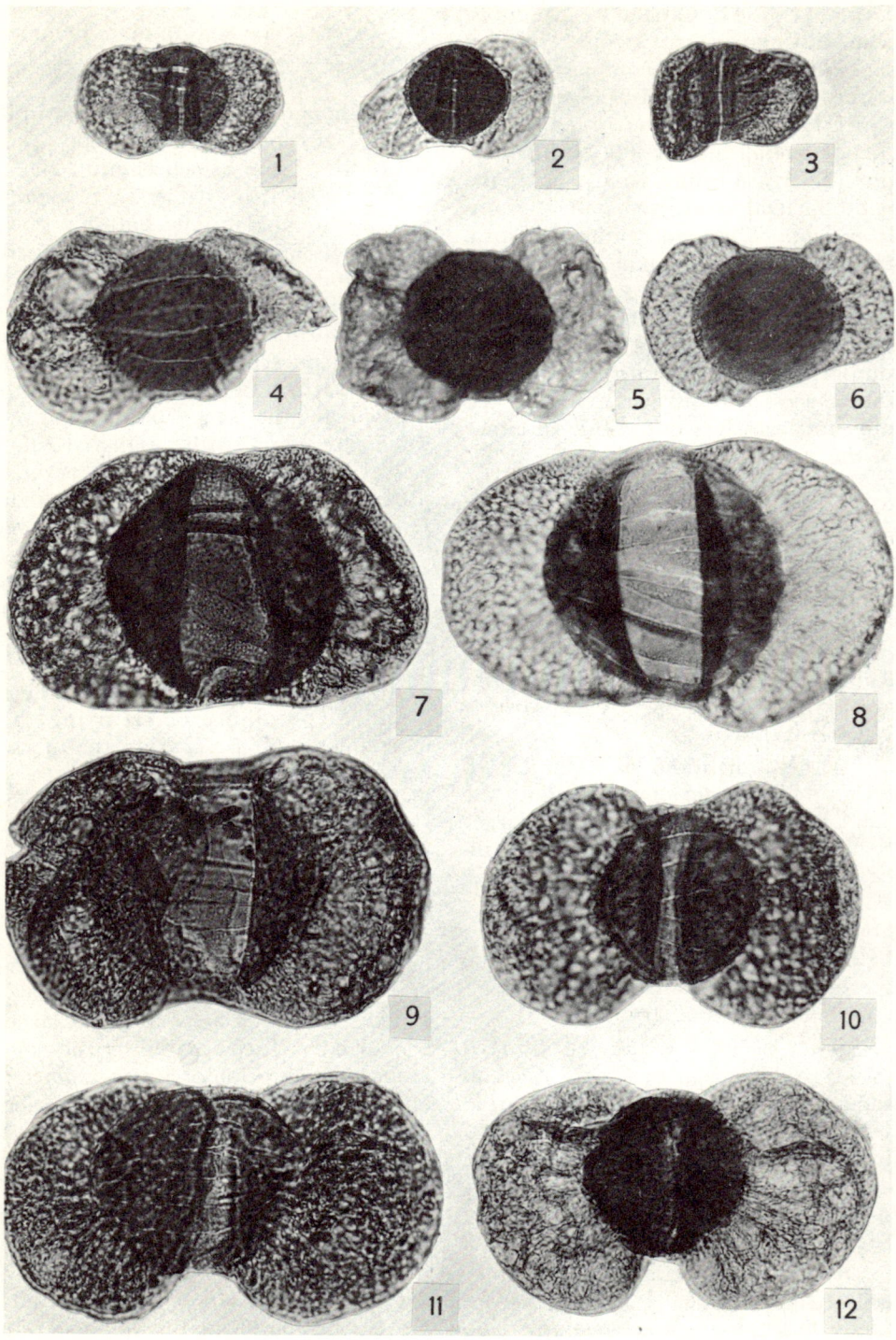

Plate 12

forms of *Taeniaesporites* is the wide spacing of their major taeniae, which are divided by expanded clefts, in which strips of thin intexine are exposed. Apart from *Lueckisporites,* with its single polar cleft, no other genus of striatitid pollen grains possesses such widely separate taeniae. Admittedly there is some inconsistency in grouping both haploxylonoid and diploxylonoid grains under the single generic name *Taeniaesporites,* in view of the systematic approach adopted towards other striatitid pollen. Diploxylonoid species of *Taeniaesporites* are, however, uncommon in assemblages from the Salt Range; and from the available data they appear, both here and elsewhere, to have the same stratigraphic distribution as the diploxylonoid forms. Apart from considerations of systematic uniformity, therefore, no strong case exists at present for subdividing *Taeniaesporites.*

Species referable to *Taeniaesporites* are widespread in Upper Permian and Triassic strata in the Northern Hemisphere. In the southern continents they have so far only been recorded from Triassic, especially Lower Triassic sediments (Balme, 1963; Goubin, 1965). To the list of generic synonymies provided by Clarke (1965b) may be added various names used, fortunately informally, by Zauer (1965). Included in these are *Paucistriatoabieites, Paucistriatoprotoconiferus* and *Paucistriatopinites* (see Zauer, 1965, pl. 16, fig. 1-5).

Taeniaesporites noviaulensis Leschik

Plate 13, figures 6-7

Taeniaesporites noviaulensis Leschik, 1956, p. 134, pl. 22, fig. 1-2.
Lueckisporites noviaulensis (Leschik), Grebe and Schweitzer, 1962, p. 11-12, pl. 5, fig. 7-8.
Taeniaesporites novimundi Jansonius, 1962, *partim,* pl. 13, fig. 25.
Taeniaesporites noviaulensis Leschik, Orlowska-Zwolinska, 1962, p. 290, pl. 2, fig. 2, ? fig. 3-4.
Taeniaesporites sp. cf. *T. noviaulensis* Leschik, Balme, 1963, *partim,* p. 28, pl. 6, fig. 6.
Striatites noviaulensis (Leschik) Schaarschmidt, 1963, p. 56, pl. 15, fig. 6, ? fig. 5, ? fig. 6-9.
Taeniaesporites noviaulensis Leschik, Freudenthal, 1964, pl. 5, fig. 2, ? pl. 4, fig. 3-5.
Taeniaesporites noviaulensis Leschik, Clarke, 1956b, p. 333, pl. 42, fig. 6-7.
Taeniaesporites noviaulensis Leschik, Goubin, 1965, p. 1422, pl. 2, fig. 1.

Occurrence: Chhidru Formation: very rare. Mianwali Formation: rare to dominant. Landa Member, Tredian Formation: very rare to common.

Representative Specimens: U.W.A. 57903, 57904, 57905, 57906, 57907. Field no. T61-226, Wargal, Salt Range, Kathwai Member, Mianwali Formation.

Description: Disaccate, slightly diploxylonoid. Corpus circular to transversely elongate oval, sometimes rather angular in polar view. Cappula dissected by three broad clefts into four major taeniae about 8-10μ wide. Exoexine of cappa 1-2μ thick, finely intrareticulate, intexine thin, translucent, bearing, in some specimens, a fine transverse scar within the polar cleft. Sacci large with marked distal inclination, saccus exoexine about 2μ thick, intrareticulate, brochi 1-3μ in diameter with a tendency to radial elongation, especially near the sacci bases. Cappula

---

PLATE 12

Permian and Triassic Striatiti: Diploxylonoid forms. All ×600.

FIGURES

1-3. *Striatopodocarpites cancellatus* (Balme and Hennelly). Lower to Upper Permian.—— 1. Specimen 57891.——2. Specimen 57892.——3. Specimen 57893.
4-6. *Taeniaesporties* sp. cf. *T. transversundatus* Jansonius. Lower to lower Middle Triassic. ——4. Unnumbered specimen from sample 54625.——5. Specimen 57910.——6. Specimen 57909.
7-9. *Striatopodocarpites pantii* Jansonius, comb. nov. Lower to Upper Permian.——7. Specimen 57901.——8. Specimen 57902.——9. Specimen 57899.
10-12. *Striatopodocarpites rarus* Bharadwaj and Salujha, comb. nov. Lower to Upper Permian. ——10. Specimen 57896.——11. Specimen 57897.——12. Specimen 57898.

breadth about half that of corpus, rectangular to elongate oval, usually bordered by longitudinal, crescentic, intexinal folds lying below the sacci bases. Exine of cappula thin and transparent.

Dimensions (15 specimens): Total breadth: 87 (96) 106μ. Corpus breadth: 40 (51) 59μ. Breadth of cappula: 18 (20) 25μ. Corpus height: 37 (47) 54μ. Sacci height: 51 (61) 69μ.

Remarks: Although Leschik's description of *Taeniaesporites noviaulensis* was sketchy, the morphology of the species is clear from his excellent illustrations (Leschik, 1956, pl. 22, fig. 1-2). Its distinguishing characters are a slightly diploxylonoid outline, broad cappula, four intrareticulate taeniae, transverse monolete scar, and fairly large sacci with a well-defined intrareticulum. The total breadth of Leschik's holotype is about 100μ.

Taeniaesporites novimundi Jansonius appears to be at least partly synonymous with *Taeniaesporites noviaulensis* and has been placed in direct synonymy by some authors (e.g., Freudenthal, 1964). Jansonius has allowed a wide range of variation within his species, however, and it is possible that some of his forms should be regarded as distinct from *T. noviaulensis*. Unfortunately, the morphology of the holotype is not as clearly illustrated as that of other specimens assigned by Jansonius to the taxon, which is consequently difficult to apply. Other published species which may also prove hard to distinguish from *T. noviaulensis* are *Striatopinites raricostatus* Romanovskaya from the Lower Triassic of West Kazakhstan and *Taeniaesporites ovatus* Goubin from the Scythian of Madagascar (Goubin, 1965).

Taeniaesporites sp. cf.
T. transversundatus Jansonius

Plate 12, figures 4-6

Taeniaesporites transversundatus Jansonius, 1962, p. 64-65, pl. 14, fig. 3-4.

Occurrence: Mianwali Formation: very rare to common. Tredian Formation: very rare to rare.

Representative Specimens: U.W.A. 57908, 57909, 57910. Field no. K6-8D, Nammal Gorge, Salt Range, Mittiwali Member, Mianwali Formation.

Description: Disaccate, diploxylonoid. Corpus circular to slightly transversely elongate oval. Cappa dissected by 3-4 usually narrow clefts into 4-5 taeniae, which are usually constricted at their equatorial extremities. Exoexine of cappa 3-4μ thick, dark in color, and apparently structureless; taeniae bearing roughly parallel, longitudinal plications. Sacci broadly crescentic in polar view with pronounced distal inclination, saccus exoexine 1-2μ thick, intrareticulate, brochi 1-2μ in diameter with slight radial elongation near the distal roots of the sacci. Cappula with convex margins, breadth about one-third that of the corpus, exine of cappula thin, translucent, unstructured or faintly punctate.

Dimensions (20 specimens): Total breadth: 60 (71) 85μ. Corpus breadth 34 (40) 47μ. Breadth of cappula: 10-20μ. Corpus height: 31 (35) 42μ. Sacci height: 34 (44) 51μ.

Remarks: A fairly wide range of variation has been allowed in specimens attributed here to *Taeniaesporites* sp. cf. *T. transversundatus*, especially in the form and arrangement of the proximal taeniae. In most specimens these are separated by narrow fairly regular clefts, but in some the clefts are wide with rather ragged margins. The essential characters of the species are considered to be its markedly diploxylonoid outline, heavily thickened cappa, and the plications of the taeniae. Specimens of *T.* sp. cf. *T. transversundatus* from the Salt Range are in general much larger than the forms described as *T. transversundatus* by Jansonius from the Scythian of Canada. Apart from this, there seems no significant difference between the two populations.

Taeniaesporites pellucidus Goubin, comb. nov.

Plate 13, figures 8-10

Taeniaesporites sp. cf. *T. noviaulensis* Leschik, Balme, 1963, *partim*, p. 28-29, pl. 6, fig. 5.
Protohaploxypinus pellucidus Goubin, 1965, p. 1423, pl. 2, fig. 4-6.

Occurrence: Mianwali Formation: rare to dominant. Landa Member, Tredian Formation: very rare to rare.

Representative Specimens: U.W.A. 57911, 57912, 57913, 57914. Field no. K6-8D, Nammal Gorge, Mittiwali Member, Mianwali Formation.

Description: Disaccate, haploxylonoid. Corpus poorly defined, transversely elongate oval. Cappa about 1μ thick, divided by three wide transverse clefts into four major taeniae, polar taeniae sometimes further subdivided by narrow clefts into subsidiary taeniae. Exoexine of cappa distinctly intrareticulate, intexine thin, translucent, absent in many specimens. Sacci hemispherical to broadly crescentic in polar view with a marked distal inclination, occasionally joined by equatorial strips of detached exoexine. Exoexine of sacci $1-2\mu$ thick, intrareticulate, diameter of brochi $1-4\mu$ with a tendency to radial elongation near the proximal and distal sacci bases. Cappula about one-third as broad as corpus, rounded quadrilateral to oval, faintly structured adjacent to the distal sacci bases, elsewhere thin and translucent.

Dimensions (40 specimens): Total breadth: 79 (88) 102μ. Corpus breadth: 50 (57) 62μ. Breadth of cappula: 14 (19) 28μ. Corpus height: 47 (53) 60μ.

Remarks: Despite the abundance of *Taeniaesporites pellucidus* in the lower part of the Mianwali Formation, it was difficult to find sufficient specimens adequately to illustrate the species, because of the generally poor preservation of the Scythian assemblages. Goubin's type specimens from the Lower Triassic of Madagascar are also poorly preserved and show evidence of shrivelling and pyrite corrosion. Obviously the choice of a form genus to accommodate pollen of the type assigned here to *T. pellucidus* is rather subjective; and, according to the usage of some authors, some of the pollen grains that I have regarded as belonging to a homogeneous taxon should be referred to *Protohaploxypinus,* because they have more than five taeniae. I believe, however, that *T. pellucidus* has essentially four major taeniae, although those bordering the polar cleft are not infrequently subdivided by narrower, subsidiary clefts so that five or even six taeniae may be present.

Taeniaesporites pellucidus is common and usually better preserved in marine Scythian strata from Western Australia (Balme, 1963), and similar pollen grains have been reported from Lower Triassic strata in Queensland and New South Wales by Evans (1966). Playford (1965), on the other hand, failed to find any species of *Taeniaesporites* in Lower Triassic sediments from Tasmania.

Genus LUECKISPORITES Potonié and Klaus emend. Klaus, 1963

Type species (original designation): *Lueckisporites virkkiae* Potonié and Klaus.

Lueckisporites, which was restricted by Klaus (1963) to disaccate pollen with the exoexine of the cappa dissected by a single, transverse, polar cleft, is one of the most interesting of late Paleozoic pollen grains. In western Europe it characterizes Zechstein sediments and is often the dominant component of Late Permian microfloras (Grebe and Schweitzer, 1962; Clarke, 1965b). Some European species range into the Triassic (Reinhardt, 1964; Klaus, 1965), but in most areas the genus is an effective Upper Permian index (Klaus, 1963, p. 355).

Records of *Lueckisporites* from Russia are less frequent or less unequivocal than those from western and central Europe, although a clear example was illustrated by Zoricheva and Sedova (1954, pl. 10, fig. 7) from the Upper Permian of the northern European part of the USSR. In the American midcontinent region, however, Wilson (1962) reported an abundance of *Lueckisporites virkkiae* in assemblages from the Guadalupian Flowerpot Formation, and Jansonius (1962) described a species of *Lueckisporites* from the Scythian of Canada. Other possible occurrences in the Northern Hempishere include the Permian of Iraq (Singh, 1964) and, less convincingly the Carboniferous-Permian of Egypt (Schurman, Burger, and Dijkstra, 1963). Virkki's Spore 64 from the Speckled sandstone at Kathwai (Virkki, 1946, fig. 33, pl. 5, fig. 5) possibly belongs to *Lueckisporites* (cf. Hart, 1960) but the assignment is dubious.

No species of *Lueckisporites* has been found in the now fairly thoroughly investigated Permian successions of peninsular India and Australia. In Africa, however, Hart (1960, 1963) has discovered representatives of the genus in Lower Permian strata from Tanzania and the Orange Free State, and Goubin (1965) states that *Lueckisporites virkkiae* occurs in both Upper Permian and Triassic sediments in Madagascar.

Lueckisporites virkkiae Potonié and Klaus

Plate 13, figures 4-5

Lueckisporites virkkiae Klaus, 1953, p. 54 (nom. nud.).
Lueckisporites virkkiae Potonié and Klaus, 1954, p. 534, pl. 10, fig. 3.
Cedripites gen. nov. Zoricheva and Sedova, 1954, pl. 10, fig. 7.
Lueckisporites virkkiae Potonié and Klaus, 1963, p. 302-303, pl. 11, fig. 50-51; pl. 12, fig. 52-55.
Lueckisporites virkkiae Potonié and Klaus emend., Clarke, 1965b, p. 331-333, fig. 8, pl. 43, fig. 3, 6-11.
?*Lueckisporites virkkiae* Potonié and Klaus, Goubin, 1965, p. 1422, pl. 1, fig. 7.

Occurrence: Wargal Limestone: very rare to rare. Chhidru Formation: very rare to rare.

Representative Specimens: U.W.A. 57915, 57916, 57917, 57918. Field no. K11-6A, Wargal, Salt Range, Chhidru Formation.

Description: Disaccate, slightly diploxylonoid. Corpus transversely elongate oval. Cappa about 2μ thick, divided into two by a wide, transverse, polar cleft in which the thin, translucent intexine is exposed. A faint transverse monolete mark is visible on the intexine at the proximal pole in some specimens. Exoexine of cappa finely but distinctly intrareticulate. Sacci almost hemispherical in polar view with a pronounced distal inclination, saccus exoexine 1-2μ thick, densely intrareticulate, brochi 1-2μ in diameter, radially elongate near the proximal sacci roots, otherwise more or less equidimensional. Cappula about one-third breadth of corpus, almost rectangular, faintly structured near its lateral margins, becoming structureless towards the distal pole.

Dimensions (20 specimens): Total breadth: 58 (66) 74μ. Corpus breadth: 40 (46) 54μ. Breadth of cappula: 8 (15)

PLATE 13
Permian and Triassic Striatiti. All ×600.

FIGURES
1-3. *Lueckisporites singhii* Balme, n. sp. Lower to Upper Permian.——1. Specimen 57921.——2. Holotype, specimen 57919.——3. Specimen 57920.
4-5. *Lueckisporites virkkiae* Potonié and Klaus. Upper Permian.——4. Specimen 57915.——5. Specimen 57917.
6-7. *Taeniaesporites noviaulensis* Leschik. Upper Permian to lower Middle Triassic.——6. Specimen 57906, proximal focus, with incipient subsidiary cleft in upper polar taenia.——7. Specimen 57903.
8-10. *Taeniaesporites pellucidus* Goubin, comb. nov. Lower to lower Middle Triassic.——8. Specimen 57913.——9. Specimen 57911, with narrow subsidiary cleft in lower polar taenia.——10. Specimen 57914.

Palynology of Permian and Triassic 375

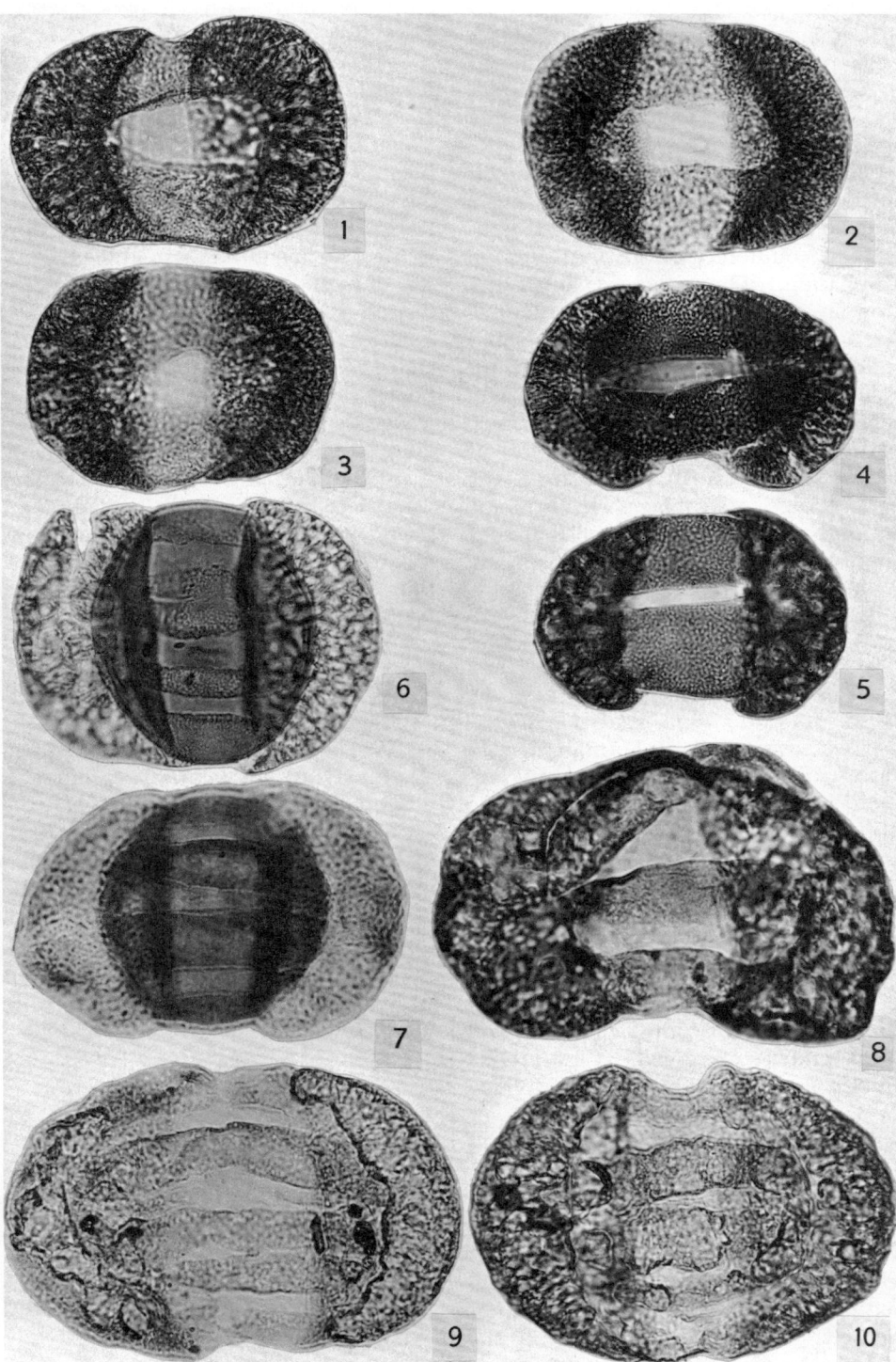

Plate 13

25μ. Corpus height: 32 (39) 48μ. Sacci height: 36 (43) 48μ.

Remarks: Agreement has not yet been reached on the acceptable limits of variation of the species *Lueckisporites virkkiae,* and the concept of the species adopted here is that of Clarke (1965b). *L. virkkiae* was not common in any material from the Salt Range, and no more than 40 or 50 specimens were encountered throughout the investigation. These showed less variation in morphology than is usual in the assemblages of western Europe, and most specimens were close to Variant A of Clarke (1965b, pl. 43, fig. 3,8,9). This, in turn, resembles the holotype of *L. virkkiae,* and the absence of Clarke's Variants B and C from the Salt Range assemblages lends some support to Klaus's contention that they should be recognized as distinct species.

Lueckisporites singhii Balme, n. sp.

Plate 13, figures 1-3

Lueckisporites sp. Singh, 1964, p. 257, pl. 46, fig. 4.

Occurrence: Amb Formation: very rare to rare. Wargal Limestone: common. Chhidru Formation: rare to abundant.

Representative Specimens: Holotype: U.W.A. 57919. Others: U.W.A. 57920, 57921, 57922, 57923. Field no. K11-6D, Wargal, Salt Range, Chhidru Formation.

Description: Disaccate, slightly diploxylonoid. Corpus circular, oval or rounded polygonal. Cappa 1-2μ thick, partially or completely divided by a broad, transverse, fusiform or wedge-shaped cleft in which the thin translucent intexine is exposed. Margins of cleft rather ragged. Exoexine of cappa finely intrareticulate or intravermiculate. Sacci crescentic in polar view with a pronounced distal inclination, saccus exoexine about 3μ thick, densely intrareticulate with strong radial elongation of the brochi near the proximal sacci bases. Cappula breadth about half that of corpus, exine of cappula faintly structured, almost invariably with a ragged rupture extending its full length.

Dimensions (30 specimens): Total breadth: 60 (72) 86μ. Corpus breadth: 40 (48) 59μ. Breadth of cappula: 14 (22) 35μ. Corpus height: 33 (46) 62μ. Saccus height: 41 (50) 62μ.

Derivation of name: After Dr. H. P. Singh, who described pollen grains of this type from the Permian of Iraq.

Remarks: Lueckisporites singhii is distinguished from *Lueckisporites virkkiae* by the form of its polar proximal cleft, which, in the former species, is poorly defined, rather ragged and wedge-shaped, so that the cappa is seldom completely and never sharply divided. The species is ubiquitous and usually common in Permian sediments from the Salt Range, and maintains such constant morphological characters that its identification is never a problem.

Genus GUTTULAPOLLENITES Goubin, 1965

Type species (original designation): *Guttulapollenites hannonicus* Goubin.

De Jekhowsky and Goubin (1964, fig. 5) illustrated a number of pollen grains of singular morphology from the Upper Permian of Madagascar, although they provided no descriptions, referring to them simply as "globular-striated saccates." Subsequently, Goubin (1965) created the genus *Gut-*

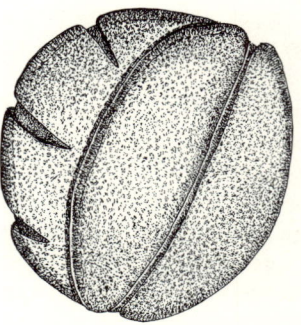

Fig. 6. *Guttulapollenites hannonicus* Goubin. Reconstruction in semilateral view, ×600.

tulapollenites, which she assigned to the Subturma Polysaccites Cookson, to accommodate these unusual forms. As it lacks analogues among living or fossil pollen grains, the morphology of *Guttulapollenites* is not easy to interpret or describe, and one of the main problems is an uncertainty concerning the direction of its polar axis. Alternative interpretations of the morphology of *Guttulapollenites* are possible, and Venkatachala, Goubin, and Kar (1967) have recently modified the generic diagnosis. They consider *Guttulapollenites* to be a tetrasaccate pollen grain. Some specimens can certainly be regarded as tetrasaccate, but the majority have more than four areas of separate exoexine and resemble the specimen illustrated in Plate 14, figure 5. I believe, therefore, that it is more appropriate to regard *Guttulapollenites* as a disaccate, taeniate, pollen grain, in which the sacci are not markedly inflated. It is accordingly referred here to the Subturma Disaccites (Striatiti).

Guttulapollenites hannonicus Goubin

Figure 6; Plate 14, figures 4-7

Form 1034, de Jekhowsky and Goubin, 1964, fig. 5, 1034.
Guttulapollenites hannonicus Goubin, 1965, p. 1431, fig. 1-2, pl. 5, fig. 5-8.

Occurrence: Amb Formation: very rare to rare. Wargal Limestone: abundant. Chhidru Formation: very rare to common. Mittiwali Member, Mianwali Formation: very rare.

Representative Specimens: U.W.A. 57924, 57925, 57926, 57927. Field no. T61-178, Kathwai, Salt Range, Chhidru Formation.

Description: Disaccate, amb circular. Cappa 5-6μ thick, divided by transverse clefts into 2-4 broad taeniae; some specimens show a short transverse monolete scar on the exposed intexine at the proximal pole. Exoexine of cappa densely columellate, appearing finely microreticulate in surface focus, intexine thin, translucent. Sacci almost hemispherical in polar view, detached distally and covering almost the whole of the distal surface, saccus exoexine about 5μ thick, structure finely and densely columellate, identical with that of the cappa. Sacci separated on the distal side by a narrow, slightly sinuous cappula consisting of exposed intexine. Longitudinal intexinal folds occasionally parallel the distal sacci bases, but otherwise the sacci show little evidence of inflation.

Dimensions (40 specimens): Total breadth (diameter): 42 (57) 70μ. Breadth of sacci: 19-27μ.

Remarks: Guttulapollenites hannonicus is a highly distinctive form, which in Madagascar is abundant in lower Sakamena (Upper Permian) sediments, but occasional specimens are said by Goubin to range into the late Triassic. In the Salt Range the species is common in the Wargal Limestone and lower part of the Chhidru Formation, becoming rare in the uppermost Permian, and only one or two specimens were found in the basal Triassic.

Apart from the occurrences in Madagascar and the Salt Range, *Guttulapollenites hannonicus* has been reported by Hart (in Goubin, 1965) from the Upper Permian of South Africa, and it almost certainly occurs in the upper Bunter of the Netherlands (Visscher, 1966, pl. 20, fig. 6). Venkatachala, Goubin, and Kar (1967) recorded it from the Barakar group, Barren Measures, and Raniganj Coal Measures of peninsula India, and a similar form occurs occasionally in the Middle Triassic of the Carnarvon Basin, Western Australia. Dr. E. M. Kemp (pers. comm.) has recently recorded the species from the Upper Permian of the Prince Charles Mountains, East Antartica.

Genus CORISACCITES
Venkatachala and Kar, 1966

Type species (original designation): *Corisaccites alutas* Venkatachala and Kar.

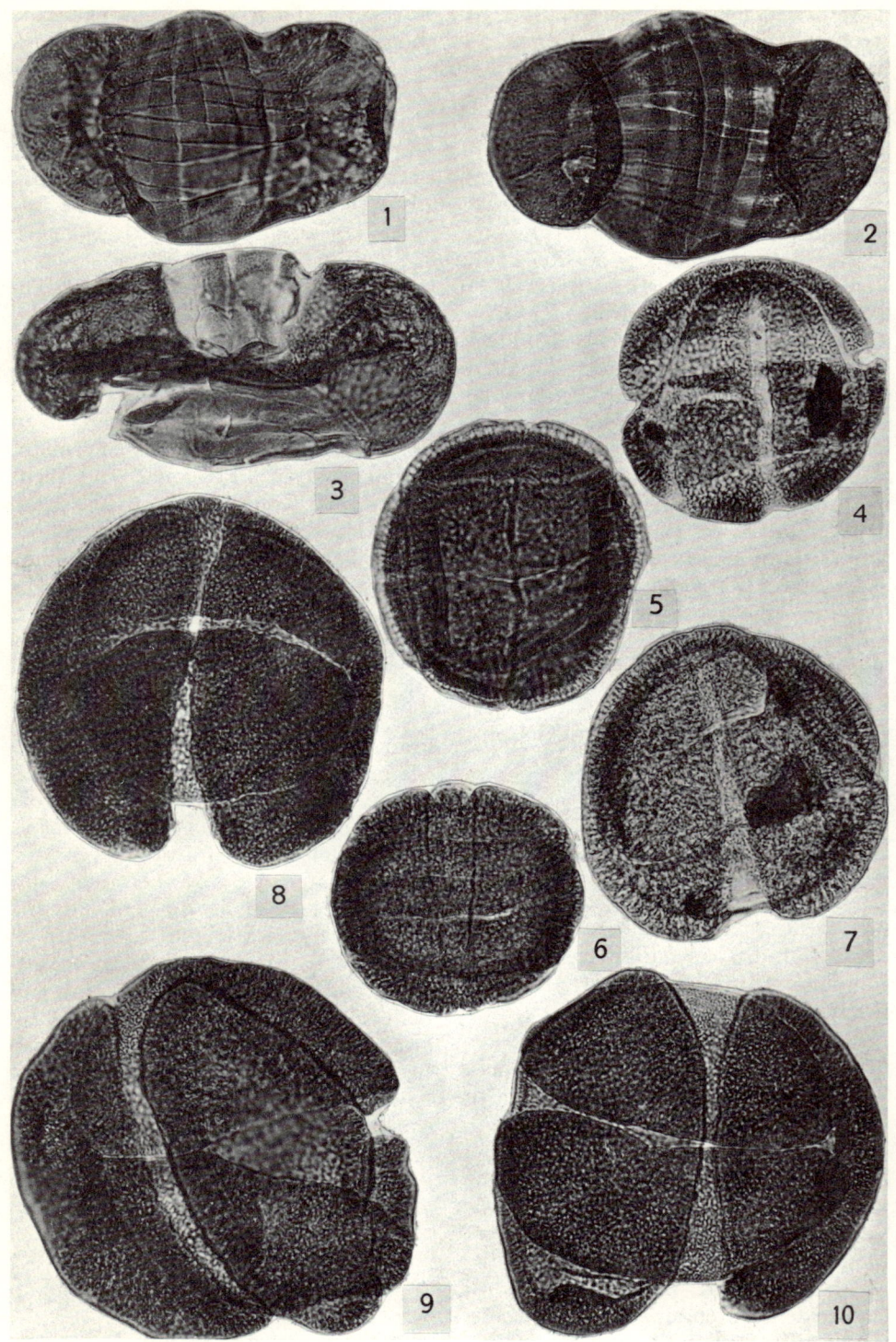

Plate 14

Corisaccites resembles *Guttulapollenites* in the sharp delimitation of its sacci margins and the structural similarity of its cappa and sacci. The two genera were not compared by Venkatachala and Kar (1966), but the most useful bases for distinguishing between them are the presence of a broad, distal cleft in *Corisaccites* together with its obviously disaccate, slightly diploxylonoid, appearance in polar view.

Corisaccites alutas Venkatachala and Kar
Figure 7; Plate 14, figures 8-10

Spore 5, Virkki, 1946, p. 126, fig. 36, pl. 6, fig. 63-64.
?*Polysaccites*-type (gen. et sp. nov.) (indet.), Ouyang, 1964, p. 518-519, pl. 8, fig. 1-4.
Corisaccites alutas Venkatachala and Kar, 1966, p. 108, fig. 1-2, pl. 1, fig. 1-7.

Occurrence: Amb Formation: rare.

Representative Specimens: U.W.A. 57928, 57929, 57930, 57931, 57932. Field no. T62-350, Warchha Water Tank, Salt Range, Amb Formation.

Description: Disaccate, slightly diploxylonoid. Corpus circular to transversely elongate oval. Cappa 4-7μ thick, finely and densely columellate with a scabrate appearance in surface focus, divided into two by a single, sharply defined, transverse, polar cleft with slightly undulate margins, in which the thin, hyaline, intexine is exposed. Sacci crescentic or longitudinally elongate oval, detached distally, little inflated, and appearing to be stuck on to the corpus, saccus exoexine 4-7μ thick with a fine structure identical with that of the cappa. Cappula narrow, usually slightly constricted at the distal pole, sharply differentiated from the sacci and consisting of thin translucent intexine.

Dimensions (30 specimens): Total breadth: 73 (86) 107μ. Corpus breadth: 51 (68) 85μ. Sacci breadth: 27 (36) 46μ. Corpus height: 43 (60) 82μ. Sacci height: 61 (73) 86μ.

Corisaccites alutas is a pollen grain of striking morphology which is so far known only from Lower Permian deposits. Virkki (1946) recorded it from two horizons in the Speckled sandstone at Kathwai, one of which also provided the type material of Venkatachala and Kar, and also from the Amb Formation (Middle Productus limestone) at Warchha. It occurs also, in small numbers, in the Artinskian Carynginia Formation in the Perth Basin, Western Australia. The forms described and illustrated by Ouyang

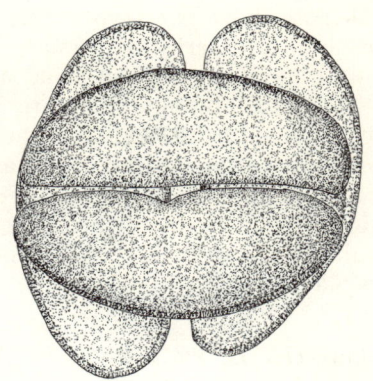

FIG. 7. *Corisaccites alutas* Venkatachala and Kar. Reconstruction in proximal view, ×600.

PLATE 14
Permian and Triassic Striatiti. All ×600.

FIGURES

1-3. *Hamiapollenites insolitus* Bharadwaj and Salujha, comb. nov. Lower Permian.——1. Specimen 57935, proximal focus.——2. Specimen 57933.——3. Specimen 57937, equatorial view distal side uppermost, showing the sacci detached laterally without distal inclination.
4-7. *Guttulapollenites hannonicus* Goubin. Lower to Upper Permian.——4. Unnumbered specimen from sample 54669.——5. Specimen 57927, polar view in proximal focus.——6. Specimen 57924, polar view.——7. Specimen 57928, semiequatorial view.
8-10. *Corisaccites alutas* Venkatachala and Kar. Lower Permian.——8. Specimen 57931, polar view.——9. Specimen 57929, semilateral view, showing gaping proximal cleft.——10. Specimen 57928, polar view.

(1964) as *Polysaccites*-type came from the Shihhotze Series of presumed late Lower Permian age and look similar to *C. alutus*, although they are larger (total breadth: 138-149μ) than any specimens so far measured and appear to have a more coarsely structured exoexine. There can nevertheless be little doubt that they are assignable to *Corisaccites*.

Genus HAMIAPOLLENITES Wilson emend. Tschudy and Kosanke, 1966

Type species (original designation): *Hamiapollenites saccatus* Wilson.

Hart (1965b), Playford and Dettmann (1965), and Tschudy and Kosanke (1966) have discussed the status of *Hamiapollenites*. The last mentioned authors widened Wilson's original diagnosis to allow forms with less than six distal taeniae to be included in the form genus. Playford and Dettmann have provided a list of synonymous generic names, which appears to be complete, apart from *Minutisaccata* Hart, which appeared while their paper was awaiting publication.

Disaccate pollen grains with taeniae on both the cappa and cappula are widely, although sporadically, distributed in Permian sediments. They have been most often reported from North America where they are known from the Late Pennsylvanian and Wolfcampian of the midcontinent (Jizba, 1962), Leonardian strata in the central United States (Jansonius, 1962; Shaffer, 1964; Hedlund, 1965), and the Guadalupian of Oklahoma (Wilson, 1962). There are no certain published occurrences from western or central Europe, but species of *Hamiapollenites*, under a variety of generic names, have been recorded frequently from Kungurian and Upper Permian strata in the USSR. A specimen undoubtedly referable to the genus was illustrated at Virkki (1946, pl. 7, fig. 90) from the Artinskian of the Salt Range, and forms from the Upper Permian of India assigned by Bharadwaj (1962) to *Distriatites* agree with Wilson's diagnosis of *Hamiapollenites*. *Striatites medius* Singh from the Permian of Iraq is also, judging from its illustration (Singh, 1964, pl. 46, fig. 6-7), more appropriately referred to *Hamiapollenites*, and excellent illustrations of the genus from the Permian of Saudi Arabia were provided by Hemer (1965). The genus is not, apparently, a prominent component of Permian microfloras in the Southern Hemisphere, and the only unequivocal published report is that of Hart (1965b) from the Artinskian of Tanzania. In Australia *Hamiapollenites* is known to occur occasionally in Artinskian strata (K. L. Segroves, unpublished data) and *Hamiapollenites insculptus* Playford and Dettmann has been illustrated from Upper Triassic sediments in South Australia (Playford and Dettmann, 1965) and Queensland (Hill, Playford, and Woods, 1965). Romanskaya (1963) described a species of *Hamiapollenites* under the name *Striatopinites reticulatus* from the Lower Triassic of Western Kazakhstan: the species is said to range into the Jurassic. A species of *Hamiapollenites* also occurs extremely rarely in the Khatkiara Member of the Tredian Formation, but the two or three specimens seen were too poorly preserved to enable it to be identified or described.

Hamiapollenites insolitus Bharadwaj and Salujha, comb. nov.

Figure 8; Plate 14, figures 1-3

Pityosporites sp. (Spore 79), Virkki, 1946, p. 134, fig. 47, pl. 7, fig. 90.
Distriatites insolitus Bharadwaj and Salujha, 1964, p. 211, pl. 12, fig. 157-158.

Occurrence: Amb Formation: rare.
Representative Specimens: U.W.A. 57933, 57935, 57936, 57937. Field no. T62-350, Warchha Water Tank, Salt Range, Amb Formation.
Description: Disaccate, weakly diploxylonoid. Corpus oval to almost circular, usually slightly elongate transversely. Cappa about 2μ thick, dis-

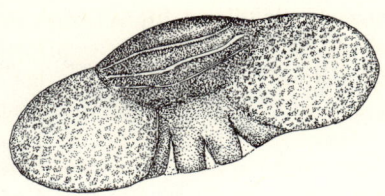

Fig. 8. *Hamiapollenites insolitus* (Bharadwaj and Salujah). Reconstruction in equatorial view, proximal face uppermost, ×600.

sected by 7-9 prominent, narrow clefts, giving rise to 8-10 transverse taeniae, which usually run the full breadth of the cappa but are sometimes discontinuous and wedge-shaped. Exoexine of cappa finely intrareticulate. Sacci crescentic to semicircular in polar view, detached laterally without any distal inclination, and often somewhat constricted at their bases. Saccus exoexine 1-2μ thick, finely intrareticulate with equidimensional brochi. Cappula breadth about four-fifths that of the corpus, cappula dissected into 3-6 longitudinal taeniae separated by clefts which are usually broader than those on the cappa and in which the intexine is exposed. Taeniae of cappa finely intrareticulate, intexine thin, transparent and structureless.

Dimensions (30 specimens): Total breadth: 76 (92) 116μ. Corpus breadth: 43 (51) 65μ. Breadth of cappula: 38 (46) 56μ. Corpus height: 43 (55) 69μ. Sacci height: 32 (42) 55μ.

Remarks: Although not a quantitatively important component of the rich and varied assemblage obtained from the Warchha Water Tank locality, *Hamiapollenites insolitus* was sufficiently plentiful to allow the examination of a hundred or so specimens. These displayed marked morphological variations, especially in such characters as the relative size of sacci and corpus and the number of longitudinal taeniae on the cappula, criteria which have both been accepted by some authors as systematically useful. Bharadwaj and Salujha (1964) gave little information on the range of variation of *Distriatites insolitus* at its type locality, and it is probable that the species was based on few specimens, for it is apparently extremely rare in the Raniganj coals (Bharadwaj and Salujha, 1965a). Similarly *Hamiapollenites insculptus* Playford and Dettmann, which may be a junior synonym of *H. insolitus,* is a rare form in the Triassic of Leigh Creek although, except in size, the specimens illustrated so far show no marked morphological variation (Playford and Dettmann, 1965, pl. 16, fig. 55-57; Hill, Playford, and Woods, 1965, pl. Txi, fig. 13). Most specimens referred here to *H. insolitus* have smaller sacci than the two illustrated by Bharadwaj and Salujha, although this is not invariably so. They have also, in the main, a lesser total breadth than the 110-140μ quoted in the original diagnosis. Neither of these divergences is considered adequate, in a form with an obviously inconstant morphology, to enable reliable specific differentiation.

b. Non-taeniate pollen grains.

Genus VITREISPORITES Leschik emend. Jansonius, 1962

Type species (original designation): *Vitreisporites pallidus* (Reissinger) (= *Vitreisporites signatus* Leschik).

Vitreisporites is the appropriate form genus for dispersed disaccate pollen grains with similar morphology to the pollen produced by the caytonialean male fructifications *Caytonanthus arberi* (Thomas), *C. kochi* Harris, and *C. oncodes* Harris. *Caytonanthus* pollen grains have been frequently described and illustrated, most recently by Townrow (1962) and Harris (1964). Their most obvious character, for a disaccate pollen, is extremely small size: according to Harris the mean total breadth of the pollen of *C. arberi* and *C. oncodes* is 22μ and 31μ respectively. In general, the shape of *Caytonanthus* pollen is haploxylonoid, although occasionally the height of the sacci is slightly greater or a little less than that of the corpus. Townrow noted that no well-defined sulcus was present on the cappula, but that a thin aperturoid area (or leptoma) was sometimes visible.

Dispersed pollen grains of the *Caytonanthus*-type *(Vitreisporites)* occur in Mesozoic sediments throughout the world and are especially abundant in Lower Jurassic deposits. From previous records they are known to range stratigraphically from Upper Permian (Balme, 1964; Medvedeva, 1960; Chalyshev, Varyukhina, and Molin, 1965) to at least Upper Cretaceous (Gray and Browning, 1959; Khlonova, 1961) and have been reported from the Eocene (Reissinger, 1950), although the Tertiary record needs confirmation and may represent specimens derived from older rocks by sedimentary processes. Rare specimens were found in Artinskian sediments during the present investigation, and these are the oldest known representatives of the genus.

Vitreisporites pallidus (Reissinger)

Plate 15, figures 5-6

Pityosporites pallidus Reissinger, 1940, p. 14 nom. nud.
Pityopollenites pallidus Reissinger, 1950, p. 109, pl. 15, fig. 1-5.
Caytonia oncodes Harris, Bolkhovitina, 1953, p. 113, pl. 20, fig. 207a, b.
Vitreisporites signatus Leschik, 1955, p. 53-54, pl. 8, fig. 10.

Coniferaletes stultulus Andreeva, 1956, p. 271, pl. 60, fig. 124.
Pityosporites pallidus (Reissinger), Balme, 1957, p. 36-37, pl. 10, fig. 112-113.
Caytonipollenites pallidus (Reissinger), Couper, 1958, p. 150, pl. 26, fig. 7-8.
Vitreisporites pallidus (Reissinger), Nilsson, 1958, p. 78, pl. 7, fig. 12-14.
Caytonipollenites subtilis de Jersey, 1959, p. 381, pl. 3, fig. 4.
Vitreisporites subtilis (de Jersey), de Jersey, 1962, p. 11, pl. 4, fig. 8-9.
Caytoniidites alaticonformis Malyavkina, 1964, p. 93, pl. 5, fig. 9; pl. 7, fig. 10.

Occurrence: Amb Formation: very rare to rare. Chhidru Formation: very rare to common. Mianwali Formation: very rare to rare. Landa Member, Tredian Formation: very rare to rare.

Representative Specimens: U.W.A. 57938, 57939, 57940. Field no. K11-4A, Wargal, Salt Range, Chhidru Formation.

Description: Disaccate, haploxylonoid to slightly diploxylonoid. Corpus longitudinally elongate oval. Cappa thin, finely intrareticulate to intrapunctate. Sacci nearly semicircular in polar view with a slight distal inclination, saccus exoexine thin, intrareticulate, brochi sharply defined and less than 1μ in diameter. Cappula nearly parallel-sided, breadth about one-fourth that of corpus, exine of cappula thin and unstructured.

Dimensions (10 specimens): Total breadth: $20\text{-}32\mu$. Corpus breadth: $9\text{-}18\mu$. Breadth of cappula: $2\text{-}5\mu$. Corpus height: $13\text{-}23\mu$.

PLATE 15

Permian and Triassic Disaccites: Non-taeniate forms. All ×600.

1-4. *Alisporites tenuicorpus* Balme, n. sp. Lower to Upper Permian.——1. Specimen 57944.——2. Unnumbered specimen from sample 54650.——3. Holotype, specimen 57941.——4. Specimen 57942.

5-6. *Vitreisporites pallidus* (Reissinger). Lower Permian to Middle Triassic.——5. Specimen 57940.——6. Specimen 57938.

7-9. *Sulcatisporites ovatus* (Balme and Hennelly). Lower to Upper Permian.——7. Specimen 57968.——8. Specimen 57967.——9. Specimen 57966.

10-14. *Pinuspollenites thoracatus* Balme, n. sp. Lower to Upper Permian.——10. Specimen 57986, equatorial view, proximal side uppermost.——11. Unnumbered specimen from sample 54673, equatorial view, proximal side uppermost.——12. Holotype, specimen 57984, proximal focus showing finely structured cappa.——13. Specimen 57985, semi-lateral view.——14. Specimen 57987, equatorial view, proximal side uppermost.

15-18. *Falcisporites nuthallensis* Clarke, comb. nov. Lower to Upper Permian.——15. Specimen 57955.——16. Specimen 57956.——17. Specimen 57958.——18. Specimen 57957.

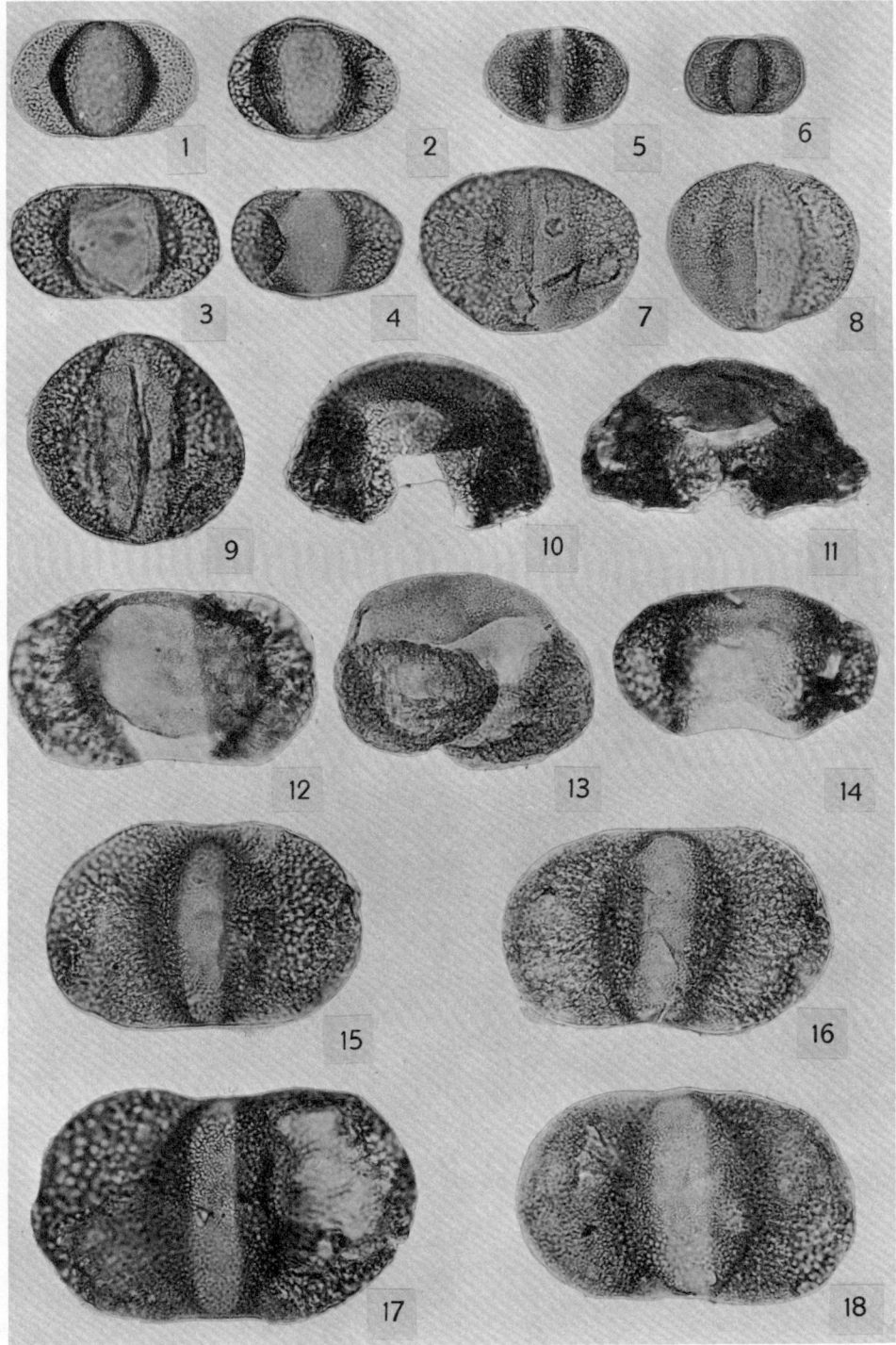

PLATE 15

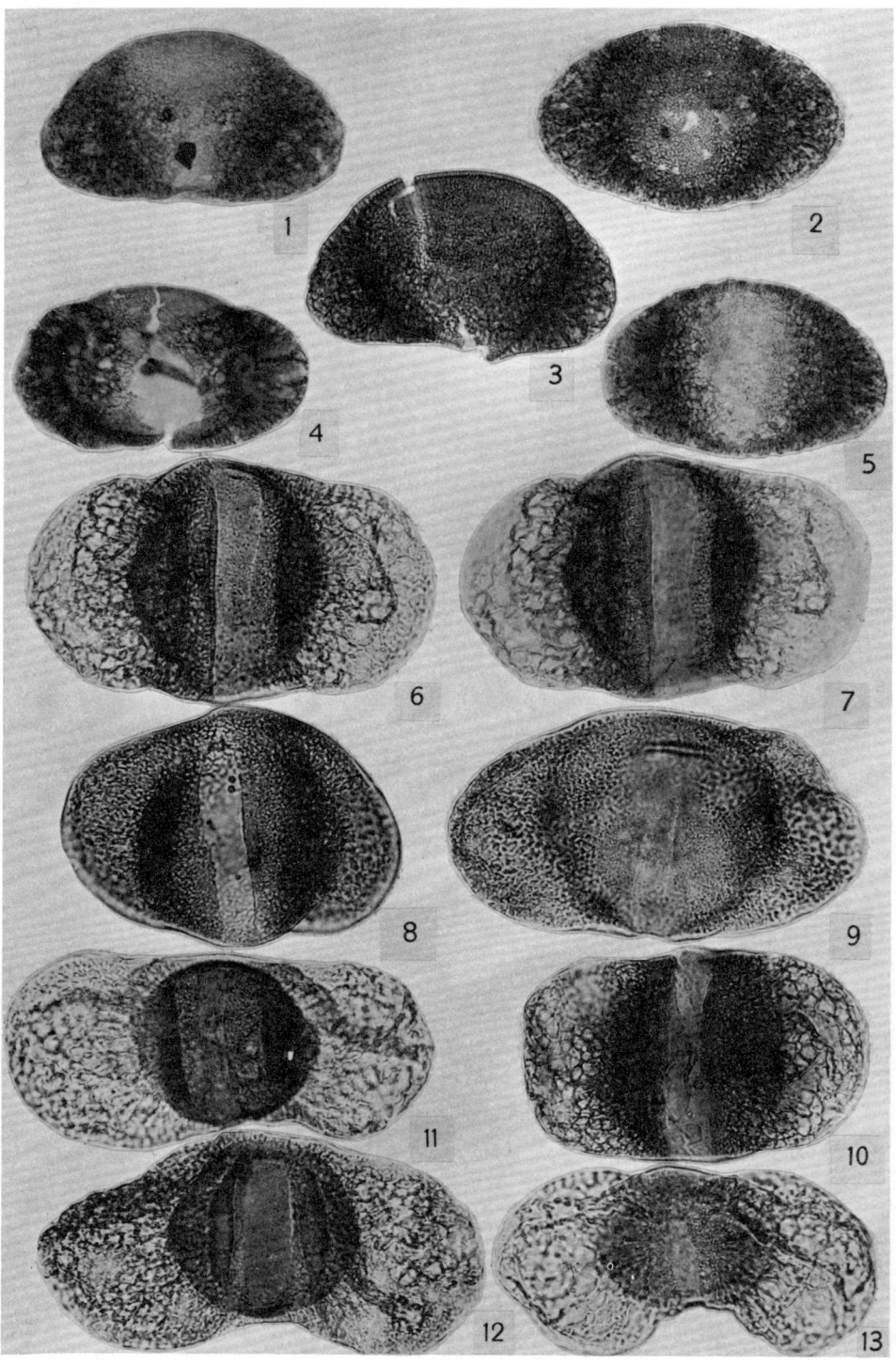

Plate 16

Remarks: Although Harris (1964, p. 14) found that the pollen grains of *Caytonanthus oncodes* could be distinguished on a statistical basis from those of *C. arberi*, there is no convincing basis for further systematic subdivision of dispersed pollen assigned to *Vitreisporites pallidus*, despite its long stratigraphic range and wide geographic dispersal. Undoubtedly it represents the pollen of many individual plants belonging to an extinct gymnosperm plexus. It is tempting to equate this with the Class Caytoniales, although both Townrow (1962) and Harris (1964) have warned against interpreting the presence of *V. pallidus* in a pollen assemblage as necessarily implying caytonialean elements in the parent flora. The most that can be said at present is that pollen grains assignable to *V. pallidus* are known to have been produced by at least three distinct caytonialean fructifications and that closely similar pollen is not known from any other group of plants.

Genus KLAUSIPOLLENITES Jansonius, 1962

Type species (original designation): *Klausipollenites schaubergeri* (Potonié and Klaus).

Klaus (1963) has clarified certain points concerning the morphology of *Klausipollenites* which were not entirely explicit in the original diagnosis. Following his commentary the generic characters of *Klausipollenites* are taken to be:

(a) Transversely (i.e., longitudinal in the sense of Klaus) elongate oval corpus.
(b) Absence of sharp differentiation between the exoexine of the saccus and that of the corpus.
(c) Gradual thinning of the exoexine towards the distal pole, giving rise to a poorly defined polar tenuitas.
(d) Small, rather rigid sacci.
(e) Tendency for sacci to be joined by equatorial strips of thickened but undetached exoexine, giving most specimens a monosaccate appearance.

Undoubted specimens of the form genus are so far known only from Permian and, rarely, Lower Triassic sediments in the Northern Hemisphere.

Klausipollenites schaubergeri (Potonié and Klaus)

Plate 16, figures 1-5

Disaccites schaubergeri Klaus, 1953, p. 54, *nom. nud.*
Pityosporites schaubergeri Potonié and Klaus, 1954, p. 536-538, fig. 8, pl. 10, fig. 7-8.
Klausipollenites schaubergeri (Potonié and Klaus), Jansonius, 1962, p. 55.
Pityosporites schaubergeri (Potonié and Klaus), Stuhl, 1962, pl. 49, fig. 1-2.
Pityosporites schaubergeri (Potonié and Klaus), Orlowska-Zwolinska, 1962, p. 292, pl. 3, fig. 8, pl. 4, fig. 1-3.
Klausipollenites schaubergeri (Potonié and Kremp), Klaus, 1963, p. 334-335, pl. 19, fig. 92-93.
Falcisporites schaubergeri (Potonié and Klaus), Schaarschmidt, 1963, p. 58, fig. 11, pl. 15, fig. 10-17.
Vesicaspora schaubergeri (Potonié and Klaus), Hart, 1965b, p. 72, fig. 168.

PLATE 16

Permian and Triassic Disaccites: Non-taeniate forms. All ×600.

FIGURES

1-5. *Klausipollenites schaubergeri* (Potonié and Klaus). Upper Permian to lower Lower Triassic.——1. Specimen 57946, equatorial view with proximal side uppermost.——2. Specimen 57945, distal view.——3. Specimen 58019, equatorial view with proximal side uppermost.——4. Specimen 57947, distal view with ruptured cappula.——5. Specimen 57949, distal view.

6-10. *Falcisporites stabilis* Balme, n. sp. Uppermost Permian to Middle Triassic.——6. Specimen 57952, in proximal focus showing fine structure of the cappa.——7. Specimen 57952, in distal focus showing the clearly defined sulcus.——8. Specimen 57954.——9. Holotype, specimen 57950.——10. Unnumbered specimen from sample 54633.

11-13. *Platysaccus queenslandi* de Jersey. Upper Lower Triassic.——11. Specimen 57982.——12. Specimen 57981.——13. Unnumbered specimen from sample 54633.

non *Vesicaspora schaubergeri* auct. non Potonié and Klaus, Jizba, 1962, p. 883-884, pl. 124, fig. 45-50.

Occurrence: Chhidru Formation: very rare to common. ? Kathwai Member, Mianwali Formation: very rare.

Representative Specimens: U.W.A. 57945, 57946, 57947, 57948, 57949, 58019. Field no. K11-6C, Wargal, Salt Range, Chhidru Formation.

Description: Disaccate, slightly diploxylonoid. Corpus transversely elongate oval to almost circular in polar view. Cappa 2-3μ thick, finely but distinctly intrareticulate, exoexine gradually becoming thicker towards the proximal sacci bases. Sacci semielliptical in polar view, slightly inclined distally with a rigid appearance. Saccus exoexine thick, densely intrareticulate, brochi with a tendency to radial elongation. Cappula not clearly delimited, with a finely structured elliptical tenuitas at the distal pole, intrareticulate structure of exoexine of the cappula becoming coarser away from pole and merging gradually into the sacci. Sacci joined equatorially by a strip of thickened exoexine in some specimens.

Dimensions (40 specimens): Total breadth: 51 (61) 72μ. Corpus breadth: 35 (43) 51μ. Corpus height: 33 (38) 49μ.

Remarks: Apart from their slightly lesser size range, specimens of *Klausipollenites schaubergeri* from the Chhidru Formation agree precisely with the description of the species given by Klaus (1963), and Dr. Klaus has, himself, confirmed the identification. The record of the typically Upper Permian *K. schaubergeri* from the Salt Range is an important extension of its known geographic range. Previously it has been most frequently described from European sediments and, with the exception of Leschik (1956), all palynologists who have studied Zechstein deposits have reported *K. schaubergeri* as a major component of their assemblages. Klaus (1955, 1963, 1965) states that, in the Southern Alps of Austria, it appears first near the top of the Grödner Sandstone and is a dominant component of most assemblages from the Bellerophon-Schichten. In the Zechsteinsalzes of Germany and the alpine Salzgebirges, *K. schaubergeri* is also abundant and may reach proportions as high as 25 percent of the spore-pollen assemblages. Occasional specimens occur in Austrian Lower Triassic strata, but according to Dr. Klaus (personal communication) these are found only in sediments associated with evaporite deposits. This apparent facies restriction suggests that *K. schaubergeri* was produced by a plant, or group of plants, which was capable of adapting to a desert environment.

Occurrences of *K. schaubergeri* from outside Europe are difficult to decipher. The North American record of Jizba (1962) is, I think, based on misidentifications, and no similar form was illustrated from the Permian of the American midcontinent by either Wilson (1962) or Shaffer (1964). It must be remembered, however, that no Late Permian microfloras have yet been described from either the United States or Canada, and it is in these that *K. schaubergeri* may be expected to occur.

No obvious representatives of *K. schaubergeri* have been illustrated by Russian workers, although Klaus (1963, p. 335) drew attention to a possible example illustrated by Samoilovich (1953, pl. 7, fig. 2) as *Protocedrus* sp. From its description this could certainly belong to *K. schaubergeri*, but according to Samoilovich the record was based on a single specimen.

Pityosporites insularis Goubin, which is said to range from Upper Permian to Upper Triassic (Goubin, 1965) of Madagascar, may be in part a synonym of *K. schaubergeri*, although it is not clearly illustrated. Otherwise, there is no record of the species, or any closely similar form, from the southern continents or peninsular India.

Genus FALCISPORITES Leschik emend. Klaus, 1963

Type species (original designation): *Falcisporites zapfei* (Potonié and Klaus).

From its emended diagnosis, *Falcisporites* is characterized by a generally haploxylonoid shape, relatively heavy rigid sacci, and a rather narrow cappula which bears a clearly delimited and well-defined longitudinal sulcus. Many specimens of *Falcisporites* also show crescentic intexinal folds bordering the cappula below the distal sacci bases. The key generic character of *Falcisporites* is its distal sulcus, a feature which is a constant character in the disaccate pollen grains of several Mesozoic "pteridosperm" fructifications investigated by Townrow (1962, 1965). These were the corystospermaceous male organs *Pteruchus africanus* Thomas, *P. dubius* Thomas, *P. simmondsi* (Shirley), and *P. petasatus* Townrow. Although, as Townrow pointed out, disaccate sulcate pollen grains occur among living and fossil conifers, none of the disaccate pollen grains known from other "pteridosperm" fructifications possesses a sulcus.

Falcisporites in this sense includes forms which have been referred by previous authors to *Pityosporites*, *Alisporites*, *Scopulisporites*, and *Pteruchipollenites*. For reasons discussed elsewhere (p. 165) I have abandoned *Pityosporites*; *Alisporites* is regarded as distinct; *Scopulisporites* is ambiguous; and *Pteruchipollenites* is accepted by most authors as a junior synonym of *Falcisporites*, although Couper (1958) did not mention a distal sulcus in his diagnosis.

Species of *Falcisporites* (i.e., *Alisporites australis* de Jersey and similar forms) are typically Triassic in Australia and are especially abundant in coals and other sediments of Middle to Late Triassic, and Rhaeto-Liassic age (de Jersey, 1962; Playford and Dettmann, 1965; Playford, 1965). They are known also from the late Triassic-Early Jurassic Ferrar Group in Antarctica (Norris, 1965), but Goubin (1965) illustrated no unequivocal examples of *Falcisporites* from the Triassic of Madagascar. In Europe *Falcisporites* has been most frequently reported from Zechstein strata; and Klaus (1965) notes only rare specimens of *F. zapfei* in Early Triassic sediments, although *Alisporites aequalis* Mädler, a late Keuper to early Liassic form, is assignable to *Falcisporites* under the interpretation accepted here. Jizba's attribution of forms from Pennsylvanian to Permian deposits in the United States to *Alisporites zapfei* (Potonié and Klaus) is incorrect, but from their illustrations (Jizba, 1962, pl. 124, fig. 54-55) the specimens may belong to *Falcisporites*. The same applies to examples of *Alisporites* depicted by McGregor (1965, pl. 3, fig. 26, 30) from the Upper Triassic of Arctic Canada.

Falcisporites stabilis Balme, n. sp.

Figure 9; Plate 16, figures 6-10

Occurrence: Chhidru Formation: very rare. Narmia Member, Mianwali Formation: common to dominant. Tredian Formation: abundant to dominant.

Representative Specimens: Holotype: U.W.A. 57950. Others: U.W.A. 57951, 57952, 57953, 57954. Field no. T62-287, Zaluch Nala, Salt Range, Khatkiara Member, Tredian Formation.

Diagnosis: Disaccate, haploxylonoid or slightly diploxylonoid. Corpus circular to transversely elongate oval. Cappa 1-2μ thick, finely but distinctly intrareticulate. Sacci almost semicircular in polar view, distally in-

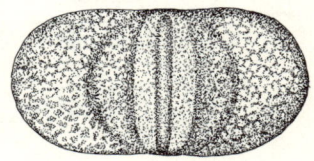

Fig. 9. *Falcisporites stabilis* Balme, n. sp. Reconstruction in distal view, $\times 600$.

clined, saccus exoexine 2-3μ thick, strongly and densely intrareticulate, brochi 1-2μ in diameter, generally equidimensional. Cappula rounded rectangular, bordered in most specimens by darkened crescentic areas lying below the distal sacci bases. Cappula finely structured towards the bases of the sacci with, extending its full length, a clearly defined longitudinal sulcus in which unstructured intexine is exposed.

Dimensions (40 specimens): Total breadth: 62 (79) 100μ. Corpus breadth: 41 (47) 63μ. Breadth of cappula: 16-24μ. Corpus length: 35 (44) 57μ.

Derivation of name: Latin: stabilis = firm.

Remarks: Falcisporites stabilis is distinguished with some hesitation from three existing species, *F. zapfei* (Potonié and Klaus), *Alisporites australis* de Jersey, and *Falcisporites snopkovae* Visscher. Klaus (personal communication) believes that *F. stabilis* has a more finely structured cappa than *F. zapfei,* and after examining a number of specimens of the latter species provided by Dr. Klaus, I agree that this character is sufficiently constant to allow the two species to be distinguished, at least in fairly well-preserved material. There is undoubtedly some overlap between *F. stabilis* and *Alisporites australis* de Jersey, and the basis for distinguishing the two species is the shape of the corpus in polar view. In *A. australis* it is elongate longitudinally and in *F. stabilis*

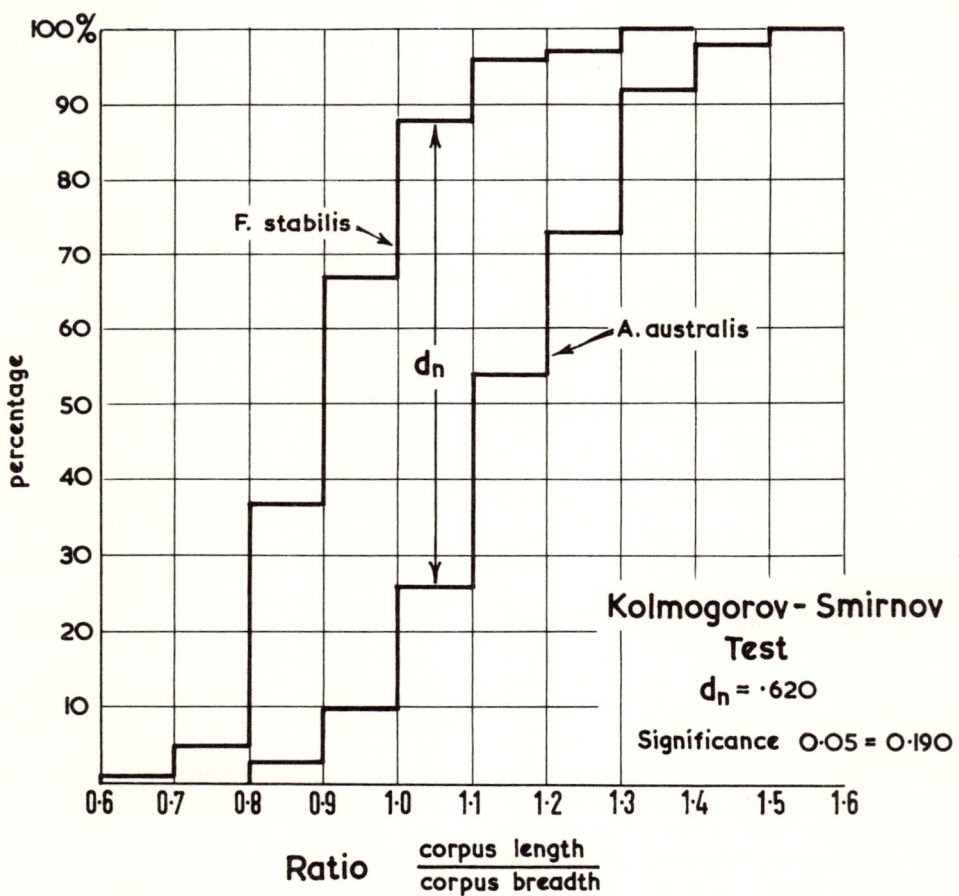

FIG. 10. Cumulative frequency distributions of ratio of corpus length to corpus breadth in 100 specimens of *Falcisporites stabilis* from the Tredian Formation, Zaluch Nala, and 100 specimens of *Alisporites australis* from Leigh Creek, South Australia.

circular or transversely elongate. The distinction emerges unequivocally when large populations are compared as in Figure 10, which compares the ratio of corpus length to corpus breadth of 100 specimens of *Alisporites australis* from Leigh Creek, South Australia, with 100 examples of *Falcisporites stabilis* from the Tredian Formation. It is equally true, however, that the two species could not be distinguished, except arbitrarily, in a mixed assemblage. *F. snopkovae* has been illustrated by only a single complete specimen, but is said to have a lemniscoid distal sulcus (Visscher, 1966).

Pant and Srivastava have probably assigned specimens of *F. stabilis* to their species *Pityosporites sakesarensis* (cf. Pant and Srivastava, 1964, fig. 3K), but this taxon is a heterogeneous one and its holotype markedly diploxylonoid.

Falcisporites nuthallensis Clarke, comb. nov.

Plate 15, figures 15-18

Spore 75, Virkki, 1946, p. 134, fig. 46, pl. 7, fig. 89.
Alisporites nuthallensis Clarke, 1965b, p. 346-347, fig. 15, pl. 43, fig. 1, 15.

Occurrence: Amb Formation: common. Wargal Limestone: very rare to rare. Chhidru Formation: very rare to rare.

Representative Specimens: U.W.A. 57955, 57956, 57957, 57958. Field no. T62-350. Warchha Water Tank, Salt Range, Amb Formation.

Description: Disaccate, haploxylonoid or slightly diploxylonoid. Corpus longitudinally elongate oval with a length/breadth ratio of about 4:3. Cappa 1-2μ thick, finely intrareticulate with sharply defined structural elements. Sacci crescentic to semicircular in polar view with a slight distal inclination sometimes joined equatorially by a narrow strip of detached exoexine. Saccus exoexine 1-2μ thick, densely intrareticulate, brochi 1-2μ in diameter with a tendency to radial elongation near the roots of the saccus. Cappula about one-third breadth of corpus with an elongate oval, transparent sulcus extending almost its full length, sulcus sometimes slightly constricted at distal pole, margins of cappula finely structured. Intexinal longitudinal folds occur beneath the distal bases of the sacci in many specimens.

Dimensions (40 specimens): Total breadth: 62 (72) 86μ. Corpus breadth: 29 (33) 40μ. Breadth of cappula: 12-16μ. Corpus height: 41 (45) 53μ. Sacci height: 39 (45) 55μ.

Remarks: There seems no doubt that *F. nuthallensis* from the British Zechstein is indistinguishable from Virkki's spore 75 from the upper part of the Amb Formation at Warchha. It also appears closely similar to *Alisporites plicatus* Jizba, which occurs in Late Pennsylvanian and probably Permian, strata in the midcontinent area of the United States. Jizba's species is, for the time being, distinguished by its more angular corpus, broader sulcus, and the constant presence of distal intexinal folds, but future studies may show that *F. nuthallensis* should be placed in synonymy with *A. plicatus*.

Genus ALISPORITES Daugherty emend. Nilsson, 1958

Type species (original monotypy): *Alisporites opii* Daugherty.

The concept of the form genus *Alisporites* has undergone many vicissitudes since it was first defined informally by Daugherty (1941, p. 98) to include "all winged spores that apparently are not related to the Abietineae or Podocarpineae." An initial difficulty lay in distinguishing it from the equally broadly characterized *Pityosporites* Seward. Manum (1960) has partly resolved the difficulties with his reexamination of Seward's type material of *Pityosporites*, although he avoided the final inference that the morphology of *Pityosporites* could never be fully understood in terms which would enable it to be effectively used as a category for dispersed pollen grains. Clarke (1965a, p. 308) reached the conclusion that *Pityosporites* was a confused name and should be abandoned. Perhaps it is fairer to regard it as analogous to a paleobotanical genus based only on an internal petrifaction, the external morphology of which is unknown and never will be known, but Clarke's recommendation is nevertheless a wise one.

The remaining question is how should *Alisporites* be restricted in relation to the proliferating taxa which have appeared since 1941 to accommodate disaccate pollen grains originally assignable to Daugherty's genus. Numerous emendations have been made to *Alisporites,* which purposively fall into two categories; those that try to preserve Daugherty's broad concept, and those that aim to define *Alisporites* as a restricted taxon on the basis of the morphology of the type species *Alisporites opii* Daugherty. Rouse (1959) adopted the first approach and has been followed with some reservations by de Jersey (1962, 1964) and to some extent by Playford and Dettmann (1965).

Attempts to restrict *Alisporites* have been based on the assumption that *Alisporites opii* possesses a distal sulcus (Keimfurche). This feature was regarded as diagnostic of the genus in the emendation of Potonié and Kremp (1954), and Klaus (1963) stated that only disaccate pollen grains with a clear fusiform sulcus should be assigned to *Alisporites*. *Falcisporites* has a sulcus with rounded ends but otherwise is closely similar to Klaus's interpretation of *Alisporites*.

The main difficulty concerns the problem of determining whether or not *A. opii* possesses a sulcus in the sense of a sharply delimited area of the cappula with a distinct and more or less constant outline. A longitudinal rupture occurs on the holotype of *A. opii* (Daugherty, 1941, pl. 34, fig. 2), but this could represent a random rent in an undifferentiated cappula rather than a sulcus. Clearly Daugherty's type material needs re-study, but through the kindness of Dr. S. A. J. Pocock of Imperial Oil Company, Calgary, I have been able to examine a slide of plant microfossils from the Chinle Formation. This contains a number of disaccate pollen grains comparable in size and gross morphology to *A. opii,* and although their preservation is imperfect, I am unable to satisfy myself that any possess a distal sulcus. Morphologically the forms I identify as *A. opii* are, except for their somewhat smaller size, close to the disaccate pollen grains of *Pamelreuthia halberfelneri* Krasser, which, according to Townrow (1962), possesses a distal leptoma rather than a sulcus. On the available evidence, therefore, I conclude that if *Alisporites* is to be restricted, forms with a clearly defined distal sulcus should be excluded from it.

Nilsson's emendation of *Alisporites* (Nilsson, 1958) is closest to the interpretation I have accepted, and essentially this agrees with Bharadwaj's concept of the genus (Bharadwaj, 1962, p. 97).

Diagnosis (slightly modified): Disaccate, haploxylonoid or slightly diploxylonoid. Corpus circular to oval. Cappa not markedly thickened, finely and uniformly structured. Sacci large, semicircular to crescentic in polar view, intrareticulate, saccus exoexine fairly thin. Cappula thin, oval, fusiform or rounded rectangular, breadth greater than about one-fourth of the corpus, lacking any clearly defined sulcus.

PLATE 17

Triassic Disaccites: Non-taeniate forms. Both ×600.

FIGURES

1-2. *Alisporites* sp. cf. *A. opii* Daugherty. Upper Lower to Middle Triassic.——1. Specimen 57961.——2. Specimen 57960.

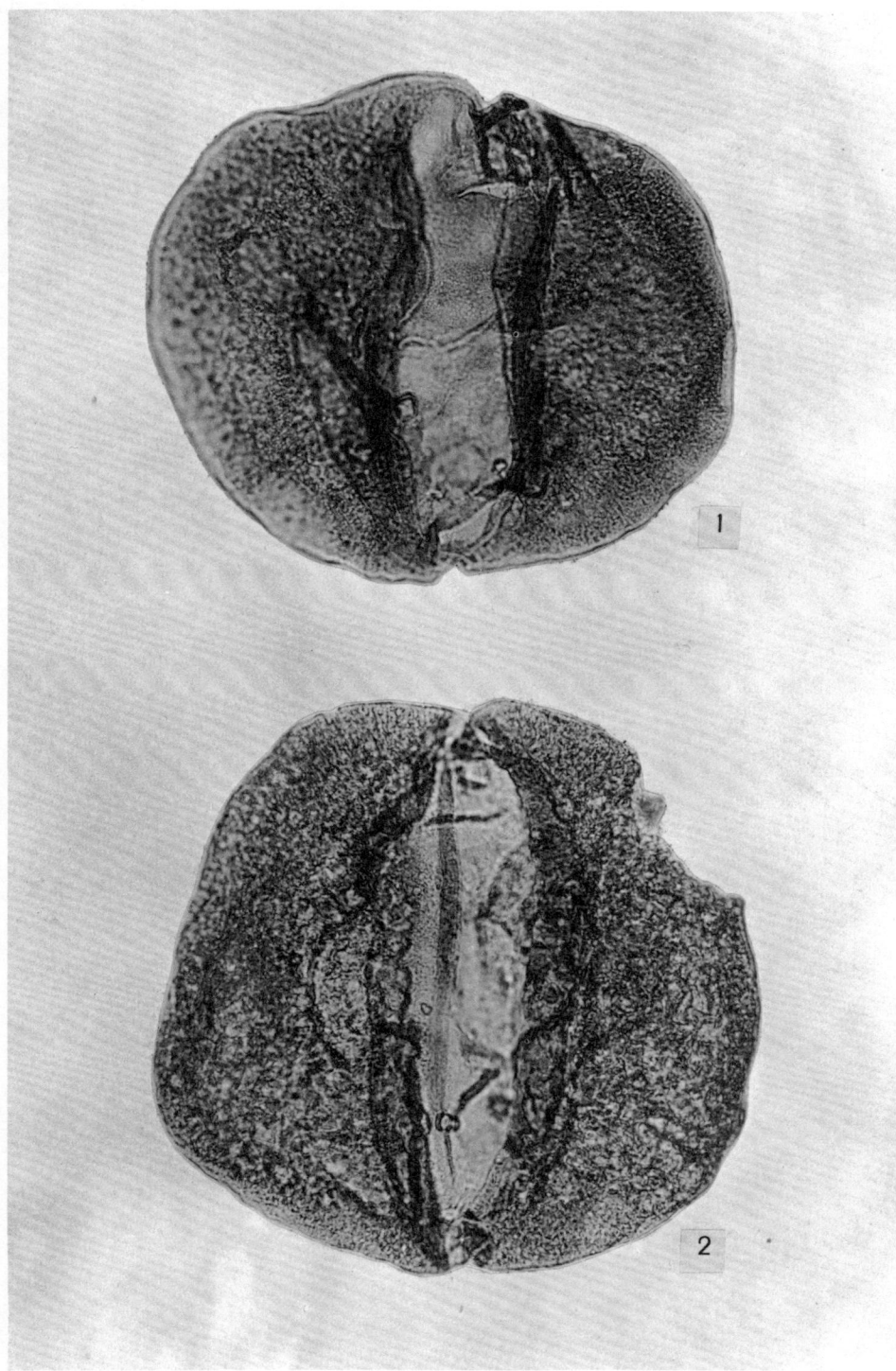

Plate 17

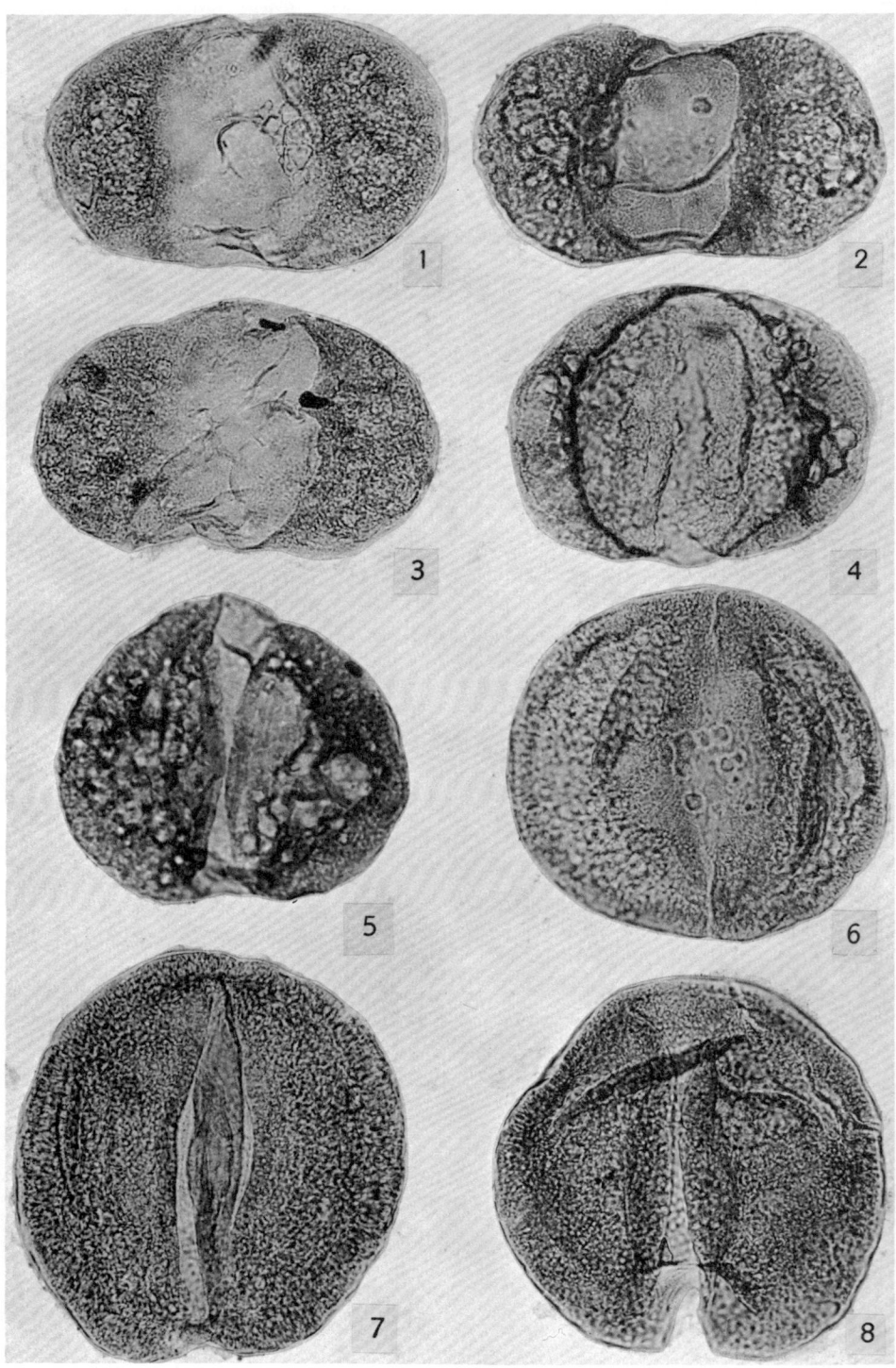

PLATE 18

Alisporites sp. cf. *A. opii* Daugherty
Plate 17, figures 1-2

Alisporites opii Daugherty, 1941, p. 98, pl. 34, fig. 2.

Occurrence: Narmia Member, Mianwali Formation: very rare to common. Tredian Formation: very rare to common.

Representative Specimens: U.W.A. 57959, 57960, 57961. Field no. K12-6, Landa Pusha, Surghar Range, Narmia Member, Mianwali Formation.

Description: Disaccate, haploxylonoid or slightly diploxylonoid. Corpus longitudinally elongate oval. Cappa 1μ or less in thickness, finely and regularly intrareticulate. Sacci semicircular or crescentic in polar view, slightly inclined distally, saccus exoexine $1\text{-}2\mu$ thick, densely and finely intrareticulate, brochi 1μ or less in diameter with weak radial elongation near the sacci bases. Maximum breadth of cappula about one-third that of corpus, outline of cappula elongate oval or nearly fusiform. Cappula thin, transparent, with minute plications and, in some specimens, an irregular longitudinal rupture.

Dimensions (15 specimens): Total breadth: 119 (132) 142μ. Corpus breadth: 57 (71) 90μ. Breadth of cappula: $21\text{-}40\mu$. Corpus height: 90 (108) 130μ.

Remarks: Alisporites sp. cf. *A. opii* is larger and better preserved than any specimens of *A. opii* so far illustrated, although its general morphological characters are similar. *Alisporites aequalis* Mädler is smaller, seems to have a distal sulcus and rigid, fairly coarsely reticulate, sacci. A specimen illustrated by Rogalska (1956, pl. 18, fig. 1) as *Piceapollenites* cf. *alatus* R. Pot. appears indistinguishable from *A.* cf. *A. opii*.

Alisporites landianus Balme, n. sp.
Plate 18, figures 1-3

Occurrence: Narmia Member, Mianwali Formation: very rare to abundant. Tredian Formation: very rare.

Representative Specimens: Holotype: U.W.A. 57962. Others: U.W.A. 57963, 57964, 57965. Field no. K12-6, Landa Pusha, Surghar Range, Narmia Member, Mianwali Formation.

Diagnosis: Disaccate, slightly diploxylonoid. Corpus outline poorly defined, circular to transversely elongate oval. Cappa about 1μ thick, translucent, finely intrareticulate to intrapunctate. Sacci semicircular to broadly crescentic in polar view with a slight distal inclination. Saccus exoexine about 1μ thick, finely and densely intrareticulate, brochi less than 1μ in diameter with a slight tendency to radial elongation. Cappula broadly oval, breadth greater than half that of corpus, thin, translucent, unstructured, and occasionally with a random rupture.

Dimensions (30 specimens): Total breadth: 82 (95) 107μ. Corpus breadth: 48 (54) 67μ. Breadth of cappula: 28 (31) 34μ. Sacci height: 47 (54) 62μ.

Remarks: Alisporites landianus is smaller and relatively more elongate transversely than *A. opii*; it also has an almost circular corpus and broad cappula. *Latosaccus latus* Mädler from the lower Muschelkalk of Ger-

---

PLATE 18

Triassic Disaccites: Non-taeniate forms. All ×600.

FIGURES

1-3. *Alisporites landianus* Balme, n. sp. Upper Lower to Middle Triassic.——1. Specimen 57963.——2. Holotype, specimen 57962.——3. Specimen 57965.

4-5. *Sulcatisporites* sp. cf. *S. kraeuseli* Mädler. Upper Lower to Middle Triassic.——4. Specimen 57975.——5. Specimen 57973.

6-8. *Sulcatisporites institatus* Balme, n. sp. Upper Lower to Middle Triassic.——6. Holotype, specimen 57976, polar view.——7. Specimen 57977, polar view.——8. Specimen 57979, semiequatorial view.

many looks generally similar to *A. landianus*, but its type specimen is too poorly preserved for detailed comparison. From Mädler's (1964, pl. 7, fig. 3-4) illustrations the brochi of the saccus intrareticulum of *L. latus* are coarse and show pronounced radial elongation. Another European Triassic species, *Scopulisporites toralis* Leschik, is also morphologically close to *A. landianus*, but has a narrower cappula and a more coarsely structured exoexine, both on the cappa and the sacci. Comparisons with disaccate species from the Russian Triassic are difficult, but it is likely that *A. landianus* falls into the rather elastic category *Pemphygaletes prolixus* Luber (cf. Chalyshev and Varyukhina, 1960, 1962).

Alisporites tenuicorpus Balme, n. sp.

Plate 15, figures 1-4

Occurrence: Amb Formation: very rare to rare: Wargal Limestone: very rare to common. Chhidru Formation: rare to abundant.

Representative Specimens: Holotype: U.W.A. 57941. Others: U.W.A. 57942, 57943, 57944. Field no. K11-6C, Wargal, Salt Range, Chhidru Formation.

Description: Disaccate, nearly haploxylonoid. Corpus circular to oval with minor transverse or longitudinal elongation. Cappa less than 1μ thick, finely intrareticulate-intrapunctate. Sacci crescentic in polar view with a slight distal inclination. Saccus exoexine about 1μ or less in thickness, finely intrareticulate, brochi equidimensional about 1μ in diameter. Cappula oval, maximum breadth about half that of the corpus, thin and translucent, sometimes bordered by weak intexinal folds below the sacci bases.

Dimensions (40 specimens): Total breadth: 28 (33) 41μ. Corpus breadth: 17 (21) 26μ. Breadth of cappula: 6 (11) 15μ. Corpus height: 17 (21) 26μ. Sacci height: 17 (20) 26μ.

Derivation of name: Latin *tenuis* = thin, *corpus* = body.

Remarks: Forms resembling *Alisporites tenuicorpus* have been assigned by some authors to *Vitreisporites* Leschik. However, in view of the close correspondence between the type species of *Vitreisporites* and the known pollen grains of the Caytoniales, I believe that the concept of that form genus should not be extended beyond the morphological variation found among the pollen of the various species of *Caytonanthus*. *A. tenuicorpus* is larger than *Vitreisporites pallidus*, has an almost circular corpus, relatively smaller sacci, and a broader cappula. For this reason it has been assigned to the more noncommittal form genus *Alisporites*.

Vitreisporites koenigswaldi Jansonius is of similar size and general morphology, but has an angular corpus, narrower cappula and, according to Jansonius (1962, p. 55), a characteristic sharp bend in the proximal sacci bases. *Klausipollenites staplinii* Jansonius also resembles *A. tenuicorpus* but possesses constant and pronounced darkened areas below the distal bases of the sacci. *Caytonipollenites latus* Mädler appears close to *K. staplinii* but is possibly distinguishable by its broader sacci and somewhat larger size. An undoubtedly similar form to *A. tenuicorpus* is *Alisporites parvus* de Jersey, which is widespread in Australian Triassic sediments. It is, however, consistently larger than *A. tenuicorpus*, having a total breadth of 42-59μ for 78 specimens measured from the original population (de Jersey, 1962). Specimens compared to *A. parvus* by Playford and Dettmann (1965) have an elliptical cappula which does not extend the full length of the corpus and on which a sulcus is sometimes developed. Forms referred by Medvedeva (1960) to *Caytonanthus tecturatus* (Luber) resemble *A. tenuicorpus*, but Medvedeva's identification seems doubtful, as the original diagnosis of *Pemphygaletes tecturatus* Luber states that the total breadth (d) of the species is 72μ (Luber and Waltz, 1941). This size would preclude the assignation of the species to *Vitreisporites*.

Genus SULCATISPORITES Leschik emend. Nilsson, 1958

Type species (original designation): *Sulcatisporites interpositus* Leschik.

Judging from the single published illustration of the type species of *Sulcatisporites* (Leschik, 1955, pl. 10, fig. 4), one is forced to agree with Potonié (1958) and other authors who doubt that it can be distinguished from *Alisporites*. It is, however, possible to interpret the morphology of *S. interpositus* in such a way that it forms the basis of a useful form category. This approach has been adopted by several authors (e.g., Nilsson, 1958; Bharadwaj, 1962; Mädler, 1964) all of whom, implicitly at least, regard the narrow cappula of *Sulcatisporites* as a sufficiently distinctive character to warrant its recognition as a separate form genus. Bharadwaj (1962) believes that the absence of a clearly defined corpus is an additional generic character of *Sulcatisporites*, but this seems too rigid, although it is true that in some Permian species the corpus is difficult or impossible to discern. In these the pollen grain appears to consist entirely of saccus exoexine separated distally by a narrow, linear, cappula. According to Mädler (1964) the sacci in *Sulcatisporites* are joined equatorially by narrow strips of detached exoexine, but again, while this is true of many specimens, I hesitate to accept as a firm systematic criterion a character which is so often gradational within a group of specimens that cannot be separated on other morphological grounds. *Sulcatisporites* was placed in synonymy with *Vesicaspora* Schemel by Hart (1965b), but following the reinterpretation of *Vesicaspora* by Wilson and Venkatachala (1963b) it seems desirable to keep the genera distinct. In *Vesicaspora wilsonii* Schemel, the type species of *Vesicaspora*, the equatorial sacci connections are pronounced and persistent and the cappula fusiform.

As I interpret *Sulcatisporites* its essential distinguishing features are a nearly elliptical amb and a very narrow cappula, which is often represented merely by a distal longitudinal parting, along which the sacci bases are almost touching. These characters also distinguish *Voltziapites* Malyavkina, a synonym, according to Malyavkina (1964, p. 137), of Sporomorph C of Lundblad (1949). *Voltziapites* is, therefore, available if future study of the type species of *Sulcatisporites* discloses that it should be properly referred to *Alisporites*.

Sulcatisporites ovatus (Balme and Hennelly)

Plate 15, figures 7-9

Florinites ovatus Balme and Hennelly, 1955, p. 96-97, pl. 5, fig. 49-52.
Pityosporites ovatus (Balme and Hennelly), Lakhanpal, Sah and Dube, 1962, p. 115, pl. 1, fig. 17.
Sulcatisporites ovatus (Balme and Hennelly), Bharadwaj, 1962, p. 97, pl. 19, fig. 249-251.
Vesicaspora ovata (Balme and Hennelly), Hart, 1965b, p. 73, fig. 170.
non *Alisporites ovatus* auct. non (Balme and Hennelly) Jansonius, 1962, p. 58-59, pl. 13, fig. 1-2.
non *Alisporites ovatus* auct. non (Balme and Hennelly), Schaarschmidt, 1963, p. 59, pl. 15, fig. 18-19; pl. 16, fig. 1-5.

Occurrence: Amb Formation: very rare to rare. Wargal Limestone: very rare to rare. Chhidru Formation: very rare to common.

Representative Specimens: U.W.A. 57966, 57967, 57968. Field no. K11-2C. Wargal, Salt Range, Chhidru Formation.

Description: Disaccate, haploxylonoid, amb oval. Corpus longitudinally elongate oval, rarely almost circular, outline poorly defined. Cappa thin, finely intrareticulate. Sacci semielliptical in polar view, saccus exoexine about 1μ thick, finely and regularly intrareticulate, brochi 1μ or less in diameter. Cappula narrow, parallel-sided and extending the full length of the corpus, often manifested as only a longitudinal line between the distal sacci bases.

Dimensions (from Balme and Hennelly, 1955): Total breadth: 46 (65) 74μ. Corpus breadth: 28-48μ. Corpus height: 36 (50) 65μ.

Remarks: Sulcatisporites ovatus is distinguished by its longitudinally elongate corpus, which although indistinct is always discernible, and its fine uniform saccus intrareticulum. It is widespread in Australian Permian sediments, in which it shows little morphological variation, and has been reported from other "Gondwanaland" countries. Specimens assigned to *S. ovatus* from North America and Europe (Jansonius, 1962; Schaarschmidt, 1963) on the other hand have a broader cappula and less sharply defined distal sacci bases and should, I consider, be regarded as distinct. *Sulcatisporites quadratus* Nilsson has a coarser saccus reticulum with radially elongate brochi (Nilsson, 1958, pl. 8, fig. 8).

Sulcatisporites nilssoni Balme, n. sp.

Plate 19, figures 1-3

Spore 76, Virkki, 1946, p. 139, pl. 7, fig. 104.

Occurrence: Amb Formation: very rare to rare.

Representative Specimens: Holotype: U.W.A. 57969. Others: U.W.A. 57970, 57971, 57972. Field no. T62-350, Warchha Water Tank, Salt Range, Amb Formation.

Description: Disaccate or pseudomonoscaccate, amb transversely oval. Corpus circular or longitudinally oval, usually clearly discernible. Cappa about 1μ thick, finely intrareticulate in area of the pole, structure becoming coarser towards the proximal base of the sacci. Sacci semielliptical in polar view, equatorially joined in many specimens by narrow strips of detached exoexine, saccus exoexine 3-4μ thick, strongly and fairly coarsely intrareticulate, brochi 1-4μ in diameter, finer near the proximal and distal bases. Cappula less than 5μ wide, parallel-sided and extending the full length of the corpus.

Dimensions (30 specimens): Total breadth: 93 (130) 155μ. Corpus breadth: 55 (75) 94μ. Corpus height: 68 (85) 108μ.

Derivation of name: After Dr. Tage Nilsson in acknowledgment of his contributions to the study of Mesozoic saccate pollen.

Remarks: Sulcatisporites nilssoni is a rare but distinctive species encountered only in the Amb Formation, from which unit it had (as Spore 76) been previously recorded at Jhallewali in the Salt Range by Virkki. A form illustrated as *Sulcatisporites* by Bharadwaj (1962, pl. 19, fig. 252) from the Raniganj Coal Measures may also belong to *S. nilssoni*. *Pityosporites potoniei* Lakhanpal, Sah, and Dube and its probable synonym *Vesicaspora maxima* Hart both have a more nearly circular amb and much less obvious corpus. Sporomorph C of Lundblad (1949, p. 13, fig. 10-13) resembles *S. nilssoni* in gross morphology, but is much bigger, with a total breadth between 204-209μ.

Sulcatisporites institatus Balme, n. sp.

Plate 18, figures 6-8

Occurrence: Narmia Member,

PLATE 19

Lower Permian Disaccites: Non-taeniate forms. All ×600.

FIGURES

1-3. *Sulcatisporites nilssoni* Balme, n. sp. Lower Permian. ——1. Specimen 57971, distal focus.——2. Specimen 57972, equatorial view, proximal side uppermost showing distal rupture between the sacci bases.——3. Holotype, specimen 57969.

PLATE 19

Mianwali Formation: rare to common. Landa Member, Tredian Formation: very rare to rare.

Representative Specimens: Holotype: U.W.A. 57976. Others: U.W.A. 57977, 57978, 57979. Field no. K6-41B, Nammal Gorge, Salt Range, Narmia Member, Mianwali Formation.

Description: Disaccate, haploxylonoid or nearly so, amb subcircular. Corpus longitudinally oval, poorly defined. Cappa 2-3μ thick, finely intrareticulate. Sacci almost semicircular in polar view, strongly inflated and distally inclined. Saccus exoexine up to 5μ thick so that the sacci have a distinct limboid margin, finely and densely intrareticulate, brochi less than 1μ in diameter. Sacci usually joined equatorially by strips of thickened exoexine. Cappula elongate fusiform, with a maximum breadth of less than about 10μ, thin and structureless, sometimes with an irregular longitudinal rupture.

Dimensions (30 specimens): Total breadth: 80 (90) 108μ. Corpus breadth: 53 (63) 81μ. Breadth of cappula: 0-12μ. Corpus height: 68 (83) 105μ.

Derivation of name: Latin *instita* = a border, referring to the limboid appearance of the sacci.

Remarks: Sulcatisporites institatus is bigger than *S. interpositus* Leschik and has a finer and denser saccus intrareticulum. *Piceapollenites fuscus* Pautsch appears from its illustration (Pautsch, 1958, pl. 1, fig. 1) to have a circular corpus, relatively smaller sacci, and thinner saccus exoexine. *Sulcatisporites kraeuseli* Mädler also resembles *S. institatus* but has a parallel-sided cappula and has less inflated and thinner sacci (vide Mädler, 1964, pl. 11, fig. 3).

Sulcatisporites sp. cf. *S. kraeuseli* Mädler

Plate 18, figures 4-5

Sulcatisporites kraeuseli Mädler, 1964, p. 65-66, pl. 4, fig. 3-4; p. 117, pl. 11, fig. 3-4.

Occurrence: Narmia Member, Mianwali Formation: very rare to rare. Tredian Formation: very rare.

Representative Specimens: U.W.A. 57973, 57974, 57975. Field no. K12-6, Landa, Surghar Range, Narmia Member, Mianwali Formation.

Description: Disaccate, haploxylonoid, amb subcircular to transversely oval. Corpus longitudinally oval. Cappa 1-2μ thick, finely intrareticulate. Sacci almost semicircular in polar view, inclined distally, saccus exoexine about 2μ thick, finely and densely intrareticulate, brochi less than 1μ in diameter. Cappula nearly parallel-sided, less than about 8μ broad, extending full length of the corpus. thin and translucent.

Dimensions (30 specimens): Total breadth: 63 (73) 94μ. Corpus breadth: 40 (49) 62μ. Breadth of cappula: 2-8μ. Corpus height: 52 (62) 69μ.

Remarks: Sulcatisporites sp. cf. *S. kraeuseli* lies on the line of demarcation accepted here between *Alisporites* and *Sulcatisporites*. It is assigned to *Sulcatisporites* because of its resemblance to Mädler's species, which appears to range throughout the German Triassic and may be the same form as *Piceapollenites fuscus* Pautsch from the Keuper of Poland. According to its diagnosis the cappa of *S. kraeuseli* is 5μ thick at the proximal pole. This is considerably thicker than any estimate made on specimens grouped here under *S.* sp. cf. *S. kraeuseli,* and for this reason I have not placed them in direct synonymy.

Genus PLATYSACCUS Naumova ex Potonié and Klaus, 1954

Type species (designated by Potonié and Klaus): *Platysaccus papilionis* Potonié and Kremp.

Platysaccus is retained here in the broad sense adopted by some previous authors (e.g., de Jersey, 1962; Playford and Dettmann, 1965; Clarke, 1965a) to embrace disaccate pollen grains with a diploxylonoid habit and a thick, almost circular corpus. The sacci may or may not be joined equatorially by thickened or detached strips of exoexine. In this sense the genus includes *Umbrosaccus* Mädler, which was instituted to incorporate *Platysaccus*-like forms with equatorially linked sacci. If the type species of *Umbrosaccus (U. hyalinus)* Mädler is considered in isolation, there appears to be a basis for separating it from *Platysaccus*, in that it is not markedly diploxylonoid and therefore lacks the "hantelförmig" amb specified in the diagnosis of *Platysaccus*. There may even be some merit in such a discrimination, for, as Clarke (1965a) has noted, the strongly diploxylonoid forms of *Platysaccus* are more common in Permian than Triassic strata. However, Mädler (1964, p. 118) specifically disclaims that the relative size of corpus and sacci can be used as a criterion to separate *Platysaccus* from *Umbrosaccus*, and as it stands, I believe the latter genus to be inadequately characterized.

Platysaccus queenslandi de Jersey
Plate 16, figures 11-13

Platysaccus queenslandi de Jersey, 1962, p. 10, pl. 4, fig. 5-6.
Platysaccus queenslandi de Jersey, Playford and Dettmann, 1965, p. 156-157, pl. 16, fig. 53-54.
Platysaccus queenslandi de Jersey, Playford, 1965, p. 198, pl. 10, fig. 5.

Occurrence: Narmia Member, Miawali Formation: very rare to common. Tredian Formation: rare to common.

Representative Specimens: U.W.A. 57980, 57981, 57982, 57983. Field no. K12-6, Landa Pusha, Surghar Range, Narmia Member, Mianwali Formation.

Description: Disaccate, diploxylonoid but not markedly so. Corpus circular or slightly transversely oval, dark in color. Cappa about 3μ thick, consisting of a structureless layer of optically dense intexine overlain by finely and densely structured (? intrabaculate) exoexine. Sacci larger than corpus, distally inclined and somewhat flaccid, sometimes joined equatorially by strips of thickened exoexine. Saccus exoexine 2-3μ thick, fairly coarsely intrareticulate, brochi diameter 1-4μ, finer towards the distal bases and with weak radial elongation. Cappula elongate oval, bordered in many specimens by longitudinal intexinal folds below the sacci bases. Cappula finely structured near its margins, bearing a sharply defined, almost rectangular sulcus running the full length of the corpus.

Dimensions (30 specimens): Total breadth: 74 (86) 97μ. Corpus breadth: 39 (43) 47μ. Breadth of cappula: 6 (13) 19μ. Corpus height: 30 (39) 46μ. Sacci height: 34 (41) 49μ.

Remarks: Specimens of *Platysaccus queenslandi* described above deviate slightly from the description given by Playford and Dettmann. They have, for example, a thicker cappa and apparently a somewhat broader cappula, although this may be a matter of interpretation, but these characters are inadequate to differentiate between the two populations.

P. queenslandi is widely distributed in Australian Triassic sediments in which it is associated with pollen of the *Falcisporites*-type (e.g., *Alisporites australis* de Jersey). Disaccate pollen grains of similar general morphology have been described occasionally from Europe; and *Platysaccus hengeloensis* Freudenthal, from the Lower Triassic of the Netherlands (Freudenthal, 1964), appears distinguishable from *P. queenslandi* only by its considerably greater size. Russian Triassic species assigned by Malyavkina (1964) to *Podocarpus (Archaeopodocarpus)* also probably include forms which could be referred to *P. queenslandi*.

Genus PINUSPOLLENITES Raatz, 1937

Type species (original monotypy): *Pinuspollenites labdacus* (Potonié).

The genus *Pinuspollenites* lacks adequate circumscription and illustration. It is used here without any phylogenetic implications as a broad form

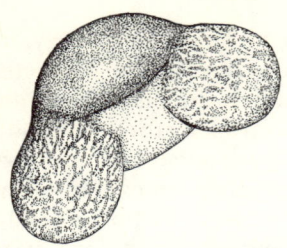

Fig. 11. *Pinuspollenites thoracatus* Balme, n. sp. Reconstruction in equatorial view, proximal side uppermost, ×600.

category to incorporate disaccate pollen grains with a similar gross morphology to that of *Pinus*.

Pinuspollenites thoracatus Balme, n. sp.

Figure 11; Plate 15, figures 10-14

Occurrence: Amb Formation: common to abundant. Wargal Limestone: common. Chhidru Formation: very rare to common.

Representative Specimens: Holotype: U.W.A. 57984. Others: U.W.A. 57985, 57986, 57987. Field no. T62-350. Warchha Water Tank, Salt Range, Amb Formation.

Description: Disaccate, more or less haploxylonoid, often preserved in off-polar compressions. Corpus transversely oval. Cappa sharply delimited, finely intravermiculate or intrareticulate, about 1μ thick, and in some specimens covering only part of the proximal side of the corpus. Sacci semicircular or slightly crescentic in polar view, abruptly differentiated from the corpus and strongly inclined distally. Saccus exoexine about 2μ thick, finely and densely intrareticulate, brochi 1μ or less in diameter, with a tendency to radial elongation. Cappula breadth about half that of corpus, consisting entirely of thin, transparent, intexine which occupies the whole area between the distal sacci bases and extends in some specimens beyond the equator on to the proximal surface.

Dimensions (30 specimens): Total breadth: 55 (62) 73μ. Corpus breadth: 38 (47) 55μ. Breadth of cappula: 16 (24) 40μ. Corpus height: 35 (39) 43μ. Saccus height: 37 (40) 46μ.

Derivation of name: Latin: *thoracatus* = armed with a breastplate.

Remarks: Pollen grains resembling *Pinuspollenites thoracatus* are uncommon in Paleozoic sediments, although they have previously been described from Mesozoic and Tertiary deposits. Some specimens of *Taedaesporites scaurus* Nilsson from the Liassic of Scania (Nilsson, 1958, pl. 7, fig. 17) appear similar to *P. thoracatus*, but have a more extensive cappa, smaller sacci, and a narrower cappula. The restricted and thin cappa also serves to distinguish *P. thoracatus* from *Granabivesiculites constrictus* Pierce from the lower Upper Cretaceous of Minnesota, a species which has been referred, at least in part, to the Pinaceae (Pierce, 1961).

Genus CEDRIPITES Wodehouse, 1933

Type species (original monotypy): *Cedripites eocenicus* Wodehouse.

Cedripites is based on a single specimen from the Eocene of Colorado and lacked an original generic diagnosis, although its morphology has been discussed subsequently by Potonié (1958). Clearly Wodehouse intended his form genus to be circumscribed by the morphological characters of modern species of *Cedrus*, although he warned explicitly that *Cedripites* was not necessarily derived from *Cedrus*. From the descriptions of Wodehouse (1959, p. 261) and Erdtman (1957, 1965), *Cedrus* pollen is characterized by its large thick-walled cappa, the structure of which merges gradually into that of the sacci. The sacci are, therefore, not sharply differentiated from the corpus; they typically have long columellate processes (endosexinous elements of Erdtman) and tend to be somewhat concave proximally so that they appear to wrap around the grain (cf. Wodehouse, 1933, p. 490).

Cedripites-type pollen grains have been most often reported from late Mesozoic and Tertiary deposits, but they are known from the Upper Permian and Triassic strata in the USSR (*vide inter alia* Takhtadzhian et al., 1963, p. 270). No forms assignable to *Cedripites* have been described from the Permian or Triassic of the southern continents, with the possible exception of forms placed in *Rimaesporites aquilonalis* Goubin by Goubin (1965) from the Upper Triassic of Madagascar. This species is said to have a distal cleft, and this character is visible on the holotype. The other illustrated specimen, however, appears to have a different morphology and may be incorrectly identified.

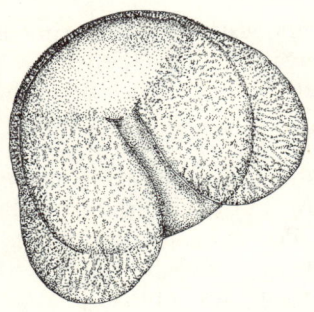

Fig. 12. *Cedripites priscus* Balme, n. sp. Reconstruction in equatorial view, proximal side uppermost, ×600.

Cedripites priscus Balme, n. sp.

Figure 12; Plate 20, figures 1-4

Occurrence: Chhidru Formation: very rare to common.

Representative Specimens: Holotype: 57988. Others: 57989, 57990, 57991. Field no. K11-6D, Wargal, Salt Range, Chhidru Formation.

Description: Disaccate with a marked tendency to preservation in off-polar compression. Corpus transversely oval to almost circular. Cappa heavily thickened, exoexine about 3μ thick at the proximal pole increasing to about 6μ near the proximal bases of the sacci, densely columellate appearing finely intrareticulate in surface focus, structure of cappa merging gradually into that of the sacci. Sacci slightly crescentic to almost semicircular in polar view with a strong distal inclination, saccus exoexine about 5μ thick, intrareticulate, brochi 1-3μ diameter becoming coarser away from proximal and distal bases, radially elongate near the sacci roots. Cappula breadth about half that of the corpus, finely structured near its lateral margins with an oval or fusiform, poorly defined tenuitas running longitudinally through the distal pole.

Dimensions (30 specimens): Total breadth: 65 (75) 83μ. Corpus breadth: 49 (56) 64μ. Breadth of cappula: 19 (29) 37μ. Corpus height: 43 (51) 61μ. Sacci height: 46 (53) 61μ.

Derivation of name: Latin *priscus* = ancient.

Remarks: Specimens resembling *Cedripites priscus* have been reported occasionally from Upper Permian deposits in the USSR (e.g., Zoricheva and Sedova, 1954, pl. 10, fig. 9), although they are apparently not common, for they do not appear in most Permian microfloral lists from Russia. Various forms have been described from Russian Jurassic and Lower Cretaceous deposits which are undoubtedly referable to *Cedripites* (cf. Bolkhovitina, 1953, 1956), but these have so far been illustrated only by sketches and detailed comparisons are impossible.

Some of the specimens assigned by Couper (1958) to *Parvisaccites* Couper bear a general resemblance to *Cedripites*, and the genera were not differentiated in the original diagnosis. Neither of Couper's two species of *Parvisaccites* is, however, likely to be confused with *C. priscus*. *Parvisaccites enigmaticus* Couper is smaller and possesses a slight marginal crest and rugulate cappa, and structural elements in the sacci of *Parvisaccites radiatus* Couper have a strongly marked radial arrangement.

Genus FIMBRIAESPORITES
Leschik, 1959

Type species (original designation): *Fimbriaesporites globosus* Leschik.

It is difficult to interpret the morphology of *Fimbriaesporites globosus* from the single illustration provided by Leschik (1959, pl. 4, fig. 29). Obviously the species is markedly diploxylonoid, and appears to have some modification of the cappa, but the "fringed" exoexine of the corpus, which Leschik regards as the main character differentiating *Fimbriaesporites* from *Platysaccus*, is not clear in the photomicrograph. Singh (1964) has provided another interpretation of *Fimbriaesporites* and believes that its reticuloid cappa distinguishes it from *Platysaccus* and other diploxylonoid form genera. It is not clear from Singh's account whether his conclusions are based on re-examination of the original material and, if not, they must be accepted with reservations, for he admits that the supposed characters are not clear in Leschik's illustration of the type. Obviously the genus still needs clarification, but Singh's usage is accepted here, as forms resembling his species of *Fimbriaesporites* from Iraq are striking, although rare, components of late Permian assemblages from the Salt Range.

Fimbriaesporites? sp.
Plate 20, figures 5-6

Occurrence: Chhidru Formation: very rare to rare. Kathwai and Mittiwali Members, Mianwali Formation: very rare.

Representative Specimens: U.W.A. 57992, 57993, 57994, 57995. Field no. T61-178, Kathwai, Salt Range, Chhidru Formation.

Description: Disaccate, diploxylonoid. Corpus circular to slightly transversely or longitudinally oval. Cappa 3-4μ thick, exoexine finely intrareticulate, dissected by irregular cracks, which give rise to islands of exoexine of various shapes and sizes. In some specimens these exoexinal islands have a crude transverse arrangement so that they simulate incipient taeniae, in others the cappa is reticuloid or rugulose. Sacci large, broadly crescentic in polar view, strongly inflated with marked distal inclination, saccus exoexine about 2μ thick, coarsely intrareticulate, brochi with a tendency to radial elongation and up to 10μ in diameter. Cappula roughly fusiform, thin, hyaline, usually with fine plications.

Dimensions (10 specimens): Total breadth: 116 (124) 137μ. Corpus breadth: 49 (52) 56μ. Breadth of cappula: 11 (19) 25μ. Sacci height: 68 (75) 82μ.

Remarks: Fimbriaesporites? sp. is a rare but striking form in uppermost Permian and lowermost Triassic strata in the Salt Range. It has a larger and less optically dense corpus than *Fimbriaesporites fimbriatus* Singh and *F. globosus* Leschik. A specimen from the Upper Permian of the Southern Alps of Austria, which was illustrated by Klaus (1963, pl. 7, fig. 31) as *Platysaccus papilionis* Potonié and Klaus, looks similar to some examples of *Fimbriaesporites?* sp. and is of comparable size. Certainly it seems morphologically far removed from the holotype of *P. papilionis*, which has a relatively small, circular and dark, almost opaque, corpus. De Jekhowsky and Goubin (1964, fig. 5, 1020) have illustrated a specimen from the Upper Permian of Madagascar that also resembles *Fimbriaesporites?* sp., but the photomicrograph does not show clearly the details of the cappa.

Turma PLICATES Naumova emend. Potonié, 1962

Subturma PRAECOLPATES Potonié and Kremp, 1954

Genus MARSUPIPOLLENITES
Balme and Hennelly, 1956, emend.

Type species (original designation): *Marsupipollenites triradiatus* Balme and Hennelly.

As originally defined, the genus

Marsupipollenites included all monosulcate pollen grains with a proximal trilete or monolete tetrad mark, excepting those large monosulcate-monolete forms which were assignable to *Schopfipollenites* Potonié and Kremp. Of the four original species of *Marsupipollenites*, *M. scutatus* and *M. fasciolatus* are now more appropriately placed in *Vittatina;* and *M. sinuous*, as Bharadwaj (1962) has pointed out, is excluded from *Marsupipollenites* even if that genus is interpreted in the initial sense of Balme and Hennelly. At the present time, therefore, *Marsupipollenites* is monotypic and consists of two forms of the type species *M. triradiatus*. To regularize its systematic status, the generic diagnosis of *Marsupipollenites* is emended here to read:

Emended diagnosis: Amb longitudinally oval or subcircular. Monosulcate, with a distal sulcus which in oval specimens is elongate and slightly narrower at the pole than at the extremities. When the amb is subcircular the sulcus appears as a broadly oval to almost circular distal opening. Proximal face bearing a small trilete tetrad scar which is invariably present but sometimes indistinct. Exine columellate with small, closely spaced, rounded structural elements. Intraexinal elements randomly disposed or arranged in roughly parallel, transverse rows simulating taeniae.

Remarks: The essential characters of *Marsupipollenites* are its distal sulcus, which may or may not be bordered by folds (cf. Potonié, 1962b), its proximal trilete scar, and its structured exine. These are not shared with any other genus of dispersed spores, but may be combined in the pollen of the presumed medullosean fructification *Potoniea adiantiformis* Zeill. and in pollen grains occurring in anther-sacs of unknown affinities from the Raniganj Coalfield which were named *Polytheca elongata* by Pant and Nautiyal (1960).

Marsupipollenites triradiatus Balme and Hennelly
Plate 21, figures 15-18

Marsupipollenites triradiatus Balme and Hennelly, 1956a, p. 60-61, pl. 2, fig. 29-35.

Occurrence: Chhidru Formation: very rare to rare.

Representative Specimens: U.W.A. 57996, 57997, 57998. Field no. T61-207, Wargal, Salt Range, Chhidru Formation.

Description: Monosulcate, amb longitudinally oval to subcircular. Sulcus extending almost full length of distal surface, slightly contracted at the pole in unexpanded specimens with a markedly oval outline, otherwise represented by an oval to circular rupture. Proximal side bearing a small triradiate tetrad mark with rays up to about 10μ long. Exine about 2μ thick, consisting of a very thin transparent intexine, which is visible only in well-preserved specimens, and a relatively coarsely columellate exoexine. Intraexinal structural elements rounded or irregular in plan, 1-3μ in diameter, closely spaced and randomly distributed.

Dimensions (20 specimens): Length: 46 (56) 65μ. Breadth: 35 (42) 51μ.

Remarks: Marsupipollenites triradiatus is widespread in Australian Permian sediments, in which it ranges stratigraphically from late Sakmarian to Upper Permian, but is in general only common in Upper Permian strata. There are also several records from the Permian of Africa, and Balme and Playford (1967) have found the species in coals and shales of Late Permian age from East Antarctica. Apart from the Raniganj occurrence noted by Pant and Nautiyal (1960), pollen grains clearly resembling *M. triradiatus* have not previously been described from sediments in the Northern Hemisphere. *Marsupipollenites tecturatus* Imgrund, from the Lower Permian of China, shows no distal sulcus in its only published illustration (Imgrund, 1960, pl. 13, fig. 3) and should be regarded as a trilete spore.

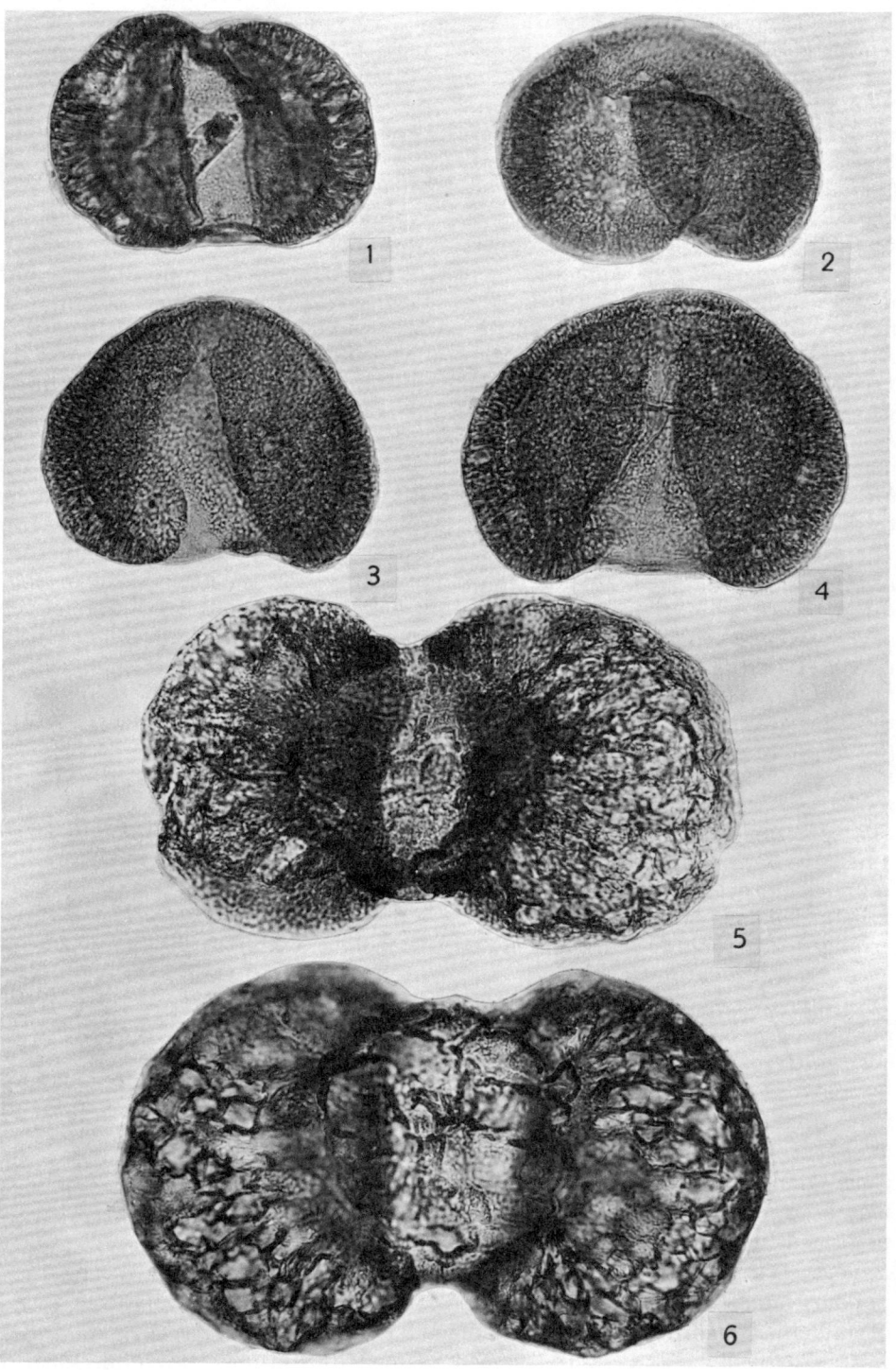

Plate 20

Genus PRETRICOLPIPOLLENITES
Danzé-Corsin and Laveine, 1963

Type species (original designation): *Pretricolpipollenites ovalis* Danzé-Corsin and Laveine.

Considering the botanical interest of the unusual type of pollen that they were describing, Danzé-Corsin and Laveine (in Briche, Danzé-Corsin, and Laveine, 1963, p. 109-110) introduced the generic name *Pretricolpipollenites* rather casually. Their generic diagnosis freely translated, runs as follows:

Pollen grains fusiform. Exine entirely smooth, color yellow-brown. Median furrow prominent, very slightly wider at the center and extremities, bordered by folds in the exine and affecting the whole length of the grain. Two less prominent lateral furrows running about two-thirds the length of the pollen and not bordered by folds. Height 25-35μ.

The genus is monotypic and figured by photomicrographs of two specimens of the type species, but only one of these shows what have been interpreted as lateral furrows. One could, therefore, have wished for a more illuminating illustration of a form which may become as controversial as *Eucommiidites troedsoni* Erdtman. According to Danzé-Corsin and Laveine, *Pretricolpipollenites ovalis* is distinguished from *E. troedsoni* by its fusiform amb, longer median sulcus, and the absence of exinal thickenings along the sulcus. If the well-documented interpretation of Hughes (1961) is accepted, however, a more basic difference lies in the arrangement of the

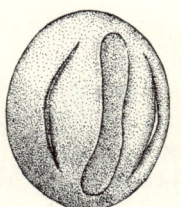

FIG. 13. *Pretricolpipollenites bharadwaji* Balme, n. sp. Reconstruction in semilateral view showing distal face with the large median sulcus and two less prominent lateral sulci, ×600.

sulci, which in *E. troedsoni* is different from that in *P. ovalis*. In the former species the main sulcus lies on the opposite face of the grain to the minor sulci, or "ring furrow" of Hughes. The major and minor sulci in *P. ovalis* are on the same face, or so one assumes from the data at present available, although it is most difficult to be sure of this in compressed specimens preserved in polar view.

Pretricolpipollenites bharadwaji
Balme, n. sp.

Figure 13; Plate 21, figures 1-7

Occurrence: Chhidru Formation: very rare to common.

Representative Specimens: Holotype: U.W.A. 57999. Others: U.W.A. 58000, 58001, 58002, 58003. Field no. K11-6C, Wargal, Salt Range, Chhidru Formation.

Description: Amb oval, trisulcate. Median sulcus clearly defined and running the full length of the grain, extremities usually slightly expanded, and margins sometimes overlapping in the vicinity of the distal pole. Two slit-

---

PLATE 20

Permian and Lower Triassic Disaccites: Non-taeniate forms. All ×600.

FIGURES
1-4. *Cedripites priscus* Balme, n. sp. Upper Permian.——1. Specimen 57989, proximal focus showing finely structured cappa and elongate columellae of the sacci.——2. Specimen 57990, semilateral view.——3. Specimen 57992, equatorial view, proximal side uppermost.——4. Holotype, specimen 57988, equatorial view, proximal side uppermost.
5-6. *Fimbriaesporites?* sp. Upper Permian to Lower Triassic.——5. Specimen 57992.——6. Specimen 57993.

like lateral sulci arranged on either side of the prominent median sulcus and often accompanied by longitudinal folds. Extremities of lateral sulci somewhat convergent, but never joining to form a continuous "ring-furrow." Exine about 1μ thick, faintly intrapunctate.

Dimensions (40 specimens): Length: 25 (28) 31μ. Breadth: 17 (20) 24μ.

Derivation of name: After Dr. D. C. Bharadwaj, Birbal Sahni Institute, Lucknow.

Remarks: Pretricolpipollenites bharadwaji was found only in the upper 12 feet or so of the Chhidru Formation, but is usually common in samples from this stratigraphic interval. It differs from *P. ovalis* in its broadly oval amb, narrow slitlike subsidiary sulci, and the general absence of longitudinal folds bordering the median sulcus.

All the several hundred specimens of *P. bharadwaji* which have been examined were preserved in polar or lateral view, so the interpretation of the positions of sulci cannot be supported by photomicrographs of specimens in equatorial view. Indeed, it is extremely difficult to be sure of the arrangement of the sulci, using only LO-analysis, when dealing with strongly compressed specimens of a form with an exine as thin as *P. bharadwaji*. This can only be convincingly demonstrated with microtome sections of individual specimens.

Subturma POLYPLICATES Erdtman, 1952

Genus GNETACEAEPOLLENITES Thiergart, 1958

Type species (original monotypy): *Gnetaceaepollenites ellipticus* Thiergart.

Pocock (1964) has recently questioned the value of the form genus *Gnetaceaepollenites*, citing the differing opinions of Krutsch (1961) and Potonié (1958) on the morphology of the holotype of its type species, as evidence that it is incapable of interpretation and therefore can typify nothing. These arguments are convincing, and

---

PLATE 21

Permian and Triassic Plicates, Polyplicates, Monocolpates, and Aletes. Fig. 26 $\times 2000$, all others $\times 600$.

FIGURES

1-7. *Pretricolpipollenites bharadwaji* Balme, n. sp. Upper Permian.——1. Specimen 58002.——2. Specimen 58000.——3. Specimen 58001.——4. Holotype, specimen 57999.——5. Unnumbered specimen from sample 54650.——6. Unnumbered specimen from sample 54650.——7. Specimen 58003.

8-11. *Cycadopites follicularis* Wilson and Webster. Lower Permian to Middle Triassic.——8. Specimen 58016, polar view.——9. Specimen 58018, semilateral view.——10. Specimen 58017, semilateral view.——11. Specimen 58015.

12-14. *Ephedripites* sp. Upper Permian to Middle Triassic.——12. Specimen 58006.——13. Specimen 58808.——14. Specimen 58007.

15-18. *Marsupipollenites triradiatus* Balme and Hennelly. Upper Permian.——15. Specimen 57997, distal focus showing sulcus outline.——16. Specimen 57997, proximal focus showing tetrad scar and fine structure of exine.——17. Unnumbered specimen from sample 54651.

19-22. *Paravittatina lucifer* Bharadwaj and Salujha, comb. nov. Lower to Upper Permian.——19. Specimen 58014.——20. Specimen 58011.——21. Specimen 58013.——22. Unnumbered specimen from sample 54650.

23-24. *Gnetaceaepollenites sinuosus* (Balme and Hennelly). Upper Permian.——23. Specimen 58004.——24. Specimen 58005.

25-27. *Inaperturopollenites nebulosus* Balme, n. sp. Upper Permian.——25. Holotype, specimen 58021.——26. Surface of unnumbered specimen from sample 54657, showing surface texture and corrosion pitting.——27. Specimen 58020.

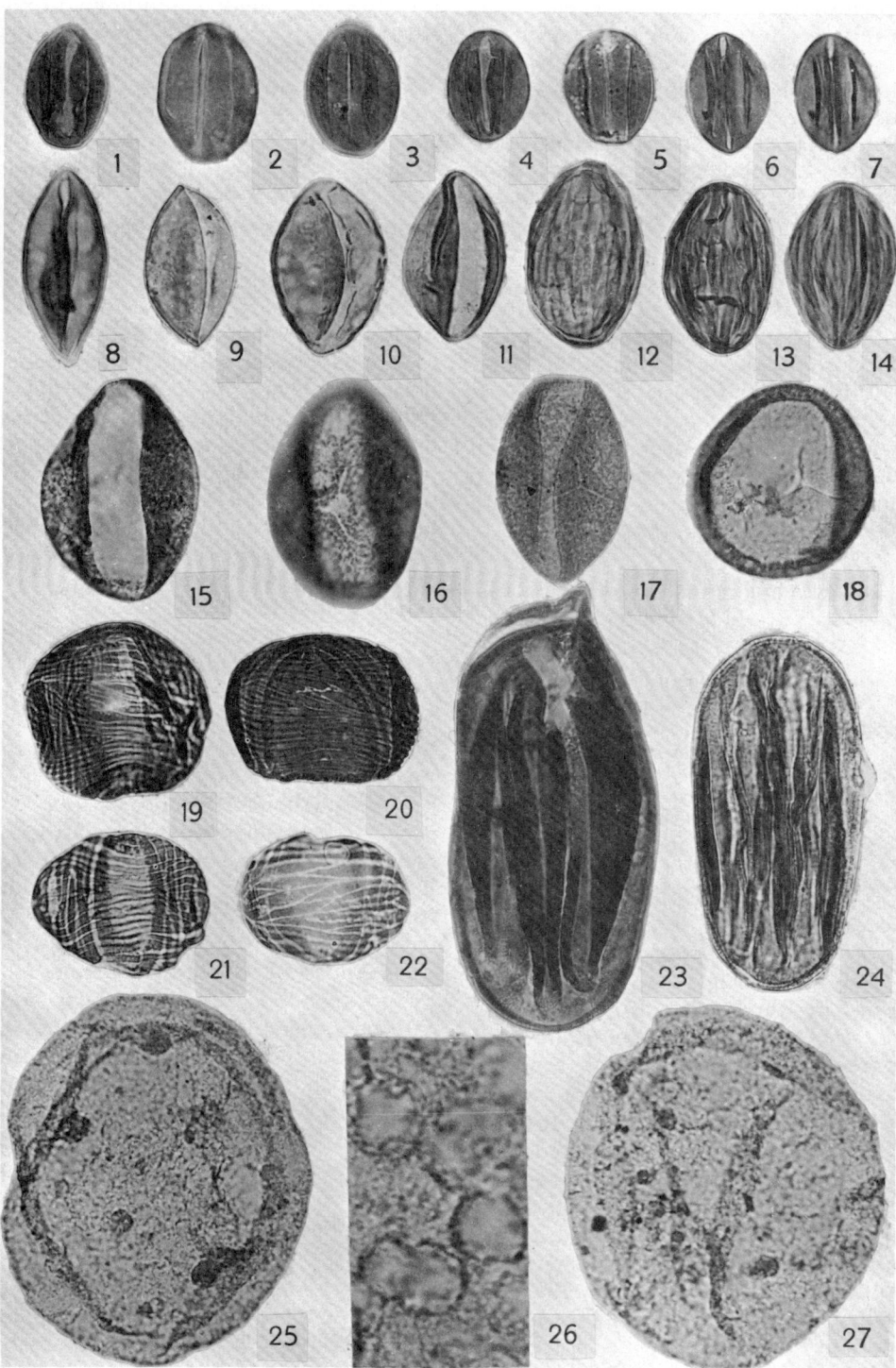

PLATE 21

it is difficult to disagree with Pocock's conclusions; but if *Gnetaceaepollenites* is rejected, further new taxa are necessary to accommodate various species that have been assigned to it. Among such species is *G. sinuosus* (Balme and Hennelly), which has a sufficiently distinctive morphology to warrant separate generic status. This species is extremely rare in Salt Range sediments, and insufficient specimens were available to form a basis for systematic revision, so the treatment of Bharadwaj (1962) is tentatively followed. The distinction made by Bharadwaj and Salujha (1964) between *Welwitschiapites* and *Gnetaceaepollenites* is unclear, particularly as no adequate differential diagnosis of *Welwitschiapites* has been published, and *prima facie* it is a synonym of *Ephedripites*.

Gnetaceaepollenites sinuosus (Balme and Hennelly)

Plate 21, figures 23-24

Marsupipollenites sinuosus Balme and Hennelly, 1956a, p. 61-62, pl. 2, fig. 25-28.
Gnetaceaepollenites sinuosus (Balme and Hennelly), Bharadwaj, 1962, p. 99, pl. 5, fig. 92.
Welwitschiapites tenuis Bharadwaj and Salujha, 1964, p. 213, pl. 12, fig. 164-165.
cf. *Gnetaceaepollenites* sp., Bharadwaj and Salujha, 1964, p. 213, pl. 12, fig. 168.

Occurrence: Chhidru Formation: very rare.

Representative Specimens: U.W.A. 58004, 58005. Field no. T61-178, Kathwai, Salt Range, Chhidru Formation.

Description: Amb elongate oval, extremities sharply to broadly rounded. Exine polyplicate with four to six heavy folds running almost the full length of the grain; associated with each fold is a narrow, linear, longitudinal cleft in the exoexine which is usually visible only after careful focussing. The clefts are arranged symmetrically round the major axes of the ellipsoidal pollen grain, and presumably one or more of them functioned as a germinal exit. In some specimens short, transverse cracks in the exoexine merge into the longitudinal clefts. Exine 2-3μ thick, consisting of a thin, transparent, intexine and a densely and finely intrabaculate exoexine which appears uniformly granulate in surface focus.

Dimensions (4 specimens): Length: 79-100μ. Breadth: 38-48μ.

Remarks: Balme and Hennelly (1956a) misinterpreted the morphology of *Gnetaceaepollenites sinuosus* in regarding it as essentially a monosulcate species with a monolete tetrad scar. It was referred to the Polyplicates by Bharadwaj (1962), who recognized that its morphology was analogous to that of *Ephedripites*, although it differs from that genus in its flaccid, finely structured exine and lack of clearly defined, regular, longitudinal ribs. There is a crude morphological resemblance between *G. sinuosus* and the pollen grains of the living *Welwitschia bainesii* (=*mirabilis*) in that both have longitudinal, and occasionally transverse, clefts in the exoexine (*vide* Bharadwaj, 1963). *Welwitschia* pollen has, however, a single major longitudinal sulcus, which is not present in *G. sinuosus,* and there is certainly no reason to infer that the latter pollen grain was gnetalean in origin.

Throughout Australia, *G. sinuosus* is chiefly found in Upper Permian sediments, although, in Queensland at least, it also occurs in Lower Permian strata (Dr. P. R. Evans, unpublished data). It is known also from the Upper Permian of Antarctica (Balme and Playford, 1967), and Bharadwaj and Salujha (1965a, 1965b) reported that forms resembling *G. sinuosus* made up about 1 percent of the assemblages from two seams in the Raniganj Coal Measures. Recently Hart (1966) has found occasional specimens in sediments from the Ecca Series of South Africa which are probably late Early Permian in age.

Genus EPHEDRIPITES Bolkhovitina ex Potonié, 1958

Type species (designated by Potonié): *Ephedripites mediolobatus* Bokhovitina.

Although the name *Ephedripites* was introduced informally by Bolkhovitina (1953), it was not validly published until 1958 when Potonié selected a type species and provided a generic diagnosis. Potonié is therefore the publishing author, and there is no force in the contention of Pocock (1964) and Singh (1964) that *Ephedripites* is invalid because of its rejection by Bolkhovitina. It is certainly possible, as Pocock and Jansonius (*in* Pocock, 1964) believe, that *Ephedripites* is a junior subjective synonym of *Equisetosporites* Daugherty, although this cannot be demonstrated with conviction. The holotype and only extant specimen of *Equisetosporites chinleana* Daugherty, the type species of the genus, is obviously preserved so imperfectly that it cannot serve satisfactorily as a basis for any taxon. In short, precisely the same objections can be levelled against the holotype of *E. chinleana* as those which led Pocock (1964, p. 144) to reject the form genus *Gnetaceaepollenites* Thiergart. *Ephedripites* also has deficiencies as a generic category, for its type species has no adequate published illustration, and the name is contrary to Recommendation 20C of the International Code of Botanical Nomenclature adopted in 1959. It has nevertheless been widely applied, and there is substantial agreement on its circumscription. There is, therefore, no convincing argument for rejecting *Ephedripites* in favor of *Equisetosporites,* a generic name against which similar but even more substantial criticisms may be levelled.

Ephedripites sp.

Plate 21, figures 12-14

Occurrence: Chhidru Formation: very rare to rare. Narmia Member, Mianwali Formation: very rare to rare. Tredian Formation: very rare to common.

Representative Specimens: U.W.A. 58006, 58007, 58008, 58009. Field no. T61-209, Wargal, Salt Range, Chhidru Formation.

Description: Polyplicate, amb longitudinally oval, extremities sharply rounded. Exine 1-2μ thick, consisting of a thin, translucent intexine overlain by a thicker, ribbed exoexine. Exoexine apparently structureless or faintly granulate, ribs 1-3μ wide and less than 1μ apart, merging at the extremities of the grain where the exoexine is continuous.

Dimensions (10 specimens): Length: 31 (37) 43μ. Breadth: 18 (23) 26μ.

Remarks: Although *Ephedripites* sp. occurs over a fairly long stratigraphic interval in the Salt Range, it was not a prominent component of any assemblage. A rather wide spectrum of morphological variants has been referred here to *Ephedripites* sp., as there were insufficient specimens to judge whether the recognition of more than one form species is justified. The commonest form is closest in gross morphology to *Equisetosporites multistriatus* Pocock from the Lower Cretaceous of Canada.

Ephedripites is widespread and often common in late Paleozoic and Mesozoic sediments from the countries of the Northern Hemisphere. It has been less frequently reported from peninsular India and the southern continents, and then only from Late Cretaceous or Kainozoic deposits. Its occurrence in Permian strata in the Salt Range hints, therefore, at the presence of northern elements which were not represented in the late Paleozoic floras of "Gondwanaland."

Genus PARAVITTATINA Balme, n. gen.[1]

Type species (here designated): *Paravittatina lucifer* Bharadwaj and Salujha, comb. nov., Bharadwaj and Salujha, 1964, p. 213, fig. 169, 171.

---

[1] In a paper published while the present manuscript was in proof, Bharadwaj and Srivastava (1969, *Palaeontographica,* Vol. 125B, p. 119-149) erected the form genus *Weylandites,* with *W. indicus* Bharadwaj and Srivastava as the type species. Al-

Pollen grains with clearly defined, more or less parallel, exoexinal ribs, but lacking properly developed sacci, are important components of Permian assemblages, and their likely botanical affinities have been the subject of a good deal of speculation. Current knowledge of the morphology and distribution of the group has been discussed by Hart (1964), Abramova and Marchenko (1964), Chaloner and Clarke (1962), Pocock (1964), and Tschudy and Kosanke (1966). From a synthesis of these contributions, it may be concluded that these ribbed pollen grains are divisible into at least three morphological categories, the characters of which are summarized below:

A: *Ephedripites*-type.

Elongate oval forms of uncertain polarity, in which the ribs are radiosymmetrically arranged and which do not possess a clearly differentiated germinal area. These are analogous to the pollen grains of *Ephedra* and have been referred in the present account to the form genus *Ephedripites*.

B. *Vittatina subsaccata*-type.

Vittatina subsaccata Samoilovich was selected by Wilson (1962) as the type species of *Vittatina,* a name customarily attributed to Luber, although it was first used in a formal publication by Samoilovich (1953). Pollen grains typified by *V. subsaccata* have prominent transverse ribs on the proximal side. These ribs terminate virtually at the proximo-distal margin. Tschudy and Kosanke (1966) discussed in detail the morphology of American Permian forms assigned to *Vittatina* and proposed an emendation of Wilson's generic diagnosis. They did not, however, discuss the status of *Aumancipollenites* Alpern, a Paleozoic form genus which also encompasses ribbed pollen grains and may be indistinguishable from *Vittatina*. This cannot be determined from available illustrations and descriptions. *Aumancisporites* was described (Alpern *et al.,* 1958) some four years before the validation of *Vittatina* and has therefore potential priority.

C. *Vittatina striata*-type.

Ribbed forms with a notably different morphology from *Vittatina subsaccata* have also been frequently assigned to *Vittatina*. In these the ribs form continuous loops from the proximal to the distal face so that they appear to run transversely on the proximal face and longitudinally on the distal, giving the impression of two sets of ribs crossing one another at right angles. The most comprehensively described and illustrated species with these characters is *Vittatina africana* Hart, from the Ecca Series of South Africa (Hart, 1966), but similar forms are obviously abundant in the Upper Permian of the USSR.

Pollen grains of the *Vittatina striata*-type form a distinctive morphological group which appears to have had phytogeographic significance during late Early and Late Permian times. They are prominent in Russian Permian sediments, especially in the Kungurian and Kazanian Stages, and are, in some assemblages, the dominant microfloral component. In Africa, also, they appear to be important constituents of late Early Permian assemblages. (Hart, 1966) and Goubin (1965) recorded them in large numbers from the Upper Permian of Madagascar. In Australia, however, they are virtually unknown and have been found only in Upper Permian sedi-

---

though Bharadwaj and Srivastava did not refer *Decussatisporites lucifer* Bharadwaj and Salujha to *Weylandites* there is little doubt that it belongs in that genus. If so, the name *Paravittatina* becomes a junior objective synonym of *Weylandites.* According to Bharadwaj and Srivastava, the strata from which *Weylandites indicus* was recovered are probably late Early Triassic in age. However, the spore assemblage considered in its entirety strongly suggests that they are Late Permian.

ments, in the southernmost part of the Perth Basin, Western Australia. Similarly they have been reported from India only once. As a very minor component of assemblages from the Raniganj coals (Bharadwaj and Salujha, 1964, pl. 12, fig. 168, 171).

On morphological, biostratigraphical, and paleogeographic grounds, I believe that there is an overwhelming case for placing pollen grains of the *Vittatina striata*-type in a separate generic category. Accordingly the form genus *Paravittatina* is proposed for dispersed pollen grains with the following characters:

Diagnosis: Monosulcate, amb transversely oval, subcircular to rounded quadrilateral. Exine with clearly defined, apparently structureless, exoexinous ribs. Individual ribs continuous and looped so that those nearer the periphery pass without interruption from the proximal to the distal face. In compressed specimens, viewed parallel to the polar axis, the ribs run transversely on the proximal face and longitudinally on the distal so that they appear to cross one another at right angles. Sulcus oval to subcircular and free of exoexine.

Differential diagnosis: Paravittatina differs from *Vittatina* in the form of its continuous looped ribs. In *Vittatina* the ribs extend only a short distance on to the distal face, and their distal extensions are transverse and parallel to the proximal ribbing. Wilson (1962), Jansonius (1962), and Tschudy and Kosanke (1966) all mention the presence of a continuous equatorial rib as a character of *Vittatina*, and this feature is certainly clear in some American species of the genus, though it is uncertain whether the type species possesses it.

Decussatisporites Leschik has a somewhat similar pattern of fine ribs, and has been used by Bharadwaj and Salujha (1964) for pollen grains of the *Paravittatina*-type. Its type species, *Decussatisporites delineatus* Leschik has, however, a longitudinally oval amb and a pronounced and strongly differentiated sulcus, and these features distinguish it from any species referred to *Paravittatina*.

Species: Apart from the type the following published species are assigned to *Paravittatina:*

Paravittatina striata Luber, comb. nov., Luber in Luber and Waltz, 1941, p. 62, pl. 13, fig. 218.
Paravittatina persecta Zauer ex Hart, comb. nov., Hart, 1964, p. 1196, fig. 64.
Paravittatina cincinnata Luber ex Hart, comb. nov., Hart, 1964, p. 1196, fig. 71.

Paravittatina lucifer Bharadwaj and Salujha, comb. nov.

Figure 14; Plate 21, figures 19-22

Decussatisporites lucifer, Bharadwaj and Salujha, 1964, *partim*, p. 213, pl. 12, fig. 169, 171.
Vittatina cincinnata Luber, Goubin, 1965, p. 1435, pl. 8, fig. 1-2.
Vittatina africana, Hart, 1966, p. 38-41, fig. 2-10.

Occurrence: Amb Formation: rare to common. Wargal Limestone: rare to common. Chhidru Formation: rare to abundant.

Representative Specimens: U.W.A. 58010, 58011, 58012, 58013, 58014. Field no. K11-6C, Wargal, Salt Range, Chhidru Formation.

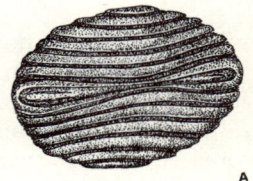

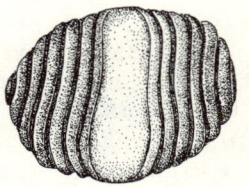

FIG. 14. *Paravittatina lucifer* (Bharadwaj and Salujha), ×750. A. Proximal face showing transverse, looped ribs. B. Distal face showing sulcus with the continuation of looped ribs running parallel to its length.

Description: Monosulcate, amb subcircular, transversely oval to rounded quadrilateral. Exine 1-2μ thick, consisting of thin, transparent intexine, overlain by ribbed, apparently structureless exoexine. Individual ribs 1-3μ wide and 1μ or less apart, forming continuous loops so that the outer ribs pass without interruption on to the distal face. In polar compressions the ribs run transversely on the proximal face and longitudinally on the distal. The number of apparent proximal ribs varied from 15-28 on the 40 specimens studied in detail. Distal sulcus oval, consisting of transparent intexine and bordered by the innermost longitudinal ribs.

Dimensions (40 specimens): Total breadth: 30 (36) 41μ. Height: 25 (31) 36μ.

Remarks: The validity of *Paravittatina lucifer* is accepted tentatively, pending clarification of several Russian species which have not yet been clearly circumscribed or illustrated by photomicrographs. Of these it appears closest to *Paravittatina cincinnata* (Hart), although none of the measured specimens of *P. lucifer* reached the lower size limit of *P. cincinnata* given by Hart (1965b, p. 54-55). There is, nevertheless, some doubt concerning the practicability of Hart's (1966) criteria for distinguishing *P. cincinnata* from *Vittatina africana* Hart, which, in turn, encompasses the holotype of *P. lucifer*.

Paravittatina striata (Luber) is also larger than *P. lucifer* and, according to Hart (1965b, p. 51-52), has a distal "keel" similar to that occurring in *Vittatina costabilis* Wilson. This is not shown in Luber's original illustration (in Luber and Waltz, 1941, pl. 13, fig. 218), nor has it been regarded as an essential character of *P. striata* by subsequent Russian authors (e.g., Samoilovich, 1953; Chalyshev and Varyukhina, 1960; Zauer, 1965).

Subturma MONOCOLPATES Iversen and Troels-Smith, 1950

Genus CYCADOPITES Wodehouse ex Wilson and Webster, 1946

Type species (original monotypy): *Cycadopites follicularis* Wilson and Webster.

Simple monosulcate pollen grains without exinal structure or sculpture have been referred to various form genera, although most recent authors have favored either *Cycadopites* Wilson and Webster or *Ginkgocycadophytus* Samoilovich. Of these two names *Cycadopites* has clear priority, and *Ginkgocycadophytus* is invalid unless it can be demonstrated that its type species, *G. caperatus* (Luber), is excluded from *Cycadopites*. From the original description of *G. caperatus* it is probable that at least some specimens assigned to the species were sculptured, and Samoilovich (1953) certainly used it for spinose forms. It may be appropriate, therefore, to retain *Ginkgocycadophytus* for certain sculptured, simple, monosulcate pollen grains.

Jansonius has reviewed the synonymy of *Cycadopites* and defined it in simple terms. His treatment is acceptable except that the genus should be explicitly restricted to forms devoid of structure and sculpture.

Cycadopites follicularis Wilson and Webster

Plate 21, figures 8-11

Cycadopites follicularis Wilson and Webster, 1946, p. 274-275, fig. 7.
Ginkgo typica (Malyavkina), Bolkhovitina, 1953, p. 62-63, pl. 10, fig. 3-4.
Monocolpopollenites acerrimus Leschik, 1955, p. 41-42, pl. 5, fig. 15.
Entylissa nitidus Balme, 1957, p. 30, fig. 78-80.
Monosulcites minimus auct. non Cookson, Couper, 1958, p. 157, pl. 26, fig. 23-25.
Ginkgocycadophytus nitidus (Balme), de Jersey, 1962, p. 12, pl. 5, fig. 1-3.
Cycadopollenites follicularis (Wils. and Web.), Danzé-Corsin and Laveine, in Briche,

Danzé-Corsin, and Laveine, 1963, p. 108, pl. 11, fig. 13.

Occurrence: Amb Formation: very rare to rare. Wargal Limestone: very rare to rare. Chhidru Formation: very rare to common. Mianwali Formation: rare to dominant. Tredian Formation: very rare to rare.

Representative Specimens: U.W.A. 58015, 58016, 58017, 58018. Field no. T61-226, Wargal, Salt Range, Kathwai Member, Mianwali Formation.

Description: Monosulcate, amb oval, uncompressed specimens fusiform. Sulcus extending full length of grain often, but not always, slightly narrower near the distal pole, margins of sulcus overlapping in some specimens, Exine about 1μ thick, unsculptured, unstructured or faintly intragranulate.

Dimensions (30 specimens): Length: 35 (45) 54μ. Breadth: 16 (24) 34μ.

Remarks: Pollen grains included in *Cycadopites follicularis* are undoubtedly of polyphyletic origin and occur in sediments of Late Carboniferous to Recent age. They are known to have been produced by members of the Peltaspermaceae (Townrow, 1960), Ginkgoales, Cycadales, and Bennettitales. Couper (1958), following his study of fossil monosulcate pollen grains of known origin, concluded that it was impossible reliably to distinguish between any of the groups mentioned, on the basis of their pollen.

Cycadopites follicularis occurs in small numbers in assemblages from most samples taken below the base of the Tredian Formation, usually as a minor component. In one sample from the Kathwai Member at Wargal (T61-226), however, it makes up 25 percent of the total assemblage.

Turma ALETES Ibrahim, 1933

Genus INAPERTUROPOLLENITES Thomson and Pflug emend. Potonié, 1958

Type species (original designation): *Inaperturopollenites dubius* (Potonié and Venitz).

Inaperturopollenites nebulosus Balme, n. sp.

Plate 21, figures 25-27

Occurrence: Chhidru Formation: very rare to common.

Representative Specimens: Holotype: U.W.A. 58021. Others: U.W.A. 58020, 58022, 58023. Field no. K11-4C, Wargal, Salt Range, Chhidru Formation.

Description: Outline circular, usually distorted by folding. Exine less than 1μ thick, scabrate with an extremely fine sculptural pattern, the elements of which are almost at the limits of optical resolution. Exine almost always folded, often with a pronounced concentric fold close to the outer margin of the compressed spore.

Dimensions (20 specimens): Diameter: 71 (77) 83μ.

Derivation of name: Latin *nebula* = a mist, referring to the fine, indeterminate, sculptural pattern.

Remarks: Inaperturopollenites is used as a noncommittal form genus for spheroidal bodies without tetrad markings or obvious germinal features, in which the exine is thin, finely structured, and easily folded. In this sense the generic name does not necessarily imply that the species is the pollen grain of a spermatophyte. *I. nebulosus* seems equally likely to have originated from a nonvascular plant.

Grebespora concentrica Jansonius, from the Lower Triassic of Western Canada (Jansonius, 1962), bears some resemblance to *I. nebulosus* but is smaller and has an unstructured exine which is characterized by a single, prominent, concentric fold.

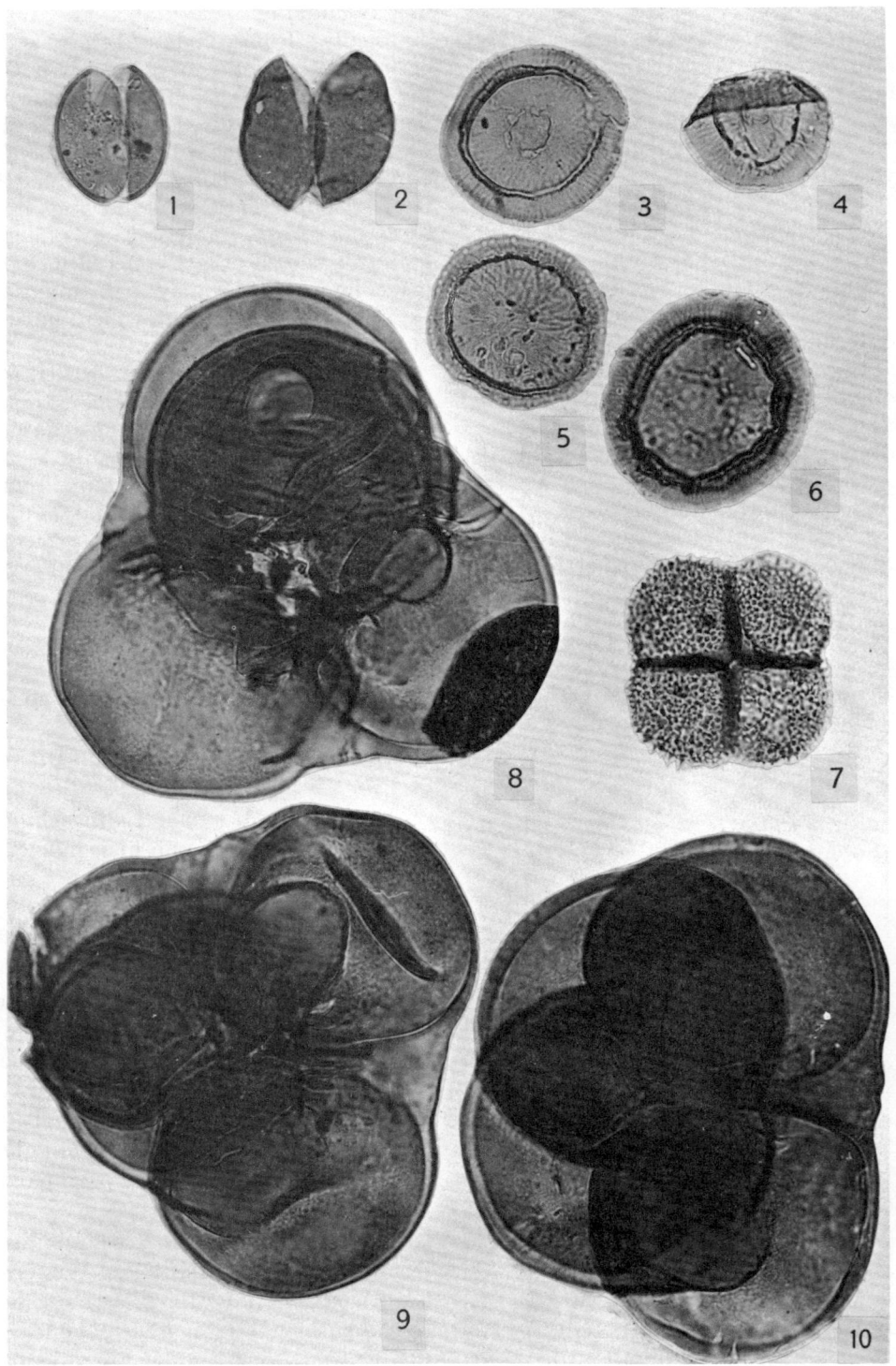

PLATE 22

INCERTAE SEDIS

Genus QUADRISPORITES Hennelly, 1958

Type species (original designation): *Quadrisporites horridus* Hennelly.

Quadrisporites horridus Hennelly

Plate 22, figure 7

Spore 23, Virkki, 1946, *partim,* p. 129, fig. 39, pl. 6, fig. 71.
Quadrisporites horridus Hennelly, 1958, p. 365-366, pl. 5, fig. 6-7.
Quadrisporites horridus (Henn.) emend., Potonié and Lele, 1961, p. 26, fig. 26-36.

Occurrence: Amb Formation: very rare. Chhidru Formation: very rare.

Representative Specimen: U.W.A. 57729. Locality K11-4C, Wargal, Salt Range, Chhidru Formation.

Remarks: Obligate, tight, tetragonal tetrads of densely spinulose "spores" occurred sporadically and rarely in the Permian assemblages. They conform to the expanded diagnosis of *Quadrisporites horridus* given by Potonié and Lele (1961) and are of uncertain origin.

Genus PYRAMIDOSPORITES Segroves, 1967

Type species (original designation): *Pyramidosporites cyathodes* Segroves.

The form genus *Pyramidosporites* was instituted to include spheroidal, spore-like bodies without sculpture and occurring in obligate, tetrahedral tetrads, in which each member of the tetrad is bound to the other three by thickened exinal strands. To date the genus is monotypic and based on *P. cyathodes,* which occurs in Upper Permian sediments of the Perth Basin, Western Australia.

Pyramidosporites racemosus Balme, n. sp.

Figure 15; Plate 22, figures 8-10

Spore 6, Virkki, 1946, p. 127-128, pl. 6, fig. 66.

Occurrence: Amb Formation, Lower Permian (Artinskian): rare.

Representative Specimens: Holotype 57730. Others: 57731, 57732. Field no. T62-350. Warchha Water Tank, Salt Range, Amb Formation.

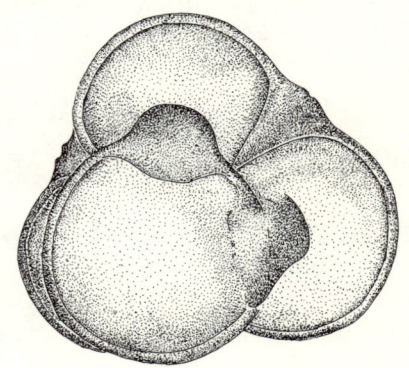

FIG. 15. *Pyramidosporites racemosus* Balme, n. sp. Reconstruction, ×400.

Description: Four spheroidal, spore-like bodies occurring as obligate, tetrahedral tetrads. No tetrad scar or germinal mechanism apparent. Each

---

PLATE 22

Permian and Triassic microfossils incertae sedis. All ×600.

FIGURES

1-2. *Schizosporis scissus* (Balme and Hennelly). Lower Permian to Lower Triassic.——1. Specimen 57731.——2. Specimen 57734.
3-6. *Peltacystia venosa* Balme and Segroves. Lower to Upper Permian.——3. Specimen 57735.——4. Specimen 57737.——5. Specimen 57736.——6. Unnumbered specimen from sample 54650.
7. *Quadrisporites horridus* Hennelly. Lower to Upper Permian. Specimen 57729.
8-10. *Pyramidisporites racemosus* Balme, n. sp. Lower Permian.——8. Specimen 57732.——9. Holotype, specimen 57730.——10. Specimen 57731.

member of tetrad bound tightly to the other three by heavy exinal strands about 20μ in diameter. Exinal strands usually homogeneous, but in some specimens (e.g., pl. 22, fig. 10) there is a trace of an incipient median split which may be a prelude to dispersal of the tetrad. Wall thickness variable, but usually 3-4μ in the distal polar areas, individual bodies faintly intragranulate but devoid of sculpture.

Dimensions (30 specimens): Total diameter: 98 (109) 124μ. Equatorial diameter of individual bodies: 60 (65) 70μ.

Derivation of name: Latin *racemosus* = bunched.

Remarks: Geometrically the tetrad form of *Pyramidosporites racemosus* is similar to that illustrated by Wodehouse (1959, p. 164) for aberrant pollen grains of chicory, although morphologically there is no analogy, for in the chicory tetrad the connecting strands are hollow and in *P. racemosus* they are solid. No basis for speculation exists as to the origin or affinities of *P. racemosus*, and the form may have derived from nonvascular or vascular plants.

There is very little doubt that Virkki's Spore 6 is identical with *P. racemosus* and that "the four large and four small lobes" of Virkki's original description were the individual tetrad members and the connecting strands respectively. Spore 6 also came from Warchha, although from a horizon just below the base of the Wargal Limestone and hence probably stratigraphically lower than T62-350.

P. racemosus is thicker walled than *P. cyathodes* and lacks the invaginated distal surface which Segroves (1967) regards as characteristic of the Australian species.

Group ACRITACHA Evitt, 1963

Sub-group SCHIZOMORPHITAE Segroves, 1967

Genus SCHIZOSPORIS Cookson and Dettmann, 1959

Type species (original designation): *Schizosporis reticulatus* Cookson and Dettmann.

Schizosporis scissus (Balme and Hennelly)

Plate 22, figures 1-2

Laevigatosporites scissus Balme and Hennelly, 1956a, p. 56, pl. 1, fig. 6-9.
Spheripollenites scissus (Balme and Hennelly), 1962, Jansonius, p. 82, pl. 16, fig. 8.
Schizosporis scissus (Balme and Hennelly), Hart, 1965, p. 14.

Occurrence: Amb Formation: very rare to rare. Chhidru Formation: very rare to rare. Mianwali Formation: very rare to common.

Representative Specimens: U.W.A. 57733, 57734. Field no. K12-6, Landa, Surghar Range, Narmia Member, Mianwali Formation.

Remarks: Schizosporis scissus is a thin-walled form which ruptures by splitting cleanly along the trace of a plane parallel to the major axis of the ellipsoid, a mode of dehiscence which recalls that of certain living members of the Chlorococcales. The species is no doubt highly artificial and has been widely reported from late Paleozoic sediments. In Australia it occurs throughout the Permian, and other records include the Lower Permian of Tanzania (Hart, 1965a), Permian of Western Canada (Jansonius, 1962), and Late Permian of Antarctica (Balme and Playford, 1967).

Genus PELTACYSTIA Balme and Segroves, 1966

Type species (original designation): *Peltacystia venosa* Balme and Segroves.

Three species from the Permian of Western Australia were included in the genus *Peltacystia* by Balme and Segroves. These are characterized by a tendency to split into two symmetrical hemispheres along a sharply defined line of equatorial weakness. Each hemisphere is divided into a polar and equatorial zone by a circumpolar ridge, or ring of processes, which encircles the body about half way between the equator and the pole.

The genus is known mainly from Upper Permian deposits in Western Australia, but also occurs in the Artinskian.

Peltacystia venosa Balme and Segroves
Plate 22, figures 3-6

Peltacystia venosa Balme and Segroves, 1966, p. 27-28, fig. 1a, b, 2a-f, 3f-k.

Occurrence: Amb Formation: very rare. Wargal Limestone: very rare. Chhidru Formation: very rare.

Representative Specimens: U.W.A. 57735, 57736, 57737. Field no. K11-6C, Wargal, Salt Range, Chhidru Formation.

Remarks: Peltacystia venosa was represented only by detached ruptured halves, which occurred sporadically in the Permian assemblages. Most specimens were found in the Upper Permian, but two or three examples were seen in spore assemblages from the Amb Formation.

ASSEMBLAGE LISTS

In the lists that follow, the microfossils from each rock unit are divided into two categories, "essential" and "accessory." Essential species were recorded in all assemblages from the stratigraphic unit under discussion; accessory forms were not identified in every assemblage, although when they did occur they were not necessarily uncommon.

AMB FORMATION

Plant microfossils were recovered from three samples from the Amb Formation, but in only one of these (Locality T62-350, Warchha) were they diverse and well preserved. Inferior preservation of material in the other two assemblages is the most likely explanation for the small number of essential species recorded in samples from the Amb Formation, and few acceptable generalizations on the composition of its microfloras can be based on the available palynological data.

Palynomorphs of Amb Formation
Anteturma: SPORITES
Turma: TRILETES

Accessory
Camptotriletes warchianus
Verrucosisporites sp. cf. *V. planiverrucatus*
Leiotriletes cf. *adnatus*
Acanthotriletes tereteangulatus
Lophotriletes novicus

Turma: MONOLETES

Accessory
Reticuloidosporites warchianus

Anteturma: POLLENITES
Turma: SACCITES

Subturma: MONOSACCATES

Accessory
Plicatipollenites indicus
Potonieisporites novicus

Subturma: DISACCITES

(a) Striatiti

Essential	*Accessory*
Protohaploxypinus limpidus	*Striatopodocarpites cancellatus*
P. goraiensis	*S. rarus*
	S. pantii
	Protohaploxypinus diagonalis
	P. varius
	Lueckisporites singhii
	Hamiapollenites insolitus
	Guttulapollenites hannonicus
	Corisaccites alutas

(b) Non-taeniate forms

Essential	*Accessory*
Pinuspollenites thoracatus	*Vitreisporites pallidus*
Falcisporites nuthallensis	*Alisporites tenuicorpus*
	Sulcatisporites ovatus
	S. nilssoni

Turma: PLICATES

Essential	*Accessory*
Paravittatina lucifer	*Cycadopites follicularis*

Incertate sedis

Accessory
Quadrisporites horridus
Pyramidosporites racemosus
Schizosporis scissus
Peltacystia venosa

WARGAL LIMESTONE

Two samples from the Wargal Limestone, both from the lower part of the section in Nammal Gorge, yielded spores, pollen grains, and acritarchs. Neither assemblage was notably satisfactory for biostratigraphic purposes as most individual specimens were corroded, and one cannot be confident of any quantitative data based on random counts of the assemblages. On the other hand, there were, in each assemblage, sufficient specimens in an adequate state of perservation to allow their reliable identification.

Palynomorphs of Wargal Limestone
Anteturma: SPORITES
Turma: TRILETES

Essential	*Accessory*
Leiotriletes cf. *adnatus*	*Lophotriletes novicus*
	Acanthotriletes tereteangulatus

Turma: MONOLETES

Essential
Punctatosporites sp. cf. *P. minutus*

Anteturma: POLLENITES
Turma: SACCITES
Subturma: MONOSACCITES

Accessory
Plicatipollenites indicus
Potonieisporites novicus

Subturma: DISACCITES
(a) Striatiti

Essential
Protohaploxypinus limpidus
Guttulapollenites hannonicus
Lueckisporites singhii
L. virkkiae

Accessory
Protohaploxypinus varius
P. diagonalis

(b) Non-taeniate

Essential
Sulcatisporites ovatus
Alisporites tenuicorpus

Accessory
Fimbriaesporites ? sp.
Pinuspollenites thoracatus
Falcisporites nuthallensis

Turma: PLICATES

Essential
Paravittatina lucifer

Accessory
Cycadopites follicularis

Incertae sedis

Accessory
Schizosporis scissus
Peltacystia venosa

CHHIDRU FORMATION

Plant microfossils were recovered from 23 samples taken from the Chhidru Formation at six localities. Sixteen of these samples represented a comprehensive, closely sampled collection from the upper part of the section exposed at Wargal in the central Salt Range. Only one or two samples were available from each of the other five localities.

In general, assemblages from the Chhidru Formation were more diverse and better preserved than those recovered from any other unit, either Permian or Triassic. This applied particularly to those from the upper part of the Formation. Acritarchs occurred in all assemblages, although they were less common than spores and pollen grains, except in the two samples from Nammal Gorge, which came from stratigraphically lower horizons than any of the material available from Wargal.

Palynomorphs of Chhidru Formation
Anteturma: SPORITES
Turma: TRILETES

Accessory
Leiotriletes cf. *adnatus*
Calamospora landiana
Acanthotriletes tereteangulatus

Lophotriletes novicus
Tigrisporites playfordi
Camptotriletes warchianus
Osmundacidites senectus
Nevesisporites fossulatus
Indospora clara
Triquitrites proratus
Iraqispora labrata
Kraeuselisporites wargalensis
K. cuspidus
K. rallus
Lundbladispora obsoleta
Densoisporites playfordi
D. complicatus
Guthoerlisporites cancellosus

Turma: MONOLETES

Essential
Polypodiisporites mutabilis

Accessory
Lunulasporites vulgaris
Laevigatosporites callosus
Reticuloidosporites warchianus
Polypodiidites sp.
Punctatosporites sp. cf. *P. minutus*

Anteturma: POLLENITES
Turma: SACCITES
Subturma: MONOSACCITES

Accessory
Plicatipollenites indicus
Densipollenites indicus

Subturma: DISACCITES
(a) Striatiti

Essential
Lueckisporites singhii

Protohaploxypinus limpidus
P. varius

Accessory
Striatopodocarpites rarus
S. pantii
S. cancellatus
Guttulapollenites hannonicus
Lueckisporites virkkiae
Protohaploxypinus diagonalis
P. goraiensis
P. microcorpus
Taeniaesporites noviaulensis

(b) Non-taeniate forms

Essential
Vitreisporites pallidus
Alisporites tenuicorpus

Accessory
Sulcatisporites ovatus
Klausipollenites schaubergeri
Cedripites priscus
Pinuspollenites thoracatus
Falcisporites nuthallensis
F. stabilis
Fimbriaesporites ? sp.

Turma: PLICATES

Essential
Paravittatina lucifer

Accessory
Gnetaceaepollenites sinuosus
Pretricolpipollenites bharadwaji
Marsupipollenites triradiatus
Cycadopites follicularis
Ephedripites sp.

Turma: ALETES

Accessory
Inaperturopollenites nebulosus

Incertae sedis

Accessory
Peltacystia venosa
Schizosporis scissus
Quadrisporites horridus

MIANWALI FORMATION

Kathwai Member

Acid-insoluble microfossils were obtained from three samples of the Kathwai Member. These assemblages were all characterized by poor preservation, lack of diversity, and enormous numbers of simple spinose acritarchs.

Palynomorphs of Kathwai Member
Anteturma: SPORITES
Turma: TRILETES

Essential
Punctatisporites fungosus
Densoisporites playfordi

Accessory
Calamospora landiana
Osmundacidites senectus
Lundbladispora obsoleta

Anteturma: POLLENITES
Turma: DISACCITES
(a) Striatiti

Essential
Taeniaesporites noviaulensis
T. pellucidus

Accessory
Taeniaesporites sp. cf. *T. transversundatus*

(b) Non-taeniate forms

Accessory
? *Klausipollenites schaubergeri*
Fimbriaesporites ? sp.

Turma: PLICATES

Essential
Cycadopites follicularis

Mittiwali Member

All five productive samples from the Mittiwali Member came from the lower half of the section exposed in Nammal Gorge. They were rich in spores, pollen grains, and especially acritarchs, but, as in the assemblages from the Kathwai Member, well-preserved individual specimens were exceptional.

Palynomorphs of Mittiwali Member
Anteturma: SPORITES
Turma: TRILETES

Essential	*Accessory*
Densoisporites playfordi	Punctatisporites fungosus
Lundbladispora obsoleta	Densoisporites nejburgii
	Calamospora landiana
	Osmundacidites senectus
	Perotrilites sp.
	Lundbladispora brevicula
	Guthoerlisporites cancellosus
	Kraeuselisporites cuspidus

Anteturma: POLLENITES
Turma: SACCITES
Subturma: MONOSACCITES

Accessory
Cordaitina gunyalensis

Subturma: DISACCITES
(a) Striatiti

Essential	*Accessory*
Taeniaesporites noviaulensis	Guttulapollenites hannonicus
T. pellucidus	Fimbriaesporites ? sp.
T. sp. cf. T. transversundatus	

(b) Non-taeniate forms

Accessory
Vitreisporites pallidus

Turma: PLICATES

Essential
Cycadopites follicularis

Incertae sedis

Accessory
Schizosporis scissus

NARMIA MEMBER

Eleven assemblages from the Narmia Member have been incorporated in this study. Seven of these came from Nammal Gorge and the remainder from Landa, in the Surghar Range. Although the sampling was not stratigraphically closely spaced at either locality, it was spread fairly well over the whole thickness of the unit. In a sense, therefore, the broad palynological trends in the Narmia Member are better documented than those in any other stratigraphic unit in the Salt Range area.

Material from the Narmia Member was usually much better preserved than that from the two lower Members of the Mianwali Formation. Acritarchs were also less abundant, although still making something over 20 percent of the total plant microfossil suite.

Palynomorphs of Narmia Member
Anteturma: SPORITES
Turma: TRILETES

Essential
Densoisporites nejburgii

Accessory
Punctatisporites fungosus
Calamospora landiana
Cyclogranisporites arenosus
Verrucosisporites narmianus
Osmundacidites senectus
Simeonospora khlonovae
Densoisporites playfordi
Lundbladispora brevicula
L. obsoleta

Turma: MONOLETES

Essential
Aratrisporites paenulatus

Accessory
Aratrisporites fischeri

Anteturma: POLLENITES
Turma: SACCITES
Subturma: MONOSACCITES

Accessory
Cordaitina gunyalensis

Subturma: DISACCITES
(a) Striatiti

Accessory
Taeniaesporites noviaulensis
T. pellucidus
T. sp. cf. *T. transversundatus*

(b) Non-taeniate forms

Essential
Falcisporites stabilis
Alisporites sp. cf. *A. opii*
A. landianus

Accessory
Sulcatisporites institatus
S. sp. cf. *S. kraeuseli*
Platysaccus queenslandi

Turma: PLICATES

Accessory
Cycadopites follicularis
Ephedripites sp.

Incertae sedis

Accessory
Schizosporis scissus

TREDIAN FORMATION

Nine samples from the Tredian Formation yielded plant microfossils. Seven of these came from the Landa Member at three localities: Landa, Zaluch Nala, and Nammal; and the other two, from the Khatkiara Member at Zaluch Nala. Essentially the assemblages from the two Members were similar, the only obviously significant difference being that those from the Landa Member contained acritarchs whereas those from the Khatkiara Member yielded only spores, pollen grains, and wood fragments. Spores and pollen were abundant and well preserved in assemblages from the Tredian Formation, although again the assemblages lacked diversity.

Palynomorphs of Tredian Formation
Anteturma: SPORITES
Turma: TRILETES

Essential	*Accessory*
Calamospora landiana	Verrucosisporites narmianus
Cyclogranisporites arenosus	Tigrisporites playfordi
	Densoisporites nejburgii
	Lundbladispora brevicula
	Guthoerlisporites cancellosus
	? Nevesisporites fossulatus

Turma: MONOLETES

Essential	*Accessory*
Aratrisporites paenulatus	Aratrisporites fischeri

Anteturma: POLLENITES
Turma: DISACCITES
(a) Striatiti

Accessory
Taeniaesporites noviaulensis
T. pellucidus
T. sp. cf. T. transversundatus

(b) Non-taeniate forms

Essential	*Accessory*
Falcisporites stabilis	Vitreisporites pallidus
Platysaccus queenslandi	Alisporites sp. cf. A. opii
	A. landianus
	Sulcatisporites institatus
	S. sp. cf. S. kraeuseli

Turma: PLICATES

Accessory
Cycadopites follicularis
Ephedripites sp.

SUCCESSION OF SPORE ASSEMBLAGES

ASSEMBLAGES OF THE PERMIAN

It is difficult to judge the validity of inferences based on palynological data from the Permian succession. The scope of sampling is certainly not comprehensive, for the entire collection comprises three samples from the Amb Formation, two from the Wargal Limestone, and 23 from the Chhidru Formation. Even the Chhidru Formation, which appears to be adequately represented, is known mainly from material collected from the uppermost part of the Wargal section. Caution and conservatism are obviously necessary in interpreting the palynological results. Certain generalizations concerning the composition and implications of the spore assemblages are nonetheless justified, and these will be treated under three sub-headings. Under the first, unifying features common to all the Permian assemblages will be discussed. The second will deal with quantitative and qualitative variations which appear to have local biostratigraphic or paleoecological significance. Finally, an attempt will be made to assess the implications of palynological data from the Chhidru Formation, relative to the phytogeography of late Permian times.

FEATURES COMMON TO THE PERMIAN ASSEMBLAGES

Figures 16 and 17 indicate that from a total of 61 Permian species only 25 occur throughout the Zaluch Group. Even so, the assemblages have an essential unity. This is manifested most

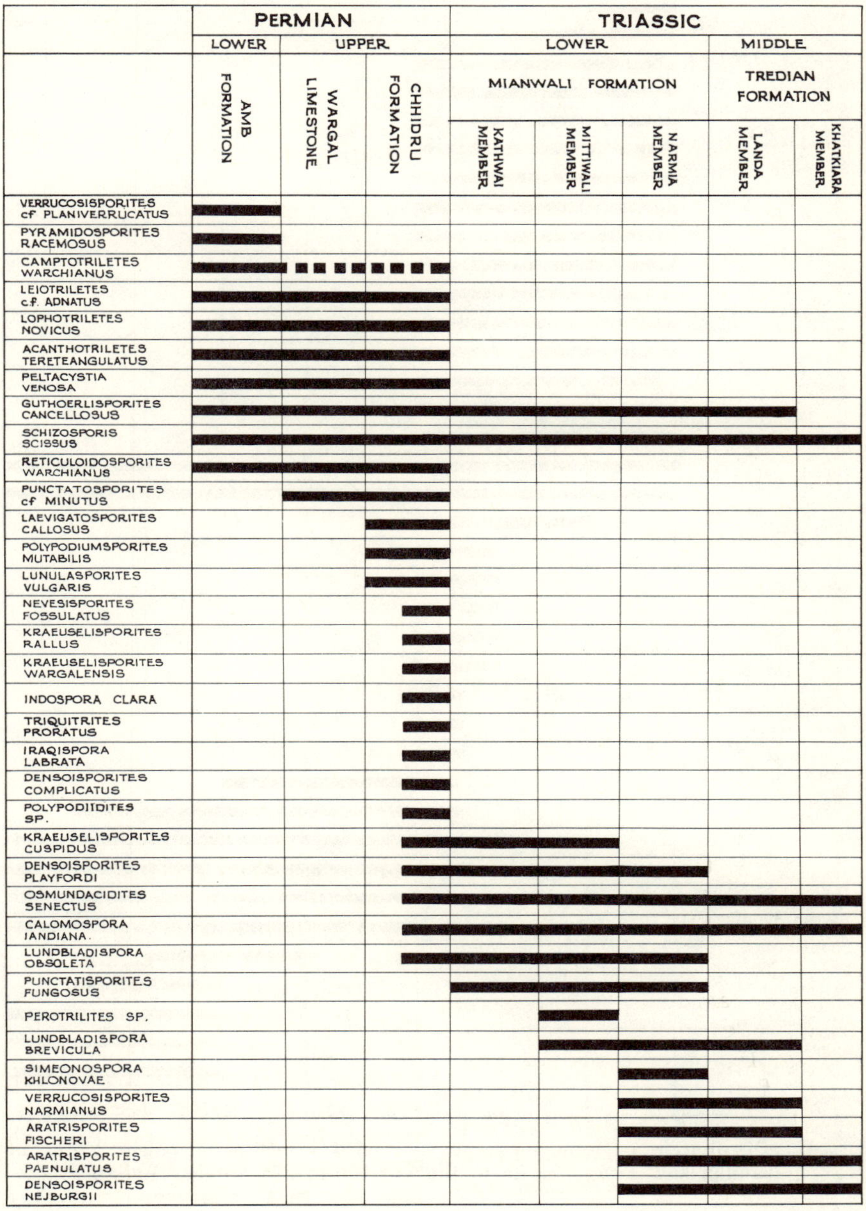

FIG. 16. Distribution of form species of presumed pteridophyte spores and plant microfossils incertae sedis in Permian and Triassic strata in the Salt Range and Surghar Range.

FIG. 17. Distribution of form species of presumed gymnosperm pollen grains in Permian and Triassic strata in the Salt Range and Surghar Range.

obviously by the constant presence of haploxylonoid, striatitid pollen grains of the *Protohaploxypinus* type. These seldom constitute less than 20 percent of any assemblage and exceed 50 percent in those from the Amb Formation.

Diploxylonoid Striatiti are associated with the haploxylonoid forms, but are much less common and their distribution is sporadic. The most unusual striatitid pollen grain is *Guttulapollenites hannonicus*, previously known from the Upper Permian and Lower Triassic of Madagascar. It occurs in all three Permian units, but quantitatively it is only important in the Wargal Limestone and the lower part of the Chhidru Formation: in the two assemblages from the former unit it was the most common single element, making up 10 and 13 percent of the total counts.

Non-taeniate disaccate pollen grains are also important components of the Permian assemblages and include the morphologically unusual form *Pinuspollenites thoracatus*. *Vitreisporites pallidus*, which closely resembles the pollen grains of *Caytonanthus*, is also a regular, although minor, constituent of the Permian microfloras. Occasional specimens of *V. pallidus* have been reported from the Upper Permian of Australia (Balme, 1960), and the species is common in the Kazanian and Tartarian of the USSR (e.g., Chalyshev, Varyukhina, and Molin, 1965). It has not, however, been previously recorded from strata as old as Artinskian.

Among the non-saccate pollen grains the most common and constantly occurring is *Paravittatina lucifer*, which was found in every sample studied and usually made up between 2 and 10 percent of the total assemblage. Pollen of similar morphology to *P. lucifer* is abundant and widespread in Kungurian and Upper Permian deposits in the cis-Ural of the USSR. It is also an important component of microfloras from certain parts of the Ecca Group in South Africa (Hart, 1966) and from the lower Sakamena Group in Madagascar (Goubin, 1965).

Striatitid and other disaccate pollen grains, together with *Paravittatina*, constitute between 75 and 90 percent of the total assemblages from the great majority of samples. Individual form species belonging to these groups range, with certain important exceptions, throughout the Permian sequence examined, and seldom display marked fluctuations in abundance from sample to sample. In the main, therefore, they are likely to represent the pollen grains of dominants in the successive Permian floras. Their gymnospermic origin is most likely, and the palynological evidence points strongly to the persistence of stable floral elements throughout most of Artinskian and Late Permian time.

Changes in the spore-pollen assemblages, which will be discussed in more detail in the next section, may reflect widespread regional floral changes or simply local responses to comparatively minor edaphic variations. Disentangling these factors is inevitably difficult, although several criteria were enumerated by Kuyl, Müller, and Waterbolk (1955, p. 65) in their review of the application of palynology to oil exploration. There is certainly evidence within the Permian succession for important changes in the regional floras, presumably resulting from extensive climatic modifications. The appearance somewhere near the base of the Chhidru Formation, of the European species *Klausipollenites schaubergeri* and the unusual disaccate pollen grains *Cedripites priscus*, may be credibly taken to indicate colonization of the Salt Range Region by certain new dominants. On the other hand, it would be dangerous to interpret the apparently spectacular microfloral changes which occur in the upper sandstone unit of the Chhidru Formation as necessarily resulting from interregional migration.

Changes in Spore Assemblages within the Permian Succession

Certain variations in the gross features of Permian and Triassic assemblages at Nammal Gorge and Wargal are shown diagrammatically in Figures 18 and 19 respectively. It should be noted that the stratigraphic sampling intervals were much closer at Wargal than at Nammal, and as a result it has been necessary to use different vertical scales on the two diagrams. Figure 19 shows, therefore, only the upper 12 feet or so of the Chhidru Formation as it is developed at Wargal.

Except for two samples from the Amb Formation (T62-350 from Warchha and T62-53 from Zaluch

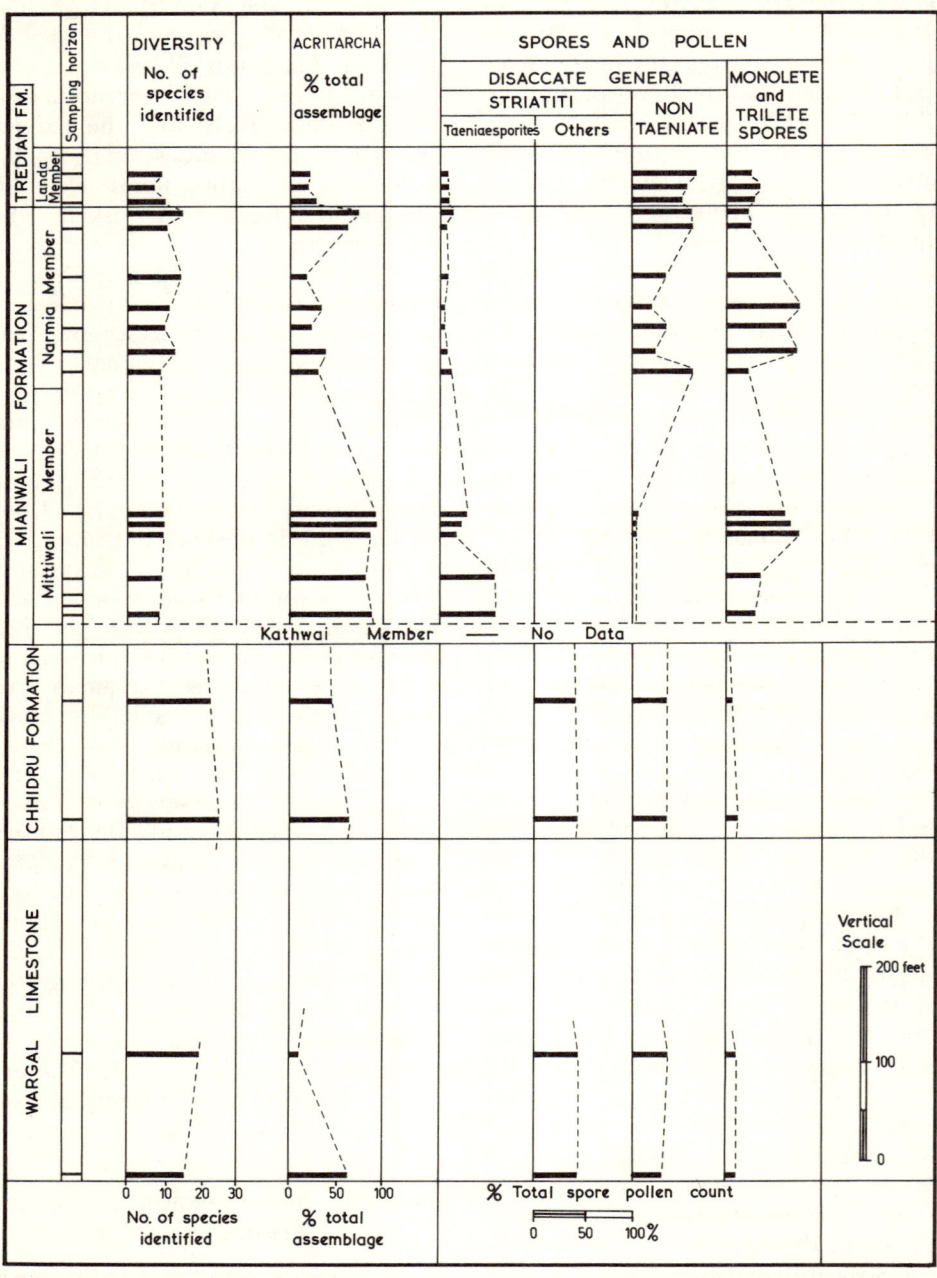

Fig. 18. Variations in certain gross characters of Permian and Triassic plant microfossil assemblages from Nammal Gorge, Salt Range.

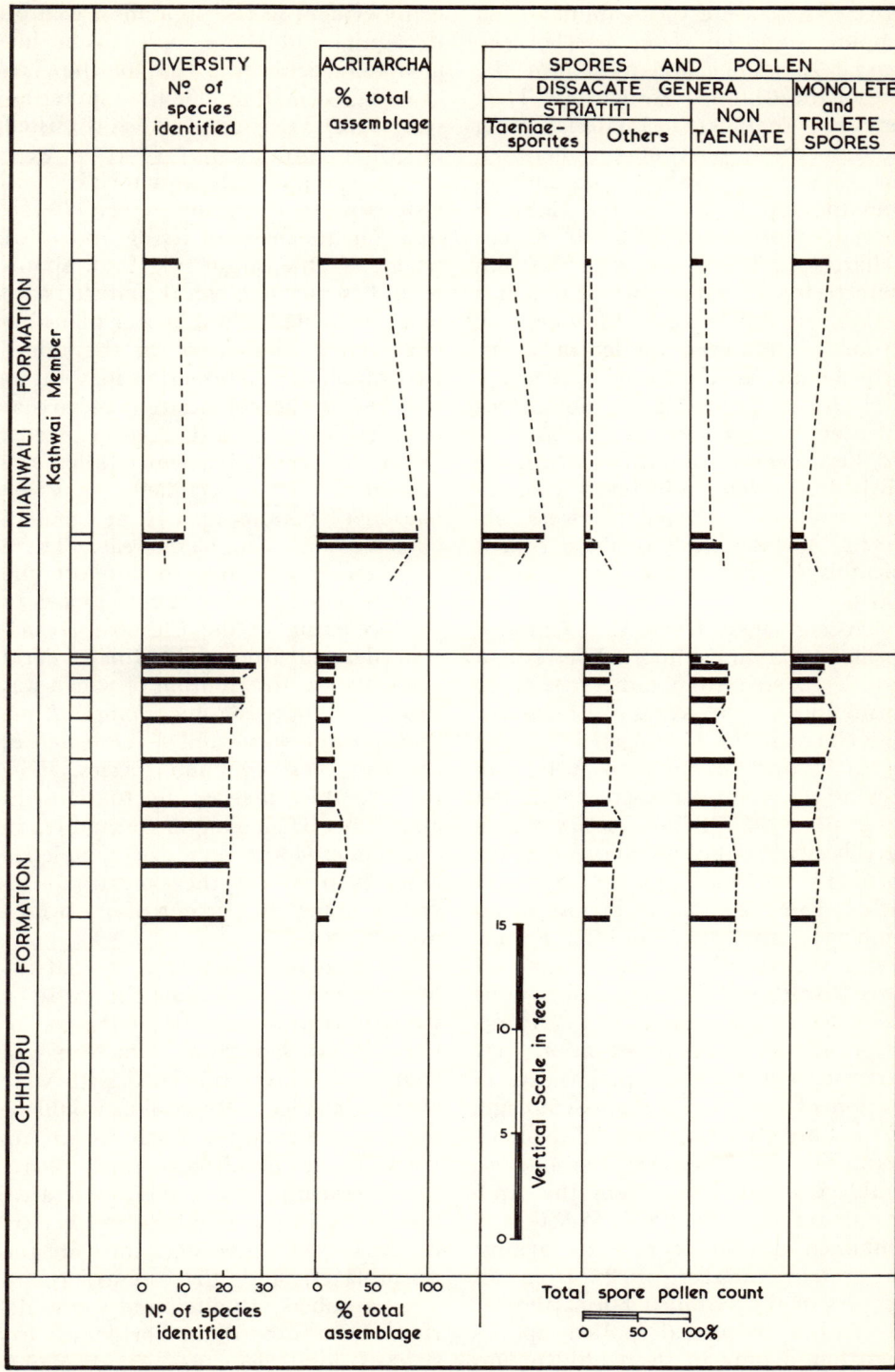

Fig. 19. Variations in certain gross characters of Permian and Early Triassic plant microfossil assemblages from Wargal, Salt Range.

Nala), all the Permian samples contained leiospheres and spinose acritarchs. These were most abundant in samples from the lower part of the Wargal Limestone and from near the base of the Chhidru Formation. They were less frequent in material from the uppermost part of the Permian succession. At Wargal, assemblages from the top 12 feet of the Chhidru Formation contained 10 to 30 percent acritarchs, and comparable proportions characterize the same part of the unit at Chhidru and Wargal. Although the acritarchs have been studied only cursorily by me, all the forms present appear to be assignable to ubiquitous Paleozoic genera such as *Leiosphaeridia, Veryhachium,* and *Micrhystridium*. No bodies which could possibly be interpreted as dinoflagellates were observed. A closer study of the acritarch assemblages was made by Sarjeant (1970).

Assemblages from the Chhidru Formation are much more diverse than those recovered from either the Amb Formation or the Wargal Limestone. This increase in heterogeneity shows up in Figure 19, although it fails to present the complete picture, in that only spore and pollen species which have been described in the section devoted to systematic palynology were included in the data. In almost all assemblages from the Chhidru Formation a variety of additional forms occurred, although individually none was sufficiently common to justify description. Superior preservation may partly account for the apparent diversification of the Chhidruan assemblages but is unlikely to be the sole explanation. The best preserved Permian assemblage of all came from the Amb Formation at Warchha (T62-350) and contained only 20 form species, against an average number of 24 from all samples of the Chhidru Formation.

Although several pollen species were found only in the Chhidru Formation, diversification of its spore-pollen assemblages is due mainly to an increase in the variety of trilete and monolete spores. This trend is especially evident in the uppermost part of the unit, and the sample taken immediately below the base of the Triassic at Wargal yielded a spore assemblage, 62 percent of which consisted of trilete forms. Most of these were cavate species with equatorial zona, assignable to the genus *Kraeuselisporites*. On the basis of living and fossil analogies, the argument that spores with these morphological characters are of lycopodiaceous origin is a plausible one. Their abundance in the upper part of Chhidru Formation may therefore reflect the colonization of coastal marshes by lycopodiaceous elements (cf. Chaloner, 1961), which filled the ecological niche occupied today by halophytic families such as the Chenopodiaceae and Plumbaginaceae. There is certainly evidence to support the view that the species of *Kraeuselisporites* occurring in the Chhidru Formation derived from subordinate floral elements, the distribution of which was subject to local edaphic factors. *Kraeuselisporites rallus* and *K. wargalensis,* the two most common species, show for example, marked fluctuations in abundance from sample to sample; in some microfloras they are completely absent; in others they make up together nearly 40 percent of the whole assemblage.

Taken in its entirety the palynological evidence points towards a strongly regressive phase at the top of the Chhidru Formation, which may of course have been associated with wide climatic changes. Regression would be expected to manifest itself by an increase in the numbers of plant microfossils reaching the depositional area, a decrease in the numbers and variety of acritarchs, diversification of the assemblages, and the occurrence in large numbers of pollen and spores derived from subordinate or locally restricted floral elements. All these features characterize assemblages from the upper 12 feet or so of the Chhidru

Formation at all localities from which it has been examined.

Although the more obvious variations in assemblages from the Chhidru Formation may well have been controlled by reversible factors, it also seems clear that important permanent changes in the regional floras occurred during Late Permian times. Twenty-eight of the 61 Permian species are confined to the Chhidru Formation, and of these 19 were seen only in assemblages from the upper part of that unit. Some of these characteristically Chhidruan species are widespread in Late Permian deposits in other parts of the world. *Klausipollenites schaubergeri* is, as we have already noted, an omnipresent form in sediments from the European Zechstein. *Marsupipollenites triradiatus* and *Gnetaceaepollenites sinuosus* are well-defined species known from the Permian of peninsular India, Australia, Antarctica, and South Africa. A curious feature, too, of the Chhidruan assemblages is their minor but persistent content of forms that are elsewhere more typical of Early Mesozoic strata. *Nevesipollenites fossulatus* and *Tigrisporites playfordi* are characteristically Triassic forms in Australia where *Guthoerlisporites cancellosus* is also known only from the Triassic and Rhaetic (Playford and Dettmann, 1965; Playford, 1965), although a similar form occurs in the European Zechstein. *Cedripites priscus* is morphologically close to certain Mesozoic disaccate pollen grains, including some specimens of *Rimaesporites aquilonalis* Goubin from the Upper Triassic and Liassic of Madagascar (Goubin, 1965, pl. 5, fig. 7). Another unusual Chhidruan pollen grain, *Pretricolpipollenites bharadwaji*, can be compared only with forms described by Briche, Danzé-Corsin, and Laveine (1963) from the Lias of France.

Future detailed collecting may enable at least two biostratigraphic units to be established within the Chhidru Formation, on a palynological basis. These cannot be delimited at present, but assemblages from the upper part of the Formation at Wargal and Chhidru are closely similar to one another and sharply differentiated from those occurring in other parts of the Permian and Triassic succession.

Comparison with Other Upper Permian Spore Assemblages

It is difficult to judge the value of phytogeographical conclusions based on palynological evidence alone. This lack of confidence stems not so much from our ignorance concerning the affinities of most fossil-dispersed spores, but more importantly, from the impossibility of determining the rank of the natural taxon represented by each form species. Certainly it would be unwise to assume that any Paleozoic spore or pollen species, however clearly characterized, denotes a natural unit lower than a family. Some, even *Vitreisporites pallidus* for example, cannot be attributed with assurance to a single natural order, and all that can be reasonably inferred about a form such as *Cycadopites follicularis* is that it is gymnospermous. It follows that many, if not most, fossil spores and pollen grains have dubious value as guides to the plant geography of the past; and, except for broad biostratigraphic purposes, there is little purpose served by comparisons based on the distributions of most form genera. Well-defined form species, combining unusual morphological characters, are likely to prove the most satisfactory units in studying the broad distributive patterns of past floras.

Twenty-one species occurring in the Chhidru Formation are considered to be sufficiently distinctive to have potential value in defining floral regions of the Late Permian. These species are listed in Table 2, which shows also the occurrence of apparently identical or closely similar forms in other parts of the world. Comparisons have usually been made only on a broad regional basis, and for the

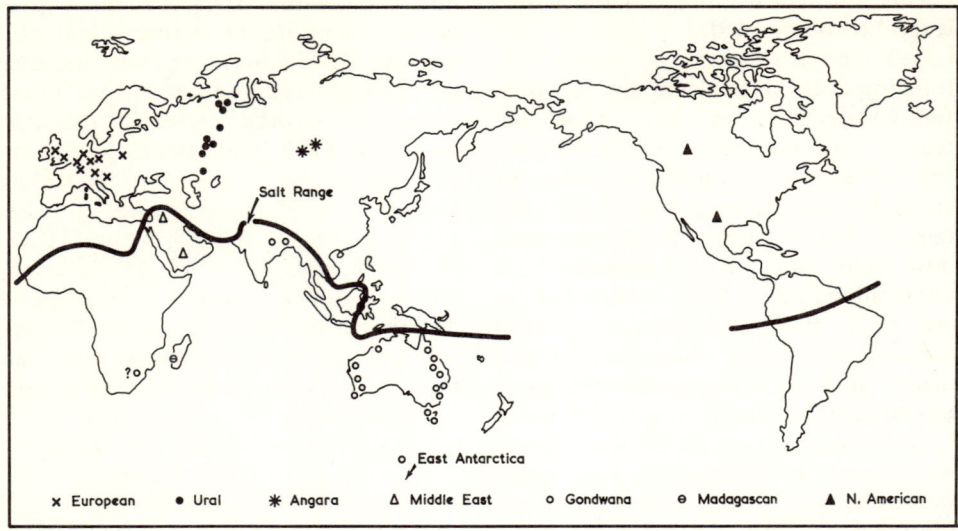

Fig. 20. Map showing the general localities of Upper Permian spore-pollen assemblages either described in published work or held in the collection of the University of Western Australia. Symbols refer to the microfloral regions discussed in the text. Heavy black line marks the approximate northern limit of the *Glossopteris*-flora (after Melville, 1966).

purposes of discussion a number of Late Permian floral provinces are tentatively recognized. The general circumscriptions of these are shown in Figure 20, and certain characteristic features of their microfloras are discussed briefly in the ensuing paragraphs.

European Region

Spore and pollen assemblages from the Upper Permian of Europe maintain constant basic characters from England in the west, throughout the Germanic Basin, and along the northern margin of the western Tethys. They have been the subject of many well-illustrated accounts, of which those of Potonié and Klaus (1954), Leschik (1956), Grebe and Schweitzer (1962), Klaus (1963), and Schaarschmidt (1963) are among the more notable. These assemblages are characterized by lack of diversity and an abundance of saccate pollen grains, derived presumably from specialized desert or semidesert vegetation, fringing the Zechstein saline lagoons. *Lueckisporites virkkiae, Taeniaesporites noviaulensis, Falcisporites zapfei, Klausipollenites schaubergeri,* and *Nuskoisporites dulhuntyi* are the most important pollen species, and disaccate grains bearing proximal tetrad markings, belonging to the *Illinites-Limitisporites-Jugasporites* suite, are sometimes prominent microfloral components.

Ural Region

Subsequent to the pioneer studies of Luber and Waltz (1941), Kungurian and Upper Permian deposits in the cis-Ural have been the subject of fairly intensive palynological investigation. Samoilovich (1953), Zoricheva and Sedova (1954), Abramova and Marchenko (1960), and Zauer (1965) have all published important accounts, although regrettably few Russian papers have been illustrated by photomicrographs. Soviet workers also often adopt a less rigorous attitude towards palynological systematics than their colleagues elsewhere, and their listed microfloras are not readily susceptible to detailed comparative analysis. Common and distinctive features are nevertheless evident in spore-pollen assemblages throughout the length of the Ural geosyncline.

Typically European species such as *Lueckisporites virkkiae, Klausipollenites schaubergeri,* and *Taeniaesporites noviaulensis* probably occur in the Russian assemblages, but quantitatively they are unimportant. Among the most important microfloral elements are striatitid pollen grains, assignable in the main to *Protohaploxypinus* and *Striatopodocarpites.* Both these genera occur in contemporaneous European assemblages as strictly minor elements; in the Upper Permian of the "Gondwanaland" countries, on the other hand, they are ubiquitous and usually abundant. Striate monocolpate pollen grains assigned by Russian workers to *Vittatina* are perhaps the most striking components of Kungurian and Upper Permian microfloras in the Ural area. In its Russian usage *Vittatina* includes the pollen type transferred in the present account to *Paravittatina;* indeed, specimens conforming to the diagnosis of *Paravittatina* are more common than those resembling *Vittatina subsaccata,* the type species of *Vittatina. Paravittatina* is often the most abundant single element in Kungurian and younger Permian sediments in the cis-Ural. Zauer (1965) reported that it composed as much as 55 percent of the total assemblage from sediments in the Solikamsk area and proportions of 70 to 90 percent of *Paravittatina* were noted by Zoricheva and Sedova (1954) in microfloras from Upper Permian deposits in the extreme northern part of the Ural Mountains. Small, disaccate pollen grains attributable to *Vitreisporites* are also prominent in assemblages from the cis-Ural. As a rule they are referred directly to the Caytoniales by Russian palynologists, who are also prepared to accept as caytonialean large disaccate forms which bear no close resemblance to pollen grains isolated from any species of *Caytonanthus.* Because of this it is often impossible to know what proportion of the grains recorded as caytonialean by Russian workers actually resemble *Vitreisporites pallidus.* There seems little doubt, however, that the species is more widespread and common in the Upper Permian of Russia than it is in Paleozoic sediments elsewhere in the world. Chalyshev, Varyukhina, and Molin (1965), for example, report that small "Caytoniales" pollen grains, which resemble *V. pallidus,* make up over 50 percent of certain Tartarian assemblages from the northern cis-Ural.

Angara Region

Permian coal deposits occurring in the Kuznetsk, Minusinsk, and Tunguska Basins are thought by Russian workers to have originated in a distinct and separate floral region from those in the area of the Ural geosyncline. Luber and Waltz (1941) recognized a Tunguska Province in which the microfloras were dominated by pteridophyte spores, monosaccate pollen grains of the *Cordaitina*-type, alete bodies, and monocolpate pollen grains. Apart from the work of Andreeva *et al.* (1956) most detailed palynological studies in the Siberian basins have been concerned with Lower Permian and Carboniferous strata. The relative importance of chronological and geographic factors in explaining discrepancies between Permian microfloras from the cis-Ural and oriental Russia is not, therefore, always clearly evident. Assemblages from the Upper Permian Erunakov deposit in the Kuznetsk Basin are highly diversified, yielding large numbers of trilete spores, monosulcate and monosaccate pollen grains, and disaccate Striatiti of the *Protohaploxypinus* type (Andreeva *et al.,* 1956). They also contain *Vitreisporites pallidus,* or a closely similar form (= *Coniferaletes impuberus* Andreeva), and striate pollen grains resembling *Vittatina* s.s. or *Ephedripites.*

No undoubted examples of *Paravittatina* have, however, been illustrated from the Angaraland successions.

Middle East

Knowledge of Upper Permian spores and pollen from the Middle East stems from two published accounts: one by Singh, (1964) which described a rich assemblage from northern Iraq, and the other by Hemer (1965) in which seven characteristic Upper Permian species from Iran were illustrated but not described. The forms illustrated by Hemer included *Lueckisporites virkkiae, Protohaploxypinus microcorpus* [as *P. jacobii* (Jansonius)], *Guthoerlisporites cancellosus,* and *Striatobietites richteri* (Klaus), all of which (or closely similar forms) are known from the Zechstein of Europe and, with the exception of *S. richteri,* from the Chhidru Formation in the Salt Range. In addition to his illustrated species Hemer also recorded a variety of species of *Vittatina* and striatitid forms, which had previously only been reported from India, Australia, or Africa.

Singh's assemblage is unusual in its comparatively low content of disaccate Striatiti, which make up only about 12 percent of the total assemblage. Although most of the species described by Singh closely match forms occurring in Permian sediments in other parts of the world, the assemblage taken as a whole has no clear-cut character. Apparently Singh's account is based on the study of a single sample, so it would be risky to generalize from what may well, from its high pteridophyte content, be a specialized association.

Gondwana Region

Upper Permian assemblages from the countries of "Gondwanaland," from which category Madagascar is excluded for the purpose of the present discussion, are known chiefly from peninsular India and Australia. Rilett (1954) published a short paper on Upper Permian spores and pollen grains from Natal, but his descriptions and illustrations are, with a few exceptions, difficult to interpret. Abundant but rather poorly preserved plant microfossils have also been recovered from Upper Permian coals and carbonaceous shales from the Prince Charles Mountains, Antarctica (Balme and Playford, 1967). The available palynological data show clear fundamental similarities between the Upper Permian microfloras of India, Australia, and Antarctica, although there well may be important differences in detail. *Dulhuntyispora,* for example, the singular form genus of trilete spores, which occurs in Upper Permian strata throughout Australia, has not yet been recorded from outside that continent. Such apparently minor differences may, of course, eventually become extremely important in testing paleogeographic hypotheses involving continental drift. In general, however, the dominant components of the Gondwana assemblages, like those of the Ural geosyncline, are disaccate striatitid pollen grains. Haploxylonoid forms are, in most places, more common than diploxylonoid species, but this is not always true. In the Amery Formation, East Antarctica, and in certain seams from the Raniganj Coalfield of India, diploxylonoid forms are especially abundant.

Apart from their general quantitative similarities, Indian, Australian, and Antarctic assemblages also have common to them a number of distinctive form species. The most important are:

Granulatisporites trisinus Balme
    and Hennelly
Marsupipollenites triradiatus
    Balme and Hennelly
Gnetaceaepollenites sinuosus
    (Balme and Hennelly)
Densipollenites indicus Bharadwaj
Anapiculatisporites ericianus
    (Balme and Hennelly)

Bascanisporites undosus Balme and Hennelly (Australia and Antarctica).

The palynological data published so far support the concept of a broad Late Permian phytogeographic region which embraced the present areas of peninsular India, Australia, and East Antarctica. Virtually no comparative material is available from South Africa or South America, so the full extent of the floral region cannot be gauged, nor can one fully assess its relationships with the traditional *Glossopteris* floral province, based on the distribution of macroscopic plant organs.

Madagascar

Spores and pollen assemblages from the lower Sakamena Group of Madagascar have been only partially described (de Jekhowsky and Goubin, 1964; Goubin, 1965). Even so there are strong indications that the Late Permian floras of Madagascar were radically distinct from those of peninsular India, Australia, and East Antarctica.

Striatitid pollen grains are by no means uncommon in sediments of the lower Sakamena Group, although the forms present are not nearly as diverse as they usually are in the Upper Permian of the Gondwana Region. According to Goubin's account, the only striatitid genus occurring in proportions of more than about 3 percent in assemblages from the lower Sakamena Group is the morphologically curious *Guttulapollenites*. *Guttulapollenites hannonicus* makes up between 5 and 40 percent of the Madagascan Late Permian assemblages and, in the northern part of the Morondava Basin, is often the most abundant single species. *G. hannonicus* has now been found in the Salt Range; the Permian of India and East Antarctica, and probably also in the Bunter of Western Europe. Apart from *G. hannonicus* the most common Madagascan pollen species is *Paravittatina* (=*Vittatina* spp. of Goubin), which occurs in proportion varying between 5 and 30 percent of the total assemblage. *Paravittatina* is known from the Gondwana Region, but only as a sporadically distributed, extremely minor element. As we have seen, it is most typical of the Ural Geosyncline Region, but is prominent also in Salt Range Permian sediments. *Lueckisporites virkkiae* and *Vitreisporites pallidus* are minor components of the lower Sakamena assemblages, and most of the other common pollen grains are nontaeniate Disaccites belonging to such genera as *Platysaccus, Alisporites,* and *Sulcatisporites*. From Goubin's account, more typical Permian striatitid from genera such as *Protohaploxypinus* and *Striatopodocarpites* are rarely encountered in Madagascan assemblages of Late Permian age.

North American Region

Only one undoubtedly Upper Permian spore assemblage has been described from North America in published literature. This came from the Flowerpot Shale of Oklahoma (Wilson, 1962), which is of early Late Permian age (Chudinov, 1965). It is, therefore, certainly older than the Chhidru Formation and probably older than most of the microfloras treated in the present discussion.

Lueckisporites virkkiae is the most abundant species in assemblages from the Flowerpot Shale which in this respect resemble certain microfloras from the European Zechstein. There are other points of resemblance between the North American and European microfloras, the presence in both forms resembling *Striatoabietites richteri* (Klaus), for example; but in detail differences outweigh similarities. The possible paleogeographic significance of these differences cannot be assessed until palynological information becomes available from late Guadalupian and Ochoan sediments.

Barss, Hacquebard, and Howie (1963) reported *Lueckisporites* from Permian sediments in the Maritime

TABLE 2. Known Distribution of Some Spore and Pollen Species Occurring in the Upper Permian of the Salt Range.

	North America	Western and Central Europe	Ural Area	Angara	Middle East	India	Australia	Antarctica	Madagascar	Salt Range
Peltacystia venosa							O			O
Guthoerlisporites cancellosus		O			O		Tr			O
Indospora clara						O	O			O
Iraqispora labrata					O					O
Plicatipollenites indicus		*?	O	O	O	O	O			O
Densipollenites indicus						*	O	*		*
Sulcatisporites ovatus			?	?		O	O	O		O
Fimbriaesporites ? sp.		O			O				O	O
Vitreisporites pallidus		*	*		O		O		*	*
Alisporites tenuicorpus										*
Pinuspollenites thoracatus										*
Cedripites priscus			?	?						*
Guttulapollenites hannonicus		Tr?				O	Tr?		*	*
Lueckisporites virkkiae	*	*	O		*				*	O
Lueckisporites singhii					O					*
Taeniaesporites noviaulensis	Tr	*	O	?	O		Tr		O	O
Paravittatina lucifer		*			O	O	O	O	*	*
Marsupipollenites triradiatus						O	*	*		O
Gnetaceaepollenites sinuosus						O	O	O		O
Pretricolpipollenites bharadwaji										*
Klausipollenites schaubergeri			*?							*

O = Rare in Upper Permian.
* = Locally common in Upper Permian.
Tr = Unknown from Permian but occurring in Triassic.

Provinces of Canada and noted broad similarities between the Canadian assemblages and those described by Wilson from Oklahoma. Possibly the sediments from which this assemblage came are Upper Permian, but the account is not illustrated and comparisons are precluded.

Salt Range Region

As has been noted already, Chhidruan assemblages from the Salt Range resemble those from the Urals and Gondwana Regions in the preponderance of Striatiti which they contain. Their subsidiary elements on the other hand are curiously mixed. Table 2 lists 21 species with well-defined morphologies which occur in the Chhidru Formation and which, with four exceptions, are known from other parts of the world. Forms listed in Table 2 fall into a number of distinct groups which are discussed in the ensuing paragraphs.

Local Elements

Four common pollen species cannot be matched closely with Permian forms known from other parts of the world. These are:
Pinuspollenites thoracatus
Cedripites priscus
Pretricolpipollenites bharadwaji
Alisporites tenuicorpus

In addition, several form species of pteridophyte spores, apart from those listed in Table 2, may not be represented in Permian assemblages elsewhere. These include *Laevigatosporites callosus*, *Kraeuselisporites rallus*, *Polypodiisporites mutabilis*, *Tigrisporites playfordi*, and *Nevesisporites fossulatus*, although the two last-mentioned species closely resemble forms known from the Australian Triassic.

Obviously inferences concerning the existence of endemics in the Salt Range flora are unjustified in the absence of detailed information on the composition and distribution of Upper Permian microfloras in adjoining areas. The species listed have, nevertheless, potential phytogeographic significance.

European Elements

Six of the forms listed are regarded as characteristically European, although most are known also to occur in other regions. These are:
Guthoerlisporites cancellosus (cf. *Endosporites velatus* Leschik)
Protohaploxypinus microcorpus
Lueckisporites virkkiae
Taeniaesporites noviaulensis
Fimbriaesporites ? sp. (cf. *Platysaccus papilionis* in Klaus, 1963, pl. 7, fig. 33)
Klausipollenites schaubergeri

The first five species are uncommon in Salt Range Permian deposits; *Taeniaesporites noviaulensis* was very rare indeed and represented by only three or four specimens among many thousands. *Klausipollenites schaubergeri*, however, was common in many Chhidruan assemblages, and its occurrence is additionally interesting as the first unequivocal record of the species from an extra-European locality. Despite the fact that specimens of *K. schaubergeri* from the Salt Range are indistinguishable from those from the Austrian type locality, the inevitable doubts persist concerning the relationships of their parent plants. *K. schaubergeri* is nevertheless a closely circumscribed form species with unusual characters, and it is not unduly incautious to infer that it represents the pollen grain of a natural taxon of familial or lower rank. If so, the distribution of such a plant group cannot have been as rigorously controlled climatically as the previously known Zechstein occurrences of *K. schaubergeri* suggested.

Finally, it may be worth noting that species belonging to the *Jugasporites-Illinites-Limitisporites* group of disaccate pollen grains were not identified in the Salt Range Permian. These genera are especially characteristic of the European Zechstein, but they also occur in the Upper Permian of "Gondwanaland" and in other parts of the world.

Ural Elements

Apart from their high content of Striatiti, the most obvious feature shared by Ural and Salt Range assemblages is the prevalence in both of *Paravittatina* and *Vitreisporites pallidus*. Comparison between microfloras of the two regions at a species level is always difficult and usually impossible,

but the following form species are probably represented in both regions:
Plicatipollenites indicus
Lueckisporites virkkiae
Taeniaesporites noviaulensis
? *Klausipollenites schaubergeri*
Vitreisporites pallidus
Paravittatina lucifer (cf. *Vittatina circinnata* Hart, in part)

Conclusions cannot be pressed, but it seems at least clear that the Salt Range microfloras have more in common with those of the Ural Region than with those of Europe.

Middle East Elements

Ten of the species listed in Table 2 occur in the Salt Range and in the Middle East. With the possible exception of *Iraqispora labrata*, however, all these forms common to both regions are known from other parts of the world and cannot be regarded as especially characteristic of the Upper Permian of the Middle East. The only justifiable conclusion from the palynological evidence is that there are certain clear resemblances between assemblages from the two regions, although the full degree of similarity cannot at present be judged.

Gondwana Elements

The following nine species are known from the Salt Range and the Gondwana Region:
Peltacystia venosa
Indospora clara
Densipollenites indicus
Plicatipollenites indicus
Sulcatisporites ovatus
Vitreisporites pallidus
Paravittatina lucifer
Marsupipollenites triradiatus
Gnetaceaepollenites sinuosus

Of these forms only *Densipollenites indicus* is known to be common in certain assemblages from both Regions. *Vitreisporites pallidus* and *Paravittatina lucifer,* both prominent in microfloras from the Salt Range, are so far known only as rarities in Permian assemblages from India, Australia, and Antarctica. Conversely the remaining six species listed, although widespread and fairly common in the Gondwana Region, were extremely rare in Salt Range microfloras.

Undoubtedly, Late Permian floras in the Salt Range Region contained Gondwanid elements, although these appear to have been subordinate components. The possibility exists, of course, that the ubiquitous species of *Protohaploxypinus* and *Striatopodocapites* reflect additional interdependence between the two floral regions. Except for very broad comparative purposes, however, the paleofloristic significance of these two genera is diminished by the difficulty of establishing sharply defined form species within them.

Madagascan Elements

In several important respects Upper Permian assemblages from the Salt Range are closer to those of Madagascar than to any other region. The resemblances may be even closer than the available data indicate, as published accounts on the palynology of the lower Sakamena Group deal only with pollen grains.

The most notable common features of the Salt Range and Madagascan assemblages are the presence of *Guttulapollenites hannonicus,* the relatively high frequency of *Paravittatina* and *Vitreisporites pallidus,* and the occurrence of *Lueckisporites virkkiae, Taeniaesporites noviaulensis,* and specimens resembling *Fimbriaesporites*? sp.

Apart from their affinities with those of the Salt Range, the lower Sakamena assemblages are also remarkable in their apparent divergence from Australian, Indian, and Antarctic microfloras of similar age. This is in marked contrast to assemblages from the Lower Permian Sakoa Group, which closely match those from equivalent strata in Africa and Australia (Doubinger and Rakotoarivelo, 1960; Rakotoarivelo, 1960). Unfortunately, virtually nothing has been published on the palynology of the upper Ecca or Beaufort Groups in southern Africa. It seems obvious, however, that relationships between the Late Permian spore assemblages of East Africa, Madagascar, and South Africa may provide a key to many problems of late Paleozoic paleogeography. Certainly there is little doubt that the Upper Permian floras of Madagascar and the Salt Range had important common elements and in both there is also evidence of a relationship with contemporaneous European and cis-Uralian floras.

Summary

Spore assemblages from the Chhidru Formation indicate that it was deposited in a shallow sea adjacent to a densely vegetated land mass. Gymnosperms were the main dominants, but the flora also contained ferns and lycopods, the latter perhaps mainly as edaphically controlled subordinates. The flora appears to have transitional characters and includes Gondwanid, European, Russian, and Madagascan elements. Taken as a whole, it appears most closely related to the Late Permian floras of Madagascar.

ASSEMBLAGES OF THE TRIASSIC

Unlike the Permian microfloras which, we have seen, maintain certain constant basic characteristics throughout the succession studied, those from the Triassic fall into two well-defined and sharply differentiated groups. The two groups have elements in common, but there can be no doubt that they represent plant associations of radically different constitution. No precise horizon can yet be established for the floral break. It lies either in the upper part of the Mittiwali Member or at the base of the Narmia Member, that is, somewhere between the Flemingitan and Prohungaritan divisions of Spath (1934, p. 27).

Early Scythian Assemblages

Palynological data on the older Scythian were obtained from the study of eight samples; three from the Kathwai Member at Wargal and five from the Mittiwali Member in Nammal Gorge. In Nammal Gorge the Mittiwali Member is 246.9 feet thick (Kummel, 1966), and the uppermost sample available for study came from the middle of the unit. No palynological information is available, therefore, from the upper part of the Mittiwali Member.

Assemblages from the Kathwai and Mittiwali Members are characterized by poor preservation, lack of diversity, and the presence of enormous numbers of small spinose acritarchs and indistinctive leiospheres. *Veryhachium reductum* Deunff and species of *Micrhystridium* and *Leiosphaeridia* are by far the most common plant microfossils and together make up between 68 and 98 percent of the total assemblage for the eight samples studied (fig. 18-19).

Nine species of spores and nine of pollen grains were recorded from the older Scythian sediments. Only one of these, *Perotrilites* sp., was not found in other strata in the Salt Range, and this form is too rare to allow much significance to be attached to its apparent restriction. Of these 18 species four, *Cycadopites follicularis, Densoisporites playfordi, Taeniaesporites noviaulensis,* and *T. pellucidus,* occurred in all eight samples. *Lundbladispora obsoleta* and *L. brevicula* were present in all assemblages from the Mittiwali Member, but occurred only sporadically in the Kathwai Member. These six species together make up from 81 to 98 percent of the total spore-pollen count in the eight assemblages under discussion. For practical purposes, therefore, the older Scythian spore assemblages are entirely composed of three form species of gymnosperm pollen grains and three spore types which are almost certainly lycopodialean in origin.

Impoverishment of the Early Triassic spore assemblages stands in marked contrast to the great diversity of those from the Late Permian, and the magnitude of the break occurring at the Permian-Triassic boundary may be gauged from Figures 16 and 17. Forty-two of the 57 form species that occur in the Chhidru Formation were not found in the Kathwai or Mittiwali Members. Indeed, of the 14 species which cross the boundary only two, *Vitreisporites pallidus* and *Cycadopites follicularis*, were other than extremely rare in the Permian, and both of these are long-ranging forms. *Taeniaesporites*, represented in the Permian by occasional specimens in the uppermost part of the Chhidru Formation, is virtually the only striatitid genus to survive into the Triassic. A few specimens of *Protohaploxypinus* and *Striatopodocarpites* were found in the Mittiwali Member, but they were not common enough to appear in the counts. Nontaeniate Disaccites were rarely identified in the lower Scythian assemblages, although the lowermost sample from the Kathwai Member contained 6 percent of a form dubiously referred to *Klausipollenites schaubergeri*. Poor preservation may be partly responsible for the failure to identify non-taeniate Disaccites. It is certainly easier, when dealing with indifferently preserved material, to be sure that a pollen grain possesses taeniae than to be confident that it does not. Even so, it is clear that apart from *Taeniaesporites* disaccate pollen grains are quantitatively unimportant components of the Early Triassic spore assemblages of the Salt Range.

Significance of the Older Scythian Assemblages

The obvious initial inference from the palynological data is that the Permian-Triassic boundary in the Salt Range is marked by a floral crisis comparable to that reflected by the invertebrate faunas. Even if the faunal break is less complete in the Salt Range than was once supposed (Kummel and Teichert, 1966), it is nevertheless notable enough, here and elsewhere, to keep the concept of catastrophism alive. Rapid, massive extinctions of marine fauna are susceptible to interpretations which do not invoke extraterrestrial causes: an ingenious recent example is Fischer's (1964) attempt to associate Late Permian extinctions with brackish oceans resulting from the abnormally large salt accumulations of that Period. If the marine crisis was accompanied by a terrestrial crisis, however, the problem defies the uniformitarian principle. Evidence from tetrapod remains, and to some extent from macroscopic plant fossils, suggests that the transition from the Permian to the Triassic on land was not accompanied by the dramatic events of the marine realm. Is it possible, then, to interpret the palynological data from the Salt Range without presuming sudden and almost total regional extinction of the Late Permian floras? The problem may be analyzed further by considering relevant information from other parts of the world.

Palynological studies of marine Scythian strata have been carried out in Canada (Jansonius, 1962; McGregor, 1965), the northern USSR (e.g., Kara-Murza, 1958; Chalyshev, Varyukhina, and Molin, 1965), Madagascar (Goubin, 1965), Western Australia (Balme, 1963), and Queensland, Australia (Evans, 1966). The Queensland assemblages have been only briefly discussed, and the Russian material is still inadequately illustrated. It is nonetheless quite obvious that plant microfossil associations from Early Triassic marine sediments in widely separated parts of the world have unmistakable common characters.

In Western Canada, West Pakistan, Western Australia, and Madagascar they contain huge numbers of small acritarchs. This may also be true in Yugoslavia and North Africa, where de Jekhowsky (1961) has reported an abundance of *Veryhachium reductum* Deunff and *Micrhystridium* spp. in strata which he referred to the "Permo-Triassic." In the Russian Arctic, acritarchs, although frequently recorded, do not appear to be as abundant as they are elsewhere. The picture is not altogether clear, however, for many forms regarded by Russian workers as alete pollen grains may well be of nonvascular origin. The microscopic algal swarms of the early Scythian are a striking phenomenon, not known to be duplicated in any other part of the geological column, at least since the beginning of the Late Paleozoic. It is difficult to provide an explanation which does not involve modification of the entire marine environment. A worldwide variation in oceanic salinity is certainly the simplest mechanism to explain the sudden multiplication of a few primitive organisms which, on modern analogies, were likely to have a wide environmental tolerance.

Broad similarities are also apparent in the spores and pollen grains occurring in marine Scythian sediments. All contain species of *Taeniaesporites* and cavate "selaginelloid" spores, although as far as one can judge, the Canadian and Russian assemblages are somewhat more diverse than those from the Salt Range and Western Australia. As a generalization, however, it is true to say that none of the Scythian assemblages so far described contains a notable variety of form species.

In places where marine Scythian strata overlie marine Upper Permian deposits, the Permian-Triassic boundary is apparently marked by a strong or complete palynological break. Among the localities so far studied, this break is most obvious in Western Australia and West Pakistan; but, even allowing for the incomplete data, it is also clear-cut in Madagascar and Arctic Russia. On the other hand, at localities where the Lower Triassic is represented by nonmarine deposits, changes in the spore assemblages are both less evident and of a different kind. In Middle Europe, Klaus (1964, 1965) is able to distinguish between Upper Permian and Lower Triassic sediments on a palynological basis, but the break at the boundary is not nearly as sharp as it is in the Salt Range. Of the 18 species taken by Klaus (1965) to characterize the Upper Permian, only three do not occur in the Triassic, although a further six are rare above the top of the Permian. In addition, the early Triassic microfloras of Middle Europe contain genera such as *Triadispora, Illinites, Voltziaceaesporites,* and *Microcachryidites* which are not obviously represented in contemporaneous assemblages from other areas. Perhaps in Western Europe floral modifications at the Permian-Triassic boundary were even less significant, for Freudenthal (1964) has described, from the Upper Bunter of Holland, an assemblage, made up almost entirely of striatitid pollen grains, which resembles quite closely German Zechstein microfloras. Indeed, from the available palynological evidence in Europe, the end of the Lower Triassic marks a more critical floral change than the beginning of the Mesozoic (cf. Reinhardt, 1964).

Continental sediments of Early Triassic age overlie the Upper Permian in eastern Australia, and here too no sharp microfloral break separates the two systems (Hennelly, 1958; Evans, 1966). In the Bowen Basin of Queensland Evans recognizes a Lower Triassic biostratigraphic unit (TR 2a) based on the presence of acritarchs,

Taeniaesporites, Densoisporites playfordi, and *Lundbladispora*; but he believes this to lie above two other Triassic palynological zones, one of which contains an abundance of *Alisporites* (= *Falcisporites*) *australis* de Jersey. The existence of unit TR2a of Evans in marine Triassic strata in Queensland emphasizes the probability that the *Taeniaesporites-Densoisporites-Lundbladispora* association is dependent as much on environmental as on strictly chronological factors.

Although it cannot be doubted that Triassic floras were distinct from those of the Permian, one is not justified, on the basis of the palynological data, in assuming massive regional extinctions at the Permian-Triassic boundary. Probably the establishment of typically Triassic floras was a less abrupt process and not complete until towards the end of Scythian time. If so, spore and pollen assemblages in marine early Scythian strata must represent specialized floras which colonized coastal areas at the end of the Permian. Their unusually high content of probable lycopodiaceous spores lends support to the view that marsh plants are strongly represented in the floras. It is still necessary to explain why the rapid spread of such specialized coastal floras was a world-wide phenomenon, and how the spores and pollen grains of upland floras were effectively screened out of marine deposits. The answer may partly lie in the exceptional paleogeographic situation which existed at the end of the Permian, when the withdrawal of the seas from the continental platforms was more complete than at any previous period of Phanerozoic time. Continental areas would then have been fringed with wide expanses of relatively low-lying land on which specialized communities could establish themselves. Because of the unusual breadth of these Early Triassic coastal belts, elements of their contemporaneous upland floras were scarcely represented in marine deposits of the time.

LATE SCYTHIAN AND MIDDLE TRIASSIC ASSEMBLAGES

The stratigraphically highest sample examined from the Mittiwali Member came from near the middle of the unit at Nammal Gorge, that is, from a horizon which probably lies within the Flemingitan division of Spath (1934, p. 27). Eleven samples from the Narmia Member yielded plant microfossils; five of these came from Landa Nala, and six from Nammal Gorge. Although not taken at closely spaced stratigraphic intervals, the samples at both localities were fairly representative of the total thickness of the unit. In a sense, therefore, palynological information on the Narmia Member is more complete than that from any other stratigraphic unit in the Salt Range. Kummel (1966) has reviewed in detail paleontological evidence bearing on the age of the Narmia Member and concludes that it is of late Scythian age, encompassing Spath's Prohungaritan division.

No invertebrate fossils occur in the Tredian Formation, which overlies the Narmia Member in the area to the west of Nammal Gorge. Fragmentary plants have been described from the unit (Sitholey, 1943; Pant, 1949), and the presence of megaspores, small spores, and pollen grains has also been reported (Pant and Srivastava, 1964). Kummel (1966) tentatively referred the Tredian Formation to the Middle Triassic, and there is little doubt, after assessing the palynological data, that this is correct. Seven samples from the Landa Member of the Tredian Formation have been studied; four from Nammal, two from Zaluch, and one from the base of the Member at Landa Pusha. Two samples only, both from Zaluch, were available from the Khatkiara Member, but these provided unusually rich and well-preserved assemblages.

Important changes take place in the plant microfossil assemblages either at, or not far below, the base of the Narmia Member. The chief of these are shown diagrammatically in Figures 18 and 21. Figure 18 illustrates certain broad quantitative trends and shows that in the Narmia Member acritarchs are generally less abundant than they are in the two lower units of the Mianwali Formation. This decline in the number of acritarchs is more marked in the Nammal section, where, except in the upper 20 feet or so of the section, they make up between 20 to 30 percent of the total assemblage. At Landa, on the other hand, spinose acritarchs and leiospheres still occur, composing between 50 to 80 percent of the total microfossil suite. Such a trend would be expected from Kummel's observation that the proportions of carbonate and shale increased in the Narmia Member towards the Surghar Range. Acritarcha are less common again (15 to 20 percent) in the Landa Member of the Tredian Formation, but they are sufficiently numerous to throw strong doubt on the assumption that the Landa Member is nonmarine. No acritarchs were found in the Khatkiara Member, however, and this fact, together with its rich suite of spores and pollen grains, indicates that the marine sedimentation initiated in early Scythian times ended with the deposition of the Landa Member.

Figure 21 shows variations in abundance of the spore and pollen species which most obviously reflect the

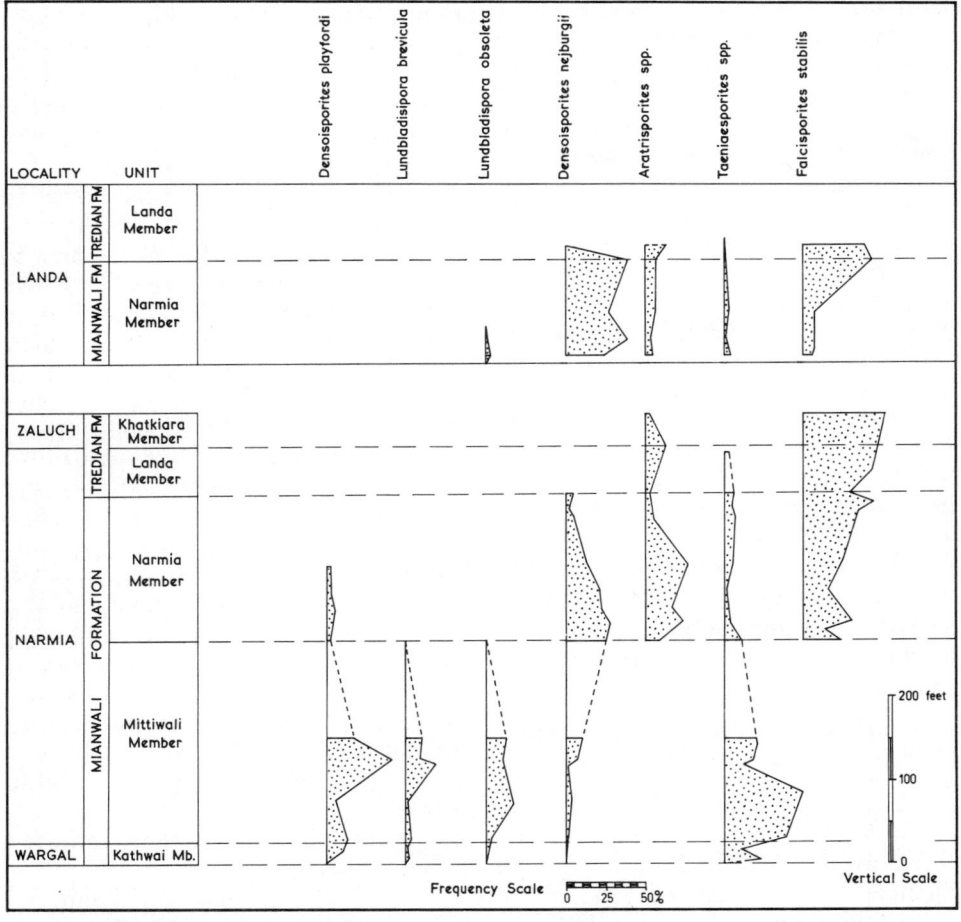

FIG. 21. Composite diagram showing quantitative variations in distribution of certain biostratigraphically important spore and pollen species in Triassic strata in the Salt Range and Surghar Range.

late Scythian floral changes. Lycopodiaceous spores are prominent in assemblages from the Narmia Member, but the common species is *Densoisporites nejburgii,* which was not found in the older Scythian sediments. *Densoisporites playfordi, Lundbladispora brevicula,* and *L. obsoleta,* the most abundant pteridophyte spores in the Mittiwali Member, are very minor components of assemblages from the Narmia Member. *Alisporites* sp. cf. *A. opii, Alisporites landianus, Sulcatisporites institatus, Verrucosisporites narmianus, Platysaccus queenslandi,* and *Cyclogranisporites arenosus* are species which occur first at the base of the Narmia Member, so that the microfloral break within the Scythian is manifested rather by the appearance of new forms than the sudden disappearance of old. Unquestionably, the most significant microfloral modifications result from the sudden appearance in large numbers of *Aratrisporites,* a lycopodialean spore of unusual morphology, and *Falcisporites stabilis,* which is almost certainly of corystospermaceous origin. The latter species, as we have seen, occurred rarely in the upper part of the Chhidru Formation, and it is interesting that other forms known from the Permian (e.g., *Tigrisporites playfordi, Nevesisporites,* and *Ephedripites* sp.) reappear in the Middle Triassic, although they were not found in the Scythian. This lends a little further support to the suggestion that elements of the Permian flora, although apparently subsidiary ones, survived into the Triassic as members of Scythian upland plant communities.

Significance of Late Scythian and Middle Triassic Assemblages

The *Taeniaesporites-Densoisporites-Lundbladisporites* association encountered in the Kathwai and Mittiwali members was comparatively short-lived, surviving only about half the Early Triassic age. Most elements of these spore assemblages range at least into the early Middle Triassic, but as strictly minor components of very different assemblages. There can be little doubt, from the prevalence of *Falcisporites,* that the corystosperms were important dominants in the Middle Triassic floras of the Salt Range. In this, and other respects, they resemble their contemporaneous floras in Australia, the only country in the Southern Hemisphere from which more than sketchy palynological data are available from the Triassic. Indeed, the succession of spore assemblages in the Salt Range Triassic strikingly resembles that in the Perth Basin, Western Australia (Balme, 1963, 1964). Here the marine Scythian Kockatea Shale, which contains a *Taeniaesporites-Densoisporites-Lundbladispora* association, is succeeded transitionally by a continental unit, the Woodada Formation. Spore assemblages from the Woodada Formation await description, but they contain *Falcisporites, Tigrisporites playfordi, Osmundacidites, Aratrisporites, Platysaccus queenslandi,* and *Densoisporites nejburgii,* all important components of assemblages from strata that occupy an analogous stratigraphic position in the Salt Range.

Middle Triassic spore assemblages from Western and Central Europe appear to reflect marked regional differentiation of their parent floras. They contain unusual saccate genera that have not been yet reported from other parts of the world (cf. Klaus, 1964; Reinhardt and Schmitz, 1965; Mädler, 1964) and in their gymnosperm component bear no close resemblance to the Salt Range assemblages. *Falcisporites zapfei,* which is morphologically similar to *F. stabilis,* is virtually confirmed to the Permian in Middle Europe, although it occurs in the Upper Bunter in Western Europe (Freudenthal, 1964).

Similarly assemblages from the Olenekian (upper Scythian) and Anisian of Arctic Russia show no obviously strong resemblances to those from the Salt Range. However, firm assertions concerning possible relationships between the Russian and Pakistani assemblages are impossible on the basis of published material at present available.

An undoubted link between Triassic floras from the Northern and Southern Hemispheres is provided by the unusual genus *Aratrisporites*, which is known to be produced by a lycopodiaceous group. *Aratrisporites* was first described from the Keuper of Switzerland (Leschik, 1955) and has been subsequently reported from many other European localities, from the USSR under a variety of generic assignations, and from all Australian States in which Triassic strata are known to be developed. Clearly the various species of *Aratrisporites* are potentially of considerable biostratigraphic importance, and the first appearance of the genus may, for practical purposes, mark a time plane which transcends broad regional floral patterns.

Aratrisporites is known from the Induan Stage of the Kolvinsk Arch in northern Russia, under the name *Zonomonoletes tschalyschevii* Varyukhina (Chalyshev, Varyukhina, and Molin, 1965). It ranges into the Middle and Upper Triassic, although it was not reported from the Olenekian by Chalyshev and Varyukhina (1962). In the Aktyubinsk district of the southern cis-Ural, however, Koptyova (1963) found a species of *Aratrisporites* (*Zonomonoletes spinosus* Koptyova) to characterize the Upper Triassic.

Schulz (1965) states that the range of *Aratrisporites* in Thuringia is Upper Buntsandstein to Keuper, but it seems to be uncommon in strata lying below the base of the Muschelkalk (Mädler, 1964). In Europe it seems not to be prominent below the Keuper, and its distribution appears to be sporadic. It was not, for example, recorded by Clarke (1965a) in his Keuper assemblages from Worcestershire in England.

Evans (1966), discussing the succession of spore assemblages in the Mesozoic of eastern Australia, regards the first appearance of *Aratrisporites* as characterizing his biostratigraphic unit TR2b. This occurs at the top of the Collaroy Claystone in the Sydney Basin, and near the top of the Rewan Formation in the Bowen Basin, Queensland. Both these units are accepted as late Early Triassic, although there is no unequivocal paleontological evidence to support such a dating. Various species of *Aratrisporites* were described by Playford (1965) from the Tiers Formation and younger Triassic strata in Tasmania. Again, the Tiers Formation cannot be independently dated, but there is no doubt that it is mainly Middle Triassic, for it overlies the Ross Formation which in turn rests on Upper Permian strata. Rare specimens of *Aratrisporites* occur in the uppermost part of the Kockatea Shale in the Perth Basin, Western Australia, and the genus is common in the overlying Woodada Formation. It occurs, too, in abundance in strata correlated with the Erskine Sandstone in the Canning Basin, Western Australia, although it has not been identified in the underlying Scythian Blina Shale.

Aratrisporites was not described from the Triassic of Madagascar (de Jekhowsky and Goubin, 1964) or the Late Triassic of Antarctica (Norris, 1965), but it would be premature to assume that it is not present in these areas.

In summary, there is a clear possibility that the first appearance of *Aratrisporites* marks a time plane of great significance in the inter-continental correlation of Triassic strata. Such a plane may lie in the Induan Stage of Russian workers and between the Gyronitan and Prohungarites divisions of Spath. It is also true, however, that the palynology of the Triassic reveals a phytogeographic pattern of enigmatic complexity.

REFERENCES

Abramova, S. A., and Marchenko, O. F., 1960, Materialy palynologicheskogo izucheniia podsolevoi tolshchi Kungurskikh otlozhenii verkhnekamskogo Kaliinogo mestorozhdeniia: Trudy Vses. Nauchno-issled. Inst. Galurgii, no. 40, p. 337-369.

——, 1964, K voprosy o klassifikatsii form roda *Vittatina* Lub. i form s rebristym telom i vozdushnymi meshkami: Izd. "Nauka," Inst. Geol. Geofiz. sib. Otd., p. 49-52.

Allen, K. C., 1965, Lower and Middle Devonian spores of North and Central Vestspitzbergen: Palaeontology, v. 8, p. 687-748.

Alpern, B., et al., 1958, Description de quelques microspores du Permocarbonifère français: Rev. Micropaleon., v. 1, p. 75-86.

Andreeva, M. O., et al., 1956, Atlas rukovodiashchikh form iskopaemykh fauny i flory permskikh otlozhenii kuznetskogo basseina: Moscow: Gos. nauchno-tekhn. izd-vo lit.-ry po geologii i okrane nedr., 409 p.

Aristova, K. E., 1963, Sporovo-pyl'tzevye kompleksy ez yurskikh i triasovykh otlozheniy vostochno-iliyskoy vpadiny yugo-vostochnogo Kazachstana: Trudy Vses. Nauchno-issled. Geol.-razv. Inst., v. 37, p. 89-92.

Arkell, W. J., 1956, The Jurassic Geology of the World: Oliver and Boyd; Edinburgh, xv + 806 p.

Balme, B. E., 1957, Spores and pollen grains from the Mesozoic of Western Australia: Comm. Sci. Indust. Research Organiz. Australia, Coal Res. Sect., T. C. 25, 48 p.

——, 1960, Some palynological evidence bearing on the development of the *Glossopteris*-Flora: in Leeper, G. W., ed., The Evolution of Living Organisms: Melbourne Univ. Press, p. 269-280.

——, 1963, Plant Microfossils from the Lower Triassic of Western Australia: *Palaeontology*, v. 6, p. 12-40.

——, 1964, The palynological record of Australian pre-Tertiary Floras: in Cranwell, L., ed., Ancient Pacific Floras: Univ. Hawaii Press, p. 49-80.

——, and Hassell, C. W., 1962, Upper Devonian spores from the Canning Basin, Western Australia: Micropaleontology, v. 8, p. 1-28.

——, and Hennelly, J. P. F., 1955, Bisaccate sporomorphs from Australian Permian coals: Austral. Jour. Botany, v. 3, p. 89-98.

——, and ——, 1956a, Monolete, monocolpate and alete sporomorphs from Australian Permian sediments: Same, v. 4, p. 54-67.

——, and ——, 1956b, Trilete sporomorphs from Australian Permian sediments: Same, v. 4, p. 240-260.

——, and Playford, G., 1967, Late Permian plant microfossils from the Prince Charles Mountains, Antarctica: Rev. Micropaléont., v. 10, p. 179-92.

——, and Segroves, K. L., 1966, *Peltacystia* gen. nov., a microfossil of uncertain affinities from the Permian of Western Australia: Jour. Royal Soc. West. Australia, v. 49, p. 26-31.

Barss, M. S., Hacquebard, P. A., and Howie, R. D., 1963, Palynology and stratigraphy of some Upper Pennsylvanian and Permian rocks in the Maritime Provinces: Pap. Geol. Survey Can., No. 63-3, 13 p.

Bharadwaj, D. C., 1962, The miospore genera in the coals of Raniganj Stage (Upper Permian), India: Palaeobotanist, v. 9, p. 68-106.

——, 1963, Pollen grains of *Ephedra* and *Welwitschia* and their probable fossil relatives: Mem. Indian Botan. Soc., No. 4, p. 125-135.

——, 1964, *Potonieisporites* Bhard., ihre Morphology, Systematik, und Stratigraphie: Fortschr. Geologie Rheinland. u. Westfalen, v. 12, p. 45-54.

——, and Salujha, S. K., 1964, Sporological study of Seam VIII in Raniganj Coalfield, Bihar (India). Part 1—Description of Sporae dispersae: Palaeobotanist, v. 12, p. 181-215.

——, and ——, 1965a, A sporological study of Seam VII (Jote Dhemo Colliery) in the Raniganj Coalfield, Bihar (India): Same, v. 13, p. 30-41.

——, and ——, 1965b, Sporological study of Seam VIII in Raniganj Coalfield, Bihar (India). Part 2—Distribution of *sporae dispersae* and correlation: Same, v. 13, p. 57-73.

——, and Singh, H. P., 1964, An Upper Triassic miospore assemblage from the coals of Lunz, Austria: Same, v. 12, p. 28-44.

Bhardwaj, D. C., 1954, Einige neue Sporengattungen des Saarkarbons: Neues Jahrb. Geologie Palaeontologie, Monatsh., v. 11, p. 512-525.

——, 1955, The spore genera from the Upper Carboniferous coals of the Saar and their value in stratigraphical studies: Palaeobotanist, v. 4, p. 119-149.

Bolkhovitina, N. A., 1953, Sporovo-pyl'tsevaia kharakteristika melovykh otlozhenii tsentral'nykh oblastei: Trudy Inst. Geol. Nauk. Moscow, No. 145, geol. ser. No. 61, 184 p.

——, 1956, Atlas spor i pyl'tsy iz yurskikh

i nizhnemelovykh otlozheniy vilyuyskoy vpadiny: Same, v. 2, 185 p.

Brenner, G., 1963, The spores and pollen of the Potomac Group of Maryland: Bull. Maryland Dept. Geology Mines, v. 27, 215 p.

Briche, P., Danzé-Corsin, P., and Laveine, J. P., 1963, Flore infraliasique du Boulonnais (Macro- et Microflore): Mém. Soc. Géol. Nord, v. 13, 143 p.

Chaloner, W. G., 1961, Palaeo-ecological data from Carboniferous spores, in Recent Advances in Botany: Univ. Toronto Press, v. 2, sect. 10, p. 980-983.

———, and Clarke, R. F. A., 1962, A new British Permian spore: Palaeontology, v. 4, p. 648-652.

Chalyshev, V. I., and Varyukhina, L. M., 1962, Stratigrafiya i sporovo-pyl'tsevye kompleksy verkhnepermskikh i triasovykh otlozheniy pechorskogo urala i gryady chernysheva: Trudy Komi Fil. Akad. Nauk, SSSR., v. 10, p. 49-58.

———, and ———, 1963, Stratigrafiya i sporovo-pyl'tsevye kompleksy verkhnetatarskikh i triasovykh otlozhenij kolvinskogo svoda: Trudy In-ta Geol. Komi Fil. Akad. Nauk, SSSR., v. 3, p. 78-96.

———, ———, and Molin, V. A., 1965, Granitsa permi i triasa v krasnotsvetiykh otlozheniiakh severnogo priural'ia: Komi Fil. Akad. Nauk SSSR., Izd. "Nauka," Moscow, Leningrad.

Chudinov, P. K., 1965, New facts about the fauna of the Upper Permian of the USSR: Jour. Geology, v. 73, p. 117-130.

Clarke, R. F. A., 1965a, Keuper miospores from Worcestershire, England: Palaeontology, v. 8, p. 294-321.

———, 1965b, British Permian saccate and monosulcate miospores: Palaeontology, v. 8, p. 322-354.

Couper, R. A., 1953, Upper Mesozoic and Cainozoic spores and pollen grains from New Zealand: Palaeont. Bull. Geol. Survey New Zealand, No. 22, 77 p.

———, 1958, British Mesozoic microspores and pollen grains: Palaeontographica, v. 103B, p. 75-179.

Cranwell, Lucy R., 1961, Coniferous pollen types of the southern hemisphere. I Aberration in *Acmopyle* and *Podocarpus dacrydioides*: Jour. Arnold Arbor., v. 42, p. 416-423.

Daugherty, L. H., 1941, The Upper Triassic flora of Arizona: Publ. Carnegie Inst., No. 526, 108 p.

de Jekhowsky, B., 1961, Sur quelques hystrichosphères permo-triasiques d'Europe et d'Afrique: Rev. Micropaléont., v. 3, p. 207-212.

———, and Goubin, N., 1964, Subsurface palynology in Madagascar: a stratigraphic sketch of the Permian, Triassic and Jurassic of the Morondava Basin, in Cross, A. T., ed., Palynology in Oil Exploration": Spec. Pap. Soc. Econ. Paleontologists and Mineralogists, No. 11, p. 116-130.

de Jersey, N. J., 1959, Jurassic spores and pollen grains from the Rosewood Coalfield: Queensland Gov. Mining Jour., v. 60, p. 346-366.

———, 1962, Triassic spores and pollen grains from the Ipswich Coalfield: Publ. Geol. Survey Queensland, No. 307, 18 p.

———, 1964, Triassic spores and pollen grains from the Bundamba Group: Same, No. 321, 21 p.

———, and Hamilton, M., 1967, Triassic spores and pollen from the Moolayember Formation: Publ. Geol. Survey Queensland, Palaeont. Pap., No. 10, p. 1-61.

———, and Paten, R. J., 1964, Jurassic spores and pollen grains from the Surat Basin: Publ. Geol. Survey Queensland, No. 322, 18 p.

Dettmann, Mary E., 1963, Upper Mesozoic microfloras from south-eastern Australia: Proc. Royal Soc. Victoria, v. 77, p. 1-148.

Doludenko, M. P., 1960, O stroyenii iskopayemykh spor *Coniopteris*: Dokl. Akad. Nauk, SSSR., v. 130, p. 627-629.

Doubinger, J., and Rakotoarivelo, H., 1960, Étude palynologique de quelques échantillons de houille du bassin de la Sakoa (Madagascar): C. R. Acad. Sci. [Paris], v. 251, p. 2758-2760.

Edgell, H. S., 1965, Techniques in the recovery of spores and pollen from surface samples: Ann. Prog. Rep. Geol. Survey West. Australia, 1964, p. 68-71.

Erdtman, G., 1957, Pollen and Spore Morphology/Plant Taxonomy. Gymnospermae, Pteridophyta, Bryophyta (illustrations): Almqvist and Wiksell, Stockholm, 151 p.

———, 1965, Pollen and Spore Morphology/ Plant Taxonomy. Gymnospermae, Bryophyta (Text): Almqvist and Wiksell, Stockholm, 191 p.

Evans, P. R., 1966, Mesozoic stratigraphic palynology in Australia: Austral. Oil Gas Jour., v. 12, p. 58-63.

Faegri, K., and Iversen, J., 1964, Textbook of Pollen Analysis, 2nd revised edit.: Hafner, New York, 161 p.

Fischer, A. G., 1964, Brackish oceans as the cause of the Permo-Triassic marine faunal crisis: in Nairn, A. E. M., ed., Problems in Palaeoclimatology: Interscience, London, p. 566-74.

Freudenthal, T., 1964, Palaeobotany of the Mesophytic. I. Palynology of Lower Triassic rock salt, Hengelo, the Netherlands: Acta botan. neerl., v. 13, p. 209-236.

Funkhouser, J. W., and Evitt, W. R., 1959, Preparation techniques for acid insoluble microfossils: Micropaleontology, v. 5, p. 369-375.

Ghosh, A. K., and Bose, A., 1947, Occurrence of microflora in the Salt pseudomorph beds, Salt Range, Punjab: Nature, v. 160, p. 796-797.

———, and ———, 1951, Permo-Triassic microflora in the Punjab Salt Range (abst.): Proc. Indian Sci. Congr., 38th, Bangalore, pt. 3, p. 126-127.

———, and ———, 1952, Spores and tracheids from the Cambrian of Kashmir: Nature, v. 169, p. 1056-1057.

———, and Sen, J., 1948, A study of the microfossils and the correlation of some productive coal seams of the Raniganj Coalfield, Bengal, India: Trans. Mining Geol. Metall. Inst. India, v. 43, p. 67-95.

———, ———, and Bose, A., 1951, Evidence bearing on the age of the Saline Series in the Salt Range of the Punjab: Geol. Mag., v. 88, p. 129-132.

Goubin, N., 1965, Description et répartition des principaux pollenites Permiens, Triasiques et Jurassiques des sondages du Bassin de Morondava (Madagascar): Revue Inst. française Pétrole, v. 20, p. 1415-1461.

Gray, J., and Browning, J. L., 1959, Caytonialean microspores from the Jurassic and Cretaceous of Alaska: Bull. Geol. Soc. America, v. 70, p. 1722.

Grebe, H., and Schweitzer, H-J., 1962, Die Sporae dispersae des niederrheinischen Zechsteins: Vorausdk. Fortschr. Geol. Rheinld. Westfalen, v. 10, p. 1-24.

Harris, T. M., 1964, The Yorkshire Jurassic Flora 2 Caytoniales, Cycadales and Pteridosperms: British Museum (Natural History), London, 191 p.

Harris, W. F., 1955, A manual of the spores of New Zealand Pteridophyta: Bull. New Zealand Dep. scient. indust. Research, No. 116, 186 p.

Hart, G. F., 1960, Microfloral investigation of the Lower Coal Measures (K2); Ketewaka-Mchuchuma Coalfield, Tanganyika: Bull. Geol. Survey Tanganyika, No. 30, 18 p.

———, 1963, A probable pre-*Glossopteris* microfloral assemblage from Lower Karroo sediments: South African Jour. Sci., v. 59, p. 135-146.

———, 1964, A review of the classification and distribution of the Permian miospore: *Disaccate Striatiti*: C. R. Ve Congr. internatl. Strat. Géol. Carb. Paris, p. 1171-1199.

———, 1965a, Microflora from the Ketewaka-Mchuchuma Coalfield, Tanganyika: Bull. Geol. Survey Tanganyika, No. 36, p. 1-27.

———, 1965b, The Systematics and Distribution of Permian Miospores: Witwatersrand Univ. Press, Johannesburg, 253 p.

———, 1966, *Vittatina africana*, a new miospore from the Lower Permian of South Africa: Micropaleontology, v. 12, p. 37-42.

Hedlund, R. W., 1965, Palynological assemblage from the Permian Wellington Formation: Oklahoma Geology Notes, v. 25, p. 236-241.

Helby, R., and Martin, A. R. H., 1965, *Cylostrobus* gen. nov., cones of lycopsidean plants from the Narrabeen Group (Triassic) of New South Wales: Austral. Jour. Botany, v. 13, p. 398-404.

Hemer, D. O., 1965, Application of palynology in Saudi Arabia: Fifth Arab Petroleum Congress, Cairo, 1965, 29 p.

Hennelly, J. P. F., 1958, Spores and pollen from a Permian-Triassic transition, N. S. W.: Proc. Linn. Soc. New South Wales, v. 83, p. 363-369.

Hill, D., Playford, G., and Woods, J. T., 1965, Triassic Fossils of Queensland: Queensland Palaeont. Soc., Brisbane, 32 p.

———, and Woods, J. T., eds., 1964, Permian Index Fossils of Queensland: Queensland Palaeont. Soc., Brisbane, 32 p.

Hughes, N. F., 1961, Further interpretation of *Eucommiidites* Erdtman 1948: Palaeontology, v. 4, p. 292-299.

———, 1964, Einige Vorschläge zur Angabe der Daten und der Klassifikation in der Sporologie: Fortschr. Geologie Rheinland. u. Westfalen, v. 12, p. 39-44.

Ibrahim, A. C., 1933, Sporenformen des Aegirhorizonts des Ruhr-Reviers: K. Triltsch., Würzburg, 46 p.

Imgrund, R., 1960, Sporae dispersae des Kaipingsbeckens, ihre paläontologische und stratigraphische Bearbeitung im Hinblick auf eine Parallelisierung mit dem Ruhrkarbon und dem Pennsylvanian von Illinois: Geol. Jahrb., v. 77, p. 143-204.

Jacob, K., Jacob, C., and Srivastava, P. N., 1953, Evidence for the existence of vascular land plants in the Cambrian: Current Sci., v. 22, p. 34-36.

Jansonius, J., 1962, Palynology of Permian and Triassic sediments, Peace River area, Western Canada: Palaeontographica, v. 110B, p. 35-98.

Jizba, Katherine M. M., 1962, Late Paleozoic bisaccate pollen from the United States midcontinent area: Jour. Paleontology, v. 36, p. 871-887.

Kara-Murza, E. N., 1958, Sporovo-pyl'tsevje kompleksy triasovylkh otlozhenij v rayone mysa Tvetkova: Sb. Stat. Paleont. Biostrat. nauchno-issled. Inst. Geol. Arktiki, v. 8, p. 33-62.

Khlonova, A. F., 1960, Vidovoy sostav pyl'tsy i spor v otlozheniyakh verkhnego mela chulymoeniseyskoy vpadiny: Trudy Inst. Geol. Geofiz. sib. Otd., No. 3, 104 p.

———, 1961, Spory i pyl'tsa verkhney poloviny verkhnego mela vostochnoy chasti zapadno siberskoy nizmennosti: Trudy Inst. Geol. Geofiz. sib. Otd., No. 7, 138 p.

Kidston, R., 1923, Fossil plants of the Carboniferous rocks of Great Britain: Mem. Geol. Survel U.K., Palaeontology, v. 2, pt. 4, p. 275-376.

Klaus, W., 1953, Alpine Salzmikropaläontologie (Sporendiagnose): Paläont. Zeitsch., v. 27, p. 52-56.

———, 1955, Über die Sporendiagnose des deutschen Zechsteinsalzes und des alpinen Salzgebirges: Zeitsch. Deutsche Geol. Gesellsch., v. 105, p. 776-788.

———, 1960, Sporen der karnischen Stufe der Ostalpinen Trias: Sonderbd. Jahrb. Geol. Bundesanst., Wien, v. 5, p. 107-183.

———, 1963, Sporen aus dem südalpinen Perm.: Jahrb. Geol. Bundesanst. Wien, v. 106, p. 229-363.

———, 1964, Zur sporenstratigraphischen Einstufung von gipsführenden Schichten in Bohrungen: Erdöl-Zeitsch. Bohr-u. Fördertechnik, v. 4, p. 119-132.

———, 1965, Zur Einstufung alpiner Salztone mittels Sporen: Zeitsch. Deutsche Geol. Gesellsch., v. 116, p. 288-292.

Koptyova, E. A., 1963, Stratigrafiya i sporovopyl'tsevye kompleksy Triasvykh otlozhenii bassiia R. Ilek (Aktyubinskoe Priural'e): Trudy Vses. nauchno-issled. Geol.-rasv. Inst., v. 37, p. 77-88.

Kosanke, R. M., 1950, Pennsylvanian spores of Illinois and their use in correlation: Bull. Illinois State Geol. Survey, v. 74, 128 p.

Kremp, G. O. W., 1965, Morphologic encyclopedia of palynology: an international collection of definitions and illustrations of spores and pollen: Univ. Arizona Press, Tucson, xiii + 185 p.

Krishnan, M. S., 1963, Geology of the salt deposits in the Punjab Salt Range, Pakistan: Spec. Pap., Geol. Soc. America, No. 73, p. 276-277.

Krutsch, W., 1961, Über Funde von ephedroiden Pollen im deutschen Tertiär: Beihefte Geologie, v. 10, No. 32, p. 15-53.

Kummel, B., 1966, The Lower Triassic formations of the Salt Range and Trans-Indus ranges, West Pakistan: Bull. Mus. Comp. Zoology Harvard Univ., v. 134, p. 361-429.

———, and Teichert, C., 1966, Relations between the Permian and Triassic formations in the Salt Range and Trans-Indus ranges, West Pakistan: Neues Jahrb. Geologie Paläontologie Abh., v. 125, p. 297-333.

———, and ———, 1970, Stratigraphy and paleontology of the Permian-Triassic boundary beds, Salt Range and Trans-Indus ranges, West Pakistan; in Kummel, B., and Teichert, C., eds., Stratigraphic Boundary Problems: Permian and Triassic of West Pakistan: Univ. Press Kansas, Dept. Geology, Univ. Kansas Spec. Publ. 4, p. 1-110.

Kuyl, O. S., Müller, J., and Waterbolk, H. T., 1955, The application of palynology to oil geology, with special reference to Western Venezuela: Geologie Mijnbouw, n. s., v. 17, p. 49-76.

Lakhanpal, R. N., Sah, S. C. D., and Dube, S. N., 1962, Further observations on plant microfossils from carbonaceous shale (Krols) near Naini Tal, with a discussion on the age of the beds: Palaeobotanist, v. 7, p. 111-120.

Lele, K. M., 1964, Studies in the Talchir flora of India: 2. Resolution of the spore genus *Nuskoisporites* Pot. and Kl.: Palaeobotanist, v. 12, p. 147-168.

Leschik, G., 1955, Die Keuperflora von Neuewelt bei Balse. 2. Die Iso-und Microsporen: Schweiz. Palaeont. Abh., v. 72, p. 5-70.

———, 1956, Sporen aus dem Salzton des Zechsteins von Neuhof (bei Fulda): Palaeontographica, v. 100B, p. 122-142.

———, 1959, Sporen aus den "Karru-Sandsteinen" von Norronaub (Südwest-Afrika): Senckenbergiana leth., v. 40, p. 51-95.

Luber, A. A., and Waltz, I. E., 1941, Atlas mikrospor i pyl'tsy paleozoya SSSR: Trudy Vses. Nauchno-issled. Geol. Inst., v. 139, 107 p.

Lundblad, B., 1948, A selaginelloid strobilus from East Greenland (Triassic): Medd. Dansk. Geol. Foren., v. 11, p. 351-363.

———, 1949, De geologiska resultaten fran borrningarna vid Höllviken. Del. 3: Microbotanical studies of cores from Höllviken, Scania: Sveriges Geol. Undersökning, Arsb. 43, No. 4, ser. C, No. 506, 16 p.

McGregor, D. C., 1964, Devonian miospores from the Ghost River Formation, Alberta: Bull. Geol. Survey Canada, v. 109, 31 p.

———, 1965, Triassic, Jurassic and Lower Cretaceous spores and pollen of Arctic Canada: Pap. Geol. Survey Canada, v. 64-65, 32 p.

Mädler, K., 1964, Die geologische Verbreitung von Sporen und Pollen in der deutschen Trias: Beihefte Geol. Jahrb., v. 65, 147 p.

Malyavkina, V. S., 1964, Spory i pyl'tsa iz Triasovykh otlozhenii zapadno-sibirskoi nizmennosti: Trudy Vses. Neft. Nauchnoissled. Geol.-razv. Inst., v. 231, 293 p.

Manum, S., 1960, On the genus *Pityosporites* Seward 1914. With a new description of *Pityosporites antarcticus* Seward: Nytt Mag. Botan., v. 8, p. 11-15.

Medvedeva, A. M., 1960, Stratigraficheskoe raschlenenie nizhnikh horizontov tunguskoi serii metodoin sporovo-pyl'tsevogo analiza: Izd. Akad. Nauk, SSSR., Inst. Geol.-razv. Gor. Isk, Moscow, 92 p.

Mehta, K. R., 1944, Microfossils from a carbonaceous shale from the Pali beds of the South Rewa Gondwana Basin: Proc. Indian Acad. Sci., v. 14B, p. 125-141.

Melville, R., 1966, Continental drift, Mesozoic continents and the migration of the angiosperms: Nature, v. 211, p. 116-120.

Nilsson, T., 1958, Über das Vorkommen eines mesozoischen Sapropelgesteins in Schonen: Acta Univ. Lundensis, n. s., pt. 2, v. 54, No. 10, 111 p.

Norris, G., 1965, Triassic and Jurassic miospores from the Beacon and Ferrar Groups, Victoria Land, Antarctica: New Zealand Jour. Geology Geophysics, v. 8, p. 236-277.

Orlowska-Zwolinska, Teresa, 1962, Pierwsze snalezisko sporomorf cechsztynskich w Polsce: Kwart. Geol., v. 6, p. 283-297.

Ouyang, S., 1962, The microspore assemblage from the Lungtan Series of Chunghsing Chekiang [Chinese, full English summary]: Acta Palaeont. Sinica, v. 10, p. 76-119.

――――, 1964, A preliminary report on sporae dispersae from the Lower Shihotse Series of Hoku District, N.W. Shansi [Chinese and English]: Same, v. 12, p. 486-519.

Pant, D. D., 1949, On some Triassic plant remains from the Salt Range in the Punjab: Nature, v. 163, p. 914.

――――, 1955, On two new disaccate spores from the Bacchus Marsh Tillite, Victoria (Australia): Ann. Mag. Nat. History, ser. 12, v. 8, p. 757-764.

――――, and Nautiyal, D. D., 1960, Some seeds and sporangia of *Glossopteris* flora from Raniganj Coalfield, India: Palaeontographica, v. 107B, p. 41-64.

――――, and Srivastava, G. K., 1964, Further observations on some Triassic plant remains from the Salt Range, Punjab: Same, v. 114B, p. 79-93.

――――, and ――――, 1965, Some Lower Gondwana miospores from Brazil: Micropaleontology, v. 11, p. 468-478.

Pautsch, Maria E., 1958, Keuper sporomorphs from Swierczyna Poland: Micropaleontology, v. 4, p. 321-325.

Peppers, R. A., 1964, Spores in strata of Late Pennsylvanian cyclothems in the Illinois Basin: Bull. Illinois State Geol. Survey, v. 90, 88 p.

Pettitt, J., and Chaloner, W. J., 1964, The ultra structure of the Mesozoic pollen *Classopollis*: Pollen et Spores, v. 6, p. 611-620.

Pierce, R. L., 1961, Lower Upper Cretaceous plant microfossils from Minnesota: Bull. Minnesota Geol. Survey, v. 42, 86 p.

Playford, G., 1965, Plant microfossils from Triassic sediments near Poatina, Tasmania: Jour. Geol. Soc. Australia, v. 12, p. 173-210.

――――, and Dettmann, Mary E., 1965, Rhaeto-Liassic plant microfossils from the Leigh Creek Coal Measures, South Australia: Senckenbergiana leth., v. 46, p. 127-181.

Pocock, S. A. J., 1964, Pollen and spores of the Chlamydospermidae and Schizaeaceae from Upper Manville strata of the Saskatoon area of Saskatchewan: Grana palynol., v. 5, p. 129-209.

Potonié, R., 1956, Synopsis der Gattungen der Sporae dispersae. 1. Teil: Sporites: Beihefte Geol. Jahrb., v. 23, 103 p.

――――, 1958, Synopsis der Gattungen der Sporae dispersae. 2. Teil: Sporites (Nachträge), Saccites, Aletes, Praecolpates, Monocolpates: Same, v. 31, 114 p.

――――, 1962a, Synopsis der Sporae in situ. Die Sporen der fossilen Fruktifikationen (Thallophyta bis Gymnospermophyta) im natürlichen System und im Vergleich mit der Sporae dispersae: Same, v. 52, 204 p.

――――, 1962b, Regeln, nach denen sich die Sekundärfalten der Sporen bilden: Paläont. Zeitsch., v. 36, p. 46-54.

――――, and Klaus, W., 1954, Einige Sporengattungen des alpinen Salzgebirges: Geol. Jahrb., v. 68, p. 517-544.

――――, and Kremp, G. O. W., 1954, Die Gattungen der paläozoischen Sporae dispersae und ihre Stratigraphie: Same, v. 69, p. 111-194.

――――, and ――――, 1955, Die *Sporae dispersae* des Ruhrkarbons, ihre Morphographie und Stratigraphie mit Ausblicken auf Arten anderer Gebiete und Zeitabschnitte; Teil 1: Palaeontographica, v. 98B, p. 1-136.

――――, and ――――, 1956, Die *sporae dispersae* des Ruhrkarbons ihre Morphographie and Stratigraphie mit Ausblicken auf Arten anderer Gebiete und Zeitabschnitte; Teil 2: Same, v. 99B, p. 85-191.

――――, and Lele, K. M., 1961, Studies in the Talchir Flora of India—1. *Sporae dispersae* from the Talchir Beds of South Rewa Gondwana Basin: Palaeobotanist, v. 8, p. 22-37.

――――, and Schweitzer, H-J., 1960, Der Pollen von *Ullmannia frumentaria:* Paläont. Zeitsch., v. 34, p. 27-39.

Rakotoarivelo, H., 1960, Étude palynologique de quelques échantillons de houille du Bassin de la Sakoa (Madagascar): Serv. Géol. Madagascar, Tanarive, 49 p.

Reed, F. R. C., Cotter, G. de P., and Lahiri, H. M., 1930, The Permo-Carboniferous succession in the Warcha Valley, western Salt Range, Punjab: Rec. Geol. Survey India, v. 62, p. 412-443.

Reinhardt, P., 1964, Über die *Sporae dispersae* der Thüringer Trias: Monatsber. Deutsche Akad. Wiss. Berlin, v. 6, p. 46-56.

――――, 1964, Einige Sporenarten aus dem

Buntsandstein Thüringens: Same, v. 6, p. 609-614.

———, and Schmitz, W., 1965, Zur Kenntnis der *Sporae dispersae* des mitteldeutschen Oberen Buntsandsteins: Freiberger Forschungeshefte, 1965, C182, Paläontologie, p. 19-36.

Reissinger, A., 1940, Die "Pollenanalyse" ausgedehnt auf alle Sedimentgesteine der geologischen Vergangenheit, 1: Palaeontographica, v. 84B, p. 1-20.

———, 1950, Die "Pollenanalyse" ausgedehnt auf alle Sedimentgesteine der geologischen Vergangenheit, 2: Same, v. 90B, p. 99-126.

Remy, R., and Remy, W., 1961, Beiträge zur Flora des Autunien IV: Monatsber. Deutsche Akad. Wiss. Berlin, V. 3, p. 489-502.

———, and ———, 1957, Durch Mazeration fertiler Farne des Paläozoikums gewonnene Sporen: Paläont. Zeitsch., v. 31, p. 55-65.

———, and ———, 1958, Die Sporen von *Dictyothalamus scrollianus* Göppert: Abh. Deutsche Akad. Wiss. Berlin, Kl. Chemie, Geologie, Biologie, 1958, No. 5, p. 3-6.

———, and ———, 1958, Beiträge zur Kenntnis der Rotliegendflora Thüringens: Sitzungber. Deutsche Akad. Wiss., Kl. Chemie, Geologie, Biologie, 1958, No. 3, 16 p.

Richardson, J. B., 1965, Middle Old Red Sandstone spore assemblages from the Orcadian Basin: Palaeontology, v. 7, p. 559-605.

Rilett, M. H. P., 1954, Plant microfossils from the coal seams near Dannhauser, Natal: Trans. Proc. Geol. Soc. South Africa, v. 57, p. 27-37.

Rogalska, M., 1956, Analiza sporowo-pylkowa liasowych osadow obszaru Mroczkow-Rozwady w powiecie Opoczynskim: Biul. Inst. Geol. Warsawa, v. 104, 89 p.

Romanskaya, G. M., 1963, Spory i pyl'tsa novykh vidor mezozoyskikh rasteniy turgayskovo progiba: Paleont. Zhurn., No. 1, 1963, p. 127-136.

Rouse, G. E., 1959, Plant microfossils from Kootenay coal-measures strata of British Columbia: Micropaleontology, v. 5, p. 303-324.

Sah, S. C. D., 1955, Plant microfossils from a Jurassic shale of Salt Range, West Punjab (Pakistan): Palaeobotanist, v. 4, p. 60-71.

Sahni, B., 1936, Wegener's theory of continental drift in the light of palaeobotanical evidence: Jour. Indian botan. Soc., v. 15, p. 319-332.

———, and Trivedi, B. S., 1943, Plant and animal remains from salt marl at Khewra in the Salt Range: Abs. Proc. Indian Acad. Sci., p. 25-26.

Samoilovich, S. R., 1953, Pyl'tsa i spory iz permskikh otlozheniy cherdynskogo i aktyubinskogo Priural'ya: Trudy Vses. nauchno-issled. Geol. razve. inst. n. ser., 75, Paleobot. Sb., p. 5-57.

Schaarschmidt, F., 1963, Sporen und Hystrichosphaerideen aus dem Zechstein von Büdingen in der Wetterau: Palaeontographica, v. 113B, p. 38-91.

Schindewolf, O. H., and Seilacher, A., 1955, Beiträge zur Kenntnis des Kambriums in der Salt Range (Pakistan): Abh. math.-naturw. Kl. Akad. Wiss. Mainz, v. 10, No. 10, p. 261-446.

Schopf, J. M., Wilson, L. R., and Bentall, R., 1944, An annotated synopsis of Paleozoic fossil spores and the definition of generic groups: Illinois Geol. Surv. Rept. Invest., no. 91, 66 p.

Schulz, E., 1962, Sporenpaläontologische Untersuchungen zur Rhät-Lias-Grenze in Thüringen und der Altmark: Geologie, v. 11, p. 308-319.

———, 1964, Sporen und Pollen aus dem Mittleren Butsandstein des germanischen Beckens: Monatsber. Deutsche Akad. Wiss. Berlin, v. 6, p. 597-606.

———, 1965, Sporae dispersae aus der Trias von Thüringen: Mitt. Zentr. Geol. Inst. Berlin, v. 1, p. 257-287.

Schurman, H. M. E., Burger, D., and Dijkstra, S. J., 1963, Permian near Wadi Araba eastern desert of Egypt: Geologie Mijnbouw, v. 42, p. 329-336.

Sellards, E. H., 1902, On the fertile fronds of *Crossotheca* and *Myriotheca*, and on the spores of other Carboniferous ferns, from Mazon Creek, Illinois: Am. Jour. Sci., 4th ser., v. 14, p. 195-202.

Sedova, M. A., 1956, Poryadok Coniferales. Mat-ry ro paleontologii Novyye semeystva i rody: Trudy Vses. Nauchno-issled. Geol. Inst., n. s., Paleont., Vyp. 12, p. 246-249.

Segroves, K. L., 1967, Cutinized microfossils of probable nonvascular origin from the Permian of Western Australia: Micropaleontology, v. 13, p. 289-305.

Selling, O. H., 1946, Studies in Hawaiian pollen statistics Part 1. The spores of the Hawaiian pteridophytes: Spec. Pub. Bernice Pauahi Bishop Mus., No. 37, 87 p.

Seward, A. C., and Smith-Woodward, A., 1905, Permo-Carboniferous plants and vertebrates from Kashmir: Mem. Geol. Survey India, Palaeont. indica, n. s., v. 2, No. 2, p. 1-13.

Shaffer, B. L., 1964, Stratigraphic and paleoecologic significance of plant microfossils in Permian evaporites of Kansas: *in* Cross, A. T., ed., Palynology in Oil Exploration, Spec. Pub. Soc. Econ. Paleontologists and Mineralogists, No. 11, p. 97-115.

Singh, H. P., 1964, A miospore assemblage from the Permian of Iraq: Palaeontology, v. 7, p. 240-265.

Sitholey, R. V., 1943, Plant remains from the Triassic of Salt Range in the Punjab: Proc. Indian Acad. Sci., v. 13, p. 300-327.

———, 1951, On the occurrence of two-winged pollen in the Triassic rocks of the Salt Range, Punjab: Current Sci., v. 20, p. 266.

Smith, A. H. V., and Butterworth, M. A., 1967, Miospores in the coal seams of the Carboniferous of Great Britain: Spec. Papers in Palaeontology, no. 1, Palaeontological Association, London, 324 p.

Spath, L. F., 1934, Catalogue of the Fossil Cephalopoda in the British Museum (Natural History). Part 4. The Ammonoidea of the Trias: British Museum (Nat. Hist.), London, 521 p.

Staplin, F. L., 1960, Upper Mississippian plant spores from the Golata Formation, Alberta, Canada: Palaeontographica, v. 107B, p. 1-40.

Stuhl, A., 1962, A Balatonfelvidek Perm Idöszaki Uledekeiben vegzett sporavizsgalatok ererd Magyar: Földtany Közl., v. 91, p. 405-412.

Sukh Dev, 1961, The fossil flora of the Jabalpur Series 3. Spores and pollen grains: Palaeobotanist, v. 8, p. 43-56.

Sullivan, H. J., 1964, Miospores from the Drybrook Sandstone and associated measures in the Forest of Dean Basin, Gloucestershire: Palaeontology, v. 7, p. 351-392.

———, and Neves, R., 1964, *Triquitrites* and related genera: C. r. Ve Congr. Internatn. Strat. Geol. Carb. Paris, p. 1079-1093.

Takhtadzhian, A. L., Vakhrameev, V. A., and Radchenko, G. P., eds., 1963, Osnovy Paleontologii-Spravochnik dla paleontologov i geologov SSSR., Golosemennye i Pokrytosemmye: Gos. Nauchno-tekhn. Izd-vo Lit-ry po Geologii i Okrane nedr., Moscow, 743 p.

Teichert, Curt, 1964, Recent German work on the Cambrian and Saline Series of the Salt Range, West Pakistan: Rec. Geol. Survey Pakistan, v. 11, p. 1-8.

———, 1966, Nomenclature and correlation of the Permian "Productus limestone," Salt Range, West Pakistan: Rec. Geol. Survey Pakistan, v. 15, p. 1-20.

Tiwari, R. S., 1964, New miospore genera in the coals of Barakar (Lower Gondwana) of India: Palaeobotanist, v. 12, p. 250-259.

Townrow, J. A., 1960, The Peltaspermaceae a pteridosperm family of Permian and Triassic age: Palaeontology, v. 3, p. 333-361.

———, 1962, On some disaccate pollen grains of Permian to Middle Jurassic age: Grana palynol., v. 3, p. 13-44.

———, 1965, A new member of the Corystospermaceae Thomas: Ann. Botany, v. 29, p. 495-511.

Tschudy, R. H., and Kosanke, R. M., 1966, Early Permian vesiculate pollen from Texas, U.S.A.: Palaeobotanist, v. 15, p. 59-71.

Venkatachala, B. S., Goubin, N., and Kar, R. K., 1967, Morphological study of *Guttulapollenites* Goubin: Pollen et Spores, v. 9, p. 357-362.

———, and Kar, R. K., 1964, Nomenclatural notes on *Striatopodocarpites* Sedova, 1956: Palaeobotanist, v. 12, p. 313-314.

———, 1966, *Corisaccites* gen. nov., a new saccate pollen genus from the Permian of Salt Range, West Pakistan: Palaeobotanist, v. 15, p. 107-109.

Virkki, C., 1937, On the occurrence of winged spores in the Lower Gondwana rocks of India and Australia: Proc. Indian Acad. Sci., sect. B, v. 6, p. 428-431.

———, 1939, On the occurrence of winged spores in a Lower Gondwana glacial tillite from Australia and in Lower Gondwana shales in India: Same, sect. B, v. 9, p. 7-12.

———, 1946, Spores from the Lower Gondwanas of India and Australia: Proc. Indian Acad. Sci., v. 15, p. 93-176.

Visscher, H., 1966, Palaeobotany of the Mesophytic. 3. Plant microfossils from the Upper Bunter of Hengelo, the Netherlands: Acta botan. neerl., v. 15, p. 316-375.

Wadia, D. N., 1934, The Cambrian-Trias sequence of N. W. Kashmir: Rec. Geol. Survey India, v. 68, p. 121-176.

Wilson, L. R., 1962, Plant microfossils Flowerpot Formation: Circ. Oklahoma Geol. Survey, No. 49, 50 p.

———, 1963, A study in the variation of *Picea glauca* (Moench) Voss pollen: Grana palynol., v. 4, p. 380-387.

———, and Venkatachala, B. S., 1963, Morphological variation of *Thymospora pseudothiessenii* (Kosanke): Oklahoma Geology Notes, v. 23, p. 125-132.

———, and ———, 1963b, A morphological study and emendation of *Vesicaspora* Schemel, 1951: Same, v. 23, p. 142-151.

———, and Webster, R. M., 1946, Plant microfossils from a Fort Union coal of Montana: Am. Jour. Botany, v. 33, p. 271-278.

Wodehouse, R. P., 1933, Tertiary pollen 2. The oil shales of the Eocene Green River Formation: Bull. Torrey botan. Club, v. 60, p. 479-524.

———, 1959, Pollen Grains: Hafner, New York, 574 p.

Zauer, V. V., 1960, O pozdne permskoy flore rayona Solikamska (po dannym sporovo-pyl'tsevogo analiza): Paleont. Zhurn., no. 4, p. 115-123.

———, 1965, Permskaia flora Solikamska:

Trudy Vses. Neft. Naucho-issled. Geol. Inst. 239, Palaeobot. Sb., p. 53-78.

Zoricheva, A. I., and Sedova, M. A., 1954, Sporovo-pyl'sevyie kompleksy verkhne- permskikh otlozhenii nekotorykh raionov severa evropeiskoi chasti SSSR.: Trudy Vses. Nauchno-issled Geol. Inst. Mosk. 1953, p. 160-201.

Indexes

[NOTE: The indexes were prepared by Penelope A. Carter and Jack D. Keim under the supervision of Curt Teichert]

I. AUTHOR INDEX

Abramova, S. A., 410, 432
Alberti, G., 295
Allen, K. C., 318
Andel, T. H. van, 54
Andreeva, M. O., 325, 364, 433
Aristova, K. E., 351
Arkell, W. J., 317
Arthaber, G. von, 144

Baksi, S. K., 293
Balme, B. E., 15, 21, 26, 27, 50, 69-72, 77, 277, 315, 321, 324, 337-339, 342, 362, 364, 367, 369, 371, 373, 382, 403, 408, 416, 427, 434, 440, 444
Barss, M. S., 435
Belousova, Z. D., 194
Bender, H., 222, 233, 239, 244, 245, 250, 253
Bentall, R., 316
Besairie, H., 173
Bharadwaj, D. C., 322, 324, 326, 332, 336, 361, 380, 381, 390, 395, 396, 403, 405, 406, 408, 409, 411
Bhardwaj, D. C., 345, 347, 351, 354, 358, 362, 365-368
Bielecka, W., 194
Bion, H. S., 179, 180
Bitterli, P., 289, 290, 297-298, 301
Bittner, A., 113, 114
Bolkhovitina, N. A., 334, 401, 409
Bose, A., 311
Böse, E., 157
Branson, C. C., 20, 69, 221
Brenner, G., 342
Briche, P., 338, 405, 431
Brosius, M., 289, 290, 297-298, 301
Browning, J. L., 382
Burger, D., 374
Butterworth, M. A., 326

Chaloner, W. J., 318, 410, 430
Chalyshev, V. I., 341, 351, 382, 394, 412, 427, 433, 440, 445
Chao, K.-K., 72, 76, 119, 143, 154, 183
Chave, K. E., 55
Chepilova, I. K., 277
Chernychev, T., See Tschernychew, T.
Chronic, J., 222
Chudinov, P. K., 435

Clark, D. L., 215, 222, 226, 239, 240, 249, 257, 258, 261, 266
Clarke, R. F. A., 355, 361, 366, 369, 371, 373, 376, 390, 398, 399, 410, 445
Collinson, C., 221, 222
Combaz, A., 299, 300
Cookson, I. C., 289, 301
Cooper, G. A., 53, 111, 118, 120, 121, 123-125, 214
Cotter, G. de P., 307
Couper, R. A., 322, 328, 335, 336, 346, 350, 387, 401, 413
Craig, G. Y., 112
Cramer, F. H., 293, 295, 297, 299, 300
Cranwell, L. R., 317
Crespin, I., 196

Dagis, A. S., 114
Danilchik, W., 5, 26
Danzé-Corsin, P., 338, 405, 431
Daugherty, L. H., 389, 390
Davidson, T., 6
Deffeyes, K. S., 55
Deflandre, G., 280, 286-289, 291, 293, 297
de Groot, K., 56
de Jersey, N. J., 331, 334, 335, 387, 390, 394, 398
Desio, A., 67
Dettmann, M. E., 318, 319, 336, 339, 342, 345-347, 351, 352, 379-381, 387, 390, 394, 398, 399, 431
Deunff, J., 286, 289-291, 293, 295, 299
Diener, C., 10-12, 131, 153, 157, 167, 172, 177-179, 181, 182, 185, 186, 189, 190, 209, 218
Dijkstra, S. J., 374
Dionne, J.-C., 42
Doludenko, M. P., 322
Doubinger, J., 439
Douglass, R. C., 63
Downie, C., 277, 287, 290, 293, 295, 297, 298, 300, 301
Dunbar, C. O., 30, 63, 117, 120-122, 157

Edgell, H. S., 315
Efremova, G. D., 277
Eisenack, A., 293, 297, 300
Ellison, S., 222, 223
Elphinstone, M., 5
Erdtman, G., 319, 400

Ethington, R. I., 223, 261
Evamy, B. D., 57
Evans, P. R., 344, 373, 440, 441, 445
Evitt, W. R., 315

Faegri, K., 320
Fager, E. W., 218, 219, 230, 237
Fairbridge, R. W., 55
Fantini Sestini, N., 128, 129, 144
Fell, H. B., 68
Fenton, C. L., 40
Fleming, A., 5, 6
Fliche, P., 40
Fox, C. S., 17
Frebold, H., 153, 157
Frech, F., 144
Frederiks, G., 145
Freudenthal, T., 372, 399, 441, 444
Funkhouser, J. W., 315
Furnish, W. M., 15, 21, 65, 72, 73, 77, 117-119, 122, 153-157, 161, 164, 165, 167, 170, 172, 184

Gansser, A., 5
Gee, E. R., 5, 12, 17, 22, 26, 27, 43, 60, 104, 105, 307, 310
Gemmellaro, G. G., 153
Gerke, A. A., 195
Ghosh, P. K., 311, 324
Ginsburg, R. N., 55
Girty, G. H., 121
Gitmez, G. U., 280, 295
Glaus, M., 129
Glenister, B. F., 15, 21, 65, 72, 73, 77, 117, 122, 154, 157, 165, 167
Goldberg, M., 57
Górka, H., 300
Goubin, N., 361, 368, 371-374, 376, 377, 386, 387, 401, 402, 410, 431, 435, 440, 445
Grabau, A. W., 12, 121, 148
Grant, R. E., 16, 21, 35, 45, 50, 53, 64-66, 72, 73, 75-77, 119, 121, 157, 158, 162, 213
Gray, J., 382
Grebe, H., 345, 369, 373, 432
Griesbach, C. L., 10, 179, 182, 185, 186, 189
Gründel, J., 196, 202, 203

Hacquebard, P. A., 435
Hamilton, M., 331

Haniel, C. A., 156, 157
Häntzschel, W., 40, 43
Harker, P., 121
Harris, T. M., 319, 335, 381, 385
Hart, G. F., 319, 324, 339, 354, 355, 361, 362, 364-367, 374, 377, 380, 395, 408, 410, 412
Hass, W. H., 221, 261
Hassell, C. W., 315
Hedlund, R. W., 380
Helby, R., 352
Hemer, D. O., 366, 380, 434
Hemphill, W. R., 4
Hennelly, J. P. F., 338, 362, 364, 367, 403, 408, 441
Heritsch, F., 63, 121, 122
Herrig, E., 202
Hess, H., 50, 67
Hill, D., 367, 380, 381
Hinde, G. J., 222
Howie, R. D., 435
Hsu, K. J., 55
Huckriede, R., 207, 222, 223, 230, 244, 249, 250, 261
Hughes, N. F., 317, 405
Husain, B. R., 18-20

Illing, L. V., 55
Imgrund, R., 332, 346, 403
Iversen, J., 320

Jacob, A., 310
Jacob, C., 311
Jansonius, J., 277, 281, 282, 286, 288, 289, 298, 300, 301, 320, 337, 338, 368, 369, 371, 372, 374, 380, 394, 396, 410, 411
Jeannet, A., 159, 161, 162, 165, 172
Jekhowsky, B. de, 277, 290, 291, 293, 295, 298, 376, 402, 435, 441, 445
Jizba, K. M. M., 380, 386, 387, 389
Johnson, J. H., 71

Kar, R. K., 105, 366, 369, 377, 379
Kara-Murza, E. N., 288, 289, 299, 301, 440
Kayser, E., 138
Khlonova, A. F., 329, 330, 382
Khouri, J., 56
Kidston, R., 322
Kindle, E. M., 42
King, R. E., 121, 161
Klaus, W., 331, 342, 351, 355, 357, 368, 369, 373, 374, 376, 385-388, 390, 402, 432, 441, 444
Kockel, C. W., 246, 253

Kohut, J. J., 218, 230, 237
Koken, E., 63, 69, 162
Kollmann, K., 195, 203
Koninck, L. de, 6
Koptyova, E. A., 445
Kosanke, R. M., 380, 410, 411
Kozlowski, R., 121
Krafft, A. von, 10, 179, 186
Kremp, G. O. W., 318, 319, 324, 350, 390
Krishnan, M. S., 310
Krutsch, W., 406
Kummel, B., 3, 12, 14-16, 22, 26, 27, 36, 49, 53, 58, 61, 66, 71, 76, 104, 105, 111, 112, 119, 122-125, 128, 130, 137, 149, 165, 177, 190, 194, 207, 208, 212-214, 278, 300, 307, 308, 311, 439, 440, 442
Kuyl, O. S., 427

Lahiri, H. M., 307
Lange, F. W., 299, 300
Laveine, J. P., 338, 405, 431
Lele, K. M., 357, 361, 364, 415
Leschik, G., 336, 337, 339, 345, 351, 362, 367, 369, 372, 386, 395, 402, 432, 445
Likharev, B. K., 138, 144
Linck, O., 68
Lloyd, R. M., 55
Lotze, F., 18
Luber, A. A., 355, 394, 410, 412, 432, 433
Lucia, F. J., 44
Lucia, J. F., 55, 56
Lundblad, B., 339, 395, 396

Mädler, K., 325, 326, 369, 393, 395, 398, 399, 444, 445
Maher, S. W., 42
Malyavkina, V. S., 395, 399
Mandwal, N. K., 215, 240
Manum, S., 389
Marchenko, O. F., 410, 432
Martin, A. R. H., 352
Martin, F., 293
Martinsson, A., 42
McGregor, D. C., 334, 338, 387, 440
Medd, A., 287, 295
Medvedeva, A. M., 277, 342, 355, 358, 382, 394
Méhes, G., 194
Mehl, M. G., 221
Mehta, K. R., 358
Merla, G., 121, 125
Meunier, A., 281
Middlemiss, C. S., 179, 186
Miklukho-Maklay, A. D., 128
Miller, A. K., 117, 122, 153, 156, 157, 164, 222
Mojsisovics, E. von, 153, 209, 218

Molin, V. A., 382, 427, 433, 445
Moore, R. C., 196-198
Mosher, L. C., 216, 226, 239, 240, 244, 245, 249, 253, 257, 258, 266
Muir-Wood, H., 118, 121
Müller, E. M., 221
Müller, J., 427
Müller, K. J., 221, 222, 235, 237, 244
Murchison, R. I., 6
Murray, F. N., 222

Nassichuk, W. W., 157, 165, 167
Nautiyal, D. D., 403
Neves, R., 332, 334
Newell, N. D., 53, 64, 116, 117
Newton, E. T., 277
Nilsson, T., 390, 395, 396, 400
Noetling, F., 10, 11, 17, 19, 30, 58, 60, 148, 157, 162
Nogami, Y., 217, 257
Norris, G., 387, 445

Öpik, A. A., 112
Ouyang, S., 332, 379

Pansart, J., 299, 300
Pant, D. D., 26, 307, 310, 311, 321, 355, 357, 367, 388, 403, 442
Pascoe, E. H., 13, 17, 20, 22, 30, 63, 65, 104, 118, 120-122, 128, 129
Paten, R. J., 334, 335
Pautsch, M. E., 398
Peppers, R. A., 332, 444
Pettijohn, F. J., 43
Pettitt, J., 318
Playford, G., 277, 321, 324, 328, 331, 334, 335, 338, 339, 341, 342, 345, 346, 351, 352, 354, 361, 362, 367, 369, 373, 379-381, 387, 390, 394, 398, 399, 403, 408, 431, 434
Plöchinger, B., 195
Plummer, H. J., 238
Pocock, S. A. J., 284, 286, 339, 406, 409, 410
Popov, Y. N., 12
Potonié, R., 317, 318, 324, 339, 342, 345, 350, 355, 361, 364, 367, 369, 390, 395, 400, 403, 406, 409, 415, 432
Potter, P. E., 43

Rahman, H., 17, 18, 22, 26
Rakotoarivelo, H., 439
Reed, F. R. C., 19, 20, 63, 111, 120, 122, 128-131, 144, 145, 170, 172, 307

Reineck, H.-E., 43
Reinhardt, P., 334, 337, 338, 341, 373, 441, 444
Reissinger, A., 382
Remy, R., 321, 322, 325, 327, 361
Remy, W., 321, 322, 325, 327, 361
Rexroad, C. B., 221, 222
Rhodes, F. H. T., 222, 239, 261
Richardson, J. B., 318, 319, 329
Rilett, M. H. P., 324, 434
Rogalska, M., 393
Ross, C. A., 158
Rothpletz, A., 153, 156
Rouse, G. E., 390
Rowell, A. J., 15, 53, 63
Ruzhentsev, V. E., 53, 69, 72, 117, 120-125, 128, 129, 132, 133, 137, 145, 154, 164, 184

Sahni, B., 307, 310
Salujha, S. K., 332, 336, 354, 365, 368, 381, 408, 409, 411
Samoilovich, S. R., 354, 355, 386, 410, 412
Sarjeant, W. A. S., 15, 21, 35, 50, 70, 77, 280, 287, 290-293, 295, 297, 298, 430
Sarycheva, T. G., 53, 69, 72, 117, 120-125, 128, 129, 132, 133, 137, 145, 154
Schaarschmidt, F., 277, 294, 296, 301, 366, 369, 396, 432
Schindewolf, O. H., 13, 14, 18, 21, 24, 33, 43, 69, 103, 111, 118, 122, 148, 153, 162, 172, 177, 184, 189, 207, 209, 249, 310
Schmidt, H., 222
Schmitz, W., 337, 338, 341, 444
Schneider, G. F., 194
Schön, M., 283, 287, 290, 301
Schopf, J. M., 310, 316
Schulz, E., 321, 334, 337, 339, 341, 445
Schurman, H. M. E., 374
Schweitzer, H. J., 317, 345, 369, 373, 432
Scott, A. J., 221
Scott, H. W., 222, 261
Sedova, M. A., 366, 368, 373, 401, 432, 433
Segroves, K. L., 277, 354, 416
Seilacher, A., 13, 310
Selling, O. H., 347
Sen, J., 310, 324
Seward, A. C., 307, 389
Shaffer, B. L., 380, 386

Shah, A., 26
Shaw, A. B., 224
Shearman, D. J., 56
Shinn, E. A., 55
Singh, H. P., 321, 324, 326, 334, 347, 351, 368, 374, 376, 380, 402, 409, 434
Sitholey, R. V., 27, 307, 310, 311, 442
Smit, D. E., 48
Smith, A. H. V., 326
Smith-Woodward, A., 307
Sohn, I. G., 15, 21, 30, 67, 76, 193-198, 203-205
Spath, L. F., 12, 58, 157, 178, 181, 182, 185, 186, 189, 190, 439, 446
Spode, F., 281, 288, 293
Srivastava, G. K., 26, 215, 240, 311, 321, 355, 357, 388, 442
Staesche, U., 214, 217, 222-224, 245, 246
Staplin, F. L., 282, 284, 286-288, 293, 322, 331
Stauffer, C. R., 238
Stepanov, D. L., 76
Stockmans, F., 281, 287, 288, 292-293, 295, 297
Stoppel, D., 222, 233, 239, 243, 250
Styk, O., 194
Sukh Dev, 354
Sullivan, H. J., 322, 324, 334
Sweet, W. C., 15, 21, 31, 35, 50, 53, 68 69, 76, 207
Swett, K., 48
Sylvester-Bradley, P. C., 203
Szczezechura, J., 202

Taha, S., 56
Takahashi, K., 288
Takhtadzhian, A. L., 401
Taraz, H., 165
Tasch, P., 277, 301
Tatarskiy, V. G., 57
Tatge, U., 230, 240, 261, 266
Taylor, J. C. M., 55
Teichert, C., 3, 14, 16, 17, 19-22, 30, 35, 49, 53, 55, 58, 65, 71-73, 76, 111, 112, 118, 119, 122-125, 128, 130, 131, 137, 149, 155, 162, 165, 172, 177, 184, 190, 194, 207, 208, 212-214, 278, 300, 307, 308, 310, 311, 440
Terry, K. D., 195
Theobald, W., 6
Thomas, G. A., 121
Thorsteinsson, R., 121
Timofeev, B. V., 289, 293
Tiwari, R. S., 338
Townrow, J. A., 319, 382,

385, 386, 390, 413
Tozer, E. T., 12, 178
Treat, I. V.-C., 153
Trivedi, B. S., 310
Tschernyschew, T., 12, 121
Tschudy, R. H., 380, 410, 411
Tucker, W. H., 195
Twenhofel, W. H., 40

Valensi, L., 280-282, 284, 287-289, 291
Varyukhina, L. M., 341, 351, 382, 394, 412, 427, 433, 440, 445
Veevers, J. J., 54
Venkatachala, B. S., 105, 347, 366, 369, 377, 379, 395
Verchère, A. M., 6
Verneuil, M. E. de, 6
Vinogradov, A. P., 55
Virkki, C., 307, 309, 310, 357, 366-368, 374, 379-389, 396, 416
Visscher, H., 377, 388

Waagen, W., 7-9, 20, 22, 26, 30, 57, 58, 63, 66-69, 72, 73, 104, 105, 118-122, 128, 129, 132, 134, 138, 139, 141, 144, 147, 153, 157, 161, 162, 183, 208, 209, 218
Wadia, D. N., 307
Walcott, C. D., 42
Wall, D., 277, 281, 286, 289, 295, 297, 300, 301
Walliser, O. H., 262
Walter, J. C., Jr., 122
Waltz, I. E., 355, 394, 412, 432, 433
Wang, Y., 143
Waterbolk, H. T., 427
Wedekind, R., 157
Wells, A. J., 55
Weyl, P. K., 55
White, D., 40
Williams, A., 118, 138
Willière, Y., 281, 287, 288, 292-293, 295, 297
Wilson, L. R., 277, 316, 317, 347, 350, 374, 380, 386, 395, 410, 411, 435
Wodehouse, R. P., 400, 416
Woods, J. T., 366, 380, 381
Wynne, A. B., 7, 8, 25, 43

Yang, T.-Y., 67
Youngquist, W., 222, 244, 246

Zoricheva, A. L., 366, 368, 373, 401, 432, 433
Zauer, V. V., 361, 366, 371, 412, 432, 433

II. SYSTEMATIC INDEX

acanthomorphid, 277
Acanthomorphitae, 278, 279, 300
Acanthotriletes, 325
Acanthotriletes tereteangulalus, 91, 324, 417-419
Acavatitriletes, 320
Acavatomonoletes, 345
Acratia?, 196
Acratia? sp., 66
Acritarcha, 52, 69, 278, 416, 440
acritarchs, 35, 277, 278, 309, 316, 423, 428, 440-442
Acrodus sp., 13, 52, 69
Aletes, 413, 421
Alexenia, 126
Algae, 52, 70
Alisporites, 310, 387, 389, 390, 394, 395, 435
Alisporites aequalis, 387, 393
A. australis, 387-389, 399
A. landianus, 393, 394, 423, 424, 444
A. nuthallensis, 389
A. opii, 389, 390, 393
A. ovatus, 395
A. parvus, 394
A. plicatus, 389
A. sp. cf. *A. opii,* 393, 423, 424, 444
A. tenuicorpus, 91, 382, 394, 418, 420, 436, 437
A. zapfei, 387
A. (=*Falcisporites*) *australis,* 441
Ambocoelia, 143
Ambocoeliidae, 141
ammonoids, 117, 118, 120
Amphissites, 196, 197, 198
Amphissites allerismoides, 197, 198
A. centronotus, 197
A. n. sp., 1, 66, 197
A. tscherdynzevi, 198
A. truncatus, 197
Amphissitidae, 197
Anapiculatisporites ericianus, 434
Anasibirites, 25
Anchignathodus, 210, 221-223
Anchignathodus coalescens, 222
A. coloradoensis, 222
A. cristulus, 222
A. isarcicus, 52, 82, 86, 96, 210, 212, 214, 219, 222-224, 271
A. minutus, 222, 223
A. penescitulus, 222
A. regularis, 221
A. scitulus, 222

A. spiculus, 221
A. typicalis, 31, 35, 52, 68, 70, 76, 77, 81, 82, 84, 86-88, 95, 96, 210, 212-214, 219, 221-224, 271
Anguisporites, 342
Anguisporites anguinus, 342
A. intonsus, 342
A. minutus, 342
Angulodus? prioniodellides, 261
Anthozoa, 62
Apatognathus, 228, 230, 235
Aplocoma cf. *A. torrii,* 44, 52, 53, 67, 88
Applanopsis, 311
Aratrisporites, 351, 352, 444, 445
Aratrisporites banksi, 351
A. fischeri, 351, 352, 423, 424
A. paenulatus, 352, 423, 424
A. parvispinosus, 351
Araxathyris, 126, 129
Araxilevis, 127
Araxoceras, 155, 165
Arcestes oldhami, 159
Arcestes priscus, 162
Archaeocidaris, 68
Archaeocidaris forbesianus, 68
Archaeohystrichosphaeridium aculeatum, 280
Articulata, 63, 131
Atatrisporites, 71
Athyrididae, 139
Athyris, 141
Aulosteges, 127
Aulsteges dalhousi, 136
A. sp., 35, 97, 120, 136
Aulostegidae, 136
Aumancipollenites, 410
Aurikirkbya, 197
Azonomonoletes, 345
Azonotriletes, 320
Azonotriletes trisulcus, 325

Bairdia, 196, 203
Bairdia armenica, 194
B. carinthiaca, 194
B. hagenowi, 194
B. intermedia, 194
B.? pseudoobuncus, 194
B. sp., 67
B? sp., 195, 203
B. subglobosa, 194
Bairdiacea, 203
Bairdiacypris, 67, 196, 204
Bairdiacypris? pannonica, 204
B.? sp., 66, 195, 204
Bairdiidae, 203
Bairdiinae, 203
Bairdiocopina, 203

Baltisphaeridium, 290, 297
Barakarites, 355
Barroisella, 114
Bascanisporites undosus, 435
Basslerella australae, 66, 196
Bellerophon, 60, 65, 79, 94, 96
Bellerophon impressus, 11
B. jonesianus, 13
B. sp., 13, 20, 31
Brachiopoda, 49, 63, 130
brachiopods, 117, 118, 120, 122, 123-125, 128, 130, 148
Brachythyrididae, 144
Bradyphyllum indicum, 63
Brunsiina? sp., 35, 63
Bryozoa, 49, 63
Bythocytheridae, 202
Bythocytherina, 202

Calamospora, 321
Calamospora hartungiana, 321
C. landiana, 52, 321, 419, 421-424
Callispirina, 127
Callispirina sp., 35, 97, 120, 145
Camptotriletes, 327
Camptotriletes corrugatus, 327
C. sp., 327
C. warchianus, 327, 417, 420
Campylites?, 90
Cancrinella, 126, 127
Carboprimitia, 67, 196
Carinaknightina, 67, 194-198
Carinaknightina carinata, 67, 198, 199
C. discarinata, 67, 199
C. tscherdynzevi, 199
Cavellina, 196
Cavellina sp., 66
Cavellinacea, 205
Cavellinidae, 205
Caytonanthus, 381, 382, 394, 427, 433
Caytonanthus arberi, 381, 385
C. kochi, 381
C. oncodes, 381, 385
C. tecturatus, 394
Caytonanthus-type, 381
Caytonia oncodes, 382
Caytoniales, 382, 394, 433
Caytoniidites alaticonformis, 382
Caytonipollenites latus, 394
C. pallidus, 382
C. subtilis, 382
Cedripites, 374, 400, 401
Cedripites eocenicus, 400
C. priscus, 90, 400, 401, 405,

Systematic Index

420, 427, 431, 436, 437
Cedripites-type pollen, 400
Cedrus, 400
Cephalopoda, 50, 65
Ceratites carbonarius, 183
Ceratobairdia, 196
Changhsingoceras, 72, 154, 155
Changhsingoceras meishanense, 154, 155
Chenopodiaceae, 430
Chlorococcales, 416
Chondrichthyes, 68
Chonetella, 127, 136
Chonetella sp., 35, 97, 120, 136
Chonetellidae, 136
Chonetid, 127, 136
Chonetidae, 136
Chonopectoides, 126
Chonostegoides, 127
Cidaris, 67
Cirratriradites africanensis, 338
C. australiensis, 339
C. splendens, 338
Cladophlebis, 307
Classopollis, 311
Cleiothyridina, 123, 127, 129, 141
Cleiothyridina cf. *C. capillata*, 35, 96, 97, 120, 130, 139
Climaccamina?, 31, 35
Climaccamina sp., 63
Clinocypris? sp., 194
Colaniella, 63
Colaniella?, 31
Colobodus sp., 13, 52, 69
Columbites, 244
Columbites parisianus, 189
Comelicania, 126
"Composita," 127, 139
Compressoproductus, 126, 127, 129
Coniferales pennatulus, 364
Coniferaletes stultulus, 382
C. impuberus, 433
conifers, upper Paleozoic, 358
Coniopteris, 322, 324
Conodontophorida, 50, 67
conodonts, 207
  L elements, 225, 226, 238
  LA elements, 225-226, 228-237
  LB elements, 225, 228-237
  LB1-elements, 229-231
  LB2-elements, 227, 229-231
  LC elements, 225-231, 234
  LD elements, 225, 233
  LE elements, 225, 232, 233
  LF elements, 225, 236, 237
  multielement skeletal apparatuses, 222

multielement species, 210, 218, 225
  U elements, 225-228, 230, 232, 235, 236, 238
  white matter, 253, 254, 259
Conularia, 17, 19
Converrucosisporites, 327
Cordaitina, 354, 355, 357, 433
Cordaitina gunyalensis, 355, 357, 422, 423
C. psiloptera, 355
Corisaccites, 377, 380
Corisaccites alutas, 377, 379, 380, 418
Coronakirkbya, 197
Corystospermaceae, 307
Costaspora, 331
Costaspora radiosa, 331
Costiferina, 127, 129
Crenispirifer, 126, 145
Crenispirifer dzhulfensis, 145
Crinoidea, 67
Crurithyris, 124, 126, 129, 142-143
Crurithyris(?), 123, 125, 126
Crurithyris? extima, 50, 64, 81, 123, 125, 130, 131, 142-144
C. speciosa, 143, 144
C. tschernyschewi, 144
Cryptobairdia intermedia, 194
C. pseudoobunca, 195
Ctenognathus conservativa, 244
C. discreta, 244
Ctenopolygnathus, 221
Cyathidites, 322
Cycadopites, 412
Cycadopites folliculares, 90, 406, 412, 413, 418, 421, 423, 424, 431, 439, 440
Cycadopollenites folliculares, 412, 422
Cyclobaculisporites indicus, 326
Cyclogranisporites, 325, 326, 336
Cyclogranisporites arenosus, 325, 423, 424, 444
C. leopoldi, 325
Cyclolobidae, 154
Cyclolobus, 21, 65, 72, 73, 117-120, 122, 123, 153-159, 162, 164
Cyclolobus astrei, 165, 170, 173
C. haydeni, 165, 167
C. insignis, 159, 162
C. krafftii, 157, 165, 167
C. kullingi, 155, 158, 161, 164, 165
C. oldhami, 65, 72, 155, 157-159, 162, 165, 172

C. cf. *oldhami*, 159
C. oldhamianus, 159
C. persulcatus, 155, 156, 158, 159, 161, 164
C. sp., 165
C. teicherti, 65, 158, 161, 162, 164, 165
C. walkeri, 158, 161, 165, 167, 170
C. walkeri madagascariensis, 165, 173
C. cf. *C. walkeri*, 65, 165, 172
Cyclolobus (Krafftoceras) haydeni, 157, 165, 167, 172
C. (K.) krafftii, 165, 167, 172
C. (K.) walkeri, 165
Cyclotriletes, 326
C. granulatus, 326
C. microgranifer, 326
C. narmiana, 326
C. oligogranifer, 326
C. pustulatus, 326
C. triassicus, 326
Cylostrobus, 351
Cypridacea, 203
Cyrolexis, 121, 127, 129
Cytheracea, 194, 202
Cythere tenera, 194
C. tubulifera, 194
Cythereis sp., 194
Cytherella sp., 194
Cytherelloidea? sp., 66
Cytherissinellidae, 202
Cytherocopina, 202

Danubites himalayanus, 185, 186
Darwinulacea, 203
dasycladacean alga, 33
Decussatisporites, 411
Decussatisporites delineatus, 411
D. lucifer, 409, 411
Dehorisporites, 352
Deltoidospora, 322
Densipollenites, 352, 354
Densipollenites indicus, 91, 352, 420, 434, 436, 438
Densoisporites, 339, 341, 342, 442, 444
Densoisporites complicatus, 90, 341, 342, 420
D. nejburgii, 341, 422, 423, 424, 444
D. playfordi, 52, 339, 341, 344, 420-423, 439, 442, 444
D. velatus, 339
Densosporites, 318
Derbyia, 118, 122, 124, 129, 131, 133, 134, 136
Derbyia?, 126, 127
Derbyia hemisphaerica, 13, 64
D. plicatella, 133, 134

D. cf. D. plicatella, 35, 96, 97, 120, 130, 134
D.? sp., 50, 64, 81, 123, 130, 131, 134
Deunffia, 70, 300
Deunffia brevispinosa, 300
D. unispinosa, 35, 279, 300
Diaphanospora, 336
Dictyoclostus indicus, 64
Dielasma, 126, 127, 129
Dielasmatids, 35, 52, 64, 78
dinoflagellates, 430
Disaccites, 361, 418-424
Discarinata, 198
Distriatites, 380
Distriatites insolitus, 380, 381
Ditomopyge fatmii, 66
Domatoceras, 95
Doroshamia, 126
Dulhuntyispora, 330, 434
Dulhuntyispora? minuta, 330
Dzhulfites, 76

Echinoconchidae, 137
Echinoconchus, 127
Echinodermata, 50, 67
Echinoidea, 67
Echinum micraster, 279, 281
Ectoprocta, 63
Edhedripites, 408-410
Edriosteges, 126, 127
Ellisonia, 210, 214, 219-221, 221, 224, 225, 229, 231, 232, 234, 235, 238
Ellisonia clarki, 220, 224-226, 230, 270
E. delicatula, 219, 220, 226-227, 230, 271
E. gradata, 52, 79, 82, 84, 86, 87, 94-96, 210, 215, 216, 218, 220, 226-232, 235, 271
E. robusta, 216, 219, 231, 232, 235, 271
E. sp., 35
E.? sp., 221, 238, 271
E. teicherti, 31, 35, 52, 81, 82, 86-88, 94, 96, 212, 219, 232-234, 271
E. torta, 220, 234, 235, 271
E. triassica, 31, 35, 52, 81-84, 86, 87, 91, 94-96, 210, 215, 216, 218, 235, 237, 238, 271
Ellivina, 144
Emphanisporites, 329
Emphanisporites erraticus, 329
Enantiognathus, 228, 230, 235
Endosporites hexareticularis, 345
E.? roeticus, 341
E. velatus, 345, 437
Endothyraceans, 33
Enoploceras, 25, 66

Enteletes, 120, 121, 124-126, 131
Enteletes conjunctus, 131
E. dzhagrensis, 125, 131
E. kayseri, 131
E. socialis, 131
E. sp., 13, 50, 64, 130, 131
E. sp. 1, 35, 96, 97, 120, 130, 131
E. sp. 2, 64, 82, 123, 125, 129
E. tschernyscheffi, 131
Enteletidae, 131
Entolium?, 52, 65
Entylissa nitidus, 412
Eocidaris, 68
Eoschizodus pinguis, 64
Ephedra, 410
Ephedripites, 408, 409, 433
Ephedripites mediolobatus, 408
E. sp., 91, 406, 409, 421, 423, 424, 444
Episageceras, 21, 65, 178, 184
Episageceras noetlingi, 156
E. wynnei, 65
Equisetales, 321
Equisetites, 307
Equisetosporites, 409
Equisetosporites chinleana, 409
E. multistriatus, 409
Ernestiodendron, 361
Eucommiidites troedsoni, 405
Eumedlicottia, 21, 65
Eumedlicottia primas, 65
Euphemites, 65
Euphemus, 65
Euphemus indicus, 10, 13, 95
Eurydesma, 17, 19

Fabilicypris oboncus, 195
Falcisporites, 70, 387, 390, 444
Falcisporites nuthallensis, 382, 389, 418, 419, 420
F. snopkovae, 388, 389
F. stabilis, 385, 387-389, 420, 423, 424
F. zapfei, 387, 388, 432, 444
Falcisporites-type pollen, 399
Faunipollenites, 362, 365
Faunipollenites varius, 365
Fibilicypris subgeinitziana, 195
Filisphaeridum, 286
Filisphaeridium setasessitante, 284
Fimbriaesporites, 401, 402
Fimbriaesporites globosus, 401, 402
F.? sp., 51, 402, 405, 419, 420, 422, 436-438
Fishes, 52
Flemingites, 24

Flemingites compressus, 261
Florinites ovatus, 395
Foraminiferida, 62
fusulinids, 117

Gangamopteris, 20, 300
Gastropoda, 50, 64
Geisiinidae, gen. indet., 66
Geisina? sp., 66
Geminospora, 318
Gemmanella parva, 194
G. schweyeri, 194
Geyerella, 127
Ginkgo typica, 412
Ginkgocycadophytus, 412
Ginkgocycadophytus caperatus, 412
G. nitidus, 412
Glassoceras, 154
Glassoceras normani, 155
Glossopteris, 20, 307, 435
Glyptophiceras, 23, 61, 66, 77, 95, 104, 178, 179, 181-185, 189, 212, 214
Glyptophiceras aequicostatum, 185, 189
G. extremum, 181, 182
G. gracile, 181, 182
G. himalayanum, 52, 181, 185, 189, 208
G. kashmiricum, 185, 186
G. minimum, 181, 182
G. minor, 181
G. nielseni, 181, 182
G. nielseni var. modesta, 182
G. ophiodes, 185, 186
G. pascoei, 181, 182, 185, 186
G. pascoei var. rotunda, 182
G. polare, 181
G. pseudellipticum, 181, 182
G. serpentinum, 181
G.? sp. ind. cf. minor, 182
G. subextremum, 181
G. triviale, 181
Glyptopleuroides, 67, 196
gnathodontids, 222
Gnathodus, 221, 222
Gnathodus sicilianus, 221
Gnetaceaepollenites, 406, 408, 409
Gnetaceaepollenites ellipticus, 406
G. sinuosus, 408, 421, 431, 434, 438
G. sp., 408
Godthaabites, 155, 157
Godthaabites kullingi, 156
Gondolella, 239, 249, 257
Gondolella carinata, 220, 240, 241
"G." milleri, 261
G. mombergensis, 239, 240
G. nevadensis, 240, 241
G. planata, 240, 241
G. sp., 258

Systematic Index

G. sp. aff. G. nevadensis, 240
G. timorensis, 256
Granabivesiculites constrictus, 400
Granulatisporites adnatus, 321
G. trisinus, 434
Graphiadactyllis, 196
Graphiadactyllis australae, 196
G.? sp. aff., 66
Grebespora concentrica, 413
Grypoceras aemulans, 66
G. bidorsatoides, 66
Gubleria, 126, 127
Guthoerlisporites, 345
Guthoerlisporites cancellosus, 70, 345, 346, 349, 420, 422, 424, 431, 436, 437
G. magnificus, 345
Guttulapollenites, 376, 377, 435
Guttulapollenites hannonicus, 70, 376, 377, 379, 418-420, 422, 427, 435, 436, 438
Gyrolepis, 69
Gyronites, 58, 82-84, 86-88, 93, 94, 215
Gyronites atavus, 13
G. frequens, 13
G. superior, 13
G. undatus, 13

Hamiapollenites, 320, 380
Hamiapollenites insculptus, 380, 381
H. insolitus, 379, 380, 381, 418
H. saccatus, 380
Hapsiphyllum indicum, 63
Haydenella, 126, 127, 129
Healdia, 196
Healdia incognita, 195
H. spp., 66
Healdiacea, 204
Healdianella doraschamensis, 195
H. splendida, 195
Hedenstroemia, 11
Helicampodus, 69
Helicampodus kokeni, 69
Helicoprion sp., 69
Heliosaccus, 355
Heliosporites, 338
Hemiprionites, 25
Hemiptychina, 127, 147, 148
Hemiptychina himalayensis, 147
H. sp., 13, 35, 64, 97, 120, 147
H. sparsiplicata, 147
Hibbardella, 225
Hibbardella acroforme, 226
H. subsymmetrica, 235, 236
Hindeodella?, 233

Hindeodella nevadensis, 235, 237
H. raridenticulata, 235, 237
H. sp., 233
H. triassica, 235, 237
Hungarella, 204
Hungarella? sp., 67, 194, 195, 204
Hustedia, 120, 127, 139
Hustedia indica, 139
H. indica var. chittidilensis, 139
H. sp., 35, 97, 120, 139
Hyattoceras subgeinitzi, 156
Hymenospora, 335
Hypotetragona, 196
Hypotetragona sp., 66
Hystrichosphaera inconspicua, 279
Hystrichosphaeridium trisulcum, 289

Idiognathodontinae, 221
Idiognathodus, 222
Illinella typica, 239
Illinites, 432, 437, 441
Inaperturopollenites, 413
Inaperturopollenites dubius, 413
I. nebulosus, 91, 406, 413, 421
Indospora, 331, 332
Indospora clara, 331, 332, 420, 436, 438
Indotheca, 307
Indotriradites, 337
Indotriradites korbaensis, 338
Invertebrate trails, 39
Iraqispora, 334
Iraqispora labrata, 93, 334, 424, 436, 438

Janiceps, 126
Judahella, 201, 202
Judahella n. sp., 194
J.? sp., 67, 195, 202
J. tsorfatia, 201
Judahellidae, 201
Jugasporites, 432, 437

Kathwaia capitorosa, 66
Kellettinidae, 196
Keyserlingina, 127
Kiangsiella, 121, 123, 127, 133
Kiangsiella sp., 35, 97, 120, 133
Kinneya, 42
Kirkbyacea, 196
Kirkbyidae, 194, 197
Kirkbyidae gen. indet., 67
Kirkbyidae? indet., 195
Klausipollenites, 385
Klausipollenites schaubergeri, 52, 70, 91, 385, 386,

420, 421, 427, 431-433, 436-438, 440
K. staplinii, 394
Knightina, 196-198, 201
Knightina? cuestaforma, 198
Knightinidae, 196, 197
Koninckites davidsonianus, 13
K. davidsonianus truncatus, 13
K. occlusus, 13
K. rotundatus, 13
K. sp., 13
Kraeuselisporites, 336-339, 430
Kraeuselisporites altmarkensis, 338
K. chamotti, 338
K. cuspidus, 338, 420, 422
K. dentatus, 336
K. rallus, 91, 337, 338, 420, 430, 437
K. vargalensis, 93
K. wargalensis, 338, 339, 420, 430
Krafftoceras, 155, 157, 165
Krotovia, 127

Laevigatosporites, 346
Laevigatosporites callosus, 88, 346, 349, 420, 437
L. scissus, 416
L. vulgaris, 346
Lahirites rarus, 368
Latosaccus latus, 393, 394
Latosporites intragranulosus, 346
L. planorbus, 346
Lebachia, 361
Leiofusa, 70, 299-301
Leiofusa banderilla, 300
L. bernesgra, 300
L. cantabrica, 299
L. jurassica, 300
L. lidae, 300
L. pumilia, 299
L. stassfurtensis, 35, 279, 299
L. tumida, 299
L. unispinosa, 300
Leiophyllites sp., 217
L. timorensis, 218
Leiosphaeridia, 279, 430, 439
Leiosphaeridia sp., 52
leiospheres, 428
Leiotriletes, 321
Leiotriletes adnatus, 321, 322
L. cf. adnatus, 417-419
L. furcatus, 335
L. sp., 322
L. sp. cf. L. adnatus, 91, 321
L. sphaerotriangulus, 321
Leptodus, 138
Leptodus richthofeni, 138
Leschikisporis, 347
Limitisporites, 432, 437

Lingula, 53, 55, 112-114, 126
Lingula anatina, 114
L. borealis, 114, 116
L. cf. *L. borealis,* 50, 63, 86, 87, 113
L. scrutata, 63, 111
L. sp., 13, 82, 111, 123, 130, 131
Lingulacea, 113
Lingulida, 113
Lingulidae, 113
lingulids, 112
Linoproductidae, 137
Linoproductus, 35, 118, 120, 125-127, 129, 131, 138
Linoproductus sp., 50, 64, 82, 97, 123, 130, 131, 138
Liroceras, 65
Lissochonetes, 136
Lochriea montanaensis, 261
Lonchodina, 232
Lonchodina discreta, 235
L. latidentata, 229, 230
L. mülleri, 235
L. sp., 235
L. triassica, 235, 237
Lonsdaleia n. sp., 63
Lophotriletes, 322, 324, 327
Lophotriletes gibbosus, 322, 324
L. novicus, 93, 322, 324, 417, 418, 420
L. sparsus, 324
L. tribulosus, 324
L. novius, 90
Loxoconcha pusilla, 194
Lueckisporites, 371, 373, 374, 435
Lueckisporites amplus, 365
L. cancellatus, 367
L. limpidus, 362
L. noviaulensis, 371
L. singhii, 91, 376, 418-420, 436
L. sp., 376
L. virkkiae, 91, 373, 374, 376, 419, 420, 432-438
Lunatisporites, 367, 369
Lunatisporites acutus, 366, 369
L. limpidus, 361
Lundbladispora, 341, 342, 442
Lundbladispora brevicula, 342, 344, 422-424, 444
L. nejburgii, 341
L. obsoleta, 52, 344, 420-423, 439, 444
L. playfordi, 339
L. simoni, 344
L. willmotti, 342, 344
Lundbladisporites, 444
L. vulgaris, 91, 350, 420
Lutkevichinella, 202, 203

Lutkevichinella bruttanae, 194, 203
L. involuta, 194, 203
L.? n. sp., 195
L.? ornata, 67, 195, 202, 203
lycopodial spores, 442-444
Lycopodiales, 337-339, 341, 342, 344, 351, 352, 430
lycopsid fructifications, Triassic, 351
Lycostrobus scotti, 351
Lygodiidites, 339
Lygodiidites balmei, 339
Lygodium, 339
Lyttonia, 125-127, 129, 138, 139
Lyttonia sp., 35, 50, 64, 82, 97, 120, 123, 130, 131, 138
Lyttoniidae, 138

Marattiaceae, 350
Marginifera, 126, 127, 129
Marginiferidae, 137
Marsupipollenites, 402, 403
Marsupipollenites fasciolatus, 403
M. scutatus, 403
M. sinuosus, 408
M. sinuous, 402
M. tecturatus, 403
M. triradiatus, 402, 403, 406, 421, 431, 434, 436, 438
M. trivadiatus, 91
Martinia, 125, 126, 129, 133, 147
Martinia?, 127
Martinia planoconvexa, 144
M. rhomboidalis, 125, 145
M. sp., 130, 131, 145
M.? sp., 35, 50, 64, 82, 97, 120, 123, 145
Martiniidae, 145
Martiniopsis, 126, 127, 129
Matomorphitae, 279
Matonisporites, 311
Medlicottia wynnei, 10, 11
M. (Episageceras), 10
M. (E.) dalailamae, 10
Medullosea, fructifications, 403
Meekella, 132
Meekellidae, 132
Meekoceras, 11, 181, 215, 217, 231, 241, 260
megaspores, 310, 311, 442
Menuthionautilus kieslingeri, 52, 66, 87
Mesolobus, 30
Metacoceras, 95
Metacoceras sp., 13
Metacopina, 204
Mexicoceras, 154
Mexicoceras guadalupense, 155
Michelinoceras, 66

Micrhystridia, 284
Micrhystridium, 70, 278, 282, 286, 290, 297, 301, 430, 439
Micrhystridium cf. *albertensis,* 286
M. alloiteaui, 286
M. angustum, 282
M. antarcticum, 289
M. bigoti, 286
M. breve, 35, 279, 281, 283
M. circulum, 35, 279, 283, 286-288
M. coronatum, 287
M. crassimuratum, 289
M. densispinum, 35, 279, 282
M. cf. *densispinum,* 283
M. fragile, 287
M. aff. *fragile,* 281
M. cf. *fragile,* 286, 288
M. inconspicuum, 35, 278, 279, 280, 281, 282
M. cf. *inconspicuum,* 279, 280
M. intromittum, 286
M. karamurzae, 35, 279, 282
M. keratoides, 288
M. aff. *M. keratoides,* 35, 279, 288
M. lejeunei, 287
M. lymense, 281
M. lymensis var. *gliscum,* 280, 281
M. microspinosum, 35, 279, 288, 289
M. minutispinum, 289
M. nannacanthum, 288, 289
M. nanum, 287
M. pachydermum, 289
M. pakistanense, 35, 279, 286
M. parinconspicuum, 288
M. parvispinum, 289
M. paryidumeti, 288
M. pascheri, 288
M. pelagicum, 288
M. polyedricum forma *reducta,* 290, 291
M. rarispinum, 289
M. recurvatum, 288
M. recurvatum forma *brevispinosa,* 281
M. resistens, 289
M. setasessitante, 35, 279, 284, 286
M. shinetonese, 287
M. singulare, 288
M. sp., 51, 280, 288, 289, 441
M. spinoglobosum, 288
M. cf. *stellatum,* 288
M. triassicum, 289
M. westphalienum, 281
Microcachryidites, 441
Microcheilinella, 67, 196, 204
Microcheilinella sp., 67, 195, 204
Microcheilus distortus, 204

Systematic Index

Minutisaccata, 380
Miocidaris, 68, 84, 86, 87, 96
Miocidaris pakistanensis, 44, 52, 53, 68
Monoceratina umbonatoides, 202
Monoceratina? sp., 67, 195, 202
Monoceratininae, 202
Monocolpates, 412
Monocolpopollenites acerrimus, 412
Monodiexodina sp., 63
Monoletes, 346, 417, 419, 420, 423, 424
Monolites, 346
Monolites major, 346
Monosaccites, 352, 418-420, 422, 423
Monosulcites minimus auct. non, 412
Myriotheca, 322

Narrabeen Group, 351
Neochonetes, 127, 129, 136
Neogondolella, 210, 215, 220, 221, 239-242, 245, 248, 249, 253, 257, 260, 262
Neogondolella aegaea, 239
N. carinata, 31, 35, 52, 68, 82-84, 86-87, 90, 94-96, 212, 214, 215, 220, 240-242, 263, 271
N. elongata, 216, 241, 242, 243, 244, 271
N. jubata, 68, 217, 220, 241-244, 271
N. mombergensis, 240, 241, 243, 257
N. nevadensis, 215
N. planata, 215
Neophricadothyris, 126, 127, 129
Neoprioniodus bicuspidatus, 243
N. bransoni, 244
N. unicornis, 235, 236
Neospathodus, 210, 221, 222, 240, 244-246, 248-250, 253-257, 260, 262, 265, 266
Neospathodus cristagalli, 68, 86-88, 216, 219, 245, 246, 248-251, 253-257, 260, 265, 271
N. dieneri, 52, 68, 79, 84, 88, 94, 215, 216, 221, 249-251, 255, 265, 271
N. dieneri n. sp., 243, 246
N. divergens, 244
N. homeri, 217, 220, 245, 246, 249, 254, 271
N. kummeli, 52, 68, 79, 84, 86, 215, 248, 251, 253, 265, 271

N. pakistanensis, 68, 216, 254, 255, 271
N. peculiaris, 216, 219, 254, 255, 256, 265, 271
N. spathi, 257, 271
N. timorensis, 67, 217, 218, 256, 257, 271
N. triangularis, 217, 220, 246, 249, 253, 254, 271
N. waageni, 68, 217, 248, 249, 260, 261, 271
Neospirifer, 124, 127, 144
Neospirifer sp., 35, 97, 120, 144, 145
N. warchensis, 144
Netromorphitae, 278, 279, 299, 300
Nevesipollenites fossulatus, 70
Nevesisporites, 334, 335, 444
Nevesisporites fossulatus, 91, 335, 420, 424, 431, 437
N. limulatus, 334
N. vallatus, 334, 335
Nordvikia, 195
Notothyris, 126, 127, 129
Non-taeniate pollen grains, 381
Nuskoisporites, 309, 355
Nuskoisporites dulhuntyi, 358, 432

Ogbinia, 126, 127
Oldhaminia, 122, 126, 129
Oligocarpia, 322
Ombonia, 124-126, 129, 130, 133
Ombonia sp., 50, 64, 82, 123, 131, 133
Ophiceras, 10, 11, 14, 23, 49, 56, 58, 61, 62, 66, 70, 76, 77, 83, 95-97, 105, 119, 126, 137, 144, 177-179, 181, 183-185, 189, 209, 212, 214
Ophiceras?, 86, 90, 93, 94
Ophiceras connectens, 13, 14, 23, 44, 46, 52, 61, 66, 86, 87, 123, 189, 190, 208
O. himalayanum, 179, 185
O. sakuntala, 177
O. tibeticum, 189
Ophioderma, 67
Ophioderma schistovertebrata, 67
Ophiuroidea, 67
Orbiculoidea, 63, 95
Orbiculoidea sp., 13, 50
Orthida, 131
Orthis crenistria?, 6
Orthothetina, 124, 126, 127, 129, 132, 133, 136
Orthothetina arakeljani, 132
O. cf. *O. arakeljani*, 50, 64, 82, 123, 125, 131, 132

O. sp., 50, 64, 82, 123, 125, 130, 131
Orthotichia, 124, 126, 127, 129, 131, 132
Orthotichia corallina, 132
O. derbyia, 132
O. sp., 96, 131
O. sp. 1, 35, 97, 120, 132
O. sp. 2, 35, 97, 120, 132
O.? sp., 130
Osmundaceae, 327
Osmundacidites, 327, 444
Osmundacidites senectus, 52, 327, 420-423
O. wellmanii, 327, 329
Osteichthyes, 68
Ostracoda, 50, 66, 195
Otoceras, 10-14, 23, 56, 105, 177-181, 183-185
Otoceras boreale, 178
O. woodwardi, 177, 179
Ottokaria, 307
Ozarkodina, 245, 262
Ozarkodina delicatula, 262
O.? sp., 235
O. tortilis, 262, 266, 268
O. typica, 261

Palaeocopida, 196
Palynomorphs, 52
Pamelreuthia halberfelneri, 390
Pandorinellina, 221
Paracalamostachys heterospora, 321
Parafusulina kattaensis, 19, 63, 117
Parametacoceras, 65
Parapronorites sp., 156
Parasaccites, 355
Paratirolites, 76, 154
Paravesicaspora, 319
Paravittatina, 409, 411, 427, 433-435, 437, 438
Paravittatina cincinnata, 411, 412
P. lucifer, 91, 406, 409, 411, 412, 418, 419, 421, 427, 436, 438
P. persecta, 411
P. striata, 411, 412
Parvisaccites, 401
Parvisaccites enigmaticus, 401
P. radiatus, 401
Parenteletes, 126, 127
Paucistriatoabieites, 371
Paucistriatopinites, 371
Paucistriatoprotoconiferus, 371
Pelecypoda, 50, 64
Peltacystia, 416
Peltacystia venosa, 91, 416-419, 421, 436, 438
Pemphygaletes prolixus, 394
P. tecturatus, 394

Perinomonolites, 351
Perinotrilites, 335
Perisaccus, 355
Perisaccus verrucalatus, 355
Permocalculus, 71, 95
Permocidaris, 68
Pernopecten?, 52, 65
Perotrilites granulatus, 335
P. sp., 335, 422, 439
Perrinites, 157
Petrolites, 318
Phylloceras oldhami, 155, 159
Piceapollenites cf. *alatus,* 393
P. fuscus, 398
Pinaceae, 400
Pinuspollenites, 399
Pinuspollenites labdacus, 399
P. thoracatus, 70, 382, 400, 418-420, 427, 436, 437
Pityosporites, 387, 389, 390
Pityosporites insularis, 386
P. ovatus, 395
P. pallidus, 382
P. potoniei, 396
P. sakesarensis, 389
P. sewardi, 367
P. sp., 380
Plagioglypta, 20, 31, 79
Plagioglypta herculea, 13, 66, 73, 95
Planetoceras, 65
Platycopida, 196, 205
Platycopina, 205
Platysaccus, 398, 399, 402, 435
Platysaccus hengeloensis, 399
P. papilionis, 398, 437
P. queenslandi, 384, 399, 423, 424, 444
Pleuromeia rossica, 339, 341
P. sp., 341
Pleuronautilus, 65
Pleuronautilus kokeni, 66
Pleuronodoceras, 155
Pleurophorus cf. *subovalis,* 13, 64
Plicates, 402, 418, 419, 421-424
Plicatipollenites, 355, 357, 358
Plicatipollenites indicus, 91, 352, 357, 418-420, 436, 438
Plicatoderbyia, 134
Plumbaginaceae, 430
Podocarpus (Archaeopodocarpus), 399
Podocopida, 194, 202
Poikilosakos, 126
Pollenites, 352, 417, 419-424
Polyedryxium, 70, 296, 298, 299
Polyedryxium deflandrei, 299
P. kraeuselianum, 295
P. sp., 35, 279, 298, 301
polygonomorphid, 277

Polygonomorphitae, 278, 289, 300
Polyplicates, 406
Polypodiidites, 349
Polypodiidites senonicus, 349
P. sp., 93, 349, 350, 420
Polypodiisporites, 346, 347, 349, 350
Polypodiisporites favus, 346
P. ipsviciensis, 347
P. mutabilis, 93, 346, 347, 349, 420, 437
Polysaccites-type, 379
Polytheca elongata, 403
Potoniea adiantiformis, 403
Potonieisporites, 358, 361
Potonieisporites bhardwaji, 361
P. bilateralis, 358
P. neglectus, 361
P. novicus, 358, 361, 417, 419
Praecolpates, 402
Prasinophyceae, 300
Prasionophycean, 277
Pretricolpipollenites, 405
Pretricolpipollenites bharadwaji, 91, 405, 406, 421, 431, 436, 437
P. ovalis, 405, 406
Prioniodella, 210, 261
"*Prioniodella*" *prioniodellides,* 216, 261, 271
Prionites, 25
Prionolobus, 58
Prismatomorphitae, 278, 298
Procarnites aff. *kokeni,* 218
Productus abichi, 121
P. cora, 6
P. costatus, 6
P. Flemingii, 6
P. (Dictyoclostus) indicus, 13
P. (Waagenoconcha) purdoni, 13
Progonocytheridae, 203
Propinacoceras? n. sp., 156
Proprisporites, 336
Proptychites, 58
Protocedrus sp., 386
Protohaploxypinus, 361, 365, 433, 435, 438, 440
Protohaploxypinus amplus, 364
P. diagonalis, 361, 364, 365, 418-420
P. goraiensis, 362, 364, 418, 420
P. jacobii, 366, 434
P. latissimus, 361
P. limpidus, 91, 362, 365
P. microcorpus, 91, 361, 366, 418-420, 434, 437
P. pellucidus, 373
P. pennatulus, 364
P. samoilovitchii samoilovitchi, 369

P. sewardi, 362
P. varius, 91, 361, 365, 366, 418-420
Prototoceras, 155
Pselioceras, 65
Pseudobythocypris? sp., 66
Pseudoceltites fortis, 13
P. radiosus, 13
Pseudogastrioceras, 21, 155
Pseudomonotis, 87
Pseudomonotis sp., 13
Pseudosageceras multilobatum, 13
Pseudostephanites, 155
Pseudotirolites, 155
Pseudotoceras, 155, 184
Pseudowellerella, 126
Psilatriletes circumundulatus, 342
Pteruchipollenites, 387
Pteruchus africanus, 387
P. dubius, 387
P. petasatus, 387
P. simmondsi, 387
Pulviella ovalis, 194
Punctatisporites, 320, 321, 336, 350
Punctatisporites fungosus, 52, 90, 320, 321, 421, 422, 423
P. minutus, 350
P. punctatus, 320
P. punjabensis, 321
P. (Marattisporites) scabratus, 350
P. sp., 350
P. sp. cf. *P. minutus,* 93, 349, 350, 419, 420
P. triassicus, 321
P. walkomi, 350
Pyramidosporites, 415
Pyramidosporites cyathodes, 415, 416
P. racemosus, 415, 416, 418

Quadrisporites, 415
Quadrisporites horridus, 415, 418, 421

Raistrickia sp., 324
Remysporites, 345
Renaultia, 322
Renngartenella avdusini, 194
R. ovata, 194
Reticulariina, 127
Reticulatia?, 126
Reticuloidosporites, 347, 349
Reticuloidosporites dentatus, 347
R. warchianus, 347, 349, 417, 420
Retusotriletes, 329
Retusotriletes bireais, 329
Retziidae, 139
Reubenella, 205

Systematic Index

Reubenella avnimelechi, 205
R.? sp., 67, 195, 196, 204
Reviya, 197
Rhipidomella, 127, 129
Rhizomospora, 319
Ricaspora, 336
Ricaspora granulata, 336
Richthofenia, 127, 129, 137
Richthofenia sp. 35, 97, 120, 137
Rimaesporites aquilonalis, 401, 431
Rivularites, 40
Rivularites permiensis, 40
Rotaraxoceras, 155
Roundya, 225, 227, 230
Roundyella, 196
Roundyella? sp., 66

Saccites, 352, 417, 419-423
Sahnites, 358
Samaropsis, 307
Saturnisporites fischeri, 351
Saurichthys?, 69
Saurichthys sp., 52
Scaphopoda, 65
Schizodus sp., 13
S. (Eoschizodus) pinguis, 13
Schizomorphitae, 416
Schizosporis, 416
Schizosporis reticulatus, 416
S. scissus, 416, 418, 419, 421, 422, 423
Schopfipollenites, 403
Schuchertellidae, 133
Scopulisporites, 387
Scopulisporites toralis, 394
Scottella typica, 262
Sehorisporites indicus, 353
Selaginellaceae, 339
Septospirigerella, 127, 129
Sibirites, 217, 244
Silenites, 196
Silenites? sp., 66
Simeonospora, 329, 330
Simeonospora khlonovae, 329, 330, 423
Simsangia, 293
Simsangia trispinosa, 290
Solenochilus, 65
Solisphaeridium, 286
Spathognathodus, 221, 222, 248
Spathognathodus bidens, 222
S. campbelli, 222
S. cristagalli, 222, 244, 246, 249
S. divergens, 222, 244, 250
S. exodentatus, 222
S. gondolelloides, 218
S. homeri, 245
S. homeris, 245, 246
S. isarcicus, 222-224
S. minutus, 223
S. minutus?, 222

S. cf. *minutus,* 222, 224
S. n. sp., 245, 246
S. pellaensis, 222
S. pulcher, 221
S. pusillus, 222
S. triangularis, 253
S. whitei, 222
Sphaeromorphitae, 279, 289
Spheripollenites scissus, 416
Sphyropteris cf. *boenischi,* 325
Spinomarginifera, 123, 124, 125, 126, 127, 129, 137
Spinomarginifera helica, 137
S. sp., 50, 64, 82, 86, 87, 123, 125, 130, 131, 137
Spiriferella, 144
Spiriferella?, 127
Spiriferella? sp., 35, 97, 120, 144
Spiriferellina, 126, 127, 129
Spiriferida, 139
Spiriferina, 25
Spiriferinidae, 145
Spirigerella, 120, 125-127, 139, 141
Spirigerella derbyi, 13, 64
S. sp., 35, 50, 64, 82, 96, 97, 120, 123, 130, 131, 139
Spores and pollen, 69
Sporites, 320, 417-419, 421, 422, 424
Stachella, 24, 60, 65
Stacheoceras, 21, 65, 155, 173
Stacheoceras antiquum, 65, 162
S. normani, 154
S. tridens, 156
Stearoceras, 65
Stenoscisma, 121, 122, 126, 129
Stenozonotriletes, 329
Stenozonotriletes exuperans, 329
S. khlonovae, 330
S. radiatus, 329, 330
S. stellatus, 329, 330
Stepanoviella, 127
Stereisportes, 335
Streptognathodus, 222
Streptorhynchus, 126, 127
Streptorhynchus pectiniformis, 121
Striatites, 367
Striatites angulistriatus, 365
S. cancellatus, 367
S. jacobii, 366
S. medius, 380
S. microcorpus, 366
S. noviaulensis, 371
S. samoilovitchii pantii, 369
S. samoilovitchii var. *pantii,* 368, 369
S. samoilovitchii var. *samoilovitchii,* 369

S. sewardi, 362
Striatobietites richteri, 434, 435
Striatohaploxypinus, 361, 373
Striatopinites raricostatus, 372
S. reticulatus, 380
Striatopodocarpites, 366-369, 433, 435, 438, 440
Striatopodocarpites cancellatus, 91, 367, 371, 418, 420
S. oblongatus, 366
S. octostriatus, 368
S. pantii, 90, 368, 369, 371, 418, 420
S. rarus, 93, 367, 368, 371, 418, 420
S. tojmensis, 366
Strigogoniatites angulatus, 156
Strobeus, 95
Strobeus avellenoides, 95
Strophomenida, 132
Strotersporites, 366, 369
Strotersporites pantii, 368
Styxisporites reissingeri, 338
Subbryantodus abstractus, 262
Subcolumbites, 217, 249
Sulcatisporites, 394, 395, 396, 435
S. institatus, 393, 396, 398, 423, 424, 444
Sulcatisporites interpositus, 395, 398
S. nilssoni, 396, 418
S. ovatus, 93, 382, 395, 396, 418-420, 436, 438
S. quadratus, 396
S. sp. cf. *S. kraeuseli,* 393, 398, 423, 424
Sundaites levis, 156
Syrdenites, 155
Syrdenites n. sp., 156

Taedaesporites scaurus, 400
Taeniaesporites, 70, 369, 371, 373, 440-442, 444
Taeniaesporites discurrens, 362
T. kraeuseli, 366, 367, 369
T. noviaulensis, 52, 90, 371, 372, 374, 420-424, 432, 433, 436-439
T. novimundi, 371, 372
T. ovatus, 372
T. pellucidus, 52, 373, 374, 421-424, 439
T. samoilovitchii pantii, 368
T. sp. cf., 372
T. sp. cf. *T. noviaulensis,* 371, 373
T. sp. cf. *T. transversundatus,* 52, 371, 372, 421-424
T. transversundatus, 372

Taenionautilus trachyceras, 66
Tainoceras, 65
Tapashanites, 155
Tasmanaceae, 300
Tasmanites, 70, 300
Tasmanites minutus, 300
T. sp., 35, 279, 300
tasmanitids, 200, 278
Temnocheilus, 65
Tenebrion, 197, 198
Terebratula, 6
Terebratula crispata?, 6
T. Royssii, 6
Terebratulida, 147
Terebratuloidea, 126, 127, 129
Thaumatocanthus blanfordi, 69
Thymospora, 346
Thymospora opaqua, 347
T. pseudothiesseni, 347
Tigrisporites, 331
Tigrisporites halleinis, 331
T. playfordi, 70, 91, 331, 420, 424, 431, 437, 441, 444
T. sp., 330, 331
Timorites, 72, 154, 156, 158, 159, 164
Timorites curvicostatus, 154, 155, 156
T. striatus, 156, 158
Tompophiceras, 178
trace fossils, 52, 70
Triadispora, 441
Triancoraesporites communis, 334
Trichonodella, 225
Triletes, 320, 417, 418, 419, 421, 422, 424
Trilobitae, 66
Tripartina, 331
Tripartites, 332
Triquitrites, 332, 334
Triquitrites arculatus, 332
T. iraqiensis, 334
T. proratus, 91, 332, 334, 420
Tschernyschewia, 121, 126, 127, 138
Tyloplecta, 126

Umbrosaccus, 399
Umbrosaccus hyalinus, 399
Uncinunellina, 126, 127, 129
Urartoceras, 155

Vediproductus, 127
Vermiporella, 35, 71
Verrucosisporites, 325-327
Verrucosisporites narmianus, 325, 326, 423, 424, 444
V. planiverrucatus, 326
V. sinensis, 326
V. sp. cf. *V. planiverrucatus*, 326, 417

V. trisecatus, 326
V. verrucosus, 325
Veryhachium, 70, 278, 290-293, 295-298, 301, 430
Veryhachium aquila, 293
V. aster, 297
V. asymmetricum, 293
V. balticum, 297
V. belgicum, 297
V. brevitrispinosum, 293
V. centrigerum, 293
V. cochinum, 293
V. delmeri, 293
V. downiei, 293
V. europaeum, 295
V. exile, 293
V. geometricum, 293
V. helenae, 293
V. hyalodermum, 293, 295
V. inflatissimum, 295
V. irregulare, 293, 295, 297
V. irregulare forma *subtetraedron*, 35, 279, 293, 295, 298
V. legrandi, 295
V. libratum, 293
V. limaciforme, 291
V. nasicum, 296
V. cf. *nasicum*, 296
V. piliferum, 293
V. reductum, 51, 290, 293, 295, 439, 441
V. rhomboidium, 297
V.? riburgense, 35, 279, 296, 297
V. scabratum, 295
V. sedecimspinosum, 296
V. sp., 297
V. tetraedron, 295
V. tetraxis, 295
V. trispininflatum, 293
V. trispinosum, 293
V. trisulcum, 289, 291, 293
V. trisulcum var. *reductum*, 291
V. valensii, 35, 278, 279, 290, 291, 292, 298
V. vandenbergheni, 297
V. wenlockium, 293
Vesicaspora, 395
Vesicaspora maxima, 396
V. ovata, 395
V. schaubergeri, 385, 386
V. wilsonii, 395
Vestigisporites, 358
Vestoceras, 155
Villozona, 197, 198
Virgaloceras, 65
Vishnuites, 179
Virkkipollenites, 355
Vitreisporites, 381, 382, 394, 395, 433
Vitreisporites koenigswaldi, 394
V. pallidus, 91, 381, 382, 385,
394, 418, 420, 422, 424, 427, 431, 433, 435, 436, 438-440
V. signatus, 381
V. subtilis, 382
Vittatina, 402, 410, 411, 433, 434
Vittatina africana, 410, 411, 412
V. circinnata, 438
V. costabilis, 412
V. spp., 435
V. striata, 410, 411
V. subsaccata, 410, 433
Voltziaceaesporites, 441
Voltziapites, 395

Waagenoceras, 154, 157
Waagenoceras cumninsi guadalupense, 154
W. girtyi, 155
W. intermedium, 156
W. mojsisovicsi, 154, 155
Waagenoconcha, 30, 121, 122, 127, 137, 138
Waagenoconcha purdoni, 64
W. sp., 96, 130
W.? sp., 35, 97, 120, 137
Waagenophyllum indicum, 63
Walchianthus, 361
Waltzispora, 334
Warthia, 65
Wellerella, 126, 127
Welwitschia, 408
Welwitschia bainesii (=*mirabilis*), 408
Welwitschiapites, 408
Welwitschiapites tenuis, 408
Wentzelella timorica, 63
Weylandites, 409
Weylandites indicus, 409
Whitspakia, 126, 127, 129, 147, 148
Whitspakia acutangula, 147, 148
W. biplex, 147
W. sp., 52
W. sp. 1, 35, 96, 97, 120, 130
W. sp. 2, 64, 82, 123, 131, 148
Wilsonastrum, 70
Wilsonastrum colonicum, 35, 279, 295, 297, 298
Wilsonites, 344

Xaniognathus, 210, 220, 239, 244, 256, 261-263, 266, 268
Xaniognathus curvatus, 52, 81-84, 86, 87, 94, 96, 212, 215, 221, 245, 260, 262, 263, 265, 266, 268, 270
X. deflectens, 52, 83, 84, 86, 94, 215, 216, 221, 245, 250, 251, 256, 263, 265, 266, 268, 271

X. elongatus, 220, 266, 268, 271
X.? sp., 268, 271
Xenaspis, 183, 184
Xenodiscus, 20, 21, 65, 183-185, 189
Xenodiscus aequicostatus, 185, 186
X. althothae, 185, 186
X. carbonarius, 65, 184
X. comptoni, 185, 186
X. cf. *ellipticus,* 185
X. himalayanus, 179, 182, 185, 186
X. cf. *lissarensis,* 185
X. cf. *ophioneus,* 185, 186
X. plicatus, 65, 183, 184
X. cf. *rotula,* 185
X. salomonii, 185, 186
X. cf. *sitala,* 185
Xestoleberis?, 194
Xystracanthus gigas, 69

Youngiella?, 196
Youngiella? sp., 66
Youngiellacea, gen. indet., 66
Youngiellacea?, gen. indet., 66

Zonomonoletes spinosus, 445
Z. tschalyschevii, 445
Zonotriletes, 332
zygopterid fructifications, 326

III. SUBJECT AND LOCALITY INDEX

[NOTE: Words such as Permian, Triassic, Salt Range, are not indexed, nor are Kummel and Teichert's (1970) localities.]

Abadeh, 165
acritarchs
  Africa, 441
  Antarctica, 277
  Assam, 293
  Australia, 440
  Belgium, 288, 297
  Canada, 277, 283, 298, 301
  Chhidru Formation, 70, 277-301, 430
  England, 277, 280, 287-288, 297, 299
  France, 280, 287, 291, 297
  Germany, 277, 288, 301
  Kansas, 277
  Kathwai Member, 70
  Landa Member, 70
  Libya, 277, 291
  Madagascar, 277, 291, 441
  Mianwali Formation, 443
  Narmia Member, 70, 443
  Oklahoma, 277
  Soviet Arctic, 283, 288, 301, 441
  Tredian Formation, 443
  Tunisia, 287, 291, 299
  Volga-Ural region, 277
  Wargal Limestone, 430
  white sandstone unit, 70
  Yugoslavia, 277, 287, 291, 441
Afghanistan, Triassic conodonts, 216, 217, 260-261
Africa, palynomorphs, 324, 362, 367, 374, 377, 380, 403, 408, 410, 416, 427, 431, 434-435, 438-441
Aktyubinsk, palynomorphs, 445
Alaska, 178
algae
  Chhidru Formation, 71, 228
  Kathwai Member, 71
Alps, Triassic ostracodes, 195
  palynomorphs, 386, 402
Amarassi, 156, 164, 170
  fauna, 156

Amarassian, 154-155, 161, 173
Amb Formation, 20-21, 66, 148-149
  ammonites, 172
  brachiopods, 111
  conodonts, 208, 214
  fusulinids, 117
  ostracodes, 66
  palynomorphs, 321-322, 324, 326-327, 347, 357-358, 361-362, 364-365, 367-368, 376, 379, 380, 382, 389, 394-396, 400, 411, 413, 415-417, 424, 426, 428, 430
Amb group, 11
Ambilobé, 158, 167-168, 170, 172
  beds, 153
Amery Formation, 367, 434
Anantnag District, Kashmir Himalayas, 215, 241
Anchignathodus typicalis Zone, 210, 212-214
Angara Region, 433, 435
Angaraland, 307, 434
Anisian, 217-218, 268, 301, 444
  lowermost, 257
Ankitohazo, 172
Antarctic
  palynomorphs, 324, 354, 362, 367, 377, 387, 403, 408, 416, 431, 434-435, 438, 445
  Prince Charles Mountains, 277, 367, 377
Anthozoa
  Zaluch Group, 63
  Wargal Limestone, 63
  Chhidru Formation, 63
apatite, in lingulid shells, 112
Araks Gorge, 165
Araks River, 54
Araksian, 72, 153-155
Arctic
  brachiopods, 121

Canadian, 178-179, 334, 337, 387
  marine plankton, 281
  Soviet
    acritarchs, 283, 288-289, 301
    palynomorphs, 72, 441, 444
Armenia, 3, 117, 121, 125, 128-130, 132, 144, 148, 153-155, 158, 164-165
Arroyo Difunta, 161
Artinskian, fusulinids, 117
  palynomorphs, 306, 307, 324, 332, 379-381, 415-416, 427
Asia, 120
  central, 128
Assa, acritarchs, 293
Astrakhan District, USSR, 194
Austria, palynomorphs, 331, 351, 364, 369, 386, 402, 437
  Upper Permian strata, 369
Australia, 70
  acritarchs, 298
  brachiopods, 121
  ostracodes, 196
  palynomorphs, 310, 321, 324, 330-332, 334, 342, 344-345, 350-351, 362, 367, 380, 387, 389, 394, 396, 399, 403, 408, 410, 415-416, 431, 434-435, 437-438, 440-441, 444-445
  tasmanitids, 277
  western, 72
Axel Heiburg Island, 179
Azerbeigjan, 3

Bad Harzburg, 325
Balaton Mountains, Hungary, 194
Baltic region, acritarchs, 297
Barakar group, 377
Barren Measures, 377
Basleo, 156, 158, 161, 170

Bear Lake Country, southeastern Idaho, 217
Beaufort Group, palynomorphs, 439
Belgium, acritarchs, 288, 297
Bellerophon impressus Zone, 11
Bellerophon Limestone, 125, 214
Bellerophon-Schichten, 386
Belloy Formation, 277
bivalve beds, 9, 22, 25
bivalve limestone, 9, 22, 25, 309
Blina Shale, 445
Bolivia, Permian brachiopods, 121
Bonaire Island, 55
Bone Springs Limestone, 221
Bowen Basin, Queensland, 331, 441, 445
Brachiopoda, 117-149
　Alps, 133
　Amb Formation, 111, 120
　Arctic, 121
　Asia, 121
　Australia, 121
　Bolivia, 121
　Chhidru Formation, 63, 117, 180
　China, 121, 138, 143
　dolomites, 121
　Dzhulfa, 117, 122, 125, 128, 133, 144, 148
　Europe, 130
　Himalayas, 131
　Iran, 128, 144
　Kathwai Member, 63-64, 111-112, 117, 214
　Mexico, 121
　reworking, 124
　Serbia, 121, 122
　Sicily, 133
　USA, 117, 120-122, 130, 133, 145
　USSR, 121
　Wargal Limestone, 63, 117
　white sandstone unit, 120
　Zaluch Group, 63, 117-118, 124
Brush Creek Limestone, Ohio, 223
Bryozoa, 63
Bunter Sandstone, 194
　acritarchs, 301
　ostracodes, 194
　palynomorphs, 321, 377, 435, 441, 444
Buntsandstein, 326, 337, 341, 445
burrows, 43

Callovian, 293
Cambrian
　acritarchs, 289

Saline Series, 310
Campiler Schichten, 217
Canada
　acritarchs, 277, 283, 298, 301, 369
　ammonoids, Permian, 155
　Arctic, 178-179
　Cretaceous, Lower, 409
　Maritime Provinces, 436
　palynomorphs, 329, 331, 334, 338, 368-369, 372, 374, 386-387, 409, 413, 416, 436, 441
　Scythian, 372-373
　western, 277, 369, 413, 416, 441
Canning Basin, 332, 445
Cape Stosch, 76
Capitan Limestone, 125, 148
Capitanian, 133, 154-157, 161
Carboniferous, 122, 132
　acritarchs, 288-289, 293
　ammonoids, 153
　brachiopods, 122, 132
　conodonts, 221-222
　European, 350
　limestone, 7
　Lower, 221-222
　Lower, Scotland, 222
　palynology, 321-322, 332, 336, 347, 350, 413, 433
Carnarvon Basin, 377
Carnian
　ostracodes, 195, 203-204
　palynomorphs, 331, 351
Carnic Alps, 133
Caryginia Formation, 379
Caucasus, 133
Cellites ? sp. Zone, 11
Cephalopoda beds, 8, 162
cephalopods
　Alaska, 178
　Arctic Islands, 178-179
　Armenia, 153, 155, 158, 164
　Canada, 155, 178
　Chhidru Formation, 65, 153
　China, South, 119, 153, 154, 155, 173
　Coahuila, 155, 156, 173
　Himalayas, 153, 157, 159, 162, 165, 167, 172, 177-180, 183-184, 189-190
　Greenland, East, 153, 155, 157-158, 178-179, 181-182, 185, 190
　Iran, 153, 158, 164-165
　Japan, 155-156
　Kalabagh Member, 65
　Kashmir, 179-182, 186, 189
　Kathwai Member, 66
　Madagascar, 153, 155, 157-158, 167, 170, 172
　Mianwali Formation, 66

Permian, 153-177
Punjab, India, 155
Siberia, 155-156, 178
Sicily, 155
Spitsbergen, 178
Timor, 155-156, 159, 164, 173
Triassic, 177-191
Ceratite beds, 7, 9, 13, 14, 22, 209, 308-310
Ceratite Formation, 10, 25
　fish, 69
Ceratite limestone, 9-11, 14, 22-24, 56-58, 60, 69, 104, 208, 215, 249
Ceratite marls, 9-11, 22-24, 58, 69, 208, 215, 249
Ceratite sandstone, 9, 11, 22-24, 208, 216
Changhsing Limestone, 72, 143-144, 154-155, 183
Changhsingian, 72, 153-155, 173, 183
Chekiang, 154-155
Chhidru Formation, 20-22, 30-31, 60-62, 113, 208
　age of, 118-119, 173
　acritarchs, 70, 277-301
　algae, 71
　anthozoa, 63
　ammonoids, 153, 157, 161-162, 165, 172-173
　brachiopods, 63, 117-128, 131-134, 136, 138, 141, 148
　bryozoa, 63
　cephalopods, 65-66, 113, 153
　conodonts, 68, 208, 212-213, 223, 234, 238, 241
　fish, 69
　foraminifers, 63
　gastropods, 65
　ostracodes, 66
　palynomorphs, 70, 308-309, 316, 321-322, 324, 327, 331, 332, 334, 335, 337-339, 341, 344-347, 349-350, 354, 357, 362, 364-368, 371, 374, 376-377, 382, 386, 387, 389, 394-395, 400-403, 405, 408-409, 411, 413, 415-417, 419, 424-425, 427-428, 430-431, 434-435, 439-440, 444
　scaphopods, 66
　tasmanitids, 277-278, 300
Chhidru Group, 11
Chhidruan, 111, 118-120, 153, 155, 173, 183, 207
Chhidruan Series, 119
Chichali Range, 8
Chideru Formation, 30

SUBJECT AND LOCALITY INDEX

Chideru Group, 11
Chideru Gruppe, 30
Chidru beds, 30
China
  ammonoids, Permian, 153-154, 173
  brachiopods, 121, 138
  polynomorphs, 326, 332, 403
  southern, 3, 119, 154-155, 332
Chinle Formation, 390
Chios, Greece, 204, 217
Chitichun, 168, 169, 172
Chitichun Limestone, 131
Chua Gorge, 12
Clavering Island, 185
Coahuila, 155-156, 161
coefficient of rank concordance (W), 230, 237
Col Lebong, 162, 172
Collaroy Claystone, 445
Colorado, 222, 400
conodonts, 207-267
  Afghanistan, 216, 260-261
  Ceratite marl, 69
  Chhidru Formation, 68, 76
  Chios, 217
  Germany, 240, 244, 250
  Greece, 246
  Greenland, 240
  Italy, 214, 217-218, 222-224, 246
  Kashmir, 240, 241
  Kathwai Member, 68, 76
  Lower Ceratite limestone, 69
  Mittiwali Member, 69
  Narmia Member, 69
  Sicily, 222, 233
  Spitsbergen, 216, 218, 239, 255
  Timor, 216-218, 257, 261
  USA, 217-218, 222-223, 231, 238-241, 244, 246, 249, 258, 261-262, 268
Conodonten-Apparat, 261
Conularia-Eurydesma beds, 19
Crinoidea
  Wargal Limestone, 67
  Kathwai Member, 67
Cretaceous
  acritarchs, 288, 289, 295, 300
  ostracodes, 202
  palynomorphs, 329-330, 339, 342, 382, 400-401, 409
  Saline Series, 310
Crittenden Ranch section, 261
Cyclolobus oldhami Zone, 11

Datta Formation, 27
dedolomitization, 48, 56
Derbyia hemisphaerica Zone, 11
Devonian, 198
  acritarchs, 283, 286, 288, 291, 293, 295, 297, 299
  conodonts, 221, 260
  ostracodes, 198
  palynomorphs, 329, 336, 355
  upper, 198, 355
Dhak Pass Formation, 22
Dhodha Wahan, 314
Dinner Springs Canyon, 261
Dinwoody Formation, 116, 214, 222-223
dolomites, 9
Dolomite unit, Kathwai Member
  ammonoids, 189-190
  brachiopods, 122-125, 130-134, 136-139, 143-145, 148-149
  conodonts, 223-224, 238, 241, 251-252, 263, 266
  fish, 69
  petrography, 43-49
dolomites, northern Italy, 214
Dzhulfa, Soviet Armenia
  ammonoids, 154, 158, 165
  brachiopods, 117, 125, 128, 132, 144
Dzhulfian, 64
  ammonoids, 153-154, 157, 164, 173, 183
  brachiopods, 117, 119-125, 128-130, 132-133, 144-145, 148-149

Ecca Group, 427
Ecca Series, 408, 410
  Upper, microflora, 439
Echinodermata
  Crinoidea, 67
  Echinoidea, 67
  Ophiuroidea, 67
Echinoidea
  Zaluch Group, 67
  Kathwai Member, 68
Egypt, Carboniferous-Permian, 374
Elburz Range, 128-130, 144
Elko County, 261
Ellesmere Island, 179
England
  acritarchs, 277, 280, 287-288, 297, 299-301
  palynology, 324, 389, 445
Eocene
  acritarchs, 289
  palynomorphs, 382, 400

Erskine Sandstone, 445
Erunakov deposit, Upper Permian, 433
Erunakova horizon, Upper Permian, 364
Ervay Tongue, 76
Euphemus indicus Zone, 11
Europe
  brachiopods, 130
  conodonts, 239-240
  palynomorphs, 332, 366, 373, 375, 380, 386, 387, 394, 396, 399, 431-435, 437-439, 441, 444-445
  western and central, 380, 435, 441, 444
extinctions, 12, 14

Ferrar Group, Antarctica, 387
fish, 69
Flemingitan Division, 439, 442
Flemingites flemingi beds, 9
Flemingites flemingianus Zone, 11
Flowerpot Formation, 277, 374
Flowerpot Shale, Oklahoma, 435
Foldvik Creek Formation, 153, 164
Fort Scott Limestone, 198
France
  acritarchs, 280, 287, 291, 297
  Faulquemont, 201
  Lessart, Poiteau, 291
  Lias, 431
  ostracodes, 201
  palynomorphs, 338, 431
Fusulina kattaensis Zone, 11
fusulinids, 117

Gabh Nala, 165
Gastropoda
  Zaluch Group, 65
  Chhidru Formation, 65
  Kathwai Member, 65
  Mittiwali Member, 65
Germanic Basin, 432
Germany
  acritarchs, 277, 287, 288, 301
  conodonts, 240, 244, 250
  palynomorphs, 324, 326, 337, 341, 386, 393, 398, 441
Glass Mountains, 133
Glossopteris floral province, 435
Godthaabian, 155

Gondwana assemblages, 307, 434
Gondwana Region, 434-436, 438
 Upper Permian, 435
 Lower Permian, 307
Gondwanaland
 palynomorphs, 307, 309, 367, 369, 396, 409, 433-434, 437-439
Greece, 204, 218, 246, 254
Greenland, 153, 155, 164, 240
 East, 153, 155, 157-158, 164, 178-179, 181-182, 185, 190
 northeast, 3
Grodner Sandstone, 386
Guadalupe Mountains, 133, 145
Guadalupian
 ammonoids, Permian, 153-154, 156-159, 173
 brachiopods, 117, 119-123, 125, 128-130, 132-133, 145, 148
 conodonts, 214
 palynomorphs, 373, 380, 435
Guryul ravine, Kashmir, 180-181
Gyronitan age, 24, 58
Gyronitan Division, 445
Gyronites-bearing limestones, 215
Gyundi River, 172

Hedenstroemia beds, 11
Hermit Shale, 40
Himalayas
 ammonoids, Permian, 157, 159, 162, 165, 167, 169, 170, 172, 182
 ammonoids, Triassic, 10, 177, 179, 180, 183-184, 189-190
 brachiopods, 131
 conodonts, Triassic, 241
Holland, 441
Holy Cross Mountains, Poland, 194
Horse Canyon, southeastern Idaho, 214, 223
Humboldt Range, 249
Hundés, 169, 172
Hungary, 194, 204
Hydaspian, 218
hyporeliefs, 43

Idaho, 217-218, 238, 243, 246
 southeastern, 214, 217, 222, 223, 243, 246
Illinois, 221
 Pennsylvanian palynomorphs, 332, 344

index of affinity (IA), 230
India
 Gondwana assemblages, 307
 northern, 155
 palynomorphs, 310, 324, 362, 364, 367, 368, 374, 377, 380, 386, 409, 411, 431, 434, 435, 438
Indiana, 204
Indonesia, 161, 164
Induan Stage, 120, 194, 445
Indus River, 5
invertebrate trails, 39
Iran
 ammonoids, Permian, 153, 164-165
 brachiopods, 128, 144
 palynomorphs, 434
Iraq
 ammonoids, Permian, 164-165
 brachiopods, 129, 144
 northern, 434
 palynomorphs, 324, 327, 334, 358, 368, 374, 376, 380, 402, 434
 Permian, 358, 368, 374
Israel, 203, 205
Italy, 246
 north, 214, 217-218, 222-224

Jabalpur Series, 354
Jabbi, 73, 158, 161, 167, 194, 197
Jabi beds, 119
Jabian Substage, 119
Japan, 155-156
Jarhanwala Nala, 307
Jhallewali, 310, 396
Jhelum River, 5
Julfa, 158
Jurassic, 203
 acritarchs, 280, 281, 284, 286-289, 291, 293, 295, 297, 301
 ostracodes, 203
 palynomorphs, 331, 334-335, 338, 339, 354, 380, 382, 387, 401
Jurgung-Tumus, Soviet Arctic, 283

Kágá, 162
Kaiping Basin, 327, 332, 346
Kala, 167
Kalabagh, 5, 162, 167
Kalabagh beds, 118, 131-134, 197
Kalabagh Member, 148-149
 age of, 119, 173
 ammonoids, 65, 153, 162, 173
 brachiopods, 118-124, 128, 131-134, 136, 139, 144-145, 148-149
 ostracodes, 66, 194, 197
 trilobites, 66
Kansas, 277
Kap Stosch, East Greenland, 178, 181
Karroo, 367
Kashmir, 3
 ammonoids, Triassic, 179-180, 182, 186, 189
 conodonts, 214, 240, 241
 Gondwana assemblages, 307
 palynology, 307, 311
Kathwai fauna, 125
Kathwai Member, 22-24, 36-37, 40-43, 58, 60-62, 117, 148-149, 208-209
 age of, 124-125, 128
 ammonoids, 177-192
 brachiopods, 114, 117, 119-120, 123, 130-133, 148-149
 conodonts, 212, 214, 223, 224, 231, 234, 238, 241, 251, 253, 266
 depositional and diagenetic environment, 53-57
 fish, 69
 lingula, from, 111
 paleontological content, 49-53, 63-71
 palynology, 309, 316, 320, 371, 374, 386, 402, 412, 421, 439, 440, 444
Katta beds, 139
Kazakhstan, western, 372, 380
Kazanian, 121, 199
 palynomorphs, 410, 427
Keuper, 325, 326, 387, 398, 445
 Lorraine, 40
Khan Zaman Nala, 64
 brachiopods, 130, 144
 conodonts, 231, 238, 241
Khasor Range, 8
Khatkiara Member, 26
 acritarchs, 443
 palynology, 308, 313, 316, 327, 331, 335, 380, 387, 423, 443
Khisor Range, 5, 8, 17, 62
 brachiopods, 119, 122, 130-134, 136-139, 141, 144, 145, 147, 148
 conodonts, 207, 214
Khoora, 7
Khúra, 58, 162
Kinderhook, 222
 uppermost, 221
Kingriali dolomites, 26-27
Kingriali Mountain, 119, 120, 130-131, 134, 138-139, 141, 148

Subject and Locality Index

Kingriali sandstone, 26, 105, 308
Kinneyan ripples, 42
kirkbyan pit, 196-198
Kockatea Formation, 320
Kockatea Shale, 321, 339, 445
Koeafeoe, 156
Kolvinsk Arch, 445
Koninckites volutus Zone, 11
Kotal-e-Tera, Afghanistan, 216, 260
Kuafeu, 156
Kuling, 157, 162, 167-170, 172
  Shale, 11, 153
Kumaon, 158, 162, 172, 179
Kunafeu, 156
Kungurian, 380
  palynomorphs, 380, 410, 427, 432, 433
  sediments, 433
Kuznetsk Basin, 364, 433
Kweichou Province, China, 67, 143

Ladinian, 201, 205
Lahore, 5
Lamar equivalent, 75, 125
Lamar Limestone, 125, 133, 145
Landa Member, 22, 26, 105
  acritarchs, 70
  palynomorphs, 307, 308, 311, 325, 339, 342, 345, 351, 355, 371, 373, 382, 397, 416, 422, 423, 442, 443
Las Delicias, 156
Lavender clay, 20
Lavender series, 20
Leigh Creek, South Australia, 381, 388
Leonardian, 301, 380
Lessart, Poitou, France, 291
Lias, 337, 387, 431
Libya, 355
  acritarchs, 277, 291
  palynomorphs, 355
Lilang, 162, 168-170, 172, 179
Limestone unit, Kathwai Member,
  ammonites, 190
  conodonts, 234, 238, 241, 263
  fish, 69
  petrography, 49
lingula, paleoecology, 112
Lockport Dolomite, 42
Loping Series, 121, 155
Lungtan Series, 332
Lyttonia nobilis Zone, 11

Madagascan assemblages, palynomorphs, 438

Madagascar, 70, 72
  acritarchs, 277, 291
  ammonoids, Permian, 153, 155, 157-158, 167-168, 170, 172
  Liassic, 431
  palynomorphs, 369, 372-373, 376-377, 386-387, 400, 402, 410, 427, 431, 434-435, 438, 439, 441, 445
  Permian, Late, floras, 435
  Permian, Upper, 410, 427
  Scythian, 372
  Triassic, Lower, 373, 427
  Triassic, Upper, 431
Madhya Pradesh, 354
Magnesian sandstone, 310
Marwat Range, 5, 8, 17
Maryland, palynomorphs, 342
Mediterranean faunas, Permian brachiopods, 125
Medlicottia wynnei Zone, 11
Meekoceras bed, 11, 215, 217, 231, 241, 260
Mersey River, Tasmania, 277
Mexico, 121, 161, 173
Mianwali Formation, 22, 26, 36, 61, 208-209
  acritarchs, 442
  cephalopods, 65, 173-191
  conodonts, 212, 253, 257
  palynomorphs, 308-309, 316, 320-321, 325, 327, 330, 336, 338-339, 341, 344-345, 351, 355, 371-373, 377, 382, 386-387, 391, 393, 398-399, 402, 409, 413, 416, 421-423
  ostracodes, 194, 197-199, 202, 204-205
  brachiopods, 117
Mianwali series, 22, 104
Midcontinent region, American, 374, 386
Middle East
  palynomorphs, 434-435, 438
  Permian, Upper, 434, 438
Minnekhata Limestone, 222
Minnesota, 400
Minusinsk Basin, 433
Mississippian
  Lower, 197
  ostracodes, 197
  palynomorphs, 331
Missouri, 198, 223
Mittiali, 11
Mittiali-Schlucht, 58
Mittiwali, 23
Mittiwali Member, 23, 24, 57, 60, 65, 66, 104-105, 208, 209
  conodonts, 68, 209, 215-216, 228, 231-232, 238, 241, 243, 249, 251, 253, 255, 256, 261, 263, 266-267
  gastropods, 65
  ostracodes, 66, 194, 198-199, 202, 205
  palynomorphs, 308-309, 314, 316, 338-339, 344-345, 355, 372, 373, 377, 402, 421-422, 439, 440, 442, 444
  pelecypods, 64
Moolayember Formation, Queensland, 331
Morondava Basin, 435
mud cracks, 40, 54
multielement species, conodonts, 218-219
Munta Nala, 60, 128, 278, 282, 286
Muntanar-Schlucht, 60
Murgabian age, 128
Murree Formation, 23, 26
Muschelkalk, 179, 240
  lower, 393, 445
  upper, 201
Muth, 172, 179

nala, 17
Nammal Gorge, 13
Narmia Member, 22, 25, 26, 66-67, 105, 208
  acritarchs, 70, 443
  conodonts, 69, 208-209, 216-217, 226, 228, 231, 235, 238, 243, 244, 246, 254, 257-258, 260, 268
  ostracodes, 194, 197, 204, 205
  palynomorphs, 70, 308, 313-314, 325, 330, 336, 341, 351-352, 355, 387, 393, 396, 398-399, 409, 416, 422, 439, 442-444
Narrabeen Group, Australia, 351
Natal, 434
Neobolus shales, 310
Neogondolella carinata Zone, 68, 214, 215
Neogondolella jubata Zone, 68, 217
Neospathodus cristagalli Zone, 68, 215-216, 219
Neospathodus dieneri Zone, 68, 215-216
Neospathodus kummeli Zone, 68, 215, 248
Neospathodus pakistanensis Zone, 68, 216
Neospathodus timorensis Zone, 68, 217-218
Neospathodus waageni Zone, 68, 216-217, 248

Netherlands, palynomorphs, 377, 399, 441
Nevada,
  conodonts, 217-218, 231, 239, 241, 244, 249, 260, 261, 268
  Triassic, Lower, 260
New South Wales, Australia, 351, 367, 373
Nilawan Group, 17, 19, 21
Nilawan Series, 17
Nipol, 156
Noe (Noil) Bunu, 156
North America, palynomorphs, 380, 386, 396, 435
  western, 240

Ochoan, 117, 122, 157, 435
Oklahoma, 277, 380, 436
Olenekian, 444-445
Oligocene
  acritarchs, 288
Ophiceras beds, 11
Ophiceras-Otoceras Zone, 56
Ophiceras Zone, 144
Ophiuroidea, 67
ostracodes, 193-205
  Amb Formation, 66
  Alps, 195
  Australia, 196
  Austria, 195
  Chhidru Formation, 66
  France, 201
  Greece (Chios), 204
  Hungary, 194, 204
  Israel, 203, 205
  Kalabagh Member, 66, 194
  Kathwai Member, 67
  Kazan, 199
  Mianwali Formation, 194
  Mittiwali Member, 67, 194
  Narmia Member, 67, 194
  Poland, 194
  Timan, 199
  USSR, 194-195, 203
  Wargal Formation, 193
  Zaluch Group, 67
Orange Free State, 367, 374
Oregon, 121
Ordovician
  acritarchs, 284, 287, 293, 297
  lingula, 112
Otoceras beds, 11-12
Otoceras Zone, 13
Otoceras-Ophiceras Zone, 23, 177, 184

Painkhanda, 177-178
paleoecology, 112
Pali beds, Rewa, 367
palynology, nomenclatural procedures, 316-320
  preservation, 316
  techniques, 315
  terminology, 319-320
palynomorphs, 306-445
  Africa, 324, 362, 367, 374, 377, 380, 403, 408, 410, 416, 427, 431, 434, 435
  Amb Formation, 417-418
  Antarctica, 324, 354, 362, 367, 377, 387, 403, 408, 416, 431, 434, 435
  Australia, 309, 324, 327, 330-332, 339, 342, 344, 345, 350-351, 354, 362, 367, 373, 377, 379, 380, 387, 389, 394, 396, 403, 408, 410, 411, 415-416, 427, 431, 434, 437, 440, 441
  Austria, 331, 351, 369, 402
  Canada, 329-331, 334, 336, 338, 368, 369, 372, 386, 387, 409, 413, 416, 436, 440, 441
  Chhidru Formation, 70, 419
  China, 332, 403
  Early Scythian assemblages, 439
  Egypt, 374
  England, 324, 389
  Europe, 332, 345, 350-351, 366, 373, 376, 387, 396, 399, 431, 432, 437, 439
  European elements, 437, 439
  France, 338
  Germany, 324-326, 337, 386, 393, 398
  Gondwana elements, 438-439
  India, 324, 336, 354, 362, 364, 367-368, 377, 380-381, 396, 403, 408-409, 411, 431, 434-435
  Iraq, 324, 327, 334, 358, 368, 374, 380, 402, 434
  Kashmir, 311
  Kathwai Member, 70, 421
  Late Scythian and Middle Triassic assemblages, 442
  Libya, 355
  Madagasgan elements, 438-439
  Madagasgar, 369, 372-374, 376-377, 386-387, 401-402, 410, 427, 431, 435, 438-440
  Middle East elements, 438
  Mittiwali Member, 70, 421-422
  Narmia Member, 70, 422-423
  Netherlands, 377, 399
  Pakistan, 369
  Permo-Triassic boundary, 442
  Poland, 398
  Saudi Arabia, 366, 380
  Tasmania, 331, 334, 339, 344, 345, 351, 369, 373
  Thuringia, 321, 334, 445
  Tredian Formation, 423-424
  Ural elements, 437-439
  USA, 342, 344, 374, 380, 386-387, 389, 396, 400, 435
  USSR, 325, 330-331, 339, 341, 351, 355, 358, 362, 364, 366, 368, 372-374, 380, 399, 401, 410, 427, 432-433, 435, 437, 439, 441
  Wargal Limestone, 418-419
  Zaluch Group, 70
Pamirian, 128
paraconformity, 54, 77, 307
Pastannah, Kashmir, 179-182
Pelecypoda
  Zaluch Group, 64
  Chhidru Formation, 64
  Kathwai Member, 64
  Mittiwali Member, 65
  Narmia Member, 65
Pennsylvanian, 201
  acritarchs, 281
  conodonts, 238
  ostracodes, 201
  palynomorphs, 342, 380, 387, 389
Perth Basin, palynomorphs, 330-331, 350, 354, 379, 411, 415, 444-445
Permian, Upper, 415
Phosphoria Formation, 76
Po, Italy, 172
Poland, 194, 398
  acritarchs, 300
  ostracodes, 194
Pomerang, 169, 172
Portuguese Timor, 218, 257
Potwar Plateau, 4, 5
Precambrian
  acritarchs, 289
Prida Formation, 249
  lower member, 249
  Middle Triassic, 258
Prince Charles Mountains, Antarctica, 277, 367, 377, 434
Prionolobus rotundus Zone, 11
Productus Limestone, 7-9, 11-12, 14, 17, 19, 196-199, 310
  ammonoids, 158, 162

Subject and Locality Index

brachiopods, 111, 117-118, 122, 128, 130, 138, 144
fish, 68
Glossopteris flora, 307
ostracodes, 193, 196, 197
palynomorphs, 309-310
Productus lineatus Zone, 11
Prohungarites Division, 209, 439, 442, 445
Prohungarites Zones, 26, 104
propareas, 114
Pufelsbach, 224
Punjab, 155, 307
Purple sandstone, 310
pusha, 17

Queensland
  Lower Triassic, 373
  palynomorphs, 331, 334, 367, 373, 380, 408, 440, 444-445

Raniganj Coalfield of India, 324, 381, 403, 411, 434
Raniganj Coal Measures, palynomorphs, 324, 332, 336, 354, 367, 377, 381, 396, 403, 408, 411
Rátang River, 172
Rawalpindi, 5
Recent
  palynology, 413
  ostracodes, 195, 203
Rewa, 367
Rewan Formation, 445
Rhaetian, 195, 334, 387, 431
ripple marks, 31, 33
  capped, 54
Roadian, 155
Rocky Mountains, northern, 155
Ross Formation, 445
Ruasnain, 156, 158, 170
Runzelmarken, 43
Russia, see USSR
Rustler Formation, 122
Ruteh Limestone, 128

Saar Basin, 358
St. Louis Limestone, 222
St. Vigel, northern Italy, 217
Saidawali Member, 20
Sakamena Group, 427, 435
  lower, 377, 427, 435, 438
Sakesar, 5
Sakesar Ridge, 310
Sakmarian, 307, 362, 403
Salem Limestone, 204
Saline Series, 309-310
Salt Range boulder bed, 18
Salzburg, 195
Sarai village, 307, 310
Sardi Formation, 20-22

Saudi Arabia, 366, 380
Scania, 400
Scaphopoda, 66
Scythian, 11, 22
  ammonoids, 179, 181-182, 189, 190
  brachiopods, 121
  conodonts, 207, 217, 218, 257
  floral changes, 444
  lingula, 116
  ostracodes, 204
  palynomorphs, 307, 351, 374, 444
Seiser Schichten (=Strati di Siusi), 214, 224
Seisian Stage, 194
Serbia, west, 121
Sheikh Budin, 5
Shihhotze Series, 380
Shingarh Range, 5
Siberia
  ammonoids, Permian, 155-156
  ammonoids, Triassic, 178
  ostracodes, 195
  palynomorphs, 329
Siberian basins, 433
Sicily
  ammonoids, Permian, 155
  brachiopods, 133
  conodonts, 222, 234
Silurian, 204
  acritarchs, 287-289, 295, 297, 299-300
  conodonts, 262
  middle, 262
  ostracodes, 204
Siran-ki-dok, 8, 58
Siwalik Group, 4
Sodhi Zairin, 162, 172
Soefa, 156, 170
Solikamsk area, USSR, 433
Sooliman Range, 6
South America, 435
South Tyrol, 122
Soviet Union, see USSR
Spain, acritarchs, 297, 300
Speckled sandstone, 19, 307
  palynomorphs, 309, 374, 379
Spirifer marcoui Zone, 11
Spiti
  ammonoids, Permian, 158, 162, 167-170, 172
  ammonoids, Triassic, 177-181, 184, 189
Spiti region, 11
Spitsbergen
  ammonoids, Triassic, 178
  conodonts, 216, 218, 240, 255
  Triassic, Lower, 255

Stachella (Bellerophon) beds, 9
Stephanian, 332, 357
Stephanites superbus Zone, 11
Stratigraphic Commission of Pakistan, 17, 22
Subcolumbites Zone, 217, 249
Sumpek (Niipol Soempek), 156
Surghar Range
  brachiopods, 130-131, 133, 136-138, 144, 145, 148
  brachiopods, Permian, 64
  conodonts, 207, 214, 257
  geography, 5
  ostracodes, 194, 197, 203-205
  Permian-Triassic boundary, 62
Switzerland
  acritarchs, 301
  palynomorphs, 445
Sydney Basin, 445

Talchir boulder beds, 17
Talchir Stage, 17
Talung Formation, 155, 183
Tanzania
  palynomorphs, 367, 374, 380, 416
  Permian, Lower, 374, 416
Tartarian, 427
Tasmania
  palynomorphs, 330, 334, 338, 339, 344, 345, 369, 373, 445
  tasmanitids, 277
  Triassic, 369
Tasmanitids
  Australia, 277
  Chhidru Formation, 277-278, 300
  Tasmania, 277
  Western Australia, 277
  USSR, 277
Tatarian, 427, 433
Tethyan Sea, 130
Tethys
  ammonoids, Triassic, 178
  geography, 56
  palynomorphs, 307, 432
Texas
  ammonoids, 156
  brachiopods, 117, 122, 125, 133, 145
  conodonts, 222
  ostracodes, 193
  west, 155-156
Thaynes Formation, 217, 244, 246
Thuringia
  palynology, 321, 334, 445

Rhaetic, 334
Tibet, ammonoids, Permian, 155, 158, 168, 172
Tiers Formation, 445
Timan, 121, 199
Timor
　ammonoids, Permian, 154-157, 159, 161, 164, 170, 173
　conodonts, 216, 218, 240, 261
　western, 156
Tobin Formation, 249
Tobra Formation, 18-19, 21, 307
Toebue Lopo, Timor, 261
Topmost beds, 8
Topmost limestone, 9, 22, 25
Tournaisian, 332
Trace fossils, 71
trails, 43
Transcaspian Lowland, 194
Transcausia, 194
Trans-Indus ranges
　brachiopods, 119, 129-130, 148
　conodonts, 207-208, 213, 214
　geography, 3
Topmost limestone dolomite beds, 309
Tredian Formation, 23, 26, 105
　acritarchs, 443
　ostracodes, 210
　palynomorphs, 307, 308, 311, 313, 316, 321, 325, 327, 331, 335, 341, 344-345, 351-352, 357, 371-373, 380, 382, 387, 393, 398-399, 409, 413, 423-424, 442, 443
Tredian Member, 26
Trilobites, 66
Tunguska Basin, 433
Tunguska Province, 433
Tunisia, 277, 291, 299

Ue Lacan, Portuguese Timor, 218
Ukraine, 329
United States
　conodonts, 241, 262
　palynomorphs, 380, 386-387, 389, 396, 410, 435
　Permian brachiopods, 120-121, 129
　western, 241, 262
Upper Brush Creek Limestone, Ohio, 223
Upper Ceratite limestone, 9, 11, 22-24, 30, 208
Uppermost richly fossiliferous unit, 31

brachiopods, 119, 138
conodonts, 31, 223, 234, 238, 241
Endothyraceons, 31
Ural Geosyncline, 433-435
Ural Mountains, 121, 433
Ural Region, 432-433, 435-438
USSR, 368
　acritarchs, 299
　ammonoids, 154-156, 164, 178
　brachiopods, 121, 124, 128
　northern, 440, 445
　ostracodes, 194, 203
　palynomorphs, 325, 330-331, 339, 341, 351, 355, 358, 362, 364, 366-368, 372-374, 380, 394, 399-401, 410, 412, 427, 432-433, 435, 439-441, 444-445
　cis-Ural, 427, 433, 439, 445
　Late Permian, 433

Virgal, 7
Virgal Group, 11
Volga-Ural region, 277

Wagina Sandstone, 332
Wahan, 16
Warcha-Gruppe, 19
Warchha, palynomorphs, 310, 327, 348, 379, 389, 416, 417, 428, 430
Warchha Formation, 307
Warchha Gorge, 124
Warchha Sandstone, 19, 22
Warchha Water Tank
　palynomorphs, 314, 316, 326, 327, 347, 358, 364, 379, 380, 381, 389, 396, 400, 415
Wargal, 7
Wargal Formation, 148-149, 208
　brachiopods, 117, 134, 137, 148
　conodonts, 208, 214
　ostracodes, 193, 194, 197
Wargal Group, 11
Wargal Limestone, 20-21, 63-64, 66-67, 148-149
　age of, 173
　acritarchs, 430
　brachiopods, 63, 118-119, 122, 133, 137, 139, 148
　crinoids, 67
　foraminifera, 63
　palynomorphs, 309, 314, 316, 321, 322, 324, 350,

358, 362, 364-366, 374, 376, 377, 389, 394-395, 400, 411, 413, 418-419, 424, 427, 430
Wellington Formation, 277
West Kazakhstan, Lower Triassic, 372
Westphalian, 281, 324, 332
White Sandstone Unit, 21, 31, 33, 35, 119
　age of, 120
　fauna, 35
　foraminifera, 63
　Acritarcha, 70
　Brachiopoda, 73, 120, 124, 130-131, 133-134, 136-139, 144-145, 147-149
　conodonts, 223, 241
white coal, 277
white matter, conodonts, 225, 227-232, 234-235, 238, 245, 254-255, 260
Winterset Limestone, 223
Wolfcampian, 380
Woodada Formation, 444-445
Worcestershire, 445
Word Formation, 120-121, 125
Wordian, 121, 125, 155, 157
Wyoming, 116, 222

Xenodiscus carbonarius Zone, 11

Yangtze Valley, 121
Yugoslavia, 441
　acritarchs, 277, 291

Zaluch Group, 17, 20, 67-68, 70, 148-149, 214
　anthozoa, 63
　brachiopods, 63, 117-118, 124, 130, 133, 139, 141, 148
　bryozoa, 63
　conodonts, 214
　echinoids, 67
　foraminifera, 63
　gastropods, 65
　ostracodes, 67
　palynomorphs, 70, 308-309, 425, 442
　pelecypods, 64
Zechstein
　acritarchs, 301
　conodonts, 222, 244, 250
　Germany, 244, 250, 277, 301, 386, 441
　palynomorphs, 345, 366, 373, 386-387, 389, 431, 434-435, 437, 441
　saline lagoons, 432
Zéwan, 179